Lecture Notes in Networks and Systems

Volume 225

The series "Lecture Notes in Networks and Systems" publishes the latest developments in Networks and Systems—quickly, informally and with high quality. Original research reported in proceedings and post-proceedings represents the core of LNNS.

Volumes published in LNNS embrace all aspects and subfields of, as well as new challenges in, Networks and Systems.

The series contains proceedings and edited volumes in systems and networks, spanning the areas of Cyber-Physical Systems, Autonomous Systems, Sensor Networks, Control Systems, Energy Systems, Automotive Systems, Biological Systems, Vehicular Networking and Connected Vehicles, Aerospace Systems, Automation, Manufacturing, Smart Grids, Nonlinear Systems, Power Systems, Robotics, Social Systems, Economic Systems and other. Of particular value to both the contributors and the readership are the short publication timeframe and the world-wide distribution and exposure which enable both a wide and rapid dissemination of research output.

The series covers the theory, applications, and perspectives on the state of the art and future developments relevant to systems and networks, decision making, control, complex processes and related areas, as embedded in the fields of interdisciplinary and applied sciences, engineering, computer science, physics, economics, social, and life sciences, as well as the paradigms and methodologies behind them.

Indexed by SCOPUS, INSPEC, WTI Frankfurt eG, zbMATH, SCImago.

All books published in the series are submitted for consideration in Web of Science.

More information about this series at http://www.springer.com/series/15179

Leonard Barolli · Isaac Woungang ·
Tomoya Enokido
Editors

Advanced Information Networking and Applications

Proceedings of the 35th International
Conference on Advanced Information
Networking and Applications (AINA-2021),
Volume 1

 Springer

Editors
Leonard Barolli
Department of Information
and Communication Engineering
Fukuoka Institute of Technology
Fukuoka, Japan

Isaac Woungang
Department of Computer Science
Ryerson University
Toronto, ON, Canada

Tomoya Enokido
Faculty of Business Administration
Rissho University
Tokyo, Japan

ISSN 2367-3370   ISSN 2367-3389 (electronic)
Lecture Notes in Networks and Systems
ISBN 978-3-030-75099-2   ISBN 978-3-030-75100-5 (eBook)
https://doi.org/10.1007/978-3-030-75100-5

This Springer imprint is published by the registered company Springer Nature Switzerland AG
The registered company address is: Gewerbestrasse 11, 6330 Cham, Switzerland

Welcome Message from AINA-2021 Organizers

Welcome to the 35th International Conference on Advanced Information Networking and Applications (AINA-2021). On behalf of AINA-2021 Organizing Committee, we would like to express to all participants our cordial welcome and high respect.

AINA is an international forum, where scientists and researchers from academia and industry working in various scientific and technical areas of networking and distributed computing systems can demonstrate new ideas and solutions in distributed computing systems. AINA was born in Asia, but it is now an international conference with high quality thanks to the great help and cooperation of many international friendly volunteers. AINA is a very open society and is always welcoming international volunteers from any country and any area in the world.

AINA international conference is a forum for sharing ideas and research work in the emerging areas of information networking and their applications. The area of advanced networking has grown very rapidly, and the applications have experienced an explosive growth especially in the areas of pervasive and mobile applications, wireless sensor networks, wireless ad hoc networks, vehicular networks, multimedia computing and social networking, semantic collaborative systems, as well as Grid, P2P, IoT, Big Data and Cloud Computing. This advanced networking revolution is transforming the way people live, work and interact with each other and is impacting the way business, education, entertainment and health care are operating. The papers included in the proceedings cover theory, design and application of computer networks, distributed computing and information systems.

Each year AINA receives a lot of paper submissions from all around the world. It has maintained high-quality accepted papers and is aspiring to be one of the main international conferences on the information networking in the world.

We are very proud and honored to have two distinguished keynote talks by Dr. Flora Amato, University of Naples "Federico II", Italy and Prof. Shahrokh Valaee, University of Toronto, Canada, who will present their recent work and will give new insights and ideas to the conference participants.

An international conference of this size requires the support and help of many people. A lot of people have helped and worked hard to produce a successful AINA-2021 technical program and conference proceedings. First, we would like to thank all authors for submitting their papers, the session chairs and distinguished keynote speakers. We are indebted to program track co-chairs, program committee members and reviewers, who carried out the most difficult work of carefully evaluating the submitted papers.

We would like to thank AINA-2021 general co-chairs, PC co-chairs, workshop co-chairs for their great efforts to make AINA-2021 a very successful event. We have special thanks to the finance chair and web administrator co-chairs.

We do hope that you will enjoy the conference proceedings and readings.

Leonard Barolli
Makoto Takizawa
AINA Steering Committee Co-chairs

Isaac Woungang
Markus Aleksy
Farookh Hussain
AINA-2021 General Co-chairs

Glaucio Carvalho
Tomoya Enokido
Flora Amato
AINA-2021 Program Committee Co-chairs

AINA-2021 Organizing Committee

General Co-chairs

Isaac Woungang	Ryerson University, Canada
Markus Aleksy	ABB Corporate Research Center, Germany
Farookh Hussain	University of Technology, Sydney, Australia

Program Committee Co-chairs

Glaucio Carvalho	Sheridan College, Canada
Tomoya Enokido	Rissho University, Japan
Flora Amato	University of Naples "Federico II", Italy

Workshops Co-chairs

Kin Fun Li	University of Victoria, Canada
Omid Ameri Sianaki	Victoria University, Australia
Yi-Jen Su	Shu-Te University, Taiwan

International Journals Special Issues Co-chairs

Fatos Xhafa	Technical University of Catalonia, Spain
David Taniar	Monash University, Australia

Award Co-chairs

Marek Ogiela	AGH University of Science and Technology, Poland
Arjan Durresi	Indiana University Purdue University in Indianapolis (IUPUI), USA
Fang-Yie Leu	Tunghai University, Taiwan

Publicity Co-chairs

Lidia Ogiela   Pedagogical University of Cracow, Poland
Minoru Uehara   Toyo University, Japan
Hsing-Chung Chen   Asia University, Taiwan

International Liaison Co-chairs

Akio Koyama   Yamagata University, Japan
Nadeem Javaid   COMSATS University Islamabad, Pakistan
Wenny Rahayu   La Trobe University, Australia

Local Arrangement Co-chairs

Mehrdad Tirandazian   Ryerson University, Canada
Glaucio Carvalho   Sheridan College, Canada

Finance Chair

Makoto Ikeda   Fukuoka Institute of Technology, Japan

Web Co-chairs

Phudit Ampririt   Fukuoka Institute of Technology, Japan
Kevin Bylykbashi   Fukuoka Institute of Technology, Japan
Ermioni Qafzezi   Fukuoka Institute of Technology, Japan

Steering Committee Chairs

Leonard Barolli   Fukuoka Institute of Technology, Japan
Makoto Takizawa   Hosei University, Japan

Tracks and Program Committee Members

1. Network Protocols and Applications

Track Co-chairs

Makoto Ikeda   Fukuoka Institute of Technology, Japan
Sanjay Kumar Dhurandher   Netaji Subhas University of Technology,
                                       New Delhi, India
Bhed Bahadur Bista   Iwate Prefectural University, Japan

TPC Members

Elis Kulla	Okayama University of Science, Japan
Keita Matsuo	Fukuoka Institute of Technology, Japan
Shinji Sakamoto	Seikei University, Japan
Akio Koyama	Yamagata University, Japan
Evjola Spaho	Polytechnic University of Tirana, Albania
Jiahong Wang	Iwate Prefectural University, Japan
Shigetomo Kimura	University of Tsukuba, Japan
Chotipat Pornavalai	King Mongkut's Institute of Technology Ladkrabang, Thailand
Danda B. Rawat	Howard University, USA
Akio Koyama	Yamagata University, Japan
Amita Malik	Deenbandhu Chhotu Ram University of Science and Technology, India
R. K. Pateriya	Maulana Azad National Institute of Technology, India
Vinesh Kumar	University of Delhi, India
Petros Nicopolitidis	Aristotle University of Thessaloniki, Greece
Satya Jyoti Borah	North Eastern Regional Institute of Science and Technology, India

2. Next Generation Wireless Networks

Track Co-chairs

Christos J. Bouras	University of Patras, Greece
Tales Heimfarth	Universidade Federal de Lavras, Brazil
Leonardo Mostarda	University of Camerino, Italy

TPC Members

Fadi Al-Turjman	Near East University, Nicosia, Cyprus
Alfredo Navarra	University of Perugia, Italy
Purav Shah	Middlesex University London, UK
Enver Ever	Middle East Technical University, Northern Cyprus Campus, Cyprus
Rosario Culmone	University of Camerino, Camerino, Italy
Antonio Alfredo F. Loureiro	Federal University of Minas Gerais, Brazil
Holger Karl	University of Paderborn, Germany
Daniel Ludovico Guidoni	Federal University of São João Del-Rei, Brazil
João Paulo Carvalho Lustosa da Costa	Hamm-Lippstadt University of Applied Sciences, Germany
Jorge Sá Silva	University of Coimbra, Portugal
Apostolos Gkamas	University Ecclesiastical Academy of Vella, Ioannina, Greece

Zoubir Mammeri   University Paul Sabatier, France
Eirini Eleni Tsiropoulou   University of New Mexico, USA
Raouf Hamzaoui   De Montfort University, UK
Miroslav Voznak   University of Ostrava, Czech Republic

3. Multimedia Systems and Applications

Track Co-chairs

Markus Aleksy   ABB Corporate Research Center, Germany
Francesco Orciuoli   University of Salerno, Italy
Tomoyuki Ishida   Fukuoka Institute of Technology, Japan

TPC Members

Tetsuro Ogi   Keio University, Japan
Yasuo Ebara   Osaka Electro-Communication University, Japan
Hideo Miyachi   Tokyo City University, Japan
Kaoru Sugita   Fukuoka Institute of Technology, Japan
Akio Doi   Iwate Prefectural University, Japan
Hadil Abukwaik   ABB Corporate Research Center, Germany
Monique Duengen   Robert Bosch GmbH, Germany
Thomas Preuss   Brandenburg University of Applied Sciences,
                                   Germany
Peter M. Rost   NOKIA Bell Labs, Germany
Lukasz Wisniewski   inIT, Germany
Hadil Abukwaik   ABB Corporate Research Center, Germany
Monique Duengen   Robert Bosch GmbH, Germany
Peter M. Rost   NOKIA Bell Labs, Germany
Lukasz Wisniewski   inIT, Germany
Angelo Gaeta   University of Salerno, Italy
Graziano Fuccio   University of Salerno, Italy
Giuseppe Fenza   University of Salerno, Italy
Maria Cristina   University of Salerno, Italy
Alberto Volpe   University of Salerno, Italy

4. Pervasive and Ubiquitous Computing

Track Co-chairs

Chih-Lin Hu   National Central University, Taiwan
Vamsi Paruchuri   University of Central Arkansas, USA
Winston Seah   Victoria University of Wellington, New Zealand

TPC Members

Hong Va Leong	Hong Kong Polytechnic University, Hong Kong
Ling-Jyh Chen	Academia Sinica, Taiwan
Jiun-Yu Tu	Southern Taiwan University of Science and Technology, Taiwan
Jiun-Long Huang	National Chiao Tung University, Taiwan
Thitinan Tantidham	Mahidol University, Thailand
Tanapat Anusas-amornkul	King Mongkut's University of Technology North Bangkok, Thailand
Xin-Mao Huang	Aletheia University, Taiwan
Hui Lin	Tamkang University, Taiwan
Eugen Dedu	Universite de Franche-Comte, France
Peng Huang	Sichuan Agricultural University, China
Wuyungerile Li	Inner Mongolia University, China
Adrian Pekar	Budapest University of Technology and Economics, Hungary
Jyoti Sahni	Victoria University of Technology, New Zealand
Normalia Samian	Universiti Putra Malaysia, Malaysia
Sriram Chellappan	University of South Florida, USA
Yu Sun	University of Central Arkansas, USA
Qiang Duan	Penn State University, USA
Han-Chieh Wei	Dallas Baptist University, USA

5. Web-Based and E-Learning Systems

Track Co-chairs

Santi Caballe	Open University of Catalonia, Spain
Kin Fun Li	University of Victoria, Canada
Nobuo Funabiki	Okayama University, Japan

TPC Members

Jordi Conesa	Open University of Catalonia, Spain
Joan Casas	Open University of Catalonia, Spain
David Gañán	Open University of Catalonia, Spain
Nicola Capuano	University of Basilicata, Italy
Antonio Sarasa	Complutense University of Madrid, Spain
Chih-Peng Fan	National Chung Hsing University, Taiwan
Nobuya Ishihara	Okayama University, Japan
Sho Yamamoto	Kindai University, Japan
Khin Khin Zaw	Yangon Technical University, Myanmar
Kaoru Fujioka	Fukuoka Women's University, Japan
Kosuke Takano	Kanagawa Institute of Technology, Japan
Shengrui Wang	University of Sherbrooke, Canada

Darshika Perera   University of Colorado at Colorado Spring, USA
Carson Leung   University of Manitoba, Canada

6. Distributed and Parallel Computing

Track Co-chairs

Naohiro Hayashibara   Kyoto Sangyo University, Japan
Minoru Uehara   Toyo University, Japan
Tomoya Enokido   Rissho University, Japan

TPC Members

Eric Pardede   La Trobe University, Australia
Lidia Ogiela   Pedagogical University of Cracow, Poland
Evjola Spaho   Polytechnic University of Tirana, Albania
Akio Koyama   Yamagata University, Japan
Omar Hussain   University of New South Wales, Australia
Hideharu Amano   Keio University, Japan
Ryuji Shioya   Toyo University, Japan
Ji Zhang   The University of Southern Queensland
Lucian Prodan   Universitatea Politehnica Timisoara, Romania
Ragib Hasan   The University of Alabama at Birmingham, USA
Young-Hoon Park   Sookmyung Women's University, Korea

7. Data Mining, Big Data Analytics and Social Networks

Track Co-chairs

Eric Pardede   La Trobe University, Australia
Alex Thomo   University of Victoria, Canada
Flora Amato   University of Naples "Frederico II", Italy

TPC Members

Ji Zhang   University of Southern Queensland, Australia
Salimur Choudhury   Lakehead University, Canada
Xiaofeng Ding   Huazhong University of Science and
                                    Technology, China
Ronaldo dos Santos Mello   Universidade Federal de Santa Catarina, Brasil
Irena Holubova   Charles University, Czech Republic
Lucian Prodan   Universitatea Politehnica Timisoara, Romania
Alex Tomy   La Trobe University, Australia
Dhomas Hatta Fudholi   Universitas Islam Indonesia, Indonesia

Saqib Ali	Sultan Qaboos University, Oman
Ahmad Alqarni	Al Baha University, Saudi Arabia
Alessandra Amato	University of Naples "Frederico II", Italy
Luigi Coppolino	Parthenope University, Italy
Giovanni Cozzolino	University of Naples "Frederico II", Italy
Giovanni Mazzeo	Parthenope University, Italy
Francesco Mercaldo	Italian National Research Council, Italy
Francesco Moscato	University of Salerno, Italy
Vincenzo Moscato	University of Naples "Frederico II", Italy
Francesco Piccialli	University of Naples "Frederico II", Italy

8. Internet of Things and Cyber-Physical Systems

Track Co-chairs

Euripides G. M. Petrakis	Technical University of Crete (TUC), Greece
Tomoki Yoshihisa	Osaka University, Japan
Mario Dantas	Federal University of Juiz de Fora (UFJF), Brazil

TPC Members

Akihiro Fujimoto	Wakayama University, Japan
Akimitsu Kanzaki	Shimane University, Japan
Kawakami Tomoya	University of Fukui, Japan
Lei Shu	University of Lincoln, UK
Naoyuki Morimoto	Mie University, Japan
Yusuke Gotoh	Okayama University, Japan
Vasilis Samolada	Technical University of Crete (TUC), Greece
Konstantinos Tsakos	Technical University of Crete (TUC), Greece
Aimilios Tzavaras	Technical University of Crete (TUC), Greece
Spanakis Manolis	Foundation for Research and Technology Hellas (FORTH), Greece
Katerina Doka	National Technical University of Athens (NTUA), Greece
Giorgos Vasiliadis	Foundation for Research and Technology Hellas (FORTH), Greece
Stefan Covaci	Technische Universität Berlin, Berlin (TUB), Germany
Stelios Sotiriadis	University of London, UK
Stefano Chessa	University of Pisa, Italy
Jean-Francois Méhaut	Université Grenoble Alpes, France
Michael Bauer	University of Western Ontario, Canada

9. Intelligent Computing and Machine Learning

Track Co-chairs

Takahiro Uchiya	Nagoya Institute of Technology, Japan
Omar Hussain	UNSW, Australia
Nadeem Javaid	COMSATS University Islamabad, Pakistan

TPC Members

Morteza Saberi	University of Technology, Sydney, Australia
Abderrahmane Leshob	University of Quebec in Montreal, Canada
Adil Hammadi	Curtin University, Australia
Naeem Janjua	Edith Cowan University, Australia
Sazia Parvin	Melbourne Polytechnic, Australia
Kazuto Sasai	Ibaraki University, Japan
Shigeru Fujita	Chiba Institute of Technology, Japan
Yuki Kaeri	Mejiro University, Japan
Zahoor Ali Khan	HCT, UAE
Muhammad Imran	King Saud University, Saudi Arabia
Ashfaq Ahmad	The University of Newcastle, Australia
Syed Hassan Ahmad	JMA Wireless, USA
Safdar Hussain Bouk	Daegu Gyeongbuk Institute of Science and Technology, Korea
Jolanta Mizera-Pietraszko	Military University of Land Forces, Poland

10. Cloud and Services Computing

Track Co-chairs

Asm Kayes	La Trobe University, Australia
Salvatore Venticinque	University of Campania "Luigi Vanvitelli", Italy
Baojiang Cui	Beijing University of Posts and Telecommunications, China

TPC Members

Shahriar Badsha	University of Nevada, USA
Abdur Rahman Bin Shahid	Concord University, USA
Iqbal H. Sarker	Chittagong University of Engineering and Technology, Bangladesh
Jabed Morshed Chowdhury	La Trobe University, Australia
Alex Ng	La Trobe University, Australia

Indika Kumara — Jheronimus Academy of Data Science, Netherlands

Tarique Anwar — Macquarie University and CSIRO's Data61, Australia

Giancarlo Fortino — University of Calabria, Italy

Massimiliano Rak — University of Campania "Luigi Vanvitelli", Italy

Jason J. Jung — Chung-Ang University, Korea

Dimosthenis Kyriazis — University of Piraeus, Greece

Geir Horn — University of Oslo, Norway

Gang Wang — Nankai University, China

Shaozhang Niu — Beijing University of Posts and Telecommunications, China

Jianxin Wang — Beijing Forestry University, China

Jie Cheng — Shandong University, China

Shaoyin Cheng — University of Science and Technology of China, China

11. Security, Privacy and Trust Computing

Track Co-chairs

Hiroaki Kikuchi — Meiji University, Japan

Xu An Wang — Engineering University of PAP, China

Lidia Ogiela — Pedagogical University of Cracow, Poland

TPC Members

Takamichi Saito — Meiji University, Japan

Kouichi Sakurai — Kyushu University, Japan

Kazumasa Omote — University of Tsukuba, Japan

Shou-Hsuan Stephen Huang — University of Houston, USA

Masakatsu Nishigaki — Shizuoka University, Japan

Mingwu Zhang — Hubei University of Technology, China

Caiquan Xiong — Hubei University of Technology, China

Wei Ren — China University of Geosciences, China

Peng Li — Nanjing University of Posts and Telecommunications, China

Guangquan Xu — Tianjing University, China

Urszula Ogiela — Pedagogical University of Cracow, Poland

Hoon Ko — Chosun University, Korea

Goreti Marreiros — Institute of Engineering of Polytechnic of Porto, Portugal

Chang Choi — Gachon University, Korea

Libor Měsíček — J.E. Purkyně University, Czech Republic

12. Software-Defined Networking and Network Virtualization

Track Co-chairs

Flavio de Oliveira Silva	Federal University of Uberlândia, Brazil
Ashutosh Bhatia	Birla Institute of Technology and Science, Pilani, India
Alaa Allakany	Kyushu University, Japan

TPC Members

Yaokai Feng	Kyushu University, Japan
Chengming Li	Chinese Academy of Science (CAS), China
Othman Othman	An-Najah National University (ANNU), Palestine
Nor-masri Bin-sahri	University Technology of MARA, Malaysia
Sanouphab Phomkeona	National University of Laos, Laos
Haribabu K.	BITS Pilani, India
Shekhavat, Virendra	BITS Pilani, India
Makoto Ikeda	Fukuoka Institute of Technology, Japan
Farookh Hussain	University of Technology Sydney, Australia
Keita Matsuo	Fukuoka Institute of Technology, Japan

AINA-2021 Reviewers

Admir Barolli
Adrian Pekar
Ahmed Elmokashfi
Akihiro Fujihara
Akihiro Fujimoto
Akimitsu Kanzaki
Akio Koyama
Alaa Allakany
Alberto Volpe
Alex Ng
Alex Thomo
Alfredo Navarra
Aneta Poniszewska-Maranda
Angelo Gaeta
Anne Kayem
Antonio Loureiro
Apostolos Gkamas
Arjan Durresi
Ashfaq Ahmad
Ashutosh Bhatia
Asm Kayes
Baojiang Cui
Beniamino Di Martino
Bhed Bista
Carson Leung
Christos Bouras
Danda Rawat
Darshika Perera
David Taniar
Dimitris Apostolou
Dimosthenis Kyriazis
Eirini Eleni Tsiropoulou
Emmanouil Spanakis
Enver Ever
Eric Pardede
Ernst Gran
Eugen Dedu
Euripides Petrakis
Fadi Al-Turjman
Farhad Daneshgar
Farookh Hussain
Fatos Xhafa
Feilong Tang

Feroz Zahid
Flavio Silva
Flora Amato
Francesco Orciuoli
Francesco Piccialli
Gang Wang
Geir Horn
Giancarlo Fortino
Giorgos Vasiliadis
Giuseppe Fenza
Guangquan Xu
Hadil Abukwaik
Hideharu Amano
Hiroaki Kikuchi
Hiroshi Maeda
Hiroyuki Fujioka
Holger Karl
Hong Va Leong
Huey-Ing Liu
Hyunhee Park
Indika Kumara
Isaac Woungang
Jabed Chowdhury
Jana Nowaková
Jason Jung
Jawwad Shamsi
Jesús Escudero-Sahuquillo
Ji Zhang
Jiun-Long Huang
Jolanta Mizera-Pietraszko
Jordi Conesa
Jörg Domaschka
Jorge Sá Silva
Juggapong Natwichai
Jyoti Sahni
K. Haribabu
Katerina Doka
Kazumasa Omote
Kazuto Sasai
Keita Matsuo
Kin Fun Li
Kiyotaka Fujisaki
Konstantinos Tsakos

Kyriakos Kritikos
Lei Shu
Leonard Barolli
Leonardo Mostarda
Libor Mesicek
Lidia Ogiela
Lin Hui
Ling-Jyh Chen
Lucian Prodan
Makoto Ikeda
Makoto Takizawa
Marek Ogiela
Mario Dantas
Markus Aleksy
Masakatsu Nishigaki
Masaki Kohana
Massimiliano Rak
Massimo Ficco
Michael Bauer
Minoru Uehara
Morteza Saberi
Nadeem Javaid
Naeem Janjua
Naohiro Hayashibara
Nicola Capuano
Nobuo Funabiki
Omar Hussain
Omid Ameri Sianaki
Paresh Saxena
Purav Shah
Qiang Duan
Quentin Jacquemart
Rajesh Pateriya
Ricardo Rodríguez Jorge
Ronaldo Mello
Rosario Culmone
Ryuji Shioya
Safdar Hussain Bouk

Salimur Choudhury
Salvatore Venticinque
Sanjay Dhurandher
Santi Caballé
Shahriar Badsha
Shigeru Fujita
Shigetomo Kimura
Sriram Chellappan
Stefan Covaci
Stefano Chessa
Stelios Sotiriadis
Stephane Maag
Takahiro Uchiya
Takamichi Saito
Tarique Anwar
Thitinan Tantidham
Thomas Dreibholz
Thomas Preuss
Tomoki Yoshihisa
Tomoya Enokido
Tomoyuki Ishida
Vamsi Paruchuri
Vasilis Samoladas
Vinesh Kumar
Virendra Shekhawat
Wang Xu An
Wei Ren
Wenny Rahayu
Wuyungerile Li
Xin-Mao Huang
Xing Zhou
Yaokai Feng
Yiannis Verginadis
Yoshihiro Okada
Yusuke Gotoh
Zahoor Khan
Zia Ullah
Zoubir Mammeri

AINA-2021 Keynote Talks

The Role of Artificial Intelligence in the Industry 4.0

Flora Amato

University of Naples "Federico II", Naples, Italy

Abstract. Artificial intelligence (AI) deals with the ability of machines to simulate human mental competences. The AI can effectively boost the manufacturing sector, changing the strategies used to implement and tune productive processes by exploiting information acquired at real time. Industry 4.0 integrates critical technologies of control and computing. In this talk is discussed the integration of knowledge representation, ontology modeling with deep learning technology with the aim of optimizing orchestration and dynamic management of resources. We review AI techniques used in Industry 4.0 and show an adaptable and extensible contextual model for creating context-aware computing infrastructures in Internet of Things (IoT). We also address deep learning techniques for optimizing manufacturing resources, assets management and dynamic scheduling. The application of this model ranges from small embedded devices to high-end service platforms. The presented deep learning techniques are designed to solve numerous critical challenges in industrial and IoT intelligence, such as application adaptation, interoperability, automatic code verification and generation of a device-specific intelligent interface.

Localization in 6G

Shahrokh Valaee

University of Toronto, Toronto, Canada

Abstract. The next generation of wireless systems will employ networking equipment mounted on mobile platforms, unmanned air vehicles (UAVs) and low-orbit satellites. As a result, the topology of the sixth-generation (6G) wireless technology will extend to the three-dimensional (3D) vertical networking. With its extended service, 6G will also give rise to new challenges which include the introduction of intelligent reflective surfaces (IRS), the mmWave spectrum, the employment of massive MIMO systems and the agility of networks. Along with the advancement in networking technology, the user devices are also evolving rapidly with the emergence of highly capable cellphones, smart IoT equipments and wearable devices. One of the key elements of 6G technology is the need for accurate positioning information. The accuracy of today's positioning systems is not acceptable for many applications of future, especially in smart environments. In this talk, we will discuss how positioning can be a key enabler of 6G and what challenges the next generation of localization technology will face when integrated within the new wireless networks.

Contents

Optimization of Task Allocations in Cloud to Fog Environment with Application to Intelligent Transportation Systems

Fatos Xhafa[1]($\boxtimes$), Alhassan Aly[1], and Angel A. Juan[2]

[1] Universitat Politècnica de Catalunya, Barcelona, Spain
fatos.xhafa@upc.cdu
[2] Universitat Oberta de Catalunya, Barcelona, Spain
ajuanp@uoc.edu

Abstract. Fog and Edge computing are opening up new opportunities to implement novel features of mobility, edge intelligence and end-user support. The successful implementation and deployment of Fog layers, as part of Cloud-to-thing-computing, largely depends on optimized allocation of tasks and applications to Fog and Edge nodes. Similarly as in other large scale distributed systems, the optimization problems that arise are computationally hard to solve. Such problems become even more challenging due to the need of application scenarios for larger computing capacity, beyond those of single nodes, requiring thus efficient resource grouping. In this paper we present some clustering techniques for creating virtual computing nodes from Fog/Edge nodes by combining semantic description of resources with semantic clustering techniques. Then, we use such clusters for optimal allocation (via heuristics and Integer Linear Programming) of applications to virtual computing nodes. Simulation results are reported to support the feasibility of the model and efficacy of the proposed approach. Applications of allocation methods to Intelligent Transportation Systems are also discussed.

1 Introduction

A number of computing paradigms and technologies have emerged after IoT and Cloud computing. Indeed, the view of IoT Cloud, in which IoT devices are directly connected to Cloud platforms and Data centers has shifted to a *continuum* view, where thereby various computing layers sit in between of IoT and Cloud. These layers referred to as Edge and Fog computing aim to retain the processing of data close to the source where the data is generated. OpenFog Consortium defines Fog computing as *"A horizontal, system-level architecture that distributes computing, storage, control and networking functions closer to the users along a Cloud-to-thing continuum"*. Besides significantly alleviating the computing burden to the Cloud, Edge and Fog layers enable faster round-trip processing and better support to end users. Most importantly, by processing the data close to the data source and to end-users, real time decision making can be designed to support a variety of use cases. This is made possible by achieving the computing and data fabric view of Edge computing [18], where different computing granularity devices can collaboratively support both real time processing and data storage requirements.

Cloud-to-thing-continuum brings the advantage of both real time processing and in-depth processing of IoT data. Real time processing enables, in particular, the detection of anomalies or abnormal events [4,19] regarding transportation fleet, etc., which would require immediate intervention by managers and local stakeholders. Further, massive offline processing enables computing of advanced descriptive, predictive and prescriptive analytics to support decision making and business intelligence and big data solutions. Cloud-to-thing-continuum enables the implementation of a plethora of application scenarios including smart logistics [8] and intelligent transportation systems [13,21]. Indeed, with computing layers of Fog and Edge computing, the premise is that latency and round trip time can be drastically reduced providing thus support to real time applications, mobility, etc.

In case of intelligent transportation systems (ITS), a new generation of ITS is being envisaged and deployed based on OBU–On Board Units– and RSU – Road Site Units– technologies deployment in road transport networks [5,11,15]. The information gathered by OBU in vehicles and RSU units in roads enables information gathering and can be used to evaluate in real time the traffic density and to predict it for short term periods as well as to compute more accurate maps of road traffics density. Through cooperative sensing and collective intelligence improved traffic efficiency can be achieved [9].

The implementation of applications in the Cloud-to-thing-continuum, however, comes with a number of challenges, among which we could distinguish the high degree of heterogeneity of resources, the variability in computing in the *continuum*, the scale and reliability. In this context, the allocation of computing resources becomes more complex than in traditional distributed systems. Many applications may require more computing resources than single Fog/Edge nodes can provide. In this paper, we address this issue by using semantic clustering of Fog resources into virtual nodes of larger computing capacity, to which applications can be allocated. Semantic categorization of resources enables a coherent clustering of resources and thus building clusters both statically and dynamically to cope with application requirements. Then, such clusters are used as input information to optimization procedures (heuristic methods and Integer Linear Programming) to optimize the allocation. Feasibility and efficacy of the proposed approach is evaluated *via* simulation.

The rest of the paper is structured as follows. In Sect. 2, we present the model for clustering of small nodes into larger virtual nodes; an application scenario from traffic load prediction is also given. In Sect. 3 we present methods for optimizing task/application allocation to clusters of nodes. Then, in in Sect. 4, we report some experimental results obtained via simulation of the Fog/Edge model and various application profiles. We end the paper in Sect. 5 with conclusions and future work.

2   Modeling of C2F Environment and Semantic Clustering

We consider the architecture depicted in Fig. 1. Fog nodes could be composed of one or more layers, however in our implementation we consider just one layer between the Cloud and the edge devices.

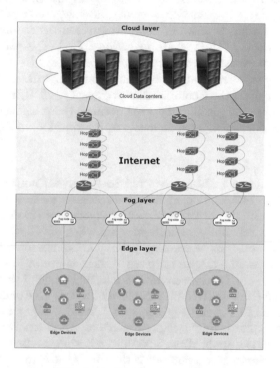

Fig. 1. Architecture overview

As shown in Fig. 1 we first have a Cloud layer which contains the Cloud infrastructure, then the Fog layer containing the Fog nodes which are connected to the Cloud through the internet. The connections between the Cloud and the Fog nodes pass through several hops. The physical distance between the Cloud data centers, and the Fog nodes affects data transfer rate, since it increases both latency and potential packet loss. The Fog nodes are directly connected to the edge devices i.e. one hop away, which makes the data transfer faster, more stable and reliable.

The Cloud node has abundant memory and processing power, while the Fog nodes are rather limited in comparison. The edge devices in our system will just have one application each that they need to run using certain specifications.

2.1 Semantic Categorization of Resources, Clustering and Offloading

The need to create groups of devices in Edge lies in exploiting capabilities of several devices simultaneously to carry out a computing intensive task collaboratively and efficiently or because the Edge devices separately cannot complete the task under the requirements of efficiency, latency, etc. The devices that are part of a group can share network computing resources among themselves to meet the requirements of the tasks/applications. To do this, we first propose to apply techniques of *semantic categorization of resources* at the Fog/Edge layer. Through semantic characterization, for example through RDF, the information of Edge devices such as their computational capacity (CPU, RAM, bandwidth) and storage capacity, the specificity of each resource (heterogeneity) and *enriched with contextual information* such as location, information on mobility, security, trust, etc.

Virtual Clusters
The clusters of devices in Edge, referred to as virtual nodes, can be formed based on semantic information of machine-readable format according to several criteria: (1) similarity of the devices (homogeneity); (2) device location; (3) computational and reliability performance analytics of a device and (4) device behavior analytics in terms of security, privacy and trust [6]. The first criterion allows similar devices in a cluster in terms of their computational characteristics in order to avoid disparity between devices and that some of them could become bottlenecks in the collaborative performance of a computational task. The second criterion allows to obtain device clusters "close to" an end user or to a data source and thus explore the "user-proximity" or "data-locality" depending on the requirements of the application. The third criterion can be taken into account when assigning tasks to groups simply to distribute the computing load. Finally, the fourth criterion would be applicable when it comes to processing sensitive data.

Static Clusters
We can calculate *a priori* static clusters, for example using the semantic similarity criterion (*semantic clustering*) or clustering based on the location (*geoclustering*), which can be useful for certain types of applications whose requirements are known *a priori*, for example to process data from a stream of fixed cameras on a highway. However, static clusters would not be suitable for all types of applications and scenarios, because finding *a priori* a suitable cluster size that fits various applications and scenarios is very difficult. In particular, static clusters cannot incorporate the current state of the device workload at run-time.

Dynamic Clusters
To satisfy the computational demands of the applications without knowing *a priori* their requirements, that is, at run-time, it is necessary to create clusters of devices dynamically and taking into account the computational load of the

devices. To do this, it is proposed to formulate optimization problems that calculate clusters of Edge devices to be assigned to computational tasks (either workflow or bag-of-tasks) in order to optimize the execution of the tasks and optimally use the Edge resources.

Offloading Algorithms

Offloading algorithms is a class of algorithms that allow to delegate or unload the computation or performance of tasks between parts of the same system or between different computer systems. The reasons for downloading the computation from an entity/system $\mathscr{A}$ to an entity/system $\mathscr{B}$ can be diverse (one or more at a time) but all of them try to take advantage of the $\mathscr{B}$ resources or features to achieve system performance, resource utilization, alleviating the burden of computations, etc.

Offloading algorithms arise very naturally in a Cloud-to-thing continuum system. In fact, the offloading algorithms are the very essence of such systems [14].

2.2 Application Scenario: Prediction of the Volume of Traffic by Images Of Mobile Crowd-Sensing

This application scenario aims to process a data flow of images sent by smart phones (*mobile crowd-sensing*) to predict the volume of traffic in real time in large cities. Traffic jams in large cities constitute a social problem and have significant economic costs to transportation systems. An alternative to calculate the congestion map of a city in real time and to make the prediction of traffic volume is to apply crowd-sensing techniques. Prediction of the volume of traffic can be very useful to optimize performance parameters in Intelligent Transportation Systems.

Specifically, thousands of drivers can send images of the state of the roads in a city in real time, forming a stream of images and corresponding positions whose real-time processing can provide valuable information on the current state and predict traffic in the city. The premise is in the large number of images that allows to achieve the prediction with high precision. This problem is defined as predicting real-time spatial information from data collected by mobile IoT sensors.

Therefore mobile crowd-sensing constitutes an alternative to the costly deployment and maintenance of video-camera infrastructures throughout a city, especially in large cities where the cost of deployment and maintenance can be very high. However, crowd-sensing has intrinsic problems and challenges:

- Driver participation is voluntary and mass participation is not guaranteed.
- The quality of the images may be poor or just *fake*, which could affect the accuracy of the prediction.
- The images can reveal information about the license plates of the cars and the information received must be anonymized.

The volunteer participation of drivers can be addressed by implementing a participation incentives scheme (a problem of interest to Vehicular Networks [10,17]). Similarly, the second point can be addressed by implementing a classifier based an image quality model of *image features* that allows deducing the contribution of an image in predicting traffic volume and thus to be able to implement an efficient classifier of the images. Finally, for the third problem, we can apply the state of the art to anonymize the license plates of cars in the images. The reader is referred to [16] for more details.

3   Optimization of Task/Application Allocation

The responsibility of the solver in our system is to perform the mapping between the tasks in an application to Fog nodes within a cluster. Its function is to find which Fog node is going to execute which tasks of an application. Thus, mapping could be considered as an optimization problem. The problem could be described as a variant to the bin-packing problem [2], where we have a number of items that we want to place in a number of bins, with a common capacity and minimize the number of used bins. Unlike the traditional problem, we will have different capacities to represent the memory of the Fog nodes. We first apply ILP –Integer Linear Programming– to get the optimum solution, then we solve the problem using a First Fit Heuristic Algorithm (FFHA) [20] approach to decrease the run-time of the optimization.

The objective is minimizing both the number of Fog nodes used, and the processing time by allocating as many tasks to Fog nodes with higher CPU as possible.

We formalize the problem as follows. Let us denote by f –a Fog node, F –the set of Fog nodes, f_{p_t}–the processing time of node f and f_used defined as equal to 1 if node f is used, 0 otherwise.

$$\min \sum_{f \in F} f_{p_t} \times f\_used$$

subject to:

$$\forall t \in T :$$
$$\sum_{f \in F} x_{tf} = 1$$

$$\forall f \in F :$$
$$\sum_{t \in T} t\_size \times x_{tf} \leq f_{RAM} \times f$$

where:

- x_{tf}: equals 1 if task t is placed in Fog node f, 0 otherwise.
- t_size: Size of task t
- f_{RAM}: RAM capacity of f.

The constraints of the ILP ensure that a task can only exist in one Fog node, and that the sum of task sizes cannot exceed the memory capacity of the Fog node they are placed in respectively.

For the ILP method, we use the *PuLP* Python package to solve the optimization problem. *PuLP* can generate Mathematical Programming System (MPS) or Linear Programming (LP) files and call GLPK [12], COIN CLP/CBC [3], CPLEX [1], and GUROBI [7] to solve linear problems. PuLP uses Coin-or Branch and Cut (CBC) as its default solver, which is an open-source mixed integer programming solver written in C++.

For the FFHA, we define our own algorithm to solve the specific problem at hand, where we have a variable number to represent Fog node's RAM. The pseudo-code of the FFHA is shown in Algorithm 1.

Algorithm 1. First Fit Heuristic Algorithm

---
for $Fogs = 1, 2, \ldots$ **do**
    for $index = 1, 2, \ldots, inRangeFogSizes.length$ **do**
        Assign $taskSize$ in sizes to $sizes[index][1]$
        if $filledFogselectedFogSize$ ... **then**
            if $taskSize <= (FogSize - filledFogSize)$ **then**
                $filledFog + taskSize$
            else
                break
            end if
        else
            return to loop
        end if
    end for
    Return assignment results
end for

As mentioned before, every time we run the simulation, the Fog nodes and the edge devices are generated in different positions with random attributes.

4 Experimental Evaluation

To assess the performance of our implementation, we benchmark the overall process by running the simulation with different sets of numbers for the Fog nodes and the devices. Since the semantic reasoning and clustering directly affects the overall performance, we can test how does the system scale by increasing the number of the Fog nodes and edge devices. Moreover, we compare between the two solvers we used by focusing on the differences in run-times, and the application response times obtained from the varied possible scenarios of how the application is being sent.

Various scenarios are considered, namely, the application is directly sent to the Cloud, the application is sent to the Cloud through a Fog node, the application is sent to a single Fog node, the application is sent to an In-range cluster and the application is sent to a neighbor cluster.

We evaluate the utility of the two proposed clustering techniques, by comparing and the response times obtained from each method of sending applications.

4.1 Experimental Setting and Results

The machine specifications used in the simulations to measure the performance of the system are as follows.

All tests where executed on a single machine with the following specifications: Windows 10 Pro, Intel Core i7-4710HQ @ 2.5 GHz, base Clock Speed up to 3.5 GHz Turbo Speed, 6 MB L3 cache, 16 GB DDR3 @ 1600 MHz, Dual channel and 500 GB NAND Sata SSD.

Since the simulation generates random attributes for the Fog nodes, the edge devices and the connections, we run the simulation for number of times and calculate the mean values to acquire statistically significant results. On the one hand, for bench-marking and comparison between the run-times of the solvers, we run the simulation using different sets of numbers for the Fog nodes and edge devices. For each set of numbers, we run the simulation 50 times with the ILP solver and 50 times with the FFHA. On the other hand, for the response time analysis, we run the simulation with the same number of Fog nodes and edge devices 50 times. In that manner, we focus on how the clustering techniques could affect the response times.

4.2 System Performance

To evaluate the optimization techniques used to solve the task allocation problem, we focus on the response times of the applications in different scenarios, for which we measure the average run-time for all the scenarios. We run the simulation with the following sets of parameters 50 times with the ILP solver and 50 times with the FFHA.

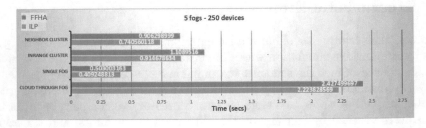

Fig. 2. Response times for 5 Fogs and 250 devices (ILP - FFHA)

The results show that the ILP solver acquires slightly better response times than that of the FFHA, in the case of running the simulation with 5 Fog nodes and 250 devices as shown in Fig. 2.

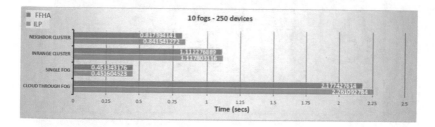

Fig. 3. Response times for 10 Fogs and 250 devices (ILP - FFHA)

When running the simulation with 10 Fog nodes and the same sets of devices, the results show that in some cases, the FFHA gets better response times compared to the ILP solver as shown in Fig. 3.

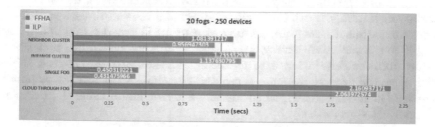

Fig. 4. Response times for 20 Fogs and 250 devices (ILP - FFHA)

The Fig. 4 shows the results when running the simulation with 20 Fog nodes. We can see that the results are similar to running the simulation with 10 Fog nodes. The ILP solver performs slightly better than the FFHA, yet the difference is not significant.

Overall, regarding the application response times, given the different methods of sending the application. The results gathered from running the simulation with the aforementioned sets of numbers for the Fog nodes and the devices, we can deduce that the ILP solver has a slight edge over the FFHA. However, the difference is rather minor. That is due to the fact that we perform the memory checks, as an initial filter, then the solvers optimize where the tasks are processed to get the least possible response time.

In contrast, when reviewing the run-times of the simulations, there is a substantial difference between the performance of the ILP solver and the FFHA, where the FFHA performs much faster than the ILP solver as shown in Fig. 5.

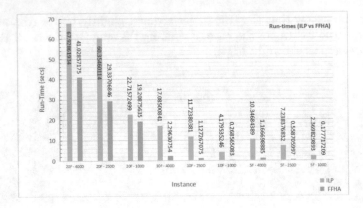

Fig. 5. Run-times (ILP *vs* FFHA)

On one hand, increasing the number of edge devices linearly affects the run-time. On the other hand, we can see an exponential increase in run-time when increasing the number of Fog nodes, which is unsurprising, since this increases the number combinations when forming either an in-range or neighbor cluster.

To measure the consistency of the optimization techniques, we calculate the standard deviation, of both the response times and the run-times by conducting another experiment, using an instance size of 10 Fog nodes and 100 devices, and 100 iterations. Additionally, we calculated the number of Fog nodes in the clusters from the same instance, to give insight on the size of the clusters. Tables 1 and 2 show the average response times and run-times with their corresponding Standard Deviations (STD), when running the simulation using the ILP solver and the FFHA respectively.

Table 1. ILP - Averages and standard deviations of response times & run-time

ILP		
	Avg response times	**Response times STD**
Cloud direct	17.1991722	8.71893098
Cloud through fog	2.18383078	0.23542268
Single fog	0.43523073	0.10711002
In-range cluster	1.09988548	0.21959514
Neighbor cluster	0.88225381	0.21159424
	Avg run-time	**Run-time STD**
	3.82993240	0.81419498

Table 2. FFHA - Averages and standard deviations of response times & run-time

FFHA		
	Avg response times	**Response times STD**
Cloud direct	18.2210832	10.5720824
Cloud through fog	2.19942492	0.26223306
Single fog	0.45561651	0.11327142
In-range cluster	1.09464597	0.19415273
Neighbor cluster	0.93392979	0.22757457
	Avg run-time	**Run-time STD**
	0.22792500	0.09110830

When comparing the run-times STD, we can see that the FFHA behaves more consistently than the ILP solver, with less deviation.

5   Conclusions and Future Work

In this paper we presented a model for Cloud to Fog computing (C2F) and allocation of tasks and applications in such environment. The aim is to evaluate the benefits of using semantic categorization of resources, semantic clustering and optimizing task allocation to such clusters. Our approach shows that there is a massive gain in response times when utilizing the Fog computing paradigm, compared to just relying on the Cloud. Based on the results, it is apparent that the application response times are understandably faster when utilizing Fog nodes or clusters. Integer Linear Programming (ILP) and an heuristic method are used for optimization purposes in this work. Various settings are considered to evaluate the scalability of the solvers. Although various application scenarios have been used in the study, more application profiles would be needed to achieve conclusive results on the allocation methods.

In our future work we would like to evaluate the proposed approach on a realistic scenario of transportation systems with V2X communication and Intelligent Road Side Units. Scenarios of safety, decreased trip times and energy savings will be considered. Also, we plan to use the iFog simulator to evaluate whether the proposed heuristic can give adequate response to mobility of Edge nodes.

Acknowledgements. This work is supported by Research Project, "Efficient & Sustainable Transport Systems in Smart Cities: Internet of Things, Transport Analytics, and Agile Algorithms" (TransAnalytics) PID2019-111100RB-C21/AEI/ 10.13039/501100011033, Ministerio de Ciencia e Innovación, Spain.

References

1. Ceselli, A., Fiore, M., Premoli, M., Secci, S.: Optimized assignment patterns in Mobile Edge Cloud networks. Comput. Oper. Res. **106**, 246–259 (2019)

2. Coffman Jr., E.G., Csirik, J., Galambos, G., Martello, S., Vigo, D.: Bin packing approximation algorithms: survey and classification. In: Pardalos, P., Du, D.Z., Graham, R. (eds.) Handbook of Combinatorial Optimization. Springer (2013)
3. COIN-OR (Common Infrastructure for Operations Research). http://www.coin-or.org
4. Corizzo, R., Ceci, M., Japkowicz, N.: Anomaly detection and repair for accurate predictions in geo-distributed big data. Big Data Res. **16**, 18–35 (2019)
5. Fouchal, H., Bourdy, E., Wilhelm, G., Ayaida, M.: A validation tool for cooperative intelligent transport systems. J. Comput. Sci. **22**, 283–288 (2017)
6. Gago, M.C.F., Moyano, F., López, J.: Modelling trust dynamics in the Internet of Things. Inf. Sci. **396**, 72–82 (2017)
7. Gurobi Optimization, LLC., Gurobi Optimizer Reference Manual (2020). http://www.gurobi.com
8. Faulin, J., Scott S.E., Juan, A.A., Hirsch, P.: Sustainable Transportation and Smart Logistics. Decision-Making Models and Solutions. Elsevier (2019)
9. Ferreira, J., et al.: Cooperative sensing for improved traffic efficiency: the highway field trial. Comput. Netw. **143**, 82–97 (2018)
10. He, Z., Cao, J., Liu, X.: High quality participant recruitment in vehicle-based crowdsourcing using predictable mobility. In: 2015 IEEE Conference on Computer Communications, INFOCOM 2015, Hong Kong, 2015, pp. 2542–2550 (2015)
11. Mfenjou, M.L., et al.: Control points deployment in an Intelligent Transportation System for monitoring inter-urban network roadway. J. King Saud Univ. Comput. Inf. Sci. In Press. https://doi.org/10.1016/j.jksuci.2019.10.005
12. Makhorin, A.: GLPK (GNU Linear Programming Kit). http://www.gnu.org/software/glpk/glpk.html
13. Hussain, M.M., Alam, M.S., Sufyan Beg, S.: Vehicular fog computing-planning and design. Procedia Comput. Sci. **167**, 2570–2580 (2020)
14. Santa, J., Fernández, P.J., Sanchez-Iborra, R., Murillo, J.O., Skarmeta, A.F.: Offloading positioning onto network edge. Wirel. Commun. Mobile Comput. **2018**, 7868796:1–7868796:13 (2018)
15. Santa, J., Bernal-Escobedo, L.: On-board unit to connect personal mobility vehicles to the IoT. FNC/MobiSPC, Ramon Sanchez-Iborra, pp. 173–180 (2020)
16. Shinkuma, R., Takagi, T., Inagaki, Y., Oki, E., Xhafa, F.: Incentive mechanism for mobile crowdsensing in spatial information prediction using machine learning. In: AINA, pp. 792–803 (2020)
17. Wang, X., Wu, W., Qi, D.: Mobility-aware participant recruitment for vehicle-based mobile crowdsensing. IEEE Trans. Veh. Technol. **67**(5), 4415–4426 (2018)
18. Xhafa, F.: The vision of edges of internet as a compute fabric. In: Advances in Edge Computing: Massive Parallel Processing and Applications. Book Series: Advances in Parallel Computing Series, Chapter 1. IOS Press (2019)
19. Xhafa, F., Kilic, B., Krause, P.: Evaluation of IoT stream processing at edge computing layer for semantic data enrichment. Future Gener. Comput. Syst. **105**, 730–736 (2020)
20. Yesodha, R., Amudha, T.: A comparative study on heuristic procedures to solve bin packing problems. Int. J. Found. Comput. Sci. Technol. **2**, 37–49 (2012)
21. Zhang, H., Lu, X.: Vehicle communication network in intelligent transportation system based on Internet of Things. Comput. Commun. **160**, 779–789 (2020). Special Issue on Internet of Things and Augmented Reality in the age of 5G

User Allocation in 5G Networks Using Machine Learning Methods for Clustering

Christos Bouras[✉] and Rafail Kalogeropoulos

Computer Engineering and Informatics Department, Patras, Greece
bouras@cti.gr, kaloger@ceid.upatras.gr

Abstract. The rapid increase in the volume of devices connected to mobile networks poses unprecedented demands on existing networking infrastructures. Machine Learning is a promising technique, already applied in various sectors of our everyday lives. It enables decision making not with the use of traditional programming but rather by using data to train models to cope with various problems without explicit programming on how to do so. The integration of Machine Learning techniques is deemed necessary in as many processes as possible to help the network face congestion and enable efficient real time decision making. In this paper we present two Machine Learning based mechanisms for improving real time user allocation on the network as well as predicting the best positioning scheme for Smallcell Base Stations to provide effective utilization of the network's resources.

1   Introduction

The introduction of more capable mobile devices has set the pace for a complete redesing of all cellular networks and their infrastructures. These new devices are able to sustain a plethora of applications. Their computing power always allows for the development of new apps while the establishment of communication standards allows them to connect with a multitude of sensors and devices. This motivates the integration of new devices such as mobile phones and sensors into the network. They can be used for a variety of purposes such as medical [9] or agricultural use [2].

Wireless networks as a result face an unprecedented rise in the number of connected devices that seek to utilize their resources, while the volume of data transfered through the network's infrastructure increases as well. In contrast to previous generations of networks that mostly consisted of a Base Stations (BSs) of the same type, namely Macro cell Base Stations (McBSs), in latest generations we saw the introduction of Smallcell Base Stations (SBSs), smaller BSs that can be further categorized based on their transmit power.

In previous generations, McBSs prevailed due to a higher transmit power, ensuring high Signal to Interference plus Noise Ratio (SINR), especially in the Downlink (DL) direction, while performance in the Uplink (UL) direction was

L. Barolli et al. (Eds.): AINA 2021, LNNS 225, pp. 13–24, 2021.
https://doi.org/10.1007/978-3-030-75100-5_2

deemed secondary. All McBSs feature the same characteristics throughout the entire network and are also capable of supporting a similar amount of users [12]. As a result, User Equipment (UE) were associated on the same BS for both directions, based solely on DL performance. These networks, called Homogeneous Networks were applicable since the required throughput for DL was traditionally higher than UL, resulting in an asymmetric utilization from the two traffic directions [10]. This approach is now completely dated, with the plethora of new devices uploading in the UL direction.

The fifth generation, considers the DL and UL direction as separate networks and does not demand users to connect to the same BS for both of them [6]. Such a technique is called Downlink and Uplink Decoupling (DUD). These networks are called heterogeneous networks (HetNets). Such networks consist of prominent McBSs and various SBSs that are scattered among a McBS's vicinity [8]. They are necessary to ensure accessibility for all connected devices and guarantee a respectable Quality of Service (QoS) in real-time.

In HetNets, SBSs become more prevalent since more traffic is generated on the UL direction. Most network connected mobile devices are battery powered with identical transmit power, though with different demands. For example autonomous cars rely on sensors that produce data needed for autonomous driving. They are extremely dependent on accuracy to function properly and demand high QoS while other applications such as music streaming usually have less QoS demands. This increases dramatically the complexity of the UL direction.

The use of Machine Learning (ML) offers a significant opportunity to refine existent network applied methods and it can even work supplementary to them to improve real time performance. So far network application of ML techniques contains the work of [1], where the authors explore an application of ML techniques to explore unknown guest user dynamics for a network of users with different demands. Self-organizing maps applied to cellular networks have been proposed to dynamically shape the connectivity of the network to the prospective demands in [4]. The survey in [3], discusses multiple open issues on ML applications for computer networks and potential problems that arise with them.

In this paper we will propose two ML based mechanisms. Our first proposal is based on Decision Trees, that can be trained on data produced by already deployed networks, pinpoint patterns on optimal user allocation and produce promising results without the need to assess all these metrics that are necessary for traditional user allocation techniques. Our second proposal can function supplementary to our first proposal, and predict the optimal positioning of SBSs based on user distribution across the network. This mechanism utilizes the k-means clustering algorithm to produce cluster centers, which we consider as candidates for positioning SBSs. Our proposals can be utilized for Efficient user allocation in networks and can be proven extremely helpful in research that aims to provide an efficient scheme for enriching network infrastructures.

In remaining sectors, we present our full proposal. Sect. 2 focuses on the system model. In Sect. 3 we present our two ML mechanisms. Section 4 presents our simulation parameters while Sect. 5 presents the simulation results. Finally, in Sect. 6 we draw our conclusions and we make suggestions on how ML can be introduced in future research.

2   System Model

With the emerge of new technologies like the Internet of Things (IoT), we expect radical changes in computer networks, such as a massive rise in the number of UEs that try to utilize the network's resources. As Small Cells become smaller we expect their deployment to be massive in current and next generation networks. These two cases create the need for solutions on how to manage such an increase in the number of connected UEs and how to position and utilize new SBSs to optimally match a UE with the appropriate BS for its UL/DL needs.

Regarding the basic network layout, we consider a Heterogeneous Network. Our network consists of BSs (McBSs and SBSs) and users (UEs). McBSs are denoted as M (M=1,...,$|M|$), SBSs are denoted as S (S=1,...,$|S|$) and the set of UEs is denoted as U (U=1,....,$|U|$). Network users attempt to transmit and receive data. Traffic can be split into two networks for transmitting and receiving data (UL Network and DL network). In HetNets, they are considered as separate channels. All BSs have limited resources, meaning they have a bound on the maximum number of users they can serve simultaneously. All BSs of the same type are considered to have the same resources.

To compute the number of RBs that a user (suppose user j) demands from a specific BS, for achieving their desired DR, we will use the following equation:

$$RB_{j,i} = [\frac{T_j}{B_{RB} * log_2(1 + SINR_{j,i})}], \tag{1}$$

where T_j denotes the UE throughput demands and DR_i the desired Data Rate for the user j, B_{RB} is the bandwidth of a RB and $SINR_{j,i}$ is the SINR between a BS and an associated user.

Next we shall define the equations for calculating Pathloss (PL) and Signal to Interference and Noise Ratio (SINR). We use the distance dependent Path loss model, to calculate pathloss, and thus we have two different equations, one for McBSs and one for SBSs. PL is used to measure the signal loss we can expect from a signal as the user receives it, relatively to when it was emitted from the BS. The equations are given below, both for McBSs and SBSs:

$$PL_M = 128.1 + 37.6 * log_{10}d, \tag{2}$$
$$PL_S = 140.7 + 36.7 * log_{10}d, \tag{3}$$

where d is the distance with a user and its serving BS.

Next we shall define SINR for DL and UL. $SINR_{j,i}$ corresponds to the SINR between a user j and its corresponding BS i. This metric represents a ratio, namely the strength of the signal received by the receiving antenna and can be calculated as:

$$SINR_{i,j}^{UL} = \frac{P_{BS}}{N + I}, \tag{4}$$

$$SINR_{i,j}^{UL} = \frac{P_{ue}}{N + I},\qquad(5)$$

Here, P_{ue} corresponds to Transmit Power (TP) of the UE, while P_{BS} is the TP of the BS, whether it is a McBS or a SBS. Regarding noise and interferences, N corresponds to the Noise power while I corresponds to the total interferences [7]. We try to simulate a typical metropolitan area network scenario. Considering that most real life networks suffer from great Non Line Of Sight (NLOS) issues, we assume that our network features more SBSs compared to McBSs. These SBSs aim to alleviate congestion in areas with high user density.

For our simulation, the main focus of applying ML is associating users and BSs, minimizing the amount of calculations necessary. Our goal is to reduce time complexity and enable better real time decision making. Network load should be efficiently distributed among all BSs, retaining an acceptable QoS level for all users. For our simulations we utilize two ML techniques that are quite different, yet incorporating them into the network can yield promising results. The two techniques are:

2.1   Decision Trees

A decision tree is a decision tool that creates a tree like presentation to model decisions and their possible outcomes. The decision tree consists a rooted tree, since it has a node called "root" with zero incoming edges while all other nodes feature exactly one incoming edge. Nodes with no outgoing edges are called leaves (and in this case they can also be called decision nodes). In a decision tree, based on a designated function each internal node splits the instance space into two or more sub-spaces.

Each leaf may be assigned to one class or it may hold a vector indicating the probability of the target attribute assigned to a certain value. The classification is produced by navigating from the tree root all the way down to a leaf [5]. In other words, following this path we can see all decisions made by the Decision Tree to come up with the depicted result (depicted on the leaf).

2.2   K-Means Clustering

Clustering is the process of grouping a set of points into "clusters", where points with a small distance belong to the same cluster, where points with a large distance are placed in different clusters. K-means algorithm assumes a Euclidean space and a predifined number of clusters. These two characteristics produce two of the most prominent issues with k-means clustering [11]. It is not always easy to know in advance the number of clusters, while it also creates similarly shaped clusters (ball-shaped), since the clusters the based on Euclidean distance.

For the algorithm itself, there are several ways to define the starting cluster centers. They can be random or we can select points from the dataset we want to process. All the points that we want to classify need to be assigned to the nearest cluster, meaning that for each point we need to calculate its Euclidean distance to the cluster center. As new points are added to the clusters, the cluster centers are constantly re-evaluated. The process can stop when we have no more points to be clustered. As an optional step, in the end we can fix the cluster centers and we can re-examine all points. Euclidean distance can be calculated as:

$$D = ||X - Z|| = \sqrt{\sum_{n=1}^{n}(x_i - z_i)^2}, \tag{6}$$

where x_i and z_i, are the coordinates of the points (user and BS) in question.

3   The Mechanisms

In this sector we present our two proposed mechanisms. These mechanisms can stand alone or they can work collectively to provide a prediction mechanism that aims to refine user allocation on BSs and provide an adequate BS positioning scheme that takes into consideration user distribution across the network.

3.1   Mechanism for Predicting User Allocation on BSs

The first mechanism is based on Decision Trees. We expect our Decision tree-based model to able to predict the best matched BS for each user based on a specific metric. We begin by modeling a network where all users are distributed uniformally across the network and allocated to a certain BS using SINR as the preferred allocation metric. For this scenario, when the network model has decided the best matching BS for each user based on our metric, the allocation results are saved on a dataset. The dataset features the coordinates of each user in the network, as well as their matching BS both on the DL as well as the UL direction. This dataset is then utilized by our ML based model for training and testing.

The size of the dataset varies with the number of users that are deployed on the network. We test our mechanisms for various dataset sizes. The produced dataset is split into two different datasets. The first one being the training dataset and the second one is the test dataset. The training dataset is fitted into a ML model that is based on the Decision Trees technique. After the model has been trained, we use the test dataset to assess its capabilities and calculate its precision. We will evaluate the model precision for a variable number of users to see how the dataset size affects the model performance.

Algorithm 1. Pseudocode for the first mechanism

U: Denoting the number of users
for $i = 1$ to U **do**
    Calculate SINR for all BSs on DL and UL;
    Create BS preference list for DL and UL over SINR;
    Associate to best matched BS for DL and UL;
end for
Produce dataset with User coordinates and DL,UL associated BS
while Precision is low **do**
    Split produced dataset into training dataset and test dataset
    Fit training dataset into Decision trees model
    Use test dataset to predict associations and calculate model precision
end while

3.2   Mechanism for Efficient Placement of SBSs

The second proposed mechanism is based on K-means Clustering and aims to provide a reliable method to optimally position BSs across the network so that we can accommodate as many users as possible to avoid congestion escalation in the network. The network consists of a set of users, distributed uniformly across the network to ensure a realistic deployment scenario. These users can be classified, based on their positions where users that are closer together, will be classified in the same class.

To complete this task we use the k-means clustering algorithm. This algorithm starts with a set of empty clusters (we begin with 42 clusters - the same number as the number of BSs in our network) where users will be allocated. The algorithm then places users into clusters based on their distance from the cluster center. As a new user is allocated into a cluster, the coordinates of the cluster center are calculated again to account for the new user that is now part of the cluster. It is essential that all users must connect to some cluster, so that we can produce accurate coordinates for the cluster centers. When the procedure is completed, for each of these final cluster centers, we calculate their distance from all BSs.

We then identify the minimum distance for all clusters and any BS. Cluster centers that share a minimum distance with McBSs, are ignored while the coordinates of cluster centers that are placed near SBSs, are considered to be the optimal coordinates for placing the SBSs in our network. Finally we compare the network performance using random coordinates for SBS positioning vs the positioning scheme that we proposed. The coordinates of cluster centers that are closer to McBSs are deleted, since McBSs are stationary and cannot be moved and thus we cannot propose a cost-efficient repositioning scheme for them.

Algorithm 2. Pseudocode for the second Mechanism

U: Denoting the number of users
Start with XX empty clusters
for $i = 1$ to U **do**
    Associate user with a specific cluster
    Calculate new cluster center
end for
Calculate cluster centers distance with all BSs
Delete cluster centers closer to McBSs
for clusters centers that are closer to SBSs **do**
    Use calculated cluster centers as the new SBS coordinates
end for

4 Simulation Setup

In this sector we will present the parameters used to model our 5G network. All
network simulations and the proposed mechanisms were produced using Python.
We used Python because it incorporates functions for a plethora of ML techniques, making it a powerful tool for ML centric simulations. Our produced
network, can be seen on Fig. 1. This network consists of 13 McBS, 29 SBSs and
200 UEs. McBS are presented as big triangles placed at the center of the network hexagons, SBSs are presented as "Y" figures, while all users are presented
as bullets.

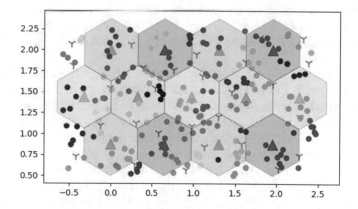

Fig. 1. The Python produced network, consisting of 13 MBSs, 29 SBs and 200 users.

In both of our simulations we have the same number of users, and they
are distributed along the network using the uniform distribution to ensure a
uniform distribution across the entirety of our network. This ensures a realistic
deployment with users wanting to exploit the entirety of our network. In the
second mechanism, we aim to produce user clusters. Using the Euclidean distance

as the distance metric for the k-means algorithm, all produced clusters seem to revolve around the cluster centers. Our mechanism is provided with the number of wanted clusters, that is equal to the number of BSs in the network, a feature that we can utilize if we want to expand the network with new BSs. All simulation parameters can be seen on Table 1.

Table 1. Simulation parameters

Parameter	Setting
Macro cell transmition power	50 dbm
Macro cell transmition power	24 dbm
User equipment transmition power	20 dbm
McBS Pathloss exponent	4
SBS Pathloss exponent	3.6
Network deployment	13 Mcells and 29 Scells
Number of users	200, 500 and 1000
Stationary user distribution	Uniform distribution

5   Results

In this section we will present the results produced from our two mechanisms. In both cases, we consider a basic network where users are static. For our first scenario, we used a dataset containing the association results from the network model based on the SINR metric, to train a ML model, based on the "Decision Trees" technique.

From the results presented on Fig. 2, we see that the model can be quite precise on the produced predictions. The dataset used is split on two parts, a training set, used to train the model and a test set, used to test its capabilities. Considering a fixed test dataset that is 20% of the size of the original dataset produced from the network model, we test our precision value. Starting from 200 users, we get a precision value of 0.625 for the UL direction and a value of 0.85 for the DL direction. When we increase the number of network users, our produced dataset size increases as well. With a dataset of 500 users, we see an improvement on the produced accuracy that now reaches 0.73 for the UL direction and 0.92 for the DL direction. Finally with a dataset consisting of the results on the allocation of 1000 users we see the accuracy rise to 0.84 for the UL direction and to 0.935 for the DL direction.

These results, show that using a ML technique can be proven extremely efficient in predicting user allocation on the network. Allocating users on BSs can be quite a complicated procedure, since it has to take into account multiple parameters for the association. As we can see from the produced results, the predictions for the DL direction are far superior to the predictions on the UL

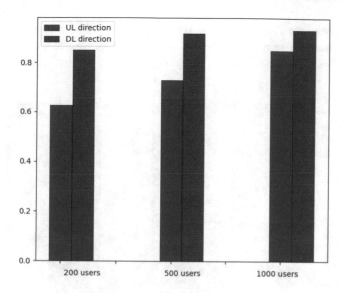

Fig. 2. Prediction accuracy.

direction. This is to be expected, since most users for the DL direction are allocated to McBSs. That means that in the produced dataset, most BSs are McBSs, making it easier for the model to predict the associated BS for each user, since it has to choose from a limited pool of BSs.

It is important though to mention that the size of our training dataset (and the test dataset) massively impacts performance. For complicated models like these, it is important that we fit them with a respectable size training dataset, so that the model can pinpoint all connections between the parameters and produce accurate results. The size of the training set should be appropriate, according to our needs, in order to avoid cases of overfitting and underfitting.

The allocation techniques that exist are quite extensive and very accurate on selecting the BS that best matches each user. This means that they can be used as a pretty good source for creating accurate datasets to use for ML models training. Especially in our case, the produced model can predict with high accuracy the best allocation for each user. In real world scenarios, this can lead to a massive improvement in real time performance for the network. Classic user allocation models take into account multiple metrics and network parameters and as a result they suffer from high complexity. Utilizing ML based models that are trained on datasets produced from classic allocation models, results in better real time decision making since a simpler model features much better time complexity, that can produce reliable results even in cases with limited network processing resources.

For our second scenario, the produced user clusters can be seen on Fig. 3, along with their respective cluster centers. The produced cluster centers are depicted as black bullets, while users that consist a cluster are depicted as bullets that share the same color. In our network we have defined 42 BSs and we also produce 42 user clusters. The position of the McBSs cannot be changed, while the positioning of the SBSs is not definite.

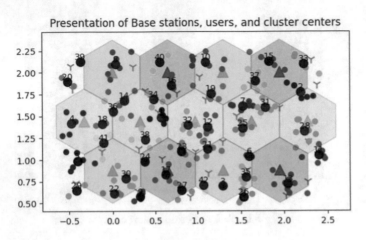

Fig. 3. Produced user clusters

Using the cluster centers as the new coordinates for the SBSs we ran our simulations. We can conclude that our proposal results in smaller distances between cluster centers and the closest available SBS. Smaller distances, result in less pathloss and signal deterioration. This means that our users will be able to enjoy higher DRs. Traditionally, UL direction demands are smaller than DL demands and are satisfied from SBSs. In our simulations, cluster centers that are close to McBSs, result in the same distance from McBSs in both results since McBSs are stationary.

Our proposal yields impressive results for the UL direction. As seen on Fig. 4, after improving the positioning scheme of SBSs, the number of associated users in the UL direction remains high and is usually better than the results with a random positioning of SBSs. This is subject to small changes, considering we can never be certain about user spawn points, and their RB needs. Since most of the associations in the DL direction are with McBSs (which remain stationary across both simulations), DL association results are the same on both cases. The difference varies with the UL direction.

All these users are matched with the BS that offers them the best SINR, meaning that they enjoy the best available QoS. Our results show that our suggestion can improve the network performance, but they ensure that our proposal remains a trustworthy method for pinpointing the best locations for placing new SBSs if we want to enrich our network's infrastructures with new SBSs. That will result in a higher offered QoS for the UL direction as well as the DL direction while the improved number of associated users results in improved total network throughput.

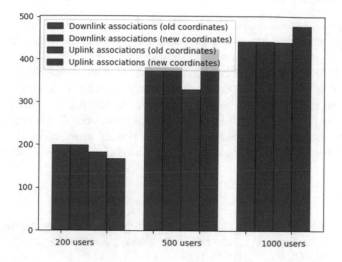

Fig. 4. Associated users with best matched BS

Considering that both pathloss and SINR are indeed heavily dependent on distance, users will see increased DRs that suffer less reduction (due to the lower pathloss). The smaller distance also means that the network resources will be better allocated. Users will enjoy better DRs, but will need less RBs from their matching BSs. That means that less RBs will be consumed by a single user, while more users will be satisfied using the same resources. As result, with our proposal we will see a massive improvement on the achieved channel throughput.

Our two proposals show that using ML we can begin to see important benefits on both user allocation as well as resource management. Our first proposal can allocate users on their best matching BSs, using minimal network resources, ensuring better real time decision making for the network. With our second proposal, we can predict the optimal positioning for SBSs. This supplements our first proposal, ensuring that more users are satisfied (especially in the UL direction) with higher DRs. This improves network performance as well as the perceived QoS, since now more users can utilize the network's resources.

6   Conclusion and Future Work

With the rise of IoT, we expect networks to face an unprecedented rise in the number of UEs that seek to use their resources. This is a crucial issue since it dictates a fruitful utilization of the existent infrastructure and the available resources. With the number of UEs expected, real time performance on networks can also be affected since massive calculations should be completed for optimal user allocation. With our two proposals, we proved that applying ML techniques can help with real time performance by predicting the optimal association between users and BSs, but it can also be used to predict the best positioning system for SBSs. This will significantly improve the existing network performance but can also complement the results produced from our first proposal.

Many ML techniques have been introduced to various scientific sectors and its introduction to networking is deemed necessary. This introduces an opportunity for massively improving existing mechanisms. In the future prediction mechanisms can be introduced to resource management (like frequency allocation), while clustering can also be used to improve the network scalability by finding an optimal system for positioning new BSs to reinforce current infrastructures.

References

1. Yang, J., Wang, C., Wang, X., Shen, C.: A Machine Learning Approach to User Association in Enterprise Small Cell Networks, pp. 850–854 (2018). https://doi.org/10.1109/ICCChina.2018.8641148
2. Mahbub, M.: A smart farming concept based on smart embedded electronics, internet of things and wireless sensor network. Internet of Things 9, 1–30 (2020). https://doi.org/10.1016/j.iot.2020.100161
3. Sun, Y., Peng, M., Zhou, Y., Huang, Y., Mao, S.: Application of machine learning in wireless networks: key techniques and open issues. IEEE Commun. Surv. Tutorials. 21(4), 3072–3108 (2019). https://doi.org/10.1109/COMST.2019.2924243
4. Mom, J.M., Ani, C.: Application of self-organizing map to intelligent analysis of cellular networks. ARPN J. Eng. Appl. Sci. 8, 407–412 (2013)
5. Rokach, L., Maimon, O.: Data mining with decision trees. Theory and Applications (2008). https://doi.org/10.1142/9789812771728-0001
6. Shi, M., Yang, K., Xing, C., Fan, R.: Decoupled heterogeneous networks with millimeter wave small cells. IEEE Trans. Wireless Commun. 17(9), 5871–5884 (2017). https://doi.org/10.1109/TWC.2018.2850897
7. Elshaer, H., Boccardi, F., Dohler, M., Irmer, R.: Downlink and uplink decoupling: a disruptive architectural design for 5G networks (2014). https://doi.org/10.1109/GLOCOM.2014.7037069
8. Feng, Z., Li, W., Chen, W.: Downlink and uplink splitting user association in two-tier heterogeneous cellular networks. In: 2014 IEEE Global Communications Conference (GLOBECOM 2014), pp. 4659–4664 (2015). https://doi.org/10.1109/GLOCOM.2014.7037543
9. Ghosh, A., Raha, A., Mukherjee, A.: Energy-efficient IoT-health monitoring system using approximate computing. Internet of Things. 9, 100166 (2020). https://doi.org/10.1016/j.iot.2020.100166
10. Sun, S., Adachi, K., Tan, P.H., Zhou, Y., Joung, J., Ho, C.K.: Heterogeneous network: an evolutionary path to 5G, pp. 174–178 (2015) https://doi.org/10.1109/APCC.2015.7412506
11. Cheung, Y.: K*-means: a new generalized k-means clustering algorithm. Pattern Recogn. Lett. 24, 2883–2893 (2003). https://doi.org/10.1016/S0167-8655(03)00146-6
12. Boostanimehr, H., Bhargava, V.: Unified and distributed QoS-driven cell association algorithms in heterogeneous networks. In: IEEE Transactions on Wireless Communications. vol. 14 (2014). https://doi.org/10.1109/TWC.2014.2371465

An Efficient Cross-Layer Design for Multi-hop Broadcast of Emergency Warning Messages in Vehicular Networks

Abir Rebei[1(✉)], Fouzi Boukhalfa[2(✉)], Haifa Touati[3(✉)], Mohamed Hadded[1(✉)], and Paul Muhlethaler[4(✉)]

[1] IResCoMath, Gabes, Tunisia
mohamed.elhadad@vedecom.fr
[2] VEDECOM, Versailles, France
fouzi.boukhalfa@vedecom.fr
[3] IResCoMath, Gabes, Tunisia
haifa.touati@cristal.rnu.tn
[4] Inria, Rennes, France
paul.muhlethaler@inria.fr

Abstract. The main objective of Vehicular ad hoc networks (VANETs) is to make road transportation systems more intelligent in order to anticipate and avoid dangerous, potentially life-threatening situations. Due to its promising safety applications, this type of network has attracted a lot of attention in the research community. The dissemination of warning messages, such as DENMs (Decentralized Environmental Notification Messages), requirse an efficient and robust routing protocol. In previous studies, the active signaling mechanism has shown its ability to prevent collisions between users trying to allocate the same resource. In this paper, we propose an original message forwarding strategy based on the active signaling mechanism. Our proposal disseminates warning messages from a source vehicle to the rest of the network while minimizing the access delay and the number of relay nodes. For this purpose, a special time slot is dedicated to forwarding emergency warning messages. To avoid access collisions on this slot, the active signaling scheme we propose favours the selection of the furthest node as the forwarder. We carry out a number of simulations and comparisons to evaluate the performances of the scheme.

Keywords: VANETs · Broadcast protocol · Cross-layer design · TDMA · Active signaling · Multi-hop communication · DENM · Network simulation

1 Introduction and Motivation

One of the most prominent technologies in the field of intelligent transportation systems (ITS) that has drawn significant interest from researchers and industry is Vehicular Ad-hoc Networks (VANETs). VANETS are essentially an extension of the unerlying principles of Mobile Ad-hoc Networks (MANETs), while featuring a number of alterations.

VANETs provide a wide range of applications covering both safety and non-safety applications. These applications aim at improving road safety traffic and assisting drivers. For example, they may warn of dangerous situations and accidents that may occur on the roads and provide internet access and infotainment services. To provide the transfer of information between a group of vehicles within range of each other or between vehicles and the infrastructure, VANETs allow for Vehicle-to-Vehicle (V2V) and Vehicle-to-Infrastructure (V2I) communications [2, 3].

Safety applications impose stringent requirements in terms of Quality of Service (QoS), given the need to achieve bounded delays and reliable communications. Guaranteeing high QoS is an major challenge in VANETs due to the high mobility of nodes, frequent changes in network topology and lack of a central control. In order to satisfy these requirements, therefore, it is necessary to have a QoS model provisioning. Moreover, different layers should cooperate to make correct routing decisions. An efficient Medium Access Control (MAC) is required to provide a broadcast service that respects the necessary requirements. Contention-free MAC protocols based on the Time Division Multiple Access (TDMA) technique provide considerable benefits, such as collision prevention [4]. Since central coordination is absent in a VANET topology, the propagation of a safety messages to the set of vehicles must be relayed through multiple relay vehicles. Furthermore, building a new multi-hop broadcast scheme of emergency messages seems difficult due to the very nature of VANETs. Many MAC-aware routing and broadcasting protocols have been designed in order to provide multi-hop communications and disseminate safety messages in a timely manner. In this paper, we present a cross-layer design for multi-hop broadcast of emergency warning messages called **AS-DTMAC Multihop**. Our approach mixes MAC and routing layers together for an efficient dissemination by selecting the best relay based on the **AS-DTMAC** [5, 6] protocol, which uses an active signaling technique to eliminate the problem of access collisions during the slot assignment process. In AS-DTMAC Multihop, we adjusted the active signaling process to let the furthest vehicle acquire the dedicated forwarding slot and thus to speed up the dissemination of warning messages.

The remainder of this paper is organized in 4 sections. The next section discusses relevant related work. In Sect. 3, we present our message broadcast cross-layer solution by presenting first the proposed MAC strategy and then the forwarding strategy. In Sect. 4, we evaluate the performance of our protocol by presenting simulations results. Finally, in Sect. 5, we conclude the paper and outline future work.

2   Related Work

Recently, several MAC-aware routing protocols have been proposed in the literature to efficiently support multi-hop communication and disseminate safety messages in vehicular networks in a timely manner. These proposals make use of different parameters from the MAC layer, like transmission time slot allocation, channel state, and collision probability, to improve data dissemination in VANETs. A detailed review of these cross-layer routing approaches for VANET is given in [25]. In this section, we focus on solutions dedicated to multi-hop emergency messages broadcast.

One of the earliest approaches to disseminating broadcast packets is called **OB-VAN** [22]. It is an opportunistic routing protocol that uses a modified 802.11 MAC layer. OB-VAN uses an acknowledgement scheme to choose relay nodes. Choosing the best relay node is performed by using an active signaling technique. Nodes that have captured the packet, transmit a short acknowledgement made up of signaling bursts, calculated based on the distance criterion just after receiving the packet. This scheme is a generalized CSMA/CA where the backoff technique is replaced by the active signaling technique. To prohibit interference on signaling bursts, OB-VAN uses the CDMA spreading code. Signaling bursts can be presented by 0 or 1. 0 denotes a listening interval and 1 denotes a transmission interval. This binary sequence is composed of two parts. The first part, is dedicated to optimizing the criterion for the best relay selection while the second is used to discriminate between nodes and permit the winner to relay the data packet.

Another TDMA-based routing protocol designed for warning message dissemination on bi-directional highways is proposed in [23]. This approach, called **Priority-Based Direction-Aware Media Access Control (PDMAC)**, classifies nodes as either cluster heads (CHs) or ordinary vehicles (OVs).

To disseminate warning messages, PDMAC develops a three-tier priority assignment process. The first tier is Direction-Based Relay Selection. A source disseminates to its neighbors a request message (REQ) that indicates its direction, destination, etc. and reserves all available time slots in this frame for itself. Neighbors respond with an acknowledgment message (ACK) that contains all free time slots and the slot to be assigned for the transmission of the message according to the severity level of the message. The node selected as the best relay is the one that is closest in distance to the destination and is moving in the direction towards it. The second tier is the Priority on the Basis of Message Type. PDMAC prioritizes warning messages over non warning messages by adding a bit in the message header to indicate the type of the message. Finally, the third tier is Priority on the Basis of Severity Levels to differentiate between different warning messages depending on their severity levels by computing the collision probability. In this case, warning messages are classified into 3 levels. In the case of a lowest priority message, the sender should wait for a free time slot to send. If it is a second level priority, it requests the release of a slot of another non-warning or warning message with lower priority. Otherwise, in the case of a highest priority level message, it is mandatory to release on the time slot of a non-warning or a lower-priority message.

A recent protocol called **Multi-Channel Token Ring Protocol (MCTRP)** is presented in [24]. MCTRP employs the multi-channel structure defined in IEEE 802.11p. The network is composed of multiple virtual rings. Nodes are classified into 5 types: Ring Founder Node (RFN), Token Holder Node (THN), Ring Member Node (RMN), Dissociative Node (DN), and Semi-Dissociative Node (SDN). There are 2 types of radio: Radio-I and Radio-II. A DN uses only Radio-I since it does not belong to any ring, but the other nodes use both of them. Also, the time system is partitioned into a control period and a data period.

The MCTRP protocol follows 3 sub-protocols. The first sub-protocol is the Ring Coordination Protocol, which manages rings and nodes and schedules Service CHannels (SCH) for each ring. First, the Ring Initialization Process consists of sending a Ring Founding Message (RFM) that includes a selected SCH number for intra-ring data communications and waiting for an invitation. After establishing a ring, a Joining Invitation Message (JIM), which includes some information such as the SCH number, the speed, etc., will be broadcasted by the RFN to the DNs. The DN will reply to the RFN with a Joining Acknowledgement Message (JAM) if the difference between its moving speed and that of the RFN is smaller than a predefined speed threshold. Other messages will be exchanged between RFN, DN and RMN such as Connection Notification Messages (CNMs), Connecting Successor Messages (CSMs), etc. using the contention-based CSMA/CA scheme. The second sub-protocol is the Emergency message exchange protocol. To efficiently deliver emergency messages, MCTRP uses Radio-I or Radio-II, depending on the case. This can be done through 4 steps. Firstly, when an RMN detects an accident, it sends an emergency message to its RFN by adopting CSMA/CA and using Radio-II. Secondly, the RFN node replies with an acknowledgement to the RMN, and then broadcasts the emergency message to all its RMNs using Radio-II. Thirdly, it also broadcasts the message to its neighboring DNs, SDNs, RFNs using Radio-I. Finally, neighboring RFNs rebroadcast the emergency message again to their RMNs using Radio-II.

The third sub-protocol is the Data Exchange Protocol. Two types of data communications exist: inter-ring data communications where packets are transmitted using CSMA/CA and intra-ring data communications where data packets are transmitted using a token based mechanism. After receiving a token, a node can transmit data during a token holding time and then pass the token to its successor.

3   Cross-Layer Solution for Emergency Messages Multi-hop Dissemination

Cross-Layer design is an emerging proposal to support flexible layer approaches in VANETs [25]. As described above, the recent ongoing research has shown increased interest in protocols that rely on interactions between different layers. In this paper, we propose to exploit the relation between MAC and network layers, in an effort to improve the performance of the Multi-hop Broadcast of DENM in Vehicular Networks. We propose the idea of combining an approach based on a MAC layer protocol named AS-DTMAC with a forwarding strategy. Figure 1 represents the general architecture of our message broadcast cross-layer solution. In this section, we first focus on the MAC protocol principle and then we describe the forwarding strategy adopted.

3.1   Mac Strategy

Time Division Multiple Access (TDMA) is a contention-free MAC[1] protocol with scheduled channel. It is mostly used for safety applications in order to satisfy real-time constraints. Nevertheless, while using TDMA with distributed schemes, the access collision[2] problem can occur (see [2] for more details). Very many protocols are susceptible to this problem, including: ADHOC MAC [15], VeMAC [18], DTMAC [4],

---

[1] According to the control scheme used to access the channel, MAC random access protocols are categorized into: contention-based or contention-free [2].

[2] The access collision problem is can occur when using distributed schemes. It happens when two or more vehicles within the same two-hop neighborhood attempt to access the same available time slot [2].

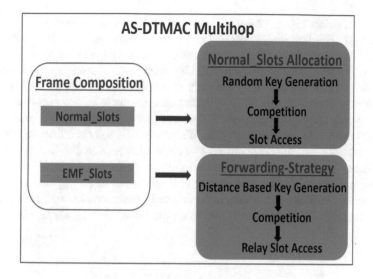

Fig. 1. AS-DTMAC Multihop Architecture

R-ALOHA [17]. To solve this problem, we have designed in [5, 8–10], an efficient mechanism named active signaling and we apply it in the DTMAC's random slot selection process.

As shown in Fig. 2, frame composition in AS-DTMAC Multihop is slightly different from that of AS-DTMAC [5]. In AS-DTMAC Multihop, each frame is composed of 100 slots and it is divided into three sets denoted as $S0$, $S1$ and $S2$, corresponding to three successive zones. AS-DTMAC Multihop defines two types of slots: *Normal_Slot* and *Emergency_Message_Forwarding_Slot* or *EMF_Slot*. Normal_slots are used by each vehicle to send data, as in AS-DTMAC. However, *EMF_Slots* are special slots dedicated, by AS-DTMAC Multihop, to forwarding emergency messages. In each frame, three slots, namely the first slot of each set S_i (i.e. $Slot_0$, $Slot_{34}$ and $Slot_{67}$), are defined as *EMF_Slots*. This choice can be explained by the fact that emergency messages are time-sensitive, hence choosing the first slot in each set to forward them will speed up warning message dissemination.

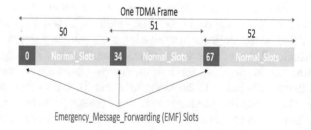

Fig. 2. AS-DTMAC Multihop Frame

As illustrated in Fig. 3, whether normal or EMF, a slot is usually formed by two time intervals. The first one, is dedicated to the selection process. However, the second is held by the winner of the competition to transmit its payload packet. In the first interval, a random binary key is generated by every node to compete for a slot. Hopefully, at the end of the time interval, only one node will remain as the winner. The key is a succession of bits (0,1). '1' means that the vehicle with a packet to send will transmits during the signaling bursts. '0' means that the vehicle with a packet to send senses the channel during this mini-slot. When a vehicle selects a listening period and senses a transmission, the competition to get the slot is over. For instance, a vehicle that draws the key '01001110' will listen during the first mini-slot and if no competing transmission is sensed during this mini-slot, it will transmit during the next mini-slot. The following two steps of the selection process in our example, will be two listening periods. The selection process continues using the same rule until the key is completely used up.

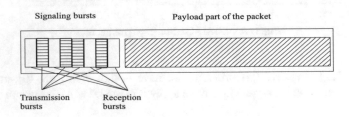

Fig. 3. Structure of the AS-DTMAC slot

3.2   Forwarding Strategy

Given the high sensitivity of emergency messages to delay constraints and collisions, it is crucial to propose an efficient multi-hop dissemination scheme that avoids these problems. Since the active signaling technique solves the collision problem, our forwarding strategy is based on an adapted version of the Active Signaling-DTMAC protocol.

As shown in Fig. 4, when a source node, S_N, detects a warning event it has to broadcast an emergency message. Vehicle S_N sends the message during its reserved *Normal_Slot*. All nodes that receive the message and that are located behind the sender: situated in the sender's zone or in the adjacent one (in the sender's transmission range), will compete to forward the received emergency message to other vehicles during the *EMF_Slot*. To avoid access collision on this slot, each vehicle generates a binary key based on the distance that separates it from the sender. The key is composed of mini-slots. As explained in the previous section, these mini-slots take the value '1' or '0'. '1' means that the node is in a transmission mode and '0' means that the node is in a listening mode. The forwarding strategy consists of selecting the best next relay from the list of vehicles that have generated keys. The winner node will forward the message to the rest of the nodes situated in the opposite direction of the sender vehicle in order to propagate the message as far as possible.

In practice, the winner is always the furthest vehicle from the transmitter. This is counted as a benefit in terms of packet propagation since the emergency message will be quickly broadcasted and the danger will be avoided.

The transmitter sends a message in its own slot reserved in the set dedicated for the zone to which it belongs. So, the forwarding will take place in the first slot of the next set.

To give a clearer idea of our forwarding strategy, we consider the example shown in Fig. 5. In this example, the sender, which is the red vehicle, sends an emergency message during its slot (slot 8). In this case, the competition and the first relay of the message will take place in the slot 34. The black vehicle (which is the furthest vehicle) is the winner of the competition and it will forward the message to the other nodes. The same process will be repeated in the next hop until all vehicles have been informed of the potentially dangerous situation.

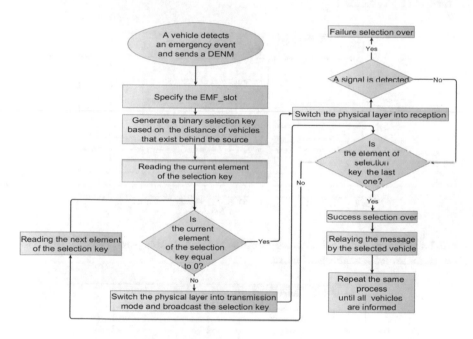

Fig. 4. Forwarding Strategy Flowchart.

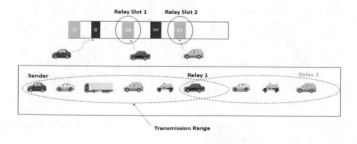

Fig. 5. Forwarding Strategy

4 Performance Evaluation

- **Latency:** this defines the time between the first broadcast of the message and the time of its reception by the last vehicle on the road (see Figs. 6 and 7).
- **Packet Loss:** as shown below in the equation, the packet loss percentage defines the number of lost packets that are not received, divided by the total number of packets that should be received. The formula to compute the packet loss is defined as follows:

$$Packet\ Loss(\%) = \frac{Losses\ Packets}{Total\ Number\ of\ Packets}$$

- **Number of Forwarders:** this metric defines the number of forwarders (relays) needed to relay a message.
- **Used Bandwidth:** the used bandwidth metric represents the total number of packets received by vehicles.

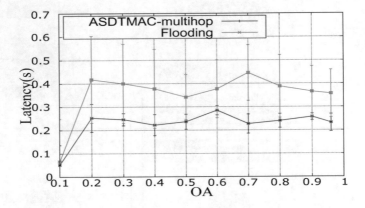

Fig. 6. Latency versus channel occupancy for AS-DTMAC Multi-hop and Flooding with $SSD =$ 20 and error bar (95% confidence interval).

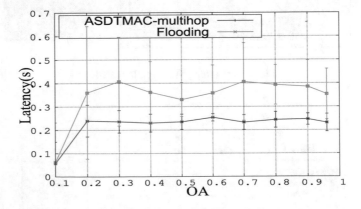

Fig. 7. Latency versus channel occupancy for AS-DTMAC Multi-hop and Flooding with $SSD =$ 30 and error bar (95% confidence interval).

Now, We move on to the **Packet Loss** metric. Figure 8 presents the packet loss versus OA in percent. The error bars are for a 95% confidence interval. It is clear that AS-DTMAC Multihop has 0% of packet losses, whereas in Flooding we find a considerable packet loss rate that can reach more than 80% in high traffic level density conditions. As we have explained above, in the Flooding mechanism, every vehicle that has received the message will attempt to forward it and this will cause a high interference in the Flooding scheme. As a result, many packets will be lost.

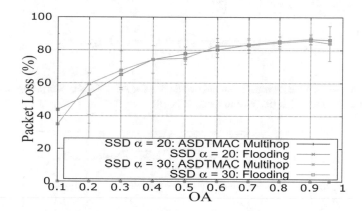

Fig. 8. Packet Loss versus channel occupancy for AS-DTMAC Multi-hop and Flooding with error bar (95% confidence interval).

We now evaluate **the Number of Forwarders** metric. In Figs. 9 and 10, we show the number of forwarders needed for each approach (AS-DTMAC Multihop and flooding) to warn of an emergency event versus OA. Figure 9 represents results with a *SSD* equal to 20 (km/h) while Fig. 10 with a *SSD* equal to 30 (km/h). As we can see, AS-DTMAC Multihop requires fewer resources, this will reduce the channel occupancy. In contrast, the flooding technique uses more resources because all the vehicles that received the packet will attempt to forward it.

We move on to **the Used Bandwidth** metric results. In Figs. 11 and 12, we plot the used bandwidth versus OA for both AS-DTMAC Multihop and flooding. This metric, represents the total number of received packets. In flooding, we notice that it provides a high value compared to AS-DTMAC Multihop. In our approach, at every hop, only one winner vehicle will relay the packet to its neighbors, whereas in flooding every receiver will relay the packet and thus vehicles could receive the packet several times.

Finally, we compare the dissemination delay achieved by AS-DTMAC Multihop and Flooding to the estimated delay. We begin by deriving an analytic expression of the estimated delay. As illustrated in Fig. 13, to deliver an emergency message from V_1 to V_4, one frame is sufficient. In fact, the message is relayed 3 times $(n_0 + n_1 + n_2 = \tau slots = one frame)$. Based on this information, we derived an analytical formula to compute **the estimated delay** needed to deliver a message from a source i to a vehicle j separated by such a distance. As defined in [7], the ED is defined in the equation as following:

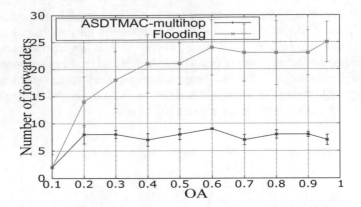

Fig. 9. Number of forwarders versus channel occupancy for AS-DTMAC Multi-hop and Flooding with $SSD = 20$ and error bar (95% confidence interval).

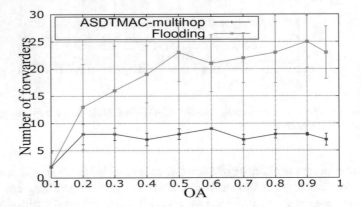

Fig. 10. Number of forwarders versus channel occupancy for AS-DTMAC Multi-hop and Flooding with $SSD = 30$ and error bar (95% confidence interval).

$$ED = \frac{Dist_{i,j}}{3*R} * \tau * s_d \tag{1}$$

where i, j, R, τ, $Dist_{i,j}$ and s_d are respectively the sender, the receiver, the transmission range, the length of the frame, the distance between the sender and the receiver and the duration of the slot.

Figure 14 presents the estimated delay to propagate a packet from a source to a receiver versus the distance between them.

We can observe from this figure that AS-DTMAC Multihop can provide a shorter delay than the estimated one. This can be explained by the fact that the forwarder will send in one of the reserved slots of forwarding ($Slot_0$ or $Slot_{34}$ or $Slot_{67}$). The delay will depend on the vehicle's position (following the AS-DMAC Multihop scheme: vehicles

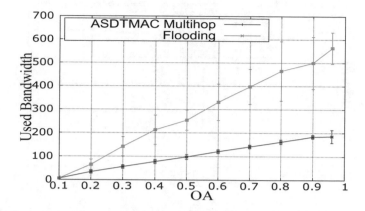

Fig. 11. Used Bandwidth versus channel occupancy for AS-DTMAC Multi-hop and Flooding with $SSD = 20$ and error bar (95% confidence interval).

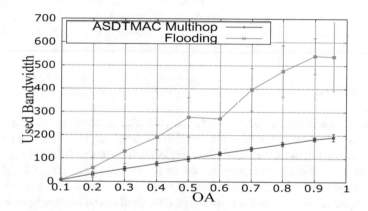

Fig. 12. Used Bandwidth versus channel occupancy for AS-DTMAC Multi-hop and Flooding with $SSD = 30$ and error bar (95% confidence interval).

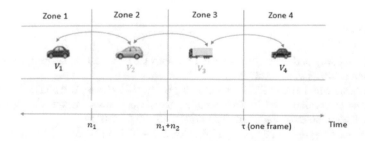

Fig. 13. Message propagation based on TDMA slot information.

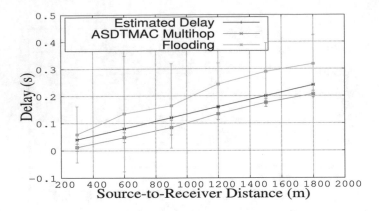

Fig. 14. Estimated Delay versus distance for AS-DTMAC Multi-hop and Flooding with *SSD* = 30 and error bar (95% confidence interval).

can access only the set of slot of their area): they can reserve at the beginning, at the middle or at the end. However, due to the interference in the flooding approach, the forwarding operation may not occur in the same frame. If this happens, a vehicle will relay on the next frame. This hypothesis can explain the results obtained.

5   Conclusion

In this paper, we have proposed to enhance DTMAC protocol by integrating active signaling. The simulation results show that AS-DTMAC drastically reduces the access collision rate and allocates slots to all the vehicles in the network in half the time it takes DTMAC to do so. We also presented a use case in the V2V for urgent and high priority traffic message like DENM, that can help to avoid an accident, all these new features are very important for the future technology described in the beginning of this paper. As future work, we will do additional simulations to compare with the standard used in V2V (IEEE 802.11p) and we plan to develop an analytical model for AS-DTMAC as well as to investigate further advanced access features that could be provided using the active signaling scheme.

References

1. Hu., F.: Security and Privacy in Internet of Things (IoTs): Models, Algorithms, and Implementations. ISBN 978-1-4987-2318-3
2. Hadded, M., Muhlethaler, P., Laouiti, A., Zagrouba, R., Saidane, L.A.: TDMA-based MAC protocols for vehicular ad hoc networks a survey, qualitative analysis and open research issues. IEEE Commun. Surv. Tutorials **17**(4), 2461–2492 (2015)
3. Laouiti, A., Qayyum, A., Saad, M.-N.-M.: Vehicular Ad-Hoc networks for smart cities. In: Proceedings of the Second International Workshop of Advances in Intelligent Systems and Computing, vol. 548, Springer (2017)

4. Hadded, M., Laouiti, A., Muhlethaler, P., Saidane, L.A.: An infrastructure-free slot assign-ment algorithm for reliable broadcast of periodic messages in vehicular Ad hoc networks. In: Vehicular Technology Conference VTC-Fall. Montreal, Canada (2016)

5. Boukhalfa, F., Hadded, M., Muhlethaler, P., Shagdar, O.: An active signaling mechanism to reduce access collisions in a distributed TDMA based MAC protocol for vehicular networks. In: Advanced Information Networking and Applications (AINA-2019), Kunibiki Messe, Matsue, Japan (2019)

6. Boukhalfa, F., Hadded, M., Muhlethaler, P., Shagdar, O.: An analytical model for perfor-mance analysis of an active signaling-based MAC protocol for vehicular networks. In: Vehic-ular Technology Conference VTC-FALL. Honolulu, Hawaii, USA (2019)

7. Hadded, M., Muhlethaler, P., Laouiti, A.: Performance evaluation of a TDMA-based multi-hop communication scheme for reliable delivery of warning messages in vehicular networks. In: Proceedings of Wireless Communication and Mobile Computing Conference (IWCMC), pp. 1029–1034 (2017)

8. Boukhalfa, F., Adjih, C., Muhlethaler, P., Hadded, M., Shagdar, O.: Physical and MAC layer design for active signaling schemes in vehicular networks. In: WiMob, pp. 86–92. Thessa-loniki, Greece (2020)

9. Boukhalfa, F., Hadded, M., Muhlethaler, P., Shagdar, O.: Coexistence of IEEE 802.11p and the TDMA-based AS-DTMAC Protocol. In: SoftCOM, pp. 1–6. Split, Croatia (2020)

10. Boukhalfa, F., Hadded, M., Muhlethaler, P., Shagdar, O.: Performance evaluation of an active signaling based time-slot scheduling scheme for connected vehicles. In: Annals of Telecom-munications, pp. 1–12 (2020)

11. Abate, J., Whitt, W.: Numerical inversion of Laplace transforms of probability distributions. ORSA J. Comput. **7**(1), 38–43 (1995)

12. Ye, F., Yim, R., Zhang, J., Roy, S.: Congestion control to achieve optimal broadcast effi-ciency in vanets. In: IEEE International Conference on Communications (ICC), pp. 1–5. South Africa, May, Cape Town (2010)

13. 802.11p-2010, IEEE standard for information technology - Telecommunications and infor-mation exchange between systems - local and metropolitan area networks - specific require-ments part 11 : Wireless LAN medium access control (MAC) and physical layer (PHY) and physical layer (PHY) specifications amendment 6 : Wireless access in vehicular environ-ments std (2010)

14. Uzcategui, R., Acosta-Marum, G.: Wave: a tutorial. IEEE Commun. Mag. **47**(5), 126–133 (2009)

15. Borgonovo, F., Capone, A., Cesana, M., Fratta, L.: Adhoc MAC: new MAC architecture for ad hoc networks providing efficient and reliable point-to-point and broadcast services. Wireless Netw. **10**(4), 359–366 (2004)

16. Naik, G., Choudhury, B., Park, J.M.: IEEE 802.11 BD & 5G NR V2X: evolution of radio access technologies for V2X communications. IEEE Access J. **7**, 70169–70184 (2019)

17. Borgonovo, F., Capone, A., Cesana, M., Fratta, L.: RR-ALOHA, a reliable R-ALOHA broadcast channel for Ad-Hoc intervehicle communication networks. In: IEEE IFIP Annual Mediterranean Ad Hoc Networking Workshop (Med-Hoc-Net). Baia Chia, Italy (2002)

18. Zhuang, W., Omar, H.A., Lio, L.: VeMAC: A novel multichannel MAC protocol for vehicu-lar ad hoc networks. In: IEEE Conference on Computer Communications Workshops (INFO-COM WKSHPS), pp. 413–418. China, Aug, Shanghai (2011)

19. Weidong, Y., Pan, L., Yan, L., Hongsong, Z.: Adaptive TDMA slot assignment protocol for vehicular ad-hoc networks. J. China Univ. Posts Telecommun. **20**(1), 11–18 (2013)

20. Ke, W., Weidong, Y., Pan, L., Hongsong, Z.: A decentralized adaptive TDMA schedul-ing strategy for VANETs. In: IEEE Wireless Communications and Networking Conference Workshops (WCNCW), pp. 216–221. China, Apr, Shanghai (2013)

21. Karnadi, F., Mo, Z., Lan, K.C.: Rapid generation of realistic mobility models for VANET. In: IEEE WCNC, pp. 2506–2511. Hong Kong, China, Mar (2007)
22. Blaszczyszyn, B., Laouiti, A., Muhlethaler, P., Toor, Y.: Opportunistic broadcast in VANETs (OB-VAN) using active signaling for relays selection. In: IEEE International Conference on ITS Telecommunications (ITST 2008), pp. 384–389, Oct (2008)
23. Abbas, G., Abbas, Z.H., Haider, S., Baker, T., Boudjit, S., Muhammad, F.: PDMAC: A priority-based enhanced TDMA protocol for warning message dissemination in VANETs. Sensors **20**, 45 (2020)
24. Bi, Y., Liu, K.H., Cai, L.X., Shen, X.: A multi-channel token ring protocol for QoS provisioning in inter-vehicle communications. IEEE Trans. Wireless Commun. **8**(11), 5621–5631 (2009)
25. Rebei, A., Hadded, M., Touati, H., Boukhalfa, F., Muhlethaler, P.: MAC-aware routing protocols for vehicular ad hoc networks: a survey. In: 2020 International Conference on Software, Telecommunications and Computer Networks (SoftCOM), pp. 1–6. Split, Hvar, Croatia (2020). https://doi.org/10.23919/SoftCOM50211.2020.9238249

Time Series Forecasting for the Number of Firefighters Interventions

Roxane Elias Mallouhy[1]([✉]), Christophe Guyeux[2], Chady Abou Jaoude[3], and Abdallah Makhoul[2]

[1] Prince Mohammad Bin Fahd University, Khobar, Kingdom of Saudi Arabia
reliasmallouhy@pmu.edu.sa
[2] University of Bourgogne Franche-Comté, Belfort, France
{christophe.guyeux,abdallah.makhoul}@univ-fcomte.fr
[3] Antonine University, Baabda, Lebanon
chady.aboujaoude@ua.edu.lb

Abstract. Time series forecasting is one of the most attractive analysis of dataset that involves a time component to extract meaningful results in economy, biology, meteorology, civil protection services, retail, etc. This paper aims to study three different time series forecasting algorithms and compare them to other models applied in previous researchers' work as well as an application of Prophet tool launched by Facebook. This work relies on an hourly real dataset of firefighters' interventions registered from 2006 till 2017 in the region of Doubs-France by the fire and rescue department. Each algorithm is explained with best fit parameters, statistical features are calculated and then compared between applied models on the same dataset.

1 Introduction

Many studies show that reaching good forecasts is vital in many activities such as industries, commerce, economy, and science. The fact of gathering a collection of observations over time will provide predictions of new observations in the future and extract meaningful characteristics of the data and statistics in different time intervals: hours, days, weeks, months, and years. The usage of data science, machine learning, and time series forecasting are feasible in the prediction of firefighters' interventions since it is logical to assume that firefighters' interventions are affected somehow by temporal, climatic, and other events such as new year's eve, snowing weather, traffic peak time, fires in summer, holidays, etc. Due to the French economic crisis (closure of small hospitals, population growth, etc.), the impact of optimizing the number of human interventions leads directly to a reduction and better control in the financial, human, and material resources. The goal is to size the number of firefighters according to the need and demand: a greater number of firefighters should be available when they are mostly used. Indeed, the number of guards available should be related to the location, number, and type of the intervention. For example, during the weekend when accidents indeed increase, the number of firefighters ready to serve the society should be greater than a regular working weekday where most of the people reside in their offices.

L. Barolli et al. (Eds.): AINA 2021, LNNS 225, pp. 39–50, 2021.
https://doi.org/10.1007/978-3-030-75100-5_4

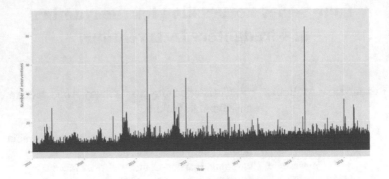

Fig. 1. Number of firefighter interventions from 2006 until 2017.

A trusted result and high prediction are affected by many factors: the algorithm used to train and test the dataset plays a big role as well as the chosen parameters. In this work, three different well-known time series algorithms have been implemented: Auto Regression (AR), Moving Average (MA), and Autoregressive Integrated Moving Average (ARIMA). Each algorithm is explained and detailed in Sect. 3. The dataset used carries information about firefighters' interventions in the region of Doubs-France from 01/01/2006 00:00:00" until 31/12/2017 23:00:00" . All the data were registered by the fire and rescue department SDIS 25 by blocks of one hour [23]. An overview of the number of firefighters interventions through the years is shown in Fig. 1.

This paper concentrates on statistical parameters calculated of three different Machine Learning algorithms to predict the number of firefighters' interventions, prophet, and comparison between the applied algorithms and related work using the same dataset.The remaining of this paper is structured as follows: literature review providing an overview about related work done by researchers on the same topic; Machine Learning algorithms for predicting firefighters' interventions explaining the three ML models used; parameters chosen showing the values used for the corresponding algorithms; building data with Prophet by applying this Facebook tool on the firefighters dataset; obtained results; results interpretation, and conclusion.

2   Literature Review

Many influential works on time series forecasting have been published in the latest years with enormous progress in this field. The idea of forecasting starts with Gardner and Snyder who boost two subsequent papers in the same year (1985) in the area of time series forecasting about exponential smoothing methods. Gardner provided a review of all existing work done to that date and extended his research to include a damped trend [2] After Gardner's work, Snyder demonstrated that simple exponential smoothing could arise from state space model innovation [3]. Most of the researches since 1985 has focused on empirical properties [10], forecasts evaluations [9], and proposition of new methods for initialization and estimation [8]. Many studies have stimulated the use of exponential smoothing methods in various areas such as air passengers [16], computer components [5], and production planning [12]. Later on, numerous variations on original methods have been proposed to deal with continuities, constraints, and renormalization at each period time [13,15]. Multivariate simple exponential smoothing was used for processing the control charts by introducing a moving average technique [21]. Moreover, Taylor [7] and Hyndman *et al.* [20] have extended basic methods and have included all 15 different exponential methods. Moreover, they have proposed models that correspond to multiplicative error cases and additive errors. However, these methods were not unique since it has been known that ARIMA models give equivalent results in forecasting, but the innovation in their work was that statistical models can lead to non-linear exponential smoothing methods.

Early studies of time series forecasting in the nineteenth century were globally based on the idea that every single time series can be seen as a realization of a stochastic process. Based on this simple proposal, many time series methods since then have been developed. Workers such as Walker *et al.* [22] formulated the concept of moving average MA model and autoregressive AR models. The conception of linear forecasting happened by the decomposition theorem. After that, many studies have appeared dealing with parameter identification, forecasting estimation, and model checking. Time series forecasting in this paper is specifically studied by running a real dataset of firefighters' interventions. This area of research in machine learning is new some-how where only a few articles are targeting this topic. A work achieved by C. Guyeux *et al.* [24] started by collecting a list of interventions and preparing the set for learning, validating, and testing. Then, by using supercomputers, the learning was carried out on an ad hoc Multi-Layer Perception. The study ends up by applying the neural architecture on a real case study with mature and encouraging results. Another work done by Couchot *et al.* [23] has shown that a machine learning tool can provide accurate results by deploying a learning process based on real and anonymized data using extreme gradient boosting to guess an accurate behavior. Ñahuis *et al.* [25,27], using the same firefighter's dataset demonstrated that machine learning is mature enough to make feasible predictions for critical events such as natural disasters. They used LSTM and XGBoost approaches to predict the number of firefighter's interventions. These investigations have been deepened following a feature-based machine learning approach in various directions [24,26,28], but never by considering the time series alone.

3 Machine Learning Algorithms for Predicting Firefighters' Interventions

Introductory books on time-series algorithms and analysis include Davis and Brockwell [19], Diggle [4], Swift and Janacek [11], Wei *et al.* [6].

3.1 Auto-Regressive Model

Auto regression is a statistical time series model that predicts an output for the near-future (number of houses sold, price of something, number of interventions, ...) based on past values. It was originated in 1920 by Udny Yule, Eugen Slutsky, and Ragnar Frisch [14]. For instance, to predict today's value based on yesterday, last week, last month, or last year one. AR models are also called Markov models, conditional models, or transition ones. Regression uses external factors which are independent as an explanatory variable for the dependent values. Autoregression model is conditioned by the product of certain lagged variables and coefficients allowing inference to be made. In reality, AR works hardly if the future predictors are unknown because it requires a set of predictor variables. On the other hand, AR is capable of adjusting the regression coefficient β and violating the assumption of uncorrelated error since the independent observations are time lagged values for the dependent observations.

In an AR model, the value of the predicted outcome variable (Y) at some time t is $Y_t = \beta_0 + \beta_1 Y_{t-1} + \varepsilon_t$ where the parameters $\beta_0 + \beta_1 Y_{t-1}$ rely on the past and ε_t which is the white noise could not be predictable from the past. It is important to mention that knowing the previous lagged values of Y_{t-2}, Y_{t-3} does not affect the prediction of Y_t because as shown in the formula, Y_t is affected only by Y_{t-1}.

3.2   Moving Average Model

Moving Average MA is a model introduced in 1921 by Hooker who considers multiple period averages to predict future output and event [1]. It is an effective and naive technique in time series forecasting used for data prediction, data preparation or feature engineering. It uses the most recent historical data values to generate a forecast. MA removed the fine-grained variation between time steps, to expose the signal. This method uses the average data periods' number. The term "moving" indicates the up and down moves of the time series done to calculate the average of a fixed number of observations. On the other hand, the process of averaging relies on the overlapping observations that create averages. Moving Average method can be used for both linear and non-linear trends. However, it is not applicable for short time series forecasting fluctuation because the trend obtained by applying the model is neither a standard curve nor a straight line. Besides, trend values are not available for some intervals at the start and the end values of time series.

The outcome value in the MA(q) model, a moving average model of order q, is presented as the following:
$y_t = c + \varepsilon_t + \theta_1 \varepsilon_{t-1} + \theta_2 \varepsilon_{t-2} + ... + \theta_q \varepsilon_{t-q}$ where ε_t is the white noise. This technique involves creating a new time series with compromised values of row observations and average in the original data set time series. Moreover, it relies on past forecast errors.

3.3   AutoRegressive Integrated Moving Average

ARIMA, also called Box-Jenkins, is a model proposed by George Box and Gwilym Jenkins in 1970 by using a mathematical approach to describe changes in the time series forecasting [18]. ARIMA is an integration of auto regression and moving average methods that use a dependent relationship between an observation and some number of lagged observations by differencing between raw observations. It subtracts an observation from the previous time step and takes into consideration the residual error. ARIMA is a powerful model as it takes into consideration history as an explanatory variable,

but in such model, the data cost is usually high due to the large observations requirement needed to build it properly. A standard notation for ARIMA being used is ARIMA
(p,d,q) where:

- p: is the auto-regressive part of the model, which means the number of lag observations that are included in the model. It helps to incorporate the effect of past values
  of the model. In other terms, it is logical to state that it is likely to need 5 firefighters
  tomorrow if the number of interventions was 5 for the past 4 d. A stationary series
  with autocorrelation can be corrected by adding enough AutoRegression terms.
- d: is the integrated part of the model. It shows the degree of differencing the number
  of times that the raw observations have been differenced. This is similar to state that
  if in the last 4 d the difference in the number of interventions has been very small, it
  is likely to be the same tomorrow. The order of differencing required is the minimum
  order needed to get a near-stationary series.
- q: order of moving average which is the size of the MA window. Autocorrelation
  graph shows the error of the lagged forecast. The ACF shows the number of MA
  terms required to remove autocorrelation in the stationaries series.

4   Parameters Chosen

Auto Regression algorithm does not need any parameters to be chosen or modified
as explained in the formula in the previous section. Nevertheless, different parameters
were tested and registered for Moving Average and ARIMA;

4.1   Window Size for Moving Average

After trying different values of window sizes for different hours and days, the best size
chosen is 3 h having the minimal values of MAE and RMSE. On the other hand, to select
the values of ARIMA, the following parameters should be taken into consideration:

1. p: the order of AR term was basically taken to be equal to the number of lags that
   crosses the significance limit in the Autocorrelation Fig. 3a. It is observed that the
   ACF lag 4 is quite significant. Then, p was fixed to 4.
2. d: let us use the Augmented Dickey Fuller (ADF) test to see if the number of interventions is stationary. The p_value found is $5.12e^{-28}$, which is lower than the significant level of 0.05. This means that no differencing is needed. Let's fix d to 0.
3. q: one lag above the significance level was found, thus q = 1 (Fig. 3b).

Figure 2 shows the actual number of interventions versus the predicted ones after
applying a moving average transformation overlaid by 3 h (Table 1).

Table 1. Different window sizes for Moving Average Algorithm.

Window Size	MAE
1 h	1.477
2 h	1.360
3 h	1.349
4 h	1.359
5 h	1.387
10 h	1.541

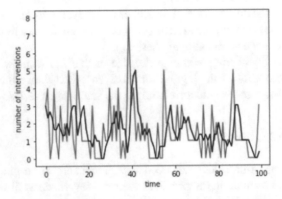

Fig. 2. Actual versus predicted number of interventions.

4.2   P, D and Q Parameters for ARIMA

After determining the values of p, d and q, ARIMA model is fitted by using order (4,0,1).

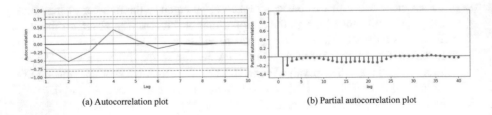

(a) Autocorrelation plot   (b) Partial autocorrelation plot

Fig. 3. p and q parameters.

5   Building Data with Prophet

A prophet forecasting model is an open-source algorithm designed by Facebook in 2017 [17] for time series having common features and intuitive parameters, where experts and non-experts in statistics and time series forecasting can use it. Prophet is based on time series models and relies on four main components: (1) Yearly and weekly seasonality, (2) Non linear trend, (3) Holidays and (4) Error. Prophet fits very well for data that have at least one year of historical inputs with daily periodicity. It is very fast in terms of fitting the model, working without converting data into time-series objects, and being robust to missing values. In addition, prophet is simpler compared to other time series forecasting algorithms, because it requires less number of parameters and models. Prophet works as following: $y(t) = g(t) + s(t) + h(t) + \varepsilon_t$ where:

- g(t): describes the increase or decrease trends in the long-term data.
- s(t): represents the impact of seasonal factors over the year on the time-series data.
- h(t): models how large events and holidays affect the data.
- ε_t: shows the non reducible error term.

For prophet's preparation, a new dataframe should be found: a new column is added to the data that emerges years, months, days, and hours. Then, this column is renamed to 'ds' and the predicted output presented in the data under the name of nbinterventions is renamed as 'y'. Figure 4 helps to visualize the forecast of the dataframe where the black dots display actual measurements, the blue line indicates Prophet's forecast and the light blue shaded line shows uncertainty intervals. It looks like the number of interventions was increasing over the years slightly (Table 2).

Table 2. New dataframe for the dataset.

Index	y	ds
0	1	2006-01-01 00:00:00
1	1	2006-01-01 01:00:00
2	0	2006-01-01 02:00:00
3	2	2006-01-01 03:00:00
4	4	2006-01-01 04:00:00

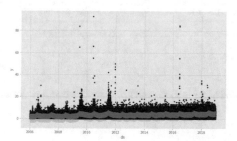

Fig. 4. Forecast for Prophet algorithm.

6   Obtained Results

Considering having 24 h per day, 7 d per week, 30 d per month, and 365 d per year, three algorithms have been implemented: AR, MA and ARIMA on the dataset in addition to the prophet. Statistical features of firemen predictions for every year from 2006 until 2017 have been registered in Table 4 and graphs that gather all statistical features are shown in Fig. 5. On the other hand, let us overview the result of the forecast by illustrating a breakdown of the former elements (Fig. 6) for daily, weekly and yearly trends using the prophet tool (see Table 3). The number of interventions during a trimmed time slot is shown in Fig. 7 where:

- yellow plot represents the actual number of interventions y.
- purple plot indicates the prediction yhat.
- blue and red plots show the upper and lower bound of prediction respectively.

Table 3. Statistical features using AutoRegression for different time slots.

Time	MAE	MSE	RMSE
1 h	0.307	0.094	0.307
2 h	0.403	0.171	0.414
12 h	1.205	1.932	1.390
1 day	1.288	2.128	1.459
5 d	1.168	1.822	1.350
1 week	1.209	2.057	1.434
1 month	1.213	2.123	1.457

(a) AR   (b) MA   (c) ARIMA

Fig. 5. Statistical features over many years using various models.

Table 4. Statistical features using AR, MA , ARIMA and prophet from 2006 until 2017.

	AR		MA		ARIMA		Prophet	
2006	1.481	2.046	1.349	1.86	1.018	1.307	2.55	3.53
2007	1.601	2.064	1.429	1.924	1.376	1.822	1.95	2.97
2008	1.496	1.952	1.385	1.868	1.263	1.644	1.31	1.63
2009	2.374	3.35	1.854	2.787	1.414	1.904	3.25	5.69
2010	2.161	3.058	1.847	2.716	1.629	2.154	2.00	2.23
2011	2.574	3.676	2.09	2.922	1.699	2.247	6.00	11.00
2012	1.99	2.5	1.84	2.415	1.642	2.031	2.44	2.87
2013	1.972	2.478	1.83	2.392	1.682	2.14	2.33	2.66
2014	2.04	2.545	1.874	2.451	1.504	1.843	2.44	2.90
2015	2.145	2.678	1.939	2.525	1.829	2.43	2.73	3.15
2016	2.223	3.137	2.026	2.807	1.898	2.31	2.45	2.99
2017	2.359	2.917	2.111	2.738	1.941	2.544	2.86	3.49
2018	2.552	3.216	2.24	2.929	2.075	2.742	1.85	2.60

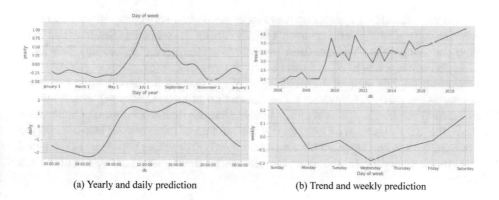

(a) Yearly and daily prediction   (b) Trend and weekly prediction

Fig. 6. Breakdown of the forecast using prophet.

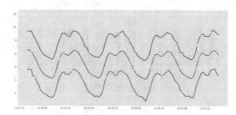

Fig. 7. Number of firefighters interventions using prophet.

7   Results Interpretation

Mean Absolute Error for AR, MA, ARIMA and prophet are compared in Fig. 8. As shown in the previous section, the best algorithm in term of fewer errors in most cases is ARIMA represented by the gray line. Among most of the years, starting from 2006 until 2018, this gray line has reached the minimum mean absolute error and root mean squared error, comparing to autoregression, moving average algorithms and prophet tool. Thus, the second, third, and fourth places are reserved respectively for Moving Average, Autoregressive, and Prophet. Generally, since ARIMA(p,d,q) stands for Auto Regressive Integrated Moving Average, it is logical to conclude that it combines AR (parameter p) and MA (parameter q) models. Other than that, ARIMA ensures the stationarity of the model (parameter d), unlike AR and MA. Therefore, by applying the components of these two models together, the probability of making errors will be reduced as shown in the experiments. It is important to mention that ARIMA is more complex than applied algorithms since it requires more time to identify the excessive number of parameters p,d and q. In contrast, when comparing eXtreme Gradient Boosting (XGBoost) and Long Short-Term Memory (LSTM) algorithms applied and experimented in [28] together with ARIMA from 2006 to 2014, it seems that ARIMA has the lowest root mean squared error values. However, XGBoost has the minimal RMSE values for 2015,2016 and 2017. This result reflects that XGBoost is better for long term forecast usage.

On the other side, by analyzing the prophet result, it is very clear that the number of interventions of firefighters increments highly during the weekend (Saturday and Sunday) and reaches the minimum during the middle of the week (Wednesday). This interpretation corresponds to official days off in France. Also, the number of interventions increments slightly starting the month of May and reaches the maximum in July, then decreases gradually till November. The fluctuation of interventions per month reflects that during summer incidents are more likely to happen than other seasons. On the other hand, the daily seasonality illustrates that the number of interventions increases during the morning starting around 5:00 am and reaches higher values between 11:00 am and 5:00 pm. It's very logical to link this variation with the departure and return time from work.

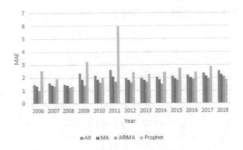

Fig. 8. Mean absotule error comparison for AR, MA and ARIMA.

8   Conclusion and Future Work

In this paper, Autoregression, Moving Average and Auto Regressive Integrated Moving Average has been implemented as well as a Facebook tool for time series forecasting called Prophet. For each model, statistical parameters have been calculated and compared between each other then compared between other results previously done. In general, as many researchers agreed, no hypothesis or rule electing a better algorithm among all-time series forecasting. The choice of the technique used depends on the specific prediction problem taking into account trends, seasonality, variables, size of the dataset, etc. In this paper, the statistical metrics indicate that ARIMA is the best model comparing to AR and MA as it combines first, the characteristics of these two algorithms and second the stationarity of the model. In contrast, XGBoost fits better than ARIMA for long term prediction. An extension to this work would be to apply different time series forecasting models to firefighters dataset.

Acknowledgement. This work has been supported by the EIPHI Graduate School (contract ANR-17-EURE-0002) and is partially funded with support from the Hubert Curien CEDRE programme n 46543ZD.

References

1. Hooker, R.H.: Forecasting the crops from the weather. Q. J. R. Meteor. Soc. **47**(198), 75–100 (1921)
2. Gardner Jr., E.S.: Exponential smoothing: the state of the art. J. forecast. **4**(1), 1–28 (1985)
3. Snyder, R.D.: Recursive estimation of dynamic linear models. J. R. Stat. Soc. Ser B (Methodological), 272–276 (1985)
4. Diggle, P.J.: Time series; a biostatistical introduction (1990) (No. 04; QA280, D5.)
5. Gardner Jr., E.S.: Forecasting the failure of component parts in computer systems: a case study. Int. J. Forecast. **9**(2), 245–253 (1993)
6. Chatfield, C.: Time-series forecasting. CRC Press, Boca Raton, USA (2000)
7. Taylor, J.W.: Exponential smoothing with a damped multiplicative trend. Int. J. Forecast. **19**(4), 715–725 (2003)
8. Ledolter, J., Abraham, B.: Some comments on the initialization of exponential smoothing. J. Forecast. **3**(1), 79–84 (1984)
9. Sweet, A.L., Wilson, J.R.: Pitfalls in simulation-based evaluation of forecast monitoring schemes. Int. J. Forecast. **4**(4), 573–579 (1988)
10. Bartolomei, S.M., Sweet, A.L.: A note on a comparison of exponential smoothing methods for forecasting seasonal series. Int. J. Forecast. **5**(1), 111–116 (1989)
11. Janacek, G.J., Swift, L.: Time series: forecasting, simulation, applications. Ellis Horwood Limited, Chichester, England (1993)
12. Miller, T., Liberatore, M.: Seasonal exponential smoothing with damped trends: an application for production planning. Int. J. Forecast. **9**(4), 509–515 (1993)
13. Rosas, A.L., Guerrero, V.M.: Restricted forecasts using exponential smoothing techniques. Int. J. Forecast. **10**(4), 515–527 (1994)
14. Klein, J.L., Klein, D.: Statistical visions in time: a history of time series analysis, pp. 1662–1938. Cambridge University Press, Cambridge, UK (1997)
15. Williams, D.W., Miller, D.: Level-adjusted exponential smoothing for modeling planned discontinuities. Int. J. Forecast. **15**(3), 273–289 (1999)

16. Grubb, H., Mason, A.: Long lead-time forecasting of UK air passengers by Holt-Winters methods with damped trend. Int. J. Forecast. **17**(1), 71–82 (2001)
17. Taylor, S.J., Letham, B.: Forecasting at scale. Am. Stat. **72**(1), 37–45 (2018)
18. Box, G.E., Jenkins, G.M. and Reinsel, G.: Time series analysis: forecasting and control Holden-day San Francisco. BoxTime Series Analysis: Forecasting and Control Holden Day (1970)
19. Brockwell, P.J., Davis, R.A., Calder, M.V.: Introduction to time series and forecasting, vol. 2, pp. 3118–3121. Springer, New York (2002)
20. Hyndman, R.J., Koehler, A.B., Snyder, R.D., Grose, S.: A state space framework for automatic forecasting using exponential smoothing methods. Int. J. Forecast. **18**(3), 439–454 (2002)
21. Gang, T.T., Yang, J., Zhao, Y.: Multivariate control chart based on the highest possibility region. J. Appl. Stat. **40**(8), 1673–1681 (2013)
22. Chen, S., Lan, X., Hu, Y., Liu, Q., Deng, Y.: The time series forecasting: from the aspect of network (2014). arXiv preprint arXiv:1403.1713
23. Couchot, J.F., Guyeux, C., Royer, G.: Anonymously forecasting the number and nature of firefighting operations. In: Proceedings of the 23rd International Database Applications & Engineering Symposium, pp. 1–8 (June 2019)
24. Guyeux, C., et al.: Firemen prediction by using neural networks: a real case study. In: Proceedings of SAI Intelligent Systems Conference, pp. 541–552. Springer, Cham (September 2019)
25. Nahuis, S.L.C., Guyeux, C., Arcolezi, H.H., Couturier, R., Royer, G. and Lotufo, A.D.P.: Long short-term memory for predicting firemen interventions. In: 2019 6th International Conference on Control, Decision and Information Technologies (CoDIT), pp. 1132–1137. IEEE (April 2019)
26. Arcolezi, H.H., Couchot, J.F., Cerna, S., Guyeux, C., Royer, G., Al Bouna, B., Xiao, X.: Forecasting the Number of Firefighters Interventions per Region with Local-Differential-Privacy-Based Data. Computers & Security, p.101888 (2020)
27. Cerna, S., Guyeux, C., Arcolezi, H.H., Couturier, R., Royer, G.: A comparison of LSTM and XGBoost for predicting firemen interventions. In: World Conference on Information Systems and Technologies, pp. 424-434. Springer, Cham (April 2020)
28. Cerna, S., Guyeux, C., Arcolezi, H.H., Royer, G.: Boosting Methods for Predicting Firemen Interventions. In 2020 11th International Conference on Information and Communication Systems (ICICS), pp. 001-006. IEEE (April 2020)

Stock Prediction Based on Adaptive Gradient Descent Deep Learning

Bo Li[1(✉)] and Li-feng Li[2]

[1] No. 32, Chengfang Street, Xicheng District, Beijing, China
lib@pbc.gov.cn
[2] 20th Floor, Block A, No. 8 Building, No. 5 Anding Road, Beijing, China

Abstract. Historical data is a permanent collection that allows people to make predictions about the future trends of stocks. With the development of machine learning especially deep learning, the prediction accuracy raised up. In this paper, we proposed feature extraction of stock data through CNN, and a stock prediction algorithm based on deep learning combined with traditional LSTM is constructed. At the same time, the third-order moment is used to improve the traditional stochastic gradient descent, and obtain a dynamic one. Experiments show that compared with other methods it has better accuracy which provides a solution for stock trend prediction by deep learning.

1 Introduction

People have diversified demands for financial management. Stocks are one of the most important financial choices for a large number of people. The aim of investments is making assets more valuable. Therefor investors prefer to buy high and sell low. But the stock market if full of uncertainty, and the return is proportional to the risk. Few investors and institutions can accurately predict stock movements, but historical data is a permanent collection that allows people to make predictions about the future trends of stocks.

Stocks and investments themselves belong to the economics. So most stock trend prediction algorithms of the early stage are based on the knowledge of economics, among which the knowledge of statistical probability theory is the most important. From the perspective of mathematics, statistically analyzing the historical trading market of stocks to obtain the laws, and then predict based on the statistical laws. However based on the category of mathematical statistics, when it comes to hypothesis testing, the difference between the predicted value and true is too large, which leads to the decline of the prediction accuracy.

With the present of machine learning especially deep learning which integrate the internal relations of training samples and build models based on them, the prediction accuracy raised up. Deep learning can not only improve many defects of traditional prediction ways but also avoid the influence of many human factors on the results because of the strong nonlinear segmentation performance of the algorithm. Most of the hundreds of senior managers in the financial field believe that machine learning and deep

L. Barolli et al. (Eds.): AINA 2021, LNNS 225, pp. 51–62, 2021.
https://doi.org/10.1007/978-3-030-75100-5_5

learning can have a substantial impact on financial industry, especially in financial risk, credit evaluation, and portfolio management.

The source of deep learning is a further exploration of neural network. For example, the framework of deep learning includes a multilayer perceptron with multiple hidden layers. In fact, deep learning is to find the internal rules among a large number of samples, with the ultimate goal of thinking and distinguish just like human beings. CNN (Convolutional Neural Network) is a feedforward neural network whose artificial neurons can respond to a part of the coverage of the surrounding units.

In this paper,we proposed feature extraction of stock data through CNN, and a stock prediction algorithm based on deep learning combined with traditional LSTM. At the same time, the third-order moment is used to improve the traditional stochastic gradient descent, and obtain a dynamic adaptive stochastic gradient descent algorithm. On this basis, experiments are carried out with actual dataset and compared with other methods. The results shows the new method is better than other algorithms which provides a solution for stock trend prediction by deep learning.

2   Literature Review

2.1   Summary of Feature Selection

The first and the most important step of stocks prediction is feature selection. If we can find the feature that affects fluctuation of stock most which other machine learning problems faced, we can model algorithm using related factors and get the target values. A Rao and YK Kwon built mathematical analysis on history stock data and used relatively mature statistical methods to get the correlation of technical indicators as a basis for stock trend prediction [1,2]. The most popular researches are on the transaction data itself, such as the mathematical statistics of the trend and relevant analysis of financial statements, which means the historical data of securities trading is taken as the research object to analyze the future trend of securities [3,4]. So most of researches would firstly select features and then confirm the orientation. Such as CF Tsai carried out a study by integrating various screening methods of impact factors [5] and JNK Liu used a completely different screening method of impact factors [6].

2.2   Summary of Deep Learning

In the last century, basic regression models were very popular, and some scholars also used classification models to predict stock trends such as Decision Tree [7], LR [8], SVM [9]. In fact some scholars started exploring securities trading by neural network. With the development of artificial intelligence, deep learning become popular and a large of researches of stock prediction based on deep learning came up. Ozbayoglu, A. M. and Bahadir, I. showed the different performance of securities trading between bayes estimation and deep learning, and came to a conclusion that the accuracy of deep learning model was better [10].

When using neural network method, the efficiency is seriously affected by the large amount of calculation, and the quality of the optimization algorithm also determines whether it can converge to the optimal solution. As an important aspect of deep learning, stochastic optimization algorithm affects the operating efficiency and final results. Adaptive optimization methods, such as Adagrad, RMSprop, and Adam are proposed for implementation a rapid training process, in which the learning rate can be flexibly adjusted. However, instability and extreme learning rates may not converge to optimal solutions (or critical points in non-convex Settings). With the development of deep learning, a variety of first-order stochastic optimization algorithm came out, of which the most widely algorithm is a stochastic gradient descent (SGD) [11]. Despite the stochastic gradient descent is relatively simple, in actual it perform well and is still one of the most common algorithm. But because of the fixed learning rate, SGD can only uniformly scaling coordinates gradient in all directions to iteratively updating parameters, which may leads slow convergence or results in loss function fluctuating around the local minimum and even divergence.

In order to solve this problem, there are many adaptive stochastic optimization method, one of the general solution is exponential moving averages (EMA) used by Momentum and Adam [12], which helps to accelerate and reduce the gradient oscillation on the gradient descent direction. And a variety of adaptive methods have been proposed, including Adagrad, AdaDelta, RMSprop, Adam et al. These methods scale the gradient by calculating the average value of the past gradient square, and due to their fast training speed, they have been successfully extended in the field of deep learning.

Although these methods have been successfully adopted in several practical applications, it has also been observed that they cannot converge in some environments. Reddi et al. provided some counter examples to prove that Adam could not converge to the optimal solution, and in some cases, the generalization ability of invisible data is not better than the non-adaptive method [13]. One of the reasons for this non-convergence is EMA. Using EMA to estimate the proportion of gradient cannot guarantee that the second-order is positively correlated with time, which will lead to the non-convergence of Adam.

Reddi et al. proposed a new Adam variant AMSGrad to solve the problem of non-convergence of the Adam, but at the expense of efficiency [13]. Zhou et al. put forward a new explanation for the problem of non-convergence [14]. An abnormal association exists between the gradient and Adam second-order moment, which may lead to the step size imbalance. Because small gradients may have larger step sizes, which can lead to larger updates, while large gradients may have smaller step sizes. Therefore, if always updates in the wrong direction, it may lead to differences.

3   Proposed Method

3.1   CNN and LSTM

Except the input layer and output layer of normal neural network, the convolutional neural network (CNN) has convolution layer, pooling layer and full connection layer. The convolution layer is mainly used to perform the convolution operation on samples by convolution verifies and then gets the input values of next layer. Pooling layer is important for CNN, its funtion is to reduce the connection inside the convolution layer and reduce the computational complexity.

The operation of convolution layer is the top priority of the whole CNN. Through a example can help us to understand the CNN. LeNet proposed by Y.LeCun [15], which structure as Fig. 1.

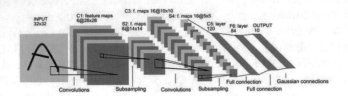

Fig. 1. Structure of LeNet

The input on the left is a hand-drawn image. In the middle, there are two pairs of pooling layers and multiple fully connected hidden layers. The number of layers C1 is 20, and layer C2 is 50 different convolution templates. The number of neurons in the next layer is 500. LeNet used only two convolutional layers at that time which is very shallow, but its effect is good. Through the experiment on MNIST dataset, the error rate of the final result is even lower than 0.95%, thus it can be seen that the depth of neural network does not need to be very deep to achieve a good effect. The model adopts gradient descent method to modify each parameter in the model layer by layer, and minimizes the loss function through constant training, so as to improve the accuracy of the model to the prediction target.

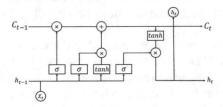

Fig. 2. LSTM neual unit

LSTM proposed by Sepp hochreiter and Jurgen Schmidhuber in 1997 [16]. It is a special RNN (Recurrent Neural Network) [17]. The main advantage of RNN is taking place in the context of information integration, and applicating in the process of the input and output. But it still has gradient disappear, gradient explosion, etc.To solve these problem, LSTM arosed. LSTM can capture long-term dependence better. By introducing a memory unit to remember values in any time interval, and simultaneously using input gate, output gate and forgetting gate to adjust the process of information entering and leaving the memory unit, the gradient disappearance or explosion problem existing in general RNN is effectively solved. The structure of each memory unit in the LSTM network is shown in Fig. 2.

Firstly, the forgetting gate determines the process of losing information. According to the output h_{t-1} of the previous memory unit and the input data x_t, the forgetting gate generates a value between 0 and 1 to determine how much information is lost in the previous long-term state C_{t-1}.

$$f_t = \sigma(W_f \cdot [h_{t-1}, x_t] + b_f), \tag{1}$$

The input gate determines which information is added. h_{t-1} and x_t are used to determine the update information through the input gate, and meanwhile, a tanh layer is used to obtain the new candidate memory unit information C_t'.

$$i_t = \sigma(W_i \cdot [h_{t-1}, x_t] + b_i), \tag{2}$$

$$C_t' = tanh(W_C \cdot [h_{t-1}, x_t] + b_C), \tag{3}$$

The previous long-term state C_{t-1} is updated to C_t through the operation of the forgetting gate and the input gate.

$$C_t = f_t * C_{t-1} + i_t * C_t', \tag{4}$$

Finally, the judgment condition is obtained from the output gate, and then a value between -1 and 1 is obtained through a tanh layer. This value is multiplied by the judgment condition to determine which state characteristics of the current memory unit to output.

$$o_t = \sigma(W_o[h_{t-1}, x_t] + b_o), \tag{5}$$

$$h_t = o_t * tanh(C_t), \tag{6}$$

3.2 Improved Adam

We proposed an improving algorithm based on Adam in which applying third-order moment. First-order moment optimization algorithm: SGDM algorithm introduces the idea of first-order moment to help accelerate and reduce the gradient oscillation in the descending direction of gradient. The first-order moment is the exponential moving average (EMA) of the gradient. It's equal to the average of the sum of the last $1/(1-\beta_1)$ gradient vectors. The mark of the era of "adaptive learning rate" optimization algorithm is the birth of the second-order moment. The second-order moment is the sum of the squares of all gradients so far. For example, compared with the learning rate $\alpha/\sqrt{v_t}$ in SGD, Adagrad effectively uses the idea of moderate learning rate attenuation, and the learning rate is $\alpha/\sqrt{\sum_{i=1}^{t} g_i^2}$. It has a good convergence effect in case of the gradient is sparse. We usually add a small smoothing term to the denominator, otherwise there will be no sense such as the denominator is 0, so the denominator is always greater than 0. According to the definition of second-order moment, we can know that the faster the parameter changes, the greater the second-order moment is, and then the learning rate decreases. When the data distribution is sparse, the effect of this method is significant.

But at the moment, we find that increasing $\sqrt{v_t}$ will cause the learning rate decreasing rapidly and approach to zero infinitely. Leading to the problem of ending the training process in advance, and there is no way to learn well for the subsequent data, which means giving up the later learning process directly. AdaDelt focus only on the descent gradient in the past sliding window. Previous algorithms often take all cumulative gradients into account. In this case, the second-order moment cumulative momentum $v_t = \beta_2 v_{t-1} + (1 - \beta_2)$ is the gradient average of the past. The advantage of this method is effectively preventing the second-order moment from increasing all the time which result in the premature stop of the training process.

Combining the advantages of all the methods described above, Adam is very popular in the field of deep learning which is to make full use of the first-order moment $m_t = \beta_1 m_{t-1} + (1 - \beta_1) \cdot g_t^2$ and the second-order moment $v_t = \beta_2 v_{t-1} + (1 - \beta_2) \cdot g_t^2$. So it can not only fully consider the acceleration and reduce the gradient oscillation in the direction of gradient descent, but also avoid the rapid decline of learning rate and end the training in advance. Although Adam is an advanced optimization algorithm, many scholars have mentioned that sometimes the performance of adaptive methods is even worse than SGD [18]. Reddi et al. pointed out the shortcomings of Adam and provided some counter examples to prove that Adam cannot converge to the optimal solution, and in some cases, the generalization ability of invisible data is not better than non-adaptive method [13].

We propose a new variant of the optimization method, which applies the third-order moment to Adam. Our goal is to propose a new method that will benefit from adaptive methods such as Adam and adadelta, as well as the excellent final performance of SGD. Based on the previous discussion, we conclude that the main reason that Adam may not be able to converge may be the oscillatory learning speed. We use the third-order moment to reduce the fluctuation of learning rate. We first propose the following definition of the third moment.

Definition 1. (skewness) The skewness of random variable x is defined as the third-order moment γ_1 defined as follows:

$$\gamma_1 = E\left[\left(\frac{X - \mu}{\sigma}\right)^3\right] = \frac{\mu_3}{\sigma_3} = \frac{E[(X - \mu)^3]}{(E[(X - \mu)^2])^{\frac{3}{2}}}$$

Where μ is the mean, σ is the standard deviation and E is the expected operator.

The concept of skewness is actually used to describe the asymmetry of random variables. We know that the probability distribution of random variables is usually irregular and asymmetric, and it will shift to a certain direction in the actual graph. So the skewness can be positive or negative. In some special cases, it can even be undefined. Negative skewness generally means that most of the values in the probability distribution are greater than the average, which results in the tail on the left side of the function longer than that on the right side. The conclusion of positive skewness is opposite to that of negative skewness.

As we all know that $\alpha / \sqrt{v_t}$ is the learning rate and $v_t = \beta_2 v_{t-1} + (1 - \beta_2) \cdot g_t^2$. Now, let's set the learning rate to $\alpha / (\sqrt{v_t} + \sqrt[3]{s_t})$, in order to make our method behave as adaptive method in the early stage of training and behave like SGD algorithm at the end of training.

Based on the third-order moment, we gave the pseudo code of adaptive gradient method.

```
Algorithm 1:
Input:
α:step size
β₁,β₂,β₃ ∈ [0,1):exponential decay rates for the moment estimates.
η₁:lower bound function,ηᵤ:upper bound function
f(θ):Stochastic objective function with parameters
θ₀:Initial parameter vector
m₀ ← 0:initialize 1st moment vector
v₀ ← 0:initialize 2nd moment vector
t ← 0:initialize time step
While θₜ not converged do
    t ← t + 1
    gₜ ← ∇θfₜ(θₜ₋₁) stochastic objective gradients at time t
    mₜ ← β₂vₜ₋₁ + (1 − β₁)·gₜ update biased first moment estimate
    vₜ ← β₂vₜ₋₁ + (1 − β₂)·gₜ² update biased second raw moment estimate
    sₜ ← β₃sₜ₋₁ + (1 − β₃)·(gₜ − mₜ)³ update third moment estimate
    m̂ₜ ← mₜ/(1 − β₂ᵗ) compute bias-corrected first moment estimate
    v̂ₜ ← vₜ/(1 − β₂ᵗ) compute bias-corrected second raw moment estimate
    v̂ₜ ← max(v̂ₜ₋₁,v̂ₜ) make sure |v̂ₜ| > |v̂ₜ₋₁|
    ηₜ ← clip(α/√[vₜ+sqrt[3][sₜ]],η₁(t),ηᵤ(t)) clip the learn rates
    θₜ ← θₜ₋₁ − m̂ₜ·ηₜ update parameters
End while
```

4 Experiments

In the experiment, we will verifie the above methods. The experiment will input the original data, predict the rise and fall of stocks in the future through CNN+LSTM structure. CNN adopts the LeNet structure mentioned before. The LSTM uses three circulating neural network units. At the same time, dynamic adaptive stochastic gradient descent is adopted in the training, and compared with the traditional stochastic gradient descent. This experiment uses deep learning framework Keras and Python to implement the model. The operating system of the experimental environment was Ubuntu, and the hardware environment was Xeon E5-2630v4 CPU and Nvidia RTX 2080TI GPU.

For data, we choose yahoo finance [19] from 2000 to 2019, a total of 19 years of the Hang Seng Index. Indexes are a better reflection of stock market volatility than individual stocks as a result of policies or specific economic conditions, at the same time Hang Seng Index has enough volatility to verifies the algorithm we proposed. We obtained the complete non-empty data of the 4439 groups where 1300 will be set as a test set, the rest is used as training set. The input data used in this paper has the following dimensions:

Table 1. Part of test data

Date	Open	High	Low	Close
2000-3-27	17827.71094	18350.60938	17815.67989	18292.85938
2000-3-28	18272.80078	18397.57031	18188.26953	18301.68945
2000-3-29	18203.13086	18249.26953	17950.19922	18096.36914
2000-3-30	18011.14063	18038.47070	17456.56055	17467.15039
2000-3-31	17383.63080	17537.13086	17107.43945	17406.53906
2000-4-3	17444.48047	17458.06055	16872.82031	16892.92969
2000-4-5	16599.07031	16736.22070	16246.53027	16318.44043
2000-4-6	16416.48047	16528.58008	16283.20020	16491.39063
2000-4-7	16657.83984	16992.52930	16657.83984	16941.67969
2000-4-10	17038.22070	17083.26953	16676.07031	16850.74023

The opening, lowest, highest, closing price, trading volume and KDJ, RSI, DMI, BIAS, total 9 dimensional indicators.

The historical data of the first 25 d are used to predict the situation of the following day. Therefore, the input of the experiment is a 25 × 9 matrix, some of which are shown in the Table 1 below. The objective of the experiment is to predict the closing price. The historical data of the experiment is shown in the following Fig. 3 In the preprocessing of experimental data, in addition to deleting and replacing the missing data, the following formula 7 is used to standardize. The standardized data is easy to calculate and shorten the running time of the program. It is also more conducive to model training and model generalization.

$$\bar{n}_t = \frac{n_t - mean(n_t)}{std(n_t)} ,$$ (7)

Use formula 8 as historical input data to predict the close price of the following day. Mean Absolute Error (MAE) and Root-mean-square error (RMSE) are used to evaluate the training results of three models. MAE is the average of the absolute error between the observed value and the true value. RMSE refers to the square root of the ratio of the squared deviation between the observed value and the truth value to the number of observations m. The smaller the MAE and RMSE are, the better prediction model is. Training set and test set are selected randomly by Gaussian distribution. Model evaluation index formula are formula 9 and formula 10.

$$x_{(t)} = \left(x_{(t-25)}, x_{(t-24)}, ..., x_{(2)}, x_{(1)}\right)^T ,$$ (8)

$$MAE = \frac{\sum_{i=1}^{m} |(x - \bar{x})|}{m} ,$$ (9)

$$RMSE = \sqrt{\frac{\sum_{i=1}^{m} (x - \bar{x})^2}{m}} ,$$ (10)

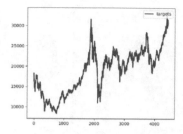

Fig. 3. Historical closing price chart

5 Discussion

SimpleRNN model and LSTM model of Keras are used for training and testing. We realizes three model under the different conditions. The first is a stock prediction model using LSTM alone, which uses traditional stochastic gradient descent in training. The second uses traditional stochastic gradient descent training and CNN+LSTM model. The third is the CNN+LSTM model using dynamic adaptive stochastic gradient descent for training. For gradient descent, we select the initial learning rate as 0.01. The Fig. 4 shows the predicted results of the above three algorithms for the rise and fall of the data set, in which the blue line represents the target value and the yellow line represents the predicted value. At the same time, in order to reflect the trend of our model more intuitively, we randomly selected a certain section of data in each experiment to predict the trend. The relevant results are shown in the Fig. 4.

Through the above experiments, we can find that for each model, the number of times that the yellow line coincides with the blue line in the pattern trend is significantly more than not coincide. At the same time, it can be seen from the trend prediction diagram that CNN + LSTM is obviously superior to the traditional LSTM under the condition of traditional stochastic gradient descent. But for CNN + LSTM using dynamic adaptive stochastic gradient descent training, it is obviously better than the traditional stochastic gradient descent training model. Next, we calculate MAE and RMSE under three model conditions, as shown in the Table 2.

Table 2. MAE and RMSE in there model

	LSTM under traditional SGD	CNN and LSTM under traditional SGD	CNN + LSTM under Dynamic adaptive SGD
MAE	0.1092	0.0279	0.0063
RMSE	0.1413	0.0396	0.0083

According to the results in the above table, the prediction errors of the three models are decreasing and the prediction accuracy is improving. The prediction accuracy of the adaptive stochastic gradient descent model is greatly improved. The MAE and RMSE of LSTM prediction model with traditional SGD training are 0.1092 and 0.1413

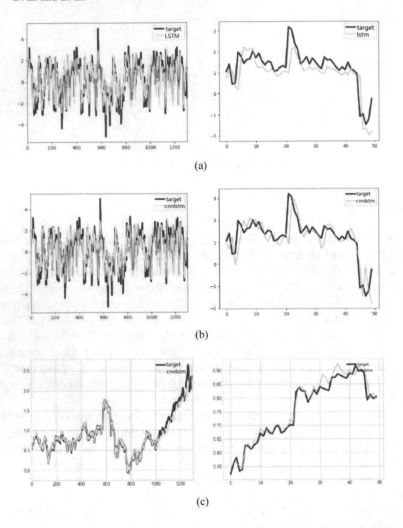

(a)

(b)

(c)

Fig. 4. Randomly selected several section of data to predict the trend

respectively. CNN +LSTM reduces the MAE by 0.0813 and the RMSE by 0.1017. After using dynamic adaptive stochastic gradient descent, the prediction MAE is reduced by 0.1029 and the RMSE is reduced by 0.133 compared with the cyclic neural network. By comparing the MAE and RMSE of the three prediction models, the prediction effect of CNN + LSTM using the adaptive stochastic gradient descent is better than that of the conventional stochastic gradient descent. The CNN + LSTM using traditional random gradient descent is superior to the single LSTM network using traditional random gradient descent. As a conclusion, the CNN + LSTM model with adaptive stochastic gradient descent has higher accuracy and more predictive value in practice.

6   Conclusions

In this paper, feature extraction of stock data through CNN is proposed, and a stock prediction algorithm based on deep learning is constructed in combination with traditional LSTM. At the same time, the third-order moment is used to improve the traditional stochastic gradient descent, and obtain a dynamic adaptive stochastic gradient descent algorithm. On this basis, the experiment is carried out with actual data and compared with other methods, and the results are better than other algorithms. It provides a solution for stock trend prediction by deep learning.

References

1. Shaikh, H., Rao, A., Hule, S.: Survey: market prediction using statistical computational methodologies and artificial neural networks (2015)
2. Kwon, Y.K., Choi, S.S., Moon, B.: Stock prediction based on financial correlation. pp. 2061–2066, June 2005
3. Lahmiri, S.: Article: a comparison of PNN and SVM for stock market trend prediction using economic and technical information. Int. J. Comput. Appl. **29**(3), 24–30 (2011)
4. Gui, B., Wei, X., Shen, Q., Qi, J., Guo, L.: Financial time series forecasting using support vector machine. In: 2014 Tenth International Conference on Computational Intelligence and Security, pp. 39–43 (2014)
5. Akita, R., Yoshihara, A., Matsubara, T., Uehara., K.: Deep learning for stock prediction using numerical and textual information. In: 2016 IEEE/ACIS 15th International Conference on Computer and Information Science (ICIS), Los Alamitos, CA, USA, pp. 1–6. IEEE Computer Society, June 2016
6. Tsai, C.F., Hsiao, Y.C.: Combining multiple feature selection methods for stock prediction: union, intersection, and multi-intersection approaches. Decis. Support Syst. **50**(1), 258–269 (2011)
7. Hunt, E.B., Marin, J., Stone, P.J.: Experiments in induction. Am. J. Psychol. **80**, 4 (1966)
8. Cramer, J.S.: The origins of logistic regression. Tinbergen Institute Discussion Papers (2002)
9. Vapnik, V.N., Chervonenkis, A.: A note on one class of perceptrons. Autom. Remote Control **25**(1) (1964)
10. Ozbayoglu, A.M., Bahadir, I.: comparison of Bayesian estimation and neural network model in stock market trading. Intelligent Engineering Systems Through Artificial Neural Networks (2008)
11. Robbins, H., Monro, S.: A stochastic approximation method. Ann. Math. Stat. **22**(3), 400–407 (1951)
12. Loshchilov, I., Hutter, F.: SGDR: Stochastic gradient descent with warm restarts, August 2016
13. Sashank, J., Reddi, S.K., Sanjiv K.: On the convergence of Adam & beyond, May 2018
14. Zhou, Z., Zhang, Q., Lu, G., Wang, H., Zhang, W., Yu, Y.: Adashift: decorrelation and convergence of adaptive learning rate methods, September 2018
15. Lecun, Y., Bottou, L., Bengio, Y., Haffner, P.: Gradient-based learning applied to document recognition. Proc. IEEE **86**(11), 2278–2324 (1998)
16. Hochreiter, S., Schmidhuber, J.: Long short-term memory. Neural Comput. **9**(8), 1735–1780 (1997)

17. Le Guillarme, N., Lerouvreur, X.: Unsupervised extraction of knowledge from S-AIS data for maritime situational awareness. In: 2013 16th International Conference on Information Fusion (FUSION) (2013)
18. Keskar, N.S., Socher, R.: Improving generalization performance by switching from Adam to SGD (2017)
19. Finance yahoo. http://table.finance.yahoo.com/

A QoS-Aware Fuzzy-Based System for Assessment of Edge Computing Resources in SDN-VANETs

Ermioni Qafzezi[1]([⊠]), Kevin Bylykbashi[1], Phudit Ampririt[1], Makoto Ikeda[2], Keita Matsuo[2], and Leonard Barolli[2]

[1] Graduate School of Engineering, Fukuoka Institute of Technology (FIT), 3-30-1 Wajiro-Higashi, Higashi-Ku, Fukuoka 811-0295, Japan
[2] Department of Information and Communication Engineering, Fukuoka Institute of Technology (FIT), 3-30-1 Wajiro-Higashi, Higashi-Ku, Fukuoka 811-0295, Japan
makoto.ikd@acm.org, {kt-matsuo,barolli}@fit.ac.jp

Abstract. In this paper, we propose a fuzzy-based system to assess the available edge computing resources in Software Defined Vehicular Ad hoc Networks (SDN-VANETs) by considering Quality of Service (QoS) of the communication link among vehicles as a new parameter. Our proposed system, called Fuzzy System for Assessment of Neighboring Vehicle Processing Capability (FS-ANVPC), determines if a neighbor vehicle is capable of helping out to complete certain tasks in terms of data processing. FS-ANVPC is implemented in vehicles and in SDN Controllers. It is used from vehicles which need additional resources to run their applications or process different application data related (but not limited) to VANETs. However, for that to happen, the vehicles should remain within the communication range of each other for a while and the communication link should satisfy certain requirements. In addition, the available resources of the neighbor vehicle are also considered by the system. The output decides the processing capability of that neighbor. We evaluate FS-ANVPC by computer simulations. A neighbor is decided as helpful when a communication session with the vehicle is maintained for a long time over a reliable link, while possessing a moderate amount of available resources.

1 Introduction

The long distances separating homes and workplaces/facilities/schools as well as the traffic present in these distances make people spend a significant amount of time in vehicles. Thus, it is important to offer drivers and passengers ease of driving, convenience, efficiency and safety. This has led to the emerging of Vehicular Ad hoc Networks (VANETs), where vehicles are able to communicate and share important information among them. VANETs are a relevant component of Intelligent Transportation System (ITS), which offers more safety and better transportation.

L. Barolli et al. (Eds.): AINA 2021, LNNS 225, pp. 63–72, 2021.
https://doi.org/10.1007/978-3-030-75100-5_6

VANETs are capable to offer numerous services such as road safety, enhanced traffic management, as well as travel convenience and comfort. To achieve road safety, emergency messages must be transmitted in real-time, which stands also for the actions that should be taken accordingly in order to avoid potential accidents. Thus, it is important for the vehicles to always have available connections to infrastructure and to other vehicles on the road. On the other hand, traffic efficiency is achieved by managing traffic dynamically according to the situation and by avoiding congested roads, whereas comfort is attained by providing in-car infotainment services.

The advances in vehicle technology have made possible for the vehicles to be equipped with various forms of smart cameras and sensors, wireless communication modules, storage and computational resources. While more and more of these smart cameras and sensors are incorporated in vehicles, massive amounts of data are generated from monitoring the on-road and in-board status. This exponential growth of generated vehicular data, together with the boost of the number of vehicles and the increasing data demands from in-vehicle users, has led to a tremendous amount of data in VANETs [8]. Moreover, applications like autonomous driving require even more storage capacity and complex computational capability. As a result, traditional VANETs face huge challenges in meeting such essential demands of the ever-increasing advancement of VANETs.

The integration of Cloud-Fog-Edge Computing in VANETs is the solution to handle complex computation, provide mobility support, low latency and high bandwidth. Each of them serves different functions, but also complements each-other in order to enhance the performance of VANETs. Even though the integration of Cloud, Fog and Edge Computing in VANETs solves significant challenges, this architecture lacks mechanisms needed for resource and connectivity management because the network is controlled in a decentralized manner. The prospective solution to solve these problems is the augmentation of Software Defined Networking (SDN) in this architecture.

The SDN is a promising choice in managing complex networks with minimal cost and providing optimal resource utilization. SDN offers a global knowledge of the network with a programmable architecture which simplifies network management in such extremely complicated and dynamic environments like VANETs [7]. In addition, it will increase flexibility and programmability in the network by simplifying the development and deployment of new protocols and by bringing awareness into the system, so that it can adapt to changing conditions and requirements, i.e., emergency services [3]. This awareness allows SDN-VANET to make better decisions based on the combined information from multiple sources, not just individual perception from each node.

In a previous work [6], we have proposed an intelligent approach to manage the cloud-fog-edge resources in SDN-VANETs using fuzzy logic. We have presented a cloud-fog-edge layered architecture which is coordinated by an intelligent system that decides the appropriate resources to be used by a particular vehicle (hereafter will be referred as *the vehicle*) in need of additional computing resources. The main objective is to achieve a better management of these resources.

In this work we focus only on the edge layer resources. We include the Quality of Service (QoS) of the communication link among vehicles to better assess their processing capability. The proposed Fuzzy System for Assessment of Neighboring Vehicle Processing Capability (FS-ANVPC) determines the processing capability for each neighboring vehicle separately, hence helpful neighboring vehicles could be discovered and a better assessment of available edge resources can be made. We see the effect of QoS on the evaluation of the neighbor vehicles capability to help *the vehicle* accomplishing different tasks.

The remainder of the paper is as follows. In Sect. 2, we present an overview of Cloud-Fog-Edge SDN-VANETs. In Sect. 3, we describe the proposed fuzzy-based system. In Sect. 4, we discuss the simulation results. Finally, conclusions and future work are given in Sect. 5.

2   Cloud-Fog-Edge SDN-VANETs

While cloud, fog and edge computing in VANETs offer scalable access to storage, networking and computing resources, SDN provides higher flexibility, programmability, scalability and global knowledge. In Fig. 1, we give a detailed structure of this novel VANET architecture. It includes the topology structure, its logical structure and the content distribution on the network. As it is shown, it consists of Cloud Computing data centers, fog servers with SDNCs, roadside units (RSUs), RSU Controllers (RSUCs), Base Stations and vehicles. We also illustrate the infrastructure-to-infrastructure (I2I), vehicle-to-infrastructure (V2I), and vehicle-to-vehicle (V2V) communication links. The fog devices (such as fog servers and RSUs) are located between vehicles and the data centers of the main cloud environments.

The safety applications data generated through in-board and on-road sensors are processed first in the vehicles as they require real-time processing. If more storing and computing resources are needed, the vehicle can request to use those of the other adjacent vehicles, assuming a connection can be established and maintained between them for a while. With the vehicles having created multiple virtual machines on other vehicles, the virtual machine migration must be achievable in order to provide continuity as one/some vehicle may move out of the communication range. However, to set-up virtual machines on the nearby vehicles, multiple requirements must be met and when these demands are not satisfied, the fog servers are used.

Cloud servers are used as a repository for software updates, control policies and for the data that need long-term analytics and are not delay-sensitive. On the other side, SDN modules which offer flexibility and programmability, are used to simplify the network management by offering mechanisms that improve the network traffic control and coordination of resources. The implementation of this architecture promises to enable and improve the VANET applications such as road and vehicle safety services, traffic optimization, video surveillance, telematics, commercial and entertainment applications.

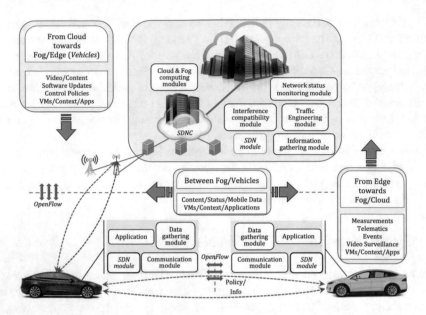

Fig. 1. Logical architecture of cloud-fog-edge SDN-VANET with content distribution.

3 Proposed Fuzzy-Based System

In this section, we present our proposed fuzzy based system. A vehicle that needs storage and computing resources for a particular application can use those of neighboring vehicles, fog servers or cloud data centers based on the application requirements. For instance, for a temporary application that needs real-time processing, the vehicle can use the resources of adjacent vehicles if the requirements to realize such operations are fulfilled. Otherwise, it will use the resources of fog servers, which offer low latency as well. Whereas real-time applications require the usage of edge and fog layer resources, for delay tolerant applications, vehicles can use the cloud resources as these applications do not require low latency.

The proposed system is implemented in the SDNC and in the vehicles which are equipped with SDN modules. If a vehicle does not have an SDN module, it sends the information to SDNC which sends back its decision. The system uses the beacon messages received from the adjacent vehicles to extract information such as their current position, velocity, direction, available computing power, available storage, and based on the received data, the processing capability of each adjacent vehicle is decided.

The structure of the proposed system is shown in Fig. 2. For the implementation of our system, we consider four input parameters: Available Computing Power (APC), Available Storage (AS), Predicted Contact Duration (PCD) and Quality of Service (QoS) to determine the Neighbor i Processing Capability (NiPC).

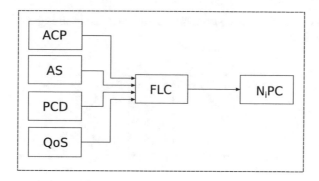

Fig. 2. Proposed system structure.

ACP: Vehicles might be using their computing power for their own applications but a reserved amount can be allocated to help other vehicles in need to complete certain tasks. Vehicles let their neighbors know that they are willing to share their resources and how much they want to share. In other words, they decide the amount of physical processor cores and the amount of memory that a particular *vehicle* can use.

AS: The neighbors should have a specific amount of storage so the *vehicle* can run the virtual machines. This storage is used also to store data after completing specific tasks of all the tasks these neighbors are asked to accomplish.

PCD: In a V2V communication, the duration of the communication session is important since it determines the amount of data to be exchanged and the services that can be performed. A *vehicle* which needs additional resources will create virtual machines on the neighbors that are willing to lend their resources, therefore the contact duration becomes even more important since much more time is needed to accomplish these tasks than just performing a data exchange.

QoS: Different VANET applications have different requirements in terms of delay, bandwidth, throughput and so on. Safety applications, for example, have strong QoS demands as they require a real time communication and a reliable link to reach their goals, that is reducing the car accidents. However, it is a challenging task to satisfy at all times the QoS requirements given the highly dynamic topology and physical obstacles and interferences on these networks.

NiPC: The output parameter values consist of values between 0 and 1, with the value 0.5 working as a border to determine if a neighbor is capable of helping out the *vehicle*. A NiPC $\geq$ 0.5 means that this neighbor i has the required conditions to help the *vehicle* to complete its tasks.

We consider fuzzy logic to implement the proposed system because our system parameters are not correlated with each other. Having three or more parameters which are not correlated with each other results in a non-deterministic polynomial-time hard (NP-hard) problem and fuzzy logic can deal with these problems. Moreover, we want our system to make decisions in real time and fuzzy systems can give very good results in decision making and control problems [1,2,4,5,9,10].

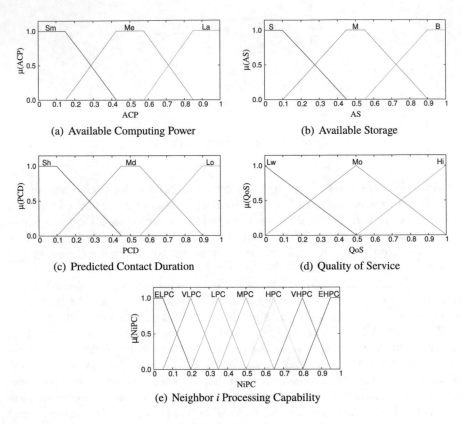

(a) Available Computing Power

(b) Available Storage

(c) Predicted Contact Duration

(d) Quality of Service

(e) Neighbor *i* Processing Capability

Fig. 3. Membership functions.

The input parameters are fuzzified using the membership functions shown in Figs. 3(a), 3(b), 3(c) and 3(d). In Fig. 3(e) are shown the membership functions used for the output parameter. We use triangular and trapezoidal membership functions because they are suitable for real-time operation. The term sets for each linguistic parameter are shown in Table 1. We decided the number of term sets by carrying out many simulations. In Table 2, we show the Fuzzy Rule Base (FRB) of our proposed system, which consists of 81 rules. The control rules have the form: IF "conditions" THEN "control action". For instance, for Rule 1: "IF ACP is Sm, AS is S, PCD is Sh and QoS is Lw, THEN NiPC is ELPC" or for Rule 50: "IF ACP is Me, AS is B, PCD is Md and QoS is Mo THEN NiPC is LPC".

Table 1. System parameters and their term sets.

Parameters	Term sets
Available Computing Power (ACP)	Small (Sm), Medium (Me), Large (La)
Available Storage (AS)	Small (S), Medium (M), Big (B)
Predicted Contact Duration (PCD)	Short (Sh), Medium (Md), Long (Lo)
Quality of Service (QoS)	Low (Lw), Moderate (Mo), High (Hi)
Neighbor i Processing Capability (NiPC)	Extremely Low Processing Capability (ELPC), Very Low Processing Capability (VLPC), Low Processing Capability (LPC), Moderate Processing Capability (MPC), High Processing Capability (HPC), Very High Processing Capability (VHPC), Extremely High Processing Capability (EHPC)

4   Simulation Results

The simulations were conducted using FuzzyC and the results are shown for three scenarios: in Fig. 4 is shown the scenario where the available computing power is small, Fig. 5 shows the results for medium available computing power and Fig. 6 the results for large available computing power. We show the relation between NiPC and QoS for different PCD while keeping constant ACP and AS values.

In Fig. 4(a), the neighboring vehicles have small processing capability and small available storage. Due to the low processing resources, we see that none of the vehicles is considered as helpful, even when a long predicted contact duration between vehicles is anticipated. However, with the increase of AS, we see that vehicles with long predicted contact duration and the highest value of QoS are considered as helpful. Although these neighbors have small available computing power, it is worth taking these vehicles into consideration as they can become potential neighbors in the case of an increase of the ACP, even if this increase might not be a big one. We see this effect in Fig. 5 where we consider the scenario with neighbors that have moderate amounts of ACP.

Table 2. FRB of FS-ANVPC.

No	ACP	AS	PCD	QoS	NiPC	No	ACP	AS	PCD	QoS	NiPC	No	ACP	AS	PCD	QoS	NiPC
1	Sm	S	Sh	Lw	ELPC	28	Me	S	Sh	Lw	ELPC	55	La	S	Sh	Lw	ELPC
2	Sm	S	Sh	Mo	ELPC	29	Me	S	Sh	Mo	ELPC	56	La	S	Sh	Mo	VLPC
3	Sm	S	Sh	Hi	ELPC	30	Me	S	Sh	Hi	VLPC	57	La	S	Sh	Hi	LPC
4	Sm	S	Md	Lw	ELPC	31	Me	S	Md	Lw	VLPC	58	La	S	Md	Lw	VLPC
5	Sm	S	Md	Mo	ELPC	32	Me	S	Md	Mo	VLPC	59	La	S	Md	Mo	LPC
6	Sm	S	Md	Hi	VLPC	33	Me	S	Md	Hi	LPC	60	La	S	Md	Hi	MPC
7	Sm	S	Lo	Lw	ELPC	34	Me	S	Lo	Lw	VLPC	61	La	S	Lo	Lw	LPC
8	Sm	S	Lo	Mo	ELPC	35	Me	S	Lo	Mo	LPC	62	La	S	Lo	Mo	MPC
9	Sm	S	Lo	Hi	LPC	36	Me	S	Lo	Hi	MPC	63	La	S	Lo	Hi	VHPC
10	Sm	M	Sh	Lw	ELPC	37	Me	M	Sh	Lw	ELPC	64	La	M	Sh	Lw	VLPC
11	Sm	M	Sh	Mo	ELPC	38	Me	M	Sh	Mo	ELPC	65	La	M	Sh	Mo	LPC
12	Sm	M	Sh	Hi	ELPC	39	Me	M	Sh	Hi	LPC	66	La	M	Sh	Hi	HPC
13	Sm	M	Md	Lw	ELPC	40	Me	M	Md	Lw	VLPC	67	La	M	Md	Lw	LPC
14	Sm	M	Md	Mo	ELPC	41	Me	M	Md	Mo	VLPC	68	La	M	Md	Mo	MPC
15	Sm	M	Md	Hi	VLPC	42	Me	M	Md	Hi	MPC	69	La	M	Md	Hi	VHPC
16	Sm	M	Lo	Lw	ELPC	43	Me	M	Lo	Lw	LPC	70	La	M	Lo	Lw	MPC
17	Sm	M	Lo	Mo	VLPC	44	Me	M	Lo	Mo	MPC	71	La	M	Lo	Mo	VHPC
18	Sm	M	Lo	Hi	LPC	45	Me	M	Lo	Hi	HPC	72	La	M	Lo	Hi	EHPC
19	Sm	B	Sh	Lw	ELPC	46	Me	B	Sh	Lw	ELPC	73	La	B	Sh	Lw	LPC
20	Sm	B	Sh	Mo	ELPC	47	Me	B	Sh	Mo	VLPC	74	La	B	Sh	Mo	MPC
21	Sm	B	Sh	Hi	ELPC	48	Me	B	Sh	Hi	LPC	75	La	B	Sh	Hi	HPC
22	Sm	B	Md	Lw	ELPC	49	Me	B	Md	Lw	VLPC	76	La	B	Md	Lw	MPC
23	Sm	B	Md	Mo	VLPC	50	Me	B	Md	Mo	LPC	77	La	B	Md	Mo	VHPC
24	Sm	B	Md	Hi	LPC	51	Me	B	Md	Hi	MPC	78	La	B	Md	Hi	VHPC
25	Sm	B	Lo	Lw	VLPC	52	Me	B	Lo	Lw	LPC	79	La	B	Lo	Lw	HPC
26	Sm	B	Lo	Mo	LPC	53	Me	B	Lo	Mo	HPC	80	La	B	Lo	Mo	EHPC
27	Sm	B	Lo	Hi	MPC	54	Me	B	Lo	Hi	HPC	81	La	B	Lo	Hi	EHPC

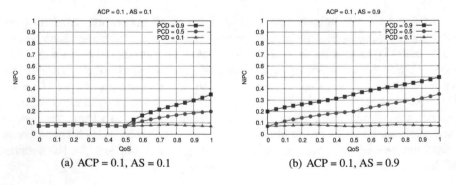

(a) ACP = 0.1, AS = 0.1   (b) ACP = 0.1, AS = 0.9

Fig. 4. Simulation results for ACP = 0.1.

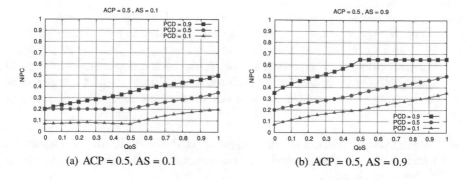

(a) ACP = 0.5, AS = 0.1   (b) ACP = 0.5, AS = 0.9

Fig. 5. Simulation results for ACP = 0.5.

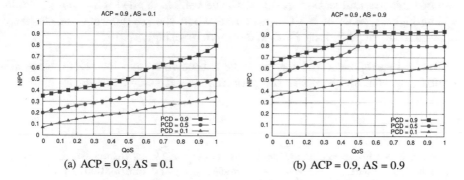

(a) ACP = 0.9, AS = 0.1   (b) ACP = 0.9, AS = 0.9

Fig. 6. Simulation results for ACP = 0.9.

When vehicles are willing to lend a large processing power but small storage (see Fig. 6(a)), we see that only the ones that will stay inside the communication range for a long period of time and offer at least a moderate value of QoS are able to help. Although these neighbors have small available storage, they can process the data while retransmitting the processed data step by step after each single sub-task/task has finished, thus successfully accomplishing the required processing tasks. However, for this to happen, the communication link between these neighbors and the *vehicle*, must assure a good value of QoS so that they could be able to retransmit the output after each completed subtask. On the other hand, from Fig. 6(b), we see that when a neighbor vehicle offers big available resources in terms of both computing power and storage, it is considered as helpful if the communication session is expected to last for a while regardless the assured QoS. In case of a bad QoS, these neighbors can store the processed data in their storage and transmit them when an improved Qos is provided.

5   Conclusions

In this paper, we proposed a fuzzy-based system to assess the available edge computing resources in a layered Cloud-Fog-Edge architecture for SDN-VANETs.

Our proposed system decides if a neighboring vehicle is capable to help a vehicle that lacks the appropriate resources to accomplish certain tasks based on ACP, AS, PCD and QoS. After assessing the processing capability for each neighbor, our previous proposed Fuzzy System for Resource Management [6] can select the appropriate layer in terms of data processing. We evaluated our proposed system by computer simulations. From the simulations results, we conclude as follows.

- The neighboring vehicles which have small ACP and small AS are not capable to help other vehicles in need regardless enabling a reliable communication link between them.
- A neighboring vehicle with a medium amount of ACP is considered as helpful, when the communication session with the vehicle in need is long and reliable.
- The highest value of NiPC is achieved when the neighboring vehicle has a large ACP, big AS, long PCD and a moderate/high QoS.

In the future, we would like to make extensive simulations to evaluate the proposed system and compare the performance with other systems.

References

1. Kandel, A.: Fuzzy expert systems. CRC Press Inc., Boca Raton, FL, USA (1992)
2. Klir, G.J., Folger, T.A.: Fuzzy sets, uncertainty, and information. Prentice Hall, Upper Saddle River, NJ, USA (1988)
3. Ku, I., Lu, Y., Gerla, M., Gomes, R.L., Ongaro, F., Cerqueira, E.: Towards software-defined VANET: Architecture and services. In: 13th Annual Mediterranean Ad Hoc Networking Workshop (MED-HOC-NET), pp. 103–110 (2014)
4. McNeill, F.M., Thro, E.: Fuzzy logic: a practical approach. Academic Press Professional Inc., San Diego, CA, USA (1994)
5. Munakata, T., Jani, Y.: Fuzzy systems: an overview. Commun. ACM **37**(3), 69–77 (1994)
6. Qafzezi, E., Bylykbashi, K., Ikeda, M., Matsuo, K., Barolli, L.: Coordination and management of cloud, fog and edge resources in SDN-VANETS using fuzzy logic: a comparison study for two fuzzy-based systems. Internet of Things **11**, 100169 (2020)
7. Truong, N.B., Lee, G.M., Ghamri-Doudane, Y.: Software defined networking-based vehicular adhoc network with fog computing. In: 2015 IFIP/IEEE International Symposium on Integrated Network Management (IM), pp. 1202–1207 (2015)
8. Xu, W., Zhou, H., Cheng, N., Lyu, F., Shi, W., Chen, J., Shen, X.: Internet of vehicles in big data era. IEEE/CAA J. Automatica Sinica **5**(1), 19–35 (2018)
9. Zadeh, L.A., Kacprzyk, J.: Fuzzy logic for the management of uncertainty. John Wiley & Sons Inc., New York, NY, USA (1992)
10. Zimmermann, H.J.: Fuzzy control. In: Fuzzy Set Theory and Its Applications, pp. 203–240. Springer (1996)

A Comparison Study of Constriction and Random Inertia Weight Router Replacement Methods for WMNs by WMN-PSOHC Hybrid Intelligent Simulation System Considering Normal Distribution of Mesh Clients

Shinji Sakamoto[1]([✉]), Leonard Barolli[2], and Shusuke Okamoto[1]

[1] Department of Computer and Information Science, Seikei University,
3-3-1 Kichijoji-Kitamachi, Musashino-shi 180-8633, Tokyo, Japan
okam@st.seikei.ac.jp
[2] Department of Information and Communication Engineering,
Fukuoka Institute of Technology, 3-30-1 Wajiro-Higashi,
Higashi-Ku, Fukuoka 811-0295, Japan
barolli@fit.ac.jp

Abstract. Wireless Mesh Networks (WMNs) are low-cost and have easy maintenance. However, WMNs have some problems e.g. node placement problems, security, transmission power and so on. To solve node placement problem in WMNs, we have implemented a hybrid simulation system based on PSO and HC called WMN-PSOHC. In this paper, we present the performance evaluation of WMNs by using WMN-PSOHC intelligent system considering Constriction Method (CM) and Random Inertia Weight Method (RIWM). The simulation results show that a better performance is achieved for CM compared with the RIWM.

1 Introduction

The wireless networks and devices are becoming increasingly popular and they provide users access to information and communication anytime and anywhere [2–4,7,8,10]. Wireless Mesh Networks (WMNs) are gaining a lot of attention because of their low cost nature that makes them attractive for providing wireless Internet connectivity. A WMN is dynamically self-organized and self-configured, with the nodes in the network automatically establishing and maintaining mesh connectivity among them-selves (creating, in effect, an ad hoc network). This feature brings many advantages to WMNs such as low up-front cost, easy network maintenance, robustness and reliable service coverage [1]. Moreover, such infrastructure can be used to deploy community networks, metropolitan area networks, municipal and corporative networks, and to support applications for urban areas, medical, transport and surveillance systems.

L. Barolli et al. (Eds.): AINA 2021, LNNS 225, pp. 73–81, 2021.
https://doi.org/10.1007/978-3-030-75100-5_7

In this work, we deal with node placement problem in WMNs. We consider the version of the mesh router nodes placement problem in which we are given a grid area where to deploy a number of mesh router nodes and a number of mesh client nodes of fixed positions (of an arbitrary distribution) in the grid area. The objective is to find a location assignment for the mesh routers to the cells of the grid area that maximizes the network connectivity and client coverage. Network connectivity is measured by Size of Giant Component (SGC) of the resulting WMN graph, while the user coverage is simply the number of mesh client nodes that fall within the radio coverage of at least one mesh router node and is measured by Number of Covered Mesh Clients (NCMC). Node placement problems are known to be computationally hard to solve [21]. In some previous works, intelligent algorithms have been recently investigated [9,15,16]. We already implemented a Particle Swarm Optimization (PSO) based simulation system, called WMN-PSO [13]. Also, we implemented a simulation system based on Hill Climbing (HC) for solving node placement problem in WMNs, called WMN-HC [12].

In our previous work [13,14], we presented a hybrid intelligent simulation system based on PSO and HC. We called this system WMN-PSOHC. In this paper, we analyze the performance of Constriction Method (CM) and Random Inertia Weight Method (RIWM) by WMN-PSOHC simulation system considering Normal distribution of mesh clients.

The rest of the paper is organized as follows. We present our designed and implemented hybrid simulation system in Sect. 2. In Sect. 3, we introduce WMN-PSOHC Web GUI tool. The simulation results are given in Sect. 4. Finally, we give conclusions and future work in Sect. 5.

2   Proposed and Implemented Simulation System

2.1   Particle Swarm Optimization

In Particle Swarm Optimization (PSO) algorithm, a number of simple entities (the particles) are placed in the search space of some problem or function and each evaluates the objective function at its current location. The objective function is often minimized and the exploration of the search space is not through evolution [11]. However, following a widespread practice of borrowing from the evolutionary computation field, in this work, we consider the bi-objective function and fitness function interchangeably. Each particle then determines its movement through the search space by combining some aspect of the history of its own current and best (best-fitness) locations with those of one or more members of the swarm, with some random perturbations. The next iteration takes place after all particles have been moved. Eventually the swarm as a whole, like a flock of birds collectively foraging for food, is likely to move close to an optimum of the fitness function.

Each individual in the particle swarm is composed of three $\mathcal{D}$-dimensional vectors, where $\mathcal{D}$ is the dimensionality of the search space. These are the current position $\vec{x}_i$, the previous best position $\vec{p}_i$ and the velocity $\vec{v}_i$.

The particle swarm is more than just a collection of particles. A particle by itself has almost no power to solve any problem; progress occurs only when the particles interact. Problem solving is a population-wide phenomenon, emerging from the individual behaviors of the particles through their interactions. In any case, populations are organized according to some sort of communication structure or topology, often thought of as a social network. The topology typically consists of bidirectional edges connecting pairs of particles, so that if j is in i's neighborhood, i is also in j's. Each particle communicates with some other particles and is affected by the best point found by any member of its topological neighborhood. This is just the vector $\vec{p}_i$ for that best neighbor, which we will denote with $\vec{p}_g$. The potential kinds of population "social networks" are hugely varied, but in practice certain types have been used more frequently.

In the PSO process, the velocity of each particle is iteratively adjusted so that the particle stochastically oscillates around $\vec{p}_i$ and $\vec{p}_g$ locations.

2.2 Hill Climbing

Hill Climbing (HC) algorithm is a heuristic algorithm. The idea of HC is simple. In HC, the solution s' is accepted as the new current solution if $\delta \leq 0$ holds, where $\delta = f(s') - f(s)$. Here, the function f is called the fitness function. The fitness function gives points to a solution so that the system can evaluate the next solution s' and the current solution s.

The most important factor in HC is to define effectively the neighbor solution. The definition of the neighbor solution affects HC performance directly. In our WMN-PSOHC system, we use the next step of particle-pattern positions as the neighbor solutions for the HC part.

2.3 WMN-PSOHC System Description

In following, we present the initialization, client distributions, particle-pattern, fitness function and router replacement methods.

Initialization
Our proposed system starts by generating an initial solution randomly, by *ad hoc* methods [22]. We decide the velocity of particles by a random process considering the area size. For instance, when the area size is $W \times H$, the velocity is decided randomly from $-\sqrt{W^2 + H^2}$ to $\sqrt{W^2 + H^2}$.

Client Distributions
Our system can generate many client distributions. In this paper, we consider Normal distribution of mesh clients. In Normal distribution, clients are located around the center of considered area as shown in Fig. 1

Particle-Pattern
A particle is a mesh router. A fitness value of a particle-pattern is computed by combination of mesh routers and mesh clients positions. In other words,

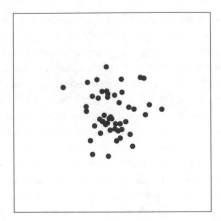

Fig. 1. Normal distribution of mesh clients.

each particle-pattern is a solution as shown is Fig. 2. Therefore, the number of particle-patterns is a number of solutions.

Fitness Function
One of most important thing is to decide the determination of an appropriate objective function and its encoding. In our case, each particle-pattern has an own fitness value and compares other particle-patterns fitness value in order to share information of global solution. The fitness function follows a hierarchical approach in which the main objective is to maximize the SGC in WMN. Thus, we use α and β weight-coefficients for the fitness function and the fitness function of this scenario is defined as:

$$\text{Fitness} = \alpha \times \text{SGC}(\boldsymbol{x}_{ij}, \boldsymbol{y}_{ij}) + \beta \times \text{NCMC}(\boldsymbol{x}_{ij}, \boldsymbol{y}_{ij}).$$

Router Replacement Methods
A mesh router has x, y positions and velocity. Mesh routers are moved based on velocities. There are many router replacement methods in PSO field [6, 18–20]. In this paper, we consider CM and RIWM.

Constriction Method (CM)
    CM is a method which PSO parameters are set to a week stable region ($\omega = 0.729$, $C_1 = C2 = 1.4955$) based on analysis of PSO by M. Clerc et al. [4–6].
Random Inertia Weight Method (RIWM)
    In RIWM, the ω parameter is changing randomly from 0.5 to 1.0. The C_1 and C_2 are kept 2.0. The ω can be estimated by the week stable region. The average of ω is 0.75 [17, 19].

3   WMN-PSOHC Web GUI Tool

The Web application follows a standard Client-Server architecture and is implemented using LAMP (Linux + Apache + MySQL + PHP) technology (see

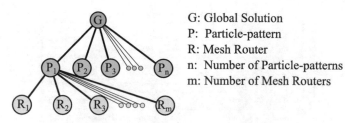

G: Global Solution
P: Particle-pattern
R: Mesh Router
n: Number of Particle-patterns
m: Number of Mesh Routers

Fig. 2. Relationship among global solution, particle-patterns and mesh routers.

Fig. 3). We show the WMN-PSOHC Web GUI tool in Fig. 4. Remote users (clients) submit their requests by completing first the parameter setting. The parameter values to be provided by the user are classified into three groups, as follows.

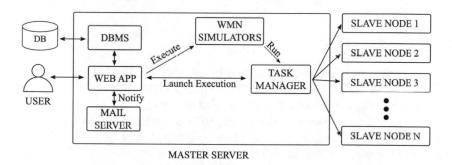

Fig. 3. System structure for web interface.

- Parameters related to the problem instance: These include parameter values that determine a problem instance to be solved and consist of number of router nodes, number of mesh client nodes, client mesh distribution, radio coverage interval and size of the deployment area.
- Parameters of the resolution method: Each method has its own parameters.
- Execution parameters: These parameters are used for stopping condition of the resolution methods and include number of iterations and number of independent runs. The former is provided as a total number of iterations and depending on the method is also divided per phase (e.g., number of iterations in a exploration). The later is used to run the same configuration for the same problem instance and parameter configuration a certain number of times.

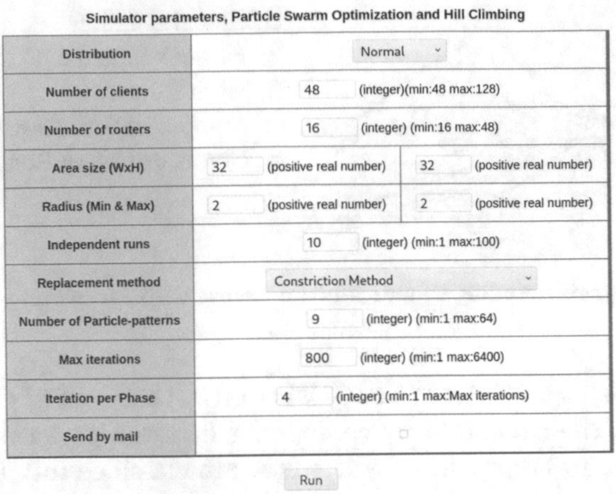

Simulator parameters, Particle Swarm Optimization and Hill Climbing

Distribution	Normal ⌄	
Number of clients	48 (integer)(min:48 max:128)	
Number of routers	16 (integer) (min:16 max:48)	
Area size (WxH)	32 (positive real number)	32 (positive real number)
Radius (Min & Max)	2 (positive real number)	2 (positive real number)
Independent runs	10 (integer) (min:1 max:100)	
Replacement method	Constriction Method ⌄	
Number of Particle-patterns	9 (integer) (min:1 max:64)	
Max iterations	800 (integer) (min:1 max:6400)	
Iteration per Phase	4 (integer) (min:1 max:Max iterations)	
Send by mail	☐	

Run

Fig. 4. WMN-PSOHC Web GUI Tool.

4   Simulation Results

In this section, we show simulation results using WMN-PSOHC system. In this work, we consider Normal distribution of mesh clients. The number of mesh routers is considered 16 and the number of mesh clients 48. We consider the number of particle-patterns 9. We conducted simulations 100 times, in order to avoid the effect of randomness and create a general view of results. The total number of iterations is considered 800 and the iterations per phase is considered 4. We show the parameter setting for WMN-PSOHC in Table 1.

Table 1. Parameter settings.

Parameters	Values
Clients distribution	Normal distribution
Area size	32.0×32.0
Number of mesh routers	16
Number of mesh clients	48
Total iterations	800
Iteration per phase	4
Number of particle-patterns	9
Radius of a mesh router	2.0
Fitness function weight-coefficients (α, β)	0.7, 0.3
Replacement method	CM, RIWM

We show the simulation results in Figs. 5 and 6. For SGC, both replacement methods reach the maximum (100%). This means that all mesh routers are connected to each other. We see that RIWM converges faster than CM for SGC.

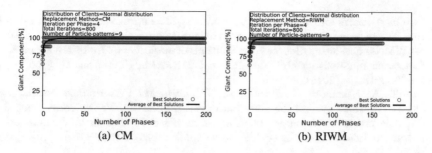

Fig. 5. Simulation results of WMN-PSOHC for SGC.

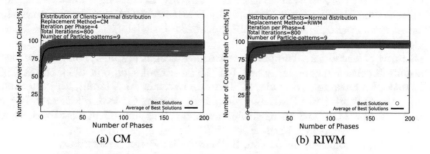

Fig. 6. Simulation results of WMN-PSOHC for NCMC.

Also, for the NCMC, RIWM performs better than CM. Therefore, we conclude that the performance of RIWM is better than CM.

5   Conclusions

In this work, we evaluated the performance of CM and RIWM router replacement methods for WMNs by WMN-PSOHC hybrid intelligent simulation system. Simulation results show that the performance of CM is better compared with RIWM.

In our future work, we would like to evaluate the performance of the proposed system for different parameters and scenarios.

References

1. Akyildiz, I.F., Wang, X., Wang, W.: Wireless mesh networks: a survey. Comput. Netw. **47**(4), 445–487 (2005)
2. Barolli, A., Sakamoto, S., Barolli, L., Takizawa, M.: A hybrid simulation system based on particle swarm optimization and distributed genetic algorithm for wmns: performance evaluation considering normal and uniform distribution of mesh clients. In: International Conference on Network-Based Information Systems, pp. 42–55. Springer (2018)

3. Barolli, A., Sakamoto, S., Barolli, L., Takizawa, M.: Performance analysis of simulation system based on particle swarm optimization and distributed genetic algorithm for WMNs considering different distributions of mesh clients. In: International Conference on Innovative Mobile and Internet Services in Ubiquitous Computing, pp 32–45. Springer (2018)
4. Barolli, A., Sakamoto, S., Barolli, L., Takizawa, M.: Performance evaluation of WMN-PSODGA system for node placement problem in WMNs considering four different crossover methods. In: The 32nd IEEE International Conference on Advanced Information Networking and Applications (AINA-2018), pp 850–857. IEEE (2018)
5. Barolli, A., Sakamoto, S., Durresi, H., Ohara, S., Barolli, L., Takizawa, M.: A Comparison Study of Constriction and Linearly Decreasing Vmax Replacement Methods for Wireless Mesh Networks by WMN-PSOHC-DGA Simulation System. In: International Conference on P2P, pp. 26–34. Parallel, Grid, Cloud and Internet Computing, Springer (2019)
6. Clerc, M., Kennedy, J.: The particle swarm-explosion, stability, and convergence in a multidimensional complex space. IEEE Trans. Evol. Comput. **6**(1), 58–73 (2002)
7. Matsuo, K., Sakamoto, S., Oda, T., Barolli, A., Ikeda, M., Barolli, L.: Performance analysis of WMNs by WMN-GA simulation system for two WMN architectures and different TCP congestion-avoidance algorithms and client distributions. Int. J. Commun. Networks Distrib. Syst. **20**(3), 335–351 (2018)
8. Ohara, S., Barolli, A., Sakamoto, S., Barolli, L.: Performance analysis of WMNs by WMN-PSODGA simulation system considering load balancing and client uniform distribution. In: International Conference on Innovative Mobile and Internet Services in Ubiquitous Computing, pp. 25–38. Springer (2019)
9. Ozera, K., Bylykbashi, K., Liu, Y., Barolli, L.: A fuzzy-based approach for cluster management in VANETs: performance evaluation for two fuzzy-based systems. Internet Things **3**, 120–133 (2018)
10. Ozera, K., Inaba, T., Bylykbashi, K., Sakamoto, S., Ikeda, M., Barolli, L.: A WLAN Triage Testbed based on fuzzy logic and its performance evaluation for different number of clients and throughput parameter. Int. J. Grid Util. Comput. **10**(2), 168–178 (2019)
11. Poli, R., Kennedy, J., Blackwell, T.: Particle swarm optimization. Swarm Intell. **1**(1), 33–57 (2007)
12. Sakamoto, S., Lala, A., Oda, T., Kolici, V., Barolli, L., Xhafa, F.: Analysis of WMN-HC simulation system data using friedman test. In: The Ninth International Conference on Complex, Intelligent, and Software Intensive Systems (CISIS-2015), pp 254–259. IEEE (2015)
13. Sakamoto, S., Oda, T., Ikeda, M., Barolli, L., Xhafa, F.: Implementation and evaluation of a simulation system based on particle swarm optimisation for node placement problem in wireless mesh networks. Int. J. Commun. Networks Distrib. Syst. **17**(1), 1–13 (2016)
14. Sakamoto, S., Ozera, K., Ikeda, M., Barolli, L.: Implementation of intelligent hybrid systems for node placement problem in WMNs considering particle swarm optimization, hill climbing and simulated annealing. Mob. Networks Appl. **23**(1), 27–33 (2018)
15. Sakamoto, S., Barolli, A., Barolli, L., Okamoto, S.: Implementation of a web interface for hybrid intelligent systems. Int. J. Web Inf. Syst. **15**(4), 420–431 (2019)
16. Sakamoto, S., Barolli, L., Okamoto, S.: WMN-PSOSA: an intelligent hybrid simulation system for WMNs and its performance evaluations. Int. J. Web Grid Serv. **15**(4), 353–366 (2019)

17. Sakamoto, S., Ohara, S., Barolli, L., Okamoto, S.: Performance evaluation of WMNs by WMN-PSOHC system considering random inertia weight and linearly decreasing vmax replacement methods. In: International Conference on Network-Based Information Systems, pp. 27–36. Springer (2019)
18. Schutte, J.F., Groenwold, A.A.: A study of global optimization using particle swarms. J. Global Optim. **31**(1), 93–108 (2005)
19. Shi, Y.: Particle swarm optimization. IEEE Connections **2**(1), 8–13 (2004)
20. Shi, Y., Eberhart, R.C.: Parameter selection in particle swarm optimization. Evolutionary programming VII, pp. 591–600 (1998)
21. Wang, J., Xie, B., Cai, K., Agrawal, D.P.: Efficient mesh router placement in wireless mesh networks. In: Proceedings of IEEE Internatonal Conference on Mobile Adhoc and Sensor Systems (MASS-2007), pp. 1–9 (2007)
22. Xhafa, F., Sanchez, C., Barolli, L.: Ad hoc and neighborhood search methods for placement of mesh routers in wireless mesh networks. In: Proceedings of 29th IEEE International Conference on Distributed Computing Systems Workshops (ICDCS-2009) pp. 400–405 (2009)

Performance Evaluation of RIWM and LDVM Router Replacement Methods for WMNs by WMN-PSOSA-DGA Considering Chi-Square Distribution

Admir Barolli[1]([✉]), Shinji Sakamoto[2], Phudit Ampririt[3], Leonard Barolli[4], and Makoto Takizawa[5]

[1] Department of Information Technology, Aleksander Moisiu University of Durres,
L.1, Rruga e Currilave, Durres, Albania
[2] Department of Computer and Information Science, Seikei University,
3-3-1 Kichijoji-Kitamachi, Musashino-shi, Tokyo 180-8633, Japan
[3] Graduate School of Engineering, Fukuoka Institute of Technology,
3-30-1 Wajiro-Higashi, Higashi-Ku, Fukuoka 811-0295, Japan
[4] Department of Information and Communication Engineering,
Fukuoka Institute of Technology, 3-30-1 Wajiro-Higashi, Higashi-Ku,
Fukuoka 811-0295, Japan
barolli@fit.ac.jp
[5] Department of Advanced Sciences, Faculty of Science and Engineering,
Hosei University, 3-7-2 Kajino-Cho, Koganei-Shi, Tokyo 184-8584, Japan
makoto.takizawa@computer.org

Abstract. Wireless Mesh Networks (WMNs) are gaining a lot of attention from researchers due to their advantages such as easy maintenance, low upfront cost, and high robustness. Connectivity and stability directly affect the performance of WMNs. However, WMNs have some problems such as node placement problem, hidden terminal problem and so on. In our previous work, we implemented a simulation system to solve the node placement problem in WMNs considering Particle Swarm Optimization (PSO), Simulated Annealing (SA) and Distributed Genetic Algorithm (DGA), called WMN-PSOSA-DGA. In this paper, we evaluate the performance of Random Inertia Weight Method (RIWM) and Linearly Decreasing Vmax Method (LDVM) for WMNs using WMN-PSOSA-DGA hybrid simulation system considering Chi-square distribution of mesh clients. Simulation results show that a good performance is achieved for LDVM compared with the case of RIWM.

1 Introduction

The wireless networks and devices are becoming increasingly popular and they provide users access to information and communication anytime and anywhere [2,11,13]. Wireless Mesh Networks (WMNs) are gaining a lot of attention because of their low cost nature that makes them attractive for providing wireless

L. Barolli et al. (Eds.): AINA 2021, LNNS 225, pp. 82–91, 2021.
https://doi.org/10.1007/978-3-030-75100-5_8

Internet connectivity. A WMN is dynamically self-organized and self-configured, with the nodes in the network automatically establishing and maintaining mesh connectivity among them-selves (creating, in effect, an ad hoc network). This feature brings many advantages to WMNs such as low up-front cost, easy network maintenance, robustness and reliable service coverage [1].

Mesh node placement in WMN can be seen as a family of problems, which are shown to be computationally hard to solve for most of the formulations [21]. We consider the version of the mesh router nodes placement problem in which we are given a grid area where to deploy a number of mesh router nodes and a number of mesh client nodes of fixed positions (of an arbitrary distribution) in the grid area. The objective is to find a location assignment for the mesh routers to the cells of the grid area that maximizes the network connectivity and client coverage. Network connectivity is measured by Size of Giant Component (SGC) of the resulting WMN graph, while the user coverage is simply the number of mesh client nodes that fall within the radio coverage of at least one mesh router node and is measured by Number of Covered Mesh Clients (NCMC). Node placement problems are known to be computationally hard to solve [9,22]. In previous works, some intelligent algorithms have been investigated for node placement problem [4,12].

In [15], we implemented a Particle Swarm Optimization (PSO) and Simulated Annealing (SA) based simulation system, called WMN-PSOSA. Also, we implemented another simulation system based on Genetic Algorithm (GA), called WMN-GA [4,10], for solving node placement problem in WMNs. Then, we designed a hybrid intelligent system based on PSO, SA and DGA, called WMN-PSOSA-DGA [14].

In this paper, we evaluate the performance of Random Inertia Weight Method (RIWM) and Linearly Decreasing Vmax Method (LDVM) for WMNs using WMN-PSOSA-DGA simulation system considering Chi-square distribution of mesh clients.

The rest of the paper is organized as follows. We present our designed and implemented hybrid simulation system in Sect. 2. The simulation results are given in Sect. 3. Finally, we give conclusions and future work in Sect. 4.

2 Proposed and Implemented Simulation System

Distributed Genetic Algorithms (DGAs) are capable of producing solutions with higher efficiency (in terms of time) and efficacy (in terms of better quality solutions). They have shown their usefulness for the resolution of many computationally hard combinatorial optimization problems. Also, Particle Swarm Optimization (PSO) and Simulated Annealing (SA) are suitable for solving NP-hard problems.

2.1 Velocities and Positions of Particles

WMN-PSOSA-DGA decides the velocity of particles by a random process considering the area size. For instance, when the area size is $W \times H$, the velocity is decided randomly from $-\sqrt{W^2 + H^2}$ to $\sqrt{W^2 + H^2}$. Each particle's velocities are updated by simple rule.

For SA mechanism, next positions of each particle are used for neighbor solution s'. The fitness function f gives points to the current solution s. If $f(s')$ is larger than $f(s)$, the s' is better than s so the s is updated to s'. However, if $f(s')$ is not larger than $f(s)$, the s may be updated by using the probability of $\exp\left[\frac{f(s')-f(s)}{T}\right]$. Where T is called the "Temperature value" which is decreased with the computation so that the probability to update will be decreased. This mechanism of SA is called a cooling schedule and the next Temperature value of computation is calculated as $T_{n+1} = \alpha \times T_n$. In this paper, we set the starting temperature, ending temperature and number of iterations. We calculate α as

$$\alpha = \left(\frac{\text{SA ending temperature}}{\text{SA starting temperature}}\right)^{1.0/\text{number of iterations}}.$$

It should be noted that the positions are not updated but the velocities are updated in the case when the solusion s is not updated.

2.2   Routers Replacement Methods

A mesh router has x, y positions and velocity. Mesh routers are moved based on velocities. There are many router replacement methods. In this paper, we use RIWM and LDVM.

Constriction Method (CM).
    CM is a method which PSO parameters are set to a week stable region ($\omega = 0.729, C_1 = C2 = 1.4955$) based on analysis of PSO by M. Clerc et al. [5,8,17].
Random Inertia Weight Method (RIWM).
    In RIWM, the ω parameter is changing randomly from 0.5 to 1.0. The C_1 and C_2 are kept 2.0. The ω can be estimated by the week stable region. The average of ω is 0.75 [7,19].
Linearly Decreasing Inertia Weight Method (LDIWM)
    In LDIWM, C_1 and C_2 are set to 2.0, constantly. On the other hand, the ω parameter is changed linearly from unstable region ($\omega = 0.9$) to stable region ($\omega = 0.4$) with increasing of iterations of computations [6,20].
Linearly Decreasing Vmax Method (LDVM)
    In LDVM, PSO parameters are set to unstable region ($\omega = 0.9, C_1 = C_2 = 2.0$). A value of V_{max} which is maximum velocity of particles is considered. With increasing of iteration of computations, the V_{max} is kept decreasing linearly [7,16,18].
Rational Decrement of Vmax Method (RDVM)
    In RDVM, PSO parameters are set to unstable region ($\omega = 0.9, C_1 = C_2 = 2.0$). The V_{max} is kept decreasing with the increasing of iterations as

$$V_{max}(x) = \sqrt{W^2 + H^2} \times \frac{T - x}{x}.$$

Where, W and H are the width and the height of the considered area, respectively. Also, T and x are the total number of iterations and a current number of iteration, respectively.

2.3   DGA Operations

Population of individuals: Unlike local search techniques that construct a path in the solution space jumping from one solution to another one through local perturbations, DGA use a population of individuals giving thus the search a larger scope and chances to find better solutions. This feature is also known as "exploration" process in difference to "exploitation" process of local search methods.

Selection: The selection of individuals to be crossed is another important aspect in DGA as it impacts on the convergence of the algorithm. Several selection schemes have been proposed in the literature for selection operators trying to cope with premature convergence of DGA. There are many selection methods in GA. In our system, we implement 2 selection methods: Random method and Roulette wheel method.

Crossover operators: Use of crossover operators is one of the most important characteristics. Crossover operator is the means of DGA to transmit best genetic features of parents to offsprings during generations of the evolution process. Many methods for crossover operators have been proposed such as Blend Crossover (BLX-α), Unimodal Normal Distribution Crossover (UNDX), Simplex Crossover (SPX).

Mutation operators: These operators intend to improve the individuals of a population by small local perturbations. They aim to provide a component of randomness in the neighborhood of the individuals of the population. In our system, we implemented two mutation methods: uniformly random mutation and boundary mutation.

Escaping from local optimal: GA itself has the ability to avoid falling prematurely into local optimal and can eventually escape from them during the search process. DGA has one more mechanism to escape from local optimal by considering some islands. Each island computes GA for optimizing and they migrate its gene to provide the ability to avoid from local optimal.

Convergence: The convergence of the algorithm is the mechanism of DGA to reach to good solutions. A premature convergence of the algorithm would cause that all individuals of the population be similar in their genetic features and thus the search would result ineffective and the algorithm getting stuck into local optimal. Maintaining the diversity of the population is therefore very important to this family of evolutionary algorithms.

In following, we present fitness function, migration function, particle pattern and gene coding.

2.4   Fitness and Migration Functions

The determination of an appropriate fitness function, together with the chromosome encoding are crucial to the performance. Therefore, one of most important thing is to decide the determination of an appropriate objective function and its encoding. In our case, each particle-pattern and gene has an own fitness value which is comparable and compares it with other fitness value in order to share information of global solution. The fitness function follows a hierarchical approach in which the main objective is to maximize the SGC in WMN. Thus, the fitness function of this scenario is defined as

$$\text{Fitness} = 0.7 \times \text{SGC}(\boldsymbol{x}_{ij}, \boldsymbol{y}_{ij}) + 0.3 \times \text{NCMC}(\boldsymbol{x}_{ij}, \boldsymbol{y}_{ij}).$$

Our implemented simulation system uses Migration function as shown in Fig. 1. The Migration function swaps solutions between PSOSA part and DGA part.

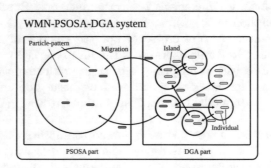

Fig. 1. Model of WMN-PSOSA-DGA migration.

2.5   Particle-Pattern and Gene Coding

In order to swap solutions, we design particle-patterns and gene coding carefully. A particle is a mesh router. Each particle has position in the considered area and velocities. A fitness value of a particle-pattern is computed by combination of mesh routers and mesh clients positions. In other words, each particle-pattern is a solution as shown is Fig. 2.

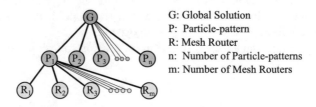

Fig. 2. Relationship among global solution, particle-patterns and mesh routers in PSOSA part.

A gene describes a WMN. Each individual has its own combination of mesh nodes. In other words, each individual has a fitness value. Therefore, the combination of mesh nodes is a solution.

3   Simulation Results

In this section, we show simulation results. In this work, we analyze the performance of RIWM and LDVM router replacement methods for WMNs by WMN-PSOSA-DGA hybrid intelligent simulation system considering Chi-square client distribution as shown in Fig. 3 [3].

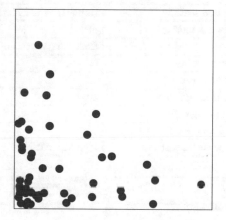

Fig. 3. Chi-square distribution.

We carried out the simulations 10 times in order to avoid the effect of randomness and create a general view of results. We show the parameter setting for WMN-PSOSA-DGA in Table 1.

We show simulation results in Figs. 4 and 5. We see that for SGC, the performance is almost the same for both replacement method. However, for NCMC, LDVM performs better than RIWM. The visualized simulation results are shown in Fig. 6. For both replacement methods, all mesh routers are connected, but some clients are not covered. We can see the number of covered mesh clients is larger for LDVM compared with the case of RIWM.

Table 1. WMN-PSOSA-DGA parameters.

Parameters	Values
Clients distribution	Chi-Square
Area size	32.0×32.0
Number of mesh routers	16
Number of mesh clients	48
Number of GA islands	16
Number of Particle-patterns	32
Number of migrations	200
Evolution steps	320
Radius of a mesh router	$2.0 - 3.5$
Selection method	Roulette wheel method
Crossover method	SPX
Mutation method	Boundary mutation
Crossover rate	0.8
Mutation rate	0.2
SA Starting value	10.0
SA Ending value	0.01
Total number of iterations	64000
Replacement method	RIWM, LDVM

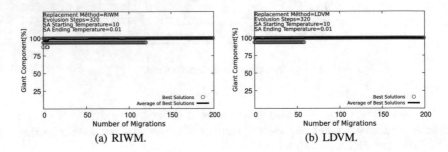

(a) RIWM.   (b) LDVM.

Fig. 4. Simulation results of WMN-PSOSA-DGA for SGC.

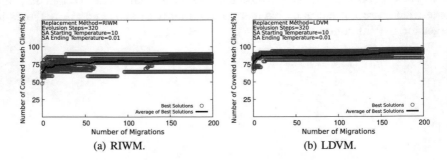

(a) RIWM.   (b) LDVM.

Fig. 5. Simulation results of WMN-PSOSA-DGA for NCMC.

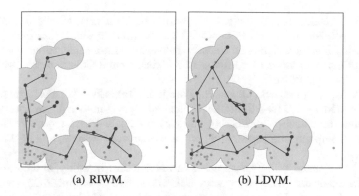

(a) RIWM.   (b) LDVM.

Fig. 6. Visualized simulation results of WMN-PSOSA-DGA for different replacement methods.

4 Conclusions

In this work, we evaluated the performance of RIWM and LDVM replacement methods for WMNs using a hybrid simulation system based on PSO, SA and DGA (called WMN-PSOSA-DGA) considering Chi-square distribution of mesh clients. Simulation results show that a good performance was achieved for LDVM compared with the case of RIWM.

In our future work, we would like to evaluate the performance of the proposed system for different parameters and patterns.

References

1. Akyildiz, I.F., Wang, X., Wang, W.: Wireless mesh networks: a survey. Comput. Networks **47**(4), 445–487 (2005)
2. Barolli, A., Sakamoto, S., Barolli, L., Takizawa, M.: A hybrid simulation system based on particle swarm optimization and distributed genetic algorithm for wmns: performance evaluation considering normal and uniform distribution of mesh clients. In: International Conference on Network-Based Information Systems, pp. 42–55. Springer (2018)
3. Barolli, A., Sakamoto, S., Barolli, L., Takizawa, M.: Performance analysis of simulation system based on particle swarm optimization and distributed genetic algorithm for WMNs considering different distributions of mesh clients. In: International Conference on Innovative Mobile and Internet Services in Ubiquitous Computing, pp 32–45. Springer (2018)
4. Barolli, A., Sakamoto, S., Ozera, K., Barolli, L., Kulla, E., Takizawa, M.: Design and Implementation of a Hybrid Intelligent System Based on Particle Swarm Optimization and Distributed Genetic Algorithm. In: International Conference on Emerging Internetworking, pp. 79–93. Springer, Data & Web Technologies (2018)
5. Barolli, A., Sakamoto, S., Durresi, H., Ohara, S., Barolli, L., Takizawa, M.: A comparison study of constriction and linearly decreasing vmax replacement methods for wireless mesh networks by WMN-PSOHC-DGA simulation system. In: International Conference on P2P Parallel, Grid, Cloud and Internet Computing, pp. 26–34. Springer (2019)

6. Barolli, A., Sakamoto, S., Ohara, S., Barolli, L., Takizawa, M.: Performance analysis of WMNs by WMN-PSOHC-DGA simulation system considering linearly decreasing inertia weight and linearly decreasing vmax replacement methods. In: International Conference on Intelligent Networking and Collaborative Systems, pp. 14–23. Springer (2019)
7. Barolli, A., Sakamoto, S., Ohara, S., Barolli, L., Takizawa, M.: Performance analysis of WMNs by WMN-PSOHC-DGA simulation system considering random inertia weight and linearly decreasing vmax router replacement methods. In: Conference on Complex, Intelligent, and Software Intensive Systems, pp. 13–21. Springer (2019)
8. Clerc, M., Kennedy, J.: The particle swarm-explosion, stability, and convergence in a multidimensional complex space. IEEE Trans. Evol. Comput. **6**(1), 58–73 (2002)
9. Maolin, T., et al.: Gateways placement in backbone wireless mesh networks. Int. J. Commun. Network Syst. Sci. **2**(1), 44 (2009)
10. Matsuo, K., Sakamoto, S., Oda, T., Barolli, A., Ikeda, M., Barolli, L.: Performance analysis of WMNs by WMN-GA simulation system for two WMN architectures and different TCP congestion-avoidance algorithms and client distributions. Int. J. Commun. Networks Distributed Syst. **20**(3), 335–351 (2018)
11. Ohara, S., Barolli, A., Sakamoto, S., Barolli, L.: Performance analysis of WMNs by WMN-PSODGA simulation system considering load balancing and client uniform distribution. In: International Conference on Innovative Mobile and Internet Services in Ubiquitous Computing, pp. 25–38. Springer (2019)
12. Ozera, K., Sakamoto, S., Elmazi, D., Bylykbashi, K., Ikeda, M., Barolli, L.: A fuzzy approach for clustering in manets: performance evaluation for different parameters. Int. J. Space-Based Situated Comput. **7**(3), 166–176 (2017)
13. Ozera, K., Inaba, T., Bylykbashi, K., Sakamoto, S., Ikeda, M., Barolli, L.: A WLAN triage testbed based on fuzzy logic and its performance evaluation for different number of clients and throughput parameter. Int. J. Grid Utility Comput. **10**(2), 168–178 (2019)
14. Sakamoto, S., Barolli, A., Barolli, L., Takizawa, M.: Design and implementation of a hybrid intelligent system based on particle swarm optimization, hill climbing and distributed genetic algorithm for node placement problem in WMNs: a comparison study. In: The 32nd IEEE International Conference on Advanced Information Networking and Applications (AINA-2018), pp 678–685. IEEE (2018)
15. Sakamoto, S., Ozera, K., Ikeda, M., Barolli, L.: Implementation of intelligent hybrid systems for node placement problem in WMNs considering particle swarm optimization, hill climbing and simulated annealing. Mob. Networks Appl. **23**(1), 27–33 (2018)
16. Sakamoto, S., Ohara, S., Barolli, L., Okamoto, S.: Performance Evaluation of WMNs by WMN-PSOHC system considering random inertia weight and linearly decreasing vmax replacement methods. In: International Conference on Network-Based Information Systems, pp. 27–36. Springer (2019)
17. Sakamoto, S., Ohara, S., Barolli, L., Okamoto, S.: Performance Evaluation of WMNs WMN-PSOHC System Considering Constriction and Linearly Decreasing Inertia Weight Replacement Methods. In: International Conference on Broadband and Wireless Computing Communication and Applications, pp. 22–31. Springer (2019)
18. Schutte, J.F., Groenwold, A.A.: A study of global optimization using particle swarms. J. Global Optim. **31**(1), 93–108 (2005)
19. Shi, Y.: Particle swarm optimization. IEEE Connections **2**(1), 8–13 (2004)

20. Shi, Y., Eberhart, R.C.: Parameter Selection in Particle Swarm Optimization. Evolutionary programming VII, pp. 591–600 (1998)
21. Vanhatupa, T., Hannikainen, M., Hamalainen, T.: Genetic algorithm to optimize node placement and configuration for WLAN panning. In: The 4th IEEE International Symposium on Wireless Communication Systems, pp. 612–616 (2007)
22. Wang, J., Xie, B., Cai, K., Agrawal, D.P.: Efficient mesh router placement in wireless mesh networks. In: Proceedings of IEEE International Conference on Mobile Adhoc and Sensor Systems (MASS-2007), pp 1–9

An Adaptive Anti-packet Recovery Method for Vehicular DTN Considering Message Possession Rate

Shota Uchimura[1], Masaya Azuma[1], Yoshiki Tada[1], Makoto Ikeda[2(✉)], and Leonard Barolli[2]

[1] Graduate School of Engineering, Fukuoka Institute of Technology, 3-30-1 Wajiro-higashi, Higashi-ku, Fukuoka 811-0295, Japan
makoto.ikd@acm.org
[2] Department of Information and Communication Engineering, Fukuoka Institute of Technology, 3-30-1 Wajiro-higashi, Higashi-ku, Fukuoka 811-0295, Japan
barolli@fit.ac.jp

Abstract. In this paper, we propose an Adaptive Anti-packet Recovery (AAR) method based on epidemic protocol. Our method can be used in Vehicular Delay Tolerant Networking (DTN). We use conventional Epidemic with recovery function and the proposed AAR method as the message delivery protocols. From the simulation results, we observed that the delivery ratio of proposed AAR is good for dense networks.

Keywords: DTN · Adaptive anti-packet · Recovery method · Epidemic

1 Introduction

During recent years, with the development of the vehicle and information industries, we see many different services via the Internet. Therefore, it is important to build a strong network that is resilient to disasters. The network should have a higher QoE, which means that it can quickly switch to an alternate link in case of a communication disconnection or long delay [10]. This research focuses on vehicular networks. We consider the vehicle as an actor in different network situations. Delay-/Disruption-/Disconnection-Tolerant Networking (DTN) technology [5] has attracted much attention as a vehicular-based communication method. However, due to the overhead, the storage resources are a problem for the conventional DTN approach. Therefore, hybrid and recovery methods have been proposed to address these problems [4,6,11].

In [4], the authors have proposed a hybrid DTN routing method which considers the Epidemic-based routing method with many replications and the SpW-based routing method with few replications. They consider storage state of the vehicles to choice the routing method.

L. Barolli et al. (Eds.): AINA 2021, LNNS 225, pp. 92–101, 2021.
https://doi.org/10.1007/978-3-030-75100-5_9

In [11], we have evaluated a message relaying method with Enhanced Dynamic Timer (EDT) considering disaster situations. The EDT is one of the recovery method for Epidemic-based routing method [6].

In this paper, we propose an Adaptive Anti-packet Recovery (AAR) method for Vehicular DTN. For evaluation, we use Epidemic with conventional anti-packet and proposed AAR method as the message delivery protocols.

The structure of the paper is as follows. In Sect. 2, we give the message relaying methods for Vehicular DTN. In Sect. 3 is described the adaptive anti-packet recovery method for Vehicular DTN. In Sect. 4, we provide the evaluation system and the simulation results. Finally, conclusions and future work are given in Sect. 5.

2   Message Relaying Methods for Vehicular DTN

DTN can provide a reliable internet-working for space tasks [3,8,13]. The space networks have possibly large latency, frequent link disruption and frequent disconnection. In Vehicular DTN, the intermediate vehicles stored messages in their storage and then transmit to others. The network architecture is specified in RFC 4838 [2].

The famous DTN protocol is Epidemic routing [7,12]. Epidemic routing is performed by using two control messages to replicate a bundle message. Each vehicle periodically broadcasts a Summary Vector (SV) in the network. The SV contains a list of stored messages of each vehicle. When the vehicles receive the SV, they compare received SV to their SV. The vehicle sends a REQUEST message if the received SV contains an unknown message.

In Epidemic routing, consumption of network resources and storage sate become a critical problem, because the vehicles replicate messages to neighbors in their communication range. Therefore, received messages remain in the storage and the messages are continuously replicated even if the end-point receives the messages. However, recovery schemes such as timer may delete the replicate messages in the network. Then, the anti-packet deletes replicate messages too late.

In the case of the conventional timer, messages have a lifetime. The messages are punctually deleted when the lifetime of the messages is expired. However, the setting of a suitable lifetime is difficult. In the case of the conventional anti-packet, the end-point broadcasts the anti-packet, which contains the list of messages that are received by the end-point. Vehicles delete the messages according to the anti-packet. Then, the vehicles replicate the anti-packet to other vehicles. However, network resources are affected by anti-packet.

3   Adaptive Anti-packet Recovery Method

In this section, we describe the proposed AAR method for Vehicular DTN. For the conventional anti-packet recovery method, end-point broadcasts the anti-packet, which contains the list of messages that are received by the end-point.

Then, the vehicles replicate the anti-packet to other vehicles. In our proposed AAR method, intermediate vehicles generate the anti-packet even if the message does not reach the end-point. The flowchart of the proposed AAR method is shown in Fig. 1.

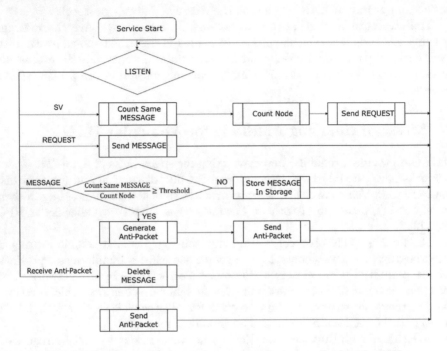

Fig. 1. Flowchart of proposed message relaying method with AAR.

In our approach, after the vehicle receives a SV, the vehicle counts the number of received SVs and measures the number of neighbors for each message. The AAR method considers the number of neighboring vehicles (Count Node) and calculates the number of Count Same Messages by comparing the Threshold. The AAR method selects the recovery mode by the condition in Eq. (1):

$$\frac{\text{Count Same MESSAGE}}{\text{Count Node}} \geq \text{Treshould.} \tag{1}$$

If the Eq. (1) is satisfied, an anti-packet is generated. On the other hand, if it is not satisfied, the message is stored in storage. In addition, the vehicle will also replicate the anti-packet continually as the conventional Epidemic.

4 Evaluation of Proposed Method

In this section, we evaluate the AAR method in Vehicular DTN. We implemented the proposed AAR method on the Scenargie simulator [9].

4.1 Evaluation Setting

We consider a grid road model for two types of scenarios (see Fig. 2). The red square in Fig. 2 indicates a vehicle. This figure shows an example of 50 vehicles/km². The following two cases are prepared for this evaluation.

- Case A: The message starting-point and end-point are fixed.
- Case B: Both starting-point and end-point are mobile.
- Common: All vehicles continue to move on the roads based on the map-based random way-point mobility model.

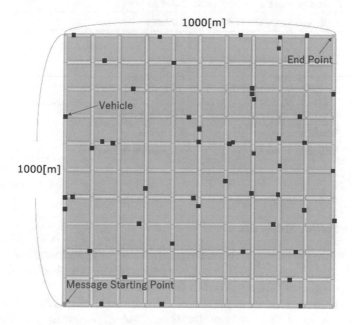

Fig. 2. Grid road model (50 vehicles).

For this evaluation, we consider four vehicular densities with 50, 100, 150 and 200 vehicles/km². Table 1 shows the simulation parameters used for the network simulator. For both scenarios, the start-point replicates 40 kinds of bundle messages to the relay vehicles and deliver the bundle message to the end-point. The simulation time is 600 s. The ITU-R P.1411 propagation model is used in this simulation [1]. We consider the interference from obstacles at 5.9 GHz radio frequency.

We evaluate the performance of delivery ratio, overhead and delay for different densities of vehicles. The delivery ratio indicates the value of the generated bundle messages divided by the delivered bundle messages to the end-point. The overhead indicates the number of times for sending replicate bundles. The delay indicates the transmission delay of the bundle to reach the end-point.

Table 1. Simulation parameters.

Parameter	Value
Simulation time (T_{max})	600 [s]
Area dimensions	$1,000$ [m] $\times$ $1,000$ [m]
Number of vehicles	50, 100, 150, 200 [vehicles/km^2]
Minimum speed	8.333 [m/s]
Maximum speed	16.666 [m/s]
Message start and end time	10–400 [sec]
Message generation interval	10 [sec]
AAR: Threshold	0.1–1.0
SV transmission interval	1 [sec]
Message size	500 [bytes]
PHY model	IEEE 802.11p
Propagation model	ITU-R P.1411
Radio frequency	5.9 GHz
Antenna model	Omni-directional

4.2   Simulation Results

We evaluate the proposed AAR method compared to Epidemic with anti-packet for Case A and Case B. In all cases, we evaluate the network performance by changing the value of Threshold from 0.1 to 1.0.

We present the simulation results of delivery ratio in Fig. 3. For 50 vehicles, the delivery ratio of Case B is better than Case A. Regardless of the number of vehicles, the results of delivery ratio improve with increasing numbers of thresholds. We have also observed that the message delivery ratio increases faster with increasing the vehicle densities. In Case A, the difference is large due to the effect of distance from starting-point to end-point. However, we found that the difference becomes smaller in Case B. This is because of the anti-packet reached the relay vehicle before the message reached the end-point.

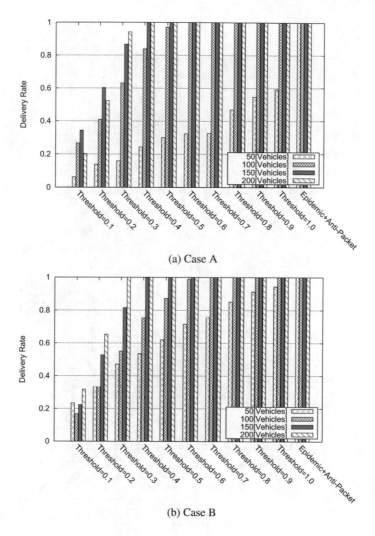

(a) Case A

(b) Case B

Fig. 3. Delivery ratio for Case A and Case B.

We present the simulation results of the overhead in Fig. 4. For 50 vehicles, the overheads of threshold 1.0 are reduced compared to the conventional Epidemic with anti-packet. For 200 vehicles, the results of overhead for Case A are higher than Case B. This is due to the high frequency of replication in some relay vehicles. For both cases with AAR method, we observed that the results of overhead are decreased with decreasing the threshold.

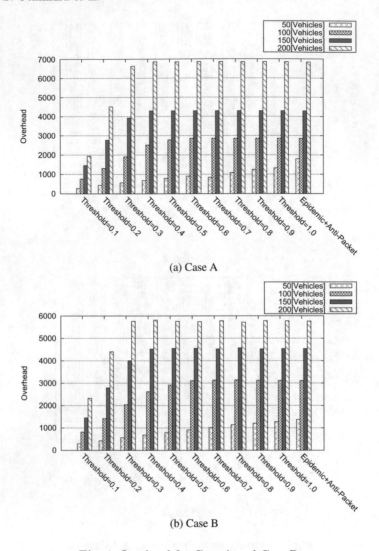

(a) Case A

(b) Case B

Fig. 4. Overhead for Case A and Case B.

We present the simulation results of delay in Fig. 5. There is little difference in the delay results compared to the conventional Epidemic with anti-packet. For Case A, the results of delay for 150 vehicles is the shortest. While for Case B, the results of delay decrease by increasing the number of vehicles. The delay of Case B is shorter than Case A. This is due to the starting-point and end-point are moving, which reduces the duration to reach the message.

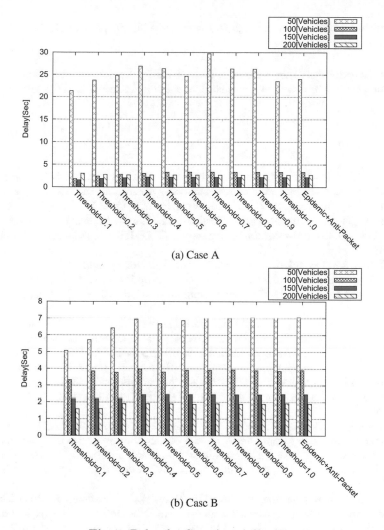

(a) Case A

(b) Case B

Fig. 5. Delay for Case A and Case B.

5   Conclusions

In this paper, we proposed an anti-packet generation method and storage management method to improve the message delivery rate in vehicle DTN. We evaluated the network performance considering the cases when starting-point and end-point are fixed and mobile. For evaluation, we used Epidemic with conventional anti-packet and AAR method as the message delivery protocols. From the simulation results, we found that the delivery ratio of AAR is good for dense networks. For sparse network, some anti-packets reached the relay vehicles before the messages reached the end-point.

In future work, we would like to investigate the impact of storage usage and the other parameters.

Acknowledgments. This work has been partially funded by the research project from Comprehensive Research Organization at Fukuoka Institute of Technology (FIT), Japan.

References

1. Rec. ITU-R P.1411-7: Propagation data and prediction methods for the planning of short-range outdoor radiocommunication systems and radio local area networks in the frequency range 300 MHz to 100 GHz. ITU (2013)
2. Cerf, V., Burleigh, S., Hooke, A., Torgerson, L., Durst, R., Scott, K., Fall, K., Weiss, H.: Delay-tolerant networking architecture. IETF RFC 4838 (Informational), April 2007
3. Fall, K.: A delay-tolerant network architecture for challenged Internets. In: Proceedings of the International Conference on Applications, Technologies, Architectures, and Protocols for Computer Communications, SIGCOMM 2003, pp. 27–34 (2003)
4. Henmi, K., Koyama, A.: Hybrid type DTN routing protocol considering storage capacity. In: Proceedings of the 8th International Conference on Emerging Internet, Data and Web Technologies (EIDWT 2020). pp. 491–502, February 2020
5. Kawabata, N., Yamasaki, Y., Ohsaki, H.: Hybrid cellular-DTN for vehicle volume data collection in rural areas. In: Proceedings of the IEEE 43rd Annual Computer Software and Applications Conference (COMPSAC-2019), vol. 2, pp. 276–284, July 2019
6. Nakasaki, S., Ikeda, M., Barolli, L.: A message relaying method with enhanced dynamic timer considering decrease rate of neighboring nodes for Vehicular-DTN. In: Proceedings of the 14th International Conference on Broad-Band Wireless Computing, Communication and Applications (BWCCA-2019), pp. 711–720, November 2019
7. Ramanathan, R., Hansen, R., Basu, P., Hain, R.R., Krishnan, R.: Prioritized epidemic routing for opportunistic networks. In: Proceedings of the 1st International MobiSys Workshop on Mobile Opportunistic Networking (MobiOpp 2007), pp. 62–66 (2007)
8. Rüsch, S., Schürmann, D., Kapitza, R., Wolf, L.: Forward secure delay-tolerant networking. In: Proceedings of the 12th Workshop on Challenged Networks (CHANTS-2017), pp. 7–12, October 2017
9. Scenargie: Space-time engineering, LLC http://www.spacetime-eng.com/
10. Solpico, D., Tan, M.I., Manalansan, E.J., Zagala, F.A., Leceta, J.A., Lanuza, D.F., Bernal, J., Ramos, R.D., Villareal, R.J., Cruz, X.M., dela Cruz, J.A., Lagazo, D.J., Honrado, J.L., Abrajano, G., Libatique, N.J., Tangonan, G.: Application of the V-HUB standard using LoRa beacons, mobile cloud, UAVs, and DTN for disaster-resilient communications. In: Proceedings of the IEEE Global Humanitarian Technology Conference (GHTC-2019), pp. 1–8, October 2019
11. Tada, Y., Ikeda, M., Barolli, L.: Performance evaluation of a message relaying method for resilient disaster networks. In: Proceedings of the 15th International Conference on Broadband and Wireless Computing, Communication and Applications (BWCCA-2020), pp. 1–10, October 2020

12. Vahdat, A., Becker, D.: Epidemic routing for partially-connected ad hoc networks. Duke University, Technical report (2000)
13. Wyatt, J., Burleigh, S., Jones, R., Torgerson, L., Wissler, S.: Disruption tolerant networking flight validation experiment on NASA's EPOXI mission. In: Proceedings of the 1st International Conference on Advances in Satellite and Space Communications (SPACOMM-2009), pp. 187–196, July 2009

From Incomplete to Complete Networks in Asynchronous Byzantine Systems

Michel Raynal[1,2(✉)] and Jiannong Cao[2]

[1] Univ Rennes IRISA, CNRS, Inria, Rennes, France
raynal@irisa.fr
[2] Department of Computing, Polytechnic University, Hung Hom, Hong Kong

Abstract. This paper presents a simple broadcast operation suited to n-process asynchronous message-passing systems in which (i) up to t processes may commit Byzantine faults, and (2), while the underlying communication network is connected (any pair of processes is connected by a path), not all the pairs of processes are directly connected by a communication channel. The algorithm proposed to implement this operation assumes (i) $t < n/3$ and (ii) $(2t+1)$-vertex connectivity of the underlying network (each pair of processes is connected by at least $(2t + 1)$ disjoint paths), requirements which are shown to be necessary. When considering incomplete networks, this abstraction can be used as the first level of a software stack on top of which, without any modifications, Byzantine-tolerant broadcast and agreement abstractions designed for fully connected networks can directly be used. The paper has also a short survey flavor from a "failure-tolerant broadcast abstractions" point of view.

Keywords: Asynchronous system · Broadcast abstraction · Byzantine process · Distributed software engineering · Fault-tolerance · Incomplete network · k-vertex connectivity · Message-passing · Modularity · Simplicity

1 Preliminary Definitions

Asynchronous Processes. The system is made up of a finite set Π of $n > 1$ asynchronous sequential processes, namely $\Pi = \{p_1, \ldots, p_n\}$. "Asynchronous" means that each process proceeds at its own speed, which can vary arbitrarily and always remains unknown to the other processes. The index i of p_i is considered to be the identity of p_i, and when there no ambiguity p_i and i are regarded as synonymous.

Communication Network (Graph). From a structural point of view, the system can be represented by an undirected graph whose vertices are the processes and whose edges are the process-to-process communication channels. From an operational point of view, the processes communicate by exchanging messages through an asynchronous network made up of bi-directional channels,

L. Barolli et al. (Eds.): AINA 2021, LNNS 225, pp. 102–112, 2021.
https://doi.org/10.1007/978-3-030-75100-5_10

each channel connecting two processes. The channels are asynchronous and reliable. "Asynchronous" means that each message that has been sent is eventually received by its destination process, but there is no bound on message transfer delays. "Reliable" means that a channel neither loose, duplicate, modify, nor create messages.

The network is connected, i.e., there is a path connecting any pair of distinct processes. It can be complete or incomplete. "Complete" (or fully connected) means that each pair of processes is connected by a channel. "Incomplete" means that not all the pairs for processes are connected by a channel.

A process p_i sends a message to a process p_j by invoking the primitive "send TAG(m) to p_j", where p_j is a neighbor of p_i, TAG is the type of the message and m its content. A process receives a message by executing the primitive "receive()".

If there is a channel connecting the processes p_i and p_j, we assume that if p_i receives a message on this channel, it knows the message is from p_j. Said differently, no process can impersonate another process.

An incomplete network (not fully connected graph) is "x-vertex connected" if it has no set of $(x-1)$ vertices (processes) whose removal disconnects it. Due to Menger's Theorem [15] if the vertex connectivity of the network is x, there exist at least x disjoint paths connecting any pair of processes[1].

Process Failure Models. A fault occurs when a process does not follow its behavior. A process that commits faults is said to be *faulty*. Otherwise, it is *correct*. The model parameter t defines an upper bound on the number of processes that can be faulty.

We consider two types of faults: crash and Byzantine behavior. A process fails by crashing when it stops executing before terminating its algorithm. The Byzantine fault model [14, 19] is more severe. A Byzantine process is a process that behaves arbitrarily: it can crash, fail to send or receive messages, send arbitrary messages, start in an arbitrary state, perform arbitrary state transitions, etc., in short a Byzantine process does not behave as specified by its algorithm. Such a behavior can be intentional (also called malicious) or the result of a transient fault which altered the content of local variables of a process, thereby modifying its intended behavior in an unpredictable way. As a simple example, a Byzantine process, which is assumed to send the same message m to all the processes, can send a message m_1 to some processes, a different message m_2 to another subset of processes, and no message at all to the other processes. Moreover, Byzantine processes can collude to "pollute" the computation. They can also control the network in the sense that they can re-order message deliveries at correct processes. It is however assumed that a Byzantine process cannot send an infinite number of messages.

---

[1] "Disjoint" means that the only vertices (processes) shared by any two of these paths are their first and last processes.

2   Broadcast in Fully Connected Crash-Prone Systems: A Brief Survey

Basic Broadcast. In a fully point-to-point connected network, the basic broadcast operation, denoted here "ba_broadcast TAG(m)", is implemented by the following loop:

$$\textbf{for } j \in \{1, \cdots, n\} \textbf{ do } \text{ send TAG}(m) \text{ to } p_j \textbf{ end } \textbf{ for.} \tag{1}$$

This non-atomic macro-operation constitutes a trivial broadcast operation. It is important to see that this operation is not reliable: if the sender crashes while executing ba_broadcast(), it is possible that only an arbitrary subset of non-crashed processes receive the message. This type of broadcast (sometimes called best effort broadcast) is clearly not reliable.

Reliable Broadcast. The previous unreliability has motivated the definition of the *Reliable broadcast* (RB), which is a communication abstraction central to fault-tolerant distributed systems. It allows each process to broadcast messages to all processes despite failures. More precisely, it guarantees that the correct processes deliver the same set M of messages. This set M includes at least all the messages they broadcast, and can also contain messages broadcast by faulty processes. A faulty process receives a subset of the messages in M.

The fundamental property of reliable broadcast lies in the fact that no two correct processes deliver different sets of messages. This communication abstraction is a basic building block used in a lot fault-tolerant distributed applications (see e.g. [2,5,17,20]).

Interestingly, RB abstraction can be built on top the unreliable broadcast operation ba_broadcast. Moreover, without enriching the model with additional computability power, RB can be extended to ensure FIFO delivery order (on each channel messages are received in their sending order) and causal delivery order (messages are delivered in their causal sending order [13]). The interested reader will find in [20] (pages 21–74) families of algorithms implementing these broadcast abstractions in fully connected asynchronous systems where up to $t < n$ processes may crash.

Total Order Broadcast. Extending RB so that all the messages be delivered in the same order at each process is impossible. This comes from the fact that, to build a common total order on message deliveries, the processes need to agree, and the kind of agreement they have to obtain requires to solve consensus, which is known to be impossible in the presence of asynchrony and (even a single) process crash [10][2].

This means that the system model (full connectivity, asynchrony, process crashes) is computationally too weak, and must be enriched with additional computational power. This can be done, for example, by adding synchrony assumptions [9], providing processes with information on failure [6], or providing the processes with randomization [1].

---

[2] Actually total order broadcast and consensus are equivalent problems in the sense any of them can be solved on top of the other one [6].

3   Byzantine Reliable Broadcast in Fully Connected Systems

Definition. Reliable broadcast has been studied in the context of Byzantine failures since the eighties. This communication abstraction (denoted BRB-broadcast) provides each process with two operations, denoted brb_broadcast() and brb_deliver(). As in [11], we use the following terminology: when a process invokes brb_broadcast(), we say that it "brb-broadcasts a message", and when it executes brb_deliver(), we say that it "brb-delivers a message". BRB-broadcast is defined by the following properties. We assume (for the moment) that a process brb-broadcasts at most one message.

– BRB-Validity. If a correct process brb-delivers a message m from a correct process p_i, then p_i brb-broadcast the message m.
– BRB-Integrity. No correct process brb-delivers twice the same message from a (correct or Byzantine) process p_i.
– BRB-Termination-1. If a correct process brb-broadcasts a message m, all correct processes eventually brb-deliver m.
– BRB-Termination-2. If a correct process brb-delivers a message m from p_i (possibly faulty) then all correct processes eventually brb-deliver m.

On the Safety Properties' Side. BRB-validity relates the output (messages brb-delivered) to the inputs (messages brb-broadcast). BRB-integrity states that there is no duplication. BRB-agreement states that there is no duplicity [20,21]: be the sender p_i correct or not, it is not possible for a correct process to brb-deliver m from p_i while another correct process brb-delivers $m' \neq m$.

On the Liveness Properties' Side. The BRB-Termination properties state guarantees on message brb-deliveries. BRB-Termination-1 states that a message brb-broadcast by a correct process is brb-delivered by all correct processes. BRB-Termination-2 is a strong delivery property, namely, be the sender correct or Byzantine, every message brb-delivered by a correct process is brb-delivered by all correct processes.

It follows that all correct processes brb-deliver the same set of messages, and this set contains at least all the messages they brb-broadcast.

The previous definition considers that each correct process RB-broadcasts at most one message. Using sequence numbers, it is easy to extend it to the case where a correct process RB-broadcasts a sequence of messages.

Byzantine-Tolerant Reliable Broadcast Algorithms. In the context of fully connected n-process asynchronous systems, a signature-free reliable broadcast algorithm has been proposed G. Bracha in [3]. This algorithm requires that at most $t < n/3$ processes be Byzantine, which is shown to be an upper bound on the number of Byzantine processes that can be tolerated to implement Byzantine-tolerant reliable broadcast on top of asynchronous fully connected reliable networks [4,20]. Hence, Bracha's algorithm is t-resilience optimal. This algorithm is based on a "double echo" mechanism of the value broadcast by the sender process. For each application message[3] it uses three types of protocol messages, requires three consecutive communication steps (one for each message type), and generates $(n-1)(2n+1)$ protocol messages.

A more recent signature-free Byzantine reliable broadcast algorithm has been introduced by D. Imbs and M. Raynal in [12]. This algorithm implements the reliable broadcast abstraction of an application message with only two consecutive communication steps ("single echo"), two message types, and $n^2 - 1$ protocol messages. The price to pay for this gain in efficiency is a weaker t-resilience, namely $t < n/5$. Hence, these two algorithms differ in their trade-offs between t-resilience and message/time efficiency.

Observation. Both the previous reliable broadcast algorithms are built on top of the ba_broadcast operation used in a Byzantine context. Hence, when considering the "for" loop implementing ba_broadcast, while a correct process sends the same message to processes, a Byzantine sender can send different fake messages, or no message at all, to different processes.

BRB-Broadcasting a Sequence of Message. The previous version of BRB-broadcast is one-shot in the sense a process invokes at most once the ba_broadcast. In this case, the identity of the message can be reduced to the identity of the invoking process. A simple way to obtain a multi-shot BBR-broadcast abstraction consists in associating an appropriate identity with each invocation of brb_broadcast() by a process. This can be easily realized with sequence numbers, a process p_i now invokes brb_broadcast(sn_i, m), where sn_i is the next sequence number value. This instance is then identified by the pair $\langle i, sn_i \rangle$.

4   Byzantine Basic Broadcast in an Incomplete Network

4.1   Aim and Broadcast Operation

Considering an n-process system including up to t Byzantine processes and where the communication network is incomplete, our aim is to build a broadcast operation, denoted bbai_broadcast() (Byzantine basic incomplete) that allows the processes to execute as if the underlying communication network was fully connected.

As shown below, this operation allows to simulate a fully connected network on top of a n-process system, where up to $t < n/3$ processes can be Byzantine, and the underlying network is $(2t+1)$-vertex connected. It follows that it is possible to build a software stack as depicted in Fig. 1. More precisely, replacing the underlying ba_broadcast operation, used at the upper level by Byzantine-tolerant algorithms designed for a fully connected network, with the bbai_broadcast()

---

[3] An *application message* is a message sent by the reliable broadcast abstraction, while a *protocol message* is a message used to implement reliable broadcast.

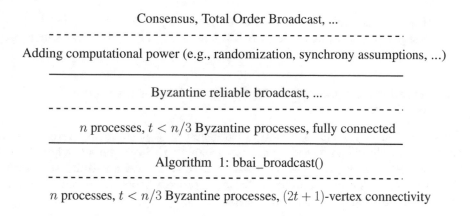

Fig. 1. A distributed software stack

operation, it follows that, without modifications, these Byzantine-tolerant algorithms can be used on top of an incomplete network. This is the case not only for the reliable algorithms presented previously [3,12], but also for consensus and total order algorithms when the computing model is enriched with additional computability power such as synchrony assumptions or randomization. An example, replacing the ba_broadcast operation it uses with the bbai_broadcast(), the randomized Byzantine binary consensus algorithm described in [16] works (without modifications) in an incomplete $(2l + 1)$-vertex connected network[4].

Byzantine Basic Broadcast for Incomplete Network. This (unreliable) broadcast is denoted BBaI-broadcast. Designed for incomplete networks, it provides the processes with two operations denoted bbai_broadcast() and bbai_deliver(). It is defined by the following properties. Without loss of generality, it is assumed that all the messages sent by a (correct) process bbai-broadcasts at most one message. It is easy to obtain a multi-shot version using the same mechanism as the one used for the multi-shot version of BRB-broadcast (see the last paragraph of Sect. 3), the identity of a message is not only the identity the sending process but a pair composed a sender identity plus a local sequence number.

- BBaI-Validity. If a correct process bbai-delivers a message m from a correct process p_i, then p_i bbai-broadcast the message m.
- BBaI-Integrity. A correct process bbai-delivers at most one message m from a (correct or Byzantine) process p_k.
- BBaI-Termination. If a correct process bbai-broadcasts a message m, then all correct processes bbai-deliver m.

Let us observe that bbai_broadcast() is pretty weak. It is nothing but ba_broadcast() in an incomplete network. As an example, if the sender p_i is

---

[4] Deterministic algorithms implementing Byzantine multivalued consensus on top of Byzantine binary consensus in asynchronous fully connected networks are described in [2,18,20].

Byzantine, it is possible that distinct sets of correct processes bbai-deliver different message from p_i, while other correct processes do not bbai-deliver a message
from p_i.

4.2   Simulating a Complete Network on Top of an Incomplete Network

The proposed algorithm is a simple extension to asynchronous systems of a
synchronous algorithm introduced by D. Dolev in [7,8]. It is assumed that $t <
n/3$, the network is $(2t + 1)$-connected, and each process knows its structure.
(Let us notice that the assumption that each process knows the communication
graph is implicit in fully connected systems.)

Notations. A path of the network from a process p_i to a process p_k is represented by the sequence $path = i, \ldots, k$ of the process identities that compose it.
Assuming $i \in path$ is not its last identity, the function $\mathsf{next}(i, path)$ returns the
identity that follows i in $path$.

Local variables at a process p_i

- The set $neighbors_i$ contains the identities of the neighbors of p_i.
- The array $dis\_paths[1..n]$ is such that, for each j, $dis\_paths[j]$ is a set containing $(2t + 1)$ disjoint paths from p_i to p_j ("disjoint" except for the first
  and the last processes p_i and p_j respectively). This implicitly means that a
  process knows the network (to be able to compute, for each other process p_j,
  the $(2t + 1)$ disjoint paths it needs to communicate with p_j).

While m denotes an application message, the protocol messages used to
implement bbai_broadcast are tagged IMPL. The ideas behind the design of the
algorithm are well-known and pretty simple. The first one can be summarized
as "discard or forward", while the second one consists in delivering a message
only when enough copies of it have been received on disjoint paths.

Broadcast of an Application Message m. When a process invokes
bbai_broadcast(m), it considers each destination process p_j (line 1), and for each
of the $(2t + 1)$ disjoint paths from it to p_j denoted $path$ (line 2), it sends the
protocol message IMPL$(m, i, path)$ to its successor p_ℓ on this path $path$ (line 3).
Hence, $path$ is the path, defined by the sender p_i, that the message m has to
follow from p_i to its destination p_j. Let us notice that if a process p_i is Byzantine, it can send different messages m, m', etc., or no message at all, to different
subset of processes.

Reception of a Message. When it receives a message IMPL$(m, k, path)$ from its
neighbor p_j, a correct process p_i first checks if the protocol message is well-formed
(line 6). "Well-formed" means that, for its receiver p_i, the message appears as it
was initially bbai-broadcast by p_k and up to p_i it traveled along the path $path$
of the network. More precisely, p_k must be the first process in $path$, which must

```
operation bbai_broadcast(m) is                                    code for p_i
(1)     for each j ∈ {1, · · · , n} \ {i} do
(2)         for each path ∈ dis_paths[j] do
(3)             let ℓ be next(i, path); send IMPL(m, i, path) to p_ℓ
(4)         end for
(5)     end for


when IMPL(m, k, path) is received from p_j ∈ neighbors_i do
(6)     if ¬ well_formed(i, j, k, path)
(7)         then discard IMPL(m, k, path)
(8)         else if i is not the last identity in path
(9)                 then let ℓ be next(i, path); send IMPL(m, k, path) to p_ℓ
(10)                else if   (m, k, −) received on (t + 1) disjoint paths)
                                ∧ ((m, k) not yet bbai-delivered)
(11)                        then bbai_deliver(m, k)
(12)                    end if
(13)            end if
(14)    end if.


function well_formed(i, j, k, path) is
(15)    return((k is the first identity in path)
                ∧ (path is an elementary path of the network)
                ∧ (j is an immediate predecessor of i in path)).
```

Algorithm 1: From an incomplete network to a fully connected network

be an elementary path of the network, and in which p_j is the predecessor of p_i (line 15). If the message is not well-formed, p_i discards it (line 7).

If the received message IMPL$(m, k, path)$ is well-formed, and p_i is not the destination process (i.e., it is not the last process in $path$, line 8), it forwards the message IMPL$(m, k, path)$ to its successor in $path$ line 9). If it is the destination of the message, p_i checks if it has received messages IMPL$(m, k, -)$ on $(t + 1)$ disjoint paths (from p_k to itself, line 10). If it is the case and additionally p_i has not bbai-delivered the pair (m, k), it does it line 11.

An Improvement. Let consider the case where the message IMPL$(m, k, path)$ received by a process p_i is such that p_k is the immediate predecessor of p_i on $path$. Hence, p_k executed "send IMPL$(m, k, path)$ to p_i" (entailed by the invocation of bbai_broadcast(m) if p_k is correct) or possibly bypassing this invocation if p_k is Byzantine. Whatever the case p_i can bbai-delivers the pair (m, k) when it receives the message IMPL$(m, k, path)$. This is due to the fact that, as p_k and p_i are neighbors, there is no Byzantine process between them which could have altered IMPL$(m, k, path)$. The algorithm must be slightly modified to take into account this observation.

4.3   Proof of the Algorithm

Theorem 1. *Algorithm 1 implements* BBaI-broadcast.

Proof. Let us first show that, while they can collude, the Byzantine processes cannot direct a correct process p_j to bbai-deliver a pair (m', k) while p_k is correct but has never invoked bbai_broadcast(m'). To this end, let p_x a Byzantine process that knows that the path $path_x = k, \ldots, x, \ldots, j$, is a path used by p_k to send a message to p_j. Process p_k can forge the well-formed protocol message IMPL$(m', k, path_x)$ and send it to its successor in $path_x$ as done in line 9. Moreover, this can be done by each Byzantine process p_y on a well-formed path $path_y$ defined similarly to $path_x$. It then follows from the predicate of line 10 that p_j can receive at most t distinct messages IMPL$(m', k, path_x)$, ..., IMPL$(m', k, path_y)$. Consequently, p_j will never bbai-deliver the pair (m', k).

The BBAI-Integrity property follows from the last predicate of line 7. To prove the BBAI-Termination property, let p_k be a correct process that invoked bbai_broadcast(m). So, p_k executed lines 1–5. Let p_i a correct process. As there are at least $(2t + 1)$ disjoint paths from p_k to p_i, and at most t Byzantine processes, it follows that at least $(t + 1)$ of these disjoint paths are made up of correct processes only. According to the forwarding mechanism used by the correct processes, it follows that p_i eventually receives $(t+1)$ well-formed messages IMPL$(m, k, -)$ and the predicate of line 5 becomes satisfied, which entails the bbai-delivered of msg by p_i.

To prove the BBAI-Validity property, let p_i a process that bbai-delivers the pair (m, k) where p_k is a correct process. In the worst case, among the $(t + 1)$ paths from p_k to p_i that make true the predicate of line 5 at p_i, at most t can include Byzantine processes. It follows that at least one of these $(t + 1)$ paths is composed of correct process only, from which we conclude that p_k invoked bbai_broadcast(m). $\square$

Theorem 2. *The condition $t < n/3$ and the $(2t + 1)$-vertex connectivity are both necessary and sufficient to implement* BBaI-broadcast.

Proof. The sufficiency results from Theorem 1. As far the necessity is concerned, we have the following.

– Let us first consider the condition $t < n/3$. On the one hand, Theorem 1 has shown that the bbai_broadcast operation simulates the basic unreliable broadcast ba_broadcast() operation used in a fully connected network. On the other hand, it has been shown (see e.g., [3,12]) that the BRB-broadcast abstraction can be implemented in a fully connected network on top of ba_broadcast() when $t < n/3$. It consequently follows that BRB-broadcast can be built on top of bbai_broadcast in an incomplete network.

  On another side, it is shown in [4,20] that $t < n/3$ is a necessary condition for BRB-broadcast in a Byzantine fully connected asynchronous message-passing system. It follows from these two observations that there is no algorithm that implements bbai_broadcast when $t \geq n/3$.

– As far as the $(2t + 1)$-vertex connectivity condition is concerned we have the following. Assuming the vertex connectivity is less than $(2t + 1)$, let us restrict the asynchronous model to its synchronous executions and consider a synchronous consensus algorithm executed on top of it (see the software stack in Fig. 1). Such an algorithm cannot exist. This is due to the fact that it is proved in [8] that, in a synchronous system where up to t processes can be Byzantine, the previous connectivity condition is a necessary requirement for consensus in the presence of up to t Byzantine processes. As synchronous systems are computationally stronger than asynchronous systems, it follows that the $(2t + 1)$-vertex connectivity condition is a necessary requirement for asynchronous systems which are not fully connected.

$\square$

5   Conclusion

This paper has presented a simple operation that provides the processes with a broadcast that simulates an unreliable broadcast on top of an $(2t + 1)$-vertex connected network in which up to $l < n/3$ processes may be Byzantine. The algorithm implementing this operation is a simple extension of a synchronous algorithm described in [8]. Its main interest lies in the fact that it allows any asynchronous algorithm (i) tolerating up to $t < n/3$ Byzantine processes, and (ii) designed to work on an underlying fully connected network, to work also on incomplete $(2t + 1)$-vertex connected networks. It has also been shown that both these conditions ($t < n/3$ and $(2t + 1)$-vertex connectivity) are necessary and sufficient, not only for synchronous systems as shown in [8], but also for asynchronous distributed systems. Last but not least, as shown by Fig. 1, the simplicity of the bbai_broadcast operation favors a modular approach when one has to cope with Byzantine processes and incomplete networks. Finding a more efficient algorithm however remains an open challenging non-trivial problem.

Acknowledgments. The authors thank the reviewers for their constructive comments. This work was partially supported by the French ANR projects DESCARTES (16-CE40-0023-03) devoted to layered and modular structures in distributed computing, and ByBLosS (20-CE25-0002-01) devoted to the design of modular building blocks for large-scale trustless multi-users applications.

References

1. Ben-Or, M.: Another advantage of free choice: completely asynchronous agreement protocols. In: Proceedings 2nd ACM Symposium on Principles of Distributed Computing (PODC'83), pp. 27–30. ACM Press (1983)
2. Ben-Or, M., Kelmer, B., Rabin, T.: Asynchronous secure computations with optimal resilience. In: Proceedings of 13th ACM Symposium on Principles of Distributed Computing (PODC'94), pp. 183–192. ACM Press (1994)

3. Bracha, G.: Asynchronous Byzantine agreement protocols. Inf. Comput. **75**(2), 130–143 (1987)
4. Bracha, G., Toueg, S.: Asynchronous consensus and broadcast protocols. J. ACM **32**(4), 824–840 (1985)
5. Cachin, Ch., Guerraoui, R., Rodrigues, L.: Reliable and Secure Distributed Programming, 367 p. Springer, ISBN 978-3-642-15259-7 (2011)
6. Chandra, T.D., Toueg, S.: Unreliable failure detectors for reliable distributed systems. J. ACM **43**(2), 225–267 (1996)
7. Dolev, D.: Unanimity in an unknown and unreliable environment. In: Proceedings of 22nd ACM Symposium on Foundations of Computer Science (FOSC'81), pp. 159–168. ACM Press (1981)
8. Dolev, D.: The Byzantine general strike again. J. Algorithms **3**, 14–30 (1982)
9. Dolev, D., Dwork, C., Stockmeyer, L.: On the minimal synchronism needed for distributed consensus. J. ACM **34**(1), 77–97 (1987)
10. Fischer, M.J., Lynch, N.A., Paterson, M.S.: Impossibility of distributed consensus with one faulty process. JACM **32**(2), 374–382 (1985)
11. Hadzilacos, V., Toueg, S.: A Modular Approach to Fault-Tolerant Broadcasts and Related Problems. Technical report 94-1425, 83 p. Cornell University (1994)
12. Imbs, D., Raynal, M.: Trading t-resilience for efficiency in asynchronous Byzantine reliable broadcast. Parallel Process. Lett. **26**(4), 8 (2016)
13. Lamport, L.: Time, clocks, and the ordering of events in a distributed system. Commun. ACM **21**(7), 558–565 (1978)
14. Lamport L., Shostack R., Pease M.: The Byzantine generals problem. ACM Trans. Programm. Lang. Syst. **4**(3), 382–401 (1982)
15. Menger, K.: Zur allgemeinen kurventheorie. Fundamentae Mathematicae **10**, 96–115 (1927)
16. Mostéfaoui, A., Moumen, H., Raynal, M., Signature-free asynchronous binary Byzantine consensus with $t<n/3$, $O(n^2)$ messages, and $O(1)$ expected time. J. ACM **62**(4), Article 31, 21 pages (2015)
17. Mostéfaoui, A., Raynal, M.: Intrusion-tolerant broadcast and agreement abstractions in the presence of Byzantine processes. IEEE Trans. Parallel Distrib. Syst. **27**(4), 1085–1098 (2016)
18. Mostéfaoui, A., Raynal, M.: Signature-free asynchronous Byzantine systems: from multivalued to binary consensus with $t<n/3$, $O(n^2)$ messages, and constant time. Acta Informatica **54**(5), 501–520 (2017)
19. Pease, M., Shostak, R., Lamport, L.: Reaching agreement in the presence of faults. J. ACM **27**, 228–234 (1980)
20. Raynal, M.: Fault-tolerant message-passing distributed systems: an algorithmic approach. Springer, 480 p, ISBN: 978-3-319-94140-0 (2018)
21. Toueg, S.: Randomized Byzantine agreement. In: Proceedings of 3rd Annual ACM Symposium on Principles of Distributed Computing (PODC'84), pp. 163–178. ACM Press (1984)

Towards the Performance Evaluation of a Trust Based Routing Protocol for VANET

Amira Kchaou[(✉)], Ryma Abassi, and Sihem Guemara El Fatmi

Digital Security Research Lab, Higher School of Communication of Tunis,
University of Carthage, Tunis, Tunisia
{Amira.Kchaou,Ryma.Abassi,Sihem.Guemara}@supcom.tn

Abstract. Vehicular Ad-hoc NETwork (VANET) is a decentralized dynamic network of vehicles that move rapidly and communicate with one another to disseminate various types of information and deliver data. However, dishonest vehicles may present in the route and may lead for changing or dropping the content of delivered messages. This latter is a challenge issue in VANET to find a trustful route for reliable data dissemination. In a previous work, we have proposed a routing protocol for VANETs based on trust (TRPV) to find an optimal route depending on the trustworthiness of the path and the number of hops. Thus, in this paper, simulations are achieved to evaluate the performance of TRPV in terms of the throughput, the Packet Data Ratio (PDR), the routing overhead, and the average end-to-end delay.

1 Introduction

Vehicular Ad-hoc NETwork (VANET) is a decentralized wireless communication network which provides internet connection and multimedia services such as video/audio streaming, and also safety services such as traffic management and road safety to avoid dangerous situations beside weather or positioning services, etc. [1]. These services depend on a timely transmission of data through messages within the network, which can achieved by the use of vehicular communications [1].

Thus, the communication links between vehicles are unreliable due to the vehicle highmobility, dynamic network, the existence of dishonest vehicles and others VANET characteristics. However, the routing is a challenging task in the VANET to find an efficient and secure route to successfully transmit the data from the source to the destination vehicles. To resolve these problems, the proposed trust and clustering mechanism are introduced in the VANET to form clusters having similar vehicle velocity, electing ClusterHeads (CHs) using the common vehicle between the neighbors of each vehicle, and to successfully transform information [2].

In previous work [3], we have proposed a trust based routing protocol for the VANET baptized TRPV through trusted CHs (i.e. CH having a trust between it

L. Barolli et al. (Eds.): AINA 2021, LNNS 225, pp. 113–124, 2021.
https://doi.org/10.1007/978-3-030-75100-5_11

and other vehicle greater than a minimum threshold value τ) that facilitates the information transmission through messages between the vehicle and the outside of its cluster.

The proposed TRPV finds a trustful route depending on the trustworthiness of the path from the source to the destination vehicles, as well as the number of hops for reliable information and message dissemination to the destination. In fact, the proposed TRPV is used for intra-cluster and inter-cluster route discovery. During the intra-cluster route discovery, the source vehicle sends a message to its corresponding CH in order to check the correctness of the message by calculating the message credibility. Afterwards, the CH directly forwards the message to the destination vehicle. On the other hand, the source vehicle sends a route request packet (RREQ) to its trusted CH. Next, the CH pursues the route discovery process via the CHs existing at the level two (i.e. having all CHs) until the routes are received. In addition, the CH forwards the routes to the requesting vehicle. Therefore, the source vehicle chooses the trustful and optimal route depending on the lowest number of hops, as well as the trustworthiness of the path.

In this paper, we evaluate the performance of TRPV through simulations of the throughput, the routing overhead, the Packet Data Ratio (PDR) and the end to end delay, and compare it to the AODV routing protocol.

The remainder of this paper is organized as follows. Section 2 reviews some related trust based routing protocols. Section 3 introduces the proposed TRPV. Section 4 presents the simulation results showing some TRPV performances compared to AODV the protocol. Finally, Sect. 5 concludes this paper.

2 Related Works

This section presents some existing works dealing with trust based routing protocol in VANETs.

In [4], the authors proposed a reliability model based routing protocol for the vehicular network, called R-AODV. The aim of the proposed is to facilitate the reliable routing between vehicles calculating the link reliability value on the basis of the location, direction and the velocity of vehicles along the route. However, R-AODV is not used trust for calculating the link reliability. In addition, R-AODV is not suited due to high payload transmission and frequent link breakage. In [5], Sahoo et al. proposed a trust based clustering with ant colony routing in VANET, called TACR in order to provide an efficient vehicular communication. In fact, VANET is organized into clusters having the RSUs as CHs and vehicles as mobile nodes. Then, CH collects the opinions of members on the vehicle, and takes a decision about the message. Thereafter, the proposed used the ant colony algorithm in order to select the efficient route among clusters using boundaries vehicles. However, TACR has two weaknesses: the slow sending decision and the using of a static CH. In [6], Kerrache et al. introduced a Trusted ROUting protocol for urban Vehicular Environment, called TROUVE in order to establish a reliable and secured route to the destination vehicle. In fact, each vehicle maintains the neighborhoods information and broadcasts it via Cooperative Awareness Messages (CAM). Next, it broadcasts the data routing through the shortest and secured route. However, TROUVE address unicast data of traffic management in urban environments. In [7], the authors proposed a trusted routing protocol for VANET on the basis of trusted metrics (the velocity, the direction, the distance, and the trust value) using trusted computing algorithm in order to provide a secured route selection and prevent malicious vehicles. However, the proposed neglects the high traffic density and the route maintenance process. In [8], the authors proposed a trust based distributed authentication method (TDA) in VANET for avoiding the collision attack, ensuring secure information transmission, and selecting the optimal path with minimal energy. In fact, honest vehicles are identified according to the channel state and the on-board unit (OBU) energy of the vehicle for establishing seamless communication. However, the proposed is focused on less energy and not based on the less number of hops to select the shortest route including in the optimal route. In [9], the authors proposed a routing protocol based on trust and minimum delay for VANET in order to select the optimal path with highest trust and minimal delay. In fact, the proposed computes the trust calculation and delay of vehicles on the basis of the participation vehicles in path using the present and previous knowledge of vehicles along with risk threshold of messages to be sent. Then, the optimal path is selected with highest trust and less delay. However, the proposed is not focused on number of hops in the selection of optimal path.

Given that the existing works benefit from trust management and routing, in this paper, we propose a new trust based routing protocol for VANET to provide an efficient and trustful route. This protocol will use the trustworthiness of the whole path from the source vehicle to the destination based the reputation of the vehicles, and the less number of hops.

3 TRPV: Trust Based Routing Protocol for VANET

This section deals with the Trust based Routing Protocol VANET baptized TRPV based on the number of hops (Hop_Count) and the trustworthiness of the path (TP) in order to select the optimal route in the network.

3.1 Basic Properties

The TRPV has several properties:

- Each CH has a CH_list containing all the other CHs in the network and its reputation value, a blacklist including the dishonest vehicle (which has a reputation value equaled to -3), and a *REPUTATION_TABLE* containing the members of the cluster and its reputation value.
- Each CH stores the TP in its routing table,which is updated for a certain time β.

- The CH having a trust between it and a vehicle higher than a predefined threshold τ is considered as trusted CH.
- The routing discovery process for inter-clusters can be achieved through trusted CHs.

3.2 TRPV Description

In VANET, each CH computes the TP to the destination vd using the trust between vehicles along the route. The TP is the product of trust between the vehicles (vehicle vi and vehicle vj) T_{ij} along the route for finding the trustful route.

The TP is formalized by the following formula (1).

$$TP = \prod_{i=1, j=1}^{k} T_{ij} \tag{1}$$

Where the T_{ij} is the trust link between vehicle vi and vehicle vj along the route.

The T_{ij} is formalized by the formula (2).

$$T_{ij} = REP_i * REP_j \; vi, vj \in route \tag{2}$$

Where the T_{ij} is equal to the product between vehicle vi and vehicle vj reputation values in the route.

When a source vehicle vs wants to transfer the data to destination vd, the process of the route discovery will be achieved according to two scenarios: the route discovery for intra-cluster and the route discovery for inter-cluster.

3.3 Route Discovery for Intra-clusters

When the source vehicle vs and the destination vehicle vd belong to the same cluster, the process of the route discovery for intra-clusters is achieved, as presented below:

- The source vehicle vs sends an URGENT_MSG to its CH to check the correctness of the message.
- After calculating the message credibility, the CH consults its CH_MEMBER_TABLE to find destination vd.
- If the destination vd exists, the CH forwards to destination vd an URGENT_MSG, and to the source vehicle vs an ACK_MSG, otherwise go to next scenario;

Table 1. Routing table

IDvd	Dest_seq_num	Routes list				
		Route_ID	Next_Hop	Hop_Count	TP	LifeTime (LT)

3.4 Route Discovery for Inter-clusters

When the source vehicle vs and the destination vehicle vd do not belong to the same cluster, the process of the route discovery for inter-clusters will be achieved.

In a previous work, when the source vehicle vs sends a RREQ to a trusted CH having T_{vsCH} higher than τ. If there is no trusted CH with an appropriate β then, the source vehicle vs resends the RREQ. Next, the CH consults its CH_list, continues the route discovery through CHs existing at level two (i.e. having all CHs while routes are received) and forwards the RREQs to the requesting vehicle. Thus, the route discovery process prohibits RREQ flooding in the VANET.

The RREQ packet header is changed as follows :

$$RREQ(\ Trusted\ CH,\ RREQ\_ID,\ Hop\_Count,\ ID_{vd},$$
$$Dest\_seq\_num,\ ID_{vs},\ S\_seq\_num,\ \tau)$$

At level two, vehicle vi receives a RREQ packet from vehicle vj, it looks up in its routing table for a valid route until destination vd. The routing table structure is depicted in Table 1. Where the route list refers to several feasible routes and to the destination ID_{vd} classified in an ascending order of Hop_Count, like $Hop\_Count_1 \leq Hop\_Count_2 \leq ... \leq Hop\_Count_n$ in the routing table. In fact, every route has a Next_Hop corresponding to the vehicle of Next_Hop in this latter. TP refers to the trust value of the overall route from the current vehicle to the destination ID_{vd}. The Lifetime (LT) represents the time of the validity.

Then, after consulting the routing table, if vehicle vi does not get a valid route to the destination vd, then it adds its address ID_{vi} to the RREQ packet, increases the Hop_Count by one, and sends it to the trusted CHs existing at level two excepting the vehicle vj.

Next, vehicle vi updates the reverse route entry in the routing table where the Next_Hop must be vehicle vj and modifies the TP of vehicle vj.

Moreover, if there is no transaction between vehicle vi and vehicle vj then, the vehicle vi forms an entry for the vehicle vj in its own local routing table. Therefore, this process will continue to destination vd.

Else if vehicle vi finds the valid route to destination vd, the dest_seq_num in the RREQ rises by one, then vehicle vi adds the route parameters: the Hop_Count and TP in the route reply (RREP), and sends it reverse to the previous vehicle vj.

Else if vehicle vj receives the RREP, then it calculates the TP to destination vd, and sends back the Current_TP_dest and Hop_Count to its previous vehicle vk using the reverse route up to the source vehicle vs. Then, the latter receives the

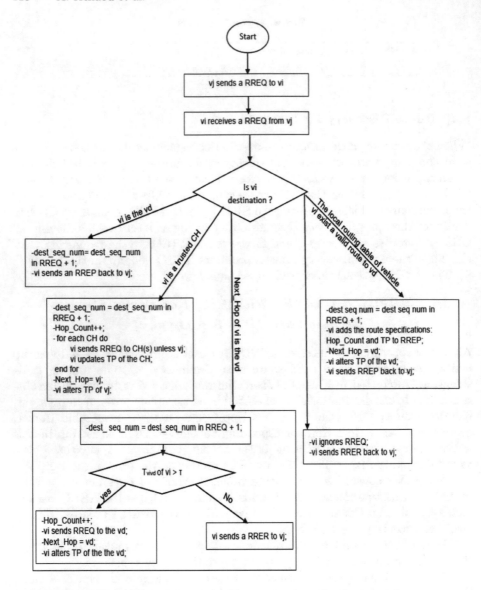

Fig. 1. Flow chart of the route discovery process for the inter-cluster.

RREP and calculates the overall *TP* up to destination vd. The RREP structure is depicted as follows:

$$RREP \ (Hop\_Count, \ Current\_TP\_dest, \ ID_{vd}, \ Dest\_seq\_ num, \\ ID_{vs}, \ LT)$$

Where the Current_TP_dest represents the current *TP* value to IDvd.

Else if the Next_Hop of vehicle *vi* represents destination *vd*, then the dest_seq_num in the RREQ increases by one and vehicle *vi* checks if the T_{vivd} of destination *vd* is higher than τ, then the Hop_Count in the RREQ rises by one. Moreover, vehicle *vi* sends the RREQ to destination *vd* and updates the reverse route entry in the local routing table where the Next_Hop, which is destination *vd*, alters the *TP* of destination *vd*. Otherwise, vehicle *vi* sends the route error (RERR) to vehicle *vj*.

After that, the vehicle source *vs* determines the optimal route depending on two metrics: the *TP* and the Hop_Count. If there are various routes having the same *TPs*, then the route having the lowest Hop_Count is elected as the optimal and trustful route. However, since several routes have the same Hop_Count, the route having the highest *TP* is elected as the optimal and trustful route. Thus, the source vehicle *vs* saves the other discovered routes to be used in some situations like broken link, the routing failure, route errors, and unreachable destinations. After a period of time, the source vehicle *vs* waits an for an ACK_MSG from the destination vehicle *vd* or for a RERR packet from the CHs.

In each step of the route discovery process, the RERR is sent back to the source *vs* when it cannot reach the destination vehicle *vd*. Then, the source vehicle *vs* and the intermediate vehicles delete the destination entry in their routing table upon receiving the RERR packet.

The RRER structure is depicted as follows :

$$RERR \; (Dest\_Count, \; Unreachable\_ID_{vd}\text{-}1,$$
$$Unreachable\_dest\_seq\_num\_1, \; Unreachable\_ID_{vd}\text{-}2,$$
$$Unreachable\_dest\_seq\_num\_2)$$

Where the Dest_Count corresponds to the computation of unreachable vehicles.

With the AODV routing protocol, if the RREP packet in our proposal is not received after a period time β, then the source *vs* will broadcast a new RREQ. Else if the failure remains, then the next RREQ will hold for $2*\beta$ seconds to be broadcast.

In this case, this process will be pursued until three broadcast RREQ packet is reached. Hence, the flow chart is illustrated in Fig. 1.

4   Simulation and Results

The aim of this section is to evaluate the performance of TRPV proposal.

4.1   Simulation Environment

We have conduct the simulations using the vehicular network simulation (Veins), the network simulator (Omnet++) and the traffic simulator: the Simulation of Urban MObility (SUMO). In our simulation, we have 50 vehicles in the network randomly placed in 2000 m * 2000 m. Additionally, each vehicle has 250 m range and moves with a maximum velocity of 30 m/s. Then, the simulation duration is 400 s and each situation is repeated 20 times for taking the average value. Table 2 illustrates the simulation parameters.

Table 2. Simulation parameters of VANET

Parameter	Value
Simulation area	2000 m * 2000 m
Mobility model	Random waypoint
Mac protocol	IEEE 802.11p
Number of vehicles	50
Number of RSUs	1
Velocity	5 m/s–30 m/s
Transmission range	250 m
Trafic type	CBR
Packet size	512 bytes
Packet rate	4 packets/s
Pause time	10 s
Bandwidth	2 Mbps
β	2.8 s
τ	0.7
Routing protocol	AODV, TRPV
Simulation time	400 s

To evaluate the performance of the proposed TRPV, we consider the following metrics:

- The throughput: is the ratio of the number of packets received by the destination vehicle per unit of time.
- The routing overhead: is the ratio between the number of the routing packets (RREQs, RERRs, RREPs and the updated packets) and the total amount of data and the routing packets.
- The Packet Data Ratio (PDR): is the ratio between the number of RREQ packets received by the destination node and the number of RREQ packets transmitted by the source node. In fact, it represents the reliability of packet transmissions in a network.
- The average end-to-end delay: is the delay experienced by the successfully forwarded packets to reach their destinations.

4.2 Simulation Results and Analysis

We assess the performance of TRPV to different velocity of vehicles. To show the performance of TRPV, we used the following metrics: throughput, routing overhead, average end-to-end delay, and Packet Data Ratio (PDR).

4.2.1   Throughput

'Figure 2 shows the variation of the throughput when varying the velocity of the vehicles for the proposed TRPV with the AODV protocol in the network. We notice that the throughput decreases as the velocity of the vehicles increases. In fact, when the vehicle velocity increases (5-10-15-20-25-30-35), the packet loss increases on the communication links as the possibility of the reliable and optimal path discovery decreases, and the network throughput should be degraded. This problem is severe in the AODV. The obtained results show that the proposed TRPV gets better throughput than AODV protocol. This can be explained by the fact that the proposed TRPV involves stable clustering and provides trustful route that improves by maximizing the selected route TP, and minimizing the number of hops.

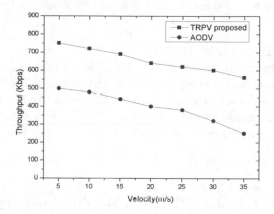

Fig. 2. Throughput according to the velocity of vehicles in the network.

4.2.2   Routing Overhead

Figure 3 shows the evolution of the routing overhead for different vehicle velocities (5-10-15-20-25 30-35). The obtained results show that the proposed TRPV is lower than AODV protocol. In fact, at high velocities, the connection between the vehicles frequently fails. However, in the proposed TRPV, the control packets (RREQ and RREP) are generated in CH configurations in order to discover and recover the route. Therefore, the proposed TRPV is achieved among CHs and avoids the routing discovery among all vehicles in the network. Moreover, the control packets are sent to qualified routes in order to meet trust in the proposed TRPV, meanwhile, the trust is not treated in the AODV protocol.

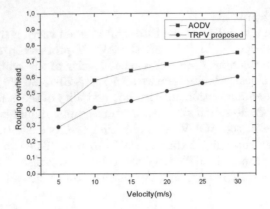

Fig. 3. Routing overhead according to the velocity of vehicles in the network.

4.2.3   Packet Data Ratio (PDR)

Figure 4 shows the variation of the PDR for different vehicle velocities (5-10-15-20-25-30-35). The obtained results show that the proposed TRPV is higher than AODV protocol. In fact, using the clustering mechanism in the proposed RREQ flooding is prevented. Moreover, the proposed TRPV excludes the malicious vehicles from the participation in network transactions as a result of malicious vehicle detection.

4.2.4   End to End Delay

Figure 5 shows the evolution of the End to End delay when varying the velocity of the vehicles for the proposed TRPV and AODV protocol in the network. The obtained results show that the proposed TRPV is lower than the AODV protocol. This can be explained by the fact that the proposed TRPV can handle alternative routes and the later alternative route can be handled when receiving an RERR packet. However, the AODV protocol has one route in routing table.

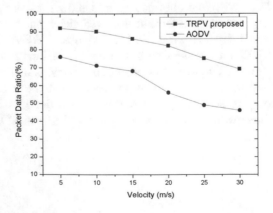

Fig. 4. Packet data ratio according to the velocity of vehicles in the network.

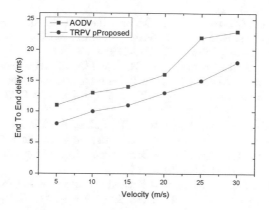

Fig. 5. End to End delay according to the velocity of vehicles in the network.

When the route failure happens, the routing step should be repeated. Thus, the packets encounter more latency when the vehicles move rapidly.

5 Conclusion

In VANET, the vehicles exchange various types of information and transmit it from the source to the destination. However, warning messages exchanged and unreliable links presented in the route pose road security problems. To mitigate this problem, we proposed a routing protocol based trust for the VANET through trusted CHs in order to find a trustful route depending on all vehicles trustworthiness along the route and the number of hops. In fact, the path trustworthiness is computed by the evaluating trust of link that must be greater than τ. In this paper, the simulation results presented the performances of the trust routing protocol for the VANET compared to the AODV in terms of throughput, Packet Data Ratio (PDR), routing overhead, and average end-to-end delay by varying the velocity of the vehicles. Hence, the proposal selects the optimal and trustful route and improve the data transmission.

References

1. Shafi, S., Ratnam, D.V.: Mobility-aware cross layer based reliable routing protocol for improved data dissemination in VANETs (2019)
2. Kchaou, A., Abassi, R., Guemara, S.: Towards a secured clustering mechanism for messages exchange in VANET. In: 2018 32nd International Conference on Advanced Information Networking and Applications Workshops (WAINA), pp. 88–93. IEEE (2018)
3. Kchaou, A., Abassi, R., El Fatmi, S.G.: A new trust based routing protocol for VANETs. In: 2018 Seventh International Conference on Communications and Networking (ComNet), pp. 1–6. IEEE (2018)

4. Eiza, M.H., Ni, Q., Owens, T., Min, G.: Investigation of routing reliability of vehicular ad hoc networks. EURASIP J. Wirel. Commun. Network. **2013**(1), 179 (2013). https://doi.org/10.1186/1687-1499-2013-179
5. Sahoo, R.R., Panda, R., Behera, D.K., Naskar, M.K.: A trust based clustering with ant colony routing in VANET. In 2012 Third International Conference on Computing, Communication and Networking Technologies (ICCCNT 2012), pp. 1–8. IEEE (2012)
6. Kerrache, C.A., Lagraa, N., Calafate, C.T., Lakas, A.: TROUVE: a trusted routing protocol for urban vehicular environments. In: 2015 IEEE 11th International Conference on Wireless and Mobile Computing, Networking and Communications (WiMob), pp. 260–267. IEEE (2015)
7. Jeyaprakash, T., Mukesh, R.: A new trusted routing protocol for vehicular ad hoc networks using trusted metrics. IJ Netw. Secur. **19**(4), 537–545 (2017)
8. Tolba, A.M.R.: Trust-based distributed authentication method for collision attack avoidance in VANETs. IEEE Access **6**, 62747–62755 (2018)
9. Sataraddi, M.J., Kakkasageri, M.S.: Trust and delay based routing for VANETs. In: 2019 IEEE International Conference on Advanced Networks and Telecommunications Systems (ANTS), pp. 1–6. IEEE (2019)

Design and Implementation of the TBOI (Time-Based Operation Interruption) Protocol to Prevent Late Information Flow in the IoT

Shigenari Nakamura[1](✉), Tomoya Enokido[2], and Makoto Takizawa[3]

[1] Tokyo Metropolitan Industrial Technology Research Institute, Tokyo, Japan
[2] Rissho University, Tokyo, Japan
eno@ris.ac.jp
[3] Hosei University, Tokyo, Japan
makoto.takizawa@computer.org

Abstract. In the CBAC (Capability-Based Access Control) model proposed for the IoT (Internet of Things), authorizers, e.g. device owners, issue capability tokens, i.e. a set of access rights on resource objects in devices, to subjects. Since data of a device are stored in another device through communication among subjects and devices, a subject sb can get data of a resource object r^1 by accessing another resource object r^2 even if the subject sb is not allowed to get the data from the resource object r^1. Here, an illegal information flow from the resource object r^1 to the subject sb occurs. In addition, each capability token is valid only for some time period. Suppose, data generated at time τ flow to a subject sb at time τ' ($> \tau$). If the subject sb is allowed to get data at time τ' but not at time τ, the subject sb should not receive the data. Here, the information flow is referred to as late. In our previous studies, the OI (Operation Interruption) and TBOI (Time-Based OI) protocols are proposed to prevent operations implying only illegal information flow and both illegal and late information flows, respectively. In this paper, we discuss the design and implementation of a device supporting the TBOI protocol and evaluate an authorization process of the TBOI protocol in terms of the execution time. In the evaluation, we show the late information flow is prevented in addition to the illegal one in the TBOI protocol although the execution time is almost the same as the OI protocol.

Keywords: IoT (Internet of Things) · Device security · CBAC (Capability-Based Access Control) model · Information flow control · Late information flow · TBOI (Time-Based Operation Interruption) protocol

1 Introduction

It is widely recognized that access control models [3] are useful to make information systems like database systems secure. For the IoT (Internet of Things) [6], the CBAC (Capability-Based Access Control) model named "CapBAC model" is proposed [5,7]. Here, subjects are issued capability tokens which are collections of access rights. Only the authorized subjects can manipulate resource objects of devices only in the authorized operations. Data are exchanged among subjects and resource objects through manipulating resource objects of devices. Therefore, a subject might obtain data via other subjects and resource objects even if the subject is granted no access right to obtain the data, i.e. illegal information flow might occur. The illegal information flow is a critical issue for not only the IoT but also other various types of information systems. In database systems with the RBAC (Role-Based Access Control) model, protocols to prevent illegal information flow are proposed [4, 8].

In order to solve this illegal information flow problem in the IoT, the OI (Operation Interruption) protocol is proposed in our previous studies [10]. Here, operations implying illegal information flow are interrupted, i.e. not performed, at devices. Since illegal information flow may occur in the CBAC model, each device can check whether or not each operation is illegal. The OI protocol is implemented on a device and evaluated in terms of the execution time to authorize access requests [11].

A capability token issued to a subject is valid only for some time period. In the OI protocol, although no illegal information flow occurs, a subject might get data generated out of validity period of a capability token to get the data. Suppose a subject sb is issued a capability token including an access right $\langle r, get \rangle$, i.e. the subject sb is allowed to get data from the resource r. The subject sb tries to get data generated at time τ from the resource r. Here, even if the time τ is not within the validity period of the capability token, the subject sb can get the data. The data are older than the subject sb expects to get, i.e. information comes to the subject sb *late*.

In our previous studies [9], the TBOI (Time-Based OI) protocol is proposed to prevent both illegal and late information flows. Here, operations implying not only illegal information flow but also late information flow are interrupted. The late information flow is defined based on the generation time of data of resources and validity periods of capability tokens issued to subjects. Each device can check whether or not each operation is late by using the time information. The TBOI protocol is evaluated in a simulation. The number of operations interrupted in the TBOI protocol is larger than the OI protocol.

In this paper, we design and implement a device supporting the TBOI protocol. Here, a Raspberry Pi3 Model B+ with Raspbian [2] is used as an IoT device. The device and a subject are a CoAP server and a CoAP client, respectively, in CoAPthon3 [14]. In the evaluation, we show that both illegal and late information flows are prevented in the TBOI protocol although the execution time of an authorization process is almost the same as the OI protocol.

In Sect. 2, we present the system model. In Sect. 3, we discuss types of information flow relations to occur in the CBAC model. In Sect. 4, we overview the TBOI protocol. In Sect. 5, we implement the TBOI protocol on a Raspberry Pi 3 Model B+. In Sect. 6, we evaluate the authorization process of the TBOI protocol in terms of the execution time.

2   System Model

In this paper, we assume that an IoT includes a pair of sets D and SB composed of the number dn of devices $d_1, \ldots, d_{dn}$ ($dn \geq 1$) and the number sbn of subjects $sb_1, \ldots, sb_{sbn}$ ($sbn \geq 1$), respectively. Each device d_k holds a set $d_k.R$ composed of the number nr_k

(≥ 1) of resource objects $r_1^k, \ldots, r_{nr_k}^k$. Notations r^k and r denote some resource object r_m^k in a device d_k and some resource object r_m^k in some device d_k, respectively. In order to make the IoT secure, the CBAC (Capability-Based Access Control) model is proposed [5,7]. Here, authorizers like device owners issue capability tokens to subjects to be able to access resource objects in the device. Each subject sb_i obtains a set CAP^i composed of the number nc_i of capability tokens $cap_1^i, \ldots, cap_{nc_i}^i$ issued. A capability token includes access rights. An access right such as a pair $\langle r, op \rangle$ indicates a subject granted the access right is allowed to manipulate a resource r in an operation op. When a subject sb_i tries to manipulate data of a resource r^k in a device d_k in an operation op, an access request and a capability token cap^i including an access right $\langle r^k, op \rangle$ are sent to the device d_k from the subject sb_i. If the capability token cap^i is valid, the access request is accepted, i.e. the operation op is performed on the resource r^k. Otherwise, the message which indicates the access request is rejected is sent to the subject sb_i from the device d_k. Since the device d_k just checks the capability token cap^i to authorize the subject sb_i, it is widely recognized that the CBAC model is more suitable for the IoT than other access control models based on the ACL (Access Control List) such as RBAC (Role-Based Access Control) [12] and ABAC (Attribute-Based Access Control) [16] models. However, it is more difficult to change capability tokens, i.e. access rights already granted to subjects, because these access rights are already distributed to subjects.

Each capability token is only valid for some period. Suppose a subject sb_i is issued a capability token cap^i including an access right $\langle r, op \rangle$ on a resource object r which is specified to be valid for time t_1 to time t_2. The subject sb_i can only access the resource object r in the operation op only in the validity period for time t_1 to time t_2. A capability token cap^i to get data from a resource r is valid from time $gt^i.st(r)$ to time $gt^i.et(r)$. Data in a resource object r are generated at some time which is referred to a generation time. Data in a resource object r may flow to a subject and another resource object. Let $minRT_m^k(r)$ be the earliest generation time of data in a resource object r which flow to another resource object r_m^k. In addition, $minSBT^i(r)$ is the earliest generation time of data in a resource object r which flow into a subject sb_i.

There are three types of devices, sensor, actuator, and hybrid devices. A sensor device just collects data by sensing events occurring around itself. An actuator device acts according to data and action requests sent from subjects. Subjects get and put data from sensor devices and to actuator devices, respectively. A hybrid device both collects data and acts such as cars and robots. Hence, subjects issue both *get* and *put* operations to hybrid devices.

3   Information Flow Relations

Through manipulating data of resource objects in devices, data are exchanged among entities, i.e. resource objects and subjects. For example, a subject sb gets data from a resource object r^1 in a device d_1 and puts the data to a resource object r^2. Here, the data of the resource object r^1 flow into the subject sb and the resource object r^2. In this paper, resource objects whose data flow into entities such as resource objects and subjects are referred to as *source* resource objects for these entities. Let $r.sR$ and $sb.sR$ are sets of *source* resource objects for a resource object r and a subject sb, respectively, which are initially ϕ. In this example, $sb.sR = r^2.sR = \{r^1\}$. In another example, a sensor device d_1 collects data by sensing events and stores the data in its resource object r^1. Here, $r^1.sR = \{r^1\}$.

Let $IN(sb)$ be a set of resource objects whose data a subject sb can get, i.e. $IN(sb_i) = \{r \mid \langle r, get \rangle \in cap^i \wedge cap^i \in CAP^i\}$. In this section, we define types of information flow relations on resource objects and subjects based on the CBAC model as follows [9,10]:

Definition 1. A resource object r_m^k *flows* to a subject $sb_i(r_m^k \rightarrow sb_i)$ iff (if and only if) $r_m^k.sR \neq \phi$ and $r_m^k \in IN(sb_i)$.

Definition 2. A resource object r_m^k *legally flows* to a subject $sb_i(r_m^k \Rightarrow sb_i)$ iff $r_m^k \rightarrow sb_i$ and $r_m^k.sR \subseteq IN(sb_i)$.

Definition 3. A resource object r_m^k *illegally flows* to a subject $sb_i(r_m^k \mapsto sb_i)$ iff $r_m^k \rightarrow sb_i$ and $r_m^k.sR \not\subseteq IN(sb_i)$.

If $r_m^k \rightarrow sb_i$ holds, data flow from the resource object r_m^k to the subject sb_i. Otherwise, no data flow from the resource object r_m^k into the subject sb_i. In the legal information flow relation ($\Rightarrow$), the condition "$r_m^k.sR \subseteq IN(sb_i)$" means that only data which the subject sb_i is allowed to get flow from the resource object r_m^k into the subject sb_i. In the illegal information flow relation ($\mapsto$), the condition "$r_m^k.sR \not\subseteq IN(sb_i)$" means that data which the subject sb_i is not allowed to get flow from the resource object r_m^k into the subject sb_i.

Definition 4. A resource object r_m^k *timely flows* to a subject $sb_i(r_m^k \Rightarrow_t sb_i)$ iff $r_m^k \Rightarrow sb_i$ and $\forall r \in r_m^k.sR(gt^i.st(r) \leq minRT_m^k(r) \leq gt^i.et(r))$.

Definition 5. A resource object r_m^k *flows late* to a subject $sb_i(r_m^k \mapsto_l sb_i)$ iff $r_m^k \Rightarrow sb_i$ and $\exists r \in r_m^k.sR(\neg(gt^i.st(r) \leq minRT_m^k(r) \leq gt^i.et(r)))$.

In the timely information flow relation ($\Rightarrow_t$), the condition "$\forall r \in r_m^k.sR(gt^i.st(r) \leq minRT_m^k(r) \leq gt^i.et(r))$" means that only the data generated while the subject sb_i is allowed to get flow from the resource object r_m^k into the subject sb_i. In the late information flow relation ($\mapsto_l$), the condition "$\exists r \in r_m^k.sR(\neg(gt^i.st(r) \leq minRT_m^k(r) \leq gt^i.et(r)))$" means that data generated while the subject sb_i is not allowed to get flow from the resource object r_m^k into the subject sb_i. This means, even if the data are not generated while the subject sb_i is allowed to get the data, the subject sb_i can get the data.

4 Protocols to Prevent Illegal and Late Information Flow

4.1 OI (Operation Interruption) Protocol

In the CBAC model for the IoT, capability tokens including access rights are issued to subjects. Data are exchanged among subjects and resource objects through manipulating resource objects in devices. If data of a resource object r^1 flowed into another resource object r^2 are furthermore brought into a subject sb which is not allowed to get the data of the resource object r^1, the data of the resource r^1 illegally flow into the subject sb. In Sect. 3, the illegal information flow relation ($\mapsto$) among subjects and resource objects is defined based on the CBAC model. In order to prevent the illegal information flow, the OI (Operation Interruption) protocol is proposed in our previous studies [10]. Here, operations implying illegal information flow are interrupted at devices. The OI protocol is shown as follows:

[OI (Operation Interruption) protocol]

1. A device d_k gets data by sensing events occurring around the device d_k.
   a. The device d_k stores the sensor data in the resource object r_m^k and $r_m^k.sR = r_m^k.sR \cup \{r_m^k\}$;
2. A subject sb_i issues a *get* operation on a resource r_m^k to a device d_k.
   a. If $r_m^k \Rightarrow sb_i$, the subject sb_i gets data from the resource r_m^k and $sb_i.sR = sb_i.sR \cup r_m^k.sR$;
   b. Otherwise, the *get* operation is interrupted at the device d_k;
3. A subject sb_i issues a *put* operation on a resource r_m^k to a device d_k.
   a. The subject sb_i puts the data to the resource r_m^k and $r_m^k.sR = sb_i.sR$;

$r_m^k.sR$ and $sb_i.sR$ are sets of *source* resource objects for a resource object r_m^k and a subject sb_i, respectively. Although $r_m^k.sR$ and $sb_i.sR$ are initially ϕ, they are updated through manipulating resource objects. $IN(sb_i)$ is a set of resource objects whose data a subject sb_i can get. The illegal information flow relation is defined based on these sets, $r_m^k.sR, sb_i.sR$, and $IN(sb_i)$, as shown in the Definition 3. If data which the subject sb_i is not allowed to get flow from the resource object r_m^k into the subject sb_i in a *get* operation of the subject sb_i, an illegal information flow ($r_m^k \mapsto sb_i$) occurs. Hence, the *get* operation is interrupted in the OI protocol.

4.2 TBOI (Time-Based OI) Protocol

In the OI protocol, although every illegal information flow is prevented, subjects might get data generated out of validity period of a capability token to get the data. For example, a subject sb_1 sends an access request and a capability token cap^1 to get data generated at time τ from the resource r. Even if the time τ is not within the validity period of the capability token cap^1, the subject sb_1 can get the data. Here, the data are older than the subject sb_1 expects to get, i.e. information comes to the subject sb_1 late. The late information flow relation is defined based on the CBAC model as shown in Sect. 3.

In order to prevent both illegal and late types of information flows, the TBOI (Time-Based OI) protocol is proposed in our previous studies [9]. Here, operations implying not only illegal information flow but also late one are interrupted. The TBOI protocol is shown as follows:

[TBOI (Time-Based OI) protocol]

1. A device d_k gets data by sensing events occurring around the device d_k at time τ.
   a. The device d_k stores the sensor data in a resource object r_m^k and $r_m^k.sR = r_m^k.sR \cup \{r_m^k\}$;
   b. If $minRT_m^k(r_m^k) = NULL, minRT_m^k(r_m^k) = \tau$;

2. A subject sb_i issues a *get* operation on a resource object r_m^k to a device d_k.
   a. If $r_m^k \Rightarrow_t sb_i$,
      i. For each resource object r such that $r \in (sb_i.sR \cap r_m^k.sR)$, $minSBT^i(r) = min(minSBT^i(r), minRT_m^k(r))$;
      ii. For each resource object r such that $r \notin sb_i.sR$ but $r \in r_m^k.sR$, $minSBT^i(r) = minRT_m^k(r)$;
      iii. The subject sb_i gets data from the resource object r_m^k and $sb_i.sR = sb_i.sR \cup r_m^k.sR$;
   b. Otherwise, the *get* operation is interrupted at the device d_k;
3. A subject sb_i issues a *put* operation on a resource object r_m^k to a device d_k.
   a. For each resource object r such that $r \in sb_i.sR$, $minRT_m^k(r) = minSBT^i(r)$;
   b. The subject sb_i puts the data to the resource object r_m^k and $r_m^k.sR = sb_i.sR$;

$minRT_m^k(r)$ and $minSBT^i(r)$ are a pair of the earliest generation times of data which are in a resource object r and flow into a resource object r_m^k and into a subject sb_i, respectively. These generation times are updated through manipulating resource objects.

If a subject sb_1 issues a get operation on a resource object r^1 to a device d_1, the device d_1 checks whether or not every data which flow from the resource object r^1 to the subject sb_1 are generated within the validity period of a capability token to get the data. The condition "$gt^1.st(r) \leq minRT^1(r) \leq gt^1.et(r)$" means that all data of the resource object r are generated within the validity period. Here, the existence of resource objects which do not hold this condition, i.e. $\exists r \in r^1.sR(\neg(gt^1.st(r) \leq minRT^1(r) \leq gt^1.et(r)))$, means that an late information flow ($r^1 \mapsto_l sb_1$) occurs from the resource object r^1 to the subject sb_1. Hence, the get operation is interrupted in the TBOI protocol.

Example 1. An example of the TBOI protocol is shown in Fig. 1. Suppose a pair of devices d_1 and d_2 and a pair of subjects sb_1 and sb_2 exist. Here, d_1 and d_2 are sensor and hybrid devices, respectively. There are resource objects r^1 and r^2 in devices d_1 and d_2, respectively. The capability token issued a subject sb_1 is composed of a pair of access rights $\langle r^1, get \rangle$ and $\langle r^2, put \rangle$ and valid from time t_1 to time t_2. In addition, another capability token issued to a subject sb_2 includes a pair of access rights $\langle r^1, get \rangle$ and $\langle r^2, get \rangle$ and is valid from time t_3 to time t_4. Here, $IN(sb_1) = \{r^1\}$ and $IN(sb_2) = \{r^1, r^2\}$.

First, the sensor d_1 gets data by sensing events occurring around itself at time τ and stores the date in a resource object r^1, i.e. $r^1.sR = \{r^1\}$ and $minRT^1(r^1) = \tau$. Next, the subject sb_1 gets data from the resource object r^1 in the device d_1 because $r^1 \Rightarrow_t sb_1$ holds. Hence, $sb_1.sR = sb_1.sR(= \phi) \cup r^1.sR(= \{r^1\}) = \{r^1\}$ and $minSBT^1(r^1) = minRT^1(r^1) = \tau$. Then, the hybrid device d_2 gets data by sensing events occurring around itself at time τ' and stores the data in a resource object r^2, i.e. $r^2.sR = \{r^2\}$ and $minRT^2(r^2) = \tau'$. After that, the subject sb_1 puts the data to the resource object r_2 in the hybrid device d_2. Here, $r^2.sR = r^2.sR(= \{r^2\}) \cup sb_1.sR(= \{r^1\}) = \{r^1, r^2\}$ and $minRT^2(r^1) = minSBT^1(r^1) = \tau$. Finally, the subject sb_2 tries to get the data from the resource object r^2. Here, a late information flow relation $r^2 \mapsto_l sb_2$ holds because $minRT^2(r^1)(= \tau) \leq gt^2.st(r^1)(= t_3)$. Hence, the get operation from the subject sb_2 is interrupted at the hybrid device d_2 to prevent late information flow.

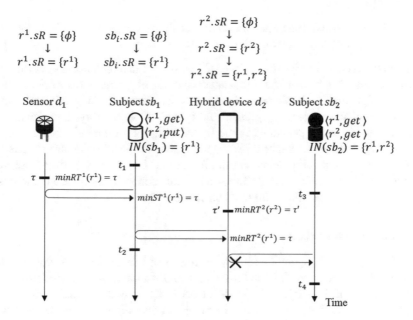

Fig. 1. TBOI protocol.

5   Implementation of the TBOI Protocol

We implement the TBOI protocol on a hybrid device d_1 which is realized on a Raspberry Pi 3 Model B+ [1] whose operating system is Raspbian [2]. A subject sb_1 is also implemented in Python on the hybrid device d_1. The hybrid device d_1 and the subject sb_1 are realized as a CoAP server and a CoAP client in CoAPthon3 [14], respectively.

5.1   Capability Token

A capability token is designed to be composed of the following fields referring to a capability token example shown in the paper [7].

- Identifier (variable size): identifier of the capability token.
- Issued-time (10 bytes): time at which the capability token is issued.
- Issuer (64 bytes): issuer of the capability token, i.e. owner of a device.
- Subject (64 bytes): subject granted the access rights in the capability token.
- Device (variable size): device whose access rights are in the capability token.
- Access rights: set of access rights granted to the subject. Each access right is composed of the following fields.
  - Operation (variable size): operation allowed, which is any CoAP (Constrained Application Protocol) [13] method, e.g. GET and PUT.
  - Resource (variable size): resource object of the device which is allowed to manipulate in the operation which is designated in the previous field.
- Not_before (10 bytes): time before which the capability token must not be valid.

- Not_after (10 bytes): time after which the capability token must not be valid.
- Signature (64 bytes): signature of the capability token.

A capability token and a signature of the subject sb_1 are included in the payload field of a CoAP request. An "Issuer" field indicates a public key of the issuer of the capability token. A "Signature" field shows a signature of the issuer. A "Subject" field contains a public key of the subject sb_1 granted the capability token. Keys and signatures of issuers and subjects are generated in the ECDSA algorithm implemented as a Python package [15]. Values of "Issuer", "Signature", and "Subject" fields are encoded into Base64 form. The capability token is valid from time "Not_before" to time "Not_after". "Issued-time", "Not_before", and "Not_after" fields denote Unix time, i.e. the number of seconds elapsed since 00:00:00 UTC on Jan. 1, 1970.

5.2   Authorization Process

In the TBOI protocol, operations implying illegal or late information flow are interrupted at devices. For this aim, the following authorization process is performed at the hybrid device d_1 if the subject sb_1 sends a CoAP request with capability tokens to the hybrid device d_1. In this paper, we consider a pair of operations, *get* and *put* on devices. If the operation op is *put*, the subject sb_1 includes its set $sb_1.sR$ of *source* resource objects in the payload field of the CoAP request.

Step 1. The hybrid device d_1 checks if the subject sb_1 is allowed to manipulate a resource object r^1 or not in the operation op. If the subject sb_1 is issued a capability token cap^1 which satisfies the following conditions, the subject sb_1 is allowed to issue the operation op. Otherwise, the operation op is interrupted.
- The operation op matches "Operation" of the capability token cap^1.
- The destination and the Resource-URI option of the CoAP request match "Device" and "Resource" of the capability token cap^1, respectively.
- The current time is larger than "Not_before" and smaller than "Not_after" of the capability token cap^1.
Step 2. If the operation op is *get*, the hybrid device d_1 checks if the subject sb_1 is allowed to get data from every resource object in $r^1.sR$ or not. If at least one resource object from which the subject sb_1 is not allowed to get data exists in $r^1.sR$, i.e. $r^1 \mapsto sb_1$ holds, the operation op is interrupted.
Step 3. If the operation op is *get*, the hybrid device d_1 checks if only the data generated while the subject sb_1 is allowed to get the data flow into the subject sb_1. If at least one resource object whose data flow late to the subject sb_1 exists in $r^1.sR$, i.e. $r^1 \mapsto_l sb_1$ holds, the operation op is interrupted.
Step 4. A public key of the subject sb_1 is in each capability token as a "Subject" filed. A signature of the subject sb_1 is in the payload field of the CoAP request. The hybrid device d_1 checks if the signature of the subject sb_1 is valid or not. If the signature is not valid, the operation op is interrupted.
Step 5. A public key and a signature of its issuer in each capability token are a "Issuer" and a "Signature" fields, respectively. The hybrid device d_1 checks if the signature of every capability token used in the above steps 1, 2 and 3 is valid or not. If at least one capability token which is not valid exists, the operation op is interrupted.

After the authorization process, sets $r^1.sR$ and $sb_1.sR$ of *source* resource objects are updated. If the authorization process is completed for a *put* operation, the hybrid device d_1 updates a set $r^1.sR$ of its *source* resource objects with $sb_1.sR$, i.e. $r^1.sR = sb_1.sR$. On the other hand, if the authorization process is completed for a *get* operation, the hybrid device d_1 sends a CoAP response to the subject sb_1. Here, the hybrid device d_1 includes its set $r^1.sR$ of *source* resource objects in the payload field of the CoAP response. The subject sb_1 adds the resource objects in the set $r^1.sR$ to its set $sb_1.sR$ of *source* resource objects, i.e. $sb_1.sR = sb_1.sR \cup r^1.sR$.

6   Evaluation

In the evaluation, first, a subject sb_1 puts data to the resource object r^1 in the hybrid device d_1. Here, the data are randomly generated. The data size is also randomly decided from 80 to 120 [Bytes]. Next, the subject sb_1 gets data in the resource object r^1 in the hybrid device d_1. We iterate the above measurement ten times, i.e. a pair of *put* and *get* operations are issued repeatedly ten times, and measure the execution time of every authorization process. Here, we generate a scenario so that every authorization process is completed because the total execution time is longest in this case. If any illegal or late information flow occurs, an authorization process aborts in the middle and the execution time is always shorter. After the measurements, we calculate the average execution time of an authorization process.

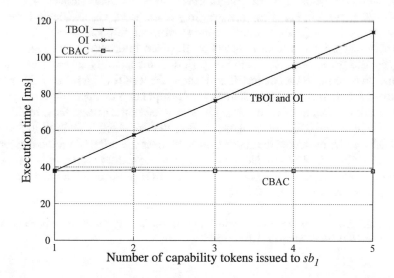

Fig. 2. Execution time for a *get* operation.

Figure 2 shows the execution time of an authorization process for a *get* operation. The horizontal axis is the number of capability tokens issued to the subject sb_1, i.e. $|CAP^1|$. In this evaluation, we assume that data are put to the resource object r^1 before each *get* operation such that $r^1.sR = \{r^m \mid m = 1, \ldots, |CAP^1|\}$. Next, we generate CAP^1 such that $CAP^1 = \{cap_m^1 \mid \langle r^m, get \rangle \in cap_m^1 (m = 1, \ldots, |CAP^1|)\}$ to complete the authorization process and the subject sb_1 issues a *get* operation on the resource object r^1 to the hybrid device d_1.

The label "CBAC" shows the execution time of an authorization process where only the CBAC model is adopted, i.e. no information flow control is supported. Here, only steps 1, 4, and 5 are performed. In the OI protocol, illegal information flow is prevented. Here, the step 2 is performed in addition to the authorization process of the CBAC model. In the TBOI protocol, both illegal and late information flows are prevented. Here, all the steps are performed. As a result, the execution time of the authorization process in the CBAC model is shorter than the OI and TBOI protocols. On the other hand, the execution time of the authorization process in the TBOI protocol is almost the same as the OI protocol. This means, in the TBOI protocol, late information flow is prevented in addition to illegal information flow although the execution time of the authorization process is almost the same as the OI protocol.

7   Concluding Remarks

The CBAC (Capability-Based Access Control) model is proposed to make the IoT (Internet of Things) secure. Here, capability tokens are issued to subjects as access rights. Since data are exchanged among subjects and resource objects through manipulating resource objects, a subject might obtain data via other subjects and resource objects even if the subject is not allowed to get the data, i.e. an illegal information flow occurs. In the OI (Operation Interruption) protocol proposed in our previous studies, operations implying illegal information flow are interrupted. However, a subject might get data generated out of validity period of a capability token to get the data. Here, the data come to the subject *late*. Hence, the TBOI (Time-Based OI) protocol was also proposed in our previous studies to interrupt not only operations implying illegal information flow but also operations implying late information flow. In this paper, we implemented the TBOI protocol on a hybrid device which is realized in Raspberry Pi 3 Model B+. We evaluated the authorization process of the TBOI protocol in terms of the execution time. In the evaluation, we showed the late information flow is prevented in addition to the illegal information flow in the TBOI protocol although the execution time is almost the same as the OI protocol.

Acknowledgements. This work was supported by Japan Society for the Promotion of Science (JSPS) KAKENHI Grant Number JP20K23336.

References

1. Raspberry pi 3 model b+. https://www.raspberrypi.org/products/raspberry-pi-3-model-b-plus/
2. Raspbian, version 10.3, 13 February 2020. https://www.raspbian.org/ (visited on 11 March 2020)
3. Denning, D.E.R.: Cryptography and Data Security. Addison Wesley, Boston (1982)

4. Enokido, T., Barolli, V., Takizawa, M.: A legal information flow (LIF) scheduler based on role-based access control model. Int. J. Comput. Stand. Interfaces **31**(5), 906–912 (2009)
5. Gusmeroli, S., Piccione, S., Rotondi, D.: A capability-based security approach to manage access control in the internet of things. Math. Comput. Model. **58**(5–6), 1189–1205 (2013)
6. Hanes, D., Salgueiro, G., Grossetete, P., Barton, R., Henry, J.: IoT Fundamentals: Networking Technologies, Protocols, and Use Cases for the Internet of Things. Cisco Press, Indianapolis (2018)
7. Hernández-Ramos, J.L., Jara, A.J., Marín, L., Skarmeta, A.F.: Distributed capability-based access control for the internet of things. J. Internet Serv. Inf. Secur. **3**(3/4), 1–16 (2013)
8. Nakamura, S., Duolikun, D., Enokido, T., Takizawa, M.: A read-write abortion protocol to prevent illegal information flow in role-based access control systems. Int. J. Space-Based Situated Comput. **6**(1), 43–53 (2016)
9. Nakamura, S., Enokido, T., Takizawa, M.: Time-based legality of information flow in the capability-based access control model for the internet of things. Concurrency Comput. Pract. Experience https://doi.org/10.1002/cpe.5944
10. Nakamura, S., Enokido, T., Takizawa, M.: Information flow control based on the CapBAC (capability-based access control) model in the IoT. Int. J. Mob. Comput. Multimed. Commun. **10**(4), 13–25 (2019)
11. Nakamura, S., Enokido, T., Ogiela, L., Takizawa, M.: Implementation of a device adopting the OI (operation interruption) protocol to prevent illegal information flow in the IoT. In: Proceedings of the 9th International Conference on Emerging Internet, Data, and Web Technologies, pp. 168–179 (2021)
12. Sandhu, R.S., Coyne, E.J., Feinstein, H.L., Youman, C.E.: Role-based access control models. IEEE Comput. **29**(2), 38–47 (1996)
13. Shelby, Z., Hartke, K., Bormann, C.: Constrained application protocol (CoAP). IFTF Internet-draft, http://tools.ietf.org/html/draft-ietf-core-coap-18 (2013)
14. Tanganelli, G., Vallati, C., Mingozzi, E.: CoAPthon: Easy development of CoAP-based IoT applications with python. In: IEEE 2nd World Forum on Internet of Things (WF-IoT 2015), pp. 63–68 (2015)
15. Warner, B.: python-ecdsa-0.11, 11 March 2014. https://github.com/ecdsa/python-ecdsa (visited on 7 July 2020)
16. Yuan, E., Tong, J.: Attributed based access control (ABAC) for web services. In: Proceedings of the IEEE International Conference on Web Services (ICWS 2005). IEEE Computer Society (2005)

How Far Should We Look Back to Achieve Effective Real-Time Time-Series Anomaly Detection?

Ming-Chang Lee[1(✉)], Jia-Chun Lin[1], and Ernst Gunnar Gran[1,2]

[1] Department of Information Security and Communication Technology, Norwegian University of Science and Technology, 2815 Gjøvik, Norway
{ming-chang.lee,jia-chun.lin,ernst.g.gran}@ntnu.no
[2] Simula Research Laboratory, 1364 Fornebu, Norway

Abstract. Anomaly detection is the process of identifying unexpected events or abnormalities in data, and it has been applied in many different areas such as system monitoring, fraud detection, healthcare, intrusion detection, etc. Providing real-time, lightweight, and proactive anomaly detection for time series with neither human intervention nor domain knowledge could be highly valuable since it reduces human effort and enables appropriate countermeasures to be undertaken before a disastrous event occurs. To our knowledge, RePAD (Real-time Proactive Anomaly Detection algorithm) is a generic approach with all above-mentioned features. To achieve real-time and lightweight detection, RePAD utilizes Long Short-Term Memory (LSTM) to detect whether or not each upcoming data point is anomalous based on short-term historical data points. However, it is unclear that how different amounts of historical data points affect the performance of RePAD. Therefore, in this paper, we investigate the impact of different amounts of historical data on RePAD by introducing a set of performance metrics that cover novel detection accuracy measures, time efficiency, readiness, and resource consumption, etc. Empirical experiments based on real-world time series datasets are conducted to evaluate RePAD in different scenarios, and the experimental results are presented and discussed.

Keywords: Anomaly detection · Time series · Unsupervised learning · Machine learning · LSTM · Look-Back and Predict-Forward strategy

1 Introduction

Anomaly detection refers to any method or technique designed to identify unexpected events or anomalies in data, and it has been used to discover anomalies in a time series, which is a series of data points generated continuously and evenly indexed in time order. During the past decade, a number of approaches and methods have been introduced for time series anomaly detections such as [2–8]. Many of them require either domain knowledge (e.g., data patterns, data distributions, pre-labeled training data) or human intervention (e.g., collecting sufficient training data, pre-labelling training data, config-uring appropriate hyperparameters, determining proper detection thresholds, etc.). These requirements consequently limit the applicability of these approaches.

It could be highly valuable to provide an anomaly detection approach that requires neither human intervention nor domain knowledge because such an approach can signif-icantly reduce human effort and that it can be easily and immediately applied to detect anomalies in any time series. In addition, offering real-time and proactive anomaly detec-tion might be also highly appreciated since it enables appropriate actions and counter-measures to be undertaken as early as possible. Last but not least, it would be beneficial to offer a lightweight anomaly detection approach because such an approach can be deployed on any commodity machine, such as desktops, laptops, and mobile phones.

To our knowledge, RePAD [1] is an approach with all above-mentioned features. RePAD is a real-time proactive anomaly detection algorithm for streaming time series based on a special type of recurrent neural networks called Long Short-Term Memory (LSTM). As soon as RePAD is employed to detect anomalies in a time series, it trains a LSTM model with short-term historical data points and automatically calculates a detection threshold to determine if each upcoming data point is anomalous or not. RePAD also decides if it should retrain its LSTM model based on the most recent short-term data points so as to adapt to minor pattern change in the time series. By utilizing LSTM with a simple network structure and training the LSTM model with short-term historical data points, RePAD demonstrates good, cost-effective, and real-time anomaly detection performance according to [1]. However, it is unclear from [1] that how different amounts of historical data points impact the performance of RePAD and that how far RePAD should look back such that it is able to provide the most effective anomaly detection. By the most effective, we mean that:

- Anomaly detect can soon start without lengthy preprocessing.
- Anomalies can be detected as early as possible.
- Low false positives and low false negatives.
- Infrequent LSTM model retraining.

In this paper, we conduct a comprehensive study to investigate the performance of RePAD. The study starts by introducing a set of metrics which cover novel detection accuracy measures that are appropriate and suitable for proactive detection approaches. The metrics also cover readiness and time efficiency since achieving lightweight and real-time anomaly detection as mentioned earlier are essential and desirable. We design two empirical experiments based on real-world time series datasets provided by the Numenta Anomaly Benchmark (NAB) [9] to evaluate RePAD in different scenarios. The experiment results show that RePAD based on long-term historical data points leads to a long preparation period, frequent LSTM retraining, and long detection time. Most importantly, RePAD loses its capability to accurately detect anomalies. On the contrary, choosing short-term historical data points is highly recommended for RePAD since it enables RePAD to provide the most effective anomaly detection.

The rest of the paper is organized as follows: Sect. 2 introduces the design of RePAD. Section 3 presents all metrics used to evaluate the performance of RePAD. Section 4 presents and discusses evaluation results based on empirical experiments and real-world time series datasets. In Sect. 5, we conclude this paper and outline future work.

2   RePAD

RePAD uses LSTM with a simple network structure (e.g., one hidden layer with only ten hidden units) to predict the values of the data points arriving at the next time point based on the data values observed at the past b continuous time points. Note that b is called the Look-Back parameter, which is set to 3 in [1]. Figure 1 illustrates the algorithm of RePAD. Let t be the current time point and t starts from 0, which is the time point when RePAD is launched. Since b is set to 3, RePAD has to collect three data points observed at time points 0, 1, and 2, denoted by v_0, v_1, and v_2, respectively. With these three data points, RePAD at time point 2 is able to train a LSTM model (called M) and uses M to predict the next data value, denoted by $\widehat{v_3}$.

When t separately advances to 3 and 4, implying that the real data-point values v_3 and v_4 have been observed respectively, RePAD is able to calculate the corresponding AARE values (i.e., $AARE_3$ and $AARE_4$) based on Eq. 1:

$$AARE_t = \frac{1}{t-b+1} \cdot \sum_{y=b}^{t} \frac{|v_y - \widehat{v_y}|}{v_y} \qquad (1)$$

where v_y is the observed data value at time point y, and $\widehat{v_y}$ is the forecast data value for time point y, and $y = b, b+1, \ldots, t$. It is clear that $AARE_3 = \frac{|v_3 - \widehat{v_3}|}{v_3}$ and $AARE_4 = \frac{1}{2} \cdot \left(\frac{|v_3 - \widehat{v_3}|}{v_3} + \frac{|v_4 - \widehat{v_4}|}{v_4} \right)$. Recall that AARE stands for Average Absolute Relative Error. A low AARE value indicates that the predicted values are close to the observed values. After deriving $AARE_3$ and $AARE_4$, RePAD replaces M with another LSTM model trained with the most three recent data points so as to predict $\widehat{v_4}$ and $\widehat{v_5}$, respectively.

When $t = 5$ (i.e., $t = 2b - 1$), RePAD calculates $AARE_5$. At this moment, RePAD is able to officially detect anomalies since it has already b AARE values (i.e., $AARE_3$, $AARE_4$, and $AARE_5$), which are sufficient to calculate detection threshold thd. In other words, RePAD requires a preparation period of $2b - 1$ time points to get its detection started. Equation 2 shows how thd is calculated based on the Three-Sigma Rule [2], which is commonly used for anomaly detection.

$$thd = \mu_{AARE} + 3 \cdot \sigma \qquad (2)$$

where $\mu_{AARE} = \frac{1}{t-b+1} \cdot \sum_{x=b}^{t} AARE_x$ (i.e., the average AARE of all previously derived AARE values), and σ is the corresponding standard deviation, i.e., $\sigma = \sqrt{\frac{\sum_{x=b}^{t}(AARE_x - \mu_{AARE})^2}{t-b+1}}$. If $AARE_5$ is lower than or equal to thd (see line 13 of Fig. 1), RePAD does not considers v_5 anomalous. Hence, it keeps using the same LSTM model (i.e., M) to predict the next data point $\widehat{v_6}$. However, if $AARE_5$ is higher than thd (i.e., line 15 holds), it means that either the data pattern of the time series has changed or an

---

RePAD algorithm
Input: Data points in a time series
Output: Anomaly notifications
Procedure:

1: **While** time has advanced {
2:      Let t be the current time point;
3:      Collect data point v_t;
4:      **if** $t = b - 1$ { // i.e., $t = 2$ if $b = 3$
5:        Train an LSTM model by taking $v_0, v_1, \dots, v_t$ as the training data;
6:        Let M be the resulting LSTM model and use M to predict $\widehat{v_{t+1}}$;}
7:      **else if** $t > b - 1$ & $t < 2b - 1$ { //i.e., $2 < t < 5$ if $b = 3$
8:        Calculate $AARE_t$ based on Equation 1;
9:        Train an LSTM model by taking $v_{t-b+1}, v_{t-b+2} \dots, v_t$ as the training data;
10:       Let M be the resulting LSTM model and use M to predict $\widehat{v_{t+1}}$;}
11:     **else if** $t \geq 2b - 1$ { //i.e., $t \geq 5$ if $b = 3$
12:       Calculate $AARE_t$ based on Equation 1 and *thd* based on Equation 2;
13:       **if** $AARE_t \leq thd$ {
14:         RePAD does <u>not</u> consider v_t an anomaly and uses M to predict $\widehat{v_{t+1}}$;}
15:       **else** {
16:         Retrain an LSTM by taking $v_{t-b}, v_{t-b+1} \dots, v_{t-1}$ as the training data;
17:         Use the new trained LSTM model to predict $\widehat{v_t}$;
18:         Re-calculate $AARE_t$ using Equation 1;
19:         **if** $AARE_t \leq thd$ {
20:           RePAD does <u>not</u> consider v_t an anomaly;
21:           Replace M with the new trained LSTM model;}
22:         **else** {
23:           RePAD reports v_t as an anomaly immediately;}}}}

Fig. 1. The algorithm of RePAD.

anomaly happens. To be sure, RePAD retrains an LSTM model with the most b recent data points (i.e., v_2, v_3, v_4) to re-predict $\widehat{v_5}$ and re-calculate $AARE_5$.

If the new $AARE_5$ is lower than or equal to *thd* (see line 19), RePAD confirms that the data pattern of the time series has changed and that v_5 is not anomalous. In this case, RePAD replaces M with this new LSTM model to accommodate the change.

On the contrary, if the new $AARE_5$ is still higher than *thd* (see line 22), RePAD considers v_5 anomalous (see line 23) since the LSTM trained with the most recent data points is still unable to accurately predict v_5. At this time point, RePAD immediately reports v_5 as an anomaly. The above detection process will repeat over and over again as time advances.

It is clear that the Look-Back parameter (i.e., b) is a critical parameter to RePAD since it controls the length of the preparation period and determines the detection performance of RePAD. In the next section, we investigate how this parameter impacts RePAD.

3 Evaluation Methodology

To comprehensively study the impact of the Look-Back parameter on RePAD, we consider the following six performance metrics:

1. Precision $\left(=\frac{\text{TP}}{\text{TP+FP}}\right)$
2. Recall $\left(=\frac{\text{TP}}{\text{TP+FN}}\right)$
3. F-score $\left(=2 \times \frac{\text{Precision}\times\text{Recall}}{\text{Precision+Recall}}\right)$
4. PP $(=2b-1$ time intervals)
5. LSTM retraining ratio $\left(=\frac{R}{N}\right)$
6. Average detection time $\mu_d \left(=\frac{\sum_t^N T_t}{N-t+1}\right)$ and standard deviation $\sigma_d \left(=\sqrt{\frac{\sum_t^N (T_t-\mu_d)^2}{N-t+1}}\right)$

Recall that RePAD is capable of proactive anomaly detection, so using traditional well-known point-wise metrics (such as precision, recall, and F-score) to measure RePAD is inappropriate. Therefore, we propose an evaluation strategy based on [10] to capture both proactive detection performance and lasting detection performance. More specifically, if any anomaly occurring at time point T can be detected within a detection time window ranging from time point $T - K$ to time point $T + K$, we say this anomaly is correctly detected, where $K \ll T$. We adopt the above-mentioned strategy together with precision, recall, and F-score to evaluate the detection performance of RePAD.

In the equations of precision and recall, TP, FP, and FN represent true positive, false positive, and false negative, respectively. Precision refers to positive predictive value, i.e., the proportion of positive/anomalous results that truly are positive/anomalous within the corresponding detection time windows. On the other hand, recall refers to the true positive rate or sensitivity, which is the ability to correctly identify positive/anomalous results within the corresponding detection time windows.

The F-score is a measure of a test's accuracy. It is defined as the weighted harmonic mean of the precision and recall of the test, i.e., F-score $= \left(\frac{2}{\text{Recall}^{-1}+\text{Precision}^{-1}}\right) = 2 \times \frac{\text{Precision}\times\text{Recall}}{\text{Precision+Recall}}$. The F-score reaches the best value at a value of 1, meaning perfect precision and recall. The worst F-score would be a value of 0, implying the lowest precision and the lowest recall.

PP, stands for Preparation Period, is a period of time required by RePAD to derive the required amount of AARE values for determining the detection threshold. More specifically, PP is defined as a time period starting when RePAD is launched and ending when RePAD officially starts its detection function. As mentioned earlier, RePAD starts at time point 0 and is able to derive the threshold at time point $2b-1$, implying that PP would be $2b-1$ time intervals. As b increases, the corresponding PP will accordingly prolong, meaning that RePAD will need more time before it can start detecting anomalies in the target time series.

Recall that whenever time advances to the next time point, RePAD might need to retrain its LSTM model. If the retraining is required, such a process will prolong the time taken by RePAD to decide whether or not the data point collected at that moment is anomalous. In addition, RePAD might require more computation resources to retrain its LSTM model. Therefore, the LSTM-retraining ratio $(=\frac{R}{N})$ is another important indicator showing the impact of Look-Back parameter on RePAD, where N is the total number of time points in the target time series, R is the total number of time points at which RePAD has to retrain its LSTM, and $R \leq N$. If the ratio is low, it indicates several things. Firstly, it means that RePAD does not need frequent LSTM retraining since its LSTM model is able to accurately predict future data points. Secondly, it further means that RePAD is able to quickly make prediction and detection whenever time advances. Thirdly, it implies that RePAD does not need extra computational resources to retrain its LSTM. Therefore, the lower the ratio, the better efficiency RePAD is able to provide.

Since the goal of RePAD is to offer anomaly detection in real-time, it is essential to evaluate how soon RePAD is able to decide the anomalousness of every single data point in the target time series. Consequently, measuring average detection time μ_d required by RePAD and the corresponding standard deviation σ_d are necessary. Hence, μ_d can be derived by equation $\frac{\sum_t^N T_t}{N-t+1}$, where T_t is the time required by RePAD to detect if v_t (i.e., the data point arrived at time point t) is anomalous or not, and N as stated earlier is the total number of time points in the target time series. More specifically, T_t is defined as a time period starting when time has advanced to t and ending when RePAD has determined the anomalousness of v_t. Note that if RePAD needs to retrain its LSTM model at time point t, T_t will also include the required LSTM retraining time. Apparently, if μ_d and σ_d are both low, it means that RePAD is able to make detection in real-time.

In the next section, we investigate how the Look-Back parameter impacts RePAD based on all the above-mentioned metrics.

4 Evaluation Results

To investigate how the Look-Back parameter impacts RePAD, we designed four scenarios as listed in Table 1. In the first scenario, the Look-Back parameter is set to 3, following the setting in [1]. In the rest three scenarios, we separately increase the value to 30, 60, and 90. Note that we did not evaluate RePAD in scenarios where the value of the Look-Back parameter is less than 3 since RePAD under the settings will not have sufficient historical data points to learn the pattern of the target time series, leading to an extremely high LSTM retraining rate according to our experience. In addition, the network structure of LSTMs in all the four considered scenarios are kept as simple as possible, i.e., one hidden layer with 10 hidden units.

Two empirical experiments were conducted based on two real-world time series datasets from the Numenta Anomaly Benchmark [11]. These datasets are called ec2-cpu-utilization-825cc2 and rds-cpu-utilization-cc0c53. They are abbreviated as CPU-cc2 and CPU-c53 in this paper, respectively. Table 2 lists the details of these two datasets. Note that the interval time between data points in all these datasets is 5 min. For this reason, RePAD in all the scenarios also followed the same time interval to perform anomaly detection. The two experiments were separately performed on a laptop running MacOS 10.15.1 with 2.6 GHz 6-Core Intel Core i7 and 16 GB DDR4 SDRAM. Note that we followed [12] and set K to 7 so as to determine the length of the detection time window. In other words, the detection time window is from time point $T - 7$ to time point $T + 7$ where T is recalled as the time point at which an anomaly occurs. The performance metrics presented in Sect. 3 were all used in the experiments.

Table 1. RePAD in four scenarios. Note that b is the look-back parameter.

Scenario	1	2	3	4
The value of b	3	30	60	90
Number of hidden layers	1	1	1	1
Number of hidden units	10	10	10	10

Table 2. Two real-world time-series datasets used in the experiments.

Dataset	Time period	# of data points
CPU-cc2	From 2014-04-10, 00:04 to 2014-04-24, 00:09	4032
CPU-c53	From 2014-02-14, 14:30 to 2014-02-28, 14:30	4032

4.1 Experiment 1

Figures 2 and 3 show the detection results of RePAD on the CPU-cc2 dataset and a close-up result, respectively. Note that this dataset contains two anomalies labeled by human experts, and these two anomalies are marked as red hollow circles in both figures. It is clear that RePAD in Scenario 1 (i.e., $b = 3$) is the only one that is able to detect the first anomaly on time and the second anomaly proactively. In the rest of the scenarios, RePAD could not detect the first anomaly at all. In addition, RePAD in Scenario 1 has more true positives and less false negatives than RePAD in the other scenarios. This phenomenon can be clearly observed from Fig. 3, and this is the reason why the recall of RePAD in Scenario 1 reached the best value of 1 (as listed in Table 3). If we further take precision and F-score into consideration (see Table 3 as well), we can see that RePAD in Scenario 1 outperforms RePAD in the other scenarios, implying that setting the Look-Back parameter to be 3 seems appropriate for RePAD.

Table 4 lists the PP required by RePAD on the CPU-cc2 dataset. In scenario 1 (i.e., $b = 3$), the required PP consists of only 5 ($=2 * 3 - 1$) time intervals according to $PP = 2b - 1$. It means that after 5 time intervals, RePAD can officially start detecting anomalies in the time series. When we increased the value of the Look-Back parameter, the corresponding PP proportionally prolonged, implying that RePAD needed more time to get itself ready. In other words, RePAD loses its capability for providing readiness anomaly detection.

Table 5 summaries the retraining performance of RePAD on the CPU-cc2 dataset. We can see that RePAD in a scenario 1 only needs to retrain its LSTM model 83 times. The corresponding LSTM retraining ratio is very low since it is around 2% (=83/4027). However, when we increased the value of the Look-Back parameter, the number of LSTM retraining required by RePAD increased. Apparently, increasing the value of the Look-Back parameter leads to more frequent LSTM retraining. In other words, training the LSTM of RePAD with long-term historical data points does not help RePAD in learning the data pattern of the CPU-cc2 dataset or making accurate predictions. In addition, as Table 6 shows, the average detection time required by RePAD to decide the

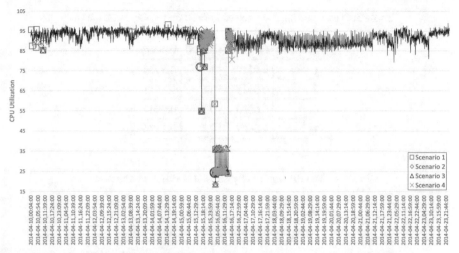

Fig. 2. The detection results of RePAD on the CPU-cc2 dataset. Note that this dataset has two labeled anomalies, marked as red hollow circles.

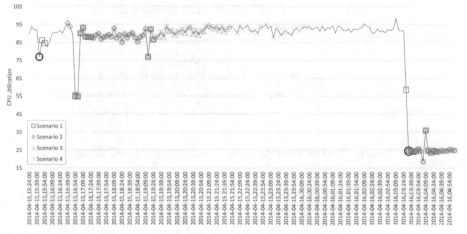

Fig. 3. A close-up of the detection results on the CPU-cc2 dataset. The two anomalies were marked as red hollow circles.

anomalousness of each data point increases when the value of the Look-Back parameter increases. The main reason is that the detection time includes both the corresponding LSTM retraining time (if the retraining is required) and the corresponding detection time. Since RePAD in Scenario 1 requires the least number of LSTM retraining and utilizes the least amount of training data (i.e., 3 data points), it leads to the shortest average detection time.

Based on the above results, we conclude that RePAD in Scenario 1 outperforms RePAD in the other three scenarios since it provides the best detection accuracy and time efficiency.

Table 3. The detection accuracy of RePAD on the CPU-cc2 dataset.

Scenario	Precision	Recall	F-score
1	0.4259	1	0.5973
2	0.0919	0.5	0.1553
3	0.0523	0.5	0.0947
4	0.0667	0.5	0.1176

Table 4. The PP required by RePAD in different scenarios.

Scenario	1	2	3	4
PP (preparation period)	5	59	119	179

Table 5. The LSTM retraining performance of RePAD on the CPU-cc2 dataset.

Scenario	# of required LSTM retraining	LSTM retraining ratio
1	83	2%
2	119	2.97%
3	199	5.01%
4	257	6.5%

Table 6. The time consumption of RePAD on the CPU-cc2 dataset.

Scenario	Average detection time (sec)	Standard deviation (sec)
1	0.015	0.027
2	0.188	0.339
3	0.385	0.691
4	0.494	0.846

4.2 Experiment 2

In the second experiment, we employ RePAD in the four scenarios to detect anomalies on the CPU-c53 dataset. The detection results and a close-up result are shown in Figs. 4 and 5, respectively. Note that this dataset contains two anomalies labeled by human experts and marked as red hollow circles. Table 7 summaries the detection accuracy of RePAD. Apparently, RePAD in Scenario 1 is the only one that is able to detect the two anomalies within the detection time windows without generating many false positives, consequently leading to higher precision, recall, and F-score than RePAD in the rest of the scenarios. Note that both Scenarios 3 and 4 led to zero precision and recall, so it is impossible to calculate the corresponding F-scores.

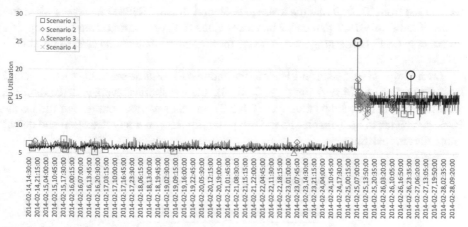

Fig. 4. The detection results of RePAD on the CPU-c53 dataset. Note that the two anomalies are marked as red hollow circles.

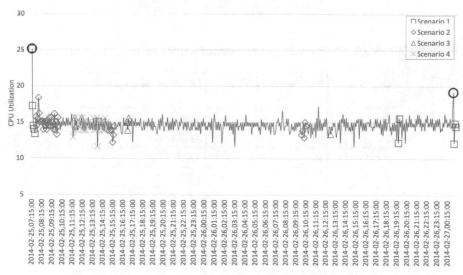

Fig. 5. A close-up of the detection results on the CPU-c53 dataset.

When it comes to PP, RePAD on the CPU-c53 dataset requires exactly the same PP as those as listed in Table 4. This is because the value of the Look-Back parameter is the only factor determining PP (i.e., $2b - 1$ time intervals). It is worth noting that the LSTM retraining ratio in this experiment does not always increase as the value of the Look-Back parameter increases. As listed in Table 8, RePAD in Scenario 3 requires the least number of LSTM retraining, whereas RePAD in Scenario 2 requires the highest number of LSTM retraining. In other words, the Look-Back parameter is not the only factor determining LSTM retraining ratio. The pattern of the target dataset is also influential. Nevertheless, we can see from Table 9 that the average detection time of RePAD increases when the value of the Look-Back parameter increases since the average detection time depends on not only LSTM retraining ratio, but also the amount of data points used to retrain the LSTM.

Overall speaking, Scenario 1 leads to the shortest PP (Please see Table 4), the second least amount of LSTM retraining (see Table 8), and the shortest average detection time with the least standard deviation (see Table 9). In other words, configuring the Look-Back parameter to be 3 enables RePAD to promptly start anomaly detection and detect anomalies in real-time.

Table 7. The detection accuracy of RePAD on the CPU-c53 dataset.

Scenario	Precision	Recall	F-score
1	0.457	1	0.627
2	0.1569	0.5	0.2388
3	0	0	n/a
4	0	0	n/a

Table 8. The LSTM retraining performance of RePAD on the CPU-c53 dataset.

Scenario	# of required LSTM retraining	LSTM retraining ratio
1	59	1.46%
2	82	2.05%
3	52	1.31%
4	73	1.85%

Table 9. The time consumption of RePAD on the CPU-c53 dataset.

Scenario	Average detection time (sec)	Standard deviation (sec)
1	0.015	0.024
2	0.122	0.199
3	0.213	0.304
4	0.321	0.522

5   Conclusion and Future Work

In this paper, we have investigated how the Look-Back parameter influences the performance of RePAD by introducing a set of performance metrics that not only evaluate detection accuracy for novel proactive anomaly detection approaches, but also evaluate readiness and time efficiency. Regardless of which real-world dataset we utilized in our experiments, setting a lower value for the Look-Back parameter always leads to better performance on all the six considered metrics. More specifically, taking short-term historic data points as online training data is an appropriate strategy for RePAD since RePAD is able to promptly learn and adapt to the changing patterns of the target time series. Furthermore, using short-term historic data points also enables RePAD to soon start detecting anomalies in the target time series and detect potential anomalies in real-time. In all the scenarios discussed in this paper, scenario 1 (i.e., setting the Look-Back parameter to 3) is the most recommended since it enables RePAD to provide real-time, lightweight, and proactive anomaly detection for time series without requiring human intervention or domain knowledge.

As future work, we plan to improve RePAD by further reducing false positives. Furthermore, we would like to extend RePAD to detect anomalies on time series data observed from the eX^3 HPC cluster [13] by referring to [14, 15, 16] and designing the methodology in a parallel and distributed way.

Acknowledgments. This work was supported by the project eX^3 - *Experimental Infrastructure for Exploration of Exascale Computing* funded by the Research Council of Norway under contract 270053 and the scholarship under project number 80430060 supported by Norwegian University of Science and Technology.

References

1. Lee, M.-C., Lin, J.-C., Gran, E.G.: RePAD: real-time proactive anomaly detection for time series. In: Proceedings of the 34th International Conference on Advanced Information Networking and Applications (AINA 2020), pp. 1291–1302 (2020)
2. Hochenbaum, J., Vallis, O.S., Kejariwal, A.: Automatic anomaly detection in the cloud via statistical learning. arXiv preprint arXiv:1704.07706 (2017)

3. Aggarwal, C.C., Yu, P.S.: Outlier detection with uncertain data. In: Proceedings of the 2008 SIAM International Conference on Data Mining. Society for Industrial and Applied Mathematics, pp. 483–493 (2008)
4. Xu, J., Shelton, C.R.: Intrusion detection using continuous time Bayesian networks. J. Artif. Intell. Res. **39**, 745–774 (2010)
5. Fisher, W.D., Camp, T.K., Krzhizhanovskaya, V.V.: Crack detection in earth dam and levee passive seismic data using support vector machines. Procedia Comput. Sci. **80**, 577–586 (2016)
6. Wu, J., Zeng, W., Yan, F.: Hierarchical temporal memory method for time-series-based anomaly detection. Neurocomputing **273**, 535–546 (2018)
7. Staudemeyer, R.C.: Applying long short-term memory recurrent neural networks to intrusion detection. South African Comput. J. **56**(1), 136–154 (2015)
8. Bontemps, L., McDermott, J., Le-Khac, N.A.: Collective anomaly detection based on long short-term memory recurrent neural networks. In: International Conference on Future Data and Security Engineering, pp. 141–152. Springer, Cham, November 2016
9. Lavin, A., Ahmad, S.: Evaluating real-time anomaly detection algorithms – the numenta anomaly benchmark. In: 14th International Conference on Machine Learning and Applications (2015).
10. Xu, H., et al.: Unsupervised anomaly detection via variational auto-encoder for seasonal KPIs in web applications. In: Proceedings of the 2018 World Wide Web Conference, pp. 187–196 (2018)
11. numenta/NAB.: The Numenta Anomaly Benchmark [Online code repository]. https://github.com/numenta/NAB Accessed 03 Mar 2021
12. Ren, H., et al.: Time-series anomaly detection service at microsoft. In: Proceedings of the 25th ACM SIGKDD International Conference on Knowledge Discovery & Data Mining, pp. 3009–3017 (2019)
13. Simula Research Laboratory: the eX^3 research infrastructure. https://www.ex3.simula.no Accessed 03 Mar 2021
14. Lin, J.-C., Lee, M.-C.: Performance evaluation of job schedulers under hadoop YARN. Concurrency Computat. Pract. Exper. (CCPE) **28**(9), 2711–2728 (2016)
15. Lee, M.-C., Lin, J.-C., Yahyapour, R.: Hybrid job-driven scheduling for virtual mapreduce clusters. IEEE Trans. Parallel Distrib. Syst. (TPDS) **27**(6), 1687–1699 (2016)
16. Lee, M.-C., Lin, J.-C., Gran, E.G.: ReRe: a lightweight real-time ready-to-go anomaly detection approach for time series. In: Proceedings of the 44th IEEE Computer Society Signature Conference on Computers, Software, and Applications (COMPSAC 2020), pp. 322–327 (2020)

A Novel Hybrid Recommendation Approach Based on Correlation and Co-occurrence Between Activities Within Social Learning Network

Sonia Souabi[✉], Asmaâ Retbi, Mohammed Khalidi Idrissi, and Samir Bennani

RIME TEAM-Networking, Modelling and e-Learning Team, MASI Laboratory, Engineering.3S Research Center, Mohammadia School of Engineers (EMI), Mohammed V University, Rabat, Morocco
soniasouabi@research.emi.ac.ma, {retbi,khalidi, sbennani}@emi.ac.ma

Abstract. Social learning is considered one of the most prevalent disciplines within e-learning. To handle learning resources within social environments, recommendation systems are gaining tremendous prominence based on a series of criteria such as the rate of learner interaction with the learning environment. On the basis of this, we highlight an overriding issue focusing on integrating the variety of activities carried out by learners within the learning environment. The challenge with most recommendation systems is that they do not address multiple activities performed by learners which may significantly affect recommendations. In our study, we focus on this point while proposing a hybrid system combining the two parameters of correlation and co-occurrence. The objective is to pinpoint the degree of connectivity between activities and their influence on recommendations. We carried out our study on an available database carried out by a Chinese team. We thus compare the results of the hybrid system with the non-hybrid system based solely on co-occurrence by measuring the performance of the two approaches. It is demonstrated that the proposed hybrid system produces highly gratifying results compared to the non-hybrid system.

Keywords: Social learning · Recommendation systems · Hybrid system · Correlation · Co-occurrence

1 Introduction

To assist the user in the information research, recommendation systems are designed to efficiently find and obtain relevant information [1]. In terms of social learning, it is noteworthy to mention that a recommendation system plays a highly significant role in the management of learning resources for learners [2]. Several approaches have been discussed: Collaborative filtering approach, hybrid approaches, content-based approaches (…) [3]. Generally, recommendation systems tend to manifest themselves in several forms: recommendation systems based on explicit data and recommendation systems

© The Author(s), under exclusive license to Springer Nature Switzerland AG 2021
L. Barolli et al. (Eds.): AINA 2021, LNNS 225, pp. 149–162, 2021.
https://doi.org/10.1007/978-3-030-75100-5_14

based on implicit data. Explicit data are collected on the basis of learner-defined annotations, while implicit data reflect the real traces of learners by collecting the frequency of events performed, as well as actions not performed [4]. Thus, implicit data are relevant in terms of recommendations as they consider the actual interactions of learners with the learning environment. In the context of learning, a recommendation approach has been advocated based on implicit learner attributes [5]. Other approaches merge implicit and explicit learner data to design a new recommendation system [6]. However, the implicit data has not been adequately addressed in the e-learning recommendation systems. Most approaches proposed still rely on the ratings assigned by learners to generate recommendations, whereas ratings are still insufficient to generate relevant recommendations. On the other hand, most recommendation systems proposed in online learning do not focus on the overall activities of learners to generate more appropriate recommendations. The focus is limited to a specific activity in many cases, except for the ratings provided by the learners. There is therefore a need to address this issue and to integrate the majority of the activities carried out by the learners into the generation of recommendations.

In order to highlight a very critical aspect that has not been addressed, we tackle an insightful issue where we can study the impact of carrying out multiple actions on the various recommendations. Implicit data approaches do not handle this point in relation to the effect of the activities undertaken on the recommendations. To further illustrate this point, we would identify the link between the mastery activity related to the recommendations and the other secondary activities of the learners recorded within the learning environment. This link is configured by two parameters, the first one, which is the correlation parameter, addresses the frequency of each action independently of another action, and the second one, the co-occurrence parameter, expresses how many times two activities appears simultaneously. Instead of considering only one parameter, we focus on both simultaneously, covering the totality of aspects that may exist between two actions performed. To prove how the combination of correlation and co-occurrence is influencing the performance of the recommendation system, we drew a comparison between the results obtained by the hybrid system and the results obtained by the non-hybrid system. The objective is to warrant that the integration of correlation and co-occurrence leads to better accuracy in terms of recommendations. For a recommendation system to be adapted for a learning social network, the dynamism of the learners must be included as well as all the actions carried out. In order for a recommendation system to be suitable for the genuine needs of learners, it is important to pay attention to the degree of learner interactivity within the learning community, including which activities were performed the most, in which categories learners are more engaged, in which categories are less engaged or not engaged at all. All these questions will play a very prominent position in generating recommendations appropriate to learners' requirements. In this respect, we propose a recommendation system that focuses on implicit data and endure a hybrid character. Two far-reaching factors are involved: correlation and co-occurrence to ensure that the calculation of recommendations includes the two factors. We carry out the test based on an online database containing a large number of activities recorded in the learning environment.

The remaining part of the document is organized as follows: Sect. 2 outlines highlights some recommendation systems proposed in the literature; Sect. 3 outlines the materials and methods we used including the recommendation strategy we proposed, Sect. 4 focuses on analysing the evaluation results and Sect. 5 draws the conclusion in association with future work.

2 Background

2.1 Social Recommender Systems

Shokeen and Rana [7] first describe social recommendation systems. They define this type of systems as the integration of recommendation systems within social media and networks. Recommendation systems, therefore, play a major role in improving the quality of recommendations. On the other hand, the article introduces the different characteristics of a social recommendation system, including: context, trust, tag, group, cross social media, temporal dynamics, heterogeneous social connections, semantic filtering. Guy [8] provide a general vision of social recommendation systems. Several types of recommendations can be defined within social media and networks: group recommendations allowing to generate recommendations for a group of people instead of individually, people recommendation, people recommendation to follow. Ahmadian et al. [9] propose a social recommendation system based on reliable implicit relationships. The proposed recommendation system consists of several steps. First, a network between users is generated based on annotations. Then, a link prediction approach is adopted to predict the implicit relationships between users, with an evaluation of prediction quality. Finally, the new predicted data are used to reconstruct a network with new neighbors. Rezvanian et al. [10] suggest analyzing two types of existing recommendations: CALA-OptMF and LTRS. The study reveals that each approach has its own advantages. CALA-OptMF improves the accuracy of recommendations in fuzzy recommendation systems. As for the LTRS, the results confirm that it can effectively manage scoring problems for users of scarce data and cold start. V. R. Revathy and Anitha S. Pillai [11] discuss the application and classification of recommendation systems within several known social networks. Social recommendation systems have been classified into two broad categories: entity-based and audience-based recommendation systems. Entity-based recommendation systems can be based on tags, context, communities, social influence. For public recommendation systems, they are generally based on friendships and groups.

2.2 Related Previous Studies

Several recommendation approaches were adapted regarding online learning, including collaborative filtering and hybrid approaches. The main purpose is to ensure the availability of pedagogical material to learners while responding to their specific needs. R. Turnip et al. [12] draw on the history of learners' behaviour and activities in order to provide them with a set of recommendations appropriate to their needs. S. P. Perumal et al. [13] propose a recommendation system based on the learner's profile based on the k-means algorithm. M. H. Mohamed et al. [14] aim to provide learners with the right courses based on the application of association rules to all transactions within the learning environment. J. Wintrode et al. [15] evaluated recommendation system techniques within the social learning platform through user-centred and data-centred evaluations. They relied on several algorithms, namely KNN, graph based algorithm and matrix factorization. N. Mustafa et al. [16] develop a recommendation system based on the study of graphs by taking into consideration the social interactions between learners. Chiefly, studies led by researchers concerning recommendation systems focus particularly on learners' profiles, without looking closely at the actions performed by learners as an entity. These studies are focusing especially on explicit data more than implicit data. In our proposition, we intend to measure the influence of activities carried out by learners on the recommendations generated and to integrate the correlation and the co-occurrence between events for calculating recommendations.

3   Our Recommendation Approach

Our recommendation system is summarized in the following figure (Fig. 1). The main components of the model are:

- The history of the learners: These are the activities and actions carried out by learners within the learning platform. These activities are divided into two categories:
- The primary activity: which is directly associated with the recommendations generated, i.e. the indicator conveying the learners' interest towards the learning objects.
- Secondary activities: which are related to the recommendations, although indirectly. These are also indicators representing the learners' interest towards the learning items, however they come in second position in respect to the primary activity.
- Data restructuring: This part is dedicated to data restructuring. Indeed, the data are structured by learner, i.e. each data expresses an activity carried out by a certain learner. The objective is to convert the data structure per learner into a data structure per learning object. The occurrence frequency of each activity will be collected for each learning object from the original database.
- Recommendation part: After restructuring our database to suit our requirements, we will generate the recommendations based on two notions:

  - The correlation between the activities performed by the learners.
  - The existing co-occurrence between the activities performed by the learners.

The top recommendations will therefore be generated from the recommendation scores based on the correlation score and the co-occurrence score.

To calculate the total recommendation score for each item, we calculate the correlation score on the one hand and the co-occurrence score on the other hand. The correlation score is obtained from the correlation matrix, while the co-occurrence score is obtained from the co-occurrence matrix.

To explain the two parameters of correlation and co-occurrence, we shall consider the following data:

M_1: *The activity or primary action*

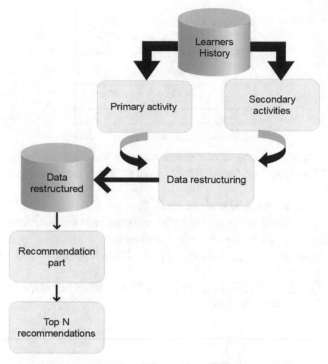

Fig. 1. The hybrid recommendation model

$\{M_2, M_3, \ldots, M_i\}$: *The other activities or actions (secondary activities)*
$\{c_1, c_2, \ldots, c_n\}$: *Items to recommend*
$\{s_{1j}, s_{2j}, \ldots, s_{ij}\}$: *The history of the activities or the number of times the activities carried out for each learning object*
$\{p_{1j}, p_{2j}, \ldots, p_{ij}\}$: *The frequency of activity appearance*

- Correlation: The correlation between two variables or two activities in our context is a notion of connection that contradicts independence. The more dependent two variables are, the higher the correlation coefficient between. We choose the Spearman correlation because it's the most appropriate measure for our case which describes a non-normal distribution.
- Co-occurrence: This notion of connection makes it possible to ascertain how many times two activities have appeared together.

	m_1	m_2
u_1	6	10
u_2	3	0

⬇

	m_1	m_2
m_1	2	1
m_2	1	1

Thus, to calculate the total recommendation score, we combine the correlation score (Eq. 1) and the co-occurrence score (Eq. 2) so as to obtain the pursuing results (Eq. 3):

$\{j_1, j_2, \ldots, j_n\}$: *Learning objects to recommend.*

$\{o_1, o_2, \ldots, o_m\}$: *Activities performed by learners such as o_1 is the primary activity and $\{o_2, \ldots, o_m\}$ are the secondary activities*

$\left\{ \left(p_{j_1 o_1}, \ldots, p_{j_1 o_m} \right), \ldots, \left(p_{j_n o_1}, \ldots, p_{j_n o_m} \right) \right\}$: *Learners historical activities towards each learning object.*

$\left\{ \left(P_{j_1 o_1}, \ldots, P_{j_1 o_m} \right), \ldots, \left(P_{j_n o_1}, \ldots, P_{j_n o_m} \right) \right\}$: *Binary history of activities performed by learners for each learning object.*

$$
\begin{aligned}
Correlation\ score\ matrix &= \begin{bmatrix} p_{j_1 o_1}\ p_{j_1 o_2} \cdots p_{j_1 o_m} \\ p_{j_2 o_1}\ p_{j_2 o_2} \cdots p_{j_2 o_m} \\ \vdots \quad \vdots \quad \vdots \end{bmatrix} \times \begin{bmatrix} 1\ cor(o_1, o_2) \ldots cor(o_1, o_m) \end{bmatrix} \\
&= \begin{pmatrix} correlation\ score(j_1) \\ \vdots \\ correlation\ score(j_n) \end{pmatrix}
\end{aligned} \tag{1}
$$

$$
\begin{aligned}
Co-occurrence\ score\ matrix &= \begin{bmatrix} P_{j_1 o_1}\ P_{j_1 o_2} \cdots P_{j_1 o_m} \\ P_{j_2 o_1}\ P_{j_2 o_2} \cdots P_{j_2 o_m} \\ \vdots \quad \vdots \quad \vdots \end{bmatrix} \\
&\times \begin{bmatrix} 1\ co-occ(o_1, o_2) \ldots co-occ(o_1, o_m) \end{bmatrix} \\
&= \begin{pmatrix} co-occurrence\ score(j_1) \\ \vdots \\ co-occurrence\ score(j_n) \end{pmatrix}
\end{aligned} \tag{2}
$$

$$
\begin{aligned}
Total\ score\ matrix &= co-occurrence\ score\ matrix + correlation\ score\ matrix \\
&= \begin{pmatrix} correlation\ score(j_1) + co-occurrence\ score(j_1) \\ \vdots \\ correlation\ score(j_n) + co-occurrence\ score(j_n) \end{pmatrix}
\end{aligned} \tag{3}
$$

3.1 The Hybrid Recommendation Algorithm

In order to generate the top recommendations for learners based on the correlation and co-occurrence between the activities carried out by the learners, we embrace the following algorithm step by step (Algorithm 1).

Algorithm	1 :	Recommendations
$(o_1, \{o_2, o_3, \dots, o_m\}, \{j_1, j_2, \dots, j_n\}$, history matrix)		

Input
Set of activities
$primary\ activity = o_1$
$the\ other\ activities = \{o_2, o_3, \dots, o_m\}$
Set of learning objects
$\{j_1, j_2, \dots, j_n\}$
History matrix
Output
Predicted scores and top N recommendations
Method
1: **for each** j_i, $i \in \{1, \dots, n\}$, do
2: Compute correlation score and co-occurrence score based on the history matrix
End for each
3 : Compute recommendation scores using (2)
4: Generate top N recommendations for learners.

4 Experiments and Evaluations

Experiments were conducted to evaluate the performance of our proposed recommendation system for recommending learning objects.

4.1 Experimental Setup

The database we will focus on in our experiment is a dataset extracted from a video-based educational experience using a social and collaborative platform[1]. The interdisciplinary learning activity is carried out between students in computer engineering and media and communication. The collaborative social network is divided into groups, with each group incorporating computer engineering and media and communication students. We opted for this database as it fits perfectly with our context and our expectations, and it handles all the activities carried out by the learners within the social network while providing them with several materials, including: documents, videos, presentations. The social network is, therefore, harnessed for a purely educational context supporting the assessment process. Students have a workspace where they can share files, images, and various resources, as well as messaging to interact with other students. The exchange, therefore, consists of sharing several types of educational materials (videos, ppt, word, images). The study involves 111 students.

---

[1] https://bera-journals.onlinelibrary.wiley.com/doi/full/10.1111/bjet.12318.

As our hybrid recommendation system is part of educational social networks, we will carry out our study on this already available database in a different perspective, namely that of recommendation systems. Our recommendation approach consists of including activities carried out by learners within a social learning network. Therefore, it is important to identify the activities included before proceeding with the operation of calculating recommendations based on correlation and co-occurrence.

In order to enhance the value of our work and based on the available data, we focus our study on the recommendations of videos contained in blogs. We will, thus, look at any action or activity directly concerned by the videos:

- The evaluation of videos by learners (fivestar).
- Creating a comment in a video (generic_comment).

Many activities have been recorded in this database. We restricted our work to relevant activities for recommending videos. Indeed, the video is a very practical support to illustrate certain notions. It is one of the strongest techniques for learning as it is supported by images and sound, and these two elements fully catch a learner's attention. Since the correlation between the primary activity associated with the recommendations and the other secondary activities, as well as co-occurrence, must be included, the primary activity must be identified in addition to the secondary activities whose relevance comes afterwards (Table 1).

Table 1. Primary activity vs secondary activities

Primary activity	Secondary activities
The evaluation of videos by learners (fivestar)	Creating a comment in a video

4.2 Experimental Case

- List of learning objects

Several learning elements are available within the database to be analyzed (Table 2).

Table 2. Learning objects

	Learning objects ID
1	3983
2	4024
3	3916
4	3887
⋮	⋮

- Original matrix

The original matrix is the matrix outlining the interaction of the different learning objects with the activities performed by learners. Here is an extract from this matrix (Table 3):

Table 3. Learning objects actions interaction

Learning object ID	Update fivestar	Create generic_comment
3983	41	48
4024	24	43
3916	29	43
3887	11	38

- Training model

Firstly, we are supposed to find the recommendation model based on our approach referring to correlation and co-occurrence. After performing the corresponding calculations, we obtain the following recommendation model (Eq. 4):

$$Total\ scores = H_1 \times M_1 + h_1 \times m_1 \qquad (4)$$

Such as:

H_1: *history matrix*
h_1: *binary history matrix*
$M_1 = correlation\ vector = \begin{pmatrix} 1 \\ 0,84 \end{pmatrix}$
$m_1 = co-occurrence\ vector = \begin{pmatrix} 1 \\ 0,8 \end{pmatrix}$

- Comparing training results to validation results

As the training model is obtained, the validation part is used to measure the performance of our recommendation system. On the latter, we compare the results obtained by the training model to the results obtained by the validation model, and then we use the required measurements. Here is an extract of the validation results obtained (Table 4). In the following section, we discuss the evaluation process under consideration.

Table 4. Total scores of some learning objects to recommend

Learning objects	Total score
Learning object 1	83,09
Learning object 2	61,9
Learning object 3	66,89
Learning object 4	44,7
Learning object 5	60,86
Learning object 6	51,37
Learning object 7	3

4.3 Experimental Results

We choose the online evaluation method based on the basic truths deduced. Thus, the basic truths deduced reveal the real preferences and orientations of a learner towards different object materials. We rely on this type of assessment to determine the point reached by our recommendation system in predicting the real preferences of the learners. In our tests, we compare the results of the assessment measures between the hybrid recommendation system and the non-hybrid recommendation system based solely on co-occurrence.

- Evaluation method
  Our analysis consists of dividing the database into two parts:

  - Part 1 (80% of the data): To test the hybrid recommendation approach by generating group recommendations for learners.
  - Part 2 (20% of the data): To validate our recommendation system by comparing the results generated in part 1 to the actual preferences of the learners in part 2.
  - Performance experiments

To calculate performance (Eq. 5), we will not include the recommendation ranking, simply whether or not the generated recommendations appear in the list of real learner preferences identified in the 20%. The more the recommendations generated in the first part of the data overlap with those in the second part of the data, the higher the performance of the recommendation system will be. Accuracy measures the ability of the system to predict learners' real preferences (Table 5).

Table 5. Interpretation of accuracy parameters

	Recommended	Not recommended
Preferred	True positive (TP)	False positive (FP)
Not preferred	False negative (FN)	True negative (TN)

The system performance is expressed by the F1 measurement. This measurement is derived from precision and recall using the following formula (Eq. 5):

$$Performance = F1 - measure = 2 \times \frac{Precision \times recall}{Precision + recall} = Precision(Precision = recall) \tag{5}$$

The accuracy is equal to the recall and the F1 measurement, so only one measurement needs to be calculated to identify the values of the other parameters. In this case we consider the traditional formula of the precision (Eq. 6):

$$Precision = \frac{TP}{TP + FP} = \frac{Number\ of\ correct\ items\ recommended}{Number\ of\ recommendations} \tag{6}$$

Table 6. Precision according to number of recommendations for hybrid and non-hybrid recommender systems

Number of recommendations	Precision of hybrid approach	Precision of non-hybrid approach
N = 4	0,75	0,5
N = 6	1	0,8
N = 5	0,83	0,66

On this basis, we obtain the following graph for the two recommendation systems (Fig. 2):

Similarly, we realize that the performance of the hybrid recommendation system outperforms that of the fully co-occurrence-based approach. The performance associated with the proposed recommendation system reaches values between 75% and 100%, while the performance associated with the non-hybrid approach is limited to 80% with a lowest value of 50% (Table 6).

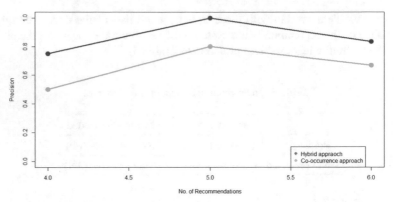

Fig. 2. Precision of the hybrid recommender system vs the co-occurrence approach based on the number of recommendations.

5 Discussion

After evaluating the results, it turns out that the performance of the hybrid recommendation system outperforms the performance of the non-hybrid recommendation system based solely on co-occurrence. This amounts to the integration of the correlation concept between the activities carried out by learners. The discrepancy between the performance of the hybrid recommendation system and the performance of the non-hybrid recommendation system stands between 25% and 30%, which reveals a significant gap in recommendation systems. Moreover, the maximal performance of the hybrid recommendation system reaches 100%, while the maximum value of the recommendation system based solely on co-occurrence is limited to 80%.

Our recommendation system achieves an accuracy rate between 75% and 100%, meaning that the recommended videos meet the real preferences of the learners with a percentage between 75% and 100%, unlike the non-hybrid recommendation system, which reaches a lower level of satisfaction than the hybrid system. Correlation allows the recommendation system to be improved and its added value to be enhanced.

This entails integrating the two matrices together: the correlation matrix and the co-occurrence matrix. On the one hand, we measure the correlation between the different activities within the learning environment, and then we ascertain how often the activities appear together from the historical data. In the literature, the hybrid recommendation system is generally still the predominant one, as it produces prominent results than non-hybrid approaches. The calculation time may be higher, but the results are actually better. Depending on the learners' activity within a social learning platform, we will be led to monitor the learner's behavior and interactivity with the pedagogical resources and in discussion forums with other learners. As the number of activities increases, the recommendation system gains efficiency. When learners are more interactive, it helps us to propose more appropriate recommendations for the learner profile and needs. We are witnessing a major technological evolution over the years, so we need to improve the quality of the recommendations provided to users, especially learners, to improve their level of knowledge and engagement. The more satisfied the learner is with the recommendations, the more engaged they will be, leading to an improvement in their level. In the literature, most research has focused on a single activity to generate recommendations. In turn, we embedded several activities simultaneously into a single recommendation system. On the other hand, we measured two key parameters to identify the connectivity between activities. Correlation is not the unique factor that can measure the relationship between two activities, but also co-occurrence, which indicates how many times two activities were performed together.

In our case, we considered only a few activities. If learners are more interactive and other activities are registered within the learning environment, our proposed recommendation system will always be worthwhile regardless of the number of activities registered.

6 Conclusion

In e-learning, particularly in social learning via social networks, recommendation systems play a very important role in managing the large amount of data available to learners. However, the recommendation systems already proposed in the literature do not support the integration of activities carried out by learners to generate more appropriate recommendations. In our approach, we propose a hybrid system that includes all learner-performed activities and combines correlation and co-occurrence between the different activities to measure the connection strength between these activities and adapt recommendations to learners' level of interaction. Experimental results show that our hybrid recommendation system generates higher accuracy for both universities compared to the non-hybrid system, proving the effectiveness of a hybrid system that merges correlation and co-occurrence into a unified system. Thus, it was possible to also integrate several types of learner activities in order to provide recommendations according to learner needs. This recommendation system therefore brings several advantages in terms of e-learning as well as the engineering education. Our future work consists of:

- Testing our recommendation system within our university.
- Integrating other learner activities to generate prominent recommendations.

References

1. Kulkarni, P.V., Rai, S., Kale, R.: Recommender system in eLearning: a survey. In: Bhalla, S., Kwan, P., Bedekar, M., Phalnikar, R., Sirsikar, S. (eds.) Proceeding of International Conference on Computational Science and Applications, pp. 119–126. Springer, Singapore (2020)
2. Abbas, M., Riaz, M.U., Rauf, A., Khan, M.T., Khalid, S.: Context-aware Youtube recommender system. In: 2017 International Conference on Information and Communication Technologies (ICICT), Karachi, December 2017, pp. 161–164 (2017). https://doi.org/10.1109/ICICT.2017.8320183
3. Lund, J., Ng, Y.-K.: Movie recommendations using the deep learning approach. In: 2018 IEEE International Conference on Information Reuse and Integration (IRI), Salt Lake City, UT, July 2018, pp. 47–54 (2018). https://doi.org/10.1109/IRI.2018.00015

4. Wang, X., Wang, C.: Recommendation system of ecommerce based on improved collaborative filtering algorithm. In: 2017 8th IEEE International Conference on Software Engineering and Service Science (2017)
5. Pujahari, A., Padmanabhan, V.: An approach to content based recommender systems using decision list based classification with k-DNF rule set. In: 2014 International Conference on Information Technology, Bhubaneswar, India, December 2014, pp. 260–263 (2014). https://doi.org/10.1109/ICIT.2014.13
6. Zhang, R., Liu, Q., Chun-Gui, Wei, J.-X., Huiyi-Ma: Collaborative filtering for recommender systems. In: 2014 Second International Conference on Advanced Cloud and Big Data, Huangshan, China, November 2014, pp. 301 308 (2014). https://doi.org/10.1109/CBD.2014.47
7. Shokeen, J., Rana, C.: A study on features of social recommender systems. Artif Intell Rev **53**, 965–988 (2020). https://doi.org/10.1007/s10462-019-09684-w
8. Guy, I.: Social recommender systems. In: Ricci, F., Rokach, L., Shapira, B. (eds.) Recommender Systems Handbook, pp. 511–543. Springer, Boston (2015). https://doi.org/10.1007/978-1-4899-7637-6_15
9. Ahmadian, S., Joorabloo, N., Jalili, M., Ren, Y., Meghdadi, M., Afsharchi, M.: A social recommender system based on reliable implicit relationships. Knowl.-Based Syst. **192**, 105371 (2020). https://doi.org/10.1016/j.knosys.2019.105371
10. Rezvanian, A., Moradabadi, B., Ghavipour, M., Daliri Khomami, M.M., Meybodi, M.R.: Social recommender systems. In: Learning Automata Approach for Social Networks. Studies in Computational Intelligence, pp. 281–313. Springer, Cham (2019). https://doi.org/10.1007/978-3-030-10767-3_8
11. Revathy, V.R., Pillai, A.S.: Classification and applications of social recommender systems. In: Abraham, A., Cherukuri, A., Madureira, A., Muda, A. (eds.) Proceedings of the Eighth International Conference on Soft Computing and Pattern Recognition (SoCPaR 2016). SoCPaR 2016. Advances in Intelligent Systems and Computing, vol. 614. Springer, Cham (2018). https://doi.org/10.1007/978-3-319-60618-7_70
12. Turnip, R., Nurjanah, D., Kusumo, D.S.: Hybrid recommender system for learning material using content-based filtering and collaborative filtering with good learners' rating. In: 2017 IEEE Conference on e-Learning, e-Management and e-Services (IC3e), Miri, Sarawak, Malaysia, November 2017, pp. 61–66 (2017). https://doi.org/10.1109/IC3e.2017.8409239
13. Perumal, S.P., Arputharaj, K., Sannasi, G.: Fuzzy family tree similarity based effective e-learning recommender system. In: 2016 Eighth International Conference on Advanced Computing (ICoAC), Chennai, India, January 2017, pp. 146–150 (2017). https://doi.org/10.1109/ICoAC.2017.7951760
14. Mohamed, M.H., Khafagy, M.H., Ibrahim, M.H.: Recommender systems challenges and solutions survey. In: 2019 International Conference on Innovative Trends in Computer Engineering (ITCE), Aswan, Egypt, February 2019, pp. 149–155 (2019). https://doi.org/10.1109/ITCE.2019.8646645
15. Wintrode, J., Sell, G., Jansen, A., Fox, M., Garcia-Romero, D., McCree, A.: Content-based recommender systems for spoken documents. In: 2015 IEEE International Conference on Acoustics, Speech and Signal Processing (ICASSP), South Brisbane, Queensland, Australia, April 2015, pp. 5201–5205 (2015). https://doi.org/10.1109/ICASSP.2015.7178963
16. Mustafa, N., Ibrahim, A.O., Ahmed, A., Abdullah, A.: Collaborative filtering: techniques and applications. In: 2017 International Conference on Communication, Control, Computing and Electronics Engineering (ICCCCEE), Khartoum, Sudan, January 2017, pp. 1–6 (2017). https://doi.org/10.1109/ICCCCEE.2017.7867668

Sentiment-Driven Breakdown Detection Model Using Contextual Embedding ElMo

Ebtesam Hussain Almansor[1,2(✉)] and Farookh Khadeer Hussain[1]

[1] School of Computer Science, Faculty of Engineering and Information Technology, University of Technology Sydney, Sydney, Australia
EbtesamHussain.Almansor@student.uts.edu.au
[2] Community College, Najran University, Najran, Saudi Arabia
http://www.elsevier.com

Abstract. The recent rapid developments in neural networks have stimulated significant performance progress in detecting breakdown in dialogue. The existing research focuses on detecting breakdown from the data label using different features such as word similarity and topic transition; however, these features are insufficient to capture contextual features. Therefore, we used ELMo language model to extract the contextual feature to help to detect the breakdown which faces a challenge due to the basis of human opinion. Another crucial challenge facing breakdown detection is the lack annotated dataset. To take this challenge we used sentiment as the data label with contextualized ELMo embedding model to detect the breakdown in dialogue. The model was evaluated with state-of-the-art models and the results shows the proposed model outperforms other models.

Keywords: Chatbot · Breakdown detection · Contextualized ELMo embedding model · Sentiment

1 Introduction

More recently, conversation agents (CAs) including chatbots, voice control interfaces and personal assistants are gaining attention in both academia and industry. The conversational agents are designed to convince humans that they are having a normal conversation with a machine or computer. Conversation agents that are based on textual data are called chatbots. Focusing on the chat-oriented system, the vital objective of the chatbot is to respond to the user's request. While the primary purpose of the chatbot is to provide a proper response to the user, sometimes the chatbot fails to provide an appropriate response provides a poor response [1]. The reason for providing an inappropriate response is

---

Fully documented templates are available in the elsarticle package on CTAN.

© The Author(s), under exclusive license to Springer Nature Switzerland AG 2021
L. Barolli et al. (Eds.): AINA 2021, LNNS 225, pp. 163–171, 2021.
https://doi.org/10.1007/978-3-030-75100-5_15

that the system could not understand user's utterance. Detecting the inappropriate responses from chatbots could enhance user experience and increase the consumer's trust in the chatbots.

More recently, a selected few studies have been focused on detecting an inappropriate response from the chatbot [2]. The point at which the responses from the chatbots are inaccurate is known as the breakdown point. The existing studies have focused on detecting the breakdown point based on chatbot response using hand features such as topic and word similarity measures. However, these feature are insufficient to capture contextual features.

Another important automated features learning process is called word embedding which uses representations of the words as vectors and uses them as an input for machine learning models. Word embedding has been used in many natural language processing tasks such as classification [3]. There are several types of word embedding such as word2vec [4], Glove [5], fastText [6]. While standard word embedding models such as word2vec are fast to train, they do not capture the context. ElMo (Embedding from language model) is an embedding model that consider the context of the words. ELMo has been used in several domains such as question answer [7] sentiment analysis [8] and name entity [9]. Due to the effectiveness of the ElMo model in NLP tasks, in this paper, we build logistic regression with contextual embedding model (ELMo) to detect breakdown.

The rest of the paper is structured as follows: Sect. 2 presents a related literature review and discussion of existing studies. In Sect. 3, the proposed framework is explained. Section 4 explains the experimental results. Section 5 presents the Conclusion and future work

2   Related Works

Recently, detecting responses that cause dialogue breakdown between a human user and a chatbot has received a lot of attention [10,11]. The key advantage of detecting a breakdown in conversation is to increase a user's trust and satisfaction. Previous studies on breakdown detection used features such as word similarities and topic differences. For example, Sugiyama et al. [12] proposed a breakdown detector model based on features such as dialogue act, sentence length and word-based similarity which improved the performance. The model used the Extra Trees Regressor to estimate the evaluation distribution. Similarly, Takayama et al. [2] developed a detection method based on an ensemble of detectors which trained data using different sets of annotators. The proposed model clustered the training data by the annotation distribution for each annotator and then built a detector for each cluster. The neural network was utilized to encode the utterance to their label. The model consists of two long short-term memory (LSTM) and two convolutional neural network encoders. They use the DBCD dataset which contains user-system dialogue which is annotated with three labels (O: Not a breakdown, T: Possible breakdown and X: Breakdown). Some annotated tasks are predictably subjective. Therefore, the annotated label could be biased due to the annotators. This model was evaluated in Dialogue Breakdown

Detection Challenge 3(DBDC3) and achieved 63.6% F-measure, outperforming the baseline by 5.6% which confirmed the effectiveness of considering the biases which occur when using annotators [13]. Also, Xie et al. [14] utilized hierarchical Bi-Direction LSTMs which are trained to map sequences of words to their breakdown labels. The model was evaluated against three dialogue models that were provided by DBDC3, and it outperformed the CRF baseline and achieved high accuracy using both classification and distributed evaluated metrics. Lee et al. [15] developed a detector model LSTM that encodes temporal utterance with memory attention. Almansor et al. [16] have proposed a word embedding model that is measuring the chatbot quality of response.

Other studies focused on inappropriate responses as a breakdown point but they did not focus on detecting these responses. For example, Higashinaka [17] proposed a taxonomy of errors in the chatbot-oriented system and classified the errors into four categories, namely, utterance-level, response-level, context-level, and environment-level errors. Furthermore, an evaluation metrics was developed for a breakdown detector [1]. These evaluation metrics were used in DBDCs which are a gold label that used classification metrics and gold distribution metrics.

While existing literature in detecting breakdown uses several models, no studies consider the context of word embedding. Therefore, in this work we take one step further and build a model that use sentiment as data label and detects the breakdown using a hybrid model that combines logistic regression and ElMo.

The next section elaborates on the proposed framework.

3   Proposed Model

We proposed a prediction model to predict the breakdown in dialogue based on sentiment score. Our main aim of this model is to overcome the limitation of the lack of label dataset. Also, the proposed model introduces the effectiveness of using ELMo as a contextual language model with logistic regression. The structure of the proposed model is shown in Fig. 1. It contains three main steps including data preparation, feature extraction and prediction model. Each step is explained in the next sub-section in more details.

3.1   Data Preparation

This step helps to prepare the data for the next step which is the extraction of features and it contains three sub-steps to clean the datasets including prepressing, lemmatization and data labelling.

Pre-processing: in this step we remove the URL links, the punctuation marks, numbers, and white spaces and convert the utterances into lower case.

Lemmatization: this step helps to reduce and returns words to their base form.

Data Labeling: we apply the Valence Aware Dictionary and sEntiment Reasoner (VADER) sentiment analyzer to score the responses. We selected VADER

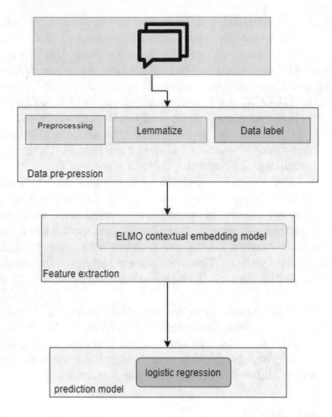

Fig. 1. The structure of the proposed model

because the literature shows it is the most accurate sentiment analyzer for social media datasets [18]. The VADER sentiment analyzer is used to score the sentiment of each response and to label it. VADER is a lexicon and rule-based sentiment analysis tool which is attuned to sentiment expression in social media [18]. VADER uses a sentiment lexicon which scores words as either positive or negative. The sentiment score is measured on a scale $[-1, 1]$, so a response with a score $[\geqslant 0.05]$ is positive; if a response has a score between -0.05 and 0.05 it is neutral; and a response with a score $[\leqslant 0.05]$ is negative.

3.2   Feature Extraction Using ELMO

ELMo provides word vectors that compute on top of two-layer bidirectional language model (biLM). The bidirectional model has two layer stacked together; each layers has two passes, one is the forward pass and the other one is the backward pass. Figure 2 shows the architecture of ElMo. This model uses character level convolutional neural network to represent the word of the data string into raw word vectors and pass on these vectors as an input to the first layer of biLM. The forward pass has information about the word and the context before this word and the backward pass has information about the word and the context after that word. The information from the forward and backward pass customs the intermediate word vectors. The intermediate word vectors fed into the next layer of biLM. The final representation is weighted sum of the raw words vectors and the two intermediate word vectors, so the input to the ELMo is computed from character level instead of words which means it captures the inner structure of the words.

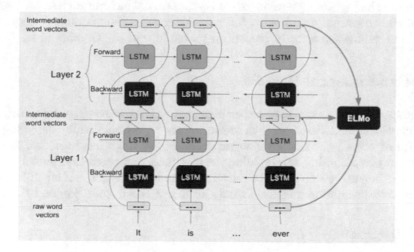

Fig. 2. The structure of ELMo model

After the preprocessing step we import the Elmo model and extract the Elmo vectors from the cleaned utterances for both train and test datasets. By this step we have ElMo vectors models for the train and test data that are ready to be used by the detection model in the next step.

3.2.1 Prediction Model

In this step we build the prediction model that is based on logistic regression with the ELMo model. So we use the vectors model to train the logistic model and make the predictions on the test data.

Logistic Regression has been used in the biological sciences and in several applications of social science [19]. Logistic regression uses a logistic function which is also known as a sigmoid function. The logistic function takes the real-value numbers and maps them between 0–1 and represents this as an S-shaped curve [19]. This function is based on Eq. 1:

$$\frac{1}{1+e^{-x}} \tag{1}$$

where x represents any number that is mapped between 0–1. The main objective behind the logistic function is to predict y for the given input x. Logistics regression predicts the probability for the first or default class based on Eq. 2:

$$p(x) = \frac{e^{\beta^T \cdot x}}{1 + e^{\beta^T \cdot x}} \tag{2}$$

where β process is followed the parameter vector and x represents the training data. The logistic regression uses an estimation algorithm such as maximum-likelihood to estimate and learn the parameter from the training dataset. The aim is to reduce the error in prediction probability in the training phase.

4   Experimental Results

In this section, we detail the experiment setup that is used to train and test the classifiers and describe their performance in automatically detecting the quality of responses. The Python programing language has been use for this study for programming and building the classifiers and prediction model. For all the experiments, we divided the data into training 85% and testing 15%. Also, we use four evaluation metrics, namely accuracy, recall, precision and F1 score.

4.1   Datasets

In this experiment we used DCDB3 datasets which is the common dataset that is used for the breakdown task. This dataset has four datasets: TKTK-150, IRIS-150, CIC-165 and YI-150 including training and testing [15].

4.2   Evaluation Metrics

For evaluation purposes, we used the standard evaluation metrics for classification tasks, which are accuracy, F1-score, recall, and precision.

$$\text{Accuracy} = \frac{TP + TN + TN}{TP + TN + TN + FN + FN + FN + FN + FP + FP} \tag{3}$$

$$\text{Precision}_N = \frac{TN}{TN + FN} \tag{4}$$

$$\text{Recall}_N = \frac{TN}{TN + FP} \tag{5}$$

$$F - \text{measure} = \frac{2 \times \text{Precision} \times \text{Recall}}{\text{Precision} + \text{Recall}} \tag{6}$$

5   Results and Discussion

We compare the proposed model with state-of-the-art models in detection break-down in dialogue. Table 1 shows the comparison of the proposed model with state-of-the art models. The comparison was based on data label so we applied the proposed model with the dataset with the original data label which was done by a human and we refer to this as the proposed model (O). Also, we apply the proposed model on the same datasets with sentiment score as data label and we refer to this as the proposed model (S). From Table 1 we can see the accuracy score by using the proposed model (O) which outperforms the state-of-the-art models by 0.67. Also, using sentiment with the proposed model obtains better results in the accuracy by around 0.84 compared with the other models.

Table 1. Comparison results of the proposed ElMo embedding model against the breakdown detect models on DBDC3 dataset

Models	Method	Accuracy	F1-score(B,NB)
CRF	CRF	0.42	0.76
Majority baseline	–	0.37	0.89
Encoded vectors from LSTM [15]	LSTM	0.47	0.74
The proposed model (O)	Logistic regression with ELMo	0.67	0.80
The proposed model (S)	Logistic regression with ELMo with sentiment score	0.84	0.73

From these results we can see using logistic regression with ELMo model has improved the accuracy of breakdown detection model. Also, using sentiment has a positive effect as the data label.

Table 2. comparison results of the proposed Elmo embedding model against other ElMo embedding model

Datasets	Model	Accuracy score
DBDC3	Elmo embedding with kersa	0.84
	Embedding model with logistic regression	0.86
MELD	Elmo embedding with kersa	0.78
	Embedding model with logistic regression	0.80

Table 2 show comparison results of Elmo embedding with kersa and with logistic regression. The accuracy results of Elmo embedding model with logistic outperform Elmo with kersa on both DBDC3 and MELD datasets as it obtain 0.86 and 0.80 accuracy score respectively.

6   Conclusion and Future Work

With rapid growth of the importance of detection breakdown in dialogue, we develop sentiment driven breakdown detection using ELMo as the contextual embedding model. The model showed that it outperformed the state-of-the-art model. From this paper, we found that sentiment has improved the accuracy of the proposed model and it has a positive effect as the data label to detect the breakdown.

References

1. Tsunomori, Y., Higashinaka, R., Takahashi, T., Inaba, M.: Evaluating dialogue breakdown detection in chat-oriented dialogue systems (2018)
2. Takayama, J., Nomoto, E., Arase, Y.: Dialogue breakdown detection considering annotation biases. In: Proceedings of the Dialog System Technology Challenge 6 Workshop (DSTC6) (2017)
3. Ge, L., Moh, T.-S.: Improving text classification with word embedding. In: 2017 IEEE International Conference on Big Data (Big Data), pp. 1796–1805. IEEE (2017)
4. Mikolov, T., Le, Q.V., Sutskever, I.: Exploiting similarities among languages for machine translation. arXiv preprint arXiv:1309.4168 (2013)
5. Pennington, J., Socher, R., Manning, C.D.: GloVe: global vectors for word representation. In: Proceedings of the 2014 Conference on Empirical Methods in Natural Language Processing (EMNLP), pp. 1532–1543 (2014)
6. Bojanowski, P., Grave, E., Joulin, A., Mikolov, T.: Enriching word vectors with subword information. Trans. Assoc. Comput. Linguist. **5**, 135–146 (2017)
7. Peters, M.E., Neumann, M., Iyyer, M., Gardner, M., Clark, C., Lee, K., Zettlemoyer, L.: Deep contextualized word representations. arXiv preprint arXiv:1802.05365 (2018)
8. Socher, R., Perelygin, A., Wu, J., Chuang, J., Manning, C.D., Ng, A.Y., Potts, C.: Recursive deep models for semantic compositionality over a sentiment treebank. In: Proceedings of the 2013 Conference on Empirical Methods in Natural Language Processing, pp. 1631–1642 (2013)
9. Sang, E.F., De Meulder, F.: Introduction to the CoNLL-2003 shared task: language-independent named entity recognition, arXiv preprint arXiv:cs/0306050 (2003)
10. Sugiyama, H.: Dialogue breakdown detection based on estimating appropriateness of topic transition. In: Dialog System Technology Challenges (DSTC6) (2017)
11. Horii, T., Mori, H., Araki, M.: Breakdown detector for chat-oriented dialogue, pp. 119–127. Springer, Singapore (2017). https://doi.org/10.1007/978-981-10-2585-3_9
12. Sugiyama, H.: Empirical feature analysis for dialogue breakdown detection. Comput. Speech Lang. **54**, 140–150 (2019)

13. Takayama, J., Nomoto, E., Arase, Y.: Dialogue breakdown detection robust to variations in annotators and dialogue systems. Comput. Speech Lang. **54**, 31–43 (2019)
14. Xie, Z., Ling, G.: Dialogue breakdown detection using hierarchical bi-directional LSTMs. In: Proceedings of the Dialog System Technology Challenges Workshop (DSTC6) (2017)
15. Lee, S., Lee, D., Hooshyar, D., Jo, J., Lim, H.: Integrating breakdown detection into dialogue systems to improve knowledge management: encoding temporal utterances with memory attention. Inf. Technol. Manag. **21**(1), 51–59 (2020)
16. Almansor, E.H., Hussain, F.K.: Modeling the chatbot quality of services (CQoS) using word embedding to intelligently detect inappropriate responses. In: International Conference on Advanced Information Networking and Applications, pp. 60–70. Springer (2020)
17. Higashinaka, R., Funakoshi, K., Araki, M., Tsukahara, H., Kobayashi, Y., Mizukami, M.: Towards taxonomy of errors in chat-oriented dialogue systems. In: Proceedings of the 16th Annual Meeting of the Special Interest Group on Discourse and Dialogue, pp. 87–95 (2015)
18. Hutto, C.J., Gilbert, E.: VADER: a parsimonious rule-based model for sentiment analysis of social media text. In: Eighth International AAAI Conference on Weblogs and Social Media (2014)
19. Brownlee, J.: Master Machine Learning Algorithms. discover how they work and implement them from scratch (2016)

The Improved Redundant Active Time-Based (IRATB) Algorithm for Process Replication

Tomoya Enokido[1]([⊠]), Dilawaer Duolikun[2], and Makoto Takizawa[2]

[1] Faculty of Business Administration, Rissho University, 4-2-16, Osaki, Shinagawa-ku, Tokyo 141-8602, Japan
eno@ris.ac.jp
[2] Department of Advanced Sciences, Faculty of Science and Engineering, Hosei University, 3-7-2, Kajino-cho, Koganei-shi, Tokyo 184-8584, Japan
dilewerdolkun@126.com, makoto.takizawa@computer.org

Abstract. Process replication can provide reliable and available distributed application services. On the other hand, a large amount of electric energy is consumed to provide replicated processes since multiple replicas of each application process have to be performed on multiple virtual machines in a server cluster system like cloud computing systems. In this paper, an improved redundant active time-based (IRATB) algorithm is newly proposed to realize energy-efficient server cluster systems for redundantly performing computation type application processes.

Keywords: Energy-efficient load balancing · Server cluster · Virtual machine · Improved redundant active time-based (IRATB) algorithm

1 Introduction

In current information systems, a large volume of data is gathered from various types of smart devices to provide various kinds of distributed application services. In order to realize distributed application services, fault-tolerant, scalable, and high performance computing systems like server cluster systems [1–4] are required since a large volume of data is manipulated and application services have to be reliable and available. Virtual machines [5] supported by server cluster systems like cloud computing systems [6] are widely used to more efficiently utilize the computation resources of a physical server. Reliable and available distributed application services can be provided by replicating [7] each application process on multiple virtual machines in a server cluster system. On the other hand, the more number of replicas of application processes are performed on multiple virtual machines than non-replication approaches. As a results, a server cluster system consumes a large amount of electric energy to perform multiple replicas of each application process on multiple virtual machines. Hence, it is necessary to realize energy-efficient server cluster systems to provide reliable and available distributed application services as discussed in Green computing [6].

L. Barolli et al. (Eds.): AINA 2021, LNNS 225, pp. 172–180, 2021.
https://doi.org/10.1007/978-3-030-75100-5_16

In our previous studies, the *redundant active time-based* (*RATB*) algorithm [7] is proposed to select multiple virtual machines for each *computation type application process* (*computation process*) so that the total electric energy consumption of a server cluster and the average response time of each computation process can be reduced. However, the RATB algorithm assumes only one thread on a CPU is allocated to each virtual machine in each physical server even if some thread is not used in the physical server. Therefore, each physical server consumes more electric energy to perform computation processes on virtual machines since it takes longer time to perform computation processes if more number of processes are performed on each virtual machine.

In this paper, the *Improved RATB* (*IRATB*) algorithm is newly proposed to furthermore reduce the total electric energy consumption of a server cluster and the response time of each process for redundantly performing each computation process. In the IRATB algorithm, the total electric energy consumption of a server cluster and the average response time of each computation process can be reduced by allocating idle threads to active virtual machines which are performing computation processes in each physical server if the total electric energy consumption of the physical server does not increase. As a result, the response time of each computation process and the total electric energy consumption of a server cluster can be more reduced in the IRATB algorithm than the RATB algorithm. The evaluation results show the total electric energy consumption of a server cluster and the response time of each computation process can be more reduced in the IRATB algorithm than the RATB algorithm.

In Sect. 2, we discuss the system model of this paper. In Sect. 3, we discuss the IRATB algorithm. In Sect. 4, we evaluate the IRATB algorithm compared with the RATB algorithm.

2 System Model

2.1 Server Cluster System

Multiple physical servers s_1, ..., s_n ($n \geq 1$) construct a server cluster S, $S = \{s_1, ..., s_n\}$. Let nc_t be the total number of cores in a server s_t ($nc_t \geq 1$) and C_t be a set of homogeneous cores c_{1t}, ..., $c_{nc_t t}$ in the server s_t. Let ct_t be the number of threads [8] on each core c_{ht} in a server s_t and nt_t ($nt_t \geq 1$) be the total number of threads in a server s_t. The total number nt_t of threads in a server s_t is $nc_t \cdot ct_t$. Each server s_t holds a set TH_t of threads th_{1t}, ..., $th_{nt_t t}$ and threads $th_{(h-1) \cdot ct_t + 1}$, ..., $th_{h \cdot ct_t}$ ($1 \leq h \leq nc_t$) are bounded to a core c_{ht}. Each server s_t supports a set V_t of virtual machines VM_{1t}, ..., $VM_{nt_t t}$. A virtual machine VM_{vt} is performed on a set $VT_{vt}(\tau)$ of threads allocated to the virtual machine VM_{vt} at time τ. $nVT_{vt}(\tau) = |VT_{vt}(\tau)|$ and $1 \leq nVT_{vt}(\tau) \leq nt_t$.

Computation processes which mainly consume CPU resources are performed on each virtual machine VM_{vt}. A term *process* stands for a computation process in this paper. Each time a load balancer K receives a request process p^i from a client cl^i, the load balancer K selects a subset VMS^i of virtual machines in a server cluster S to redundantly perform the request process p^i. The load balancer K forwards the process p^i to every virtual machine VM_{vt} in the subset VMS^i. A notation NF denotes the maximum number of physical servers which concurrently stop by fault in the cluster S. Let rd^i be the *redundancy* of a process p^i, i.e. number of replicas of a process p^i. We assume $NF + 1 \leq rd^i = |VMS^i| \leq n$. On receipt of a request process p^i, a virtual machine VM_{vt} creates and performs a replica p^i_{vt} of the process p^i. On termination of the process p^i_{vt}, the virtual machine VM_{vt} sends a reply r^i_{vt} to the load balancer K. The load balancer K takes only the first reply r^i_{vt} and ignores every other reply. Hence, a load balancer K can receive at least one reply r^i_{vt} from a virtual machine VM_{vt} even if NF servers stop by fault in a server cluster S. A virtual machine VM_{vt} is *active* iff (if and only if) at least one replica is performed on the virtual machine VM_{vt}. Otherwise, the virtual machine VM_{vt} is *idle*. A thread th_{kt} is *active* iff at least one virtual machine VM_{vt} is active on the thread th_{kt}. Otherwise, the thread th_{kt} is *idle*. A core c_{ht} is *active* iff at least one thread th_{kt} is active on the core c_{ht}. Otherwise, the core c_{ht} is *idle*. In this paper, we assume each thread th_{kt} in a server s_t is not allocated to multiple virtual machines at time τ.

2.2 Computation Model

A notation pt^i_{kt} stands for a replica of a process p^i performed on a thread th_{kt} allocated to a virtual machine VM_{vt} in a server s_t. Let Th^i_{kt} be the total computation time [msec] of a replica pt^i_{kt} where the replica pt^i_{kt} is performed on a thread th_{kt}. A notation $minTh^i_{kt}$ shows the minimum computation time of a replica pt^i_{kt} where the replica pt^i_{kt} is exclusively performed on one thread th_{kt} on a core c_{ht} and the other threads on the core c_{ht} are *idle* in a server s_t. We assume $minTh^i_{1t} = minTh^i_{2t} = \cdots = minTh^i_{rt}$ in a server s_t. $minTh^i = minTh^i_{kt}$ on the fastest server s_t. We assume one virtual computation step [vs] is performed for one time unit [tu] on one thread th_{kt} in the fastest server s_t. The maximum computation rate $Maxf_{kt}$ of a thread th_{kt} on a core c_{ht} in the fastest server s_t where only the thread th_{kt} is active on the core c_{ht} is 1 [vs/tu]. $Maxf_{ku} \leq Maxf_{kt}$ on the slower server s_u. We assume $Maxf_{1t} = Maxf_{2t} = \cdots = Maxf_{nt_t t}$ in a server s_t. $Maxf = max(Maxf_{k1}, ..., Maxf_{kn})$ which is defined to be one. A replica pt^i_{kt} is considered to be composed of VS^i_{kt} virtual computation steps. $VS^i_{kt} = minTh^i_{kt} \cdot Maxf = minTh^i_{kt}$ [vs]. The maximum computation rate $maxf^i_{kt}$ of a replica pt^i_{kt} is $VS^i_{kt}/minTh^i_{kt}$ $(0 \leq maxf^i_{kt} \leq 1)$ where the replica p^i_{kt} is exclusively performed on a thread th_{kt} on a core c_{ht} and only the thread th_{kt} is active the a core c_{ht} in a server s_t.

The computation rate $FT_{kt}(\tau)$ of a thread th_{kt} on a core c_{ht} in a server s_t at time τ is given as follows:

$$FT_{kt}(\tau) = Maxf_{kt} \cdot \beta_{kt}(at_{kt}(\tau)). \tag{1}$$

Here, $at_{kt}(\tau)$ is the number of active threads on the core c_{ht} at time τ where the thread th_{kt} is bounded to the core c_{ht}. Let $\beta_{kt}(at_{kt}(\tau))$ be the *performance degradation ratio* of a thread th_{kt} on a core c_{ht} at time τ $(0 \leq \beta_{kt}(at_{kt}(\tau)) \leq 1)$ where multiple threads are active on the same core c_{ht}. $\beta_{kt}(at_{kt}(\tau)) = 1$ if $at_{kt}(\tau) = 1$. $\beta_{kt}(at_{kt}(\tau_1)) \leq \beta_{kt}(at_{kt}(\tau_2)) \leq 1$ if $at_{kt}(\tau_1) \geq at_{kt}(\tau_2)$.

Suppose a virtual machine VM_{vt} is performed on a set $VT_{vt}(\tau)$ of threads in a server s_t at time τ. The computation rate $FV_{vt}(\tau)$ of the virtual machine VM_{vt} at time τ is given as follows:

$$FV_{vt}(\tau) = \sum\nolimits_{th_{kt} \in VT_{vt}(\tau)} FT_{kt}(\tau). \tag{2}$$

In this paper, each replica p_{vt}^i of a request process p^i is performed on a virtual machine VM_{vt} installed in each server s_t. Replicas which are being performed on a virtual machine VM_{vt} at time τ are *current*. A notation $CP_{vt}(\tau)$ shows a set of current processes on a virtual machine VM_{vt} at time τ and $NC_{vt}(\tau) = |CP_{vt}(\tau)|$. In this paper, we assume the computation rate $FV_{vt}(\tau)$ of a virtual machine VM_{vt} at time τ is uniformly allocated to every current replica on the virtual machine VM_{vt}.

The computation rate $f_{vt}^i(\tau)$ of a replica p_{vt}^i performed on a virtual machine VM_{vt} at time τ is given as follows:

$$f_{vt}^i(\tau) = \begin{cases} \alpha_{vt}(\tau) \cdot FV_{vt}(\tau) \ / \ NC_{vt}(\tau), & \text{if } NC_{vt}(\tau) > nVT_{vt}(\tau), \\ FT_{kt}(\tau), & \text{otherwise.} \end{cases} \tag{3}$$

Here, $\alpha_{vt}(\tau)$ is the *computation degradation ratio* of a virtual machine VM_{vt} at time τ ($0 \le \alpha_{vt}(\tau) \le 1$). $\alpha_{vt}(\tau_1) \le \alpha_{vt}(\tau_2) \le 1$ if $NC_{vt}(\tau_1) \ge NC_{vt}(\tau_2)$. $\alpha_{vt}(\tau) = 1$ if $NC_{vt}(\tau) \le 1$. $\alpha_{vt}(\tau)$ is assumed to be $\varepsilon_{vt}^{NC_{vt}(\tau)-1}$ where $0 \le \varepsilon_{vt} \le 1$.

Suppose that a replica p_{vt}^i starts and terminates on a virtual machine VM_{vt} at time st_{vt}^i and et_{vt}^i, respectively. Let T_{vt}^i be the total computation time of a replica p_{vt}^i performed on a virtual machine VM_{vt}. Here, $T_{vt}^i = et_{vt}^i - st_{kt}^i$ and $\sum_{\tau = st_{vt}^i}^{et_{vt}^i} f_{vt}^i(\tau) = VS_{vt}^i$. At time st_{vt}^i a replica p_{vt}^i starts, the computation laxity $lc_{vt}^i(\tau)$ is VS_{vt}^i [vs]. The computation laxity $lc_{vt}^i(\tau)$ [vs] of a replica p_{vt}^i at time τ is $VS_{vt}^i - \sum_{x=st_{vt}^i}^{\tau} f_{vt}^i(x)$.

2.3   Power Consumption Model of a Server

A notation $E_t(\tau)$ denotes the electric power [W] of a server s_t at time τ to perform replicas of computation processes on multiple virtual machines. Let $maxE_t$ and $minE_t$ be the maximum and minimum electric power [W] of a server s_t, respectively. Let $ac_t(\tau)$ be the number of active cores in a server s_t at time τ and $minC_t$ be the electric power [W] where at least one core c_{ht} is active on a server s_t. Let cE_t be the electric power [W] consumed by a server s_t to make one core active.

The *power consumption model of a server with virtual machines (PCSV model)* [9] to perform computation processes on virtual machines is proposed. According to the PCSV model, the electric power $E_t(\tau)$ [W] of a server s_t to perform replicas of computation processes on virtual machines at time τ is given as follows [9]:

$$E_t(\tau) \ = \ minE_t \ + \ \sigma_t(\tau) \ \cdot \ (minC_t \ + \ ac_t(\tau) \ \cdot \ cE_t). \tag{4}$$

Here, $\sigma_t(\tau) = 1$ if at least one core c_{ht} is active on a server s_t at time τ. Otherwise, $\sigma_t(\tau) = 0$. The electric power $E_t(\tau)$ of a server s_t depends on the number of active cores at time τ.

The processing power $PE_t(\tau)$ [W] is $E_t(\tau)$ - $minE_t$ at time τ in a server s_t. The total processing electric energy $TPE_t(\tau_1, \tau_2)$ [J] of a server s_t from time τ_1 to τ_2 is $\sum_{\tau=\tau_1}^{\tau_2} PE_t(\tau)$.

3   The Energy-Efficient Process Replication

3.1   The Redundant Active Time-Based (RATB) Algorithm

In our previous studies, the *redundant active time-based* (*RATB*) algorithm [7] is proposed to reduce the total electric energy consumption of a server cluster S and the average response time of each process for redundantly performing each computation process. The RATB algorithm assumes only one thread th_{kt} is allocated to each virtual machine VM_{vt} in a server s_t even if some other threads are idle in the server s_t. Let d_{Kt} be the delay time [msec] between a load balancer K and a server s_t. The minimum response time $minRT_{vt}^i$ of a replica p_{vt}^i where the replica p_{vt}^i is exclusively performed on a virtual machine VM_{vt} and only VM_{vt} is active on a core c_{ht} in a server s_t is calculated as $minRT_{vt}^i = 2d_{Kt} + minT_{vt}^i \cdot Maxf/Maxf_{vt} = 2d_{Kt} + 1 \cdot 1/Maxf_{vt} = 2d_{Kt} + 1/Maxf_{vt}$ [msec]. The RATB algorithm estimates the increased active time $iACT_{ht}(\tau)$ of each core c_{ht} in a server s_t at time τ based on the response time RT_{vt}^i of each replica p_{vt}^i performed on each virtual machine VM_{vt} in the server s_t. The total processing electric energy laxity $tpel_t(\tau)$ [J] shows how much electric energy a server s_t has to consume to perform every current replica on every active virtual machine in the server s_t at time τ. In the PCSV model, the electric power $E_t(\tau)$ [W] of a server s_t at time τ depends on the number $ac_t(\tau)$ of active cores in the server s_t at time τ as shown in Eq. (4). In the RATB algorithm, the total processing electric energy laxity $tpel_t(\tau)$ of a server s_t at time τ is estimated by the following Eq. (5):

$$tpel_t(\tau) = minC_t + \sum_{h=1}^{nc_t}(iATC_{ht}(\tau) \cdot cE_t). \tag{5}$$

Let $TPEL_{vt}^i(\tau)$ be the total processing electric energy laxity of a server cluster S where a replica p_{vt}^i of a request process p^i is allocated to a virtual machine VM_{vt} performed on a server s_t at time τ. In the RATB algorithm, a virtual machine VM_{vt} where the total processing electric energy laxity $TPEL_{vt}^i(\tau)$ of a server cluster S is the minimum at time τ is selected for a replica p_{vt}^i of a request process p^i.

3.2   The Improved RATB (IRATB) Algorithm

In the RATB algorithm [7], only one thread is allocated to a virtual machine in a server. Hence, some threads on a same core may be idle while the other threads are active to perform replicas. As a result, the more number of replicas are performed on each virtual machine, the more electric energy a server consumes since it takes longer time to perform replicas and the active time of each core in the server increases. In this paper, we newly propose an *Improved RATB* (*IRATB*) algorithm to furthermore reduce the total processing electric energy consumption of a server cluster S for redundantly performing each process. In the PCSV model [9], the electric power consumption $E_t(\tau)$ of a server s_t depends on the number $ac_t(\tau)$ of active cores in the server s_t at time τ. At time τ, the electric power consumption $E_t(\tau)$ of the server s_t where current replicas are performed on a virtual machine VM_{vt} with one thread th_{kt} on a core c_{ht} is the same as the electric power where current replicas are performed on the virtual machine VM_{vt} with multiple threads on the same core c_{ht} since $ac_t(\tau) = 1$. In the IRATB algorithm, the total processing electric energy consumption of a server cluster S and the average response time of each process can be reduced by allocating idle threads to active virtual machines in each server s_t. In the IRATB algorithm, if a thread $th_{k't}$ on a core c_{ht} allocated to a virtual machine $VM_{v't}$ is idle, the thread $th_{k't}$ is used for another active virtual machine VM_{vt} performed on the same core c_{ht} in each server s_t at time τ. Then, the computation time of each replica performed on the virtual machine VM_{vt} can be reduced in the IRATB algorithm than the RATB algorithm since the computation rate $FV_{vt}(\tau)$ of the virtual machine VM_{vt} increases. As a result, the total processing electric energy consumption of each server s_t can be more reduced in the IRATB algorithm than the RATB algorithm since the active time of each core can be reduced. Let $idle_{ht}(\tau)$ be a set of idle threads on a core c_{ht} in a server s_t at time τ and $avm_{ht}(\tau)$ be a set of active virtual machines on a core c_{ht} at time τ. At time τ, idle threads on a core c_{ht} are allocated to each active virtual machine VM_{vt} performed on the core c_{ht} in a server s_t by the THREAD_ALLOC procedure as shown in Algorithm 1.

4 Evaluation

The IRATB algorithm is evaluated in terms of the total processing electric energy consumption [KJ] of a homogeneous server cluster S and the response time of each process p^i compared with the RATB algorithm [7]. A homogeneous server cluster S is composed of three servers s_1, s_2, s_3 ($n = 3$). Every server s_t ($1 \le t \le 3$) and every virtual machine VM_{vt} follows the same computation model and the same power consumption model as shown in Tables 1 and 2. The parameters of each server s_t are obtained from the experiment [9]. Every server s_t is equipped with a dual-core CPU ($nc_t = 2$). Two threads are bounded for each core in a server s_t, i.e. $ct_t = 2$. The number nt_t of threads in each server s_t is four. The total number of threads in the server cluster S is twelve. Initially, each thread th_{kt} is allocated to one virtual machine VM_{vt} in a server s_t (k, $v = 1$, ..., 4 and $t = 1$, ..., 3). Hence, there are twelve virtual machines in the server cluster S. We assume the fault probability fr_t for every server s_t is the same $fr = 0.1$.

Algorithm 1. Thread_Alloc procedure

Input: s_t, τ.
Output: VT_{vt}.
  procedure THREAD_ALLOC(s_t, τ)
    for each core c_{ht} in a server s_t **do**
       $avm_{ht}(\tau)$ = a set of active VMs on a core c_{ht} at time τ;
       if $|idle_{ht}(\tau)| \geq 1$ and $|avm_{ht}(\tau)| \geq 1$ **then**
          while $|idle_{ht}(\tau)| > 0$ **do**
             $th = th_{kt} \in idle_{ht}(\tau)$;
             VM_{vt} = a virtual machine where $CP_{vt}(\tau)$ is the maximum in $avm_{ht}(\tau)$;
             $VT_{vt}(\tau) = VT_{vt}(\tau) \cup \{th\}$;
             $idle_{ht}(\tau) = idle_{ht}(\tau)$ - $\{th\}$;
             $avm_{ht}(\tau) = avm_{ht}(\tau)$ - $\{VM_{vt}\}$;
             if $avm_{ht}(\tau) = \phi$ **then**
                $avm_{ht}(\tau)$ = a set of active VMs on c_{ht} at time τ;
             end if
          end while
       end if
    end for
  end procedure

Table 1. Homogeneous cluster S.

Server	nc_t	ct_t	nt_t	$minE_t$	$minC_t$	cE_t	$maxE_t$
s_t	2	2	4	14.8 [W]	6.3 [W]	3.9 [W]	33.8 [W]

The number m of processes p^1, ..., p^m ($0 \leq m \leq 8,000$) are issued. The starting time st^i of each process p^i is randomly selected in a unit of one millisecond between 1 and 3600 [ms]. The minimum computation time $minT_{vt}^i$ of every replica p_{vt}^i is assumed to be 1 [ms]. The delay time d_{Kt} of every pair of a load balancer K and every server s_t is 1 [msec] in the server cluster S. The minimum response time $minRT_{vt}^i$ of every replica p_{vt}^i is $2d_{Kt} + minT_{vt}^i = 2 \cdot 1 + 1 = 3$ [msec].

4.1   Response Time

Figure 1 shows the average response time of each process in the IRATB and RATB algorithms. In Fig. 1, IRATB(rd) and RATB(rd) stand for the average response time of each process in the IRATB and RATB algorithms with redundancy rd (= 1, 2, 3), respectively. The average response time of each

Table 2. Parameters of virtual machine.

Virtual machine	$Maxf_{vt}$	ε_{vt}	$\beta_{vt}(1)$	$\beta_{vt}(2)$
VM_{vt}	1 [vs/msec]	1	1	0.6

process increases as the number m of processes increases in the IRATB(rd) and RATB(rd) algorithms. The average response time in the IRATB algorithm can be more reduced than the RATB algorithm. In the IRATB algorithm, idle threads in each server s_t are allocated to active virtual machines in the server s_t if the total processing electric energy consumption of the server s_t does not increase. Then, the computation rate of each active virtual machine to perform replicas of each process increases. As a result, the average response time of each process can be more reduced in the IRATB algorithm than the RATB algorithm since the computation resources in the server cluster S can be more efficiently utilized in the IRATB algorithm than the RATB algorithm.

4.2   Total Energy Consumption of a Server Cluster

Figure 2 shows the average total processing electric energy consumption $ATEC$ of the server cluster S to perform the number m of processes in the IRATB and RATB algorithms. In Fig. 2, IRATB(rd) and RATB(rd) stand for the average total processing electric energy consumption of the server cluster S in the IRATB and RATB algorithms with redundancy rd ($= 1, 2, 3$), respectively. In the IRATB algorithm, the response time of each replica can be more reduced than the RATB algorithm as shown in Fig. 1 since idle threads in each server s_t are allocated to active virtual machines in the server s_t. Then, the active time of each virtual machine can be more reduced in the IRATB algorithm than the RATB algorithm. As a result, the average total processing electric energy consumption of the server cluster S can be more reduced in the IRATB algorithm than the RATB algorithm. Following the evaluation, the IRATB algorithm is more useful than the RATB algorithm.

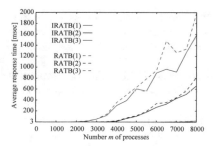
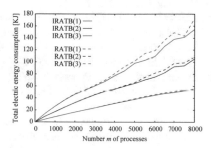

Fig. 1. Average response time.

Fig. 2. Total electric energy consumption.

5   Concluding Remarks

In this paper, the IRATB algorithm was newly proposed to reduce the total electric energy consumption of a server cluster and the response time of each process for redundantly performing computation processes by allocating idle threads in each server s_t to active virtual machines. We showed the total electric energy consumption of a server cluster and the response time of each process can be more reduced in the IRATB algorithm than the RATB algorithm. The evaluation results showed the IRATB algorithm is more useful than the RATB algorithm.

Acknowledgements. This work was supported by the Japan Society for the Promotion of Science (JSPS) KAKENHI Grant Number 19K11951.

References

1. Enokido, T., Aikebaier, A., Takizawa, M.: Process allocation algorithms for saving power consumption in peer-to-peer systems. IEEE Trans. Industr. Electron. **58**(6), 2097–2105 (2011)
2. Enokido, T., Aikebaier, A., Takizawa, M.: A model for reducing power consumption in peer-to-peer systems. IEEE Syst. J. **4**(2), 221–229 (2010)
3. Enokido, T., Aikebaier, A., Takizawa, M.: An extended simple power consumption model for selecting a server to perform computation type processes in digital ecosystems. IEEE Trans. Industr. Inform. **10**(2), 1627–1636 (2014)
4. Enokido, T., Takizawa, M.: Integrated power consumption model for distributed systems. IEEE Trans. Industr. Electron. **60**(2), 824–836 (2013)
5. KVM: Main Page - KVM (Kernel Based Virtual Machine) (2015). http://www.linux-kvm.org/page/Mainx_Page
6. Natural Resources Defense Council (NRDS): Data center efficiency assessment - scaling up energy efficiency across the data center industry: Evaluating key drivers and barriers (2014). http://www.nrdc.org/energy/files/data-center-efficiency-assessment-IP.pdf
7. Enokido, T., Duolikun, D., Takizawa, M.: An energy-efficient process replication algorithm based on the active time of cores. In: Proceedings of the 32nd IEEE International Conference on Advanced Information Networking and Applications (AINA 2018), pp. 165–172 (2018)
8. Intel: Intel Xeon Processor 5600 Series: The Next Generation of Intelligent Server Processors (2010). http://www.intel.com/content/www/us/en/processors/xeon/xeon-5600-brief.html
9. Enokido, T., Takizawa, M.: Power consumption and computation models of virtual machines to perform computation type application processes. In: Proceedings of the 9th International Conference on Complex, Intelligent and Software Intensive Systems (CISIS 2015), pp. 126–133 (2015)

The Actual Cost of Programmable SmartNICs: Diving into the Existing Limits

Pablo B. Viegas[1], Ariel G. de Castro[1], Arthur F. Lorenzon[1],
Fábio D. Rossi[2], and Marcelo C. Luizelli[1(✉)]

[1] Federal University of Pampa (UNIPAMPA), Alegrete, Brazil
{pabloviegas.aluno,arielcastro.aluno,arthurlorenzon,
marceloluizelli}@unipampa.edu.br
[2] Federal Institute Farroupilha (IFFAR), Alegrete, Brazil
fabio.rossi@iffarroupilha.edu.br

Abstract. Programmable Data Planes is a novel paradigm that enables efficient offloading of network applications. An important enabler for this paradigm is the current available SmartNICs, which should satisfy rigid network requirements such as high throughput and low latency. Despite recent research in this field, not much attention was given to understand the costs and limitations of offloading network applications into Smart-NIC devices. Existing offloading approaches either neglect the existing limitations of SmartNICs or assume that as a fixed cost – leading, therefore, to sub-optimal offloading approaches. In this work, we conduct a comprehensive evaluation of SmartNICs in order to quantify existing performance limitations. We provide insights on network performance metrics such as throughput and packet latency while considering different key building blocks of complex P4 programs (e.g., registers, cryptography functions, or packet recirculation). Results show that line-rate throughput can degrade up to 8x, while latency can increase as much as 80x when performing memory-intensive operations in the data plane.

1 Introduction

Programmable Data Planes (PDP) is a mainstream technology that has been recently redesigning the networking domain. PDP allows to (re)define the behavior of network devices (e.g., programmable routers and SmartNICs), allowing to deliver specialized packet processing mechanisms [2]. Recent advances in data plane programmability have enabled offloading typical control plane applications to the data plane (e.g., machine learning algorithms [19], routing [12], or network monitoring [4,8,9]). On shifting the operation of these applications to the data plane, it brings the benefit to process every single packet and react to network conditions in the order of nanoseconds, with minimum control plane intervention. Despite that, data plane operation might become complex – and the complexity comes at a price: *lower throughput and higher latency*.

Current SmartNIC architectures (e.g., [11]) do not limit the number of operations performed by the data plane in a single pipeline stage. For example, a PDP application could trigger the read and write of an unbounded number of registers in a given stage of the packet processing pipeline. Yet, the same application could recirculate the ingress packet multiple times in order to mimic a loop-based mechanism. These are straightforward examples of simple operations commonly used by more complex PDP applications (e.g., in-network clustering [19]). Therefore, understanding the current performance limitations of existing SmartNICs is paramount to the design of efficient PDP applications.

A recent study [7] has made the first effort to understand the existing limitations of SmartNICs. Harkous et al. [7] have focused on evaluating the performance of general PDP metrics such as parsers, control blocks, and header modifications in P4 programs. Despite this effort, no study has yet thoroughly evaluated key building blocks of complex P4 programs (e.g., registers, cryptography functions, or packet recirculation) – which are essential for most recent P4 applications. To fill in this gap, we perform an extensive performance evaluation of SmartNICs to understand and quantify PDP application primitives and existing limitations. We focus our evaluation on measuring the performance in terms of latency and throughput for a variety of packet sizes (from 64B to 1500B) when (i) operating multiple registers of different sizes/widths (e.g., used to implement bloom filter alike structures [1]), (ii) matching on multiple tables in the ingress and egress pipeline (e.g., used to implement machine-learning algorithms [19]), (iii) performing packet recirculation (e.g., used to implement IoT data desegregation [17]), and (iv) using cryptography functions and arithmetic operations. Results show that network throughput can degrade up to 8X, while latency can increase as much as 80X when performing memory-intensive operations in the data plane. The main contributions of this paper can be summarized as:

- an in-depth performance evaluation of SmartNICs;
- a discussion of current limitations in SmartNIC architectures; and
- an open-source code of all experiments in order to foster reproducibility.

The remainder of this paper is organized as follows. In Sect. 2, we describe the SmartNIC architecture used in this work. In Sect. 3, we overview the recent literature regarding PDP applications and performance evaluation. In Sect. 4, we present and discuss the obtained results and, in Sect. 5, we conclude this paper with final remarks.

2   SmartNIC Architecture

The hardware techniques used to deliver high-speed network packet processing require context switching on the order of nanoseconds and very high degrees of processing parallelism to scale the performance of P4-based programs to Gb/s of throughput and beyond [11]. This is particularly challenging given the ever-increasing complexity of offloaded codes to network adapters. Current programmable NICs (also referred to as SmartNICs) rely their architectures on multi-threaded, multicore flow processor units to cope with this increasing and stringent demand. Next, we focus our discussion on the general architectural details of the Netronome SmartNIC architecture [11] – which is used later in our experiments.

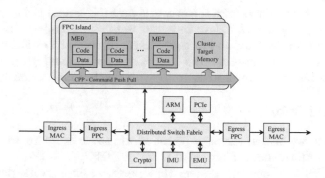

Fig. 1. An overview of the Netronome SmartNIC architecture.

The SmartNIC Netronome NFP4000 architecture is illustrated in Fig. 1 and organizes its flow processing cores (FPC) in multiple islands. FPC is a 32-bit machine, and therefore, all of its internal registers and local memory consists of 32-bit words. Each FPC contains eight Micro Engines (MEs) – a separate processor with its own instruction store (*code*) and local memory (*data*). Because of that, every ME can run in parallel with all other MEs. Each ME has 8 threads that can be used for co-operative multithreading, in the way that at most one thread is executing code from the same program at any given moment. Hence, each FPC runs at most 8 parallel threads at 1.2 GHz (one thread per ME). FPCs follow a Harvard Architecture, and therefore code and data occupy different memories – 4K bytes are shared between threads for data and private memory, while 8K instructions for the coding store.

In each FPC, local memory is composed of a set of 32-bit registers, shared between all 8 threads. These registers are divided into: (*i*) general-purpose registers (256 32-bits registers) – used by default to store any register of up to 32-bits size; (*ii*) transfer registers (512 32-bits registers) – used for copying register over the interconnection bus (e.g., from or to other FPCs or memories); (*iii*) next-neighbor registers (128 32-bits registers) – used mainly to communicate with neighboring FPCs; and last (*iv*) local memory (1024 32-bit registers) – which is a little bit slower than general register (3 cycles access, instead of just 1 cycle). When there is a need for more memory than available space in local FPC registers, variables are automatically and statically allocated to other in-chip memory hierarchies.

There are four other kinds of memory which are available to FPCs: (*i*) Cluster Local Scratch (CLS) (20–50 cycles); (*ii*) Cluster Target Memory (CTM) (50–100 cycles); (*iii*) Internal Memory (IMEM) (120–250 cycles); and (*iv*) External Memory (EMEM) (150–590 cycles). In summary, the local memory register is used for data that is used in every packet. The CLS is used for data, which is needed for most packets and small shared tables. The CTM is used for packet headers and coordination between other sub-systems. Then, IMEM is used for packet bodies and medium-sized shared tables. Finally, EMEM is used for large shared tables.

As packets are received from the network, an FPC thread picks up the packet from the *Distributed Switch Fabric* and processes it (i.e., on-path SmartNIC [6]). Additional threads are allocated to new packets as they arrive. For instance, the SmartNIC NFP-4000 supports up to 60 FPCs, each supporting up to 8 threads. Then, the device is able to process up to 480 packets simultaneously.

3   Related Work

In this section, we discuss the most recent efforts towards P4 SmartNIC offloading and the performance evaluation of them in PDP.

DAIET [15] is a network system that performs in-network data aggregation. It uses Machine Learning (ML) to judiciously decide which partition of the application (e.g., MapReduce) is deployed into PDP to reduce network traffic while maintaining the correctness of the overall computation. FairNIC [6] utilizes SmartNICs to decrease CPU host utilization while providing performance isolation in a multi-tenant environment. It provides an abstraction that allows network applications to access NIC resources. In turn, Clara [13] provides performance clarity for SmartNIC offloading. It analyzes network functions (NFs) and predicts their performance when ported to SmartNIC targets. It uses a logical SmartNIC model to capture SmartNIC architecture. Then, an intermediate representation identifies the code blocks and maps them onto the logical model while optimizing for a performance objective. Finally, it outputs the performance profile for the original NF input on a particular workload. Similarly, SmartChain [18] minimizes the redundant packet transmission by analyzing service function chaining (SFC) forwarding paths and reducing the packet transmissions between the SmartNIC and the host CPU. In the same direction, iPipe [10] allows to offload distributed applications onto SmartNICs. At its core, a scheduler combines first-come & first-serve strategy with deficient round-robin to tolerate applications with variable execution costs.

In addition to the efforts mentioned above towards enabling efficient offloading of network applications to SmartNICs, there are also recent initiatives to offload ML algorithms to PDPs. In this context, N2Net [16] and Sanvito et al. [14] represent the first steps toward in-network inference. N2Net [16] is a compiler that generates a P4 program configuration for an RMT-like switch pipeline [3] based on a binary neural network (BNN) model description while Sanvito et al. [14] introduce BaNaNa SPLIT, a system capable of offloading BNNs from CPUs to SmartNICs through a quantization process that transforms the NN model into a format that can be appropriately executed on PDPs. More recently, Xiong et al. [19] propose to deploy trained ML classification algorithms into PDPs. The proposed approach relies on multiple match-action tables and, therefore, is portable between different PDP targets.

As one can observe, most of the existing efforts are still restricted to offloading mechanisms to PDP. However, to the best of our knowledge, these studies do not take into account the current limitations of existing SmartNICs on the offloading process. Therefore, these solutions might either lead to infeasible solutions (e.g., using more resources than available) or suffer high penalties on the expected performance. One noticeable exception is the recent study conducted by Harkous et al. [7]. They evaluate different P4 programs and their impact on the packet processing latency. They gradually increase the complexity of a SmartNIC pipeline (i.e., including parser, control blocks, and deparser) to identify the most influential variables for predicting packet latency. Despite this effort, the work conducted by [7] is still full of gaps considering key building blocks of complex P4 programs (e.g., registers, cryptography functions, packet recirculation, or multiple tables) – which are essential in P4 applications (especially ML applications). In this work, we take a step further into thoroughly understanding the performance of SmartNICs and their existing limitations.

4   Deployment Evaluation

In this section, we evaluate the performance of Netronome SmartNIC for P4 programs concerning their properties and achieved throughput and latency. We start describing our environment setup and performance metrics, followed by the discussion of results.

4.1   Setup

Our environment setup consists of two high-end servers. Each server has an Intel Xeon 4214R processor with 32 GB RAM. One server is our Device Under Test (DUT) – i.e., the server in which P4 programs are loaded – and the other is our traffic generator. Both servers have a Netronome SmartNIC Agilio CX 10 Gbit/s network device with two network interfaces, which are physically connected. We use MoonGen [5] as our DPDK[1] traffic generator. We instruct MoonGen using the Netronome Packet Generator[2]. In our experiments, we send IPv4 packets at line rate (i.e., 10Gbit/s) with random source and destination prefixes. For our evaluation, we varied the packet size from 64B to 1500B. All experiments were run at least 30 times to ensure a confidence level higher than 90%.

4.2   Metrics

In our evaluation, we aim to measure the performance of P4 programs with respect to the achieved throughput and latency, and identify existing hardware limitations. We evaluate the impact on those metrics regarding (i) the number of operations on registers, (ii) the number of access to match+action tables,

---

[1] https://www.dpdk.org/.

[2] https://github.com/emmericp/MoonGen/tree/master/examples/netronome-packetgen.

(*iii*) the number of packet recirculation, (*iv*) the number of applied cryptography functions, and (*v*) the number of performed arithmetic operations. To evaluate these metrics, we automatically generate P4 codes with the properties to be analyzed. All P4 codes have at least one match+action table used to perform IP forwarding between physical ports. After generating the P4 source codes, we compiled them using the Netronome compiler and loaded the generated firmware into the physical SmartNIC. Then, we pump network traffic with MoonGen and collect the obtained throughput and latency. To measure data plane latency (i.e., the amount of time a packet stays on PDP), all P4 programs have at least one register which keeps that information (i.e., the difference between the ingress and egress timestamps). During our experiments, we read that register data using the Netronome CLI. The obtained throughput (measured in packets per second) is obtained directly from MoonGen. In order to foster reproducibility, our experimental codes are public available[3]. It is important to note that other SmartNICs and compilers can be easily adapted to our experimental codes.

4.3   Results

4.3.1   The Cost of Reading and Writing at Multiple Registers

We start by analyzing the cost of performing multiple register operations in the P4 pipeline. Register operations are one of the main building blocks of recent P4 applications (e.g., [1,17,19]). In the experiments, we varied the number of register operations performed sequentially by the P4 program from 10 to 200 registers while varying the register width from 32- to 512-bit words. We consider that registers are placed in the ingress pipeline and are either read, write, or read & write. Our goal is to quantify the impact on throughput and latency, as well as to quantify the existing limitation of the current architecture. Figure 2 illustrates the measured throughput (in packets per second) and latency (in nanoseconds).

Packet Intense Line-Rate Network Throughput. Figures 2(a) and 2(b) depict the measured throughput and latency, respectively, for small packets (64 Bytes) and register width of 32 bits. As the number of registers increases (and consequently the number of P4 instructions), there is a sharp performance degradation on observed throughput and latency. The line rate for 10 Gbit/s (i.e., 14.88 million packets per second – Mpps) can be sustained for reading operations to only 10 registers (Fig. 2(a)). Even with 10 registers, the bandwidth degradation for writing operations is 30% (and it reaches 50% for read & write) – as they demand more micro engine cycles to be performed. After, we observe a linear decrease of up to 87% (i.e., 2 Mpps) considering 200 registers. In Fig. 2(b), the latency increases linearly as the number of operations performed by each processed packet in the pipeline. For a small number of registers (i.e., 10), there is an acceptable overhead of 8650 ns (reading), 20296 ns (writing), and 35839 ns (both operations). However, this overhead can be as high as 0.12 and 0.24 ms for 50 and 200 registers, respectively.

---

[3] https://github.com/mcluizelli/performanceSmartNIC

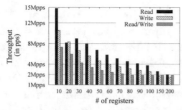

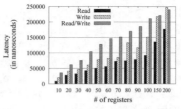

(a) Measured throughput for different register operations (packet size 64B).

(b) Measured latency for different register operations (packet size 64B).

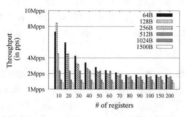

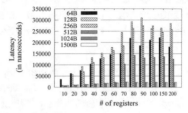

(c) Measured throughput for different packet sizes.

(d) Measured latency for different packet sizes.

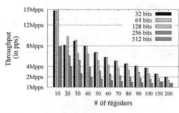

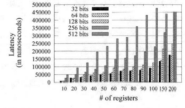

(e) Measured throughput for different register width.

(f) Measured latency for different register width.

Fig. 2. Measured throughput and latency for register operations.

Line Rate Network Throughput for Different Packet Sizes. Figures 2(c) and 2(d) depict the measured throughput and latency, respectively, for different packet sizes (from 64B to 1500B). We fixed the register width to 32 bits and the register operation to read & write (as it is the most resource-consuming one). Observe that the larger the packet size (and, consequently, the less the number of packets per second to achieve 10Gbit/s), the more register operations can be sustained at line rate. For instance, to 512B-size packets, the line rate throughput can be sustained up to 60 registers being read and write (2Mpppps). For 1024B-size packets and higher, there is no throughput degradation – even for 200 registers. Although there is no throughput degradation, Fig. 2(d) illustrates that per-packet latency increases substantially. For instance, for 1500B-size packet, the latency increases up to 3X (from 7983 ns to 22172 ns), while for small packet sizes (and, consequently, higher packet throughput), this latency overhead can be as high as 0.25 ms per packet (i.e., a 34X increase).

Line Rate Network Throughput for Different Register Width.
Figure 2(e) and Fig. 2(f) illustrate the throughput and the latency for a varying
width of registers. We fixed the packet size to 64B and the operation to read
– as the goal is to quantify the performance degradation with respect to the
achieved line rate. As one can observe, the line-rate operation is kept for a register width of up to 128 bits (and 10 registers). Larger register width demands
more cycles to fetch the data from memory. As discussed in Sect. 2, Netronome
SmartNIC follows a 32-bit architecture. Therefore, any register width wider than
that requires extra cycles to be fetched. In addition to that, it is important to
mention the memory hierarchy. The more register is needed, the more external
memories are used – which directly affects the time to fetch data.

Current Limitations. *Is there any limitation on operating registers in a
SmartNIC?* In our experiments, we were able to define at most six arrays of 32-
bit registers, each having 130M positions. This limitation is due to the amount
of memory available on the board (see Section 2). Although we could instan-
tiate such a large number of registers, we could not access all of them in a
single pipeline pass because micro engines have a limited code space to store
the instruction set (i.e., at most 8K instruction). Differently from traditional
languages, P4 does not have go-to primitives, and therefore all instructions are
defined at compiling time. For that reason, we were able to operate at most 200
registers in a single P4 program (considering that our P4 program has also other
instructions to perform the forwarding).

4.3.2 Impact of Packet Recirculation

Next, we evaluate the impact of performing packet recirculation in the P4
pipeline. As previously mentioned, the P4 language does not support iteration-
based structures, and therefore, packet recirculation has been used as a way to
circumvent such limitations. In short, packet recirculation consists of sending a
packet back to the ingress pipeline after processing it (and therefore, mimic a
loop-based structure). In the experiments, we varied the number of packet recirculation made (from 0 to 50) in each packet while varying the packet size (from
64B to 1500B). We consider that our P4 program is just forwarding network
traffic from physical interfaces. Figure 3 illustrates the measured throughput
and latency.

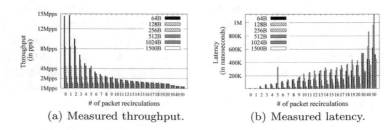

(a) Measured throughput.   (b) Measured latency.

Fig. 3. Analysing the effect of packet recirculation in the P4 pipeline.

Figure 3(a) illustrates the throughput behavior as the number of packet recirculations increases (it decreases in a super-linear manner). For small-packet network traffic, we observe that fewer packet recirculations are sustained at line rate (e.g., three packet recirculations for 128B size packets). In contrast, for larger packets (e.g., 1024B), it can maintain line rate even for up to two dozen packet recirculations. As the packets are recirculated, more packets are being pushed into the data plane – which enqueue them, and eventually dropped occurs (reducing throughput). In turn, Fig. 3(b) depicts the observed latency. As one can observe, there is a non-negligible increase in latency as packets recirculate. For instance, even for large packets (e.g., 1500B) – in which we observe little or no throughput degradation, the per-packet latency doubles on performing just 3 packet recirculations (from 5610 ns to 10688 ns). This overhead is even sharper as the number of recirculations increases. For example, there is a 6x latency increase (32762 ns) for doing 10 recirculations, and an 80x latency increase (450062 ns) for 50 recirculations. As one can observe, this overhead is even greater for small packets.

Current Limitations. In our experiments, we observe that the Netronome architecture does not allow for recirculating custom-made metadata structure. This limits substantially the ability to write complex P4 programs – especially the ones using packet recirculation to circumvent the lack of iteration-based structure.

4.3.3 Impact of Using Multiple Tables

We evaluate the impact of using multiple match-action tables. Unlike traditional forwarding devices that use match+action tables exclusively for routing (i.e., to look up network addresses), the P4 language has opened up new possibilities for this construction type. For instance, Xiong et al. [19] have used multiple tables to implement data plane clustering approaches (e.g., k-means with a table per cluster). Similar to the work conducted by Harkous et al. [7], we also observe that the performance of P4 programs is not affected by the size of tables – as a hash-based data structure implements them. Usually, large match+action tables are already placed on larger and slow memories (e.g., DRAM). Here, instead, we aim at analyzing the impact of using multiple match+action tables at different stages of the pipeline. In the experiments, we varied the number of existing tables in our P4 programs (from 1 to 10), and we ensure that every packet is always matched sequentially on all tables. An action is invoked to read a single 32-bit data from the table and store it in a metadata structure on a packet matching. We varied the packet size (from 64B to 1500B) and the number of tables per pipeline (either on the ingress or egress pipeline). Figure 4 illustrates the measured throughput and latency.

Figure 4(a) illustrates the measured throughput for an increasing number of match+action tables. As observed, the throughput for 64B packets (most packet intensive network traffic) is almost negligible for up to 5 match+action tables (i.e., it keeps the line rate). However, we observe an abrupt decay after 5 tables, followed by a constant throughput behavior (up to 10). In the Netronome architecture, a P4 program can only have 5 tables in each pipeline (ingress/egress). Therefore, tables 1–5 are located in the ingress pipeline, while 6–10 in the egress. As all the memory is statically allocated for a P4 program, the Netronome compiler tends to use faster, closer available memory to micro engines to allocate ingress tables. Even when defining tables only in the egress pipeline, the compiler tends not to use faster memory. We empirically show this behavior in Fig. 4(c). We incrementally place 5 tables either in the ingress or egress pipeline. We observe that the ingress pipeline is always faster to use available tables (w.r.t. latency) – even in the cases where no tables are used in the ingress. Last, Fig. 4(b) illustrates the per-packet latency for an increasing number of tables (both in the ingress/egress pipeline). On average, there in an increase of 40–50% in the latency in the ingress pipeline (between 1 and 5 tables).

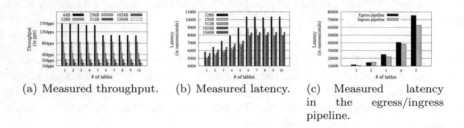

(a) Measured throughput.   (b) Measured latency.   (c)   Measured   latency in   the   egress/ingress pipeline.

Fig. 4. Analysing the effect of using multiple tables on the P4 pipeline.

Current Limitations. As mentioned, the Netronome architecture poses a limit on the number of match+action tables one can use in a P4 program (i.e., at most 5 in each pipeline). That limits the applicability of more complex algorithms (e.g., [7]) in SmartNICs. Further, we also observe performance differences w.r.t. to the incurred latency (i.e., tables placed in the ingress behave 20% faster).

4.3.4   Impact of Using Cryptography Functions and Arithmetic Operations

Next, we evaluate the impact of using cryptography functions in our P4 programs. Bloom filters or hash-based data structures widely use them (e.g., to identify heavy-hitters [1]). Cryptography functions are target-dependent (i.e., the implementation depends on the hardware), and in the Netronome architecture, they are implemented by specific micro engines (named Crypto in Fig. 1). Netronome implements eight hash functions: (*i*) crc32, (*ii*) crc16, (*iii*) identity, (*iv*) csum16, (*v*) crc32custom, (*vi*) crc16custom, (*vii*) random, and (*viii*) xor16. In the experiments, we analyse the impact of applying consecutive calls to these cryptography functions (from 0 to 30) in each packet being processed. For the purpose of this experiment, we keep the packet size in 64B and consider that our P4 program just forward network traffic between physical interfaces. Figure 5 illustrates the measured throughput and latency.

Figures 5(a) and 5(b) illustrate the throughput and latency, respectively. As one can observe, only three out of the eight cryptography functions (`crc32-custom`, `random`, and `csum16`) do not lead to performance degradation w.r.t. throughput and latency when increasing the number of calls to them. The remaining ones (i.e., `crc32`, `crc16`, `identity`, `crc16-custom`, and `xor16`) lead to some performance degradation from applying 10 cryptography functions (e.g., 23% of throughput degradation for applying `xor16`). This overhead is even higher for 30 cryptography functions (up to 55% overhead for cryptography function `xor16`). In turn, Fig. 5(b) illustrates the incurred per-packet latency. For up to 10 cryptography functions, the per-packet latency remains acceptable (i.e., below 10000 ns). However, on applying higher number of cryptography functions (e.g., from 15–20 and on), the latency cost grows exponentially. For instance, the `xor16` function reaches up to 7X higher latency (applying 30 functions) in comparison to the simple forwarding (the case of 0 cryptography functions). We further evaluate the impact of using arithmetic operations in our P4 programs (i.e., $+$, $-$, $*$, $/$, $\%$, and $<<$ (bit-shifting)). In the experiments, we analyse the impact of applying consecutive arithmetic operations (varying from 10 to 10000 operations) for intensive packet processing (i.e., 64B packets). We also consider that our P4 program just forwarding network traffic between physical interfaces. In this experiment, we do not observe any statistically significant latency or throughput degradation.

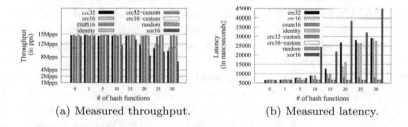

(a) Measured throughput.   (b) Measured latency.

Fig. 5. Analysing the effect of applying multiple hash functions on the P4 pipeline.

Current Limitations. Netronome architecture poses a few limitations restricting the design of more complex P4 programs to its SmartNICs. For instance, multiplication and division are only performed over an integer. Further, the architecture restricts multiplication and division operations of fixed-size operands (i.e., at most 32 bits operands). This limits, for instance, the implementation of precise fixed-point representation for real numbers. Another limitation is the bit shifting operation. The current architecture requires bit-shifting to be static-compiled with predefined values – limiting its applicability. Last, the number of arithmetic operations is limited to 10000 operations per pipeline – related to the number of instructions a micro engine can store.

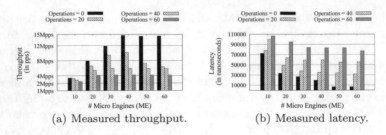

<table>
<tr><td>(a) Measured throughput.</td><td>(b) Measured latency.</td></tr>
</table>

Fig. 6. Analysing the effect of using multiple Micro Engines (CPU cores) to process network traffic.

4.3.5   Analysing Used Cores and Energy Consumption

Last, we evaluate the impact of varying the number of micro engines used by the SmartNIC (from 10 to 60 ME). The goal is to verify whether or not it affects the obtained performance. We keep the packet size at 64B for this experiment and consider that our P4 program forwards network traffic between physical interfaces. We varied the number of reading operations (from 0 to 60) in existing 32-bit registers to stress out the hardware. Figure 6(a) illustrates the measured throughput. As expected, the more ME is available, the more throughput is achieved (in general). However, we can observe that the SmartNIC does not need all ME working in parallel for the evaluated workload. For instance, for 0 read operations, the line rate throughput is achieved using 40 ME. When considering the case of 60 read operations, we observe that more than 20 ME does not affect the performance. Yet, we also observe that allocating a higher number of ME is not always the best strategy. In some cases (e.g., for 20 read operations), the performance is worsened by increasing the number of MEs from 40 to 50–60. Figure 6(b) depicts the measured latency. As one can observe, there is always a latency reduction when increasing the available ME (even when there is no improvement in the throughput). Finally, we evaluated the energy consumption. In our experiments, the energy consumption varied 0.2 W between using 10 ME and 60 ME.

5   Final Remarks

In this paper, we performed an extensive performance evaluation of SmartNICs to understand and quantify existing limitations. We focus our evaluation on measuring the performance in terms of latency and throughput for a plethora of packet memory-intensive scenarios. We showed that the line-rate throughput is bounded by (i) the number of register operations (up to 10 operations), (ii) the number of multiple match+action tables user in the pipeline (up to 5), (iii) the number of cryptography operations (up to 10). As future work, we intend to build an analytical model that can accurately estimate the performance of P4 applications executing on SmartNICs.

Acknowledgements. This work was partially funded by National Council for Scientific and Technological Development (CNPq) (grant 427814/2018-9), São Paulo Research Foundation (FAPESP) (grant 2018/23092-1), Rio Grande do Sul Research Foundation (FAPERGS) (grants 19/2551-0001266-7,20/2551-000483-0, 19/2551-0001224-1).

References

1. Ben Basat, R., Einziger, G., Friedman, R., Luizelli, M.C., Waisbard, E.: Constant time updates in hierarchical heavy hitters. In: Proceedings of the ACM SIGCOMM, SIGCOMM 2017, pp. 127–140. ACM, New York (2017)
2. Bosshart, P., Daly, D., Gibb, G., Izzard, M., McKeown, N., Rexford, J., Schlesinger, C., Talayco, D., Vahdat, A., Varghese, G., Walker, D.: P4: programming protocol-independent packet processors. ACM SIGCOMM **44**(3), 87–95 (2014)
3. Bosshart, P., Gibb, G., Kim, H.S., Varghese, G., McKeown, N., Izzard, M., Mujica, F., Horowitz, M.: Forwarding metamorphosis: fast programmable match-action processing in hardware for SDN. ACM SIGCOMM CCR **43**(4), 99–110 (2013)
4. Castro, A.G., Lorenzon, A.F., Rossi, F.D., Da Costa Filho, R.I.T., Ramos, F.M.V., Rothenberg, C.E., Luizelli, M.C.: Near-optimal probing planning for in-band network telemetry. IEEE Commun. Lett. 1 (2021). https://ieeexplore.ieee.org/document/9330755/keywords#keywords
5. Emmerich, P., Gallenmüller, S., Raumer, D., Wohlfart, F., Carle, G.: MoonGen: a scriptable high-speed packet generator. In: Proceedings of the ACM IMC, IMC 2015, pp. 275–287. ACM, New York (2015)
6. Grant, S., Yelam, A., Bland, M., Snoeren, A.C.: SmartNIC performance isolation with FairNIC: programmable networking for the cloud. In: Proceedings of the ACM SIGCOMM, pp. 681–693 (2020)
7. Harkous, H., Jarschel, M., He, M., Priest, R., Kellerer, W.: Towards understanding the performance of p4 programmable hardware. In: ACM/IEEE Symposium on Architectures for Networking and Communications Systems, pp. 1–6. IEEE (2019)
8. Hohemberger, R., Castro, A.G., Vogt, F.G., Mansilha, R.B., Lorenzon, A.F., Rossi, F.D., Luizelli, M.C.: Orchestrating in-band data plane telemetry with machine learning. IEEE Commun. Lett. **23**(12), 2247–2251 (2019)
9. Hohemberger, R., Lorenzon, A.F., Rossi, F.D., Luizelli, M.C.: A heuristic approach for large-scale orchestration of the in-band data plane telemetry problem. In: Barolli, L., Amato, F., Moscato, F., Enokido, T., Takizawa, M. (eds.) Advanced Information Networking and Applications, pp. 381–392. Springer International Publishing (2020)
10. Liu, M., Cui, T., Schuh, H., Krishnamurthy, A., Peter, S., Gupta, K.: Offloading distributed applications onto SmartNICs using iPipe. In: Proceedings of the ACM Special Interest Group on Data Communication, pp. 318–333 (2019)
11. Netronome: Internet (2020). https://www.netronome.com/static/app/img/products/silicon-solutions/WP_NFP4000_TOO.pdf
12. Pizzutti, M., Schaeffer-Filho, A.E.: Adaptive multipath routing based on hybrid data and control plane operation. In: IEEE INFOCOM, pp. 730–738 (2019)
13. Qiu, Y., Kang, Q., Liu, M., Chen, A.: Clara: performance clarity for SmartNIC offloading. In: Proceedings of the ACM Hot Topics in Networks, pp. 16–22 (2020)
14. Sanvito, D., Siracusano, G., Bifulco, R.: Can the network be the AI accelerator? In: Proceedings of the Workshop on In-Network Computing, pp. 20–25 (2018)

15. Sapio, A., Abdelaziz, I., Aldilaijan, A., Canini, M., Kalnis, P.: In-network computation is a dumb idea whose time has come. In: Proceedings of the 16th ACM Workshop on Hot Topics in Networks, pp. 150–156 (2017)
16. Siracusano, G., Bifulco, R.: In-network neural networks. arXiv preprint arXiv:1801.05731 (2018)
17. Wang, S.Y., Wu, C.M., Lin, Y.B., Huang, C.C.: High-speed data-plane packet aggregation and disaggregation by p4 switches. J. Netw. Comput. Appl. **142**, 98–110 (2019)
18. Wang, S., Meng, Z., Sun, C., Wang, M., Xu, M., Bi, J., Yang, T., Huang, Q., Hu, H.: SmartChain: enabling high-performance service chain partition between SmartNIC and CPU. In: IEEE International Conference on Communications, pp. 1–7. IEEE (2020)
19. Xiong, Z., Zilberman, N.: Do switches dream of machine learning? Toward in-network classification. In: Proceedings of the 18th ACM Workshop on Hot Topics in Networks, pp. 25–33 (2019)

On the Adaptation of Bitcoin-Like Blockchains for the Secure Storage of Traffic-Related Events

El-hacen Diallo[1(✉)], Omar DIB[2], and Khaldoun Al Agha[1]

[1] CNRS, Laboratoire Interdisciplinaire des Sciences du Numérique,
Université Paris-Saclay, 91190 Gif-sur-Yvette, France
{diallo,alagha}@lisn.fr
[2] Department of Computer Science, Wenzhou-Kean University, Wenzhou, China
odib@kean.edu

Abstract. This paper investigates Bitcoin-like blockchains enabled traffic efficiency applications in VANETs. Prior works do not provide a thorough study on the adaptation of the Proof-of-Work (PoW) mechanism in the context of traffic records management, nor a detailed study of performance. Therefore, in this work, we study a PoW based blockchain for secure traffic-related records in the VANETs' context. We introduce a blockchain simulator to assess our protocol's performance (i.e., throughput and latency). Results show that the proposed scheme ensures the integrity and traceability of the shared traffic-related records and point out trade offs between security and performance.

1 Introduction

The rapid development of modern devices, and the evolution of wireless technology, have extended Vehicular Ad-hoc NETworks' (VANETs) applications to the next level. Autonomous vehicles are equipped with smart devices such as OnBoard Units (OBUs), sensors, Global Positioning System (GPS) receivers. They are also supplied with WAVE (Wireless Access in Vehicular Environments) module, easing their interaction with nearby vehicles as well as with infrastructures deployed on the road, i.e., Road-Side-Units (RSUs). A huge amount of traffic-related data is collected throughout vehicles. And this data is used to enhance traffic safety and efficiency by minimizing accidents and providing optimized trip planning, parking services, real-time streaming, etc. Traditionally, collected traffic records are uploaded and managed relying on a central cloud [1]. However, such centralized architectures face privacy issues and high bandwidth and storage cost which can lead to severe network congestion [2]; they are also highly prone to *a single point of failure* vulnerability.

The Edge computing paradigm is a potential solution to cope with the aforementioned drawbacks of centralized schemes. The notion of Edge computing consists of splitting the cloud to Edge devices and bringing it to the edge of the network [2], in other words, bringing facilities closer to vehicles. For example,

L. Barolli et al. (Eds.): AINA 2021, LNNS 225, pp. 195–207, 2021.
https://doi.org/10.1007/978-3-030-75100-5_18

Edges nodes could be embedded on RSUs. Doing so will minimize network congestion, bandwidth usage and, more importantly, help support real time data processing. As a result, the QoS related to traffic management will be improved.

Nevertheless, although the Edge cloud paradigm will offer better QoS by tackling latency issues in cloud-centric systems, it has brought along daunting security challenges. Ensuring secure and transparent communication between Edge devices is a major challenge because of the Edge computing's distributed nature. Moreover, the network has become more vulnerable as it is difficult to protect Edge devices from intrusion, unauthorized access, and various attacks [3].

We believe that blockchain technology, and its exciting capacity of maintaining a decentralized and distributed database between untrusted entities without relying on a central authority, can meet the aforementioned security issues. The Blockchain technology consists of a peer-to-peer network exploiting a consensus algorithm to agree on a distributed database state. From a data structure point of view, a blockchain is a linked list of blocks, where each block stores the hash of the previous block. As the core of any blockchain platform, the consensus mechanism is the main component that differentiates blockchain platforms. Introduced as the underlying consensus mechanism with Bitcoin [4], the first blockchain platform, the Poof-of-Work (PoW) consists of dedicated processors trying to solve a difficult but easy-to-verify cryptographic puzzle. PoW based blockchains may witness inconsistencies that cause branches in the chain. This phenomenon is known as *forks*, where more than one valid block is broadcasted to the network simultaneously [5]. In Bitcoin, the most extended branch is considered as the main chain. And blocks that do not belong to the main chain are called *stale blocks*.

In this work, we take advantage of the decentralization, data integrity, and public variability properties of the blockchain, aiming to secure related traffic records. We focus on investigating and studying the adoption of the Proof-of-Work consensus for efficient traffic-related records management.

2   Related Work

A considerable number of applications in VANET's context relied on the Proof-of-Work mechanism. Such PoW-based applications can be categorized as follows:

Vehicles as Blockchain Nodes: In [6], authors proposed a blockchain-based anonymous reputation system (BARS), aiming to prevent the distribution of forged messages. The global consensus mechanism was provided by the vehicles assuming they have enough amount of resources to act as miners. Also, in [7], basic concepts of blockchain were used for data sharing among intelligent vehicles while relying on the PoW algorithm. Unfortunately, no details were provided about the blockchain design, nor the protocol's performance metrics, thus making the assessment of their work very hard.

Indeed, the idea of using vehicles as blockchain nodes seems interesting from a decentralization point of view. However, such a scheme faces challenges regarding the scalability, latency, and global blockchain update [8]. Because of the storage, communication, and computing load resulting from the usage of the PoW as consensus protocol between vehicles, guaranteeing the reliability of latency-sensitive applications (e.g., safety messages dissemination) is a challenging task [9]. To mitigate the storage overhead on vehicles, in this work [10], the authors suggested a light blockchain. However, a scalability evaluation is still needed.

RSUs as Blockchain Nodes: On the other hand, RSUs have been used as blockchain miners. For example, in [11] the PoW mechanism is executed by the RSUs to build a secure and trustworthy event message delivery in VANETs. Also, in [12] RSUs were relied on as blockchain nodes, which compute to solve the PoW puzzle. The authors' goal is to propose a blockchain scheme for secure announcement messages sharing in IoV (Internet of Vehicles). Furthermore, in [13], the authors proposed a blockchain-based protocol for distributed trust management, where a fixed number of RSUs validate events using fuzzy logic and participate in the block creation process using PoW.

3 Limitations of Related Works

All the works mentioned above lack metrics related to their system performance. Indeed, VANETs applications are usually delay-sensitive; it is, therefore, crucial to provide metrics such as message confirmation time and the number of transactions that can be processed by the blockchain in a unit of time.

Furthermore, data validation in VANETs applications differs from Bitcoin and other coin based blockchains. The latter relies only on the history (i.e., committed data) for transaction validation. In contrast, traffic-related records need to be approved by the witnesses of the concerned events. Therefore, it is essential to investigate and study the PoW mechanism's suitability in VANETs applications; and particularly for implementing a secure traffic-related records sharing system.

Despite PoW, other consensus algorithms were introduced in the context of VANETs. We cite: Proof-of-Event (PoE) [14] and Proof-of-Driving (PoD) [7]; however, further discussions regarding their security models are required so that such protocols can be applied reliably and securely. PBFT (Practical Byzantine Fault Tolerance) [15], and Proof-of-Authority (PoA) [16] were also adopted; nevertheless, such protocols seem to be centralized since a pre-defined subgroup of RSUs is designated by a central trust authority to maintain the blockchain. Furthermore, stake-based protocols like Proof-of-Stake (PoS) [17] was also introduced. Although this approach mitigates the CPU demands, new issues arise that were not present in PoW-based blockchains, such as *Nothing at stake attack* and *Grinding attack* [18].

4 Contributions

To the best of our knowledge, the state-of-the-art lacks a more throughout study about adoption of PoW based blockchain in the context of traffic-related records management, followed by a rigorous performance evaluation. The main contributions of this paper are as follows:

- We detail how the Proof-of-Work (PoW) can be adapted to secure the traffic messages in the VANETs context.
- We analyze and discuss PoW contribution regarding traffic-events validation.
- We present a benchmark framework whereby we assess the performance of PoW enabled traffic events management in terms of throughput (i.e., number of confirmed events per second) and latency (event confirmation delay).
- We also assess the impact of the PoW puzzle difficulty on the performance of the system (i.e., throughput and latency).
- Finally, we highlight and discuss trade-offs between security and performance.

The remainder of this article is organized as follows. In Sect. 5, we present the proposed architecture. Section 6 describes the protocol, which is a blockchain-based scheme for traffic records storage. And Sect. 7 describes the implemented simulation environment. In Sect. 8, the performance of the proposed system is assessed. Finally, Sect. 10 concludes this paper and discusses future works.

5   System Model

In this section, we describe the system model, where a blockchain is enabled traffic records management. We first describe the system architecture, then highlight VANETs applications and finally discuss traffic events validation techniques (Fig. 1).

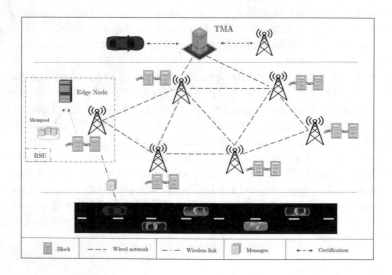

Fig. 1. System architecture

5.1   System Components

The proposed architecture can be separated into three layers. The first layer represents the Ad-hoc domain, which represents intelligent vehicles. Through their dedicated communication modules and embedded sensors, cars record relevant traffic events (e.g., accidents, traffic congestion data, safety warnings). These records are signed by vehicles with their private keys to ensure the communications' authenticity and reliability before being uploaded to RSUs.

The second layer of the proposed architecture illustrates RSUs connections, which form a peer-to-peer network. Each RSU is supplied with an Edge node, providing the required resources (i.e., computation and storage) to maintain a blockchain and solve the PoW puzzle. Received traffic records are stored into RSU's *mempool* under a pending state. Pending events wait for a threshold of confirmations before turning to valid.

The last layer represents the Traffic Management Authority (TMA); it is a trusted organization generating cryptographic credentials to newly joining vehicles and RSUs. The provided certificates must be anonymous to guarantee vehicle privacy. Besides, the TMA monitors the blockchain and inflicts punishment to misbehavior actions.

We recall here that the paper's scope is to investigate and study the adoption of PoW towards traffic-related records management.

5.2   Traffic-Related Events Validation

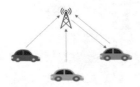

Fig. 2. RSU-centric events validation

Fig. 3. Vehicle-centric events validation

Messages validation is the first step toward trustworthy traffic records sharing and storage since data issuers could be reporting fictitious events. We classify existing VANETs protocols for events validation into two main groups: Rsu-centric events validation and Vehicle-centric events validation methodologies.

RSU-Centric Events Validation Protocol: In this category, RSUs validate messages uploaded by vehicles. More precisely, after receiving a message reporting a traffic event, a RSU waits for a threshold of confirmation messages from vehicles before considering the event to be plausible. If necessary, RSUs would request confirmations from nearby vehicles as illustrated in Fig. 2.

Vehicle-Centric Events Validation: Differently, in this approach, events are validated and signed by vehicles through threshold of signatures or a consensus algorithm as suggested in [12]. As we can notice, no need for RSUs intervention as illustrated in Fig. 3. However, reaching a consensus between highly mobile vehicles could be an obstacle. Indeed, such a scheme may add a substantial load to the vehicle-to-vehicle communications (V2V), hence engendering a significant delay to the event validation time.

Therefore, in this paper, we will be considering the RSU-centric events validation methodology when designing our model. More precisely, in our protocol, the RSUs will act as blockchain nodes; they will receive traffic events from vehicles and use a PoW to validate the events according to a threshold and then update their maintained blockchain nodes accordingly. As such, a RSU-based blockchain will be built to achieve a decentralized, transparent, immutable, and secure traffic-related system in the VANETs context. In the following, we will describe in detail how the blockchain is enabled traffic message storage.

6   Protocol Description

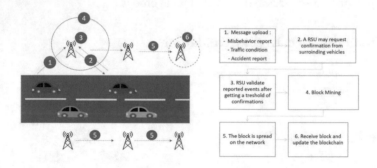

Fig. 4. Model flow

This section describes the proposed scheme for secure traffic-related data validation, storage, and sharing. In Fig. 4, we present a diagram flow of the proposed model, which consist of 5 steps. We also provide pseudo-codes describing the block creation and confirmation processes.

The first step of our protocol is the collection of traffic-related messages. This phase consists of vehicles reporting hazardous traffic events such as accidents, attacks performed by other vehicles, traffic jams, ice on the road, etc. Collected data are used to optimize road-trips, traffic flow, and hence reduce the CO_2 emissions and fuel consumption; they can also help report vehicle's misbehavior and help reduce accidents caused by traffic jams. The immutability property of the blockchain ensures received data integrity. Nevertheless, collected messages have to be validated in the first place. So, in steps 2 and 3 of the protocol, RSUs will gather enough confirmations from surrounding vehicles. After a threshold is reached, the corresponding event will be valid. Step 4 consists of block mining, whereas RSUs take advantage of their CPUs to solve the Proof-of-Work puzzle. Algorithm 1 depicts how RSUs mine blocks. Next, in step 5, the block is spread through the network. Finally, step 6 consists of mined block verification. When a RSU receives a new block, it verifies if the block has been correctly signed, the PoW was correctly solved, and finally, if the required threshold of confirmations was reached for each event within the block. After verification, the RSU will stop trying to mine the same block (in PoW, RSUs are parallelly trying to mine the next block) and add the block to its local blockchain. See Algorithm 2.

Algorithm 1. Block Mining

Input : mempool : *mempool*;
blockchain : *chain* ;
peers : *peers*;
block time out : $block_{time}$;
PoW difficulty : d

1:   $b \leftarrow createBlock()$
2:   $nonce \leftarrow 0$
3:   $tries \leftarrow 10000$
4: **repeat**
5:     $b.setBlockNonce(nonce)$
6:     $b.calculateBlockHash()$;
7:     **if** $b.getHash().getLeadingZeros() \geq d$ **then**
8:       break
9:     **else if** $chain.getHeight() \leq b.getHeight()$ **then**
10:      break
11:    **end if**
12:    $nonce \leftarrow nonce + 1$
13: **until** $nonce \leq tries$
14: **if** $nonce \leq tries$ **then**
15:    $b.shuffleTransactions()$
16: **else**
17:    $chain.addBlock(b)$
18:    $mempool.update()$
19: **end if**
20: $wait(block_{time})$ % wait for block spreading delay %
21: $Repeat()$ % Repeat the mining process %

6.1   Security Model

PoW security lies down on the global computing power within the system; for example, Bitcoin is considered secure as long as adversary miners do not hold 51% of the total CPU power [4]. In Bitcoin, over 1 million miners maintain the

---

Algorithm 2. Receive block

Input : mempool : *mempool*;
blockchain : *chain* ;
block : *block*;
peers : *peers*;
PoW difficulty : *d*

 1: **if** *chain.HasBlock(block.getHash())* **then**
 2:      **return**
 3: **end if**
 4: *height ← block.getHeight()*
 5: **if** IsVALIDBLOCK(*block*) **then**
 6:      *stopMinig(height)*
 7:      *chain.addBlock()*
 8:      *chain.update()*
 9:      *sendTo(peers)*
10:      *Mine(height + 1)* % Start mining the next block %
11: **end if**
12: **function** IsVALIDBLOCK(*block*)
13:      *hash ← block.concAttributes()* %   Concatenate all block atrributes except block hash %
14:      **if** *SHA256(hash) <> hash* **then**
15:          **return** *FALSE*
16:      **else if** *hash.getLeadingZeros() < d* **then**
17:          **return** *FALSE*
18:      **else if** *notblock.hasThreshold()* **then** % Test if has enough confirmations from vehicles %
19:          **return** *FALSE*
20:      **end if**
21:      **return** *TRUE*
22: **end function**

---

blockchain, hence the challenge of getting the majority of the computation power. In our context, the blockchain might be for a given zone, where the number of RSUs is a lot lower than Bitcoin. Miners are RSUs; therefore, it is easier to obtain the majority of RSUs global CPUs. However, RSUs are authenticated in advance; thus, only known RSUs can join the network. Hence, the network is less subject to external attacks or RSUs conspiring to attack the system. Besides the 51% attack, forks increase the advantage of the adversary in the network [19]. Essentially, the more the number of forks is important, the less secure the blockchain becomes. Therefore, in the evaluation section, we assess the impact of forks on the overall performance of the adopted PoW consensus protocol.

7   Simulation Environment

We have designed our simulation environment using NS-3 [20], a discrete events simulator. RSUs represent nodes in NS-3; each of them runs an instance of the blockchain. The positioning of RSUs is settled following a grid topology. RSUs are connected by a point to point channel to form a peer-to-peer network. The average distance between RSUs is 8–11 km. Each RSU has between 2 to $\sqrt{N}$ connections (peers) generated randomly, where N designate the number of RSUs. And a TCP connection is settled between RSUs. The network connection speed is 100 Mbps. Moreover, communications are encrypted using Schnorr signature [21] implemented using the OpenSSL library. Finally, SHA-256 is used as hash function.

The presented simulator is used to picture the life cycle of traffic-events storage in VANETs, from the collection, reception, and validation of such events to the blockchain confirmation and final persistence in the ledger.

The evaluated metrics are the average number of events that can be confirmed per second, i.e., the throughput of the blockchain (event/s), the average delay between traffic event creation time to event confirmation, i.e., the latency (s). Also, we assess the impact of the PoW puzzle difficulty (d), and the number of RSUs involved in the consensus protocol (N), as well as events arrival rate, i.e., f (event/s) on the performance of the system (i.e., throughput and latency). We suppose that the traffic is dense enough and that the fixed threshold of needed confirmation messages to validate an event will be reached in less than $\lambda = 500$ ms. Some parameters of the simulator have been fixed, such as the event size is set to 800 *bytes* as defined in [15,17], and the period to generate events is set to 60 s (i.e., if the event arrival rate (f) is equal to 50 event/s, there will be 50×60 generated events).

The above-mentioned parameters include real-world configurations. For instance, we have analyzed a dataset of alert messages from San Franciso's city for the year 2019 [22]. The traffic warnings arrival rate is 60 events per day, which is not that much. Indeed, the frequency of traffic alert arrival will increase with the number of autonomous vehicles. Therefore, when running simulations, we vary the events' arrival rate from 1–500 event/s to capture respectively low and heavy traffics. On the other hand, regarding the number of RSUs and the network speed. The RSUs can be communicating using WiMAX [23], as the converge rate can reach 15 km with a communication bandwidth of 100 Mbps [23]. Thus, less than 24 RSUs is enough to cover San Francisco, whose area is $121{,}4\,\mathrm{km}^2$.

8   Performance Evaluation

This section evaluates various instances of the PoW enabled traffic-related data management protocol using a server with the following settings: Dell R640 server, Intel(R) Xeon(R) Silver 4112; CPU 2.60 GHz; 8 core CPU; 64 GB RAM, and running Ubuntu 18.04. The plotted results represent end-to-end measurements from all RSUs. The throughput, i.e., the number of confirmed events per second (event/s), is examined by dividing the total confirmed events by the simulation time. As for the latency, we subtract the creation timestamp from the generated event's confirmation timestamp for each event; then, the average delay is computed. Each experiment is repeated for 10× with different seeds, and the mean is plotted with errors, using the 95% confidence interval.

Table 1. PoW solving delay.

PoW difficulty (leading zeros)	Delay (s)
2	0.0014
4	0.2635
6	91.5003

We evaluate the PoW puzzle-solving delay by launching multiple instances of PoW with various difficulties (2, 4, and 6). For each difficulty (d), we repeat the simulation 100× to obtain enough granularity. All those 100 PoW puzzle-solving delays are stored in a file through which the RSUs read the PoW delays. Table 1 represents the mean of the filtered results from the file mentioned above. Results shows that PoW delay increases exponentially with d.

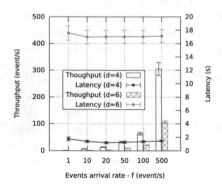

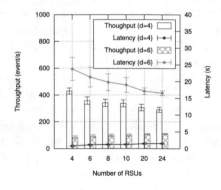

Fig. 5. Performance (throughput, latency) vs. Events arrival rate - f

Fig. 6. Performance (throughput, latency) vs. Number of RSUs

Figure 5 depicts the system' performance of with the increasing events arrival rate (f). Results show that reducing the PoW difficulty (d) from 6 to 4 enhances the performance. When d is set to 4, the system can process up to 300 event/s out of 500 event/s generated, and on average, the event confirmation delay does not exceed 2 s. Lower performance is depicted when $d = 6$ as the throughput does not exceed 106 event/s and the latency can attain 18 s. This significant difference of performance comes from PoW solving' high delay when d = 6.

Figure 6 shows the impact of the number of RSUs on the system's performance. As can be seen, the performance decreases with the increasing number of RSUs. For example, in the worst case, if $d = 4$, increasing the number of RSUs from 4 to 24 decreases the throughput by 141 event/s. This drop in performance is partially due to the block spreading delay, which increases with the network size; nevertheless, it is not the leading cause. Because when the number of RSUs is important, there are more collisions, i.e., RSUs solving the PoW puzzle simultaneously; hence, worsening the system performance. As shown in Fig. 7, the number of stale blocks increases with the number of RSUs. For instance, increasing the number of RSUs from 4 to 24 induces 39% of mined blocks to be stale. Hence, affecting the system performance.

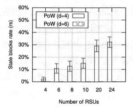 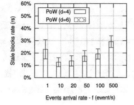 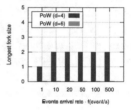

Fig. 7. Stale blocks (%) vs. Number of RSUs

Fig. 8. Stale blocks (%) vs. Events arrival rate-f.

Fig. 9. Forks size vs. Events arrival rate-f.

In Figure 8, the rate of stale blocks (rs) versus events arrival rate (f) is assessed. Results show that if $d = 4$, rs can attain 39%. i.e., RSUs waste 39% of their resources (i.e., computation and bandwidth) on processing blocks that won't be part of the blockchain. That's because, with low difficulty, many RSUs will be likely to solve the puzzle at the same time. However, with $d = 6$, no resources are wasted. Nevertheless, in return, the gained time will be exhausted on PoW solving.

To determine the required number of blocks before considering a block to be persistent in the blockchain, in Fig. 9, we measure the most extended fork size, i.e., the longest branch that does not belong to the blockchain. The obtained results show that implicitly reducing PoW puzzle difficulty (d) increases the forks' size because the RSUs are likely to solve the PoW at the same time. In case $d = 4$, the most extended fork is equal to 2 blocks, while for $d = 6$, there are no forks.

9 Discussion

Although reducing the Proof-of-Work performance would minimize the events' confirmation delay, it may imply a significant number of forks, which disturbs the blockchain's immutability. Moreover, forks cause a huge waste of computation and bandwidth. On the other hand, a greater PoW difficulty induces a considerable delay due to the PoW solving task. Hence, raising a trade-off between security and performance.

10   Conclusion

We investigated and adapted PoW-of-Work (PoW) based blockchain for traffic-related events securing in VANETs' context. We showed how the PoW adaptation differs from the Bitcoin-like use cases. We also presented a benchmark framework whereby we evaluated different instances of the studied protocol; the assessed metrics are the number of events processed by the blockchain (throughput) and events' confirmation time (latency). Results show that PoW can be adapted to validate and track traffic-messages in VANET's. From a performance point of view, it was shown that reducing PoW difficulty engenders a significant number of stale blocks; thus, weakening the security and inducing extra resource costs (communication and computation load). In future works, we have planned to study the impact of vehicle mobility on blockchain performance and to assessing the efficiency of the protocol under malicious vehicles.

References

1. Lee, E., Lee, E.-K., Gerla, M., Oh, S.Y.: Vehicular cloud networking: architecture and design principles. IEEE Commun. Mag. **52**(2), 148–155 (2014)
2. Kaur, K., Garg, S., Aujla, G.S., Kumar, N., Rodrigues, J.J.P.C., Guizani, M.: Edge computing in the industrial internet of things environment: software-defined-networks-based edge-cloud interplay. IEEE Commun. Mag. **56**(2), 44–51 (2018)
3. Khan, W.Z., Ahmed, E., Hakak, S., Yaqoob, I., Ahmed, A.: Edge computing: a survey. Future Gener. Comput. Syst. **97**, 219–235 (2019)
4. Nakamoto, S., et al.: Bitcoin: A peer-to-peer electronic cash system (2008)
5. Shahsavari, Y., Zhang, K., Talhi, C.: A theoretical model for fork analysis in the bitcoin network. In: 2019 IEEE International Conference on Blockchain (Blockchain), pp. 237–244. IEEE (2019)
6. Zhaojun, L., Liu, W., Wang, Q., Gang, Q., Liu, Z.: A privacy-preserving trust model based on blockchain for VANETs. IEEE Access **6**, 45655–45664 (2018)
7. Singh, M., Kim, S.: Blockchain based intelligent vehicle data sharing framework. arXiv preprint arXiv:1708.09721 (2017)
8. Mendiboure, L., Chalouf, M.A., Krief, F.: Survey on blockchain-based applications in internet of vehicles. Comput. Electr. Eng. **84**, 106646 (2020)
9. Elagin, V., Spirkina, A., Buinevich, M., Vladyko, A.: Technological aspects of blockchain application for vehicle-to-network. Information **11**(10), 465 (2020)
10. Mostafa, A.: VANET blockchain: a general framework for detecting malicious vehicles. J. Commun. **14**(5), 356–362 (2019)
11. Shrestha, R., Bajracharya, R., Shrestha, A.P., Nam, S.Y.: A new-type of blockchain for secure message exchange in VANET. Digit. Commun. Netw. **6**, 177–186 (2019)
12. Zhang, L., Luo, M., Li, J., Au, M.H., Choo, K.-K.R., Chen, T., Tian, S.: Blockchain based secure data sharing system for internet of vehicles: a position paper. Veh. Commun. **16**, 85–93 (2019)

13. Kchaou, A., Abassi, R., Guemara, S.: Toward a distributed trust management scheme for VANET. In: Proceedings of the 13th International Conference on Availability, Reliability and Security, p. 53. ACM (2018)
14. Yang, Y., Chou, L., Tseng, C., Tseng, F., Liu, C.: Blockchain-based traffic event validation and trust verification for VANETs. IEEE Access **7**, 30868–30877 (2019)
15. Zhang, X., Chen, X.: Data security sharing and storage based on a consortium blockchain in a vehicular Adhoc network. IEEE Access **7**, 58241–58254 (2019)
16. De Angelis, S., Aniello, L., Baldoni, R., Lombardi, F., Margheri, A., Sassone, V.: PBFT vs proof-of-authority: applying the CAP theorem to permissioned blockchain (2018)
17. Yang, Z., Yang, K., Lei, L., Zheng, K., Leung, V.C.M.: Blockchain-based decentralized trust management in vehicular networks. IEEE Internet Things J. **6**(2), 1495–1505 (2018)
18. Siim, J.: Proof-of-stake. In: Research Seminar in Cryptography (2017)
19. Gervais, A., Karame, G.O., Wüst, K., Glykantzis, V., Ritzdorf, H., Capkun, S.: On the security and performance of proof of work blockchains. In: Proceedings of the 2016 ACM SIGSAC Conference on Computer and Communications Security, pp. 3–16 (2016)
20. https://www.nsnam.org . Network simulator ns-3
21. Schnorr, C.-P.: Efficient identification and signatures for smart cards. In: Conference on the Theory and Application of Cryptology, pp. 239–252. Springer (1989)
22. Moosavi, S., Samavatian, M.H., Nandi, A., Parthasarathy, S., Ramnath, R.: Short and long-term pattern discovery over large-scale geo-spatiotemporal data. In: Proceedings of the 25th ACM SIGKDD International Conference on Knowledge Discovery & Data Mining, pp. 2905–2913 (2019)
23. Vaughan-Nichols, S.J.: Achieving wireless broadband with WiMAX IEEE Comput. **37**(6), 10–13 (2004)

Trustworthy Acceptance: A New Metric for Trustworthy Artificial Intelligence Used in Decision Making in Food–Energy–Water Sectors

Suleyman Uslu[1], Davinder Kaur[1], Samuel J. Rivera[2], Arjan Durresi[1(✉)],
Mimoza Durresi[3], and Meghna Babbar-Sebens[2]

[1] Indiana University-Purdue University Indianapolis, Indianapolis, IN, USA
{suslu,davikaur}@iu.edu, adurresi@iupui.edu
[2] Oregon State University, Corvallis, OR, USA
{sammy.rivera,meghna}@oregonstate.edu
[3] European University of Tirana, Tirana, Albania
mimoza.durresi@uet.edu.al

Abstract. We propose, for the first time, a *trustworthy acceptance metric* and its measurement methodology to evaluate the trustworthiness of AI-based systems used in decision making in Food Energy Water (FEW) management. The proposed metric is a significant step forward in the standardization process of AI systems. It is essential to standardize the AI systems' trustworthiness, but until now, the standardization efforts remain at the level of high-level principles. The measurement methodology of the proposed includes human experts in the loop, and it is based on our trust management system. Our metric captures and quantifies the system's transparent evaluation by field experts on as many control points as desirable by the users. We illustrate the *trustworthy acceptance metric* and its measurement methodology using AI in decision-making scenarios of Food-Energy-Water sectors. However, the proposed metric and its methodology can be easily adapted to other fields of AI applications. We show that our metric successfully captures the aggregated acceptance of any number of experts, can be used to do multiple measurements on various points of the system, and provides confidence values for the measured acceptance.

1 Introduction

Despite so many advantages of AI systems and their uses, these systems sometimes directly or indirectly harm the users and society. It has become essential to make these systems safe, reliable, and trustworthy. Lately, trustworthy AI has been gaining increasing attention from governments, organizations, and scientific communities. So European Union (EU) has proposed ethnic guidelines and laws [50] for trustworthy AI to govern and facilitate the development and working of AI systems [13]. DARPA [15] also launched an XAI program, whose motive

L. Barolli et al. (Eds.): AINA 2021, LNNS 225, pp. 208–219, 2021.
https://doi.org/10.1007/978-3-030-75100-5_19

was to make these AI systems explainable and trustworthy. Garter estimates that 30% of all the digital products that use AI will require a trustworthy AI framework by 2025 [6], and 86% of users will trust and remain loyal to companies that use ethical AI principles [4]. So, developing AI systems using a trustworthy framework has become a necessity for today's society. Furthermore, there are various recent studies on developing trustworthy and explainable algorithms and AI [19, 28, 29, 40–42, 49, 55]. Kaur et al. [23] surveyed similar approaches that aimed to create trustworthy and explainable AI systems.

And as AI technologies mature, they have to follow the natural process that all established technologies have gone through, standardization. Among many advantages, standards allow manufacturers and users to speak the same language, enable the users to check the quality of products on the market, reduce the legal liability of manufacturers and providers, etc. Already ISO, an organization that deals with standardization, has presented different approaches to establish trust in AI systems using fairness, transparency, accountability, and controllability [20]. However, such works remain at the level of high-level principles. What is needed is to develop metrics and measurement procedures to establish standards for trustworthy AI [27]. We envision that in the future, various agencies will use such metrics to certify AI-based solutions, similarly as FDA certifies medications and treatments. For this reason, in this paper, for the first time, we propose a concrete metric, *trustworthy acceptance*, and its measurement methodology. Our metric captures and quantifies the system's transparent evaluation by field experts.

In this paper, we use the following definition: *Trustworthy AI is a framework to ensure that a system is worthy of being trusted concerning its stated requirements based on the evidence. It makes sure that the users' and stakeholders' expectations are met in a verifiable way* [20]. Furthermore, for the time being, AI lacks many aspects of human intelligence, including meaning, multidimensional data beyond the set used for algorithm training, meaningful causality, ethics, etc. Therefore, AI systems should complement and empower humans without replacing them. This is the essential requirement to make AI trustworthy. And for this reason, when developing the proposed metric and corresponding measurement procedures, we include human experts in the loop. Our contributions can be summarized as follows:

- We propose in Sect. 4 a *trustworthy acceptance* metric for the evaluation of the acceptance of AI-based systems by field experts.
- The measurement procedure for the proposed metric is described in Sect. 4 and is based on the concept of a distance acceptance approach that is adaptable to a wide range of systems. In addition to the acceptance value, our metric provides the confidence of the acceptance.
- Our metric utilizes the trust of the experts in the given context, managed by our trust system, summarized in Sect. 3.
- Our metric can be measured in many points of the system in order to reach an assessment of the whole system, as discussed in Sects. 4 and 5.

- Finally, in Sect. 5, we illustrate the application of our trustworthy acceptance metric and its measurement methodology using three systems for environmental decision making.

2   Related Work

As part of our daily lives, decision-making is also an essential element of processes of the most significant fields such as economics, finance, healthcare, and the environment. Kambiz [21] explained the importance of decision makings in such fields by giving examples of the crucial results caused by erroneous or dissatisfactory decisions such as the world economic crisis. He indicated that the difficulty of making such critical decisions relies on their complex nature and the involvement of multiple stakeholders who can have different expertise, background, perspectives, or maybe even competing for objectives.

As it is possible and common in some areas to have experts as stakeholders, there is a need to have a consensus mechanism for such decision makings. Dong and Xu [11] proposed approaches to minimize the modifications to the solutions that the experts propose at each round of the decision making. Furthermore, Hegselmann and Krause [17] investigated the consensus reaching mechanisms of decision makings involving agents with different behaviors in both mathematical modelings and computer simulations. Similarly, Babbar-Sebens and Minsker [3] proposed a design employing an algorithm for the utilization of expert feedback for an improved optimization in the environmental field. These studies clearly show a considerable need for methods and frameworks for advanced decision-making, especially for the ones where humans and machines need to collaborate for superior results.

Trust has long been a concept that is believed to be an essential part of the decision-making process especially involving multiple stakeholders. It has been shown that the utilization of trust contributed to the integrative behavior and helped disruptive activities to decline [14,18,25]. There has been multiple studies [30–32,52] and surveys [10,33,53] discussing trust management and its frameworks. In [39], a trust management framework based on measurement theory is proposed for online social communities. There are several applications of this framework, such as stock market prediction with Twitter data [37]. Other examples include trust management in social networks [54], cloud computing [34,35], internet of things [36,38], healthcare [8,9], emergency communications [12], and detection of crime [24] and fake users [22].

In the environmental field, Alfantoukh et al. [1,2] proposed a model for water allocation problem and a more generic model for consensus reaching problem involving trust. We introduced a trust-based decision support system for natural resource sharing problems in Food-Energy-Water sectors [48]. Also, we presented the enhanced versions of our system utilizing discrete and precomputed solutions [43], game-theoretical approach [44], and scenarios with different evaluation criteria [45]. Furthermore, we presented the role of trust sensitivity of actors in environmental trust-based decision-making scenarios [47].

When computers make decisions, the liability of the decisions becomes a significant issue. Therefore, it requires a comprehensive testing process before deploying and utilizing such systems, which brings the acceptance of a system in the picture. Although there are studies to reach a desired level of acceptance, proposals could be concentrated on revealing and satisfying the needs of the users [26]. Also, there are studies to model users' desire to use such AI-based systems [16]. One of the most critical areas that we need the high acceptance of is healthcare, where trust between doctors and computerized systems is highly crucial [51]. In [24], it has been shown that in decision makings involving both humans and computers, there is no exact winner when considering all scenarios of crime detection, which opens the door for collaboration for more significant results. Similarly, in [22], AI-based fake user detection gave better results when it utilized the community's preferences.

3 Trust Management Framework

We summarize here our measurement theory-based trust management framework [39], because the new metric and its methodology are based on this framework. Trust defined in this framework has two main parameters: impression, m, and confidence, c. Impression represents the level of trust from one party to another, while confidence is the degree of certainty of the impression. Impression calculation is done by averaging the measurements, as shown in Eq. 1 whereas the confidence is calculated related to the standard error of the mean as shown in Eq. 2. In these formulas, $m^{A:B}$, $c^{A:B}$, and $r_i^{A:B}$ are the impression, confidence, and a measurement from A to B.

$$m^{A:B} = \frac{\sum_{i=1}^{N} r_i^{A:B}}{N} \tag{1}$$

$$c^{A:B} = 1 - 2\sqrt{\frac{\sum_{i=1}^{N}(m^{A:B} - r_i^{A:B})^2}{N(N-1)}} \tag{2}$$

Another essential feature of this framework is to anticipate the trust even without communication between two parties. It supports the propagation methods, namely transition and aggregation, to assess trust between two entities that are connected through third-party nodes. Ruan et al. [39] proposed several trust propagation methods and provided their error propagation functions. For aggregation, we selected the averaging method, as shown in Eq. 3, and provided its error formula in Eq. 4. In these formulas, S represents the source, D is the destination, and Ts are the transitive nodes.

$$m_{T_1}^{SD} \oplus m_{T_2}^{SD} = \frac{m_{T_1}^{SD} + m_{T_2}^{SD}}{2} \tag{3}$$

$$e_{T_1}^{SD} \oplus e_{T_2}^{SD} = \sqrt{\frac{1}{2^2}((e_{T_1}^{SD})^2 + (e_{T_2}^{SD})^2)} \tag{4}$$

4 Trustworthy Acceptance Metric and Its Methodology

This section describes the methodology of our trustworthy acceptance metric. We assume that an AI based system generates sets of solutions that a group of experts will evaluate. Each expert has its own set of solutions to evaluate. Part of each set of solution is a reference, optimal solution and some sub-optimal solutions, based on the trade-offs and the criteria applied by the system. Experts might chose the optimal solution or another sub-optimal solutions, based on their expertise, as shown in Fig. 1.

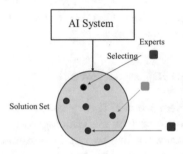

Fig. 1. Experts evaluate and select solutions from the set generated by the AI system.

Acceptance is based on and inversely related to the distance between the proposals of two parties. Distance is measured using the Euclidean distance formula, where each parameter of a proposed solution becomes a dimension. In other words, solutions can be considered as vectors when measuring the distance. After normalizing the each dimension, we also normalize the final distance where the maximum distance becomes 1. Also, distances are always non-negative. An example of the distance between n-dimensional solutions S and T, represented as d_S^T, is shown in Eq. 5 where S_i and T_i are the values of the dimension i of each solution.

$$d_S^T = \sqrt{\frac{\sum_{i=1}^n (S_i - T_i)^2}{n}} \tag{5}$$

Acceptance is measured as shown in Eq. 6 where A_e is the acceptance of the expert e, S is the solution selected by the expert, and T is the reference point, optimal solution. In this case, acceptance is also in the same range of distance which is [0,1]. Although the distance is bidirectionally the same, we usually stress the acceptance of the system evaluated by the expert. Therefore, if a system fails to provide appropriate solution alternatives close enough to the reference point, it is appropriate for experts to reject the system or rate it with lower acceptance.

$$A_e = 1 - d_S^T \tag{6}$$

We consider each acceptance as a trust assessment. Since it is possible to evaluate the acceptance of a system by multiple experts, we use our trust system, see Sect. 3, to aggregate the individual acceptances of experts weighted by their trust as shown in Eq. 7 where k is the number of experts. Similarly, we find the confidence of the acceptance measurements by calculating the population standard error of the mean as shown in Eqs. 8 and 9. We used the regular standard deviation formula instead of the weighted one for simplicity.

$$Tw_A = \frac{\sum_e A_e T_e}{k} \qquad (7)$$

$$SE_{Tw_A} = \frac{\sqrt{\sum_e (Tw_A - A_e)^2}}{n} \qquad (8)$$

$$c_{Tw_A} = 1 - 2(SE_{Tw_A}) \qquad (9)$$

With these two values, the weighted mean acceptance and the confidence, we have our trustworthy acceptance metric, (Tw_A, c_{Tw_A}). It is also possible to have multiple sample measurements to evaluate a system's acceptance by the same group of experts. In this case, again we use our trust system for aggregation and calculate the aggregated mean acceptance of the system as shown in Eq. 10 where n represents the sample size. Furthermore, we update the standard error of the mean using Eq. 11 which is a generalized version of Eq. 4 that we use in our trust measurements explained in detail in [5, 39].

$$System_{Tw_A} = \frac{\sum_n Tw_A}{n} \qquad (10)$$

$$SE_{System_{Tw_A}} = \sqrt{\frac{1}{n^2} \sum_n (SE_{Tw_A})^2} \qquad (11)$$

5   Simulated Measurement Results

To illustrate the evaluation of the proposed acceptance metric, we used the data set that consists of environmental solutions for over 200 fields in a region as explained in [47]. The genetic algorithmic system [7], generates near-optimal solutions while having constraints on the resource usage. The tests were run using three systems that generate solutions. System 1 is the default system where there is no constraint on the number of solutions generated for a field. In System 2 and 3, we reduced the number of available solutions to 7 and 4, respectively. Five agent experts were assigned to evaluate these systems by selecting the most appropriate solution for the specific field which could represent a neighborhood of farmers. Trust of such experts is assumed to be high because it makes sense that the organization which is testing the system deploy top experts. However, if needed, their trust can be dynamically adjusted by techniques described in our previous work [45–47]. Furthermore, a reference solution is also determined and assumed to be legitimate by the user. In our experiments, for simplicity, we used only the environmental protection values in the distance function.

An evaluation of a system starts by assigning each expert a field and presenting them the solutions prepared by the system for that specific field. After we completed the assignment, the experts selected the fittest solution based on their expertise. For an expert who didn't select the reference solution, a nonzero distance is measured between the expert and the reference solution, which is calculated using Eq. 5. Then, we calculated the individual acceptance rates using Eq. 6. After having individual acceptance rates, we averaged the acceptance rates weighted by their trust, as in Eq. 7, and also calculated the confidence of the acceptance, shown in Eq. 9. We call the pair of trust-weighted average acceptance and its confidence as our trustworthy acceptance metric. To increase the confidence in our metric, we repeated this process two more times for each system. We named the selected fieldsets A, B, and C. Then, we aggregated the trustworthy acceptance measurements for each fieldset using Eq. 10. After completing the evaluation for selected samples and generating the system's trustworthy acceptance, we performed the same tasks for System 2 and 3 and measured the acceptance of each system.

Figure 2 shows the trustworthy acceptance of fields A, B, and C for Systems 1, 2, and 3. Compared to System 1, the acceptance declined in System 2 because of limiting the available solutions to 7 which eliminated some solutions closer to the optimal point. Similarly, a stricter constraint, having only four available solutions, resulted in even less acceptance of System 3. These results could be evident for a person who understands and can compare the systems. For example, such results could be used by USDA to certify only Systems 1 and 2 but not System 3. Also, the aggregation of samples' acceptances could be essential in real

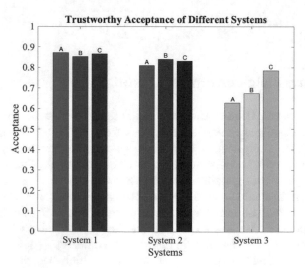

Fig. 2. Trustworthy acceptance of system 1, 2, and 3 measured over the sampled fields A, B, and C from the whole region is presented.

scenarios to reduce the bias of uneven sampling. As shown in Fig. 3, we aggregated the results of fields A, B, and C and presented the trustworthy acceptance of each system measured by our experts. As figures illustrate the acceptance of each system, Table 1 shows both the acceptance values and their confidence which together form our *trustworthy acceptance metric*.

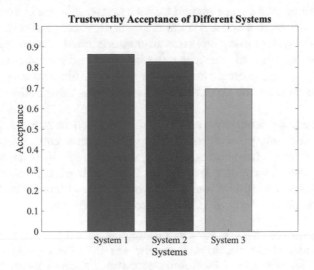

Fig. 3. Measured acceptances for fields A, B, and C are aggregated to find the trustworthy acceptance of system 1, 2, and 3.

Table 1. Trustworthy acceptances, which consist of acceptance and its confidence, are presented for fields A, B, and C and also their aggregation for system 1, 2, and 3.

System 1	Tw_Acceptance	Confidence
Field A	0.872	0.956
Field B	0.853	0.960
Field C	0.866	0.964
Tw_Acceptance_S1	0.863	0.977
System 2	Tw_Acceptance	Confidence
Field A	0.809	0.856
Field B	0.841	0.948
Field C	0.831	0.918
Tw_Acceptance_S2	0.827	0.942
System 3	Tw_Acceptance	Confidence
Field A	0.627	0.726
Field B	0.673	0.719
Field C	0.785	0.857
Tw_Acceptance_S3	0.695	0.861

6 Conclusions

We presented a new *trustworthy acceptance metric* and its measurement methodology for evaluating the approval of an AI system that generate solutions for environmental decisions for a region. Our metric can be used for standardization of Trustworthy AI. The human experts assigned for the evaluation selected the most appropriate solutions presented by each system. We measured the distance from these solutions to a optimal, reference point and calculated the trustworthy acceptance of a system using our trust framework. Furthermore, using our trust framework, we aggregated multiple measurements and provided the confidence of the acceptance using error propagation methods. Finally, we calculated and compared the trustworthy acceptance of each system measured by the assigned experts.

Our *trustworthy acceptance metric* can be applied to many AI applications that can use the concept of distance-based acceptance. Our approach to confidence measurement is based on our trust system that considers trust assessments as measurements. Lastly, the experts' inclusion of trust helped build a metric that can differentiate experts and is even more appropriate for scenarios where there is a possibility to engage different groups of experts.

Acknowledgements. This work was partially supported by the National Science Foundation under Grant No.1547411 and by the U.S. Department of Agriculture (USDA) National Institute of Food and Agriculture (NIFA) (Award Number 2017-67003-26057) via an interagency partnership between USDA-NIFA and the National Science Foundation (NSF) on the research program Innovations at the Nexus of Food, Energy and Water Systems.

References

1. Alfantoukh, L., Ruan, Y., Durresi, A.: Trust-based multi-stakeholder decision making in water allocation system. In: International Conference on Broadband and Wireless Computing, Communication and Applications, pp. 314–327. Springer, Cham (2017)
2. Alfantoukh, L., Ruan, Y., Durresi, A.: Multi-stakeholder consensus decision-making framework based on trust: a generic framework. In: 2018 IEEE 4th International Conference on Collaboration and Internet Computing (CIC), pp. 472–479. IEEE (2018)
3. Babbar-Sebens, M., Minsker, B.S.: Interactive genetic algorithm with mixed initiative interaction for multi-criteria ground water monitoring design. Appl. Soft Comput. **12**(1), 182–195 (2012)
4. Barometer, E.T.: Edelman trust barometer global report. Edelman (2019). https://www.edelman.com/sites/g/files/aatuss191/files/2019-02/2019_Edelman_Trust_Barometer_Global_Report_2.pdf
5. Berendsen, H.J.: A Student's Guide to Data and Error Analysis. Cambridge University Press, Cambridge (2011)
6. Burke, B., Cearley, D., Jones, N., Smith, D., Chandrasekaran, A., Lu, C., Panetta, K.: Gartner top 10 strategic technology trends for 2020-smarter with gartner (2019)

7. Chipperfield, A., Fleming, P.: The MATLAB genetic algorithm toolbox (1995)
8. Chomphoosang, P., Durresi, A., Durresi, M., Barolli, L.: Trust management of social networks in health care. In: 2012 15th International Conference on Network-Based Information Systems, pp. 392–396. IEEE (2012)
9. Chomphoosang, P., Ruan, Y., Durresi, A., Durresi, M., Barolli, L.: Trust management of health care information in social networks. In: Complex, Intelligent, and Software Intensive Systems (CISIS), 2013 Seventh International Conference on, pp. 228–235. IEEE (2013)
10. Chomphoosang, P., Zhang, P., Durresi, A., Barolli, L.: Survey of trust based communications in social networks. In: 2011 14th International Conference on Network-Based Information Systems, pp. 663–666. IEEE (2011)
11. Dong, Y., Xu, J.: Consensus Building in Group Decision Making. Springer, Singapore (2016)
12. Durresi, A., Durresi, M., Paruchuri, V., Barolli, L.: Trust management in emergency networks. In: 2009 International Conference on Advanced Information Networking and Applications, pp. 167–174. IEEE (2009)
13. EC: Ethics guidelines for trustworthy AI (2018). https://ec.europa.eu/digital-single-market/en/news/ethics-guidelines-trustworthy-ai
14. Gunia, B.C., Brett, J.M., Nandkeolyar, A.K., Kamdar, D.: Paying a price: culture, trust, and negotiation consequences. J. Appl. Psychol. **96**(4), 774 (2011)
15. Gunning, D.: Explainable artificial intelligence (xai). Defense Advanced Research Projects Agency (DARPA), nd Web **2**, 2 (2017)
16. Gursoy, D., Chi, O.H., Lu, L., Nunkoo, R.: Consumers acceptance of artificially intelligent (AI) device use in service delivery. Int. J. Inf. Manag. **49**, 157–169 (2019)
17. Hegselmann, R., Krause, U., et al.: Opinion dynamics and bounded confidence models, analysis, and simulation. J. Artif. Soc. Soc. Simul. **5**(3), 1–2 (2002)
18. Hüffmeier, J., Freund, P.A., Zerres, A., Backhaus, K., Hertel, G.: Being tough or being nice? a meta analysis on the impact of hard-and softline strategies in distributive negotiations. J. Manag. **40**(3), 866–892 (2014)
19. Hurlburt, G.: How much to trust artificial intelligence? IT Prof. **19**(4), 7–11 (2017)
20. Information technology – Artificial intelligence – Overview of trustworthiness in artificial intelligence. Standard, International Organization for Standardization, Geneva, CH (2020)
21. Kambiz, M.: Multi-Stakeholder Decision Making for Complex Problems: A Systems Thinking Approach with Cases. World Scientific, Singapore (2016)
22. Kaur, D., Uslu, S., Durresi, A.: Trust-based security mechanism for detecting clusters of fake users in social networks. In: Workshops of the International Conference on Advanced Information Networking and Applications, pp. 641–650. Springer, Cham (2019)
23. Kaur, D., Uslu, S., Durresi, A.: Requirements for trustworthy artificial intelligence–a review. In: International Conference on Network-Based Information Systems, pp. 105–115. Springer, Cham (2020)
24. Kaur, D., Uslu, S., Durresi, A., Mohler, G., Carter, J.G.: Trust-based human-machine collaboration mechanism for predicting crimes. In: International Conference on Advanced Information Networking and Applications, pp. 603–616. Springer, Cham (2020)
25. Kimmel, M.J., Pruitt, D.G., Magenau, J.M., Konar-Goldband, E., Carnevale, P.J.: Effects of trust, aspiration, and gender on negotiation tactics. J. Pers. Soc. Psychol. **38**(1), 9 (1980)

26. Kocielnik, R., Amershi, S., Bennett, P.N.: Will you accept an imperfect AI? exploring designs for adjusting end-user expectations of AI systems. In: Proceedings of the 2019 CHI Conference on Human Factors in Computing Systems, pp. 1–14 (2019)
27. Lakkaraju, S., Adebayo, J.: Neurips (2020) tutorial. In: Tutorial: (track2) explaining machine learning predictions: State-of-the-art, challenges, and opportunities (2020)
28. Mueller, S.T., Hoffman, R.R., Clancey, W., Emrey, A., Klein, G.: Explanation in human-AI systems: A literature meta-review, synopsis of key ideas and publications, and bibliography for explainable AI. arXiv preprint arXiv:1902.01876 (2019)
29. Rossi, F.: Building trust in artificial intelligence. J. Int. Aff. **72**(1), 127–134 (2018)
30. Ruan, Y., Alfantoukh, L., Durresi, A.: Exploring stock market using Twitter trust network. In: Advanced Information Networking and Applications (AINA), 2015 IEEE 29th International Conference on, pp. 428–433. IEEE (2015)
31. Ruan, Y., Alfantoukh, L., Fang, A., Durresi, A.: Exploring trust propagation behaviors in online communities. In: Network-Based Information Systems (NBiS), 2014 17th International Conference on, pp. 361–367. IEEE (2014)
32. Ruan, Y., Durresi, A.: Trust management for social networks. In: Proceedings of the 14th Annual Information Security Symposium, p. 24. CERIAS-Purdue University (2013)
33. Ruan, Y., Durresi, A.: A survey of trust management systems for online social communities-trust modeling, trust inference and attacks. Knowl. Based Syst. **106**, 150–163 (2016)
34. Ruan, Y., Durresi, A.: A trust management framework for cloud computing platforms. In: Advanced Information Networking and Applications (AINA), 2017 IEEE 31st International Conference on, pp. 1146–1153. IEEE (2017)
35. Ruan, Y., Durresi, A.: A trust management framework for clouds. Comput. Commun. **144**, 124–131 (2019)
36. Ruan, Y., Durresi, A., Alfantoukh, L.: Trust management framework for internet of things. In: Advanced Information Networking and Applications (AINA), 2016 IEEE 30th International Conference on, pp. 1013–1019. IEEE (2016)
37. Ruan, Y., Durresi, A., Alfantoukh, L.: Using Twitter trust network for stock market analysis. Knowl. Based Syst. **145**, 207–218 (2018)
38. Ruan, Y., Durresi, A., Uslu, S.: Trust assessment for internet of things in multi-access edge computing. In: 2018 IEEE 32nd International Conference on Advanced Information Networking and Applications (AINA), pp. 1155–1161. IEEE (2018)
39. Ruan, Y., Zhang, P., Alfantoukh, L., Durresi, A.: Measurement theory-based trust management framework for online social communities. ACM Trans. Internet Technol. (TOIT) **17**(2), 16 (2017)
40. Smith, C.J.: Designing trustworthy AI: A human-machine teaming framework to guide development. arXiv preprint arXiv:1910.03515 (2019)
41. Smuha, N.A.: The EU approach to ethics guidelines for trustworthy artificial intelligence. CRi-Comput. Law Rev. Int. **20**(4), 97–106 (2019)
42. Sutrop, M.: Should we trust artificial intelligence? TRAMES: J. Humanit. Soc. Sci. **23**(4), 499–522 (2019)
43. Uslu, S., Kaur, D., Rivera, S.J., Durresi, A., Babbar-Sebens, M.: Decision support system using trust planning among food-energy-water actors. In: International Conference on Advanced Information Networking and Applications, pp. 1169–1180. Springer, Cham (2019)

44. Uslu, S., Kaur, D., Rivera, S.J., Durresi, A., Babbar-Sebens, M.: Trust-based game-theoretical decision making for food-energy-water management. In: International Conference on Broadband and Wireless Computing, Communication and Applications, pp. 125–136. Springer, Cham (2019)

45. Uslu, S., Kaur, D., Rivera, S.J., Durresi, A., Babbar-Sebens, M.: Trust-based decision making for food-energy-water actors. In: International Conference on Advanced Information Networking and Applications, pp. 591–602. Springer, Cham (2020)

46. Uslu, S., Kaur, D., Rivera, S.J., Durresi, A., Babbar-Sebens, M., Tilt, J.H.: Control theoretical modeling of trust-based decision making in food-energy-water management. In: Conference on Complex, Intelligent, and Software Intensive Systems, pp. 97–107. Springer, Cham (2020)

47. Uslu, S., Kaur, D., Rivera, S.J., Durresi, A., Babbar-Sebens, M., Tilt, J.H.: A trustworthy human-machine framework for collective decision making in food-energy-water management: the role of trust sensitivity. Knowl. Based Syst. **213**, 106683 (2021)

48. Uslu, S., Ruan, Y., Durresi, A.: Trust-based decision support system for planning among food-energy-water actors. In: Conference on Complex, Intelligent, and Software Intensive Systems, pp. 440–451. Springer, Cham (2018)

49. Varshney, K.R.: Trustworthy machine learning and artificial intelligence. XRDS: Crossroads, ACM Mag. Stud. **25**(3), 26–29 (2019)

50. Wachter, S., Mittelstadt, B., Russell, C.: Counterfactual explanations without opening the black box: automated decisions and the GDPR. Harv. JL Tech. **31**, 841 (2017)

51. Wang, W., Siau, K.: Trusting artificial intelligence in healthcare (2018)

52. Zhang, P., Durresi, A.: Trust management framework for social networks. In: 2012 IEEE International Conference on Communications (ICC), pp. 1042–1047. IEEE (2012)

53. Zhang, P., Durresi, A., Barolli, L.: Survey of trust management on various networks. In: Complex, Intelligent and Software Intensive Systems (CISIS), 2011 International Conference on, pp. 219–226. IEEE (2011)

54. Zhang, P., Durresi, A., Ruan, Y., Durresi, M.: Trust based security mechanisms for social networks. In: Broadband, Wireless Computing, Communication and Applications (BWCCA), 2012 Seventh International Conference on, pp. 264–270. IEEE (2012)

55. Zhang, Y., Liao, Q.V., Bellamy, R.K.: Effect of confidence and explanation on accuracy and trust calibration in AI-assisted decision making. arXiv preprint arXiv:2001.02114 (2020)

Classifying Holes, Voids, Negative Objects and Nothing

Katrina Hooper$^{(\boxtimes)}$, Alex Ferworn, and Fatima Hussain

Ryerson University, Toronto, Canada
{katrina.hooper,aferworn}@ryerson.ca, fatima.hussain@ryerson.com

Abstract. In the field of Urban Search and Rescue (USAR), searching for people trapped in the rubble of a collapsed building largely assumes the presence of many objects that are not the trapped victims, but may provide clues as to the location of them. However, the lack of objects can be just as informative. Unlike positive objects (POs), *negative objects* extend into the surface of interest [1], and may contain people. These negative objects (NOs) are commonly known as "holes" or "voids".

In this paper we discuss the necessity for creating a lexicon and classifier for negative objects or holes. We review current efforts to detect and classify them, and finally begin building a vocabulary to effectively discuss negative objects. This paper aims to spark a discussion about negative object research and suggest a starting point for a novel research area.

Keywords: Negative objects · Holes · Voids · USAR

1 Introduction

1.1 What Is a Hole?

Machine vision research has largely focused on detecting and classifying physical objects [2]. Though, the lack of objects can be just as important as the presence of them [3]. An example of this is the assumption in robotic path planning, that the absence of an object means "safe space". A similar argument can be made for the operation of Unmanned Aerial Systems (UASs) where "up" is often assumed to be safe because "down" often has objects restricting travel, like the ground.

This logic of positive objects as a guide, breaks down when a robot must move through a space with no objects, i.e. negative objects. As NOs are clearly involved in questions of traversal and detection, we argue, it would be useful to have a lexicon describing them and their characteristics.

1.2 Types of Holes

The classic definitions of holes typically only emphasize absence, which can be misleading. Holes require additional characteristics to be well defined and thus

L. Barolli et al. (Eds.): AINA 2021, LNNS 225, pp. 220–232, 2021.
https://doi.org/10.1007/978-3-030-75100-5_20

distinguishable from one another. For example, the hole in a donut is very different from the hole in a cup, both can be filled by a PO, but the size, shape, surrounding material and functionality differ greatly. We include Table 1, providing definitions for a wide variety of holes discussed in various research works from Sect. 2, where the absence of something is a key component of the definition as interpreted by the researchers. We see that many NOs are identified by function or by a PO that can occupy the space of a NO. For example, "manhole" defines a hole that is designed to provide access for a human but uses the absence of a human as a kind of abstraction based on filled space (space the size of a "man") rather than the characteristics of empty space–for which there is no lexicon.

1.3 Importance of Detection and Classification in USAR

In today's society, the likelihood of a disaster is much higher than in the past due to the effects of urban development, climate change and other factors making the world more dangerous than it used to be [4,5]. In addition to natural calamity, it is known that heavily populated areas are at risk of man-made threats like industrial accidents [6] and terrorist attacks [7]. These threats make it more likely that urban structures may collapse and trap people inside. It is the role of Urban Search and Rescue (USAR) teams to respond to such events in a timely manner. The purpose of their response is to stabilize the new structures caused by the collapse of a building and locate, medically stabilize and extricate trapped people as quickly as possible because any delay in extrication reduces the probability that these people will survive [8]. One of the characteristics of collapsing buildings is that new openings are created throughout what will become the new structure of rubble. Usually, the fastest method of access to the people trapped inside rubble will be through new "holes". Therefore, we argue, it is important for first responders to be able to identify and classify the type of entry ways into the rubble to determine which holes are more likely to lead to someone trapped. The complexity of rubble piles is also an accessibility issue as rubble can be comprised of many chaotically placed (and potentially unstable) materials of random shape [9]. Being able to classify the structure of completely contained and semi-contained NOs can provide responders better insight into the rubble pile's structural integrity, in addition to helping responders and structural engineers devise a plan on how to access the internal spaces potentially containing victims.

As the use of robotic traversal and interaction on and within disaster rubble piles is an ongoing research pursuit to increase human safety [10], hole or NO classification can help in decision-making with respect to autonomous navigation tasks and with determining traversability. For USAR robots on the ground, detecting NOs (especially at long range) is challenging because many sensors are not in a position to detect the signs of a NO until the robot is very close to it [3]. Imaging can be distorted by external lighting making this task even harder [11]. A non-trivial problem is that NOs cannot be detected by sensors, since sensors sense the presence of things. Therefore the presence of NOs must be inferred from the lack of sensor data [2].

2   Related Work and Hole Specifications

In this section we will discuss methods researchers have used to identify broad and specific types of NOs. We concentrate on how authors approached their definitions of NOs to gain better insight into creating a lexicon for all NO types.

2.1   Potholes and Cracks

Eriksson et al. [12] develops an algorithm to differentiate between road imperfections using readings from GPS and accelerometer sensors. The authors aim to classify potholes, pothole clusters, manholes, railroad crossings as well as some POs. In this paper the authors define potholes as being areas of missing pavement from the road surface, significantly sunken manhole covers, as well as protruding manhole covers and other notable road surface irregularities [12], shown in Fig. 1. This definition of pothole, a type of NO, holds some contradictions: what is commonly known as a pothole does not include sunken-in manhole covers and especially not protrusions in the road surface, shown in the centre and right-most images in Fig. 1. It is within this definition that Eriksson et al. combine the search for a NO with a PO.

Fig. 1. These are examples of road surface anomalies that generated a *pothole* detection in the Pothole Patrol algorithm developed by [12]. The left picture is what Eriksson et al. call a typical small pothole, the centre picture is a sunken in manhole cover and the right picture is a bridge expansion joint.

Wang et al., [13], and Mednis et al., [14], propose computationally lightweight, real-time algorithms for detecting and marking the location of potholes using GPS and accelerometer sensors from a smartphone, by bettering the work of Mohan et al. [15]. [14] expanded upon the accelerometer approach from [15], it maintained the need for a restricted smartphone position and instead introduced the combination of 4 detection functions based on available accelerometer data to give an accurate final classification. The 4 functions are determined by if the reading surpasses a vertical threshold (Z-$THRESH$), a derivative threshold (Z-$DIFF$), has a standard deviation value threshold ($STDEV$-Z), and when 3 readings of G-force all reach around zero Gs (G-$ZERO$). [13] proposes an algorithm to reduce errors in all classifiers from [15], noting that Z-DIFF and STDEV-Z are time and frequency dependent, and that G-ZERO and Z-THRESH would be influenced by their respective peak measurements. The authors incorporate a time reading between when Z-DIFF and STDEV-Z functions returned positive readouts to ensure they weren't false positives, and combined the Z-THRESH and G-ZERO readings to also reduce false positives. In the above papers the definition of a pothole is based on sensor readings and not explicitly defined. At most, an example of a pothole is shown in Fig. 2 of the pothole [13] used as a case study. Comparing this to Eriksson et al., it is clear that at minimum, Wang et al. does not agree with the definition given in [12] and because of this non-uniform definition we can question what [15] and [14] were detecting and classifying.

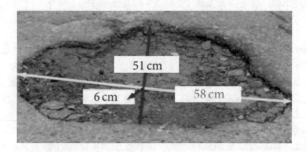

Fig. 2. This is the case study pothole used in Wang et al. 2015 [13]

Lin et al., [16], uses a non-linear support vector machine (SVM) algorithm on segmented gray-scale images to identify the "sink piece" of a pothole. The sink piece is the region along the wall of the pothole where the composition is much more granular than road top [16]. With the SVM sorting the image segments by granularity to find if a sink piece is present, the authors can classify if a pothole is in the image. The authors show the beginning of a clear definition of a pothole in saying that it must contain a sink piece, though authors do not restrict the size of a pothole and their algorithm could not classify a pothole that was filled with muddy water. With the interference of weather conditions, we see that a sink piece cannot be the only characteristic of a pothole. Since a pothole can exist in all weather conditions, and further more it is possible that other types of NOs may also contain a sink piece, such as the NO shown in Fig. 3. Here, we see the need to refine the definition of a pothole from a visual perspective.

Fig. 3. This shows another type of NO that contains a sink piece but should not be classified as a pothole. Image taken from The Denver Post [17].

Yu et al. propose a method for autonomous pothole detection using a ground robot to collect the data. [18] use a strong laser to project a line perpendicular to the direction of travel along the ground and a camera to capture the image of the laser line on the target surface. The authors create a template of the ground (potentially containing a NO) based on the deformation of the laser beam. The authors then create a distress image based on depths calculated from the deformation readings. Using parameters of depth distress, horizontal distress, vertical distress and size, the authors classify potholes, vertical cracks and transversal cracks, as well as give a rating to the severity of the pothole after it has been classified. Since their algorithm is a threshold based algorithm, there is no upper limit as to how large a pothole can be. As for the scale of severity, if we were to compare this analysis of a pothole to one by [14], we would expect discrepancies in the classification of a pothole corresponding to a pothole that is small in size but has high severity by the algorithm in [18]. We expect these discrepancies as running over a severe pothole will result in greater accelerometer data and thus surpassing set thresholds and classifying it as a larger pothole than it is.

Sy et al. use image processing to identify cracks in the pavement in real-time. The authors define two criteria that cracks must meet, that the crack must be photometrically darker than its surroundings as it is harder for light to reach its inner portions. The authors also restrict cracks to being continuous regions in the image [19]. We notice that among many algorithms using computer vision as a means of data analysis, that photometric brightness is a widely agreed upon quality of NOs.

2.2   USAR

Both [20] and [21], implement algorithms, proposed by [11], to analyze image pairs with data types RGB and depth to determine if the full colour-depth image contains a hole large enough to support a human, this is known as an *access hole*.

Their algorithm checks if any part of the image surpasses all predefined thresholds, representing the access hole qualities of image darkness, depth disparity and aspect ratio. In addition, authors define an access hole as having clearly defined edges on an RGB image, a noticeable depth change from its surroundings and being poorly illuminated.

Hu et al. focus on isolating inner-structure voids in images produced by ground penetrating radar (GPR) data, radargrams, on disaster rubble [22]. The authors isolate that Lean-To Voids must have a triangular shape with clearly marked boundaries. The authors explain that their algorithm can also classify V-Shape and A-Frame Voids because they are similar in shape, since their algorithm focuses more on the internal structure of the void as opposed to the components making up the void itself. The authors also explain that voids with thickness' much smaller than the depth scanned to or is around the same size as the material surrounding the void becomes essentially undetectable due to noise. An example of these undetectable voids are Pancake Voids, which are pockets created by layers of heterogeneous material commonly found in rubble piles.

2.3   Robotic Navigation and Path Planning

[23] and [24] work on a broad idea of NOs, in which authors aim to identify edges of any area of a ground-level path that is below path level. Both Coughlan and Murarka use stereo cameras to create a depth disparity map to represent the surrounding ground space around the robot. Coughlan uses discontinuities in the disparity map to identify edges of NOs. The authors absorb everything beyond the edges as a NO and deem the area as *unsafe for travel* [23], whereas Murarka uses labeling within the disparity map to first classify all sections of the surrounding area. The authors label sections based on if the surrounding ground is level with its current position, if the ground is inclined, if there is a drop-off expected, if there are areas that are "non-ground"-such as walls- or unknown if the [24] doesn't have enough data to classify within the other labels. Unknown areas could be due to occlusions caused by NOs [24].

[25] presents a reversed outlook on the NO problem, where instead of disregarding incomplete information caused by a NO or simply labeling the area as unsafe as a whole, the authors focus on trying to fill in the missing data. Using a laser range scanner the authors produce a 3-D map representing the robot's surroundings. Using ray tracing to label which areas of the map are considered to be occluded and those laying between the sensor and the occluded area are labeled visible. Then, Heckman et al. create a 2-D map in the same shape as the ground plane of the 3-D map used for marking where potential NOs lie. To find NOs the authors search each 3-D column and mark if a transition between visible and occluded occurs, if it is entirely occluded or if it is the end of the data structure. Finally, based on the type of occlusion, the authors can infer additional data and determine if there is a NO present. While this algorithm does not categorize detected NOs, the authors bring to light the behaviour of NOs from the perspective of unmanned ground vehicles (UGVs).

[2,3,25,26] use laser range sensors to identify both potential NOs and confirmed NOs. All authors approach the presence of a NO as being a gap in the point cloud data mapping the UGV surroundings. The work to decipher the difference between sparse data, caused by non-uniform point spacing, and missing data caused by occlusions that meet the criteria shown in [25]. Larson et al. classify potential NOs based on how the angles of the points surrounding the point cloud gap (PCG) relate to each other, recognizing that an occlusion will skew the way the points are located on the ground. [3] uses the angle difference to attempt to determine the edges of the NO. [26] and [2] focus more on finding patterns in point density surrounding the PCG as opposed to vertical depth changes.

[27] uses stereo cameras and computer vision tactics to classify the areas of a UGV's surroundings. Hu et al. establish characteristics of photometric appearance, similar to what is determined for access holes from [11], and further the condition to account for a brightness change based on distance and angle of image approach. The authors also adjust the theory of occlusion from [25] to work for stereo cameras. Like Hu et al., [28], uses stereo cameras for detecting and classifying NOs, though Hergehelegiu et al. use the paired images to create a 3-D full colour disparity map of the surroundings. Herghelegiu et al. extend the conditions defined in [27], however the condition of photometric brightness is used but only as an aid to the assumptions that the NOs will have disparity values lower than the ground plane or will not produce disparity values at all. [28] furthers their classification to identifying if the NO has a gentler extension into the PO of the ground or if it more abruptly cuts into ground thereby occluding data points and causing data gaps.

3   Definitions

We notice the definition of *pothole* has many different interpretations and, in some cases, even includes POs as part of its definition [12]. We also notice that the definition of a pothole was implied but not explicitly defined in [14–16]. This can cause contradictions as well as inaccuracies when comparing and extending research.

Another example of this is shown in the definition of *gaps*, where it can be taken as a type of NO [15,25] as well as a term to represent the lack of points in a point cloud caused by a NO [11]. Since definitions are the backbone of object recognition, we must work to create terms with clear and unique definitions to standardize research and make accurate comparisons in algorithm performance.

Table 1. This table contains terms representing sub-classifications of NOs and their definitions as authors are used in research papers discussed in Sect. 2

Term	Definitions	Term	Definitions
Pothole	• [12]: Missing chunks of pavement, severely sunk in or protruding manhole covers, other significant road surface anomalies • [13]: Not explicitly defined, example shown in Fig. 2 • [14]: Not explicitly defined, further separated into *large potholes* and *small potholes*, neither is explicitly defined • [15]: Not explicitly defined • [16]: Contains a sink piece • [18]: If both depth, horizontal and vertical distress measures have a large value, the distress, which indicates the deviation is significant in both directions	Access Hole	• [11,20,21]: – defined through its intended use – sufficiently large to permit human entry – potential for access into a collapsed structure, i.e., its function – a hole must be deeper in the interior than the surrounding terrain – clearly marked edges – strong change in depth – strong change in pixel darkness – a hole must be large enough to support entry by a searcher, such as a human, dog, or robot – has three attributes that characterize a hole and allows us to perform access hole detection: (i) depth disparity, (ii) hole size, and (iii) photometric brightness
Pothole Cluster	• [12]: a series of a minimum of k pothole detections with high enough confidence in a localized area • [14]: Not explicitly defined	Manhole	• [12]: Manhole covers and other equipment in the road that are nearly flush with the road surface. Moderate cracking, sinking or bulging
Crack	• [19] – region that is darker than its surrounding – is a continuous region	Drain Pits	• [14]: Not explicitly defined
Transversal Crack	• [16]: Not explicitly defined • [18]: The vertical distress measure should have a low value, which indicates that there is little deviation between any of the columns for an image • [19]: Not explicitly defined	Longitudinal Crack	• [16]: Not explicitly defined, sample image provided in paper
Alligator Crack	• [16]: Not explicitly defined, sample image provided in paper • [19]: Not explicitly defined	Gaps	• [14]: These objects are located on the main multi-lane street where the total road smoothness is better than average and the transitions between several road segments have only minor impact on the suspension of the vehicles • [27]: data missing that is not caused by PO occlusion • [2]: Not explicitly defined • [3]: An absence of data
Pancake Void	• [22]: – are the most complicated void type because of the complex geometry – are often very small and covered by several layers of heterogeneous rubble	A-Frame/Lean-To/V-shape Void	• [22]: Have well defined boundaries
Hazardous negative obstacles	• [28] Areas below ground that emerge due to natural or human factors, e.g., holes in the ground due to cracks in the pavement/asphalt, missing sewer cap	Man-made negative obstacles	• [28]: Areas below ground purposely existing in the environment, e.g., edge of a railway platform, stairs down

4   Framework for Moving Forward

What we hope to do in this paper is to create a framework for further developing sub-classifications of NOs with clearly defined boundaries between them. Creating a lexicon for NOs will provide a method to have an informed and widely understood discussion about NOs. In many fields such as USAR, mining and autonomous path planning, discussions about NOs can be of great value.

In USAR, NOs house the areas in which trapped victims can be located and they can also provide the location for potential instabilities. When a NO is completely contained within collapsed rubble, it leaves the opportunity for a PO to occupy the space. *Lean-To Voids*, a sub-classification of NOs, shown in Fig. 4, is an example of a type of NO contained in a rubble pile. We notice that the walls of the void precariously balanced against each other and when under the weight and stress of building rubble and debris, it can cause this void to collapse in on itself. Having an informed way to speak about the void and classify it allows responders to quickly assess the stability of the void and how much the collapse of the void can impact the rest of the rubble pile, which a vital part of victim rescue and public safety precautions. NOs also provide access into the rubble to get responders closer to voids that can contain people [11]. Making the distinction between shallow NOs that rest on the surface of the rubble and NOs that are large enough to support a responder and extend into the rubble to provide a means of access to voids contained within the rubble is a crucial part of the initial analysis of a rubble pile. Having a vocabulary to discuss aspects of NOs and efficiently and effectively communicate features of targeted NOs may greatly increase response team productivity.

Essentially, mining is the construction of NOs to safely contain humans and withstand vibrations caused by heavy machinery, explosives and working within these NOs to retrieve resources. Similar to USAR, understanding weak points in the shape of the constructed NOs can help increase miner safety. Furthermore, having the ability to effectively communicate about mine structure and necessary updates to modify mine configuration (potentially transitioning to a different type of NO) would increase the efficiency and safety of mine construction and maintenance.

Autonomous path planning for robotics extends across all versions of unmanned vehicles. Frequently in path planning, any ground with a lower depth than the current path is labeled unsafe for travel. It would be beneficial for NOs to be classified by features, as some may be traversable. For autonomous driving, the ability to classify holes in the road (called potholes) and sunken manholes could allow cities to track and record the locations that require road maintenance. Further classification of potholes could allow for priority rankings based on size[1], depth, impact on car, and frequency of being driven over, to increase the efficiency of road repairs to maintain safe roads. NOs can also provide useful information on choosing paths, instead of focusing on them as objects to avoid, they can be used as guides towards targeted locations.

---

[1] **Size** - area on the surface plane the NO rests in.

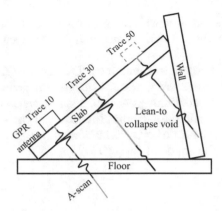

Fig. 4. Example of the structure of a Lean-To Void. Taken from [22]

In [28], the authors work to develop headgear to aid visually impaired persons with walking safety. We can suggest an example of a visually impaired person looking for a cross-walk. On either side of the slope that connects a sidewalk to a road is a type of NO with a strong depth change, commonly known as a curb.

5   Challenges

NO research is a challenging field as it is frequently based on the absence of data or the inability to obtain the necessary data. We see in [11,20,21] how obtaining the data and maintaining appearance constraints poses to be a challenge as external lighting and image point of view skew the assumption that access holes are poorly illuminated compared to its surroundings. Furthermore, in order to accurately determine if the access hole leads into the rubble you would need to be able to "trace the path" that the NO takes into the rubble and ensure there are no blockages restricting further access and finally you need to determine a threshold for how deep into the rubble an access hole must stretch in order to be considered an access hole by the responders who created the term.

As discussed in Sect. 3, without having clear categories and definitions of NOs, research centred around specific types of NOs becomes ambiguous. It is necessary to explicitly define the criteria of the NO that is being studied, especially when using a direct comparison between different algorithms as a method of verification. This is shown in [26], where the authors compare their method of detecting NOs to several other algorithms including [2] and [3]. Zhong et al. propose an algorithm to detect NOs with respect to a small autonomous SUV while Sinha et al. work on detecting NOs with respect to a small UGV. NOs detected by [2], are the subset of NOs that are considered hazardous to the UGV. This means that some of the NOs detected may be considered NOs in the algorithm by Zhong et al., however due to the vehicle size difference, many NOs detected by Sinha et al. would not be detected by Zhong et al. Furthermore, the vehicle's means of travel differs as one is a 4-wheeled UGV and the other is a tracked

UGV again affecting what is considered a NO to an algorithm. This stresses the need for defining criteria for NOs so researchers can accurately evaluate their algorithms.

Since sensors are limited to detecting POs and are frequently limited by occlusion to collect further data, we are limited on gaining informative data about NOs. The idea of combining sensors, such as those that are and are not limited by occlusion to gain data around the external surrounding of the NO as well as the internal structure of NOs. This idea offers to give more data about the surface bounding the shape of the NO.

6   Conclusion

Developing a lexicon for NOs is the first step towards understanding and appropriately classifying them. NOs make up all of the space that is not taken up by POs and in many fields that space is much more valuable. In all variations of path planning it is labeled safe, unless it impacts traversability and in USAR it is the location of instabilities in a rubble pile and the space that can enclose trapped human beings.

Further research should include creating clear definitions for classes of NOs. Classes could be determined by the types of POs that could fill the NO or by way of defining new classes unique to NOs. Furthermore, a vocabulary of terms to allow for effective discussions about NOs can be explored.

This research can be extended into the fields of USAR to classify NOs contained in or on the surface of rubble piles and evaluating stability and weak points in the POs surrounding the NO. It can also be extended into the field of robotics for autonomous piloting of UASs and UGVs to further classify surroundings to make better decisions regarding path planning. Finally, it can be used to better communities by classifying road imperfections and prioritizing maintenance to ensure safety and reduce the risk of emergency situations.

Glossary

Base - the region of the NO located the deepest inside of the NO
Depth - the average length of the wall creating the NO, measured perpendicular to the surface the NO resides in
Hole - there exists at dimensionally most a 2-dimensional cross section such that the NO is entirely surrounded
Negative Object (NO) - An object that can be occupied by a positive object
Negative Obstacle - A NO that cannot be traversed
Sink Piece - The region along the wall of the pothole where the composition is much more granular than road top
Tunnel - A NO that extends entirely through a positive object (does not have a base)
Wall - the surface of the NO shared with its bounding positive object
Void - NO completely contained inside an positive object

References

1. Matthies, L. et al.: Obstacle detection for unmanned ground vehicles: a progress report. In: Robotics Research, pp. 475–486 (1996). https://doi.org/10.1007/978-1-4471-1021-7_52
2. Sinha, A., Papadakis, P.: Mind the gap: detection and traversability analysis of terrain gaps using LIDAR for safe robot navigation. Robotica **31**(7), 1085–1101 (2013). https://doi.org/10.1017/S0263574713000349. ISSN 02635747
3. Larson, J., Trivedi, M.: Lidar based off-road negative obstacle detection and analysis. In: IEEE Conference on Intelligent Transportation Systems, Proceedings, ITSC, pp. 192–197 (2011). https://doi.org/10.1109/ITSC.2011.6083105
4. Anderson, J., Bausch, C.: Climate change and natural disasters: scientific evidence of a possible relation between recent natural disasters and climate change. In: Policy Department Economic and Scientific Policy 2 (2006)
5. Banholzer, S., Kossin, J., Donner, S.: The impact of climate change on natural disasters. In: Reducing Disaster: Early Warning Systems for Cimate Change, pp. 21–49. Springer, Dordrecht (2014)
6. Joseph, G., Kaszniak, M., Long, L.: Lessons after Bhopal: CSB a catalyst for change. J. Loss Prev. Process Ind. **18**(4–6), 537–548 (2005). https://doi.org/10.1016/j.jlp.2005.07.009. ISSN 09504230
7. Perry, W.L., et al.: Predicting Suicide Attacks: Integrating Spatial, Temporal, and Social Features of Terrorist Attack Targets. Rand Corporation, Santa Monica (2013)
8. Edwards, F.L., Steinhäusler, F.: NATO and Terrorism: On Scene: New Challenges for First Responders and Civil Protection. Springer Science & Business Media, Berlin (2007)
9. Ferworn, A., et al.: Interacting with a virtual destroyed environment constructed from real disaster data. In: 12th IEEE International Symposium on Safety, Security and Rescue Robotics, SSRR 2014 - Symposium Proceedings, pp. 24–27 (2014). https://doi.org/10.1109/SSRR.2014.7017654
10. Krotkov, E., et al.: The DARPA robotics challenge finals: results and perspectives. J. Field Rob. **34**(2), 229–240 (2017)
11. Kong, C., et al.: Toward the automatic detection of access holes in disaster rubble. In: IEEE International Symposium on Safety, Security, and Rescue Robotics, SSRR 2013 (2013). https://doi.org/10.1109/SSRR.2013.6719364
12. Eriksson, J., et al.: The pothole patrol : using a mobile sensor network for road surface monitoring. In: Proceedings of the 6th International Conference on Mobile Systems, Applications, and Services, pp. 29–39 (2008)
13. Wang, H.W., et al.: A real-time pothole detection approach for intelligent transportation system. Mathematical Problems in Engineering 2015 (2015). https://doi.org/10.1155/2015/869627. ISSN 15635147
14. Mednis, A., et al.: Real time pothole detection using android smartphones with accelerometers. In: 2011 International Conference on Distributed Computing in Sensor Systems and Workshops, DCOSS'11, pp. 1–6 (2011). https://doi.org/10.1109/DCOSS.2011.5982206
15. Mohan, P., Padmanabhan, V.N., Ramjee, R.: Nericell: rich monitoring of road and traffic conditions using mobile smartphones. In: Proceedings of the 6th ACM Conference on Embedded Network Sensor Systems, pp. 323–336 (2008). https://doi.org/10.1145/1460412.1460450

16. Lin, J., Liu, Y.: Potholes detection based on SVM in the pavement distress image. In: Proceedings - 9th International Symposium on Distributed Computing and Applications to Business, Engineering and Science, DCABES 2010, pp. 544–547. IEEE (2010). https://doi.org/10.1109/DCABES.2010.115. ISBN 9780769541105

17. Parker, R.: Xcel refilling big holes in Mayfair streets, but may need to dig more. In: The Denver Post (2012)

18. Yu, X., Salari, E.: Pavement pothole detection and severity measurement using laser imaging. In: 2011 IEEE International Conference on Electro/Information Technology, pp. 1–5. IEEE (2011). https://doi.org/10.1109/EIT.2011.5978573

19. Sy, N.T., et al.: Detection of defects in road surface by a vision system. In: Proceedings of the Mediterranean Electrotechnical Conference - MELECON, pp. 847–851 (2008). https://doi.org/10.1109/MELCON.2008.4618541

20. Kong, C., et al.: What is a hole? discovering access holes in disaster rubble with functional and photometric attributes. J. Field Rob. **33**(6), 825–836 (2015). http://ncart.scs.ryerson.ca/wp-content/uploads/2015/10/What-is-a-Hole-Discovering-Access-Holes-in-Disaster-Rubblewith-Functional-and-Photometric-Attributes.pdf

21. Waismark, B., Ferwom, A., Tran, J.: Enhancing autonomous access hole detection. In: IHTC 2017 - IEEE Canada International Humanitarian Technology Conference 2017, pp. 51–54 (2017). https://doi.org/10.1109/IHTC.2017.8058198. ISBN 9781509062645

22. Hu, D., et al.: Detecting, locating, and characterizing voids in disaster rubble for search and rescue. Adv. Eng. Inf. **42**(March) (2019). https://doi.org/10.1016/j.aei.2019.100974. ISSN 14740346

23. Coughlan, J., Shen, H.: Terrain analysis for blind wheelchair users: computer vision algorithms for finding curbs and other negative obstacles. In: CVHI (2007)

24. Murarka, A., Kuipers, B.: A stereo vision based mapping algorithm for detecting inclines, drop-offs, and obstacles for safe local navigation. In: 2009 IEEE/RSJ International Conference on Intelligent Robots and Systems, IROS 2009, pp. 1646–1653 (2009). https://doi.org/10.1109/IROS.2009.5354253

25. Heckman, N., et al.: Potential negative obstacle detection by occlusion labeling. In: IEEE International Conference on Intelligent Robots and Systems, pp. 2168–2173 (2007). https://doi.org/10.1109/IROS.2007.4398970

26. Zhong, Z., et al.: Robust negative obstacle detection in off-road environments using multiple LiDARs. In: 2020 6th International Conference on Control, Automation and Robotics, ICCAR 2020, pp. 700–705 (2020). https://doi.org/10.1109/ICCAR49639.2020.9108058

27. Hu, T., et al.: Negative obstacle detection from image sequences. In: Third International Conference on Digital Image Processing (ICDIP 2011), vol. 8009, p. 80090Y, July 2011 (2011). https://doi.org/10.1117/12.896288. ISSN 0277786X

28. Herghelegiu, P., Burlacu, A., Caraiman, S.: Negative obstacle detection for wearable assistive devices for visually impaired. In: 2017 21st International Conference on System Theory, Control and Computing, ICSTCC 2017, pp. 564–570 (2017). https://doi.org/10.1109/ICSTCC.2017.8107095

Empirical Study of Software Adoption Process in the Bitcoin Network

Mitsuyoshi Imamura[1(✉)] and Kazumasa Omote[1,2]

[1] University of Tsukuba, Tennodai 1–1–1, Tsukuba 305-8573, Japan
s1730148@s.tsukuba.ac.jp, omote@risk.tsukuba.ac.jp
[2] National Institute of Information and Communications Technology,
4-2-1 Nukui-Kitamachi, Koganei, Tokyo 184-8795, Japan

Abstract. Managing the flood of devices is a modern challenge in distributed networks as it is becoming increasingly difficult to manage the software updates needed to protect the devices from malicious attacks. Bitcoin, which is a blockchain-based system that relies on approximately 10,000 volunteer nodes to manage its ledger, faces similar problems. A centralized system mitigates this problem by enforcing update policies through one authority. In contrast, a distributed system with different update policies for different users could enter a worst-case scenario of sparse deployment due to forgotten updates or refusal to apply them. Nevertheless, we find that Bitcoin peer-software updates have been surprisingly successful in terms of evading this scenario. Therefore, we investigate the modeling of the impact caused by software version history and the remaining challenges, such as the inability to work well with update cycles. Consequently, we report that previous vulnerable software versions are still being used, and the share growth decelerates as the number of versions increases. Moreover, we clarify that the probability of recovering the adoption rate to previous levels is small, while surviving older versions increases the vulnerability of the network.

1 Introduction

The research of Satoshi Nakamoto [7] is widely accepted as the starting point of the academic movement of blockchains. Moreover, Satoshi Nakamoto is also considered an open source software (OSS) developer, as he first proposed the concept of Bitcoin and uploaded a prototype version to SOURCEFORGE[1] on January 9, 2009[2]. Since then, Bitcoin has been developed with contributions from many anonymous volunteers and, as of August 2020, it has been released as 20 major versions as Bitcoin Core. In other words, Nakamoto was the initial developer of the well-known Bitcoin Core, software that has been continuously maintained for over 10 years and has grown to become one of the most used programs in the blockchain network. According to information from BitcoinNodes[3], approximately 10,000 nodes currently contribute to the maintenance of

---

[1] https://sourceforge.net/.
[2] https://www.mail-archive.com/cryptography@metzdowd.com/msg10142.html.
[3] https://bitnodes.io/.

L. Barolli et al. (Eds.): AINA 2021, LNNS 225, pp. 233–244, 2021.
https://doi.org/10.1007/978-3-030-75100-5_21

the blockchain network, as of the author's writing. Although the ledger in a blockchain network is a critical component that all users must reference in a trustless network, there is no economic incentive from sharing it to the network. This suggests that there are still many volunteers in the network, despite the growing storage costs of maintaining the ledger.

Hence, the interesting point of the Bitcoin network is that its success for over 10 years and the maintenance of such a widely distributed network with no economic incentives is being completed by volunteers.

Herein, we study the management entity of volunteers in the network, assuming that the latest released software must be used to protect devices from attackers and keep networks secure. It becomes increasingly challenging to support software updates as the number of managed devices in a distributed network increases [8]. Furthermore, the update policy differs from one volunteer to another in a blockchain network because of the decentralized management structures, whereas centralized systems have unified update policies. Unlike storage costs, which are considered at launch, maintenance costs reduce the motivation of volunteers who have no economic incentive because they continuously check the latest release status and update the installed software. In this context, it is unclear whether the release and widespread adoption of software are compatible in the network.

Therefore, this research focuses on volunteers who maintain the ledger and investigates the trends in software managed by volunteers in the Bitcoin network. These trends in software update cycles and adoption rates provide insight into volunteer management. The contributions of this work are summarized in the following points:

- Empirical trends are identified by analyzing the software update cycles in Bitcoin networks.
- Potential challenges caused by the usage of older versions are determined based on the software adoption rate in the network.
- The software adoption rates in networks are modelled using the Gompertz curve model.

The remainder of this paper is organized as follows. Section 2 introduces the evolution of the major software of the Bitcoin network, which is the main subject of this research. Section 3 presents previous works that report software trends in the network and case studies related to releases in general software. In Sect. 4, the adoption rate of software in the network is analyzed, while in Sect. 5, the issues and impact of the adoption rate in the network are discussed on the basis of the performed analysis. Finally, Sect. 6 concludes this paper.

2   Background

In this section, we review the evolution of the most major Bitcoin peer software, Bitcoin Core. Herein, we introduce four generations of software based on its historical background: 1st generation (from 0.1 to 0.3), 2nd generation (from 0.4 to 0.8), 3rd generation (from 0.9 to 0.15), and 4th generation (from 0.16 to 0.20).

2.1 Version History and Implementation

2.1.1 1st Generation (Versions 0.1–0.3)

The first major version of Bitcoin Core, version 0.1, was released by Satoshi Nakamoto on January 9, 2009. The implementation at that time mainly provided initial functionality and could support only Windows; Linux has become available since version 0.2. As a classic countermeasure to the 51% attack on the blockchain, 11111, 33333, and 68555 block hash values were hard-coded as a checkpoint in version 0.3. Bitcoin Core versions 0.1 to 0.3.20 were released before the community management was moved to bitcoin.org. From version 0.3.21 onwards, the community has moved from SOURCEFORGE to GitHub[4], which is currently used for version management.

2.1.2 2nd Generation (Versions 0.4–0.8)

Versions 0.4 to 0.8 started branding as Bitcoin-Qt from the Qt-based user interface, which was implemented in version 0.5. Since initial implementation, the UI/UX has been improved, growing closer to the current prototype. In this generation, wallet encryption and multi-signature transactions on a preliminary began being supported. Additionally, as it relates to network interfaces, support for the Tor hidden service was introduced. A specification change to switch the database from Berkeley DB to Level DB was also implemented for storing transactions and block indexes.

2.1.3 3rd Generation (Version 0.9–0.15)

Versions 0.9 to 0.15 started rebranding as Bitcoin Core and implemented usability enhancements, including the introduction of automatic estimation to improve hard-coded transaction fees. Additionally, the replace-by-fee feature, which allows sent transactions to change to new transactions for an additional fee before storage in the block, was implemented. Performance was improved by the block file pruning mode, which automatically discards old block data and transactions instead of keeping the complete block data. However, storage costs were increased by changing the chain state DB management method used for unspent transaction (UTXO) tracing from per transaction to per UTXO to speed up the validation and re-indexing of the blockchain in the initial block download (IBD). Other implementations included parallel block downloading after block header synchronization and performance improvements to speed up verification by changing OpenSSL to libsecp256k1 and retiring mempool memory. Finally, Segregated Witness (SegWitt) was implemented in the test net.

2.1.4 4th Generation (Version 0.16–0.20)

Since the release of 0.16, SegWit has been introduced to improve scalability [10], and the network block size limit has been gradually relaxed. Segwit fully supports

---

[4] https://github.com/bitcoin/bitcoin.

more transactions into blocks on the main network packs by removing the signature part from the transaction ID and signing it in a transaction-independent area. This change was also associated with the failure to call mining protocols that do not define SegWit rules, which were implemented to encourage SegWit compatibility.

Nevertheless, the new features could be removed in some cases because of operational issues. Based on the BPI61 proposal, reject messages were implemented in version 0.9 to improve interoperability between peers. Reject messages were sent directly to peers to inform them a block or transaction was rejected or give feedback on errors in the protocol. However, this feature was removed because it was mainly used for debugging and resulted in security vulnerabilities from DoS attacks targeting memory pools, and bandwidth issues.

2.2 Handling Vulnerabilities

While Sect. 2.1 describes the implementation and modification of the main features, it should be noted that fixes for known security issues were also introduced. As the software was being updated, developers encountered not only dependency library issues but also critical bugs in the software, as detailed in the unique common vulnerabilities and exposures (CVE) catalog of Bitcoin Core.

Vulnerabilities related to OpenSSL, such as CVE-2014-0076, CVE-2014-0160, and CVE-2014-0224, were announced in minor updates of version 0.9, whereas vulnerabilities related to the OS version, such as CVE-2017-8798 and CVE-2017-1000494, were announced in version 0.19.0.1; these are not directly related because they are not software-specific vulnerabilities.

Direct vulnerabilities include CVE-2011-4447, CVE-2012-3789, and CVE-2013-4165. The former reported an issue with a private key leak in a wallet encryption function, which was fixed in version 0.5. CVE-2012-3789 was related to denial of service (DoS) attacks, which caused Bitcoin processes to become unresponsive and was fixed in version 0.6.3. CVE-2013-4165 also reported problems with the Bloom filter processing code, which was fixed in version 0.8.4.

Further, CVE-2018-17144 was a particularly serious issue that not only caused DoS attacks on full nodes stored by the ledger but also triggered coin inflation issues. This issue was announced by the community soon after it was found and affected versions 0.14.x, 0.15.x, and 0.16.x before versions 0.14.3, 0.15.2, and 0.16.3, respectively.

Besides the issues reported in the CVE, the recently released version 20 also embedded an experimental implementation called Asmap to counter the large-scale EREBUS attack, which was reported in previous research [9] as a potential disruption to the Bitcoin network.

3   Related Work

In this section, we briefly review related works that perform software-based Bitcoin network analysis and report general software release trends to understand the empirical trends of software releases in the Bitcoin network.

3.1 Network Analysis

Although many studies on Bitcoin network analysis showcase network information, such as IP addresses, regions, and autonomous systems (AS), few related works report software information.

Pappalardo et al. [5] included information on the software version used in an investigation of the timing when transactions broadcasted to the Bitcoin network are stored in blocks. Particularly, they reported on the type of software and number of nodes in the process of analyzing the dynamics of transactions and blocks in the network. Their observations from May 4 to May 11, 2016 demonstrate that users mostly used Bitcoin Core, Bitcoin Classic, Bitcoin XT, BTCC, and BitcoinUnlimited. Regarding Bitcoin Core, versions 12 and 8 were the most current and oldest versions available, respectively.

Park et al. [6] analyzed the differences between the recent Bitcoin network structure in 2018 and that investigated in previous studies from 2013 to 2016 [5]. Their results highlighted the different characteristics between the observed network and the various previously used software, taking into account that Bitcoin Core version 15 had the majority share in the network. This study considered Bitcoin Core and Bitcoin ABC, BitcoinUnlimited, BUCash, Bitcoin Classic, Bitcoin XT, and BTCC; however, users used only a few of the abovementioned networks. The oldest version of Bitcoin Core, version 8, was still in use by a few users.

Neudecker [4] reported on network characterization based on measurements of the Bitcoin network between 2015 and 2018 and showed the long-term software distribution in the Bitcoin network in the long term. The measurement results suggested the dynamism of the distribution of Bitcoin Core and Bitcoin ABC, as users preferred the most common software during the observation period from April, 2016 to April, 2018.

Nevertheless, this study was limited to researching the characteristics of the Bitcoin network, which is insufficient for demonstrating any correlation between software release and adoption.

3.2 Software Releases

Bosu et al. [1] collected questionnaires from blockchain-based software developers to understand the development motivations, challenges, and needs and investigated the differences from other OSS. Their research points regarding the difficulty of post-release updates in non-blockchain software can be easily upgraded for functionality. Contrarily, blockchain software needs to be backward compatible to allow the validation of previous transactions and anticipates worldwide nodes for upgrading functionality. In the community level, the dissemination of software releases to users was found to be difficult.

Dong et al. [2] studied the impact of release timing on user behavior. Particularly, they analyzed the impact between the initial software release rate, the update rate at which new versions and patches are released, and the user's interest to download the OSS. Their results indicated that, although a high initial release or update speed is a factor in increasing user interest, too high of an initial release or update speed may have the opposite effect.

The abovementioned studies suggest that promoting software in general can be challenging and that the blockchain community has low expectations. The analysis in Sect. 4 explores various points, such as the release timing.

4   Empirical Analysis

As mentioned in Sect. 3, the reality of software management in the network is unclear because it is only considered in part of the network analysis. To address this issue, in this section, we focus on the usage of Bitcoin software versions in the network.

4.1   Data

First, the data used in this study include the containing node information of the Bitcoin network collected by the Bitcoin Network Monitoring and Research project, which is run by the DSN research group[5]. The data are obtained using a peering protocol feature and include the operating system version, transaction arrival speed, geographic location, IP version, and autonomous system number (ASN) of the nodes on the Bitcoin network. The software version data cover the period from January 29, 2016 to January 11, 2021; the analysis uses data from this observation period, unless otherwise noted. Corrupted data or data with obvious gaps owing to maintenance actions or system failures, such as network outages and system restarts, were excluded from the analysis. Our results also face the same potential limitations as those of previous studies, such as the inability to reach nodes hiding on top of firewalls and network address translation (NAT), which is a common problem.

Next, the software version to be analyzed in this study was determined by the identifier defined by the string in the payload of the version message in the Bitcoin protocol. However, there are users in the network that use software with modified identifiers, so the name identification process is necessary. The preprocessing of the analysis conducted in this study is as follows:

- If the software does not have the same name as the first work, it is considered a unique software version.
- Clustering is performed by the longest match from the top.
- A version that has been modified to a non-released version is considered different software, even if the version control identifiers match (e.g., Satoshi:10.0.1).

Of course, this match of identifiers and protocol versions does not guarantee the assumed software name. Users presumably do not modify the software version unless there are special circumstances.

---

[5] https://dsn.tm.kit.edu/bitcoin/index.html.

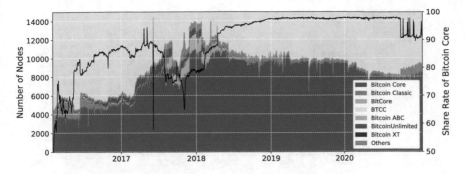

Fig. 1. Software share in the Bitcoin network

4.2 Software Distribution in Networks

To investigate the usage of software version updates, it is important to know the distribution of major software and consider its impact on the network. Therefore, the position of the Bitcoin Core in the network in terms of usage should be clarified before software version analysis is performed.

Figure 1 shows the distribution of software used by the nodes in the network during the observation period. The colored areas corresponding to different software represent the number of nodes, and the black line shows the share of Bitcoin Core in the network. In addition to Bitcoin Core, we checked the distribution of Bitcoin Classic, BitCore, BTCC, Bitcoin ABC, BitcoinUnlimited, and Bitcoin XT, which were introduced in previous research [5,6]. All other software are classified as "Other"; Specifically, this category included BitcoinNodes, dsn.tm.kit.edu [4], and the Falcon Relay Network [3]. A few emerging software programs, such as btcd(btcwire)[6], have been updated recently, whereas others, such as bitprim[7], have not been updated for several years.

Confirming the dynamics of the nodes on which the software depends, we found that the overall number of nodes in the network nearly tripled between 2016 and 2018; in fact, it reached a peak in 2018 and held steady until 2020, before declining slowly. We also confirmed the Bitcoin Core had the dominant share over the observation period. Bitcoin ABC, which launched in the Bitcoin Cash divide in late 2017, affected the share of Bitcoin Core for a short time, but Bitcoin Core maintained a share of over 80% long-term. Over 90% of users were using Bitcoin Core, making it the dominant software in the network and the evident recent trend since 2018.

This implies that Bitcoin Core provides more information about installation, is easier to install compared to newer software, enables easy first-time access, and is highly recognized because it is maintained based on Satoshi's original software. Moreover, when we consider the overall reduction in the number of nodes, a software with an existing major share of the market is more advantageous.

---

[6] https://github.com/btcsuite/btcd.
[7] https://github.com/bitprim/bitprim-core.

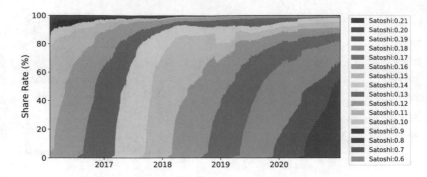

Fig. 2. Trends in Bitcoin Core network share by version

4.3   Version Distribution

We assumed that the behavior of users in the network in terms of software updating is reflected in Bitcoin Core, as it is the proprietary software of the network. Therefore, we categorized the nodes that use Bitcoin Core based on major versions to explore the corresponding changes in the share of each version.

Figure 2 illustrates the post-release share of each software version in the market share of Bitcoin Core. Despite the fact that software updates in the network depend on user behavior, new versions were found to consistently disseminate in the network after release. This implies that users are more interested in the implementation of new features and the security of the network than expected. Additionally, we confirmed that previously released versions do not override the market share of the latest version, and the network as a whole is not being divided in user support for the software version.

While this is not the worst case of unregulated and unhealthy software updates in the network, two concerns remain.

First, the unsupported scope is not at the same level as before because growth decelerates with the release of new versions. To confirm the peak share dynamism of each software version, Table 1 lists the maximum share on the network from Satoshi:0.12 to Satoshi:0.19 released during the observation period. The decreasing trend of maximum values has continued since Satoshi:0.14, which was released March 8, 2017 and has since gained the largest share of the market. The peak adoption percentage of the version of Satoshi:0.19 is below 50%, which means that the majority of the network does not benefit from the software update.

The second concern stems from the fact that a few users continue using very old software versions, some of which are no longer supported. The oldest version we found was Satoshi:0.6, which was released on March 30, 2012. Approximately 0.2% of users use versions from Satoshi:0.6 to Satoshi:0.9, while Satoshi:0.10 is now considered the mainstream version.

Table 1. Peak adoption rate and release interval in released software versions

Release version	Peak adoption rate		Peak reach days	Released interval	NewReleased - Peak
Satoshi:0.12	71.78%	(N/A)	175 days	226 days	7 days
Satoshi:0.13	72.38%	(+0.59)	188 days	182 days	9 days
Satoshi:0.14	79.71%	(+7.34)	174 days	197 days	16 days
Satoshi:0.15	78.55%	(−1.16)	157 days	190 days	8 days
Satoshi:0.16	64.26%	(−14.29)	217 days	165 days	−59 days
Satoshi:0.17	60.56%	(−3.70)	256 days	158 days	−45 days
Satoshi:0.18	59.74%	(−0.82)	262 days	211 days	−102 days
Satoshi:0.19	45.39%	(−14.35)	295 days	160 days	2 days

4.4   Analysis of Adoption Process Model

Assuming that the decline in software adoption in the network follows the Gompertz curve model, we introduce Eq. (1) using two contradictory factors: k, which accelerates the progression of adoption as the adoption rate, S, increases; and b^t, which inhibits the progression of adoption over time. It should be noted that k and b are constants, and $0 < b < 1$.

$$\frac{dS}{dt} = kSb^t \tag{1}$$

Equation (1) was solved for an adoption rate, S; then, assuming $\alpha = e^\beta$, $\beta = k/\log b$, and $C = S(0)/\alpha$. Equation (2) was obtained.

$$S(t) = C\alpha^{b^t} \tag{2}$$

In summary, $S(t)$ equals C when t becomes ∞, and C in turn equals the value of the adoption rate, S_{max}. Then, the inflection point, t_i, at which the adoption rate starts to decelerate, is computed as:

$$t_i = \frac{-\ln[-\ln \alpha]}{\ln b}, \quad S(t_i) = \frac{C}{e} \tag{3}$$

Figure 3 shows an estimate of the acceleration constant value, k, and the decay constant value, b, for versions 12 to 19, for which the peak values during the observation period are clear. The blue solid, orange solid, and black dashed lines correspond to the raw values, estimated values, and inflection points, respectively. The x-axis and y-axis of each figure correspond to the number of days and the adoption rate (%), respectively.

For versions 12 to 16, a trend of immediate adoption after release can be observed, and the inflection point appears within two weeks to one month. In contrast to this trend, for versions 17 to 19, the inflection point is delayed 3 to 5 months, as the adoption of the software starts some time after release. Finally, comparing the decay constant values for each version, we notice that the decay constant values after version 16 are larger than those before version 15, which corresponds to the timing of the sharp drop in the market share peak.

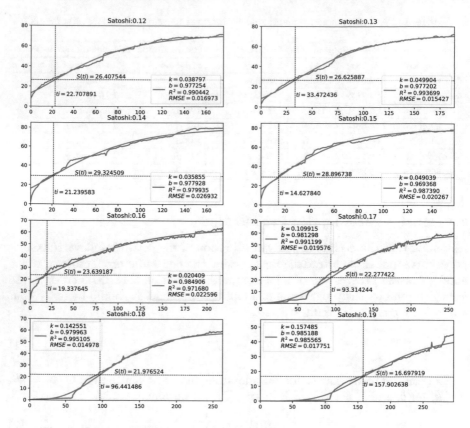

Fig. 3. Gompertz model fitting of the growing share of software versions

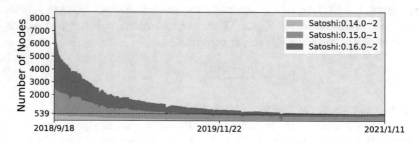

Fig. 4. Changes in the number of nodes containing vulnerabilities after the release of a fixed version

5 Discussion

In this section, the security risks in the network are discussed based on the results of the analysis described in Sect. 4.

Despite the improvements implemented in the latest version, some users in the network continue to use older versions of Bitcoin Core, which, as described in

Sect. 2, include security risks. Here, we focus on CVE-2018-17144, which affected various versions, to assess software usage. Figure 4 shows the software usage for versions 14.0 to 16.2, where vulnerability was reported, based on data collected on September 18, 2018, when CVE-2018-17144 was published.

The results show that 539 (7%) Bitcoin Core users continue to use software that poses security risks. These users presumably have different user characteristics because they do not correlate with the overall adoption trend. The present research classifies users according to their intentions as follows.

If users unintentionally use vulnerable versions, we assume that they are unaware of the security risk information or have abandoned maintenance actions. In any case, the cost of volunteering to run the node with no economic incentive is found to be high. However, the reason for not quitting on running nodes is unknown and needs to be investigated.

If users intentionally use vulnerable versions, we assume that they run a honeypot in the network. Because this is a node crash-type vulnerability, it is useful to set up a honeypot to verify that a malicious user actually exists in the network.

We expect that these nodes are non-maintenance nodes abandoned for maintenance by users because of the large number of nodes and the historical trend confirming the survival of old software versions.

6   Conclusion

In this research, we analyzed the software update cycles and adoption rates of Bitcoin, which is a blockchain-based software, to determine the reality of volunteer management in a trustless and distributed volunteer-dependent network.

As a result, we found that software update cycles in the Bitcoin network correspond to the release of new software; thus, the software adoption rate increases steadily. Therefore, the assumption holds that once a version is released, it does not override the prevalence of previous versions, and the network overall is not divided in user support for the software version.

However, older versions of software remain in use. Software with vulnerabilities reported in the CVE affect not only the individual user but the whole network, and should be considered a potential security risk. Furthermore, we should note that the peak of software adoption has been declining every year. This trend confutes the assumption that the currently established adoption rate is not overridden. Additionally, by modeling empirical trends in adoption rates, we found that there are greater time-dependent impediments to adoption compared with past versions. This means if the latest software is released at constant intervals and no fork is formed, an increase in the number of days to reach the adoption rate peak consequently contributes to a reduced adoption rate peak.

In future work, we intend to clarify the mechanisms behind the adoption model through detailed analysis. Although the release and adoption of software operates better than expected, our research does not identify the factors and mechanisms behind the observed behavior. Moreover, a simplified model was deployed in this study; thus, there is room for improvement. Future studies should focus on improving the fit of a better growth curve model. In this paper, we optimized the model for each software version, but the model can be further optimized for all versions present in the network.

Acknowledgements. This work was supported by Grant-in-Aid for Scientific Research (B) (19H04107).

References

1. Bosu, A., Iqbal, A., Shahriyar, R., Chakraborty, P.: Understanding the motivations, challenges and needs of blockchain software developers: a survey. Empirical Softw. Eng. **24**(4), 2636–2673 (2019)
2. Dong, J.Q., Wu, W., Zhang, Y.S.: The faster the better? innovation speed and user interest in open source software. Inf. Manage. **56**(5), 669–680 (2019)
3. Gencer, A.E., Basu, S., Eyal, I., Van Renesse, R., Sirer, E.G.: Decentralization in bitcoin and ethereum networks. In: International Conference on Financial Cryptography and Data Security, pp. 439–457. Springer, Heidelberg (2018)
4. Neudecker, T.: Characterization of the bitcoin peer-to-peer network (2015–2018) (2019). http://dsn.tm.kit.edu/bitcoin/publications/bitcoin_network_characterization.pdf
5. Pappalardo, G., Di Matteo, T., Caldarelli, G., Aste, T.: Blockchain inefficiency in the bitcoin peers network. EPJ Data Sci. **7**(1), 30 (2018)
6. Park, S., Im, S., Seol, Y., Paek, J.: Nodes in the bitcoin network: comparative measurement study and survey. IEEE Access **7**, 57009–57022 (2019)
7. Satoshi, N.: Bitcoin: a peer-to-peer electronic cash system (2008). http://www.bitcoin.org/bitcoin.pdf
8. Taivalsaari, A., Mikkonen, T.: A roadmap to the programmable world: software challenges in the IoT era. IEEE Softw. **34**(1), 72–80 (2017)
9. Tran, M., Choi, I., Moon, G.J., Vu, A.V., Kang, M.S.: A stealthier partitioning attack against bitcoin peer-to-peer network. In: IEEE Symposium on Security and Privacy (S&P) (2020)
10. Xie, J., Yu, F.R., Huang, T., Xie, R., Liu, J., Liu, Y.: A survey on the scalability of blockchain systems. IEEE Netw. **33**(5), 166–173 (2019)

Investigation and Analysis of Features in Decentralized Network Management of Minor Cryptocurrencies

Mitsuyoshi Imamura[1(✉)] and Kazumasa Omote[1,2]

[1] University of Tsukuba, Tennodai 1–1–1, Tsukuba 305-8573, Japan
s1730148@s.tsukuba.ac.jp, omote@risk.tsukuba.ac.jp
[2] National Institute of Information and Communications Technology,
4-2-1 Nukui-Kitamachi, Koganei, Tokyo 184-8795, Japan

Abstract. Blockchain is a trustless interoperability platform for users and is managed in a decentralized structure. There are no controllers in this platform, and service continuity depends on the users present in the network. Surveys of such network have mostly focused on miners who are authorized to write to the ledger; however, the non-mining nodes that maintain the ledger and are operated by volunteers with no direct incentive are also important indicators of stability. Although there are several projects that are inspired by or forked from Bitcoin, it is not intuitive to assume that projects not pertaining to Bitcoin are running in the same manner. Compared with major currencies, networks running minor currency nodes, despite having the same infrastructure, are more challenging to focus owing to their unclear operational realities. Thus, learning from the characteristics of failed projects is as important as learning from successful ones. Therefore, in this study, we investigate the unique features in the network by analyzing blockchain-based minor currencies. We analyze 317 minor cryptocurrency nodes and report that they contain locality, we also report on the issues of blacklisted nodes.

1 Introduction

Blockchain is a trustless interoperability platform for peer-to-peer (P2P) networks, and the infrastructure was used in the original paper published by Satoshi Nakamoto [15]. One of the major differences from traditional systems is that it has a decentralized structure with no management, and service continuity is dependent on the users in the P2P network. Bitcoin has been developed by several volunteers, and this aspect is well demonstrated in this technology, which has been operating for more than 10 years since January 9, 2009[1], since its released to SOURCEFORGE[2]. It provides a baseline for not only cryptocurrency but also for technical discussions in the blockchain.

---

[1] https://www.mail-archive.com/cryptography@metzdowd.com/msg10142.html.
[2] https://sourceforge.net/.

L. Barolli et al. (Eds.): AINA 2021, LNNS 225, pp. 245–258, 2021.
https://doi.org/10.1007/978-3-030-75100-5_22

In an observational review focusing on the distributed network aspect of blockchain, the volunteer nodes that contribute to ledger maintenance in the network were found to be an interesting indicator. According to a previous classification study [8], there are three types of activities that take place in the network: P2P transactions, which constitute the transaction network, mining, which records the transactions in a ledger, and decentralized management, which maintains the ledger. Transaction networks and mining, which are activities authorized to write to the ledgers, are the most popular topics to focus on in view of the attack on privacy and consensus, owing to their strong relationships with direct incentives. Decentralized management is different from the other aforementioned activities in the sense that it is a volunteer activity with no direct incentives. Therefore, the economic cost of storing the ledger can easily lead to the user leaving the network and is an important indicator of network stability, including its availability and robustness. The measurement methods and network status for major cryptocurrencies, such as Bitcoin and Ethereum, have been reported in previous studies [5, 11].

Minor currencies, in addition to major ones, must also be analyzed for understanding the conditions during the provisioning of services. It is difficult to understand the trends and instability in the network when a service is abandoned due to project failure as well as when it moves toward convergence. Further, it has been noted that Bitcoin and Ethereum bubbles burst in 2017; however, it is challenging to identify the signs of service termination, even in 2020. Nevertheless, we assume that minor currencies cannot survive in the same history. Furthermore, they would not have survived under the same conditions as those of Bitcoin and Ethereum.

There are more than 2,000 projects inspired by or forked from the Bitcoin initiative. However, it is not intuitive to assume that minor projects, other than those that grow into major ones such as Ethereum, continue to run in the same manner. They receive less attention than major currencies and their operational realities are not as clear. Thus, we can expect that the networks operating the nodes of these minor currencies have the same infrastructure and features that successful projects do not have.

Therefore, in this study, we analyze the distribution of nodes in the networks of these minor currencies to identify their similarities with the major currencies and the characteristics of those who remain in the network at the end of the service. We also investigate the status of nodes through security scans, assuming that unpopular services potentially contain nodes that have either given up on maintenance or have been abandoned.

The main contributions of this study can be summarized as follows:

- We identify the characteristics of minor currencies in the network that differ from those of major currencies.
- We identify blacklisted nodes that are included in the operational nodes.
- We report on the empirical successes and availability risks in decentralized structures, as nodes continue to operate despite the lack of economic incentives.

The remainder of this paper is organized as follows. Section 2 presents related works on dynamism in blockchain networks. Section 3 reports the data used in this study and the results of the analysis. Finally, Sect. 4 concludes this paper.

2 Related Work

Observational research leads to the discovery of potential challenges in the network, providing important feedback to improve the system and contributing to modeling the empirical rules that historical changes clear the dynamism behind the user's behavior. In this section, we introduce related work on the well-known Bitcoin and Ethereum, as well as other cryptocurrencies.

Donet et al. [5] reported an early observational study that provided methods and empirical measurements regarding data collection in a Bitcoin network. They focused on the network information that was provided as a reference in previous research on transactions and block propagation [4] and presented detailed information, such as the size of the P2P network, the geographic distribution of nodes, network stability, and propagation time. They also reported on the trend for networks with many nodes located in the United States and China; such nodes were rare in sparsely populated areas and developing countries and evenly distributed in Western Europe and the United States. Most previous studies refer to this trend as a baseline trend.

Feld et al. [6] reported a network-level classification using autonomous system numbers (ASNs), in addition to reporting a country-level classification using geolocation and analyzing Bitcoin anonymity using IP addresses. In their work, the networks of the most deployed nodes included AS7922 (Comcast), AS701 (Verizon), AS4134 (China Telecom), AS5089 (Virgin Media, UK), and AS22773 (Cox Communications Inc. US).

Gencer et al. [7] conducted an analysis of multiple infrastructures covering not only Bitcoin, which is a major platform, but also Ethereum, which receives considerable attention. They reported improved provisioning bandwidth compared with that in previous research [2] and clusters of geographical proximity between the nodes.

Park et al. [14] reported on the evolution of the Bitcoin network structure. The problem of decentralization in the Bitcoin network was highlighted in another study [8,12].

As Bitcoin and Ethereum are widely known technologies, there is a significant amount of historical background and several reported features of these networks. In contrast, minor currencies have received less attention and only a few studies have been conducted on this topic.

Padmavathi et al. [13] extended the previous investigation of Bitcoin to include Litecoin, which was created via forking from the Bitcoin project. They reported on the characteristics of the Litecoin network, such as the relatively small number of routable nodes as compared with Bitcoin for alternative coins.

Daniel et al. [3] proposed a topology inference model that relies on a passive timing analysis of measured block arrival times for Zcash, which is a highly anonymous cryptocurrency, according to network size and node distribution.

To characterize the network, Cao et al. [1] measured the size, node distribution, and connectivity of Monero, a well-known anonymous cryptocurrency similar to Zcash.

Kim et al. [10] investigated the distribution of nodes in the network to analyze network security in Stellar.

These investigated cryptocurrencies are inspired by or forked from Bitcoin, and they use enhanced versions of anonymity and consensus algorithms. Further, the community behind these technologies is different from that of the initial one. However, these are relatively well-known cryptocurrencies, and the features they implement are too popular to be classified as minor. The focus of this study is on currencies that have not been discussed in these related works.

3   Analysis

3.1   Data

First, we use the nodes on the network to analyze the network information for 317 cryptocurrencies listed on cryptoID[3] and Trittium[4]. We excluded from our analysis 11 currencies that are among the top 200 currencies in terms of market capitalization according to Coinmarketcap at the time of our observation because our study targets only minor cryptocurrencies. We also excluded 7 currencies that consist only of the onion router (Tor) servers. As a result, 301 currencies are included in our analysis. We obtain the network information every 60 min and summarize it as a daily statistic. Note that we merge and aggregate currencies that are duplicated in the reference data site.

Next, we use VirusTotal[5], which is a well-known online scanning engine, to investigate security risks after obtaining the node information. The risk results are "Clean," "Unrated," "Malware," "Phishing," "Malicious," "Suspicious," and "Spam." Then, we consider results other than "Clean" and "Unrated" to be security risks unless otherwise stated.

The observation period is from May 1 to August 31, 2020. There are potential limitations to our results owing to our reliance on website information and deviations from real-time information. Furthermore, it should be noted that because the target cryptocurrencies are minor, there may be a sudden suspension of distribution due to a decline in popularity or difficulty in survival. Therefore, investigating minor cryptocurrencies of this type, unlike major cryptocurrencies, makes it difficult to collect data on the network.

---

[3] https://chainz.cryptoid.info/.

[4] https://chains.trittium.cc/.

[5] https://www.virustotal.com/gui/.

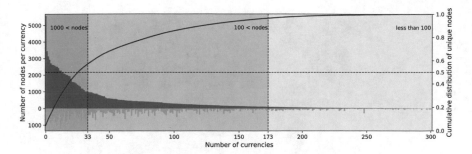

Fig. 1. Node and cumulative distributions of cryptocurrencies

3.2   Node Information

We summarize the basic information regarding the collected nodes of the cryptocurrency to investigate the security risk.

First, in Fig. 1 we show the node and cumulative distributions per currency obtained during the observation period. The green bars correspond to the number of unique nodes per currency, and the orange bars denote the number of nodes overlapping between currencies. The red line denotes the cumulative distribution of the number of unique nodes. The distribution of the collected nodes is color-coded according to the area in the order of 1000 or more, 100 or more, and less than 100.

The number of unique nodes collected during the observation period is 140,385, of which 16,949 nodes (12.07%) are duplicates across currencies. If these duplicate nodes are excluded, the total number of currencies with unique nodes is estimated at 300. Unless otherwise noted, we calculate the distribution and share ratio based on this value.

Second, we focus on the distribution of the number of currencies as well as the cumulative distribution. We report that there are 33 currencies with more than 1,000 nodes, 140 currencies with more than 100 nodes, and 128 currencies with less than 100 nodes. The number of unique nodes collected in approximately 90% of the currencies is less than 1,000, classified roughly in half of the currencies with a baseline of 100 nodes.

With regard to most of the collected node distributions, the intersection of the cumulative distribution line and the 50% dotted line is located in the > 1000-node area. In other words, a few currencies still tend to use IP addresses in the network relatively well, even though they are minor currencies. In our analysis, half of the nodes are collected in 33 currencies, and we find that the contribution of a particular currency to the collected data is significant.

In terms of currency, the maximum number of unique nodes is 5,578 and the maximum number of duplicate nodes is 1,236. As the number of nodes to be discovered increases, the number of duplicate nodes also tends to increase; however, because the identity of the nodes is not identified, it is necessary to consider duplicate counting using dynamically assigned IP addresses.

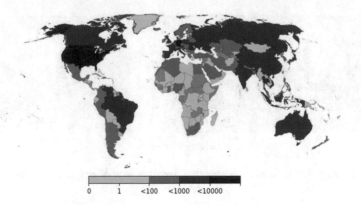

Fig. 2. Node distribution according to country

3.3 Regional and Network Areas

Next, we classify the regions where the nodes are built to identify trends in the regional characteristics of the collected data. Figure 2 shows the number of nodes distributed according to country. The countries are classified according to the IP addresses assigned using the whois API[6] and IP2Location[7]. The colors in the figure represent the number of nodes; regions with zero nodes are shown in gray, and regions with only one node are shown in blue. Regions with more than one node are indicated in red, with the higher number of nodes shown in darker colors.

Ranking the number of nodes in Fig. 2, there are 164 countries with nodes. The countries with more than 10,000 nodes, which account for a majority of nodes in the network, are the United States (20,425 nodes) and Germany (18,821 nodes), which together account for approximately 31.79% of the total nodes in the considered countries. Further, there are 21 countries with the next largest number of nodes, in the range of 1,000–10,000 nodes; these countries include European nations such as Russia (8,585 nodes), the United Kingdom (7,068 nodes), and France (3,373 nodes), and Asian nations such as China (8,051 nodes) and Thailand (3,144 nodes), and Australia (2,742 nodes). Finally, 50 countries are in the 100–1,000 node range, 73 countries are in the 1–100 node range, and only 18 countries have only one node.

Our results are similar to those presented in most recent research on Bitcoin and Ethereum networks [8,9,12,14]. Most nodes were found to be distributed in North America and Europe, followed by a strong presence in Asia. One unique trend that was noted in our study was that there are more nodes in countries such as Nigeria (985 nodes) and South Africa (892 nodes), which are located in a minor regions in terms of Bitcoin and Ethereum presence. Overall, the distribution trend is the same as that of major currencies; however, we find different trends

---

[6] https://whois.whoisxmlapi.com/.

[7] https://lite.ip2location.com/.

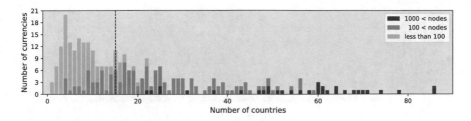

Fig. 3. Number of currencies per number of countries

in some regions. We attribute this to the fact that minor currencies have different user usage trends as compared to that of major currencies.

Moreover, half of the minor currencies were found to be distributed in a few countries, whereas the major currencies are distributed in different areas. This implies a locality property that is operationalized in a particular community when it loses popularity.

Furthermore, the distribution of the number of currencies per number of countries is shown in Fig. 3 to confirm the regional dispersion trend of the currencies. The bar is color-coded according to the number of nodes distribution, as in Fig. 1. The vertical dotted line indicates the position where the number of currencies is half.

There are two currencies with nodes distributed across 86 countries. In contrast, there are three currencies with nodes built in only one country. The largest number of cryptocurrencies have nodes distributed in four countries. There are 20 currencies in total. In addition, half of the currencies are distributed in less than 14 countries. Furthermore, there are 34 currencies that have more than 100 nodes and 111 currencies that have less than 100 nodes. Intuitively, we know that a small number of nodes denotes a limited number of regions to distribute; accordingly, this result provides a concrete criterion to classify minor currencies.

In addition, to verify the trend at a finer granularity, we confirm the distribution at the ASN level at which the IP addresses are managed. Figure 4 classifies the nodes into ASNs by referring to the whoisAPI and IP2Location, similar to the country classification, and shows the top 15 ASNs, which are approximately 30% of the total network nodes. The bar is color-coded according to the number of nodes distribution, as in Fig. 1.

In the ranking of ASNs, telecommunication companies such as AS3320 (Deutsche Telekom) and AS4134 (China Telecom) are at the top. Some of the most popular cloud providers reported in previous research [12], such as AS24940 (Hetzner), AS16276 (OVH), and AS14061 (DigitalOcean), are listed, whereas the well-known service AS16509 (Amazon) is not the most popular and ranks 26th.

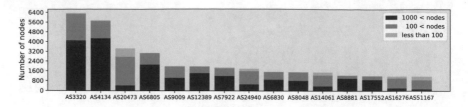

Fig. 4. Top 15 ASNs in the distribution of ASN levels

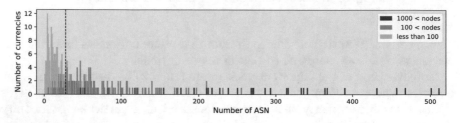

Fig. 5. Number of currencies per number of ASNs

Then, we focus on the distribution of currencies, and the results identified that AS20473 (CHOOPA), AS24940 (Hetzner), AS14061 (DigitalOcean), AS16276 (OVH), and AS51167 (Contabo) have an interesting feature. This feature is that all these ASNs have less than half of the currencies with more than 1000 nodes, contradictory to the case of the other top ASNs. Additionally, these networks are provided by cloud providers that supply hosts with relatively low prices.

This result fits with the intuition that the group of nodes that users eventually abandon are concentrated in cloud providers that have relatively low prices. However, we note the contradictory behavior of users; that is, they are not economically rational when it comes to managing nodes despite their unpopularity, while they use cloud providers with relatively low prices.

The number of currencies per number of ASNs, as well as number of countries, is shown in Fig. 5. The bar is color-coded according to the number of nodes distribution, as in Fig. 1. The vertical dotted line indicates the position where the number of currencies is half. This figure shows the distribution at a more detailed granularity, as compared to the distribution per country shown in Fig. 3.

There is only one currency that was identified with the largest number of nodes distributed in most ASNs, and this currency is distributed in 511 ASNs. In contrast, there are three currencies that were identified as being distributed in only one ASN. Further, it was found that the largest number of currencies is distributed when the nodes are distributed in four ASNs, with 12 currencies. In addition, half of the currencies were found to be distributed among lesser than 28 ASNs.

It is well known that users of nodes in major currencies tend to prefer those nodes that are offered by major cloud providers, as these providers have low maintenance costs. We find that the most recent reports [8, 12] show that this tendency is becoming increasing pronounced year by year. Therefore, we assume that hosting nodes in a region without a maintenance cost advantage has a different siting tendency for minor currencies than for major currencies, such as being dependent on the community's area of operation and the location of the developer, similar to the analysis at the country level.

The distribution of individual currencies shows that half of the currencies operate in networks with less than 30 ASNs. In other words, network availability is extremely low and based on the national distribution results, we can see that it is operated by a very narrow community.

In summary, we find that minor currencies are characterized by locality. This is based on our observation that such currencies are managed in regions and specific networks that are rare and therefore differ from those for major currencies. Moreover, if we assume the scenario of users abandoning nodes, we find that the results agree with the intuition that nodes that require a relatively low maintenance price remain in the network. However, there are inconsistencies in the rationality in the behavior of users.

3.4   Security Risk Scanning

In this section, we describe our investigation of the security risks of 140,385 nodes discovered during the observation period using the scanning engines via the VirusTotal API. As a result, 5,203 unique nodes were reported to have security risks, which is 3.71% of the total nodes that were identified.

Here, we consider the results of the scanning engine to reflect the type of cryptocurrency, and the nodes are located in the region and network identified in the previous section.

The nodes where security risks are reported by the scanning engine are categorized by cryptocurrency labels, and we find that they are distributed to 238 (79.07%) different cryptocurrencies. However, this classification includes nodes that are duplicated in multiple cryptocurrencies, so we need to consider the results of unique nodes to avoid overestimation. There are 1,242 nodes with duplicate runs across currencies in nodes that security risks are reported by scanning engines. Excluding these nodes, we confirm the cryptocurrency labels of the nodes that run with unique cryptocurrencies. We find that there are 206 different cryptocurrency labels. This means that 68.44% of all minor cryptocurrencies have a security risk warning, excluding duplicate nodes. Although there is a cryptocurrency bias in the collected nodes, we find that this result does not reflect the specific cryptocurrency results that we were concerned about. This conclusion remains the same even when the nodes located in different countries and networks are investigated.

Table 1. The number of nodes that the scanning engine predicted to have a security risk

No.	Engine	Nodes	No.	Engine	Nodes	No.	Engine	Nodes
1	ADMINUSLabs	126	34	EonScope	0	67	Quttera	3
2	AICC (MONITORAPP)	0	35	Feodo Tracker	0	68	Rising	0
3	AegisLab WebGuard	25	36	Forcepoint ThreatSeeker	46	69	SCUMWARE.org	0
4	AlienVault	177	37	Fortinet	66	70	Sangfor	2
5	Antiy-AVL	0	38	FraudScore	0	71	SecureBrain	0
6	Armis	0	39	FraudSense	0	72	Segasec	0
7	Artists Against 419	0	40	G-Data	3	73	Snort IP sample list	15
8	AutoShun	0	41	Google Safebrowsing	0	74	Sophos	4
9	Avira	4	42	GreenSnow	464	75	Spam404	0
10	BADWARE.INFO	0	43	Hoplite Industries	0	76	Spamhaus	3911
11	Baidu-International	0	44	IPsum	292	77	StopBadware	0
12	BitDefender	59	45	K7AntiVirus	0	78	StopForumSpam	0
13	BlockList	663	46	Kaspersky	18	79	Sucuri SiteCheck	0
14	Blueliv	1	47	Lumu	0	80	Tencent	0
15	Botvrij.eu	0	48	MalBeacon	0	81	ThreatHive	0
16	CINS Army	99	49	MalSilo	0	82	Threatsourcing	277
17	CLEAN MX	27	50	Malc0de Database	0	83	Trustwave	4
18	CMC Threat Intelligence	0	51	Malekal	0	84	URLQuery	0
19	CRDF	630	52	Malware Domain Blocklist	0	85	URLhaus	0
20	Certego	11	53	Malware DomainList	0	86	VX Vault	0
21	Cisco Talos IP Blacklist	96	54	MalwarePatrol	1	87	Virusdie External Site Scan	0
22	Comodo Site Inspector	0	55	Malwarebytes hpHosts	0	88	Web Security Guard	0
23	Comodo Valkyrie Verdict	1073	56	Malwared	0	89	Yandex Safebrowsing	0
24	CyRadar	221	57	Netcraft	4	90	ZCloudsec	0
25	Cyan	0	58	NotMining	0	91	ZDB Zeus	0
26	CyberCrime	0	59	Nucleon	0	92	ZeroCERT	0
27	Cyren	0	60	OpenPhish	0	93	Zerofox	0
28	DNS8	0	61	Opera	0	94	ZeusTracker	0
29	Dr.Web	16	62	PREBYTES	0	95	desenmascara.me	0
30	ESET	5	63	PhishLabs	0	96	malwares.com URL checker	0
31	ESTsecurity-Threat Inside	0	64	Phishing Database	1	97	securolytics	0
32	EmergingThreats	91	65	Phishtank	0	98	zvelo	5
33	Emsisoft	2	66	Quick Heal	32			

All nodes, including the unique and duplicate nodes, are distributed in 114 countries, with the highest number of nodes being in the UK (960 nodes). Further, the number of nodes in the United States does not reflect a distributional bias. We also find that this number is not dependent on the regional location and the scale of the network space, based on the result of 905 different ASe distributions.

In summary, the scanning engine results are not weighted by the level of the number of nodes collected, and we ignore the trends in the node distribution and analyze them as characteristics of minor cryptocurrencies. In the following sections, we will elaborate on the results of the scanning engine.

First, regarding the scanning engines, we find that not all engines reported security risks, and that there exists a trend of engines that are more likely to report a security risk. Table 1 presents the number of nodes classified as having a security risk for the 98 different scanning engines.

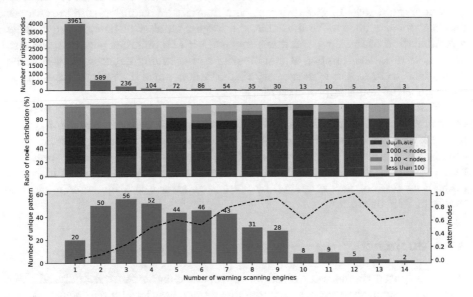

Fig. 6. Number of scanning engines that reported a security risk per node

A total of 36 different scanning engines have reported security risks, and the scanning engine with the greatest number of alerts is Spamhaus[8], which classified 3,911 nodes (75.17%). Other scanning engines that were classified as having several nodes with security risks include Blocklist, Comodo Valkyrie Verdict[9], CRDF[10], and GreenSnow[11]. Conversely, there are other scanning engines, such as

---

[8] https://www.spamhaus.org/.
[9] https://verdict.valkyrie.comodo.com/.
[10] https://www.crdf.fr/.
[11] https://greensnow.co/.

the Blueliv[12], MalwarePatrol[13], and Phishing Database[14], that respond uniquely to a small number of nodes.

Next, focusing on the node-level, we find a trend in the node distribution of the scanning engine that reported security risks. Figure 6 shows the relationship between the nodes and scanning engines that reported security risks. The top panel depicts the number of nodes per scanning engine that reported a security risk, the middle panel depicts the ratio of the number of nodes using the distribution of nodes per currency, and the bottom panel illustrates the number of unique patterns of the scanning engines.

Our results show that 3,961 nodes have the highest number of alerts from a single scanning engine, and a total of 1,242 nodes have reported security risks from two or more scanning engines. As more scanning engines report security risks, the number of nodes decreases. The maximum number of scanning engines in this investigation is 14. However, with the increase in the number of alerts, the ratio of the currencies with fewer and duplicate nodes also tends to increase.

There are a total of 397 different combinations of scanning engine response patterns for the nodes. We find that the largest number of patterns represents a combination of the three scanning engines, with 56 different patterns. We also assume that as the number of combinations of scanning engines increases, the nodes tend to have a unique security risk criterion, since the ratio of patterns to the number of nodes approaches unity.

With regard to user behavior, we assume the establishment of a backing community based on the trends that the nodes have locality in the country and network area level presented in Sect. 3.3. In contrast, we assume that the variability that appears in the security of nodes reflects a microscopic trend for management to depend on individuals. It also emphasizes a scenario in which nodes that are not well managed by volunteers after operation remain as fragments of P2P network.

4 Conclusion

We investigated the network of minor cryptocurrency nodes and analyzed their running status in the network to understand the operational differences between major and minor currencies. We surveyed 317 minor cryptocurrency nodes and found that locality is a feature that differentiates them from major currencies. We also identified features that support the overall trend of node abandonment, that is, low-cost nodes tend to remain in the network, and over 75% of the currencies have nodes that are blacklisted.

According to the results of this analysis, we assume that there are only certain individuals and small communities that have worked toward building the network of minor currencies in the last stage, and that it is a sham operation involving

---

[12] https://www.blueliv.com/.

[13] https://www.malwarepatrol.net/.

[14] https://github.com/mitchellkrogza/Phishing.Database.

abandoned nodes. If the major currencies that are currently in the spotlight also have the same network tendencies, they will essentially shift to minor currencies.

Furthermore, several nodes run without economic rationality, and we assume that they are used by malicious users without the node owners being aware of it. Although the countermeasure is a simple method that involves stopping the nodes, it is necessary to wait for a natural shutdown in a distributed network because the blockchain protocol cannot be stopped. This decentralized management functions in the network until the service ends, similar to the case of major currencies. It acts as a powerful feature when the system needs to be started; however, it becomes a bottleneck if the need to shut down the system arises.

In the future, we must discuss the design of blockchains including the case of stopping service. We believe that our analysis results could be useful for developing countermeasures in the future.

Acknowledgements. This work was supported by Grant-in-Aid for Scientific Research (B) (19H04107).

References

1. Cao, T., Yu, J., Decouchant, J., Luo, X., Verissimo, P.: Exploring the monero peer-to-peer network. In: Financial Cryptography and Data Security 2020, Sabah, 10-14 Feb 2020. Springer, Cham (2020)
2. Croman, K., Decker, C., Eyal, I., Gencer, A.E., Juels, A., Kosba, A., Miller, A., Saxena, P., Shi, E., Sirer, E.G., et al.: On scaling decentralized blockchains. In: International Conference on Financial Cryptography and Data Security, pp. 106–125. Springer, Heidelberg (2016)
3. Daniel, E., Rohrer, E., Tschorsch, F.: Map-z: exposing the zcash network in times of transition. In: 2019 IEEE 44th Conference on Local Computer Networks (LCN), pp. 84–92. IEEE (2019)
4. Decker, C., Wattenhofer, R.: Information propagation in the bitcoin network. In: IEEE P2P 2013 Proceedings, pp. 1–10. IEEE (2013)
5. Donet, J.A.D., Pérez-Sola, C., Herrera-Joancomartí, J.: The bitcoin P2P network. In: International Conference on Financial Cryptography and Data Security, pp. 87–102. Springer, Heidelberg (2014)
6. Feld, S., Schönfeld, M., Werner, M.: Traversing bitcoin's P2P network: insights into the structure of a decentralised currency. Int. J. Comput. Sci. Eng. **13**(2), 122–131 (2016)
7. Gencer, A.E., Basu, S., Eyal, I., Van Renesse, R., Sirer, E.G.: Decentralization in bitcoin and ethereum networks. In: International Conference on Financial Cryptography and Data Security, pp. 439–457. Springer, Heidelberg (2018)
8. Imamura, M., Omote, K.: Difficulty of decentralized structure due to rational user behavior on blockchain. In: International Conference on Network and System Security, pp. 504–519. Springer, Cham (2019)
9. Imtiaz, M.A., Starobinski, D., Trachtenberg, A., Younis, N.: Churn in the bitcoin network: characterization and impact. In: 2019 IEEE International Conference on Blockchain and Cryptocurrency (ICBC), pp. 431–439. IEEE (2019)
10. Kim, M., Kwon, Y., Kim, Y.: Is stellar as secure as you think? In: 2019 IEEE European Symposium on Security and Privacy Workshops (EuroS&PW), pp. 377–385. IEEE (2019)

11. Kim, S.K., Ma, Z., Murali, S., Mason, J., Miller, A., Bailey, M.: Measuring ethereum network peers. In: Proceedings of the Internet Measurement Conference 2018, pp. 91–104 (2018)
12. Mariem, S.B., Casas, P., Romiti, M., Donnet, B., Stütz, R., Haslhofer, B.: All that glitters is not bitcoin–unveiling the centralized nature of the BTC (IP) network. In: NOMS 2020-2020 IEEE/IFIP Network Operations and Management Symposium, pp. 1–9. IEEE (2020)
13. Padmavathi, M., Suresh, R.: Secure P2P intelligent network transaction using lite-coin. Mobile Netw. Appl. **24**(2), 318–326 (2019)
14. Park, S., Im, S., Seol, Y., Paek, J.: Nodes in the bitcoin network: comparative measurement study and survey. IEEE Access **7**, 57009–57022 (2019)
15. Satoshi, N.: Bitcoin: a peer-to-peer electronic cash system (2008). http://www.bitcoin.org/bitcoin.pdf

DynAgentX Framework: Extending the Internet Management to the VCPE in the SD-WAN Context

Maxwell E. Monteiro[✉]

Federal Technology Institue of Espirito Santo, 6, 5KM Rd. ES-010, Serra, Brazil
maxmonte@ifes.edu.br

Abstract. The virtual customer premises equipment (VCPE) is a virtual Computing platform, capable of to host simultaneous and dynamically provisioned virtual network functions (VNFs), the VCPE has become an important element in SD-WAN and Edge Computing. Notwithstanding the SDWAN points to new management solutions (such as, gRPC-Network Management Interface, or RestConf), the basic management tasks, namely monitoring, and simple configuration, still the same. In this interim of coexistence between traditional WAN and SD-WAN, the maturity, the past investments, and dominance of Internet management solutions, suggest that it would be quite interesting for the operators to extend their legacy Internet management solutions till VCPE and SD-WAN get their maturity. This paper presents the DynagentX, a VCPE internal management instrumentation's architecture, aiming to adapt the Internet management framework to the SD-WAN, and edge computing scenario, focusing the VCPE. In this work I show results that substantiate the proposal feasibility, performance, and security.

Keywords: CPE · Management · Internet management framework · OSGi · Agent X

1 Introduction

The Software Defined WAN (SD-WAN) [12] has risen as a new paradigm for Wide Area Networks. The idea of an overlay software defined network upon a mix of well known WAN technologies was enthusiastically received by network operators and vendors. Besides the network programmability, the network function virtualization (NFV) has become an important SD-WAN, and the edge computing cornerstone [15]. SD-WAN has coined what has been called the Virtual Customer Premises Equipment (vCPE) [10]. The VCPE is a virtual Computing platform capable of to host simultaneous and dynamically provisioned virtual network functions (nfvs). This innovative Wide Area Network interpretation has imposed some conflicts with the legacy WAN technologies, among which

L. Barolli et al. (Eds.): AINA 2021, LNNS 225, pp. 259–271, 2021.
https://doi.org/10.1007/978-3-030-75100-5_23

the most notorious is the Internet [2] management framework. Notwithstanding the SD-WAN points to new management solutions (such as, gRPC-Network Management Interface, or RestConf), the basic management tasks, namely monitoring, and simple configuration, still the same. In this interim of coexistence between traditional WAN and SD-WAN, the maturity, the past investments, and the Internet management solutions dominance, suggest that it would be quite interesting for the operators [10] to extend their legacy Internet management solutions till VCPE and SD-WAN get their maturity, in terms of standards. This is especially true for those operators that are migrating from traditional WAN to SD-WAN technology. Unfortunately, due to the clear difference between to manage a single function physical boxes (CPEs) and a dynamically provisioned multi-functional virtual CPE (vCPE), the Internet framework in its basic operation mode seems to be an inadequate technology. The challenge resides on the dynamism of the Virtual network functions deployed inside the VCPE. This dynamic composition involves a variety of processing technologies, such as containers, virtual machines, and traditional network services, all of them enchained to fulfill the costumer services contract. Furthermore, each virtual network function may be provided by different vendors. Besides, some services providers have a mix of physical and virtual CPEs. The Network Management System (NMS)has to treat the CPEs and VCPE as units, although the VCPE could be conformed by a bunch of Virtual network Functions. For the legacy NMS is extreamely important a certain abstraction from the internal VCPE structure, in order to treat it as a unit.

Motivated by the above mentioned challenge and the advantages of extending the Internet management framework life-cycle, while the SD-WAN and VCPE come to maturity, we have investigated how to adapt the Internet management framework to the VCPE scenarios.

As the result of the aforementioned investigation, this paper presents the DynagentX an VCPE internal management's instrumentantion architecture, aiming to adapt the Internet management framework to the SD-WAN VCPE scenario. Capturing pros and cons of Different VCPES internal structures and virtualization technologies, we have figured out an approach based on the conservation, as much as possible, of the traditional Internet Network management ecosystem. Which implies the use of the well-known and mature protocols and technologies. DynagentX uses the standard AgentX architecture and functionalities [5], along with the modern dynamically provisioned, and service oriented OSGi technology. This combination adequates the Internet framework for the VCPE management task in the SD-WAN context. This work does not advocate that Internet management is the best solution for the VCPE management, but an alternative to adapt the legacy Internet NMS to the facing the transition from traditional WAN to SD-WAN. The comparison among SNMP, gRPCNMI, and Restconf for the VCPE is out of the paper's scope.

Concerning the paper organization, Sect. 2 presents our contribution claims, Sect. 3 shows the related works, while Sect. 4 describe the fundamental concepts and technologies involved in DynAgentX. Finally the Sect. 5 presents and describes the DynAgentX architecture and implementation. The Validation and conclusion are respectively in Sects. 6 and 7.

2   Our Contributions

In our investigation and development aiming the adaptation of the Internet Management solutions toward the new SD-WAN paradigm, we believe have made contributions in three directions: **(a) the vision of the most adequate Internet Management standards for the Virtual Custumer Premise Equipment**. Having in mind the VCPE management challenge described in the Introduction, we have investigated among the SNMP standard documents those with the potential to face it. We have identified the AgentX (RFC2741) [5] as the most relevant for the VCPE case, though its well-known implementation presents some practical issues for the VCPE scenario; **(b) the design of an adaptable and dynamic VCPE's internal management instrumentation architecture**. Considering the practical limitations of the AGentX's well-known implementation the net-snmp [4] for the VCPE case, we had the intuition that a technology like OSGi [1] would help to minimize them. Thus we have designed the VCPE's internal management instrumentation and communication architecture, giving it the management flexibility and adaptability needed; **(c) the design and development of a OSGi-JAVA API, aiming to reach the behavior demanded in contribution (b)**. Since there wasn't a well-known implementation of the agentX using OSGi technology.

3   Related Works

Once our contribution claims were divided into three directions, the related works were organized in the same way:

- **1. SD-WAN and VCPE management proposals:** most of the works involving Software Defined WAN management are focused on the network survivability [17], or on the distributed Control Plan performance [8]. To the extent of our knowledge we couldn't find works addressing the VCPE internal management architecture, and technology.
- **2. Agent X, and its well-know implementations:** As we have identified the Agent X as the most relevant SNMP standard for the VCPE management. Here we cite its well-known implementation (the net-snmp) [3]. The Agent X description is provided in the Background section, as well as the motives for which we consider net-snmp basic operation mode inadequate for the VCPE case. Even alternative Agent X implementation such as PyAgentX, SNMP4J and AgentC++ present similar limitations, since they either have limitations to identify master and sub-agents process at run-time.
- **3. Implementations of Internet Management using OSGi:** Finally as we have identified the OSGi Technology as the dynamism provider for our

Agent X implementation. We have investigated OSGi SNMP agents implementations. The OSGI SNMP Agent implementation is not a new idea, as can be seen in [14]. However, the main contribution of Merino's et al. was a SNMP agent aiming the OSGI platform management, not a network devices or functions. Although the Merino's agent internal architecture seems close to the Agent X proposal, the work didn't mention or implement the rfc2741.

This work distinguishes itself from those above by proposing a VCPE management instrumentation architecture, which allows a variety of VCPEs configuration behaving as a single device.

4   Background

4.1   The Software-Defined WAN (SD-WAN) and the vCPE

The Software defined Wide Area Network (SD-WAN), shown in Fig. 1, is a set of technologies organized in order to orchestrate a overlay WAN upon a diversity of communication technologies [12]. The software defined control allows the construction of a complex, and dynamic network, supposedly increasing availability, and performance, besides decreasing costs. An important component of the SD-WAN is the Virtual Customer Primesi equipment (vCPE) [10], illustrated by the black equipment at the branch office in the Fig. 1. First, it's important to emphasize that the VCPE hasn't a standard or consolidated definition. Some authors claim that VCPEs are based upon virtual entities, comprising a chain of network functions (NFVs) [9] allocated in the Cloud or inside a bare metal (X86 or ARM) box at the edge of the S-D WAN. These two VCPE's definitions open room for a third one, the hybrid, joining both.In this paper we consider the three VCPE's definitions, admitting that in all cases there is a main computing Container either physical or virtual, wrapping the network functions. In

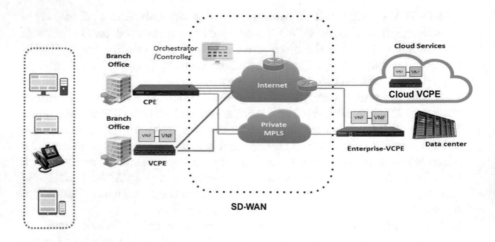

Fig. 1. S-D WAN and VCPE

the Fig. 1 the Virtual Customer premise Equipment (VCPE) is shown as a bare metal (X86 or ARM) box (black equipment) or as a virtual Computer (inside a Cloud provider), in which Virtual network functions run.

4.2   AgentX

The SNMP extensible Agent Agent X is an IETF initiative, aiming the hability of a SNMP agent to be composed by a group of dynamically aggregated auxiliary agents. Thus, The RFC 2741 describes an architecture in which there are two roles:

(a) **Master Agent:** The principal entity, which behaves as a single SNMP agent for the external Network management System (NMS) an as a coordinator for the internal auxiliary sub-agents.
(b) **Sub Agents:** They are Auxiliary agents which contribute with the instrumentation of part of Management information Base (MIB), but are not directly accessible from the NMS. A sub agent is accessed via Master Agent, only when a MIB object under its control is required.

The main Idea behind Agent X is the coordination among SNMP agents forming a single point of access for multiple MIB instrumentation providers. The Master Agent is the coordinator of the group, exposing a regular external SNMP access port. A internal group control port is used by the sub-agents to register and unregister their contributions to the MIB instrumentation.

Besides the roles, the Agent X standard defines a protocol to rule the relationship between Master Agent and Sub Agents. Among the Agent X protocol primitives the most relevants are:

1. **Open and Close Session:** primitives used to organize the dialog of each pair Master-SubAgent;
2. **register and unregister MIB responsibility:** primitives used by the sub agents to inform the master agent, which part of the MIB they are responsible for;
3. **internal management data exchange (GET, GET BULK, GET NExt, SET):** primitives used to move management data between a sub agent and the Master agent;

The most well-known implementation of AgentX is the net-snmp framework [4]. Unfortunately In its basic operating mode, net-snmp runs each agent (master and sub-agents) as a instance of the snmpd deamon. This makes hard to some one, at run-time, to know which snmpd is the master and which is a specific sub-agent. This control must be done collecting the PID of each one upon the start command. this PID collection procedure imposes tailor made scripts or software, out of net-snmp API scope. Without this capacity of easily start and stop sub-agents, net-snmp is not adequate to fuflil the requirements of a internal VCPE management framework. It is important to remind that in a VCPE services and functions are deployed, started and stoped dynamically, which requires from the internal management framework the same capacity.

4.3 OSGi and Distributed OSGi

OSGI is a set of specifications, and reference implementations of a middleware for the dynamically deployable java modules.

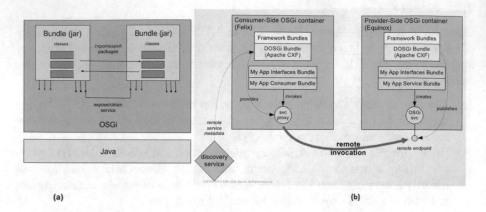

(a)   (b)

Fig. 2. OSGI and D OSGI reference architecture by Richard S. Hall

Figure 2(a) shows the standard OSGI system Stack. The OSGi middleware runs on top of the JVM, offering a set of services to loc ate, and dynamically deploy bundles (the term that OSGi uses to name its modules). A bundle is a self-contained bunch of computing artifacts able to offer and consume services. Typically a bundle is a   *.jar* artifact. Bundles receive unique identities in the OSGi middleware's register service, turning possible to operate each one, individually, at run-time. Furthermore, bundles can interact with the middleware, requesting: the deployment of a new bundle, the stopping, or updating of the active ones. Since the OSGi specification release 5, bundles upon OSGi middleware platforms in different hosts can access services from each other, as presented in Fig. 2(b). This feature is called Distributed OSGi (DOSGi) [13].

5   The DynagentX Framework

The DynagentX framework is an AgentX standard adaptation for fulfill the requirements of the SD-WAN, or edge computing VCPE. The framework starts from a VCPE internal management instrumentation architecture, and goes through the OSGi agentX implementation. It is important to highlight the flexibility and adaptability provided by the DynagentX, the main requirement of the VCPE's management.

5.1   The VCPE Internal Management Instrumentation and Communication Architecture

The Fig. 3 shows the DynAgentX proposed architecture for a VCPE's internal management instrumentation. Its components are:

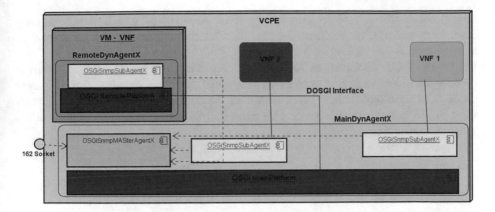

Fig. 3. DynAgentX architecture

- **The Main DynAgentX platform:** It is the architecture's foundation piece, in which the main OSGi platform is instantiated. Upon this main OSGi platform the Snmp AgentX and the auxiliary bundles are deployed.
- **The OSGiSnmpMasterAgentX:** This is the bundle playing the Master AngentX role, which exposes the external SNMP (udp 161) interface, and the internal AgentX control services. On this same main OSGi platform.
- **The OSGiSnmpSubAgentX:** These are the bundles that consume Master's internal services, and offer internal management data exchange services (GET, GET BULK, GET NExt, SET). Each of them is bonded with a VNF (either implemented as a container, or a process).
- **The Remote DynagentX:** It represents the secondary OSGi platforms deployed into a separated and self-contained virtual machines or container. All the OSGi Remote platforms connect to the Main DynAgentX OSGi Platform via DistributedOSGi (DOSGi) interface. After DOSGi connection, OSGiSnmpSubAgentX bundles in remote OSGi platforms can interact with the OSGiSnmpMasterAgentX as they would do if they were in the Main DynAgentX platform.

It's fundamental to emphasize the dynamism provided by the OSGi platform, allowing run-time deployment, and shutdown of the OSGiSnmpSubAgentX bundles.

5.2   The OSGiSnmpAgentX Bundle Implementation

Figure 4(a) shows the OSGiSnmpAgentX Class diagram, in which can be noticed that both Master and Sub-agents derive from the same superclass, which is a extension of the SNMP4j library [7]. The SNMP4J was used intending to avoid the insertion of bugs if we had to re-code the SNMP protocol from scratch. Besides the OSGiSnmAgentX, the DynagentX has two other auxiliary bundles as shown in Fig. 4(b). The *Monitor Bundle:* is responsible for to detect VCPE's new functionalities, and to load the **OGSGiSnmpSubAgenX** bundle associated with each OSGi platform has its own Monitor. The *Topology Bundle*: is responsible for to keep track of all VCPE's remote OSGi platforms. Only the main OSGi platform has its Topology bundle. The remote OSGi platforms update their connection status using the main platform topologyBundle's services. The topology bundle keeps a unified view of all OSGi platforms involved in a VCPE's instance. It also can be extended to accommodate VCPEs with a more complex internal architectures.

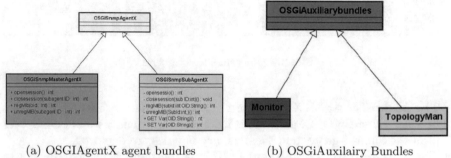

(a) OSGIAgentX agent bundles   (b) OSGiAuxilairy Bundles

Fig. 4. The implemented bundle UML class diagram

The interaction diagram shown in Fig. 5(a) (left side of the figure) shows how a Network Management System (NMS) accesses a nfv sub-agent through the Agent Master, while the (b) (right side of the figure) highlights how the Monitor bundle starts a new OSGiSnmpSubAgentX bundle when a new virtual network function is detected. As well as also shuts it down when the functionality is no longer needed. Whenever a new OSGiSnmpSubAgentX is needed, the monitor bundle requests the download from the secure https service-provider's site. A table associating nfvs and associated OSGiSnmpSubAgentX Bundle. Which is regularly updated by the monitor bundle. This mechanism is relevant to DynAgentX's self-adaptable behavior.

5.3   OSGiSnmpAgentX Implementation Notes

Security Note: DynAgentX's security relies on two main points:

(i) As a SNMP's derived technology, its security level depends on the utilized protocol's version 3.
(ii) Once the second external interaction happens between the Monitor bundle and the service provider secure site. DynAgentX's security depends on the provider site's security level.

It is capital to mark that, although possible, no other interactions are especified to the DynAgentX at this moment. So there are no other security risks mapped to DynAgentX.

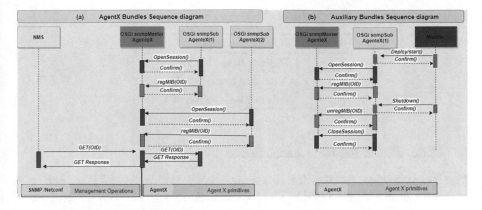

Fig. 5. OSGi bundles interaction diagrams

Third Part Libraries Used: Our implementation has used some third part open code presented bellow:

(i) the Apache karaf Cellar 4.1.5 running over the Apache Felix OSGi platform 1.18.0. Karaf Cellar provides all DOSGi features required by the DynAgentX.

(ii) We have used the SNMP4J2.2.0. within the JVM 1.8, as mentioned before

Developed Sub-agents and Respective MIBS: In order to permit some validation setups we have developed the following OSGisnmpSubagents:

(a) for DNS, Vnf1 : DNS-Bind9 - MIB : DNS ServerMIB (rfcs 1611, and 1612)
(b) for DHCP, vnf2 : DHCP-dhcp3-serve -MIB : IETF DHCP-ServerMIB (draft)
(c) for HAProxy, vnf3 : HAProxy-1.8.- MIB : the HA Proxy MIB (OID1.3.6.1.4.1.29385.106.1.0.0)

The OSGiSnmpSubAgentX for those vnfs were based on the respective MIBS. Once developed, the subagents can be loaded on the Main Platform or on the remote platform, depending on the VCPE composition and the orchestration decision. Orchestration is beyond the scope of this paper.

6   Validation Tests and Results

The validation experiments were mainly designed to verify the DynAgentX feasibility, and its performance behaviour. The DynagentX was functionally validated in Three VCPE composition scenarios, as shown in Table 1. All of them were setup on the bare metal VCPE, but only scenarios 1 and 2 were setup on the Cloud VCPE. For the two tested setups we simulated the dynamic deployment of vnfs described in Sect. 5.

Table 1. Validation scenarios

Scenario	VCPE composition
1	Process based VNFs
2	Container based VNFs
3	Virtual machine based VNFs

6.1 The VCPE Experimental Setup

The Bare Metal Setup: for this setup we've used a typical X86 box, a i3 processor, 8GRAM, 500G HD. The VCPE's HOST OS was the Ubuntu server 18.04.1 LTS. The Hypervisor was the qemu-kvm. The Container manager was the Docker-ce 18.03.1. The NMS station was a notebook i3 4G, With peppermint9 OS, using the net-snmp as SNMP client. A python script was created in order to lunch snmp PDUs and register the enlepsed time. The NMS station and the VCPE was connected by a Ethernet switch. **The Cloud based VCPE Experimental Setup:** for the VCPE Cloud based setup we've used the amazon EC2 service. Aiming to create a as fair as possible comparison, we've used a m4.large instance (2 vcpus, 8Gi RAM), which costed us 24 dollars cents per 24 h of tests. The NMS station was the same used in the bare metal VCPE setup.

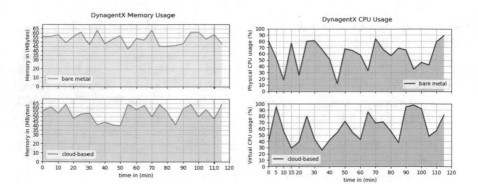

Fig. 6. OSGi Agent X JVM resource usage

The Performance Results: the performance test was made considering the 5 minuts EC2 monitoring data consolidation. The NMS station was configured to send SNMP PDUS separated by 1 min intervals, during approximately 2 h. The graph shown in Fig. 6 (left side) presents the worst case for JVM memory usage in MBytes, among scenarios 1, 2, and 3 for the both setups. While the Fig. 6 (right side) presents the worst case for the CPU (bare-metalVCPE), among scenarios 1, 2, and 3; and VCPU usage (cloud-based VCPE), between scenarios 1 and 2. **The response time Test Results** Unfortunately We had not space to present The response time graph with the collected data. Even though, the typical response time for Bare Metal and Cloud-based VCPEs (among scenarios 1, 2 and 3) was mean = 25 ms, min = 5 ms, max = 45 ms. In order to take in account the Internet delay between our lab and the AWS instances, we've measured the RTT from the NMS station to the VCPE Amazon instance, which was mean = 145 ms, min = 140 ms, max = 156 ms. So the worse response time for the Cloud-based yagentX Framework (considering the RTT was mean = 170 ms, min = 145 ms, max = 201 ms.

Note: The graphs presented in this section do not consider the memory amount and CPU percentage used by the VNFs. The measures were made focusing the resource usage by the DynagentX's components. The performance tests have confirmed that DynagentX is not greedy for resources, being suitable to deployment over low-end cloud virtual machines or low-end bare-metal boxes. Observing the Dynagentx response time, it is possible to affirm that, for a few vnfs (three in our tests) its internal infrastructure does not compromise its responsiveness. Certainly, Dynagentx must be tested on a more computing intensive scenario, where vnfs would be heavily demanded, provoking high competition for memory and cpu. The Dynamic SubAgent Deployment Tests have shown that the DynagentX is prone to automation and migth be suitable to operate within an autonomic SD-WAN system.

The Dynamic SubAgent Deploymnent Validation Results. Additionally to performance tests, we conducted a limited validation to the dynamic deploy of the following virtual network function: DNS and DHCP. In order to do this validation, we programmed the Monitor Bundle to find the OSGiSNMPSub-Agents at a https secure url. We have made a packet trace to verify de subagent upload, but we did not collect performance measures for these sub-agents transfer. We also did not register the elapsed time till the sub-agent has became fully operational.

7   Conclusion

This paper presented the Dynagentx a framework aiming to extend the Internet management technology into the SD-WAN and edge computing scenarios.

The framework was based on OSGi Technology and its validation results have proved its adequacy to the VCPE management, either in the X86 bare-metal boxes, as well as in the cloud-based VCPE. The performance tests have shown that Dynagentx framework is responsive, even though doesn't consume too much computing resources.

The D-OSGi proved to be the right technology to give the flexibility and adaptability to the SNMP AgentX standard. The presented VCPE internal management instrumentation and communication architecture is the framework highlight, allowing different VNFs' implementation technology scenarios.

As future works, we intend to evaluate the DynAgentX in a more complex cloud-based VCPE internal-structure. Also we intend to find the minimum X86 bare metal hardware and EC2 instance for each VCPE scenario in Table 1. As mentioned before, we intend to perform a broader performance test involving a heavy vnfs demand. The VCPE orchestration and provisioning, not covered by the DynagentX framework, seems to be the natural extension for this work. Other interesting future path is to include the NetConf [6] into the presented management instrumentation and communication architecture. Furthermore, a massive security test is also on our radar. We think that a vCPE orchestration service compatible with some Management platfors such as OSM [11] and ONAP [16] would be helpful.

References

1. Alliance, O.: OSGI-the dynamic module system for Java (2009)
2. Case, J.D., Fedor, M., Schoffstall, M.L., Davin, J.: Simple network management protocol (SNMP). Technical report (1990)
3. Chavan, S.S., Madanagopal, R.: Generic SNMP proxy agent framework for management of heterogeneous network elements. In: 2009 First International Communication Systems and Networks and Workshops. COMSNETS 2009, pp. 1–6 (2009)
4. group at CMU, N.G.: the cmu-snmp suite (2002). http://www.net-snmp.org/
5. Daniele, M., Wijnen, B., Ellison, M., Francisco, D.: Agent extensibility (AgentX) protocol version 1. Technical report (1999)
6. Enns, R.: NETCONF configuration protocol. Technical report (2006)
7. Fock, F., Katz, J.: SNMP4J-The Object Oriented SNMP API for Java Managers and Agents, August 2008. http://snmp4j.org/index.html
8. Görkemli, B., Parlakışık, A.M., Civanlar, S., Ulaş, A., Tekalp, A.M.: Dynamic management of control plane performance in software-defined networks. In: NetSoft Conference and Workshops (NetSoft), 2016 IEEE, pp. 68–72 (2016)
9. Herrera, J.G., Botero, J.F.: Resource allocation in NFV: a comprehensive survey. IEEE Trans. Netw. Serv. Manage. **13**(3), 518–532 (2016)
10. Kerpez, K., Ginis, G.: Software-defined access network (SDAN), pp. 1–6 (2014)
11. Kourtis, M.A., McGrath, M.J., Gardikis, G., Xilouris, G., Riccobene, V., Papadimitriou, P., Trouva, E., Liberati, F., Trubian, M., Batalle, J., et al.: T-NOVA: an open-source MANO stack for NFV infrastructures. IEEE Trans. Netw. Serv. Manage. **14**(3), 586–602 (2017)
12. Michel, O., Keller, E.: SDN in wide-area networks: a survey. IEEE Internet Things J. 37–42 (2017). https://doi.org/10.1109/SDS.2017.7939138
13. Rellermeyer, J.S., Alonso, G., Roscoe, T.: R-OSGI: distributed applications through software modularization. In: ACM/IFIP/USENIX International Conference on Distributed Systems Platforms and Open Distributed Processing, pp. 1–20 (2007)
14. Rodero-Merino, L., Vaquero, L.M., Caron, E., Muresan, A., Desprez, F.: Building safe PaaS clouds: a survey on security in multitenant software platforms. Comput. Secur. **31**(1), 96–108 (2012)
15. Shi, W., Cao, J., Zhang, Q., Li, Y., Xu, L.: Edge computing: vision and challenges. IEEE Internet Things J. **3**(5), 637–646 (2016)

16. Slim, F., Guillemin, F., Gravey, A., Hadjadj-Aoul, Y.: Towards a dynamic adaptive placement of virtual network functions under ONAP. In: 2017 IEEE Conference on Network Function Virtualization and Software Defined Networks (NFV-SDN), pp. 210–215. IEEE (2017)
17. Vdovin, L., Likin, P., Vilchinskii, A.: Network utilization optimizer for SD-WAN. In: 2014 First International Science and Technology Conference (Modern Networking Technologies) (MoNeTeC), pp. 1–4 (2014)

k-Level Contact Tracing Using Mesh Block-Based Trajectories for Infectious Diseases

Kiki Adhinugraha[1]([⊠]), Wenny Rahayu[2], and David Taniar[3]

[1] Department of Computer Science and Information Technology,
La Trobe University, Melbourne, Australia
k.adhinugraha@latrobe.edu.au
[2] School of Engineering and Mathematical Sciences, La Trobe University,
Melbourne, Australia
w.rahayu@latrobe.edu.au
[3] Faculty of Information Technology, Monash University, Melbourne, Australia
david.taniar@monash.edu

Abstract. Contact tracing is a method of identifying people who may have come into contact with an infected person during a virus incubation time. Standard contact tracing is performed to identify only the first level of exposures, whereas in reality, exposures can occur across multiple levels. Multi-level contact tracing can reveal a wider range of objects which contribute to the spread of the virus. This paper proposes a k-Level Contact Tracing Query (kL-CTQ) to reveal a broader range of objects which have possibly been exposed to pathogens. To minimize manual location tracing while preserving the user's privacy, we propose a mesh block sequence (MBS) method where the trajectories are transformed into an MBS before being shared with health authorities. While our simulation uses an Australian administrative region structure, this method is applicable in countries which implement similar administrative hierarchical building blocks.

1 Introduction

Pathogens that cause infectious diseases can spread directly from one person to another through respiratory droplets. When a pathogen infects a person, the time it takes for the symptoms to appear is called the incubation time or subclinical stage. During this incubation time, infected hosts might spread the virus to the nearby objects while travelling from one place to another [7]. Identifying these newly exposed objects as a result of being in close contact with an infected person is referred to as contact tracing [11].

Contact tracing has become critical in understanding how and where the virus has spread. Several contact tracing strategies have been applied to assist the government and health authorities to fight the COVID-19 pandemic such as COVIDsafe [4], PEPP-PT [14], and SafePaths [16]. In Australia, despite the high

© The Author(s), under exclusive license to Springer Nature Switzerland AG 2021
L. Barolli et al. (Eds.): AINA 2021, LNNS 225, pp. 272–284, 2021.
https://doi.org/10.1007/978-3-030-75100-5_24

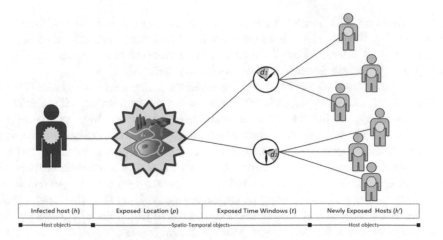

Infected host (h)	Exposed Location (p)	Exposed Time Windows (t)	Newly Exposed Hosts (h')

●——Host objects——●————————Spatio-Temporal objects————————●——Host objects——●

(a) Object Exposure

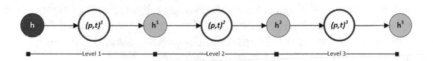

●——————Level 1——————●——————Level 2——————●——————Level 3——————●

(b) 3-Level Exposures

Fig. 1. Pathogen transmission

number of COVID-19 cases, the COVIDsafe app has not provided much assistance to health authorities, and the number of users who utilize this application is still below the minimum required number [13].

Figure 1 shows how objects can be exposed to pathogens which can then spread from one host to the broader community. There are two types of objects which can possibly be exposed which need to be identified during the contact tracing process. The first type is the spatiotemporal objects, and the second one is the host objects h, as shown in Fig. 1a. The spatiotemporal objects compromise two inseparable objects, which are location and time window as (p, t). The spatiotemporal object is the location p where the infected host visited at a certain time window t. The spatiotemporal objects can be obtained by examining the hosts' historical trajectories.

When an infected host visits a certain location p at a specific time t, this location is considered to be an exposed location. If a group of people visited this exposed location (p, t), these people will have a high chance of having been exposed to the virus and will be the newly infected hosts h'. The location, time, and new hosts which have had direct interaction with the infected host h are considered to be the level-1 exposed objects. Common contact tracing queries aim to identify the first level objects. A k-level exposure is a condition where the newly exposed hosts h^n infect other spatiotemporal objects and hosts. As shown in Fig. 1b, the newly exposed hosts h^{n+1} has one intermediate connection to the original infected host h. As the level number increases, the number of newly infected hosts h^n and exposed locations increases.

It is important to understand the object's movement in contact tracing queries since the virus spreads with the urban movement [3]. In relation to contact tracing, an infected host is considered to have a high probability of spreading the virus if this host comes into close contact with other people while walking. Identifying a close contact event requires two people or trajectories to be located close to each other while walking, so there is no confusion between the trajectories of people who are travelling in separate vehicles. A user is considered to be walking if the velocity of trajectory point $V(x)$ is within the range of walking velocity, which is around 5–8 km/h. Due to the issues of inaccuracy and incompleteness in a trajectory and also the privacy issue, trajectory data is not suitable for contact tracing queries without undertaking a prior preprocessing step [2].

This paper proposes a k-Level Contact Tracing Query (kL-CTQ) to identify possible spatiotemporal objects and newly exposed hosts on a certain level of exposure using the users' trajectories. The trajectories will be transformed into a mesh block sequence (MBS) to minimisethe privacy issue [1]. In summary, we describe the contribution of our works as follows:

1. We propose a k-Level Contact Tracing Query (kL-CTQ) to identify wider object exposures.
2. We demonstrate the effectiveness of kL-CTQ in identifying wider object exposure than common CTQ using crowdsourced trajectory data.

This paper is organized as follows: Sect. 2 discusses about the related works and Sect. 3 discusses about the proposed model. The simulation and evaluation are presented in Sect. 4 and Sect. 5 concludes our work.

2   Related Works

During a pandemic, contact tracing is the most critical step in identifying the places which have been visited by an infected person and the people with whom they have come in contact during the incubation period. The COVID-19 virus can be transmitted to another person located within a 1.5 m–2 m range of a COVID-19 infected person. Therefore, a proximity-based detection approach using Bluetooth Low Energy (BLE) from a smartphone is utilized for many contact tracing methods and apps [15]. Most proximity-based contact tracing apps do not record the locations that have been visited due to privacy issues [17].

Another contact tracing approach is to use the person's historical trajectory data extracted from their mobile phone. Trajectory data have been widely used for spatial analysis problems, such as data mining for movement patterns [5,19], traffic modelling [22] and contact tracing [8].

One of the main aims of contact tracing is to identify all the places visited by an infected person during a trip. The geolocation problem has attracted many researchers who have proposed and developed new methods, especially in mobility environments [6,21]. Since revealing a person's geolocation for contact tracing might raise privacy issues, several methods have been developed to hide the person's real identity or location such as FakeMask to generate fake context [23], KAT to replace real locations with dummy locations [10], PLQP [9] and Landmark Tree [18].

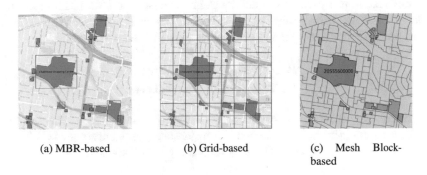

(a) MBR-based   (b) Grid-based   (c) Mesh Block-based

Fig. 2. Spatial object representation

In the contact tracing procedure, the coordinates of a trajectory point can be substituted with the name of the area in which the point is located. The area is considered an exposed spatiotemporal object instead of the coordinates themselves, as shown in Fig. 2. A minimum bounding rectangle (MBR) is used to represent a commercial building on a map in this figure. While an MBR might appropriately represent a commercial place, MBR coverage is limited to the object itself. Therefore, not all commercial locations will be considered. For example, if the infected host visited a park, the location will not be recognized.

The gap problem in the MBR approach can be solved using a grid model, where the idea is to split the area into uniform cells to cover the entire map [12,20] as shown in Fig. 2b. In a grid model, the entire map is covered without any visible gaps. However, an important location might be split into several cells. A mesh block-based approach is superior since this method does not show leave gaps or overlapping areas, and one crucial object is mapped into exactly one mesh block.

3   k-Level Contact Tracing

3.1   Preliminaries

Trajectory or GPS traces are a collection of s user's position x in a time interval or speed interval where $T = \{x_1, x_2, ..., x_n\}$. A trajectory can be obtained by utilizing a GPS-enabled device, such as a smartphone, to record movement and store the information either in local or cloud storage. A trajectory file may contain numerous trajectory points, where each point has several attributes or tags to describe the geolocation condition[1]. The most common tags are coordinates (latitude, longitude), elevation and time. Some trajectories have a time zone indicator in the tag such as +11:00 for AEST during daylight savings, while some trajectories denote the time in UTC format.

Definition 1. Movement mode $M(x_i)$ represents the estimated transportation mode used at a certain trajectory point x_i based on $V(x_i)$. The movement mode is the walk point when $V(x_i)$ is within the walking speed threshold v_w. The movement mode is vehicle point when $V(x_i)$ exceeds v_w.

$$M(x_i) = \begin{cases} WalkPoint(WP), & \text{if } 0 \leq V(x_i) \leq v_w \\ VehiclePoint(VP), & \text{otherwise} \end{cases} \quad (1)$$

Definition 2. An **MBS** is the set of visited MBs that represent the trajectory path in chronological order. The time a trajectory point enters an MB is called *entry time* t_e while the time the last trajectory point stays in the same MB is called *exit time* t_x.

Definition 3. Reduced Mesh Block Sequence (RMBS) is a subset of MBS where for each MB, the occurrences of WP must be higher than the walk point number threshold σ_{WP}, where $\sum WP \geq \sigma_{WP}$.

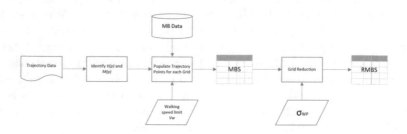

Fig. 3. Trajectory to RMBS transformation

The process to transform trajectory to RMBS is shown in Fig. 3. The first step is to identify the velocity point $V(x)$ and movement mode $M(x)$ for each trajectory point in the trajectory file. Using the grid map layer, the number of walk points WP and vehicle points VP are calculated and kept in the MBS table. The grids that have $\sum WP \geq \sigma_{WP}$ will be removed from the MBS table,

---

[1] https://www.topografix.com/GPX/1/1/.

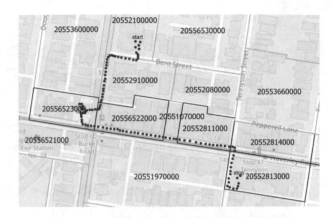

Fig. 4. Example of a mesh block trajectory

Table 1. Example of MBS to RMBS

MB	t_e	t_x	$\sum WP$	$\sum VP$	$\overline{V(p)}$
20552100000	.	.	7	2	4
20552910000	.	.	2	24	28
20556522000	.	.	3	1	12
...	.	.	...	...	...
20552813000	.	.	8	8	17

(a) MBS table

MB	t_e	t_x	$\sum WP$	$\sum VP$	$\overline{V(p)}$
20552100000	.	.	7	2	4
20556523000	.	.	16	6	3
20552813000	.	.	8	8	17

(b) Reduced MBS table $\sigma_{WP} = 5$

and the final result will be the RMBS table. Since an RMBS only contains MBs with a high number of walking points, RMBS is considered to be a list of exposed spatiotemporal objects.

We demonstrate the transformation of MBS to RMBS in Table 1. As shown in the example in Fig. 4, the MBS for trajectory T is the set of MBs in chronological order as follows: $MBS(T) = \{20552100000, 20552910000, 2055652 2000, 20556523000, 20556522000, 20551070000, 20552811000, 20552814000, 20552 813000\}$. We then populate the statistical data in each MB and present the MBS in Table 1a. The average speed $\overline{V(p)}$ for each MB is set to km/h. We disregard the entry time and exit time for the sake of simplicity. As shown in this table, Mesh Block 20556522000 occurs twice as the user visits the same MB at different times.

3.2 k-Level Contact Tracing Query

A contact tracing query is a method of obtaining exposed objects from the infected host by observing the host's historical trajectories at a particular time. The k-Level Contact Tracing Query (kL-CTQ) is used to obtain all the exposed objects at level k from an infected host.

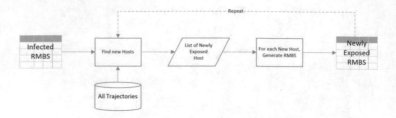

Fig. 5. Contact tracing query in RMBS

In kL-CTQ, there are two objectives that need to be achieved, which are:

1. To identify all exposed spatiotemporal objects.
2. To identify all new hosts and their trajectories.

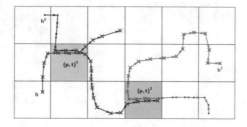

Fig. 6. 2-level contact tracing

The kL-CTQ is based on MBS where the exposed spatiotemporal objects are obtained from RMBS since the RMBS table only contains all mesh blocks with a high number of walking points. The MBS-based CTQ is shown in Fig. 5. The first step is to identify new hosts based on the infected RMBS. The infected RMBS is a table that contains all the spatiotemporal objects for the infected host. Once all newly exposed hosts have been identified, each exposed new host generates RMBS from each host trajectory. For k-Level CTQ, this process is repeated $k - 1$ times to obtain all k-level exposed objects.

Figure 6 shows the 2-level CTQ where the infected host is identified by the red trajectory h. Let $(p, t)^1$ be the level-1 exposed spatiotemporal object when the trajectory from host h^1 goes through this exposed spatiotemporal object. This host becomes the level-1 exposed host. When host h^1 could expose other users, the location and time is called the spatiotemporal object level-2 $(p, t)^2$. When the green trajectory from host h^2 goes through this location and becomes exposed due to host h^1, host h^2 will become the newly exposed host level-2. The 2L-CTQ(h) will obtain $(p, t)^1, (p, t)^2$ as two spatiotemporal objects, and exposed hosts h^1, h^2.

4 Experiments and Evaluation

This section evaluates the efficiency of kL-CTQ in identifying the k-level of exposed objects, which can be spatiotemporal objects or other hosts. We modify the GPX dataset from planet.gpx[2] to ensure all the trajectories share a similar travelling time to simulate the massive movements in the $850\,\mathrm{km}^2$ of the Melbourne metropolitan area. For these experiments, we processed 300+ trajectories from various contributors. We transformed the trajectories into RMBS for a day trip. Our methods are implemented in a system running Java with PostGIS as the database engine[3]. For the visualization tool, we use QGIS[4]. The system uses Intel i7-8665U CPU with 16 GB of RAM.

4.1 Level-1 Object Exposure

In this evaluation, as an example, we assume trajectory number 000509760 belongs to an exposed host. As the average walking speed is around 5 Km/h, we set the walking speed threshold to $V(p) = 8\,\mathrm{km/h}$ to accommodate a faster walking speed. Therefore, all the velocity points beyond this limit are considered a non-walking mode, while all the points below this velocity limit are considered a walking mode. Common contact tracing queries are focused on obtaining the exposed level-1 objects.

The example of obtaining the 1st level of exposure from an infected host is illustrated in Fig. 7. Figure 7a shows the trajectory of the infected host and the exposed spatial-temporal objects $(p, t)^1$ along the trajectory path indicated by the red polygon, where the average trajectory speed is within the walking speed of 8 km/h. Figure 7b shows the table of newly exposed hosts in a mesh block 20400630000, which has two-time windows at 8 AM and 10 AM. During time windows, eight possible newly exposed hosts visited these exposed spatiotemporal objects. The trajectory routes from these newly exposed hosts are shown in Fig. 7c.

The number of newly exposed hosts from all the exposed spatiotemporal objects from this infected host is shown in Fig. 8, where some exposed locations might have no host exposure at all. There might be no other users or trajectories at these spatiotemporal objects, or the trajectories moved faster than the walking speed limit.

---

[2] https://planet.openstreetmap.org/gps/.
[3] https://postgis.net.
[4] https://www.qgis.org.

(a) Spatiotemporal Objects Level-1

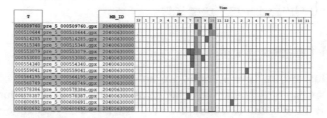

(b) Exposed Hosts list in (204006300000,08:00-11:00)

(c) Newly Exposed Trajectories from Exposed MB 204006300000

Fig. 7. 1st level exposure

4.2 Level-k Exposed Objects

For the level-k object exposure, we limit our evaluation to level-3 object exposure. From the same infected host, the results of our evaluation are shown in Fig. 9. The exposed locations or MBs are shown in Fig. 9a, where the number of distinct mesh blocks increases significantly from level 1 to level 2. The same trend can be seen for the number of spatiotemporal objects shown in Fig. 9b. While the number of exposed locations and spatiotemporal objects increases significantly from level 1 to level 2, the exposed area expansion goes from level 2 to level 3 as shown in Fig. 9c. The exposed coverage area has increased 10-fold, from only 22 km^2 to 229 km^2 at level 3 as shown in Fig. 10.

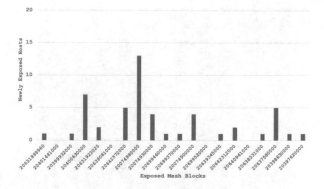

Fig. 8. Newly exposed hosts for each exposed MB

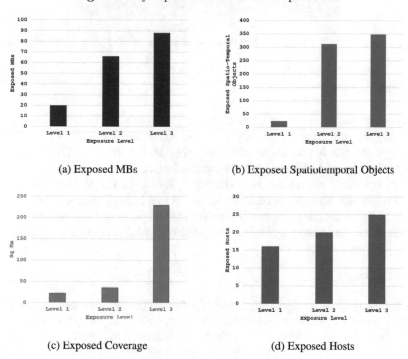

(a) Exposed MBs

(b) Exposed Spatiotemporal Objects

(c) Exposed Coverage

(d) Exposed Hosts

Fig. 9. *k*-level exposure

Although the exposed region is vast, the crowdsourced data does not provide sufficient trajectories for contact tracing evaluation. Therefore, there is no significant increase in the number of newly exposed hosts from level 1 to level 3, as shown in Fig. 9d.

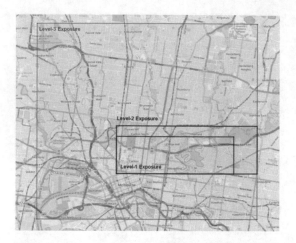

Fig. 10. Exposed locations on level 2

5   Conclusion

In this paper, we proposed k-Level Contact Tracing Query (CTQ) with the aim of identifying two objects, namely exposed spatiotemporal objects and newly exposed hosts for a certain level. The value of k can show the relationship between the exposed objects and the infected host, where $k > 1$ indicates there are no direct relationships between the exposed objects and the original infected host. Our simulation using a crowdsourced dataset in Melbourne shows that kL-CTQ might play an essential role in identifying the potential spread of pathogens for contagious diseases and prevent unnecessary lockdown during a pandemic.

References

1. Adhinugraha, K., Rahayu, W., Hara, T., Taniar, D.: Contact Tracing with Mesh Block-based Trajectories. ACM Trans. Spat. Algorithms Syst. (submitted) (2020)
2. Adhinugraha, K., Rahayu, W., Hara, T., Taniar, D.: Dealing with noise in crowdsourced GPS human trajectory logging data. In: Concurrency Computation, September, pp. 1–18 (2020). https://doi.org/10.1002/cpe.6139
3. Alamri, S., Taniar, D., Safar, M.: A taxonomy for moving object queries in spatial databases. Future Gener. Comput. Syst. **37**, 232–242 (2014). https://doi.org/10.1016/j.future.2014.02.007. http://dx.doi.org/10.1016/j.future.2014.02.007
4. Australia, D.O.H.: COVIDSafe. https://www.health.gov.au/resources/apps-and-tools/covidsafe-app
5. Goh, J., Taniar, D.: Mining frequency pattern from mobile users. Lecture Notes in Computer Science (including subseries Lecture Notes in Artificial Intelligence and Lecture Notes in Bioinformatics), vol. 3215, 795–801 (2004). https://doi.org/10.1007/978-3-540-30134-9_106

6. Goh, J.Y., Taniar, D.: Mobile data mining by location dependencies. Lecture Notes in Computer Science (including subseries Lecture Notes in Artificial Intelligence and Lecture Notes in Bioinformatics), vol. 3177, 225–231 (2004). https://doi.org/10.1007/978-3-540-28651-6_33

7. Hellewell, J., Abbott, S., Gimma, A., Bosse, N.I., Jarvis, C.I., Russell, T.W., Munday, J.D., Kucharski, A.J., Edmunds, W.J., Sun, F., Flasche, S., Quilty, B.J., Davies, N., Liu, Y., Clifford, S., Klepac, P., Jit, M., Diamond, C., Gibbs, H., van Zandvoort, K., Funk, S., Eggo, R.M.: Feasibility of controlling COVID-19 outbreaks by isolation of cases and contacts. Lancet Global Health **8**(4), e488–e496 (2020). https://doi.org/10.1016/S2214-109X(20)30074-7

8. Kim, W., Lee, H., Chung, Y.D.: Safe contact tracing for COVID-19: a method without privacy breach using functional encryption techniques based-on spatiotemporal trajectory data, pp. 1–16 (2020). https://doi.org/10.1101/2020.06.29.20143180

9. Li, X.Y., Jung, T.: Search me if you can: privacy-preserving location query service. In: Proceedings - IEEE INFOCOM, pp. 2760–2768 (2013). https://doi.org/10.1109/INFCOM.2013.6567085

10. Liao, D., Li, H., Sun, G., Anand, V.: Protecting user trajectory in location-based services. In: 2015 IEEE Global Communications Conference, GLOBECOM 2015 (2015). https://doi.org/10.1109/GLOCOM.2014.7417512

11. Marcel, S., Christian, A.L., Richard, N., Silvia, S., Emma, H., Jacques, F., Marcel, Z., Gabriela, S., Manuel, B., Annelies, W.S., Isabella, E., Matthias, E., Nicola, L.: COVID-19 epidemic in Switzerland: on the importance of testing, contact tracing and isolation. Swiss Med. Wkly. **150**(11–12), 4–6 (2020). https://doi.org/10.4414/smw.2020.20225

12. Mariescu-Istodor, R., Fränti, P.: Grid-based method for GPS route analysis for retrieval. ACM Trans. Spat. Algorithms Syst. **3**(3), 1–28 (2017). https://doi.org/10.1145/3125634

13. Meixner, S.: How many people have downloaded the COVIDSafe app and how central has it been to Australia's coronavirus response? (2020). https://www.abc.net.au/news/2020-06-02/coronavirus-covid19-covidsafe-app-how-many-downloads-greg-hunt/12295130

14. PEPP-PT: Pan-European Privacy-Preserving Proximity Tracing (2020). https://www.pepp-pt.org, https://github.com/pepp-pt/pepp-pt-documentation/blob/master/PEPP-PT-high-level-overview.pdf

15. Pietrzak, K.: Delayed Authentication Replay and Relay Attacks on DP-3T, pp. 1–6 (2020). https://eprint.iacr.org/2020/418.pdf

16. Raskar, R.M.: Private Kit: Safe Paths; Privacy-by-Design (2020). https://safepaths.mit.edu/

17. Reichert, L., Brack, S., Scheuermann, B.: Privacy-Preserving Contact Tracing of COVID-19 Patients. IACR Cryptol, pp. 4–5 (2020). http://arxiv.org/abs/2004.06818, https://eprint.iacr.org/2020/375.pdf

18. Shao, Z., Taniar, D., Adhinugraha, K.M.: Range-kNN queries with privacy protection in a mobile environment. Pervasive Mob. Comput. **24**, 30–49 (2015). https://doi.org/10.1016/j.pmcj.2015.05.004

19. Taniar, D., Goh, J.: On mining movement pattern from mobile users. Int. J. Distrib. Sens. Netw. **3**(1), 69–86 (2007). https://doi.org/10.1080/15501320601069499

20. Thomas, S.A., Gharge, S.: Enhanced security for military grid reference system using visual cryptography. In: 2018 9th International Conference on Computing, Communication and Networking Technologies, ICCCNT 2018 (2018). https://doi.org/10.1109/ICCCNT.2018.8494139

21. Waluyo, A.B., Srinivasan, B., Taniar, D.: Research on location-dependent queries in mobile databases. Comput. Syst. Sci. Eng. **20**(2), 79–95 (2005)
22. Xu, F., Huang, Z., Dai, H., Zhu, X., Wu, H., Zhang, J.: A real-time traffic index model for expressways. Concurrency Comput. **31**(24), 1–20 (2019). https://doi.org/10.1002/cpe.5155
23. Zhang, L., Cai, Z., Wang, X.: FakeMask: a novel privacy preserving approach for smartphones. IEEE Trans. Netw. Serv. Manage. **13**(2), 335–348 (2016). https://doi.org/10.1109/TNSM.2016.2559448

Architecture for Real-Time Visualizing Arabic Words with Diacritics Using Augmented Reality for Visually Impaired People

Imene Ouali[1(✉)], Mohamed Saifeddine Hadj Sassi[2], Mohamed Ben Halima[1], and Ali Wali[1]

[1] REGIM-Lab: Research Groups in Intelligent Machines, National Engineering School of Sfax, University of Sfax, BP 1173, 3038 Sfax, Tunisia
{mohamed.benhlima,ali.wali}@ieee.org
[2] Digital Research Center of Sfax (CRNS), Laboratory of Signals, systeMs, aRtificial Intelligence and neTworkS (SM@RTS), Sfax, Tunisia
mohamed-saifeddine.hadjsassi@enis.tn

Abstract. The number of visually impaired people is continuously increasing around the world due to the aging of the population. Several methods of communication are needed as solutions to perform their daily activities as Arabic text which appear everywhere and contains several diacritical marks such as i'jām, consonant score and tashkil. Accordingly, several architectures have adopted different techniques to detect and recognize Arabic words without diacritics and few are capable of reading the Arabic word with diacritics. In this context, a new architecture based on Augmented Reality (AR) is proposed in order to detect and recognize Arabic letters with diacritical marks in real-time. The proposed architecture shows great potential for using AR engine to detect Arabic words with diacritics within orientation, writing style and complex background. In addition, it improves the visualization by reading the detected Arabic words with diacritics through a created dataset. Our work aims to improve the user's experience and simplicity for partially sighted and visually impaired people.

1 Introduction

Nowadays, the use of Immersive Virtual Technologies (IVT) leads to a more efficient and creative future. Augmented Reality (AR) provides an ideal interface to smart devices in order to show information such as the text displayed on natural Arabic scenes in real time. Accordingly, text is used everywhere in our environment like advertising posters, business cards and driving directions. Thus, it is very important to understand it to interact with our environment. However, this is not the case for visually impaired people. The information can be treated incorrectly in case of unclear text. Visually impaired people may find difficulties to understand texts in case of complex background or bizarre text

L. Barolli et al. (Eds.): AINA 2021, LNNS 225, pp. 285–296, 2021.
https://doi.org/10.1007/978-3-030-75100-5_25

type or small size text. Moreover, the complex script of Arabic text requires big effort and attention to ensure an accurate text detection. The Arabic script has numerous diacritics, including i'jam, consonant pointing, and tashkil, supplementary diacritics. Thus, the detection and recognition of words with diacritics is a complex process. Even though several architectures exist for detection and recognition of Arabic text based on deep learning (DL), few systems are implemented in intelligent devices capable of detecting, recognizing and pronouncing Arabic words with vowels. Hence, a new architecture has to be created in order to detect and recognize Arabic words with diacritics. In this context, we propose a new architecture based on AR for detection, recognition and reading of the Arabic with diacritics and their vowels. Consequently, our architecture uses AR should have a transition from supporting communication among text on natural scene, and human to enabled by data visualization methods through speeches for people with visually impaired. In the current work, our contributions can be summarized as follows:

- Several works [1, 26] related to text detection and recognition recently published by the research community have been introduced. The problem of detection and recognition has been studied for many years. Scientific researchers indicate that the small fonts and the complex background and adjustments like color contrasts are big problems facing detection and recognition tasks. It is therefore necessary to continue testing and researching with different technologies to resolve such problems. This will enhance the selection of appropriate technologies for our architecture that deal with needs of visually impaired people.
- We provide an architecture of detection and recognition Arabic text with vowels. Several systems based on Deep Learning (DL) for Arabic text detection. However, some of such systems focus on the issue related to detecting Arabic vowels. For this reason, we have studied methods of text with vowels detecting based on DL. The limitation of these works allows us to study other technologies like AR in order to detect correctly Arabic text with vowels.
- We present the detection process method using AR engine and show its effectiveness on detecting Arabic Text in natural scenes through different devices cited in related work section. This can help future researcher to understand how to choose the suitable engine.

The remainder of this paper is organized as follows: Sect. 2 defines several related works. In Sect. 3, the framework of the proposed method and system is described. Section 4 present the evaluation of our method. Section 5 present the result and the discussion. Finally, Sect. 6 is reserved for the conclusion and future work.

2   Related Work

In this section, we present related work focused on the development of new correspondence methods and systems. Several works of text detection and recognition from videos such as [6–8] and from images such [4,5,10–14,17] and [29] were proposed. A comparative study of different methods and systems of detection and recognition of text from images is presented in Table 1. Among these methods of text detection and recognition, there are systems based on AR as a device such as [1–3,16,18] and DL such as [16,19,20,23,24,27] and [26].

Table 1. Comparison between text detection and recognition systems

	Ref	Engine	Device	Databases	Language
AR	[1]	Vuforia	Smartphone or tablet	Vuforia Word List (VWL)	Arabic - English
	[2]	Unreal Engine 4	Android tablets	Photo gallery of Jesi (Italy)	Italy
	[3]	Vuforia	Android smartphone or tablet	Image data of Tajweeed Qur'an	Arabic
	[16]	Vuforia	Android smartphone or tablet	Vuforia Word List (VWL)	Banjar - Indonesian
	[18]	Vuforia	Android smartphone	Vuforia Word List (VWL)	Aceh - Indonesian
DL	[19]	OCR	Mobile android	Data from Web	Latin language
	[20]	OCR	Smartphone	Database created by authors	Latin language
	[24]	OCR	Microsoft HoloLens	Database created by authors	English
	[9]	OCR- FCN	Computer	ACTIV dataset	Arabic
	[23]	OCR	Optical glasses	Database created by authors	Latin language
	[26]	Google Vision API - OCR	Microsoft's HoloLens	Web	Latin language

Mostafa, et al. [1] has proposed an intelligent educational application based on AR technology. The proposed system is simple and useful and dedicated mainly for children by teaching them the alphabets, words and numbers in two languages (Arabic and English). It contains image recognition, text to speech recognition, fun games and other ideas that are appealing to children. In addition, Mori, et al. [2] show a virtual interactive experience created for the photo gallery of Jesi (Italy). Namely, three interactive works produced with Unreal Engine 4 to give the viewer a greater immersion in the immortal images of Lorenzo Lotto. The goal was achieved by creating three choreographies with audio supervised by a historian, a recreation of the works with threedimensional graphics and a soundtrack specially composed by Tecla Zorzi. Furthermore, Andriyandi, et al. [3] define an application of AR as one of the multimedia technologies that can be used as interactive educational support to help study the tajweed of the Koran. The method used in this research is Accelerated segment test corner detection (FAST) features. Detecting Arabic text with vowels consider as an open research issue since there are Arabic language have numerous diacritics, including i'jam, consonant pointing, and tashkil, supplementary diacritics. Moreover, Syahidi, et al. [16] present a mobile application (DewataAR) providing information on tourist objects in Bali. This application uses AR to display three-dimensional objects, video and audio information of the temples on these tourist objects. DewataAR works by scanning the brochure of a tourist object using an Android device such as: a tablet or a smartphone.

In addition, Ardian, et al. [18] describe a translation application of the Indonesian language Aceh (ARgot). This application is able to operate as a real-time application without the need for an input medium such as text that must type using a good quality camera and having the characteristics of the autofocus in the camera. This research focuses on the process of detecting and tracing text available in the company. This research also aims to see how AR technology can be applied to unmarked text and have a different font. The result of the search is an Android smartphone software, which enables real-time multimedia translation. It can therefore be used almost in all types of smartphone. Based on the test and the results of the respondent, this application is able to arouse the interest of the user and can also be applied in real life.

3   System Architecture

In this section, we present a new architecture to detect, recognize and read vowelized Arabic text. This work is inspired and based on our previous work [15]. We aim to detect the Arabic word using smart camera features such as a camera. In addition, the data collected will be analyzed using a tool. It detects the word in the image and verify it in a database. This tool establishes connections between the letters stored in the Vuforia dataset and the audioclip dataset. In addition, the analyzed data will be transmitted to an Audioclip data stored for reading and viewing using appropriate techniques and technologies. This makes the results of the Arabic text available to the user (Fig. 1). A detection and recognition method based on ontology such as [28] is proposed as a Future work.

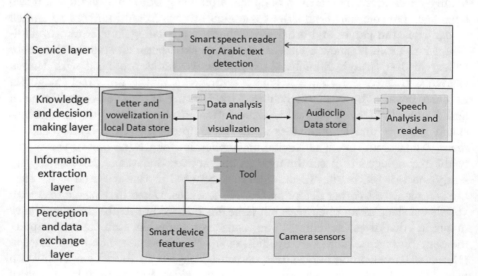

Fig. 1. Layers of our architectures

3.1 Mapping Between Schema Architecture and Existing Technologies

Based on a knowledge of the data processing [15] and [25] within proposed existing AR technologies at the related work section, we map the system components to more concrete oriented system components (Fig. 2). Firstly, we transmit collected data from camera sensors of smart devices to structured data that can be analyzed as this method in [21] and [22]. Accordingly, we define a tool that can extract and recognize data from unstructured data. Android fulfill requirements of the smart building system thanks to its availability, popularity, and portability. Android is not designed to be a platform only for mobile phones, it is a comprehensive platform for smart building devices such as a smartwatch, smart glasses, and so on. It can collect data from smart devices features that have capabilities like Wi-Fi, camera. Besides, the output data from the tool will be stored into a Comma Separated Values (CSV) file and loaded into a MySQL database through JSON Web service. The study of IVT technologies in the second section permit to select the appropriate AR toolkits and technologies for our architecture. We have decided to use Vuforia AR software development kit. This engine is more accessible for our study. Furthermore, we can use the local database of Vuforia to store data of holders. Besides, this engine helps to recognize Arabic words in a few seconds through the letter and vowelization in local data store of Vuforia. Moreover, Vuforia AR rendering engine delivers reliable vision-based experiences across a range of conditions and ensures that users have a great UX. Furthermore, audioclip data store aims to analyse and read the detected letters of the word. Finally, a smart speech reader for Arabic text available to end-to-end user.

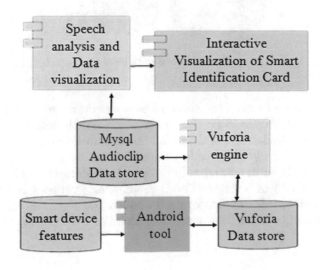

Fig. 2. Detailed level of our architecture

3.2   Used AR Engine and Database

Vuforia performs the detection and tracking of text elements by performing a camera display scan called the Region of Interest (ROI). This technology uses two different ROI, one used for detection and one used for tracking. Vuforia can detect words included in the list of predefined words. You can also specify a list of special words and filters that prevent word recognition. In our system, we add a database for the Arabic language. Text recognition is useful for applications that need to recognize individual words and word series. Text recognition may be used as a stand-alone feature or in combination with a target.

Vuforia relies on UTF-8 character encoding standards and can recognize all listed characters. It detects and tracks text elements by scanning the camera display called the Region Of Interest (ROI). A word database containing the words used to process the translator (Fig. 3). Word lists can be loaded from files and can be extended with additional words specific to application use cases. It is also possible to filter word lists using a filter list to exclude the detection of certain words (using a black list filter) or allow only the detection of certain words (using a white list filter). Word lists represent a basic set of words that Vuforia can use as a reference to appropriate text elements appearing in the camera's field of view.

Fig. 3. Used databases

The system process contains three phases: detection, recognition and concatenation (Fig. 4). In the detection phase, each Arabic letter with vowels is stored in the Vuforia dataset. A detection function for each letter is used in the system background to identify each vowelized letter by it correspond Identity (Id). Besides, an algorithm searches in the audio clip dataset on the corresponding audio of each Id of a letter. The concatenation synthesis is based on the concatenation of recorded speech segments from audio clip database. Thus, each Id is concatenated in order to generate the sound of the detected word.

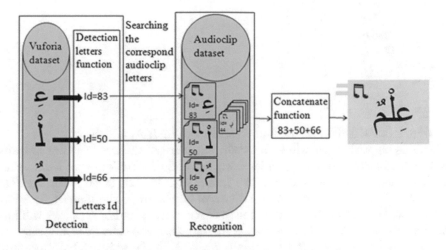

Fig. 4. Process of a word detecting and reading letters with vowels

3.3   Specification of Our Architecture

We have identified architectural requirements that are interdependent on the existence of the word in the dataset. Therefore, in this section, we will illustrate an algorithm for finding words *name* in the Vuforia dataset BD_v and the audioclup dataset BD_a. The new architecture is defined as:

$$\sum_{i=0}^{length(name)} (id(l)_i) \qquad (1)$$

- name: the name of the word.
- i: between 0 and length word.
- l: the letter i of the word.
- id: the identifier of each letter of the word.

To facilitate the tasks and clarify the principle of our idea, we will define this algorithm which allows to read Arabic word in the dataset. After giving the specification of our architecture, we define an algorithm to find the word in BD_v and BD_a.

```
Algorithm 1: Read Arabic word with vowels
res=""

For (i = 0, i ++, i < length(name))
    For (j = 0, j ++, j < length((BD_v))
        If (name[i] == (BD_v) [j])
            For (k = 0, k ++, k < length((BD_a))
                If (name[i] == (BD_a) [k])
                    res = res + id(name[i])
```

```
            ENDIF
         ENDFOR
      ENDIF
   ENDFOR
ENDFOR

return res
```

Algorithm 1 is used to read an Arabic word in BD_v and BD_a.

In this algorithm, we will look for the existence of the word (name) captured by camera in BD_v and BD_a. The word goes segmented letter by letter. Each letter (name[i]) has a specific id. The same letter has the same id for the BD_v and BD_a. We will look for the existence of the word in the dataset, letter by letter until the end of the word. If the id of the word found in BD_v, we will look for the same id in BD_a. In this case, if we find the same id in BD_v and BD_a, we will add this id in the result variable (res). This initialized variable was empty before starting the execution of the algorithm.

4 Evaluation

In this section, we assess the contribution of our proposed approach. First, we explain the experimental configuration for the evaluation of the proposed system. Then, we focus on the evaluation of this system (Fig. 6). The purpose of this evaluation is to test the potential of an AR device for the general public to improve the functional vision of the visually impaired and people with almost complete vision loss and to successfully detect, recognize and read Arabic word with vowels. During the evaluation of the functionalities of our AR application which exploit information from reality to facilitate large tasks, the participants commented positively and confirmed their usefulness in real life. In this part, we will detail the configuration steps to evaluate our proposed system. We have tested our system on 25 participants. The latter are 9 women and 16 men of different ages (between 20 and 50). All of participants were visually impaired peoples. Participants were identified as P1–P25, respectively. Each participant has tested the application and after answering a short questionnaire (Q1–Q4).

- Q1 (User Experience): How was your experience?
- Q2 (Satisfaction): Was the system comfort and easy to use?
- Q3 (Efficiency): Were the best results achieved for Arabic text visualization?
- Q4 (Effectiveness): Were the intended results achieved as error rates?

The provided mean value for Usability and User Experience of AR system for Arabic text with diacritics are significantly higher than other as shown in Fig. 6. Based on the test and the results of the respondent, this application is able to arouse the interest of the user and can also be applied in real life. The overall interface is considered to be powerful enough and flexible in order to detect and recognize the Arabic words with vowels. Furthermore, the detection method using the Vuforia engine with the ability to detect the word regardless of their character size, writing style, complex background (Fig. 5).

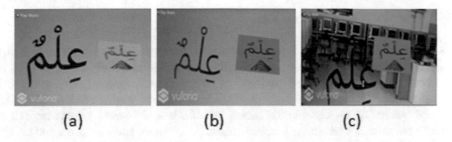

(a)   **(b)**   **(c)**

Fig. 5. Arabic word detection based on size (a), orientation and writing style (b), and complex background (c).

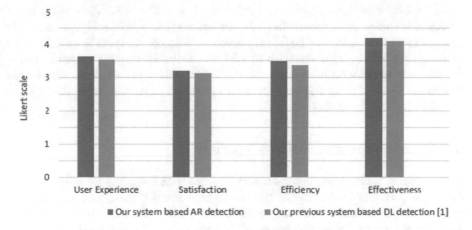

■ Our system based AR detection   ■ Our previous system based DL detection [1]

Fig. 6. Mean values per question for measuring usability and security on Likert scale

Table 2. Median standard deviation (SD) values per question for measuring usability and UX on Likertscale.

	Users of	Median	SD
UX	Our system	3.6	0.6
	System [9]	3.2	1.2
Satisfaction	Our system	3.6	1
	System [9]	3.5	1.1
Efficiency	Our system	3.2	1
	System [9]	4.2	0.2
Effectiveness	Our system	3.5	1.2
	System [9]	2.7	1.4

5   Result and Discussion

Based on the analysis performed on the test, the results confirm that our proposed system has no problems on the different styles of Arabic word, as long as the text still has a standard pattern on paper. The complex background, orientation and the size are solved through the Vuforia engine. Arabic language has 28 letters of the Arabic alphabet. Arabic has six vowel phonemes (three short vowels and three long vowels). Most of systems such as proposed in the work [9], cannot detect the Arabic language with vowels. We have created the database that can treat all the letters and vowels in Arabic language. Vuforia engine deals with the detection phase. We have created another database that can deal with recognition phase and solve the issue of vowels. Our system based on Android-based Smartphone in real-time unlike [16,19,20,23,24] systems. The used architecture is similar to [3]. However, a different method is used in order to recognize letters with vowels and add an audioclip dataset to read vowels. As we have mentioned before, we have used two datasets that can handle the different letters with vowels and solve the problem of writing style that can be found in [1,16,18] (Table 2).

6   Conclusion and Future Work

In this paper, we have presented an architecture that helps the visually impaired to read Arabic word with vowels. Hence, we have studied and analyzed the existing architectures under two categories based on the AR and DL technologies. This helped us to define the appropriate engine, device and database for our architecture. Furthermore, we have described the most important limitation of existing works. Accordingly, we have designed an AR based architecture for Arabic text detection to assist visually impaired people. Besides, we have mapped between schema architecture and suitable existing technologies. This allows to deal with problems that was discussed in the related works section. Our architecture provides a straightforward mobile application that uses Vuforia as an AR engine to bring in a real-time element with intuitively comprehensible interfaces. Our paper has focused on solving the limitations of existing works for Arabic text detection and recognition by taking into consideration visually impaired people needs. Several works have used deep learning to detect Arabic text without vowels. However, our system can detect and recognize Arabic words with vowels. In the future, we will try to explore the prospects of extending the data store by improving the system for a bigger data store of audio-clip. In addition, we intend to implement Microsoft Hololens display system. This could be used to improve the AR experience to be more comfortable.

References

1. Mostafa, A., Elsayed, A., Ahmed, M., Mohamed, R., Adel, M., Ashraf, Y.: Smart educational game based on augmented reality. Technical report, EasyChair (2020
2. Mori, B., Giovent'u, C.: An augmented reality (ar) experience forlorenzo lotto. In: Virtual and Augmented Reality in Education, Art, and Museums, pp. 324–332. IGI Global (2020)
3. Andriyandi, A.P., Darmalaksana, W., Maylawati, D.S., Irwansyah, F.S., Mantoro, T., Ramdhani, M.A.: Augmented reality using featuresaccelerated segment test for learning ta. Telkomnika **18**(1), 208–216 (2020)
4. Wei, Z.-H., Du, X.-P., Cao, L.: Research on the text detection and recognition in natural scene images. In: Eleventh International Conference on Graphics and Image Processing (ICGIP 2019), vol. 11373, p. 113730F. International Society for Optics and Photonics (2020)
5. Ahmed, S.B., Razzak, M.I., Yusof, R.: Progress in cursive wild text recognition. In: Cursive Script Text Recognition in Natural Scene Images, pp. 85–92. Springer, Heidelberg (2020)
6. Elleuch, H., Wali, A., Samet, A., Alimi, A.M.: A real-time eye gesture recognition system based on fuzzy inference system for mobile devices monitoring. In: International Conference on Advanced Concepts for Intelligent Vision Systems, pp. 172–180. Springer, Heidelberg (2016)
7. Elleuch, H., Wali, A., Samet, A., Alimi, A.M.: Interacting with mobile devices by fusion eye and hand gestures recognition systems based on decision tree approach. In: Ninth International Conference on Machine Vision (ICMV 2016), vol. 10341, p. 103410D. International Society for Optics and Photonics (2017)
8. Abdelhedi, S., Wali, A., Alimi, A M.: Human activity recognition based on mid-level representations in video surveillance applications. In: 2016 International Joint Conference on Neural Networks (IJCNN), pp. 3984–3989. IEEE (2016)
9. Beltaief, I., Halima, M.B.: Deep fcn for Arabic scene text detection. In: 2018 IEEE 2nd International Workshop on Arabic and Derived Script Analysis and Recognition (ASAR), pp. 129–134. IEEE (2018)
10. Turki, H., Halima, M.B., Alimi, A.M.: Text detection in natural scene images using two masks filtering. In: 2016 IEEE/ACS 13th International Conference of Computer Systems and Applications (AICCSA), pp. 1–6. IEEE (2016)
11. Ayed, A.B., Halima, M.B., Alimi, A.M.: Map reduce based text detection in big data natural scene videos. In: INNS Conference on BigData, pp. 216–223 (2015)
12. Turki, H., Halima, M.B., Alimi, A.M.: Scene text detection images with pyramid image and mser enhanced. In: 2015 15th International Conference on Intelligent Systems Design and Applications (ISDA), pp. 301–306. IEEE (2015)
13. Selmi, Z., Halima, M.B., Wali, A., Alimi, A.M.: A framework of text detection and recognition from natural images for mobile device. In: Ninth International Conference on Machine Vision (ICMV 2016), vol. 10341, p. 1034127. International Society for Optics and Photonics (2017)
14. Turki, H., Halima, M.B., Alimi, A.M.: A hybrid method of natural scene text detection using msers masks in hsv space color. In: Ninth International Conference on Machine Vision (ICMV 2016), vol. 10341, p. 1034111. International Society for Optics and Photonics (2017)
15. Sassi, M.S.H., Jedidi, F.G., Fourati, L.C.: A new architecture for cognitive internet of things and big data. Procedia Comput. Sci. **159**, 534–543 (2019)

16. Syahidi, A.A., Tolle, H., Supianto, A.A., Arai, K.: Bandoar: real-time text based detection system using augmented reality for media translator banjar language to Indonesian with smartphone. In: 2018 IEEE 5th International Conference on Engineering Technologies and Applied Sciences (ICETAS), pp. 1–6. IEEE (2018)

17. Sassi, M.S.H., Fourati, L.C.: Architecture for visualizing indoor air quality data with augmented reality based cognitive internet of things. In: International Conference on Advanced Information Networking and Applications, pp. 405–418. Springer, Cham (2020)

18. Ardian, Z., Santoso, P.I., Hantono, B.S.: Argot: text-based detection systems in real time using augmented reality for media translator aceh-Indonesia with android-based smartphones. In: Journal of Physics: Conference Series, vol. 1019, p. 012074. IOP Publishing (2018)

19. Rajendran, P.S., Christian, I.S., Shedge, M.S.: Aredai augmented reality based educational artificial intelligence system (2019)

20. Gupta, T., Sisodia, M., Fazulbhoy, S., Raju, M., Agrawal, S.: Improving accessibility for dyslexic impairments using augmented reality. In: 2019 International Conference on Computer Communication and Informatics (ICCCI), pp. 1–4. IEEE (2019)

21. Sassi, M.S.H., Fourati, L.C., Jedidi, F.G.: Business information architecture for big data and Internet of Things. In: 2019 15th International Wireless Communications Mobile Computing Conference (IWCMC), pp. 1749–1756. IEEE (2019)

22. Sassi, M.S.H., Jedidi, F.G., Fourati, L.C.: A new architecture for cognitive internet of things and big data. Procedia Comput. Sci. **159**, 534–543 (2019)

23. Zhao, Y., Hu, M., Hashash, S., Azenkot, S.: Understanding low visionpeople's visual perception on commercial augmented reality glasses. In: Proceedings of the 2017 CHI Conference on Human Factors in Computing Systems, pp. 4170–4181 (2017)

24. Huang, J., Kinateder, M., Dunn, M.J., Jarosz, W., Yang, X.-D., Cooper, E.A.: An augmented reality sign-reading assistant for users with reduced vision. PloS one **14**(1), e0210630 (2019)

25. Sassi, M.S.H., Jedidi, F.G., Fourati, L.C.: Computer-aided software engineering (CASE) tool for big data and IoT architecture. In: 2019 15th International Wireless Communications Mobile Computing Conference (IWCMC), pp. 1403–1410. IEEE (2019)

26. Huang, J.: A hololens application to aid people who are visually impaired in navigation tasks (2017)

27. Sassi, M.S.H., Fourati, L.C.: Investigation on deep learning methods for privacy and security challenges of cognitive IoV. In: 2020 International Wireless Communications and Mobile Computing (IWCMC), pp. 714–720. IEEE (2020)

28. Ouali, I., Ghozzi, F., Taktak, R., Sassi, M.S.H.: Ontology alignment using stable matching. Procedia Comput. Sci. **159**, 746–755 (2019)

29. Ouali, I., Sassi, M.S.H., Halima, M.B., Ali, W.: A new architecture based ar for detection and recognition of objects and text to enhance navigation of visually impaired people. Procedia Comput. Sci. **176**, 602–611 (2020)

Progressive Mobile Web Application Subresource Tampering During Penetration Testing

Tobiasz Wróbel[1] ⓘ, Michał Kędziora[1] ⓘ, Michał Szczepanik[1] ⓘ, Piotr P. Jóźwiak[1] ⓘ, Alicja M. Jóźwiak[2] ⓘ, and Jolanta Mizera–Pietraszko[3](✉) ⓘ

[1] Faculty of Computer Science and Management, Wroclaw University of Science and Technology, Wrocław, Poland
`{tobiasz.wrobel,michal.kedziora,michal.szczepanik,`
`piotr.jozwiak}@pwr.edu.pl`
[2] Faculty of Architecture, Wroclaw University of Science and Technology, Wrocław, Poland
`239963@student.pwr.edu.pl`
[3] Department of Innovative Projects Management, Military University of Land Forces, Wrocław, Poland
`jolanta.mizera-pietraszko@awl.edu.pl`

Abstract. Since the boost of mobile devices popularity, both operating systems and mobile applications have become more complex which in turn transfers into greater vulnerability to hackers' attacks. Penetration testing is aimed at detection of security gaps in mobile systems. On the other hand, Progressive Web uses Web browser API to enhance the range of functionalities to cross-platform. Thus, this paper focuses on mobile Web application penetration tests of Progressive Web. First, some new functionalities were evaluated for vulnerabilities, then an in-depth analysis of the Web push functionalities was carried out. External resources, which deliver Web push services, were explored for the libraries security. Then, Man-in-the-Middle attack on Subresource Integrity Mechanism (SIM) was analyzed to exploit the vulnerabilities detected.

1 Introduction

Progressive Web Applications (PWA) are becoming a viable option when it comes to targeting the mobile market. The main benefit from a business standpoint is cost reduction. PWA is a type of application software delivered through the web and built using common web technologies including HTML, CSS and JavaScript. It is intended to work on any platform that uses a standards-compliant browser, including desktop and mobile devices. Based on that this solution also helps to increase audience and improve user experience on all popular platforms. Instead of creating separate applications for each mobile operating system, one application can be used for all of them. All that comes with almost no requirements of the user's experience on the contrary to the native applications. The user has the application's icon on his device's home screen, receives notifications and the application is not fully relying on the internet connection. But since Progressive Web applications have become in fact the only Web applications in usage with some extra functionalities, that modern Web browsers APIs implement, their vulnerability to attacks is comparable to the native apps [1].

This paper performs analysis of the Progressive Web application standard functionalities. We study Web push functionalities, external Web push services, Man-in-the-Middle attacks, subresource integrity mechanism, and subresource tampering. Based on it, we propose a tool to support penetration tests against the vulnerability detected. The proposed solution was then implemented and tested both in an artificial simulated environment and a real environment.

1.1 Progressive Web Applications

A Progressive Web application or PWA is a loose term that describes a type of Web application built using Web technologies but adjusted to give the user a native app [7]. There were three technical approaches that had a big impact on PWA as found an innovative and very popular solutions. First of all, there was a need to store data on the devices to make the applications less dependent on the constant network connection. After Application Cache or AppCache has become a part of the HTML standard [17], Websites and applications started informing the user's browser to save the files needed. They used a Manifest file, which contained a list of files, paths or URLs to be stored on the device's memory. Then, they can be accessible, even in case of the network unavailability. However, the question is how to check the cached data for validity. This is where Service Workers [6] bring their functionalities. They are event-driven JavaScript processes that work separately from the main browser thread. Such that it enables them to work in the background even if the user closes the browser. They can perform the operations such as data synchronization, or they can handle caching resources, intercept requests, receive updates for the application used by the service worker, and handle push notifications. The last functionality was defined by the World Wide Web Consortium in their recommendations [15]. The document describes an API that uses HTTP/2 to display notifications outside the context of a Web page. Under such circumstances, even in case when the user does not have his or her browser open, the notification will be in the background displayed to the user (see Fig. 1).

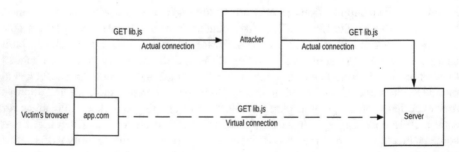

Fig. 1. The application is submitting requests for a library – first step of MitM Attack

When developing a Web application, we usually need some standard functionalities such as authentication, managing user accounts, logins, passwords, and access control. We expect these functionalities to be reliable, work fast, and be secure. Just security is the most crucial points. Diego Poza from Auth0, an authentication, and authorization service provider, describes, why it is not advised to manage user identities on your own [16]. Building the identity management component of an application may be a challenge. Integrating the application with the various login options may cause problems even with a very straightforward design. Another one is the user input validation. Custom user identity management comes with a great risk. Any vulnerabilities pose a threat and any data breach result in finding product unreliable. Apart from security issues, the service must be reliable as well. For this reason, according to a W3Techs survey [9] only 23.8% developers prefer to use frameworks and libraries, while others avoid them. Library usage means acquiring the source code from the application's server or accessing a content delivery network. In both cases, the library is transported over HTTP that can be either secure or insecure.

1.2 Subresource Tampering via MITMM

The transport phase of acquiring external resources might be vulnerable to a Man-in-the-Middle attack. In computer security, it is an act of sniffing by connecting both the victim and the destination, that the victim is trying to contact, and modifying messages between them without their knowledge [3]. A successful all-out attack gives the hacker a full knowledge about the content of the messages sent between the user and the server and even the ability to modify the contents or to insert messages into the communication channel. Man-in-the-Middle (MITM) is just a model of an attack and can be executed on different communication layers and channels.

Some attacks performed on HTTPS [3], cellular network UMTS [14], or Bluetooth [11] are reported in numerous resources. Nonetheless, the most dangerous out of these three mentioned is the attack on HTTPS, because of the common belief, that the lock symbol in the browser, which indicates a HTTPS connection to the destination, means their communication is secure. In practice, the security relies on the validity of the certificate used to encrypt the communication. The authors mention that a hacker may successfully perform the attack using his own certificate to secure the communication between him and the user, and the server's certificate for the communication between the attacker and the server simultaneously. Then the lock symbol misleads the user giving him a false sense of security. After introducing the HTTP Strict Transport Security [10] the encryption certificates are checked for trustworthiness that is whether or not it has been issued by a valid certificate authority. If HSTS (HTTP Strict Transport Security) is set in the website header, the connection is encrypted and consequently, any other connection that uses unverifiable certificates SSL is blocked such that it cannot be modified, and the user is alarmed about the incident. Once a hacker has successfully placed himself between the victim and the server that it is trying to contact him, there's a lot of harm to the victim. One of the methods called libraries tampering is while the victim's application is working the data can be downloaded in the background without the victim's awareness. Tampering means inserting, changing, or deleting any contents

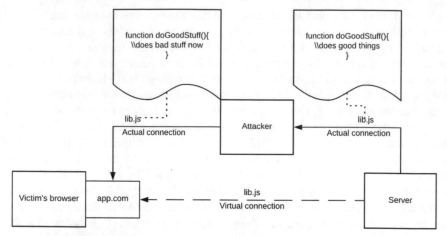

Fig. 2. When receiving a resource that can be tampered, the attacker changes it and then returns to the victim – final step of MitM Attack

of the library but only in the case when the resources reach the victim because the attacker redirects the traffic to it. A simplified subresource tampering attack is shown in Fig. 2.

After the main HTML file of the Website has been downloaded onto the victim's device, the browser starts executing the code. Without the the victim's awareness, the browser starts requesting for any external resources he needs. A well-performed attack does not have any impact on the application's standard behavior. Instead, it performs malicious actions in the background. For example, the authors of the "Pride and Prejudice in Progressive Web Apps: Abusing Native App-like Features in Web Applications" paper were able to change the URL of the third-party Web push server, that the attacked application tried to use. They replaced it with a self-controlled server were able to send self-crafted push notifications to the victim.

2 Related Research

With the rise of the popularity of Progressive Web Applications, the concerns about security related to it have arisen [19, 20]. A quite thorough study was conducted by a group of researchers at the Korea Advanced Institute of Science and Technology. The results were presented at the 2018 ACM SIGSAC Conference on Computer and Communications Security [12]. The authors presented how to abuse some of the new PWA features. One of them is capturing the subscription object, an object that enables the third-party Web-Push service to send push notifications to the application user. Given that the notifications resemble the original, the user may not notice it and become a victim of this attack. Even if the object is transported over a secure layer, an attacker can still access it. Some third-party Web-Push services need inclusion of a library to run the application service. Providing the library sent uses an insecure medium (plain HTTP for example), it can easily be intercepted and changed. As described in "6.2 - Reflected Transmission of Subscription Objects over HTTPS" [12] some libraries were holding the push service's server address, which when redirected to a hacker-controlled server, results in the user receiving attacker-crafted notifications. According to the "Involvement of Third Par-ties" paragraph in the paper "How the Web Tangled Itself: Uncovering the History of Client-Side Web (In)Security" [21] there are two major risks for the applications with third-party libraries in-built. The first one is the usage of some outdated versions having all security vulnerabilities identified. The second one is an insecure library transfer to the application requesting this library. So is the role of SIM. According to the recommendation from the W3 Consortium [2], in order to assure SIM of the requested resources, application developers are recommended to use the "integrity" attribute which contains a cryptographic hash representation of the resources. Upon receiving the server response with the resources, the application computes the hash to check for any inconsistency between the present and expected data.

3   Implementation of the Solution

The behavior of SIM attack can be approximated for testing purposes with a HTTP or HTTPS proxy server. As described in the paper about the CERN httpd proxy server [13], the server waits for requests from the clients. When he receives a request, he sends it to the destination server. The response is handled by the proxy server, which then forwards it directly to the client. Having the same rights as a MITM attacker the proxy performs additional tasks. One of the key functionalities of httpd [13] was caching resources. The server stores the resources to his own memory. Then, on receiving a request for a resource stored in his memory, the server returns it immediately, which gives a significant performance boost for the client and reduces the traffic on the channel to the destination server. Another proxy function can monitor and filter the resources. Having access to the destination URL, the proxy can manage transfer to the resource that the URL points to.

The search for an extendable proxy starts at the beginning of the implementation process. At first proxy.py [18] was considered. It does not need a lot of resource to work. The author describes its background for mobile developers to test their applications with server-side software in-built in another network. The proxy server allows to redirect the traffic. It is released under the BSD 3-Clause license. Proxy.py uses custom plugins, but the user cannot relate a particular request to a response. For this reason, the mitmproxy [5] was chosen as our proxy server to build upon. It provides easy-to-use interfaces for submitting requests and receiving responses, quite a straightforward filtering and a content replacement in the traffic. In addition, it handles relations between the requests and responses, so that it seems rather unlikely that the proxy user may receive the wrong response. Creating the plugin was also simplified by the mitmproxy developers. Mitproxy server provides an event-based interface that allows defining callbacks for many Web events, like request, response, TCP, or Websocket connection. The solution proposed is a plugin dedicated for PWA type of apps which provides an option to modify server requests and responses in a way that can detect some typical for PWA vulnerabilities.

3.1 Rule Based Replacements

The created tool (see Fig. 3) is an extension for the mitmproxy. The user can define the content replacements in HTTP response messages. When the response event in mitmproxy appears, it means that the whole HTTP response has been received by the proxy server. From now on the user gets an access to the request and the response. Also the replacements can be easily made.

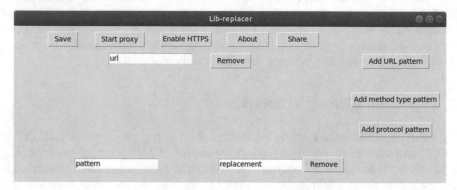

Fig. 3. The user interface of the proxy tool

At first, the request URL is matched against a set of patterns defined by the user. Each set is related to the pairs of a pattern and the replacement in case the pattern is matched. After all the modifications have been made to the response's content, it is forwarded to the proxy user. This way the same set of rules on various functions of the application being tested can be defined. After the first matching the URL pattern changes the content of the message. However, the rule applies that the changes defined by the user are minimal because another traffic is kept when received by the proxy and sent to the its user.

3.2 Transfer Rules Solution

The rules defined by the user can be saved to a file. The JavaScript Object Notation was chosen as the destination file format. In our opinion, it simplifies handling the read & write commands from and to the json library for Python [8]. Also, it automatically handles the mapping of the data from the file to a Python dictionary object being a set of the key & value pairs. The text-based file with the set of the rules is integrated with the version control systems. When comparing two versions of the rules file, the user may have some preferences such as his favorite text editor, the integrated development environment (IDE), or the version control system, because most of them implement an interface for highlighting differences between the files. The fundamental structure of the file containing the rules is shown in Fig. 4. The character sequence in line 2 is the pattern that the request URL is matched against. In case the pattern matches the URL, the tool starts processing the key value pairs that are related to this URL pattern.

4 Runtime Environment and Testing

In order to run the background proxy process, which is in fact mitmproxy's mitmdump [4], and the user interface, Python environment version at least 3.6 is required. To quickly setup the environment we will also need pip, a package installer for Python.

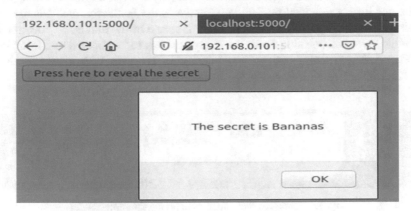

Fig. 4. The resulting alert with the changed text message

To separate the environment, we use a virtual environment, which allows creating individual Python environments with their own packages isolated from other environments or the operating system's (OS) packages. Venv is a module that comes as standard with Python. Before running the proxy we have to establish a set of rules to be in use. As the tool uses regular expressions, we can define our own rules to be matched. After that, we want to define the content's regular expression and the character sequences for replacements. On having that done the rules in the graphical user interface (GUI) are saved to the JSON file. From then on, no work will be lost. After setting up all the URL patterns and the replacement rules, we start the proxy server. By default, it runs on port 8899, but it can be changed in the "config.conf" file in order to filter and tamper only the tested application's traffic instead of our whole computer's internet traffic.

The mock application is a simple Web application that has only a single button. On clicking a button, it executes the "myFunction" method whose body is located in the "lib.js" library, that the Web browser requests after it receives the HTML file. The method displays an alert. In the test for the tool, we change the library content and make the Web application display an alternative alert (see Fig. 4). We have to match against "lib.js" in the request URL and then replace the strings.

For the regular expressions usage to match and replace character sequences, we use "lib[.]js" which matches the desired request URL, as the dot is a special sign in regular expressions representing any character. Adding the brackets indicates that we want to match the dot character rather than any character. Then we put our strings for the content pattern to match and replace it. After refreshing the page by running the proxy and pressing the same button again, we can see an alert as in Fig. 5. The researchers behind the "Pride and Prejudice in Progressive Web Apps: Abusing Native App-like Features in

Web Applications" paper [12] have proved the "AFN.az" news portal to be vulnerable, so we want to test our tool on it. When accessing the portal using the URL "afn.az" we are redirected to a non-secure HTTP page. There is a HTTPS version, but the user would have to explicitly type in the full URL including the right protocol. In this test, we try to replace the standard footer of the dialog shown at startup, which asks the user if he would like to receive notifications from the portal. After a brief research, we can see that the library provided by SendPulse contains the JavaScript methods to display the subscription dialog. To replace the footer message of the dialog, we identified the part of the library responsible for it. We change the text to "Tampered by LibReplacer" (see Fig. 5). After running the proxy and accessing the tested Website, we receive the following dialog that encourages us to subscribe to the service.

Fig. 5. The tested Website after applying the replacement

As for quantity evaluation, we have tested the user inputs and data presented in this solution on 132 Progressive Web Apps altogether. Our analysis indicated that around 83% of the server data were vulnerable and 41% of user inputs data were affected by this attack, but they are usually additionally verified on server-side before processing. Environment settings needed for performing this kind of attack decrease the possibility that this kind of attack can be performed in real scenario secure environment on the user's site.

5 Conclusion and Further Work

The goal of this paper was to analyze some attacks on subresource integrity SIM using the Man-in-the-Middle method. Vulnerabilities were assessed in the new Web application standard, called Progressive Web application. After a more in-depth analysis of Web push, a new functionality brought by the PWA standard, external Web push services were assessed for their security. On detecting an attack against an insecure Web push library transport, we created an attack vector. Man-in-the-Middle attack was analyzed for its impact on the attacked application's activity. A Man-in-the-Middle attack type on insecure resource transport was chosen as the main attack vector that represents the developed tool. After establishing the attack method, mitmproxy was chosen as the proxy server to build the tool upon. We developed an extension for mitmproxy that allows the user to modify the responses received by the proxy before they were relayed to the client. The tool was tested on a mock Web application and a real news portal application. In both cases the tool successfully modified the contents of the libraries required by the Web applications, but it did not change their functionality nor impacted their stability.

Defense strategy against the Man-in-the-Middle based attack on subresource integrity SIM consists of two stages. The first stage is a defense against a Man-in-the-Middle attack, which is a matter of network security. A rule of thumb is to avoid connections to a suspicious public Wi-Fi network, always encrypt the home Wi-Fi network, never access sensitive information while being connected to a suspicious network, also, to do his best to avoid being attacked by a Man-in-the-Middle hacker.

As for the second stage of defense, the security of the transport must be assured. The Web application server should use at least HTTPS as the communication protocol to the application users and subresource integrity SIM is a must. Subresource integrity serves for validating whether or not the content of the requested resource, usually provided by a third-party service, is what was expected. SIM is achieved by using a cryptographic hash of the resource content. When the application receives the resource, it computes the cryptographic hash of the received resources and compares it to the hash of the expected one. Any violation means that the resource was somehow changed and may suggest a security breach.

Our future work on the developed tool goes in two directions. One of them is to study some altered HTTPS traffics in order to target the applications that implement secure connections but not necessarily in a fully secure way. Another one is automation of the detection. Currently, the tool allows using the knowledge about the application through manual analysis. This part could be potentially done by the computer to allow fully automatic detection of such vulnerabilities.

References

1. Chalupnik, R., Kedziora, M., Jozwiak, P., Jozwiak, I.: Correspondent sensitive encryption standard (CSES) algorithm in insecure communication channel. In: International Conference on Dependability and Complex Systems, pp. 90–98. Springer, Heidelberg (2019)
2. Braun, F., Akhawe, D., Weinberger, J., West, M.: Subresourceintegrity. In: W3C Working Draft, vol. 3, pp. 3655–3682 (2014)
3. Callegati, F., Cerroni, W., Ramilli, M.: Man-in-the-Middle attack to the https protocol. IEEE Secur. Priv. 7(1), 78–81 (2009)
4. Cortesi, A., Hils, M., Kriechbaumer, T.: Contributors. Mitmdump (2010)
5. Cortesi, A., Hils, M., Kriechbaumer, T.: Contributors. Mitmproxy: A free and open source interactive HTTPS proxy, 2010–. [Version 5.0] (2010)
6. GoogleDevelopers.Introductiontoserviceworker.   https://developers.google.com/Web/ilt/pwa/introduction-to-service-worker, Accessed 19 Dec 2019
7. GoogleDevelopers.ProgressiveWebappchecklist. https://developers.google.com/Web/Progressive-Web-apps/checklist, Accessed 17 Dec 2019
8. Python Software Foundation. json—json encoder and decoder. Accessed 23 Dec 2019
9. W3Techs. Usage of javascript libraries for Websites. https://w3techs.com/technologies/overview/javascript_library, Accessed 20 Dec 2019
10. Hodges, J., Jackson, C., Barth, A.: Rfc 6797: Http strict transport security (hsts). IETF (2012). https://tools.ietf.org/html/rfc6797
11. Kügler, D.: Man in the Middle attacks on bluetooth. In: International Conference on Financial Cryptography, pp. 149–161. Springer, Heidelberg (2003)
12. Lee, J., Kim, H., Park, J., Shin, I., Son, S.: Pride and prejudice in progressive web apps: abusing native app-like features in Web applications. In: Proceedings of the 2018 ACM SIGSAC Conference on Computer and Communications Security, pp. 1731–1746. ACM (2018)

13. Luotonen, A., Altis, K.: World-wide Web proxies. Comput. Netw. ISDN Syst. **27**(2), 147–154 (1994)
14. Meyer, U., Wetzel, S.: A Man-in-the-middle attack on umts. In: Proceedings of the 3rd ACM workshop on Wireless security, pp. 90–97. ACM (2004)
15. W3C. W3c recommendation (2015). https://www.w3.org/TR/2015/REC-notifications-201 51022/, Accessed 19 Dec 2019
16. Poza, D.: Why you should not Manage your users' identities. https://auth0.com/blog/why-you-should-not-Manage-your-users-identities/, Accessed 20 Dec 2019
17. W3C. Application cache as part of the w3c recommendation. https://html.spec.whatwg.org/multipage/offline.html#appcache, Accessed 17 Dec 2019
18. Singh, A.: Proxy.py – a lightweight, single file http proxy server in python (2018). Accessed 23 Dec 2019
19. Kołaczek, G., Mizera-Pietraszko, J.: Security framework for dynamic service-oriented IT systems. J. Inf. Telecommun. **2**, 428–448 (2018). https://doi.org/10.1080/24751839.2018.1479926
20. Kołaczek, G., Mizera-Pietraszko, J.: Analysis of dynamic service oriented systems for security related problems detection. In: 2017 IEEE International Conference on Innovations in Intelligent Systems and Applications (INISTA), Gdynia, pp. 472–477 (2017). https://doi.org/10.1109/INISTA.2017.8001206
21. Stock, B., Johns, M., Steffens, M., Backes, M.: How the Web tangled itself: uncovering the history of client-side Web (in) security. In: 26th {USENIX} Security Symposium ({USENIX} Security 17), pp. 971–987 (2017)

CO-BPG: A Centralized Optimizer for Routing in BGP-Based Data Center Networks

Chen Duan, Wei Peng, and Baosheng Wang[✉]

College of Computer, National University of Defense Technology,
Changsha 410073, Hunan, China
{duanchen,wpeng,bswang}@nudt.edu.cn

Abstract. Load-balancing routing is an important and challenging issue in data center networks. Traditional internet routing protocols like OSPF and BGP cannot provide satisfiable network throughput because they are traffic-unaware and cannot utilize the rich connections and topology regularity in data center networks. New routing approaches proposed for data center networks can achieve good load balance, but cannot deploy widely because they often introduced in special switch/host hardware or software. In this paper, we propose a centralized routing optimizer, CO-BGP, for data center networks using BGP as a base routing protocol. It provides routing optimization with minimal deployment cost with off-the-shelf data center switches and servers. A routing optimization algorithm is proposed to optimize the routing performance by taking into account of network traffic load. The algorithm uses Bayes learning to cope with the dynamic of network traffic. Experiment results have shown that the proposed algorithm has better performance than the well-known equal cost multi-path (ECMP) approach.

1 Introduction

Data center networks have some important characteristics which differ from the Internet. Firstly, they normally have a regular topology structure with rich connections like fat tree, spine-leaf structure [1], which can provide full bisection bandwidth and multiple paths between any source-destination pairs. Secondly, centralized control is feasible since devices in a data center are normally concentrated in a local area. Thirdly, data centers could offer really high bandwidth even upon 40 Gbps using current Ethernet techniques.

Routing in data center networks is a well-known yet complicated research topic for decades. It greatly influences the performance of a data center network [2]. There are two main categories for routing in data center networks. One is to use the Internet routing protocols as they are mature techniques. Another is to devise new routing schemes considering the unique characteristics of data center networks.

Traditional Internet routing protocols include intra-domain routing protocols like OSPF and IS-IS, and inter-domain routing protocol, e.g. BGP. It is argued that IS-IS and OSPF work well if the network scale is small, but BGP is better for large-scale data center networks [3]. These distributed routing protocols provide a single path for a destination and are traffic-unaware, so they are inefficient and cannot fully take advantage of rich connections in data center networks. To alleviate the problem, the equal cost multi-path (ECMP) technique has often been used to distribute data traffic upon multiple paths.

Some new routing protocols are specifically designed for data center networks. There are two main categories. The first category adopts the software-defined networking (SDN) framework which uses a central controller to collect both the network topology and traffic load information, and then find the optimal flow table configuration. The OpenFlow protocol is necessary for communication between the SDN controller and SDN switches. The second category uses programmable switches or smart network interface controllers (NICs) to realize customized load balancing functions based on local decisions. Both categories rely on new techniques and do not take advantage of cost-effective traditional switches.

In this paper, we try to achieve load balancing in a data center network with commodity switches running BGP routing protocol. A centralized optimizer, CO-BGP, is proposed which collects routing and traffic information from switches and reroutes traffic on congested links to avoid packet losses. The Bayes learning technique is used to make rerouting decisions. The optimizer can be immediately deployed in data centers without any hardware improvement. We have done experiments on 3-tier CLOS topology to validate the performance of the proposed scheme. The experimental results have shown that CO-BGP can achieve better performance than ECMP with least updating cost on data center networks.

The rest of the paper is organized as follows. Section 2 introduces the related research work. The problem formulation is described in Sect. 3. The design of the central optimizer CO-BGP is illustrated in Sect. 4. Section 5 presents the experiment results and a brief conclusion is given in Sect. 6.

2 Related Work

To realize load balancing and improve resource utilization ratio of data center networks, ECMP is often applied as its principle is simple and easy to implement. In ECMP, packets are forwarded to different interfaces based on the calculation of a hash value using some selected fields of packet header. Despite of its simplicity, it is believed that ECMP has three disadvantages. First, it is traffic-unaware, just like traditional routing protocols used in the Internet. Secondly, it does not support using non-shortest paths. Thirdly, it is not very efficient in the scenarios with asymmetric network topology or non-uniform network traffic.

Within the SDN framework, Kannagevlu et al. [4] proposed an edge-to-edge rerouting scheme with flow granularity for data center networks which have a mixing of elephant flows and mice flows. Only elephant flows are rerouted to alternative paths to achieve load balancing. Similarly, FastPass [5] is also a centralized approach which can realize zero or extra-shallow queues in data center switches. It uses a centralized controller to assign a transmission path and timeslots for each data flow.

CONGA [6] is a distributed congestion-aware load balancing scheme for data centers which use global congestion information by feedback and make routing decisions on a flowlet granularity on access switches. The load information of a path is collected by piggybacking the information on packets traversing the path. HULA [7] is a hop-by-hop traffic-aware algorithm on flowlet granularity too. Each switch sends probing packets to disseminate its load information periodically. Fan et al. [8] proposed a new method based on them. It divides a network into multiple individual routing domains. Core switches can carry load information of the links. Edge switches can make optimized routing decisions.

3 Problem Formulation

3.1 Macro vs. Micro Routing Optimization

The main purpose of routing optimization in a data center network is to balance the distribution of traffic in the network so that high network bandwidth utilization ratio, low transmission delay and low packet loss rate can be achieved. To this end, a routing optimization method needs to collect network states and make optimizing decisions. Network states usually include the network topology and traffic distribution information. In addition, the routing information can also be collected by the optimizer.

We classify the optimization into microscopic optimization and macroscopic optimization based on the available network status information. Micro optimization provides fine-grained traffic scheduling functions. It usually uses programmable switches to collect information of flows [9, 10] or utilize servers' buffer to predict the traffic amount of flows [11].

Macro optimization schedule the traffic with coarse granularity. Commodity switches count the transmitted bytes and discarded packets of the links. However, there is no available method to calculate the precise traffic matrix of the server pairs. It is hard to calculate the size of the traffic going to each destination. Researchers [12] adopted deep learning like GNN to estimate the traffic matrix, but the effect is not good.

Current commodity data centers deploy traditional routing protocols and use ECMP. They couldn't afford the high deployment cost of updating switch hardware. Obviously, in data center networks, there is a gap in routing optimization technology between academic research and industry. We aim to analyze the macro optimization problem and propose a deployable optimizer to fill in this gap.

3.2 Problem Definitions

Formally, a data center network is represented by a directed graph $G = (V, E)$, where V is the set of switch and server nodes, and E is the set of links connecting different nodes. The set V consists of server nodes and switch nodes. The switch nodes are commodity switches that support SNMP and run the BGP. E represents physical links and each element in E could be identified by a switch's interface or a tuple $<u, v>$ which bandwidth is denoted by $B_{[v,i]}$ or $B(u, v)$.

In addition, a controller is deployed to compute the optimization decisions, and it communicates with switch nodes through a separate control network. The controller also plays as a dumb BGP router which establishes a BGP connection with each switch and collect routing update messages from switches.

Assumption 1. According to RFC 7938 [1], we assume all the switches running an eBGP process and each switch is configured with a unique AS (Autonomous System) number. A directed link in E also can be identified by $[as_1, as_2]$.

Assumption 2. The controller queries each switch about its interface statistics in a period of T, including: 1) TX (Transmit) dropped count *lossN*. 2) TX packets count *txpktN*. 3) TX bytes count *txB*.

The distributed routing protocol BGP selects the best path to a destination based on routing update messages from neighbors. It means that for a switch v, there is only one valid route item from s to t, which causes wasting of the bandwidth of idle links and couldn't utilize the inherent multi-path advantage in data centers. We aim to realize the load balance by periodically rerouting some traffic from the saturated links to the underutilized links. Specifically, we filter out the links needed optimize, and reroute part of the traffic to available underutilized links by configuring the static routes.

Definition 1. In the n-th period, the loss rate of the interface i of node v, *lossr(n,v,i)* is defined as formula 1. It also represents the loss rate of the link which connects with interface i in the output direction and the value is in [0, 1].

$$lossr(n, v, i) = lossN(n, v, i) \, / \, demandN(n, v, i) \tag{1}$$

$$demandN(n, v, i) = txN(n, v, i) + lossN(n, v, i) \tag{2}$$

where *demandN(n,v,i)* represents the number of packets which are forwarded to interface i. It may include packets to different destinations and from different input interfaces. *demandB(n,v,i)* represent the total bytes of these packets.

Definition 2. The bandwidth utilization of the link in the n-th period is defined as formula 3.

$$util(n, v, i) = \frac{txB(n, v, i)}{B_{[v,i]}}. \tag{3}$$

Definition 3. In the n-th period, congestion links are the links whose loss rates are greater than zero. Let *Int(v)* be the set of switch v's interfaces, in the n-th period, the set of congestion links $L_{con}(n)$ is defined as formula 4.

$$L_{con}(n) = \{[v, i] | loss(v, i) > 0, v \in V_{switch}, i \in Int(v)\} \tag{4}$$

Definition 4. In the n-th period, average loss rate of network links reflects congestion status. Because only a few links are congested, the *lossr(n,v,i)* of most links are equal to zero which causing the average loss rate really small. We define the true average loss rate which only calculate the average loss rate of congestion links by formula 5.

$$TlossR(n) = \frac{1}{|L_{con}|} \sum_{[v,i] \in L_{con}} lossr(n, v, i), n \in N^+ \tag{5}$$

Definition 5. In the n-th period, the load deviation of the bandwidth utilization among links could briefly reveal the load balance status in data center networks. The load-deviation of the bandwidth utilization is defined as formula 6.

$$LB(n) = (1/N_{interface}) \sum_{v \in Vswitch} \sum_{i \in Int(v)} (1 + |util(n, v, i) - avgU(n)|)^2, n \in N^+ \tag{6}$$

$$avgU(n) = 1\bigg/ (|V_{switch}| * n) * \sum_{v \in Vswitch} \sum_{i \in Int(v)} util(n, v, i), n \in N^+ \quad (7)$$

where $avgU(n)$ is the average load of links, and $N_{interface}$ is the total number of links which is equal to the twice of physical link number because edges are bidirectional. Due to $0 <= |util(n, v, i) - avgU(n)| <= 1$, we add an offset to make it greater than 1. Obviously, when traffic loads become more unbalanced, $LB(n)$ increases accordingly. In addition, it could be proved that $1 <= LB(n) <= 2.25$.

Definition 6. In the route table of a switch v, $routeinfo(v,dst)$ represents the set of available output interfaces to destination dst.

3.3 Problem Formulation

For a data centers network $G = (V, E)$, in the n-th period, the routing optimization is a multi-objective optimization problem. It is formulated as formula 8.

$$\begin{aligned} &\text{Minimize TlossR (n), } LB(n), n \in N^+ \\ &\text{s.t. TlossR(n)} \geq 0, 1 \leq LB(n) \leq 2.25 \end{aligned} \quad (8)$$

These objective functions are non-deterministic relationship and difficult to perform regression analysis. Therefore, we discuss how to transform this multi-objective optimization problem to a single objective function. Given a link, the relation of loss rate and bandwidth utilization is concluded as formula 9.

$$util(n, v, i) = \begin{cases} \rightarrow 1, loss(n, v, i) > 0 \\ < 1, loss(n, v, i) = 0 \end{cases} \quad (9)$$

In other words, if packet loss happens, the link is saturated and the bandwidth utilization tends to be 1. Under such condition, the multi-objective optimization problem could be simplified as:

$$Minimize\ TlossR(n),\ if\ TlossR(n) > 0 \quad (10)$$

$$Minimize\ LB(n),\ if\ TlossR(n) = 0 \quad (11)$$

4 Design and Algorithm

4.1 CO-BGP Overview

Our goal is to design a centralized routing optimizer CO-BGP to alleviate network congestion and realize load balancing. Figure 1 shows the architecture of CO-BGP, which is implemented in C language and deployed on the controller. CO-BGP can be immediately deployed in data centers without any hardware improvement. The controller may be a Linux server and communicates with switches by separate control network. The data networks carry business traffic. In this way, it provides high bandwidth and ultra-low latency between the controller and switches which brings reliability and efficiency to CO-BGP.

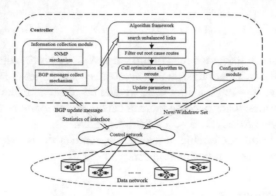

Fig. 1. CO-BGP architecture.

Information Collection Module. It consists of SNMP mechanism and dumb BGP router mechanism which collect interfaces statistics such as ifInDiscards, ifInUcastPkts and ifInOctets and global routes forwarding tables. They are implemented by SNMP protocol and a dumb BGP router that establishes connection with each switch by control network, and only sends HELLO and KEEPLIVE messages. So it learns the global routes and hand these information over to route optimization module.

Route Optimization Module. According to the current network states uploaded by the information collection module, it needs to search the unbalanced links and adjust routes to achieve load balancing which are implemented by the heuristic algorithm and the Bayes learning-based rerouting mechanism. The details are described in Sect. 4.2, 4.3. This module hands over the routing update set to Configuration module.

Configure Module. It is responsible for updating static route entries in RIBs. It may add or delete route entries which is performed by Netconf or Telnet protocol.

4.2 Routing Optimization Module

It is the main body of CO-BGP. It uses heuristic algorithm to divide the global optimization problem into multiple sub-problems corresponding to unbalanced saturated links. The main idea of it is rerouting part of traffic traversing saturated links to underutilized links so as to satisfy the formula 12.

In the n-th period, the process of this module is described as follow.

```
Input   A connected network G=(V,E)
        Initial traffic matrix txBMatrix
        Optimize parameter opN, q
Ouput   NewSet: Set of new route items
        WithdrawSet: Set of route items to be withdrawed
step1   If TlossR(n)>0
           sort congestion links based on the loss rate
        else
           sort unbalanced links based on the utilization
step2   Select the top opN links to optimize
        2.1 Search the routes through the link, sort them
and select q ratio of them
        2.2 call optimization algorithm to reroute the
selected routes
        2.3 optimization algorithm return Newset WithdrawSet
step3   After the n-th period
        3.1 calculate the optimization traffic matrix
        3.2 update the opN, q for the heuristic algorithm
```

step1 it determines the optimization object according to the formula 10 and 11, and sorts the links based on the network conditions.

Step2 it chooses the top opN links and for each link <u, v> or [as_u, as_v], it finds out the route, that is as-path [as_1, as_2, …, as_n], traversing this link by matching AS numbers. It prefers to reroute routes that have a smaller impact on global network. Therefore, it sorts them based on the index of as_u in as-path which also represents the distance between the source of as-path and the congestioned link. It just chooses the top q ratio of these sroutes, then in step 2.2 it hands over the chosen links and routes to rerouting mechanism. The rerouting mechanism outputs the routing updates.

step3 it collects the network information in the $n+1$-th period and updates the parameters as formula 12. The main idea of updating parameters is that the worse network condition, the larger parameters. In addition, the updating policies of rerouting mechanism are shown in Sect. 4.3.

$$(opN, q) = \begin{cases} (opN_{n-1} + 1, q * 1.2) & if \ TlossR(n) > TlossR(n-1) \geq 0 \\ (opN_{n-1}, q) & if \ TlossR(n-1) > TlossR(n) > 0 \\ (opN_{n-1} - 1, q * 0.8) & if \ TlossR(n-1) > TlossR(n) = 0 \end{cases}$$

$$else \ if \ TloosR(n) = TLossR(n-1) = 0$$

$$(opN, q) = \begin{cases} (opN_{n-1} + 1, q * 1.2) & if \ loadVar(n) > loadVar(n-1) \\ (opN_{n-1} - 1, q * 0.8) & if \ loadVar(n) \leq loadVar(n-1) \end{cases} \quad (12)$$

4.3 Bayes Learning-Based Rerouting Mechanism

The main idea of the mechanism is rerouting part of the traffic traversed congested links to more appropriate paths, while the rerouting behavior wouldn't cause congestion in new paths. Therefore, it is critical to know the size of traffic to each destination on switches. In SDN, researchers tend to estimate the traffic matrix of end to end in flow granularity. In commodity switches, the collected corse-grained statistics in Assume 2 are too rough to identify the flows, that is difficult to estimate, so macro routing optimization cannot reserve bandwidth for each flow like FastPass [3]. None the less, in order to make reasonable optimization decisions and avoid routing oscillation, we propose a novel Bayes learning-based rerouting mechanism.

We use Bayes learning to build a mapping from statistic data matrix to the score of each destination traffic routing on switches. We reroute the selected network prefix to the path with highest score and according to the optimization result matrix fine tune scores online. It could converge quickly and doesn't need pre-trained models. The details of the rerouting mechanism are shown below.

We use Bayes learning to estimate the traffic. The posterior probability is defined as formula 13, where $P(c|S)$ represents the probability of routing traffic to interface c in the current network state. $P(c)$ represents the prior probability and $P(S|c)$ represents the likelihood. $P(S)$ could be regarded as the normalization factor. Then ignoring $P(S)$ we could infer that the formula 14 is true.

$$P(c|S) = P(c)\,P(S|c)\,/\,P(S) \tag{13}$$

$$P(c|S) \propto P(c)\,P(S|c) \tag{14}$$

Because the Beta distribution could represent many kinds of distribution by changing parameters, we assume that the likelihood $P(S|c)$ submits to the Beta distribution. In addition, the prior probability $P(c)$ submits to the Binomial distribution obviously shown in formula 15. Therefore, because of the conjugacy of the Beta distribution and the Binomial distribution, it is easier to infer that the posterior probability also submits to the Beta distribution in formula 16.

$$P(c) \sim B(1, 1\,/\,|\,routeinfo(v,\ dst)\,|) \tag{15}$$

$$P(c|S) \sim Beta(a,\ b) \tag{16}$$

In the n-th period, given a route entry to destination dst on switch v, if the $|routeinfo(v,dst)|>1$, rerouting mechanism would calculate the mean of each posterior distribution and choose the largest one as the interface to reroute. The mean of the Beta distribution is defined as formula 17.

$$u = a\,/\,(a + b) \tag{17}$$

After the n-th period, the heuristic algorithm would generate optimization traffic matrix as the initial traffic matrix in the $n + 1$-th period. It also updates the posterior probability in rerouting mechanism as formula 18, where *reward* represents the gains of rerouting and *regret* represents the loss. The detailed policy is shown in formula 19.

$$P(c|S) \sim Beta(a + reward,\ b + regret) \tag{18}$$

$$(reward, regret) = \begin{cases} (0, 2) \ if \ TlossR(n) > TlossR(n-1) \ and \ LB(n) > LB(n-1) \\ (1, 2) \ if \ TlossR(n) > TlossR(n-1) \ and \ LB(n) < LB(n-1) \\ (2, 1) \ if \ TlossR(n) < TlossR(n-1) \ and \ LB(n) > LB(n-1) \\ (2, 0) \ if \ TlossR(n) < TlossR(n-1) \ and \ LB(n) < LB(n-1) \end{cases}$$

$$(19)$$

5 Evaluation

In this section, we evaluate CO-BGP's performance with real TCP/UDP traffic as well as large-scale simulations. The real traffic experiments are established on the testbed which consists of docker containers. They illustrate CO-BGP's good performance with realistic workload. Our detailed simulations confirm that CO-BGP scales to large topologies and give suggestions on the algorithm parameters. We compare our optimization algorithm with the benchmark ECMP.

5.1 Experiment with Realistic Traffic

Experiment Settings. We evaluate the performance of proposed algorithms on the 3-tier clos topology network in Fig. 2 and compare with ECMP. We use frr-routing to simulate data centers switches and ubuntu containers to simulate servers. The switches are connected by vcth (Virtual Ethernet Device) and we modify the driver code of veth to realize differentiated bandwidth. The optimization period of CO-BGP is 10 s.

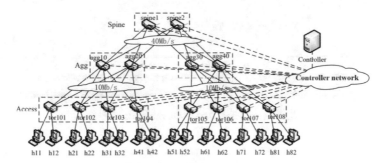

Fig. 2. Experiment topology.

The traffic model is relatively stable and mixing with some burst traffic. We send TCP or UDP flows at a rate of 5 Mbit/s across network, every time for 120 s and send multiple times. The following metrics are used to measure the performance:

1) true average loss count: In order to show the result more directly, we use the absolute value of the average of TX dropped count instead of the percentage of loss rate and define it as TlossA.

2) load-Deviation of bandwidth utilization: It shows the load balance status of the global network. We calculate the load-deviation of bandwidth utilization of the links according to formula 7.

Experimental Results. Figure 3 shows a sampling results of the optimization performance in constant period. In fact, we tested it for many times and they all have significant results.

Figure 3(a) shows the *true average loss count* and it could be partitioned into three stages corresponding to three traffic periods. It is shown that in each stage the *TlossA* of CO-BGP decreases significantly in three or five period, and it could reach to zero. ECMP performs better than no-optimize but because of traffic-unaware, it keeps l oss packets continuously and the TlossA is above the horizontal dotted line closed to 70.

Figure 3(b) shows the load-deviation of the bandwidth utilization in two traffic periods. It shows that the LB(n) of CO-BGP decreases significantly. Although ECMP performs better than no-optimize, it oscillates between two horizontal dotted lines. It shows that CO-BGP could achieve load balancing meanwhile achieve congestion control.

According to above experiment results, we could conclude that CO-BGP shows significant effect. In the constant optimizing condition, it could achieve better result as expected because of estimating the traffic.

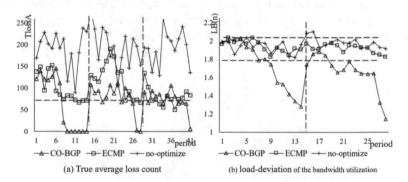

(a) True average loss count   (b) load-deviation of the bandwidth utilization

Fig. 3. Comparison of two evaluation metrics.

5.2 Large-Scale Simulations

Our experimental results are for a comparatively small topology. In order to explore the performance in large scale topology, we simulate the route optimization module on the topology of 24 nodes (6 spine and 18 leaf), 120 nodes (10 spine and 110 leaf) and 256 nodes (16 spine and 240 leaf). We implement it in Python. We mainly test the parameter space and the optimization performance on different scale topologies.

Parameter Space. The opN determines the number of links which needed to be optimized. The initial parameter influence the convergence time of the algorithm. If it was set too small, the algorithms would need lots of rounds to converge. On the contrary, if it was set to large, in each period, it would reroute too many flows and cause congestion so that it may cause route oscillation in the following periods.

Therefore, we set it to 5%, 30%, 50% of the total number of links, and test them. According to the simulation results in Fig. 4, we find that when setting opN to 30% of the total number of links, no matter the true average loss count or load-Deviation of bandwidth utilization it performs better than 5% and 50%.

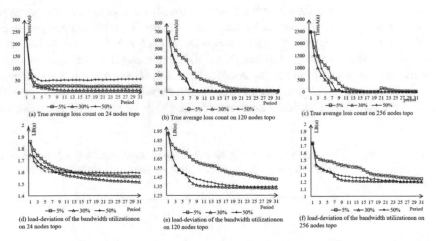

Fig. 4. Comparison of different parameters in simulations.

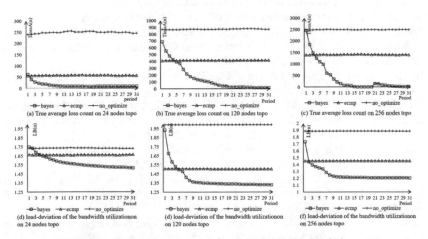

Fig. 5. Comparison of two evaluation metrics in simulations.

Optimization Performance. According to the simulations above, we set opN to 30% of the total number of links and test the optimization performance on different scale topologies compared with ECMP. The results in Fig. 5 show that our algorithm could decrease the Tloss(n) and LB(n) rapidly in each topology. In most instances, after three periods the Tloss A(n) and LB(n) of the proposed algorithm is less than ECMP.

We conclude that proposed algorithm could perform well in large scale topology, and the optimal opN configuration should be around 30% of the total number of links.

6   Conclusion

In this paper, we analyze the centralized routing optimization problem. We survey the related works and conclude our motivation. Then the routing optimization problem is classified, formulated and the problem properties are analyzed. We proposed an easily deployed centralized routing optimizer, CO-BGP, and a novel Bayes learning-based rerouting mechanism to fill in the gap between academic research and industry. Finally, the experiment and simulation results show that with minimal deployment cost with off-the-shelf data center switches and servers, the CO-BGP also achieves significant effect.

Acknowledgments. This work is supported by National Key Research and Development Program of China (No. 2018YFB0204301) and a project from National Science Foundation of China (No. 61972412).

References

1. Bao, J., Dong, D., Zhao, B.: DETOUR: a large-scale non-blocking optical data center fabric. In: Asian Conference on Supercomputing Frontiers, pp. 30–50. Springer, Cham (2018)
2. Wei, Z., Dong, D., Huang, S., Xiao, L.Q.: EC4: ECN and credit-reservation converged congestion control. In: 2019 IEEE 25th International Conference on Parallel and Distributed Systems (ICPADS), 209–216. IEEE (2019)
3. Internet Engineering Task Force. Use of BGP for Routing in Large-Scale Data Centers: RFC 7938. The Internet Society, New York (2016)
4. Kanagavelu, R., Mingjie, L.N., Mi, K.M., Lee, F.B.H.: OpenFlow based control for rerouting with differentiated flows in Data Center Networks. In: 2012 18th IEEE International Conference on Networks (ICON), Singapore, pp. 228–233 (2012)
5. Perry, J., Ousterhout, A., Balakrishnan, H., Shah, D., Fugal, H.: Fastpass: a centralized zero-queue datacenter network. In: SIGCOMM, pp. 307–318 (2014)
6. Alizadeh, M., Edsall, T., Dharmapurikar, S., Vaidyanathan, R., Chu, K., Fingerhut, A., et al.: CONGA: distributed congestion-aware load balancing for datacenters. In: Proceedings of the ACM conference on SIGCOMM, pp. 503–514 (2014)
7. Katta, N., Hira, M., Kim, C., Sivaraman, A., Rexford, J.H.: Scalable load balancing using programmable data planes. In: Proceedings of the Symposium on SDN Research, pp. 1–12 (2016)
8. Fan, F., Hu, B., Yeung, K.L.: Routing in black box: modularized load balancing for multipath data center networks. In: INFOCOM on Computer Communications, pp. 1639–1647. IEEE (2019)
9. Zhang, J., Ren, F., Huang, T., Tang, L., Liu, Y.: Congestion-aware adaptive forwarding in datacenter networks. Comput. Commun. **62**, 34–46 (2015)

10. Ghorbani, S., Yang, Z., Godfrey, P.B., et al.: Drill: micro load balancing for low-latency data center networks. In: Proceedings of the Conference of the ACM Special Interest Group on Data Communication, pp. 225–238 (2017)
11. Li, Z., Zhang, Y., Li, D., Chen, K., Peng, Y.: OPTAS: decentralized flow monitoring and scheduling for tiny tasks. In: IEEE INFOCOM 2016, pp. 1–9. IEEE (2016)
12. Rusek, K., Suarez-Varela, J., Almasan, P., Barlet-Ros, P., Cabellos-Aparicio, A.: RouteNet: leveraging graph neural networks for network modeling and optimization in SDN. IEEE J. Sel. Areas Commun. **38**(10), 2260–2270 (2020)

Joint Channel and Power Allocation Algorithm for Flying Ad Hoc Networks Based on Bayesian Optimization

Pengcheng Wang, Wei Peng[✉], Wenxin Zhang, and Gaofeng Lv

College of Computer Science, National University of Defense Technology,
Changsha 410073, China
{pcwang,wpeng}@nudt.edu.cn

Abstract. Flying ad hoc networks (FANETs) formed with unmanned aerial vehicles (UAVs) have wide applications in military and civilian fields, including patrolling, search and rescue operations, monitoring. Due to limited bandwidths and the fragility of wireless communication, channel and power allocation has become a key issue which significantly affects the performance of a FANET. This problem is usually formulated as a mixed-integer nonlinear programming problem and some intelligent approximation algorithms like particle swarm optimization algorithms have been proposed in wireless mesh network (WMN) and D2D (Device-to-Device) network scenarios. In order to improve the overall network throughput and spectrum utilization ratio, a joint channel selection and power allocation algorithm based on Bayesian Optimization is proposed, which uses the conjugacy property of Beta distribution to obtain approximate optimal allocation results. Simulation experiment results have shown that the proposed algorithm has better performance in network throughput and normalized network satisfaction ratio than the compared algorithms.

Keywords: Channel allocation · Transmitting power control · Bayesian Optimization · Flying ad hoc network

1 Introduction

Owing to the explosive expansion of wireless communication and UAV technologies, UAVs have been widely used in civilian and military fields to carry out the missions including patrolling, search and rescue operations, monitoring, and other military missions. The cooperative UAVs can be used as wireless relays and flying sensors covering a wider range, enhancing the multitasking ability and scalability. In order to meet the increasing network bandwidth and transmission throughout requirements, channel and power allocation have become a key issue in FANETs.

In wireless mesh network (WMN), paper [1] proposed a swarm intelligence algorithm named PSO to solve the channel allocation. And paper [2] proposed a joint channel and power allocation algorithm based on graph theory. In the cognitive radio network, paper [3] proposed a channel and power allocation algorithm based on random game theory. In

Device-To-Device (D2D) networks, paper [4] employs a secondary allocation strategy to solve the joint channel and power allocation problem. Whether these algorithms are applicable to FANETs is still a question that needs to be verified. Paper [6] proposed a slot-based channel allocation and power optimization algorithm under the network organization relying UAV ground station [5]. Paper [7] considers difference of priority between UAVs, and makes tendentious allocation of network resources according to the priority. Paper [8, 9] apply deep reinforcement learning to the channel selection of wireless networks, but they have not taken care of the interference between different channels.

Different from the above research work, this paper proposes a heterogeneous multi-channel network structure (Fig. 1). In this network structure, each node has two sets of radio frequency equipment, which work in a high frequency band and a low frequency band respectively. Different frequency bands use different encoding, decoding and modulation methods. The low frequency bands have strong anti-interference ability called Highly-Protected Band. The high frequency bands called Primary Band here have higher data rate. In the network, the core node can communicate with all nodes at different time slots in the Highly-Protected Band, and allocates the channel in the Primary Band according to the communication demand generated by other nodes. We research the power selection and channel allocation of the Primary Band under such a network structure, and proposes an algorithm based on Bayesian Optimization. Depending on the good conjugacy of the Beta Distribution, the probability of a better-performing allocation strategy is continuously increased in the iterative process. Experimental results show that the algorithm we proposed has better performance in network throughput and normalized network satisfaction when compared with random allocation algorithm, PSO, and greedy algorithm.

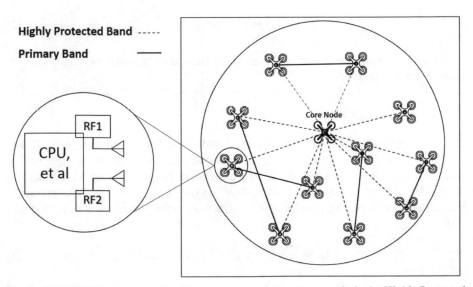

Fig. 1. Each UAV has two sets of radio frequency equipment, one works in the Highly Protected Band, the other works in the Primary Band.

2   Network Model and Problem Definition

2.1   Network Model

In a multi-node FANET network, we define $N = \{1,2, 3, ...N\}$ as the set of all nodes, which are peer nodes. Each node has two sets of radio frequency equipment. One set of radio frequency works in Highly-Protected Band, and all nodes share a channel. This channel is the control channel that completes the communication between the core node and the others. The other set of radio frequency works in Primary Band, which have k channels to choose. These channels are called data channel. Define channel selection strategy $c_i = \{c_i^1, c_i^2, c_i^3, \ldots c_i^k\}$, when user i selects k channel c_i^k is 1, otherwise it is 0. The power of the control channel remains unchanged. Power P of the data channel can be selected between levels $\{p_1, p_2, p_3 \ldots p_n\}$.

2.2   Propagation Model

In this paper, we select the free space propagation model. The minimum transmit power level of the free space propagation model is calculated by:

$$p_{ij} = \left[\frac{R_{\text{thresh}}(4\pi d_{ij})^2}{G_t G_r \lambda^2} / \frac{P_{max}}{p_n}\right] \tag{1}$$

R_{thresh} is the threshold at which the receiving node can normally receive signals, d_{ij} is the Euclidean distance between nodes i and j, and G_t and G_r are the gains of the transmitting antenna and the receiving antenna. The power received by the receiving node is calculated by Eq. (2):

$$P_{ij}^{r,c,p} = P_{ij}^{t,c,p} d(j, i)^{-\theta} \tag{2}$$

Θ is the path loss coefficient, and usually the value is a constant between (2, 4), $P_{ij}^{t,c,p}$ is the transmission power.

2.3   Interference Model

As shown in Fig. 2, different communication node pairs (such as node pairs A, B, and C, D) select different channels to minimize interference and improve communication quality. Relative to the node B, the communication between node C and D is a source of interference. In the network, when node N_i and node N_j are in communication, the interference that it receives can be defined according to the physical interference model [10]:

$$I_{ij}^c = \sum_{y \in N} P_y^{t,c,p} d(j, y)^{-\theta} \varepsilon_{j,y} (y \neq i, j) \tag{3}$$

It is the sum of the received signals from other nodes at the receiving point N_j. Where $P_y^{t,c,p}$ is the transmission power of the node y, $d(j, y)^{-\theta}$ is the path loss, $d(j, y)$ is the

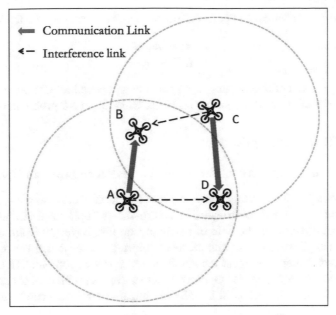

Fig. 2. Relative to the node B, the communication between node C and D is a source of interference.

Euclidean distance between nodes j and y. ε_{jy} is the weight of the interference which is related to the channel selected and its value range is [0,1]. The closer the two channels you choose are, the larger the value is.

Define the signal to interference noise ratio (SINR) based on the physical interference model:

$$\gamma_{ij}^{c,p} = \frac{P_{ij}^{t,c,p} \cdot d(j,i)^{-\theta}}{P_N + I_{ij}^c} \tag{4}$$

Where $d(j,y)^{-\theta}$ is the path loss, P_N is the noise, and I_{ij}^c is the interference received from other nodes.

2.4 Problem Definition

According to Shannon's formula, the channel capacity is:

$$C_{ij} = B \cdot \log_2\left(1 + \gamma_{ij}^{c,p}\right) \tag{5}$$

Where B is the channel bandwidth. When the channel capacity in the network communication link is greater than the network constant packet rate v, $C_{ij} = v$. The total throughput rate in the network is the sum of all transmission link channel capacities:

$$U = \sum_{\forall i,j \in N\, i \neq j} C_{ij} \tag{6}$$

Formally, the joint channel and power allocation problem can be described as:

$$x^* = \text{arg} maxU$$
$$\text{s.t. } (1) - (6)$$

Solving the problem of maximizing U by channel and power allocation in FANET is a mixed integer nonlinear programming problem, which is an NP problem.

3 Algorithm Description

3.1 Channel and Power Allocation Algorithm Based on Bayesian Optimization

Bayesian Optimization has attracted widespread attention in the field of artificial intelligence by virtue of its good performance in hyperparameter optimization, and is currently a hot research direction in the field of machine learning. Bayesian Optimization has a wide range of applications in environmental monitoring and sensor networks, natural language text processing, intelligent recommendation systems, combination optimization and other fields. we use Bayesian Optimization to solve the problem of channel selection and power allocation in FANETs. Usually optimization problems can be expressed by the following equation:

$$x^* = \text{arg} \max_{x \in \chi} f(x) \tag{7}$$

Where x represents the selected decision vector, χ represents the decision space, and f(x) represents the objective function. Specific to the optimization problem in this article, x represents the vector of the channel and power selected by the communication pair in the network, and $f(x)$ is the network throughput. The framework of Bayesian optimization is composed of two parts: model which can be updated and queried to drive optimization decisions and acquisition function. The model can be divided into a priori probability and a posterior distribution. The prior probability refers to the prior setting of the unknown problem before the observation. And the posterior distribution refers to the result of Bayesian analysis, reflecting all our knowledge of the problem under the given data and model. An appropriate acquisition function can balance the relationship between exploration and exploitation while searching for the optimal decision. We can infer the posterior distribution by the Bayes' rule:

$$p(H \mid D) = \frac{p(D \mid H)p(H)}{p(D)} \tag{8}$$

Among them, $p(H)$ is the prior distribution, $p(D|H)$ represents the likelihood, which reflects the credibility of a certain set of data under a given parameter, $p(H|D)$ is the posterior distribution, $p\ (D)$ is the marginal likelihood, which refers to the average of the probability of the specified observation value obtained under the condition of taking all possible values of the model parameters [11]. We choose Beta-Bernoulli model to match the prior and the likelihood. Under the good benefit of its conjugacy, we calculate the posterior by the Eq. (9):

$$p(H \mid \alpha, \beta) = \prod \text{Beta}(H_i \mid \alpha, \beta) \tag{9}$$

After one interaction, we update the distribution:

$$p(H|D) = \prod \text{Beta}(H_i|\alpha + \text{Reward}, \beta + \text{Regret}) \tag{10}$$

In the algorithm, we define Reward as the ratio of the current total network throughput U_n to the first throughput (U_0) after network initialization. Regret is the absolute value of the difference between 1 and U_n/U_0. The specific update strategy is as follows:

$$\alpha = \alpha + \rho \frac{U_n}{U_0} \text{ if } U_n > U_{n-1} \tag{11}$$

$$\beta = \beta + \rho \left| 1 - \frac{U_n}{U_0} \right| \text{if } U_n < U_{n-1} \tag{12}$$

Where ρ is the amplification factor, and the value is a constant greater than 1 to improve the convergence speed of the algorithm. The algorithm is encapsulated in the Algorithm 1.

Algorithm 1 Bayesian Optimization

```
1: Initialize each element of the hyperparameter variable
matrix α(i, j), β(i, j);
2: for i = 1 to T do
3:    for i = 1 to m do
4:       for j = 1 to (k + pₙ) do
5:          samples the expected value of the channel or
power level of the link of i : θ(i, j) ~ Beta(α(i, j),
β(i, j));
6:          end for
7:       end for
8:    Choose the optimal channel and power allocation
strategy;
9:    According to the selected channel and power
allocation strategy, calculate the network throughput,
and then calculate the Reward and Regret;
10:   update α(i, j) and β(i, j);
11: end for
12: end
```

3.2 Other Algorithm Description

(1) Particle Swarm Optimization. The Particle Swarm Optimization is a typical swarm intelligence algorithm. Similar to a flock of birds searching for food and notifying each other to find the best place, PSO finds the optimal solution through continuous exploration of particles iteratively. In the problem of channel selection and power allocation, PSO first initialize the solution vector $X_i(x_1, x_2 \ldots x_n)$ of each particle (the solution vector here includes the channel and power level) and the velocity vector $V_i(v_1, v_2 \ldots v_n)$. The iterative equation of the velocity vector is as follows:

$$V_i = \omega V_i + c_1 r_1 (Pbest_i - X_i) + c_2 r_2 (Gbest - X_i) \tag{13}$$

Where ω is the weight of inertia, generally between [0,1]. The parameters $c1$ and $c2$ are called learning factors, which are random values between [0,1]. The iterative formula of the solution vector is $X_i = X_i + V_i$. The specific algorithm is described as follows according to paper [1]:

Algorithm 2 Particle Swarm Optimization

```
1: Initialize the particles, calculate the total network
throughput under the individual solution vector, select
the individual local optimal solution vector Pbest_i and the
global optimal Gbest of the population;
2: for 1 to gm (gm is the number of iterations)
3:   Update the individual velocity vector V_i according to
the velocity vector update equation;
4:   Update the solution vector X_i according to the
solution vector update equation;
5:   Calculate the total network throughput under the
solution vector, update the local optimal solution vector
Pbest_i and the global optimal Gbest of the population;
6: end for
7: output the Gbest;
8: end
```

(2) Greedy Algorithm. According to the joint power and channel allocation algorithm based on secondary allocation mentioned in the paper [4], for the network model of this paper, the algorithm is now described as follows:

Algorithm 3 Greedy Algorithm

```
1: Sort the communication node pairs in the network in
ascending order of distance;
2: for i in k (k represents the total number of
communication links)
3:   Set the power of the transmitter to the maximum;
4:   Calculate the total throughput of the network when
different channels are selected respectively, and choose
the one with the largest throughput as the final channel;
5:   Calculate the total throughput of the network under
different power levels in turn, and choose the one with
the largest throughput as the final power level;
6: end for
7: end
```

(3) Random Channel Allocation Algorithm. In this allocation algorithm, the probability of the channel being selected is the same. According to the communication requirements in the network, the channel and power level are randomly assigned to each pair of communication chains in an optional range with a uniform distribution model.

3.3 Algorithm Complexity Analysis

The time complexity of the algorithm in this paper is $O(T*N*(k + p_n))$ while carrying out channel selection and power allocation. Where T is the number of iterations, N is the channel need to be allocated. k is the total number of channels that can be allocated, and p_n represents the total number of power levels. The time complexity when calculating throughput is $O(T*N*N^2)$, so the total time complexity of the algorithm in this paper is $O(T*N^3)$.

4 Evaluation

4.1 Environment Model

Assume that UAVs are distributed in the geographic area of $H*W$, where H is the length of the area, and W is the width of the area. N communication nodes are randomly deployed in the area. Supposing that the communication in the network is evenly distributed, there are k random communication pairs in the network at a certain moment. The network is updated by random movement of the node position at regular intervals, and the communication requirements are updated at the same time. The traffic of nodes in the communication is assumed to be constant bit rate (CBR, Constant Bit Rate) traffic, and the communication packet rate is assumed to be v. The parameters used in the simulation are listed in Table 1.

Table 1. Simulation parameters

Symbol	Meaning	Value
H (m)	Length of the area	1000
W (m)	Width of the area	1000
N	Communication nodes	40
v (KB/s)	Communication packet rate	128
R_{th} (dBm)	Receive power threshold	−40

The evaluation indicators of algorithm performance include network throughput rate *(THt)* and normalized network satisfaction *(St)*. Calculate the channel capacity of each link according to Shannon's formula and compare it with the packet sending rate of the communication nodes. If the link channel capacity is greater than the packet sending rate, the packet sending rate is recorded as the throughput rate of the link, otherwise the actual channel capacity is recorded. Calculate the total throughput rate of communication node pairs in the network. The calculation is shown as Eq. (14).

$$TH_t = \begin{cases} C_{ij} \text{ if } C_{ij} < v \\ v \text{ if } C_{ij} > v \end{cases} \quad C_{ij} = B \cdot \log_2\left(1 + \gamma_{ij}^{c,p}\right) \tag{14}$$

Calculate the percentage of the number greater than the packet rate v to the total number of communication requirements, and record it as the normalized network satisfaction.

$$S_t = \frac{m_t}{k_t} \tag{15}$$

As the Eq. 15 shows, m_t is the number of communication nodes greater than the packet sending rate v until the time period t, and k_t is the total number of communication requirements.

4.2   Evaluation Result

(1)   Evaluation of Different Algorithm

Network Throughput. Update the node position in the range of 1000 *1000 m, and then randomly generate communication requirements, calculate the total network throughput rate after 100 update cycles. The network throughput rate of each algorithm is shown in Fig. 3:

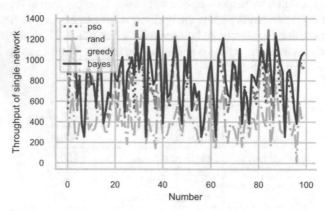

Fig. 3. This picture shows the throughput of the network in each update period.

It can be seen from the Fig. 3 that the Bayesian Optimization algorithm is better than the other three algorithms in most cases in the randomly generated network. We accumulate the throughput obtained by these 100 network updates to obtain Fig. 4. It can be seen from the Fig. 4 that the Bayesian Optimization algorithm is significantly better than the other three algorithms.

Normalize Network Satisfaction. Count the number of communication links in the network whose channel capacity is greater than the packet sending rate, and compare with the number of communication links. The network satisfaction of various algorithms is shown in Fig. 5. As can be seen, the bayesian network algorithms have higher satisfaction, and significantly better than the other three algorithms.

(2)   Evaluation of Different Network Settings

Number of Network Nodes. The number of nodes in the network increases from 40 to 90 sequentially, calculating the average throughput and the average normalized satisfaction after 20 update period (Fig. 6).

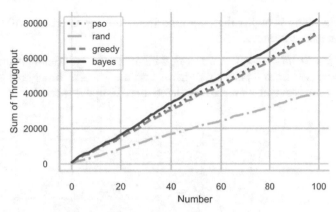

Fig. 4. The sum of the network throughput in the 100 period is plotted in the figure.

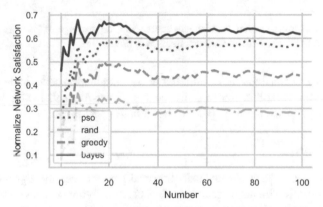

Fig. 5. The cumulative normalize network satisfaction is shown above.

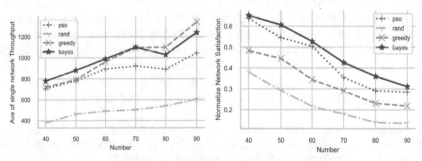

Fig. 6. With the increasing nodes, the average of the network throughput is rising, and the normalize network satisfaction is descending.

According to the experimental results, the total network throughput shows an increasing trend, but the interference increases, and the network satisfaction shows a downward trend. Although when nodes increasing to a certain extent, the throughput rate of the greedy algorithm is higher than our algorithm, but our algorithm still has obvious advantages in network satisfaction.

Communication Requirement. In a network with 40 nodes, the communication requirements are increased from 8 pairs to 19 pairs, and the average throughput and average normalized network satisfaction are calculated after 20 update period.

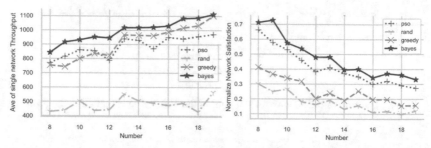

Fig. 7. With the communication requirement increasing, the average of network throughput is incremental, and the normalize network satisfaction is downward.

As can be seen from Fig. 7, as the number of communication nodes increases, the total network throughput is still increasing, and the overall network satisfaction is decreasing. The algorithm proposed in this paper still has a better performance.

5   Conclusion

This paper proposes a joint channel selection and power allocation algorithm based on Bayesian optimization for heterogeneous multi-channel flying wireless ad hoc networks. The algorithm uses the Beta distribution as the prior distribution, and iteratively changes the distribution parameters to maximize the network throughput. The Bayesian optimization algorithm is compared with the traditional swarm intelligence algorithm PSO [1], greedy algorithm [4] and the simplest random allocation algorithm. Experimental results show that the Bayesian Optimization algorithm is better than the other three algorithms in terms of network throughput and normalized network satisfaction. The study in this paper is based on a single-hop network model and proposes a centralized channel and power allocation algorithm. Realizing distributed channel and power allocation algorithms in a multi-hop network model is the direction of future research.

Acknowledgments. This work was supported in part by the National Natural Science Foundation of China (No. 61601483), Hunan Young Talents Grant (No. 2020RC3027) and the Training Program for Excellent Young Innovators of Changsha (No. kq2009027).

References

1. Zhang, Y.C., Wang, Y.J, Yao, S.W.: A PSO-based channel assignment algorithm in wireless mesh networks (in Chinese). J. Univ. Electron. Sci. Technol. China, **05**(v.46), 90–95+108 (2017)

2. Chen, X., Zhang, Z.Y., Luo, H.Y.: Joint optimization of power control, channel assignment andscheduling in wireless mesh network (in Chinese). J. Zhejiang Univ. (Eng. Sci.) **43**(8), 1406–11 (2009)
3. Wang, Z.Y., Zhang, H.Y., Xu, N.: Channel assignmentand power control based on stochastic learning game in cognitive radio networks (in Chinese). J. Acta Electronica Sinica **046**(12), 2870–7 (2018)
4. Hsu, C.S., Chen, W.C.: Joint power control and channel assignment for green device-to-device communication. In: DASC/PiCom/DataCom/CyberSciTech 2018, pp. 881–884 (2018)
5. Bacco, M., Pietro Cassarà, C.M., Gotta, A., Marchese, M., Patrone, F.: A survey on network architectures and applications for nanosat and UAV swarms. In: WISATS 2017, pp. 75–85 (2017)
6. Xue, Z., Wang, J., Shi, Q., Ding, G., Wu, Q.: Time-frequency scheduling and power optimization for reliable multiple uav communications. J. IEEE Access **6**, 3992–4005 (2018)
7. Zhou, L., Chen, X., Hong, M., Jin, S., Shi, Q.: Joint channel assignment and power allocation for multi-UAV communication. In: SPAWC 2020, pp. 1–5 (2020)
8. Naparstek, O., Cohen, K.: Deep multi-user reinforcement learning for dynamic spectrum access in multichannel wireless networks. In: GLOBECOM 2017, pp. 1–7 (2017)
9. Wang, S., Liu, H., Gomes, P.H., Krishnamachari, B.: Deep Reinforcement Learning for Dynamic Multichannel Access. IEEE Trans. Cogn. Commun. Netw. **4**(2), 257–265 (2018)
10. Cardieri, P.: Modeling interference in wireless ad hoc networks. J. IEEE Commun. Surv. Tutorials **12**(4), 551–572 (2010)
11. Shahriari, B., Swersky, K., Wang, Z., Adams, R.P., De Freitas, N.: Taking the human out of the loop: a review of bayesian optimization. Proc. IEEE **104**(1), 148–175 (2016)

Cluster-Based Distributed Self-reconfiguration Algorithm for Modular Robots

Mohamad Moussa, Benoit Piranda, Abdallah Makhoul$^{(\boxtimes)}$, and Julien Bourgeois

FEMTO-ST Institute, University Bourgogne Franche-Comté, CNRS, Montbéliard, France
{mohamad.moussa,benoit.piranda,abdallah.makhoul, julien.bourgeois}@univ-fcomte.fr

Abstract. Modular robots are automated modules that can change their morphology self-sufficiently and progressively for control or reconfiguration purposes. Self-reconfiguration is a very challenging problem in modular robots systems. Existing algorithms are complex and not suitable for low resources devices. In this paper, we propose a parallel and fully distributed cluster-based algorithm to convert a set of blocks/modules in a new geometrical configuration. We proposed a cluster based self-reconfiguration algorithm. The main idea is to study the impact of clustering on the self-reconfiguration problem. The modules in each cluster remain together and try to move in order to find the final configuration. Based on this concept, our algorithm operates in parallel in each cluster to fasten reconfiguration process and ultimately the set of blocks will reach the desirable shape. We evaluate our algorithm on the centimeter-scale sliding blocks, developed in the Smart Blocks project in a simulator showing the entire reconfiguration process in real-time. We show the effectiveness of our algorithm, especially the benefits of clustering in self-reconfiguration task by comparing performance of both approaches (with and without clustering) while varying the number of clusters.

1 Introduction

Programmable matter consists of tactile materials that have physical form dynamically modifiable. They are able to transform on its own, depending on program controlled. It is essentially composed from materials and components that have capabilities to be programmed and change their shape property on demand. The main approach to implement programmable matter is to build a huge self-reconfigurable robot consisting of a bunch of modules. Each of these

This work has been supported by the EIPHI Graduate School (contract ANR-17-EURE-0002).

modules has some computation, sensing and communication capabilities. The modules can detach, move independently and connect to each other to form new configurations.

Modular robots are distributed systems made up of identical connected modules which are able to coordinate together in order to perform complex tasks that can revolutionize everything starting from surgical applications, to transportation applications or space exploration [1]. The main challenge in modular robots is self-reconfiguration problem which enables the modules to modify their structure from one arbitrary initial shape to another (goal shape) [15]. This problem is very challenging because of the very high number of possible configurations and a fixed set of rules does not work for all possible situations. It poses several challenges. Firstly, this process is very slow due to the limited number of modules which can move simultaneously without interfering which results in limited parallel possible motions and other system constraints. Secondly, self-reconfiguration requires a coordination between modules during motion in order to avoid any collision by adapting a set of rules that does not work for all situations since some motions requires cooperation outfaced by motion algorithms [11]. However, existing solutions for the self-reconfiguration problem propose several algorithms using search-based and control-based techniques [5,14]. The basic concept is to search for a goal configuration in the modular robots configuration space. These solutions are still very complex and not suitable for modular robots with low computation and energetic resources. One solution to fasten self-reconfiguration process is clustering. Grouping modules into clusters and carrying out some tasks in cluster-based approaches can improve the effectiveness in terms of execution time and communication cost. Therefore, the partitioning of the modules into clusters and the parallel displacement of the modules in each partition will accelerate the convergence of the modules to the goal shape [2]. In fact, nodes would care only for executing tasks at intra-cluster level and will not be affected by inter-cluster changes, thus reducing scope of inter-cluster interactions. Clustering can yield improvements to the execution of tasks related to modular robots' programmable matter such as self-reconfiguration where modules forming specific parts of the initial shape can form one cluster. Therefore, it requires less reconfiguration adjustments to attain a more or less similar parts of the goal shape. Hence, here lies the importance and effectiveness of clustering.

Our objective in this paper is to show the impact and the efficiency of clustering and parallelization on the self-reconfiguration problem. We propose a cluster-based distributed and parallel self-reconfiguration algorithm that scales to large modular robot systems in order to speed up the reconfiguration of the modular robot systems from an initial shape to a goal one. Our aim in this paper is not to propose a clustering technique, but to provide a new cluster-based distributed self-reconfiguration algorithm while using our clustering techniques as described in [2]. The number of clusters is specified based on the goal configuration and based on the density cut algorithm [13]. The proposed algorithm consists in sliding the not well-placed robots in the interior the current configuration in order to fill target free cells. For each not well-placed robot, the method compute a gradient distance from free goal cells, then construct a path for each not well-placed robot to a free cell which does not cross the path of another robot, and then moves successively horizontal and vertical lines of robots from the end to the begin of the path. These steps are repeated until all the goal cells are filled. The remainder of this paper is organized as follows. In Sect. 2 we provide a short description of the related work. Section 3 presents the system model and some assumptions. Section 4 provides the details of our self-reconfiguration algorithm. The experimental results are presented in Sect. 5 and Sect. 6 concludes our paper.

2   Related Works

In this paper, we consider modular self-reconfigurable robot made from centimeter-size cubes (blocks) able to move [10]. Works with sliding cubes was widely used in the literature [6–8]. In [6], it is showed that using homogeneous Limited Sliding-Cube modules could be performed in quadratic time using meta-modules and tunneling motions. The method's advantage is a high degree of motion parallelism. It was then further investigated in [7] and a novel approach was proposed for heterogeneous modules. Moreover, a tunneling-based recon-figuration algorithm was later proposed in [8] to remove the limitation on the arrangement of start and goal configurations by employing tunneling techniques.

Self-reconfiguration is one of the most critical challenges for modular robots. It may seem easy to solve, however, it is one of the most challenging issues. Several solutions have been proposed in the literature and there is no general solution has been appeared yet. For instance, in [4] the authors propose a self-reconfiguration in two dimensions. They propose the idea of module substraction, where unwanted modules remove themselves from the system without external intervention to attend the desired configuration. The algorithm is composed of two parts, sequential one for unwanted modules selection and parallel one for modules movements. The algorithm has quadratic complexity and does not work in all situations especially in hollow structures. In [11], the authors propose a distributed algorithm for reconfiguration of lattice-based modular robots systems made from the same cubic modules used in this paper. This iterative algorithm can transform various large modular robots, beginning from a filled shape towards a target one by using motion rules and creating a train of modules made up of the border of the configuration, which will be moved gradually in order to reach the target shape. In cite [3] the authors introduced geometric algorithms for copying, rotating, scaling and reflecting a given polyomino and they also provided an algorithm for constructing a bounding box surrounding a polyomino.

Most of existing Self-reconfiguration algorithms are based on sequential movements which results in dramatically increasing the reconfiguration process's duration. In fact, there exists an obvious trade-off between a high degree of motion parallelism and full convergence to the goal shape due to collisions and deadlocks that may appear with increasing number of modules moving simulta-neously. In the literature, existing methods with parallel motions suffer from a limitation in the number of parallel motions depending on the approach used. In [9], a scaffolding technique is proposed which allows this parallelism while avoiding collisions. The contribution of this paper is to propose a fully parallel and distributed self-reconfiguration algorithm using clustering approaches with increase in motion parallelism for accelerating self-reconfiguration that can manage almost to build any goal shape given the initial shape divided into a pre-defined number of clusters.

3 System Model and Assumptions

In this paper, we consider modular self-reconfigurable robot made from centimeter-size cubes (blocks) able to move [10]. Blocks are equipped wih a micro-electro-mechanical-system (MEMS) actuator array in the upper face to move the objects and they are powered externally. A block can slide along the border of one of its neighbors. Each block can connect to up to 4 neighbors positioned on each of its faces. They can communicate together by using neighbor-to-neighbor message passing.

We assume that these blocks are placed in a simple 2D Cartesian coordinate system along $\vec{x}$ and $\vec{y}$ axes. Each cell can either be occupied by a block or empty. It is assumed that blocks can recognize their attached neighbors at any given time and moving blocks cannot communicate with any other ones. We assume that, at any time, a block can detect if there exists a surrounding cell $\exists\, f \subset \{East; West; North; South\}$ that should be filled.

Moreover, each cluster is given its goal shape G, also know by cluster members. We consider that this target shape is fixed during self-reconfiguration process, so no new modules can join or quit any cluster.

Furthermore, the number of blocks in each cluster must be sufficient enough (greater or at least equal) in order to be able to build the cluster's target shape.

4 Cluster Based Self-reconfiguration

4.1 Agents

Throughout this paper, we consider the *Cluster Head* agent for explaining the algorithm. The *Cluster Head* is a special treatment to manage a cluster, the module with this agent acts as a local coordinator guiding the transformation process to form the target shape of its own cluster. The *Cluster Head* agent is dynamic and can change throughout the algorithm. A new *Cluster Head* module can be elected in each cluster during reconfiguration, it is defined as:

Definition 1. (Cluster Head Coordinator)
Let $N(M, D)$ be the neighbor cell of M in the direction D, and G be the goal configuration. M is a *Cluster Head* coordinator, if it exists: D verifying $(N(M, D) \in G) \wedge (N(M, D)$ is empty$)$.

4.2   Principle

In previous work [10], a module could block all the others by the path it defined to move to an empty cell. Our algorithm allows reconfiguring a set of connected blocks divided into clusters considering that each cluster is self-reconfigured to form its corresponding part of the overall goal shape. This algorithm consists in exchanging messages with data transmission in the network made up of connected blocks. Each block is considered as an independent robot equipped with its own processing unit and memory able to execute the program instructions. A first stage affects a position to each block in the grid and the goal shape for its cluster. Algorithm 2 illustrates how cluster-based self-reconfiguration is handled by the *Cluster Head* block.

At each iteration, one *Cluster Head* module is elected in each cluster among potential ones. This block will trigger cluster building (see Algorithm 2) and then propagate useful information, i.e. distance, to cluster member blocks in order to start the motion and self-reconfigure into cluster's goal shape. Also, it will schedule activities in the cluster and will be in charge of handling intra-cluster interactions. This *Cluster Head* block is not static during the self-reconfiguration algorithm. It can change dynamically at the end of each blocks' movement. It is the case when the previous *Cluster Head* module does not verify the Definition 1 anymore, in other words it no longer has any empty surrounding cell to fill. *Cluster Head* block will take in charge of finding the motion path starting from the block not belonging to goal shape with maximum distance in order to start self-reconfiguration into cluster's target configuration.

The algorithm is detailed in the following. It allows to start the motion events once all the computation is done on all blocks. Mainly, our algorithm consists of 3 major steps. As shown in Fig. 1, these three steps will be executed in parallel and repeated in each cluster until the cluster's goal shape is reached.

1. Set blocks distance in each cluster.
2. Find the block B_{max} with maximum distance not in the goal shape.
3. Displacement of blocks by establishing motion path following decreasing distance order starting from B_{max}.

4.3   Self-reconfiguration Algorithm

We use message passing system to communicate and transmit data between connected blocks. As aforementioned, our reconfiguration approach follows three main steps we detail bellow.

4.3.1   Defining Distances to Blocks in Each Cluster

- *CLUSTER_GO*: Sent by *Cluster Head* block to its connected neighbors to start construction of its own cluster's tree while distributing distance. When received by a cluster member blocks, it defines its distance as being that received in the message plus 1.

Algorithm 1: Distributed algorithm for the construction of a *cluster's tree.*

    $myCH$ // Cluster Id
    $parentCluster$
    $subTreeSize$ // per cluster
    $neighbors$ // set of neighbours
    $childrenCluster$ // set of children
    $nbExpectedAnswers$
    $initialized$ // boolean value

1   **if** $isHead()$ **then**
2   TriggerCluster()

3   **Msg Handler** $CLUSTER\_GO(msgCH, msgDistance)$:
4      $initialized \leftarrow false$
5   **if** $myCH = msgCH$ **then**
6          $distance \leftarrow msgDistance + 1$
7   **if** $parentCluster = null$ **then**
8              $parentCluster \leftarrow sender$
9              $nbExpectedAnswers \leftarrow neighbours.size - 1$
10   **if** $nbExpectedAnswers = 0$ **then**
11                  $subTreeSize \leftarrow 1$
12   **send** CLUSTER_BACK($myCH$,1) **to** *sender*
13   **else**
14   **foreach** $nId \in neighbours$ **do**
15   **if** $nid \neq M_j.id$ **then**
16   **send** CLUSTER_GO($myCH$) **to** nId
17   **else**
18   **send** CLUSTER_BACK($myCH$,−1) **to** *sender*
19   **else**
20   **send** CLUSTER_BACK($myCH$,−1) **to** *sender*

21   **Msg Handler** $CLUSTER\_BACK(msgCH, msgSize)$:
22      $nbExpectedAnswers \leftarrow nbExpectedAnswers - 1$
23   **if** $msgSize \neq -1$ **and** $myCH = msgCH$ **then**
24   childrenCluster.add(*sender*)
25          $subTreeSize \leftarrow subTreeSize + msgSize$
26   **if** $nbExpectedAnswers = 0$ **then**
27          $subTreeSize \leftarrow subTreeSize + 1$
28   **if** $parentCluster \neq null$ **then**
29   **send** CLUSTER_BACK($myCH$,$subTreeSize$) **to** *sender*
30   **else**
             // head
31   **foreach** $child \in childrenCluster$ **do**
32   **send** REQUEST_MAX($myCH$) **to** $child$

- *CLUSTER_BACK*: Response to *CLUSTER_GO* sent by a block to either acknowledge receipt of distance and that all its neighbors have successfully set their distance, or to notify the sender that it does not belong to the same cluster.

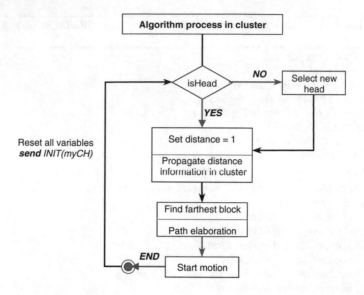

Fig. 1. Algorithm of cluster-based reconfiguration.

Algorithm 2: Distributed control algorithm for the *Cluster Head* agent.

---

    neighbours // maximum 4 neighbours
    head ← *true*
    bmaxId // farthest block's id not in goal shape
    bmaxDistance // bmaxId's distance
    myCH // Cluster Id

1 **while** *head* **do**
2 | *TriggerCluster()*
3 | *isHead()*

4 **Function** *TriggerCluster()*:
5 | setNeighbours()
6 | $distance \leftarrow 1$
7 | $nbExpectedAnswers \leftarrow neighbours.size()$
8 | **foreach** $nId \in neighbours$ **do**
9 | **send** CLUSTER_GO($myCH$,$distance$) to nId

10 **Event Handler** *TRANSLATION_END*:
11 | setNeighbours()
12 | $nbExpectedAnswers \leftarrow neighbours.size$
13 | **foreach** $nId \in neighbours$ **do**
14 | **reset** all variables
15 | **send** INIT($myCH$) to nId

16 **Function** *isHead()*:
17 | return ($\exists emptyCell \in \{West, North, East, South\} \wedge emptyCell \in goal$)

Once cluster tree is built, the block with maximum distance and not in goal shape must be found, then the motion starts by following the decreasing path order of distance. Algorithm 1 illustrates the first step in details. Cluster tree construction is done by calling the helper function *TriggerCluster*. Starting from a *Cluster Head* module in each cluster, a structure tree rooted at the *Cluster Head* node is built with distance assignment using two messages *CLUSTER_GO* and *CLUSTER_BACK*.

4.3.2 Finding B_{max}

B_{max} is defined as the block with maximum distance and not belonging to cluster's goal shape. The root (*Cluster Head*) in each cluster will then launch the search for B_{max} by sending *REQUEST_MAX* to its children through its cluster tree. This message will be transmitted locally across all cluster members. When received by a block already in the cluster's goal shape this message is propagated to children blocks until leaf ones, otherwise block is not in cluster's goal shape and should send its *id* and *distance* back to tree's root using *SEND_MAX* message which is progressively returned to the parent until it reaches the root. Thus, the root will maintain the block's id (*bmaxId*) not in goal shape having the maximum distance (*bmaxDistance*) value.

4.3.3 Blocks Displacement

The motion path can be easily established. Starting from B_{max} block and following decreasing order distance by gradually crossing *parentCluster* blocks benefiting from the tree structure. This way back leads us to the *Cluster Head* block that has empty spot to fill, once *Cluster Head* is reached the displacement of the blocks starts where each block has to replace and move to its *parentCluster*'s position.

Using programming, the blocks are able to deal with state changes locally and to handle different events, such as connection/disconnection of a new block, or the end of a motion. When a *Cluster Head* block finishes motion, i.e., all the blocks on the path succeeded in moving to its *parentCluster*'s position, it resets all its variables and sends a message of type *INIT(myCH)* to all neighbors in order to notify the blocks to reinitialize all their variables with a new iteration taking place. Then, this *Cluster Head* must verify if it can still acts as *Cluster Head* block or not. If it is the case, old *Cluster Head* triggers a new iteration and repeat the process, otherwise a new *Cluster Head* takes over. This reconfiguration algorithm is being repeated in each cluster, until all cluster's blocks converge towards goal shape, hence cluster's target shape is well formed.

4.4 Message Evaluation

The proposed cluster-based self-reconfiguration algorithm operates by using a message–passing model. Our algorithm involves only local and intra-cluster interactions between blocks. The messages are used in the system for communication purposes, where a block sends a message to a neighbor one asking for data. We

---

Algorithm 3: Message Handler for any module M_i.

---

wellPlaced // module $\in$ goal shape or not

1 **Msg Handler** *REQUEST_MAX(msgCH)*:

2   **if** $\neg wellPlaced$ **and** $myCH = msgCH$ **then**

3   **send** SEND_MAX($M_i.id$,*distance*) to *parentCluster*

4   **foreach** *child* $\in$ *childrenCluster* **do**

5   **send** REQUEST_MAX(*msgCH*) to *child*

6 **Msg Handler** *SEND_MAX(msgId,msgDistance)*:

7   **if** *parentCluster* $\neq null$ **then**

8   **send** SEND_MAX (*msgId,msgDistance*) to *parentCluster*

9   **else**

10   **if** $bmaxDistance < msgDistance$ **then**

11          $bmaxDistance \leftarrow msgDistance$

12          $bmaxId \leftarrow msgId$

13   **start motion**

14 **Msg Handler** *INIT(msgCH)*:

15   **if** $myCH \neq msgCH$ **then**

16   **send** INIT_ACK() to *sender*

17   **else**

18   **if** *!initialized* **then**

19   **reset** all variables

20          $toAnswer \leftarrow sender$

21          $nbExpectedAnswers \leftarrow neighbours.size - 1$

22   **if** $nbExpectedAnswers = 0$ **then**

23   **send** INIT_ACK () to *sender*

24   **else**

25   **foreach** $nId \in neighbours$ **do**

26   **send** INIT(*myCH*) to *nId*

27   **else**

28   **send** INIT_ACK () to *sender*

29 **Msg Handler** *INIT_ACK()*:

30    $nbExpectedAnswers \leftarrow nbExpectedAnswers - 1$

31   **if** $nbExpectedAnswers = 0$ **and** $toAnswer \neq null$ **then**

32   **send** INIT_ACK () to *toAnswer*

33   **else**

34   **if** *isHead()* **then**

35   TriggerCluster()

36   **else**

37   **foreach** *nid in neighbours* **do**

38   **send** NEW_HEAD (id) to *nid*

39 **Msg Handler** *NEW_HEAD(msgId)*:

40   **if** $M_i.id = msgId$ **then**

41   TriggerCluster(); // new Head

42   **foreach** $nid \in neighbours$ **do**

43   **if** $nid \neq M_j.id$ **then**

44   **send** NEW_HEAD (*msgId*) to *nid*; // except sender

---

note m the maximum number of connections a module can have, n the number of blocks, and k the number of clusters. During the first step, except for the *Cluster Head*, when a block receives a *CLUSTER_GO* message, it will send it back to all its neighbours except the sender hence the number of *CLUSTER_GO* messages exchanged is $O(k \times n \times (m-1))$. Each reception of a *CLUSTER_GO* is followed by a response of *CLUSTER_BACK* message. Therefore, the total number of messages exchanged during step 1 is $O(2k \times n(m-1))$. The total number of messages exchanged during the last two steps, by using $\langle REQUEST\_MAX,$ $SEND\_MAX \rangle$ and $\langle INIT, INIT\_ACK \rangle$ is $O(2k \times m)$ at each step. Then, the new *Cluster Head* selection requires $O(k \times m)$ messages. Therefore, the total number of messages exchanged at each iteration's execution is $O(k \times n \times m)$.

5 Simulation and Results

We implemented our proposed algorithm in *VisibleSim* [12], a simulation and development environment that supports multi-target large scale modular robot systems (*Smart Blocks*, *Blinky Blocks*, *Claytronics*...). We conducted extensive simulations to evaluate the performance of the proposed algorithm. In order to visualize clustering impact, we focus on comparing cluster-based and without clustering algorithms with varying number of clusters and ensemble sizes. Two types of complex reconfiguration scenarios have been carried out on several shapes while varying the number of modules in the ensemble. For each scenario, we generated different versions of the goal configurations using different scales ranging from hundreds to ten thousands of blocks. The first scenario formed by the text: 'smart' with various number of blocks. 'Smart' word is composed from five letters, one idea is to let each cluster form one letter as shown in Fig. 2a. In fact, we have five clusters, each corresponding to one letter with a specific color. We consider that blocks are initially divided into 5 clusters using [2]. Each block belongs to a cluster and the cluster's goal shape is stored in all cluster members.

(a) Smart word formed by 5 clusters.   (b) Large comb formed by 2 clusters.

Fig. 2. Two self-reconfiguration scenarios.

The second scenario is to build a large comb. We test our algorithm to self-reconfigure large networks of more than 10,000 blocks. In this experiment (cf. Fig. 2a), a large set of *Smart Blocks* is assembled presenting very regular volumes of blocks with symmetries. For each experiment, blocks can glow in different colors depending on running program in order to show a state upon receipt of

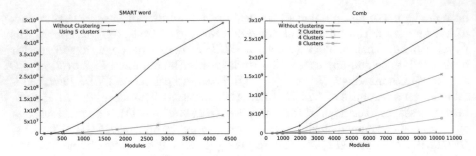

Fig. 3. Number of messages exchanged for 'smart' word using 5 clusters and to build a comb using 2,4 and 8 clusters.

a message, detection of an event or to indicate that it has successfully reached the goal shape. They are also able to display value upon their front face, used in our case to display useful information such as distance and cluster identifiers. In Fig. 3, we studied the number of messages exchanged in the two scenarios according to the size of the goal shape. Firstly, using five clusters to form the smart' word with various network sizes (up to 4,800 blocks) and compare it with results obtained without clustering. The total messages exchanged by both algorithms linearly increases with the number of blocks. Obviously, the number of messages exchanged using 5 clusters is much lower without clustering. Secondly, Fig. 3 also shows the total number of messages sent to construct a comb. It displays the messages exchanged for a dividing into 2,4 and 8 clusters, versus those exchanged without clustering. Without clustering approach requires the highest number of messages exchanged due to the fact that the whole system is considered as a single cluster, hence a higher number of connections and messages exchanged between blocks. When no cluster is used, the number of messages increases exponentially, while it decreases gradually and becomes linear with the increase in the number of clusters. Hence, in each cluster a lot of messages will be dropped, especially by modules at the boundaries between clusters. As a consequence, this confirms that our algorithm only involves local intra-cluster interactions as announced in previous sections. Fig. 3 enables one to easily notice that with the increase of both number of clusters and ensemble size, also the gap between the two approaches is largely increasing.

6   Conclusion and Future Work

In this paper, we proposed a cluster based self-reconfiguration algorithm in a fully distributed fashion. We conducted two types of complex reconfiguration experiments, and we evaluated the algorithm on several shapes while varying the number of clusters and the number of modules in the ensemble. Our proposal has shown to be effective, i.e., self-reconfiguration algorithms can be aided by clustering approaches to speed up reconfiguration process. From obtained results, one can easily notice that the speed of algorithm's convergence with clustering is much way better than without clustering. The results also show that clustering has many benefits including energy-efficiency by reducing communication load (messages exchanged), scalability. As a future work, we aim to improve the proposed algorithm by making many possible motions within each cluster while avoiding collisions, thus achieving a high degree of motion parallelism per cluster. This will surely affect the reconfiguration speed and reduce complexity, but it is a complex task since it involves more interactions at the inter-cluster level.

References

1. Alattas, R.J., Patel, S., Sobh, T.M.: Evolutionary modular robotics: survey and analysis. J. Intell. Rob. Syst. **95**(3–4), 815–828 (2019)
2. Bassil, J., Moussa, M., Makhoul, A., Piranda, B., Bourgeois, J.: Linear distributed clustering algorithm for modular robots based programmable matter. In: IEEE/RSJ International Conference on Intelligent Robots and Systems, IROS (2020)
3. Fekete, S.P., Gmyr, R., Hugo, S., Keldenich, P., Scheffer, C., Schmidt, A.: Algorithmic aspects of manipulating programmable matter with finite automata. Algorithmica **83**, 1–26 (2020)
4. Hall, M.D., Ozdemir, A., Groß, R.: Self-reconfiguration in two-dimensions via active subtraction with modular robots. In: Robotics: Science and System XVI (2020)
5. Hauser, S., Mutlu, M., Léziart, P.-A., Khodr, H., Bernardino, A., Ijspeert, A.J.: Roombots extended: challenges in the next generation of self-reconfigurable modular robots and their application in adaptive and assistive furniture. Rob. Auton. Syst. **127**, 103467 (2020)
6. Kawano, H.: Complete reconfiguration algorithm for sliding cube-shaped modular robots with only sliding motion primitive. In: IEEE/RSJ International Conference on Intelligent Robots and Systems (IROS), pp. 3276–3283 (2015)
7. Kawano, H.: Tunneling-based self-reconfiguration of heterogeneous sliding cube-shaped modular robots in environments with obstacles. In: 2017 IEEE International Conference on Robotics and Automation (ICRA), pp. 825–832 (2017)
8. Kawano, H.: Distributed Tunneling Reconfiguration of Sliding Cubic Modular Robots in Severe Space Requirements, p. 14 (2018)
9. Lengiewicz, J., Hołobut, P.: Efficient collective shape shifting and locomotion of massively-modular robotic structures. Auton. Rob. **43**(1), 97–122 (2019)
10. Piranda, B., Laurent, G., Bourgeois, J., et al.: A new concept of planar self-reconfigurable modular robot for conveying microparts. Mechatronics **23**(7), 906–915 (2013)
11. Piranda, B., Bourgeois, J.: A distributed algorithm for reconfiguration of lattice-based modular self-reconfigurable robots. In: 2016 24th Euromicro International Conference on Parallel, Distributed, and Network-Based Processing (PDP), pp. 1–9. IEEE (2016)
12. Piranda, B., Fekete, S., Richa, A., Römer, K., Scheideler, C.: Your simulator for programmable matter. In: Algorithmic Foundations of Programmable Matter, Visiblesim (2016)
13. Shao, J., Yang, Q., Liu, J.,Kramer, S.: Graph clustering with density-cut. arXiv preprint (2016)

14. Thalamy, P., Piranda, B., Bourgeois, J.: A survey of autonomous self-reconfiguration methods for robot-based programmable matter. Rob. Auton. Syst. **120**, 103242 (2019)
15. Yim, M., Shen, W.M., Salemi, B., Rus, D., Moll, M., Lipson, H., Klavins, E., Chirikjian, G.S.: Modular self-reconfigurable robot systems [grand challenges of robotics]. IEEE Rob. Autom. Mag **14**(1), 43–52 (2007)

Prediction of the Response Time in Web Services Based on Matrix Factorization

Maksymilian Iwanow[1](✉) ⓘ and Jolanta Wrzuszczak-Noga[2] ⓘ

[1] Faculty of Management, Computer Science and Finance, Wroclaw University of Economics and Business, ul. Komandorska 118-120, 53-334 Wrocław, Poland
Maksymilian.Iwanow@ue.wroc.pl
[2] Faculty of Computer Science and Management,
Wroclaw University of Science and Technology,
Wybrzeże Wyspiańskiego 27, 50-370 Wrocław, Poland
Jolanta.Wrzuszczak-Noga@pwr.edu.pl

Abstract. This paper concerns the prediction of the response time of internet resources depending on a group of various factors, such as agent location, web browser and cookies, using algorithms related to matrix factorization techniques. The experiment was based on testing ten university websites from five different agent locations. The collected data was used to predict the load time of pages using the SVD (Singular Value Decomposition), SVD++ (Singular Value Decomposition++) and NMF (Non-negative Matrix Factorization) algorithms. The quality of the prediction was evaluated using RMSE (Root Mean Square Error), MSE (Mean Square Error) and MAE (Mean Absolute Error) indexes.

1 Introduction

The Internet has a fundamental impact on people's lives. It allows to communicate on such a scale, that was not possible in the time before it was invented, regardless of gender, nationality, race, religion and from every point in the world. The Internet gives new opportunities to share sound, images and video. Due to the faster and faster technology expansion, there is a need to download resources from various places in the world. Time is money, so every second of the response time matters. It is related to the development of new online enterprises, start-ups. It creates new e-business opportunities. It is clear that further progress of e-services will continue making human life easier. Therefore, load time issues will become more and more important.

To ensure the network ability to provide the necessary services to given traffic within a certain technological framework, QoS – Quality of Service was introduced. There are different criteria related to the QoS. The most important for users are metrics that generally refer to the response time. Resources need to be available as fast as possible – it is said, that a speed of HTTP (Hypertext Transfer Protocol) load should be less than 2–3 s [1]. On the other hand, there is a server viewpoint. Providers should ensure reasonable and precise QoS expectations – parameters are related to the latency, the minimal number of failures, bit rate, throughput and many others [2].

This work aims to give insides referring to the prediction of the pages load time using algorithms that are based on matrix factorization techniques. The word 'prediction' relates to the performance, i.e. downloading resources according to a set of factors like agent location, browser. Forecast approaches are correlated with the NMF (Non-negative Matrix Factorization) and SVD (Singular Value Decomposition) factorization techniques. Three prediction algorithms (SVD, SVD++, NMF) were evaluated by prediction performance indexes like RMSE (Root Mean Square Error), MSE (Mean Square Error) and MAE (Mean Absolute Error).

This paper provides the comparison of prediction algorithms based on matrix factorization methods, used to predict load time for different agent locations, browsers and cookies (caching) policy.

Our experiment contributes to the intelligent measurements, suggests which agent (agent type, agent localization) and which forecasting method could be used to predict the future behaviour of the monitored websites.

This paper is organized as follows. In the second section, we present the background information, including related work. In the third section, we describe the implementation, setup and scenarios of our real-world Internet experiment. In the fourth section, we present the results of our study. The final section presents conclusions.

2   Background. Related Work

Prediction is discussed within many areas because it can provide the information that complements the limited data. In telecommunication and energy networks forecasting is very important in the context of serving customers as the aspects of the business processes. There are many different methods used for predicting. Recently collaborative filtering with flagship matrix factorization techniques has become very important. Based on the literature, matrix factorization algorithms are usually applied in the QoS field for predicting metrics related to Web Services, runtime service adaptations, predicting distances in large-scale networks. Those methods are superior techniques for product and social or community network recommendations as well.

In [3] there is presented a model for large-scale networks forecasting by matrix factorization. The authors proposed a scalable system, called Internet Distance Estimation Service (IDES), for predicting large numbers of network distances from limited measurements. There were applied two algorithms for representing a matrix of network distances: SVD and NMF. Simulations carried out on the real-world datasets confirmed that IDES prediction was efficient and robust.

The power of matrix factorization was demonstrated by the Netflix Prize competition [4]: it turned out, that such based models are superior techniques to formulate product recommendations. Those techniques have become popular in recent years because of combining a scalability aspect with predictive accuracy, they offer flexibility in modelling real-life problems.

In [5] few matrix factorization based algorithms were employed to improve the Social Recommendation. Nowadays such user-specific applications seem more and more important. The authors describe the most inconvenient elements in such systems (data sparsity, cold-start problems).

In [6] an adaptive approach based on matrix factorization is proposed for performing online QoS prediction of response time (RT) and throughput (TP). The presented model is based on the stored data stream applying data transformation and online learning techniques. The algorithm was evaluated within the prototype implementation for flight orders. The results showed, that the online QoS prediction approach can be successfully applied in the runtime service adaptation.

Another model related to the QoS forecast was proposed in the paper [7]. The time-aware matrix factorization framework (TMF) integrates QoS time series to provide two-phase QoS predictions in the field of cloud service recommendation.

One more idea of applying matrix factorization was offered in the paper [8] for the links prediction. The framework based on NMF takes into account not only the graph structure, but also both internal and external information (extracted from network). The model solves the link prediction problems basically with local information about a node attribute and global information related to the topological structure.

[9] deals with the community networks where clients exert random or nearest proxy Internet access. The situation clearly showed that some of the proxies will be saturated while the others will be idle. There was proposed a solution based on three matrix factorization algorithms for estimating the response time: Probabilistic Matrix Factorization, Non-negative Tensor Factorization, Adaptive Matrix Factorization. The results were compared with the baseline algorithm.

More and more Web Services are published every day, therefore there is a need to rate them basing on the non-functional features. The forecast of QoS metrics plays very important role in the service selection field. In [10] an efficient approach is proposed that allows to predict the missing QoS parameters values in terms of Location and Reputation aware Matrix Factorization. They show that matrix factorization outperforms many other existing prediction algorithms.

[11] presents very similar approach. It concerns the forecast of the response time of downloading web resources, depending on the location of the client and the browser with the use of the regression based on Fine Tree algorithm and the regression based on Gaussian 5/2 Metric.

In [12] the application of data mining algorithms for the prediction of Web performance is presented. Researchers evaluate four different predictive models by different algorithms: neural networks, decision trees, time series, and regression algorithms. Authors compare features of two general data mining systems, i.e. Microsoft SQL Server and IBM Intelligent Miner.

3 Experimental Setup

The initial step of the experiment was to gather the real response time measurements for the group of ten websites. The process of downloading resources lasted almost four months. It involved launching five WebPageTest agents and gathering results to the text-file-structured database. Collected data was exploited for prediction of the pages loading performance for algorithms SVD, SVD++ and NMF. This section describes the details of the experimental phase.

3.1 Data Collection

Tests were conducted according to the following plan:

- 10 tested websites,
- 5 tested locations,
- tests requested 2 times per day (in the morning and in the evening, every website in each agent location),
- first and repeat view.

The following websites were tested:

- Wrocław University of Science and Technology – https://pwr.edu.pl/
- University of Wrocław – https://uni.wroc.pl/
- AGH University of Science and Technology – https://www.agh.edu.pl/
- Jagiellonian University in Kraków – https://www.uj.edu.pl/
- Warsaw University of Technology – https://www.pw.edu.pl/
- Nanyang Technological University – https://www.ntu.edu.sg/Pages/home.aspx
- University of Tokyo – https://www.u-tokyo.ac.jp/en/index.html
- University of Oxford – https://www.ox.ac.uk/
- Harvard University – https://www.harvard.edu/
- Lomonosov Moscow State University – https://www.msu.ru/

The following agents were chosen:

- Dulles:Chrome – Dulles, USA, Chrome browser
- London_EC2:Chrome – East Central London, England, Chrome browser
- ec2-ap-northeast-1:Firefox – East Central Tokyo, Japan, Firefox browser
- ec2-eu-central-1:Firefox – East Central Frankfurt, Germany, Firefox browser
- ec2-ap-southeast-2:Chrome – East Central Sydney, Australia, Chrome browser

To gather the initial dataset, necessary for conducting the analysis process, there was used Node.js platform and JavaScript files to reach WebPageTest API. The considered data was divided into 5 files – each file per one agent (2.000 records per file).

3.2 Algorithms Implementation and Evaluation Details

For the prediction process, three matrix factorization-based methods were tested and compared with the Normal Predictor and Nearest Neighbour algorithms: SVD, SVD++, NMF. Prediction with simple SVD is calculated in the following way:

$$\hat{r}_{ui} = \mu + b_u + b_i + q_i^T p_u \tag{1}$$

If the user u is unknown, then the bias b_u and the factors p_u are assumed to be zero. The same applies to item i with b_i and q_i. The SVD++ extends the basic method with a new set of item factors that capture implicit ratings. The NMF predictions refer to the simple equation $\hat{r}_{ui} = q_i^T p_u$ where both (user and item) factors are kept positive.

Algorithms were implemented in Python with the great help of Surprise Library, It was designed for the prediction systems with support in the form of few packages like prediction algorithm package, dataset module, accuracy module. All three algorithms: SVD, SVD++ and NMF belonged to the Matrix Factorization Prediction algorithms package, Normal Predictor was related to the Basic Prediction algorithms package and KNN – to the KNN-inspired Prediction algorithms package.

Algorithms were ranked with three indexes:

- Mean squared error (MSE)

$$MSE = \frac{\sum_{i=1}^{N}(y_i - \hat{y}_i)^2}{N} \tag{2}$$

- Mean absolute error (MAE)

$$MAE = \sum_{i=1}^{N}|y_i - \hat{y}_i| \tag{3}$$

- Root mean square error (RMSE)

$$RMSE = \sqrt{\frac{\sum_{i=1}^{N}(y_i - \hat{y}_i)^2}{N}} \tag{4}$$

Where y_i – actual value, $\hat{y}_i$ – predicted value, N – number of observations. The following scenarios were tested:

- first view measurements prediction
- repeat view measurements prediction
- all-measurements, browser prediction
- median load time for agents.

4   Results and Discussion

This paragraph presents the experiment results involving three matrix factorization prediction algorithms: SVD, SVD++, NMF and two other popular ones: KNN and Normal Predictor (Tables 1, 2 and 3), (Fig. 1). There were used three indexes for accuracy evaluation: RMSE, MSE and MAE. Results highlighted with bold font are considered as the most significant.

Results: Scenario 1 (First View Measurements Prediction)

Table 1. Quality prediction indexes for SVD, SVD ++, NMF, KNN and Normal Predictor for five agents – first view measurements prediction.

Algorithm	Index	Dulles, USA	London, England	Tokyo, Japan	Frankfurt, Germany	Sydney, Australia
SVD	MSE	**0.097**	0.132	**0.463**	0.253	0.211
	MAE	0.155	0.169	**0.349**	0.208	0.245
	RMSE	**0.308**	0.364	**0.678**	0.498	0.460
	RMSE-MAE	0.153	0.195	**0.329**	0.29	0.215
SVD++	MSE	0.159	0.136	0.452	0.257	0.217
	MAE	0.248	0.174	0.347	0.210	0.256
	RMSE	0.396	0.369	0.672	0.505	0.457
	RMSE-MAE	**0.148**	0.195	0.325	0.295	0.201
NMF	MSE	0.099	0.137	0.456	0.255	0.215
	MAE	**0.151**	0.165	0.347	0.193	0.245
	RMSE	0.314	0.366	0.675	0.495	0.459
	RMSE-MAE	0.163	0.201	0.328	0.302	0.214
KNN	MSE	0.097	0.133	0.464	0.250	0.214
	MAE	0.155	0.171	0.351	0.195	0.252
	RMSE	0.311	0.360	0.681	0.499	0.461
	RMSE-MAE	0.156	0.189	0.33	0.304	0.209
Normal predictor	MSE	0.863	1.118	1.452	1.425	0.993
	MAE	0.730	0.809	0.908	0.891	0.779
	RMSE	0.929	1.057	1.205	1.194	0.997
	RMSE-MAE	0.199	0.248	0.297	0.303	0.218

Results: Scenario 2 (Repeat View Measurements Prediction)

Table 2. Quality prediction indexes for SVD, SVD ++, NMF, KNN and Normal Predictor for five agents–repeat view measurements prediction.

Algorithm	Index	Dulles, USA	London, England	Tokyo, Japan	Frankfurt, Germany	Sydney, Australia
SVD	MSE	**0.030**	0.037	0.249	0.060	0.140
	MAE	**0.039**	0.052	**0.159**	0.059	0.069
	RMSE	**0.170**	0.187	**0.499**	0.243	0.374
	RMSE-MAE	0.131	0.135	0.34	0.184	0.305

(*continued*)

Table 2. (*continued*)

Algorithm	Index	Dulles, USA	London, England	Tokyo, Japan	Frankfurt, Germany	Sydney, Australia
SVD++	MSE	0.032	0.035	**0.252**	0.061	0.141
	MAE	0.051	**0.042**	0.152	0.058	0.063
	RMSE	**0.170**	0.187	**0.499**	0.244	0.332
	RMSE-MAE	**0.119**	0.145	0.347	0.186	0.269
NMF	MSE	0.033	0.039	0.250	0.064	0.145
	MAE	0.060	0.059	0.131	0.063	0.076
	RMSE	0.180	0.195	**0.499**	0.247	0.381
	RMSE-MAE	0.12	0.136	**0.368**	0.184	0.305
KNN	MSE	0.030	0.036	0.250	0.061	0.143
	MAE	0.040	0.047	0.152	0.056	0.074
	RMSE	0.171	0.189	0.500	0.240	0.348
	RMSE-MAE	0.131	0.142	0.348	0.184	0.274
Normal predictor	MSE	0.064	0.066	0.355	0.108	0.208
	MAE	0.156	0.152	0.332	0.187	0.233
	RMSE	0.253	0.257	**0.595**	0.329	0.443
	RMSE-MAE	**0.097**	0.105	0.263	0.142	0.21

Results: Scenario 3 (All-Measurements, Browser Prediction)

Table 3. Quality prediction indexes for SVD, SVD++, NMF, KNN and Normal Predictor due to the browser: Chrome and Firefox.

Algorithm	Index	Chrome	Firefox
SVD	MSE	**0.403**	0.602
	MAE	0.456	**0.511**
	RMSE	**0.628**	0.771
	RMSE-MAE	0.172	0.26
SVD++	MSE	0.409	0.598
	MAE	0.459	**0.511**
	RMSE	0.633	0.768
	RMSE-MAE	0.174	0.257
NMF	MSE	0.406	**0.604**
	MAE	**0.455**	0.510
	RMSE	0.631	**0.772**
	RMSE-MAE	0.176	0.262

(*continued*)

Table 3. (*continued*)

Algorithm	Index	Chrome	Firefox
KNN	MSE	0.414	**0.612**
	MAE	0.461	0.514
	RMSE	0.636	0.776
	RMSE-MAE	0.175	0.262
Normal predictor	MSE	0.875	1.268
	MAE	0.699	0.815
	RMSE	0.932	1.122
	RMSE-MAE	0.233	0.307

Referring to the first view results (Table 1) the best outcomes provided the SVD algorithm (minimal errors in most cases). Common out-turns delivered NMF and KNN, a little bit worse was SVD++. In general, MSE and MAE index values for NP algorithm were about 4 times worse comparing to SVD.

Repeat view outcomes (Table 2) succeeded, in general, similarly: the best was SVD, a little bit worse: SVD++, NMF and KNN. MSE and MAE index values for NP were about 2–4 worse comparing to the SVD algorithm. It is pretty clear that repeat view results (Table 2) are around 3–4 times better (in case of MSE, MAE for SVD, SVD++, NMF and KNN) that first view outcomes.

The Tokyo and Sydney agent locations delivered the biggest differences between RMSE and MAE indexes providing the biggest fluctuation errors (Table 1 and Table 2).

The average inaccuracies for two tested browsers (Chrome, Firefox) are presented in (Table 3). Firefox loses in every scenario with about 10–30% worse performance than Chrome.

Results: Scenario 4 (Median Load Time)

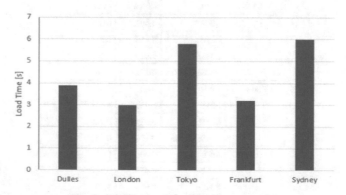

Fig. 1. Median load time for five agents [s].

According to the median load time for all of the agents, Sydney is the one with the biggest response time (~6 s), while London with the smallest one (~3 s) (Fig. 1). Tokyo is the worst after Sydney, but Dulles in that case lose even with Frankfurt. All in all, Tokyo and Sydney statistically perform the worst in all prediction scenarios.

5 Conclusion

The experiment shows, that all tested factors like agent location, browser, cookies/cache policy influence the prediction of resources response time.

The first, general remark is that in every scenario case, per every agent localization, appropriate index values are in the one, similar result range for SVD, SVD++, NMF and KNN algorithms, however Normal Predictor always deliver the biggest errors.

When analyzing the agent localization, Tokyo (Firefox browser) generates the biggest errors. On the other hand, location that statistically appears the best, is Dulles (Chrome browser). Very evident is the difference between prediction errors for the browsers. In fact, in every case Chrome browser generates smaller mistakes than Firefox. According to the median load time for all of the agents, Sydney is the one with the biggest response time (~6 s), while London with the smallest one (~3 s).

Generally, the smallest differences between RMSE and MAE errors generate the SVD++. NMF algorithm very often outputs quite similar to SVD.

The future work related to this experiment could be connected with the choice of other prediction algorithms group, for example regression algorithms. Another idea is to check the same locations with other browsers. There could be developed other factors to compare the impact on the websites load time. As page loading is a process, there might be a point in comparing other metrics that refers to it, for example start rendering or first byte of server response.

References

1. Schwartz, B.: Google keep site speed below 2–3 seconds (2016). https://www.seroundtable. com/google-site-speed-recommendation-23057.html, Accessed 10 Oct 2020
2. Cignoli, C.: Why managing QoS matters more than ever (2017). https://www.appneta.com/ blog/why-managing-qos-matters-more-than-ever/, Accessed 10 Oct 2020
3. Mao, Y., Saul, L.K.: Modeling distances in large-scale networks by matrix factorization. In: Proceedings of the 4th ACM SIGCOMM conference on Internet Measurement, pp. 278–287 (2004)
4. Koren, Y., Bell, R., Volinsky, C.: Matrix factorization techniques for recommender systems. Computer **42**(8), 30–37 (2009)
5. https://labs.engineering.asu.edu/sarwat/wp-content/uploads/sites/26/2014/11/a17-forsati. pdf, Accessed 30 Oct 2020
6. Zhu, J.M., He, P.J., Zheng, Z.B., Lyu, M.R.: Online QoS prediction for runtime service adaptation via adaptive matrix factorization. IEEE Trans. Parallel Distrib. Syst. **28**(10), 2911–2924 (2017)
7. Shun, L., et al.: Time-aware QoS prediction for cloud service recommendation based on matrix factorization. IEEE Access **6**, 77716–77724 (2018)
8. Wang, W., Tang, M., Jiao, P.: A unified framework for link prediction based on non-negative matrix factorization with coupling multivariate information. PloS one **13**(11), e0208185 (2018). https://journals.plos.org/plosone/article?id=10.1371/journal.pone. 0208185, Accessed 30 Oct 2020

9. Gómez-Sánchez, E., et al.: Estimation of web proxy response times in community networks using matrix factorization algorithms. Electronics **9**(1), 88 (2020)
10. Li, S., Wen, J., Luo, F., Cheng, T., Xiong, Q.: A location and reputationaware matrix factorization approach for personalized quality of serviceprediction. In: Proceedings IEEE ICWS, pp. 652–659 (2017)
11. Wrzuszczak-Noga, J., Borzemski, L.: Prediction of load time for different agent locations and browsers. In: Workshops of the International Conference on Advanced Information Networking and Applications, pp. 1298–1307. Springer Cham (2020)
12. Borzemski, L., Kliber, M., Nowak, Z.: Using data mining algorithms in web performance prediction. Cybern. Syst. Int. J. **40**(2), 176–187 (2009)

BO-RL: Buffer Optimization in Data Plane Using Reinforcement Learning

M. Sneha[1(✉)], Kotaro Kataoka[2], and G. Shobha[1]

[1] R V College of Engineering, Bengaluru, Karnataka, India
sneham@rvce.edu.in, shobhag@rvce.edu.in
[2] Indian Institute of Technology Hyderabad, Sangareddy, Telangana, India
kotaro@cse.iith.ac.in

Abstract. Fine tuning of the buffer size is well known technique to improve the latency and throughput in the network. However, it is difficult to achieve because the microscopic traffic pattern changes dynamically and affected by many factors in the network, and fine grained information to predict the optimum buffer size for the upcoming moment is difficult to calculate. To address this problem, this paper proposes a new approach, Buffer Optimization using Reinforcement Learning (BO-RL), which can dynamically adjust the buffer size of routers based on the observations on the network environment including routers and end devices. The proof of concept implementation was developed using NS-3, OpenAI Gym and TensorFlow to integrate the Reinforcement Learning (RL) agent and the router to dynamically adjust its buffer size. This paper reports the working of BO-RL, and the results of preliminary experiments in a network topology with limited number of nodes. Significant improvements are observed in the end to end delay and the average throughput by applying BO-RL on a router.

1 Introduction

Low latency and high throughput are the global demands for any of modern networked services. Many optimization techniques have been explored to improve latency and throughput at the same time including Quality of Service (QoS) [1,2] transport layer protocol extensions [3,4] and traffic engineering in controlled environment such as Data Center Networks [5–8] dynamic buffer-size adjustment for packet forwarding in the data plane of the network has also been actively explored [9–12]. Packet buffering in the data plane determines how many packets can wait before they are transmitted to the link or discarded. Consequently, it significantly affects the latency and packet loss rate caused by the devices in the data plane. However, algorithm based solutions cannot exhibit their best performance if the traffic pattern goes out of their model. When a new destructive protocol emerges, such a protocol may introduce new factors and parameters that are not supported by those algorithm based solutions.

In order to achieve the resilience to the dynamic changes of traffic trend on the buffer size optimization in the data plane, this paper proposes Buffer Optimization using Reinforcement Learning (BO-RL). BO-RL uses Deep Q Learning to implement the RL agent, which continuously makes the decision to change the buffer size based on the

L. Barolli et al. (Eds.): AINA 2021, LNNS 225, pp. 355–369, 2021.
https://doi.org/10.1007/978-3-030-75100-5_31

constant monitoring results of end to end latency and throughput in the network. This paper developed a proof of concept (PoC) implementation using NS-3, OpenAI Gym, and TensorFlow to optimize the buffer size of one router in the network. The experimentation results showed the significant improvement on both latency and throughput, and left some lessons for both further improvement and real-world deployment. The main contributions of this research work are: 1) designing and implementing a practical use case of RL to optimize the buffer size, 2) experimentally proving the effectiveness of the proposed approach in the small and controlled environment, 3) sharing the lessons for the real world deployment.

The rest of the paper is organized as follows: Related work is reviewed in Sect. 2. BO-RL system design is described in Sect. 3 and system implementation in Sect. 4. Section 5 shows the simulation results with limitations and potential improvements to the study. Finally, the work is concluded in Sect. 6.

2   Related Work

Quang Tran Anh Pham et al. [1] used RL with DDPG (Deep Deterministic Policy Gradient) algorithm for Software Defined Networks (SDN) which takes traffic patterns as input and incorporates QoS into the reward function. This work explored a better routing configuration as the action of the agent, which also improves the latency. H. Yao et al. [3] proposed load balance routing schemes for SDN considering queue utilization (QU). The routing process was divided into three steps, the dimension reduction, the QU prediction using Neural Networks, and the load balance routing. H. Yao et al. assumed that if the worst throughput is higher, then the load balancing is better. Yiming Kong et al. [4] proposed TCP Congestion Control using RL based Loss Predictors where the static buffer is altered manually at router to check different scenarios. RL-TCP showed a 7–8% decrease in RTT and 9% increase in throughput on an extremely under-buffered bottleneck link.

M. Elsayed et al. [13] proposed a Deep Reinforcement Learning (DRL) based dynamic resource allocation algorithm to reduce the latency of devices offering mission-critical services for small cell networks. Both the base station and the end nodes were running the DRL agent and the results showed 30% improvement in latency for dense scenarios with 10% throughput reduction. N. N. Krishnan et al. [14] proposed a DRL framework to learn the channel assignment for the Distributed Multiple Input Multiple Output (D-MIMO) groups for WiFi network in order to maximize the user throughput performance. They achieved 20% improvement in the user throughput performance compared to other benchmarks.

N. Bouacida et al. [9] proposed LearnQueue which is an AQM (Active Queue Management) technique using ML. It dynamically tunes the buffer size of Wi-Fi Access Point to decrease the time that packets spend in the queues and thus to reduce the latency. N. Bouacida et al. observed from the experimental results that LearnQueue reduces queuing latency drastically while maintaining the similar throughput in most of the cases. Minsu Kim proposed [10] a DRL based AQM which selects a packet drop or non-drop action at the packet departure stage depending on the current state consisting of current queue length, dequeue rate, and queuing delay. After an action is selected, a reward is calculated based on the queuing delay and packet drop-rate. Dropping or non-dropping action of a packet is not taken by the agent rather it is left to the queuing discipline employed on the node.

3   System Design

3.1   The Overall Architecture

This paper proposes a system to improve end-to-end delay and throughput in a network by optimizing the buffer size on the routers using RL algorithm as shown in Fig. 1. The proposed architecture introduces RL Agent, which is an additional node in the network to run the RL operation. Network nodes and end hosts are extended to interact with RL Agent to allow it to give the instruction of buffer size adjustment on the network. There are four different steps in the RL operation: 1) Observing 2) Training & Learning 3) Taking the Action on the environment 4) Getting the reward. RL Agent observes the network state from both router and end hosts from the environment, takes actions based on the total reward earned for the previous action, gets the reward for current action.

3.2   Formulating the State Space s_t

Four parameters are considered to form a state space $s_t = \{x_1, x_2, x_3, x_4\}$ as shown in Fig. 1 The parameters x_1 = the number of packets dropped because of queue length and x_2 = average qsize at the router are observed from the switch/router. The parameters x_3 =average per packet delay and x_4 = average throughput are observed from the destination node.

3.2.1   Observing the Router for the Parameters x_1 and x_2

At every time step t the agent observes the average queue size and the packets dropped due to queue size from the router in the network environment. Here queue refers to the QueueDisc [15] of traffic control layer in Linux systems. Figure 2(a) illustrates the flow chart of the algorithm used for converting this network information from the router to agent's observation. *CheckQueueDiscSize* procedure is called every 0.1 ms to check the average queue size and the number of packets dropped by exceeding the queue size within the time step t.

3.2.2   Observing the Destination for the Parameters x_3 and x_4

At every time step t the agent observes the network environment to read the average throughput and average per packet delay from the destination. Figure 2(b) illustrates the procedure to calculate the average end-to-end delay and throughput within the time step t as the observation parameters for the agent. Whenever a packet is received at the destination node the *FeedDelayThroughput* procedure is called, to record the delay experienced by that packet and the size of the packet (in bytes) received. The packet's transmission time is attached to the packet at the sender when it's transmitted, which is extracted at the destination node for calculating the delay.

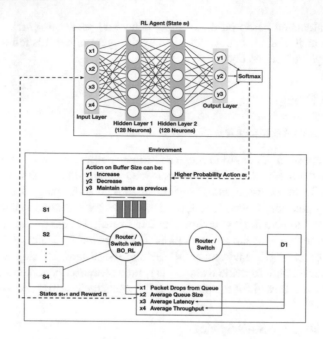

Fig. 1. Overall architecture of the proposed system

3.3   Choosing the Algorithm for RL Agent

Once the state space is formulated, the agent algorithm must be identified, which would best suit the requirements for BO-RL. The agent learning algorithms can be value-based, policy-based or model based [16]. The first two categories are model free algorithms and are suitable for networking research because the agent will not be having access to a model of the network environment.

BO-RL implements the DQN algorithm proposed by Mnih et al. [18] with experience replay as the agent algorithm because; 1) DQN can accurately approximate the desired value function for the environments with large state space. Since we have network as the environment, it is impossible that the agent would observe a previously observed state 2) DQN's function approximation based value function can scale very well, which is a natural choice for more complicated large networks.

In the value based algorithm, the goal of the algorithm is to build a value function which eventually gets defined into a policy. The simplest algorithm in this category is Q-Learning[17]. Q-learning builds a lookup table of Q(s,a) values, with one entry for each state-action pair. Algorithm is well suited for the environments where the states and the actions are small and finite. Q-learning algorithm cannot be used for BO-RL, because the environment in the proposed system is a network and is highly dynamic in nature.

The state as defined in the Sect. 3.B has 4 observation parameters and even a small change in one of the parameters will result in a different state, increasing the number of states to infinite. The Q-learning algorithm will start storing the Q-values for each such

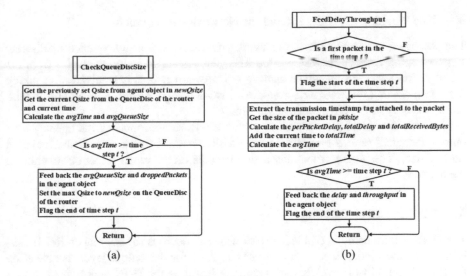

Fig. 2. (a) FlowChart for CheckQueueDiscSize(), and (b) FlowChart for FeedDelayThroughput()

state-action pair into the Q-table and the convergence will be very slow. The memory requirement of the agent will also increase as the table size increases. But the only limitation of the DQN is, because of the deep neural networks, which takes more time to train compared to Q-Learning. The DQN algorithm leverages the advantages of Deep Neural Networks (DNN) as a Q-function that takes the state as input and outputs Q-values for each possible action.

These Q-values are optimized using mean squared error loss function (L), which is defined as in equation below:

$$L = \frac{1}{2}[r + \gamma max_{a'}Q(s',a') - Q(s,a)]^2. \tag{1}$$

Where r is the total reward, s is current state, a is the action taken in current state, s' next state where the agent might move to by taking the action a' and γ is the discount factor to balance between immediate and future rewards: Every time, the network predicts: the Q-values for all the actions, by feed forward pass for the current state s, calculates the $max_{a'}Q(s',a')$ by doing a feed forward to state s', sets the Q-value for action a as $r + \gamma max_{a'}Q(s',a')$, update the weights using the back propagation. The agent tries to minimize this error by learning. But convergence is still an issue. To overcome this problem the authors in [18] have proposed two methods: 1) Experience replay 2) Target network. In experience replay method, instead of training the network and updating the weights for every transition, agent stores the last N experiences defined as (s,a,r,s') in a replay memory. When the number of experiences in the replay memory is exceeding the minibatch size, agent selects the samples of the experiences randomly from the replay memory which will help in taking proper actions. This helps in breaking the strong correlation between consecutive samples and reduces the variance of the updates.

3.4 The Working of RL Agent and the Network Environment

The RL agent interacts with the Network environment E in a sequence of discrete steps $= 1,2,3....$ At each time step t, the RL agent selects an action $a_t \in A(s)$ from the set of action space on the given state $s_t \in S$. Reward $r_t \in R$ is received after taking the action a_t, and observes a new state s_{t+1}. BO-RL implements DQN algorithm with replay memory as the RL agent's algorithm.

At each time step, the state along with the reward are inputs to the agent. As an output, the agent returns one of the three possible actions (increase/decrease/maintain same Qsize). The higher probability action is taken on the router under observation in the network.

3.5 The Process of Training

The DNN is built using two hidden layers with 128 neurons for each layer. ReLU [19] is used for the hidden layers and softmax is used for the output layer as activation functions. The network weights are gradually learnt as the DNN gets trained. Adam Optimizer [20], an optimization algorithm which is an extension to stochastic gradient descent is used to make the learning faster. Different hyper parameter of the RL agent are listed in Table 1.

Table 1. Hyper parameters of the RL agent

Parameter	Brief description
ε	Percentage you want to explore (selecting a random action) or exploit (selecting an action with max future reward)
Epsilon_min	Minimum ε value
Epsilon_decay	Rate at which ε is decreased from 1.0 to Epsilon_min
Batch_size	Replay memory batch size
Gamma (γ)	Discount factor

For every time step t the RL agent learns the queue size in n number of episodes and each episode has m number of steps, so totally in $n \times m$ number of steps before an action is taken. Game over condition should be defined, so that the RL agent stops learning infinitely.

3.6 The Action a_t

The RL agent can take one of the actions(Increase/decrease/remain the same). The output layer of the DNN assigns probabilities to different actions and the Softmax function picks the action with higher probability. The selected action will be implemented back on the router in the network environment.

Every time the future actions which an RL agent takes are influenced by the reward which it had earned by taking the previous action. While choosing the action RL agent must decide whether to take the action randomly (exploration) or to take the same action which had got better reward in the past (exploit). This is a difficult choice and depends on the environment as well. Greedy policies can be used to balance between exploration and exploitation. BO-RL implements decaying ε – greedy policy to choose between exploration and exploitation while choosing the actions. After each time step t, ε is updated using the equation below:

$$\varepsilon = \varepsilon \times Epsilon\_decay. \tag{2}$$

3.7 The Reward r_t

After the RL agent takes an action, it waits for the next state s_{t+1} to see the impact on the network. This impact is evaluated by a reward function to optimize the queue size further to achieve better delay and throughput. The reward can be on a single parameter or can be cumulative based on the requirement. A penalty will be given if the delay is higher and the throughput is decreased. The RL agent always tries to achieve higher rewards by which it automatically achieves better delay and throughput. The Gamma (γ), is used to balance between immediate and future reward; value near to 0.1 for future rewards and value near to 0.9 for immediate rewards. Since BO-RL is an online learning setting, only the immediate rewards have been considered for taking the action decisions.

4   System Implementation

4.1   Implementation Details

BO-RL is implemented using four main components;Environment Network, OpenAIGym, ns3-gym and RL agent as shown in Fig. 3.

Environment Network: Environment Network is implemented using ns3 network simulator [15]. The simulation scenario script contains the internet topology and serves as the environment for the RL agent. The procedures (*Queue size Monitoring Function and Delay, throughput monitoring function*) in the script collects the observation parameter data as defined in Sect. 3 and sends them to the RL agent through the *Environment gateway*. The procedure *Queue size Monitoring Function* executes the received action from the RL Agent. All the routers in the topology built, implement the DropTail queue as the device queue. The traffic control layer between the netdevice and the IP layer implements FIFO QueueDisc [15], which is manipulated by the RL Agent.

OpenAIGym: *gym* is a toolkit for developing RL algorithms. It is a collection of environments, which can be directly used to implement RL algorithms. Custom environments can also be added to this library according to the requirements of the problem definition. *gym* is independent of the structure of the RL agent implemented. The main purpose of *gym* is to have a standardized interface allowing to access the state and execute actions in the environment.

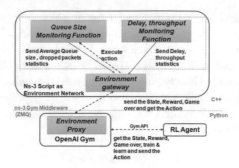

Fig. 3. System block diagram

ns3-gym Middleware: It interconnects ns-3 simulator with OpenAI Gym framework. It provides communication between the components over the *ZMQ sockets* (ZeroMQ is a library used to implement messaging and communication between two applications) using *Protocol Buffers* (library for serialization of messages). ns3-gym has two parts; *Environment Gateway and Environment Proxy*. The *Environment Gateway* created by *OpenGymGateway* object resides inside the simulator and is responsible for:

1. Collecting environment state into structured numerical data as RL agent would expect using callback functions *GetObservationspace(), GetAction-Space(), GetObservation(), GetReward, GetGameOver()*
2. Translating the numerical actions received from the RL agent to proper function calls with proper arguments using the callback function *ExecuteActions()*

Environment Proxy created by the standard *Gym::make('ns3-v0')* receives environment state and exposes it towards the RL agent through the *Gym API* written in Python. The *Environment Proxy* translates the *gym* function calls into messages and sends them towards the *Environment Gateway* over ZMQ socket.

RL Agent: The RL agent is implemented with Python 3.5 using TensorFlow [21] library for Deep Learning Networks.

4.2 Experimental Setup

Network Configuration and Flow Generation: In order to evaluate the effect of BO-RL, network topologies are created on NS3. Table 2 summarises the parameters and their values used for traffic generation and the RL agent implementation.

TCP flows are generated using the BulkSend application of ns3 with randomized senders and receivers. The generation of flows follow a Poisson distribution while the start time follows an exponential distribution. The choice of initial queue size is important to allow the RL agent perform reasonably. The initial queue size was set to 300 packets that does not unnecessarily degrade the performance of the RL agent after performing preliminary experiments with different values between the minimum of 100 packets and maximum of 500 packets. If the initial buffer size is too small, it causes

Table 2. Simulation parameter setting

Parameter	Values
Max size of the queue on all the routers	500 packets
Total size of data to transmit over TCP	1 MB
Total simulation time for testing	10 s
Agents observation time interval t	2 ms
Exploration rate ε	1.0
Epsilon_min	0.01
Epsilon_decay	0.999
Batch_size	32
Discount factor Gamma(γ)	0.7
Learning rate	0.001
Total number of training episodes n	200
Total number of steps m in each episode n	100

lots of packet drops and the throughput gets affected. If the initial queue size is too large, it affects the delay.

RL Agent Configuration: The possible actions to be taken by the RL agent are:

1. increase the queue size by 1,
2. decrease the queue size by 1, or
3. maintain the same queue size as previous.

Initially the value of ε is 1.0 so that the RL agent will explore more by taking random actions and as the time passes the ε is eventually decreased by the epsilon decay factor to reach 0.01 where the RL agent tries to exploit by picking the actions from its stored experiences. The game over condition is met, when the RL agent achieves 95% of the minimum possible delay (link delay). The below equation is used to check the game over condition:

$$delay = min\_delay * 1.05. \tag{3}$$

The reward function is designed to give more priority to delay and throughput parameters than packet drops. RL agent gets cumulative rewards/penalty according to the following conditions.

1. if the *cur_delay*
    a. is lesser than the *prev_delay* a +ve reward
    b. is greater than the *prev_delay* a −ve reward
    c. is same as the *prev_delay* then a 0 reward
2. if the *cur_throughput*
    a. is lesser than the *prev_throughput* a +ve reward
    b. is greater than the *prev_throughput* a −ve reward
    c. is same as the *prev_throughput* then a 0 reward

5 Evaluation

We evaluate the BO-RL proof of concept implementation based on the latency and throughput of the end-to-end communication in NS-3 based testbed.

5.1 Choice of Reward Function

Figure 4 plots the CDF of average reward earned by the RL Agent for 200 episodes during training. The average reward and the standard deviation for different delay and throughput ratios were 1270 and 156 for 40:60, 1183 and 168 for 50:50, 1079 and 148 for 60:40 respectively. Hence 40:60 reward ratio was considered in the conduction of all the experiments.

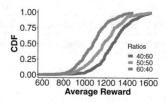

Fig. 4. CDF of average reward for 200 training episodes

5.2 Buffer Optimization in Single Destination Environment

This experiment observes the baseline performance of the BO-RL RL by enabling it on Router 1 (R1) in the network given in Fig. 5 on the basis of delay and throughput of the flows from multiple sources to a single destination. We compare the results with R1 without BO-RL, which works as an ordinary router same as R2. In this setup for enabling BO-RL, the RL agent receives the observation parameters from R1 and Destination 1 (D1). A total of 16 flows were generated by the sources to D1 to transfer 1 MB data in each flow.

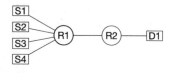

Fig. 5. The network configuration where the link between R1 and R2 is supposed to be congested due to the concurrent flows from 4 Sources to 1 Destination

Figure 6(a) plots the CDF of the average end-to-end delay of packets from the sources to D1 through the corresponding routers, and also shows that R1 with BO-RL successfully reduced the overall latency compared with R1 without it. The average delay and its standard deviation for packets through R1 with BO-RL were 53.43 ms and 9.780 ms respectively, which out perform those of R1 without BO-RL 64.264 ms and 19.347 ms. The maximum latency with BO-RL was also significantly shorter than the case without BO-RL.

Figure 6(b) plots the CDF of the end-to-end throughput of flows observed in the same setup. The average throughput and the standard deviation were 2.215 Mbps and 1.534 Mbps with BO-RL, and 1.940 Mbps and 1.113 Mbps without BO-RL respectively.

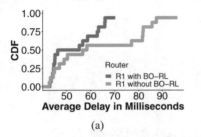
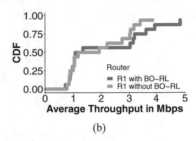

(a)   (b)

Fig. 6. (a) Comparison of CDF End-to-end Average delay at D1 for the flows routed through R1 without BO-RL and R1 with BO-RL, and (b) the corresponding throughput comparison

Figure 7(a) shows the queue size variation, and the number of packets dropped at R1 because of queue limits exceeded, and Fig. 7(b) the corresponding rewards gained by RL agent in the same test phase. Figure 7(a) shows that the queue size is decreasing irrespective of packet drops observed at R1 since the throughput and delay is still improving and the reward gained is increasing as shown in Fig. 7(b). This is because of the choice of RL agent's reward function is more biased towards improving the delay and throughput rather than decreasing the packet drops.

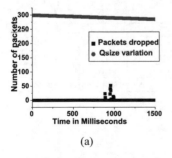
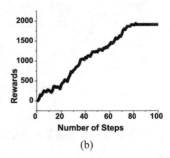

(a)   (b)

Fig. 7. (a) Dynamic adjustment of queue size Vs number of packet dropped at R1 by RL agent, and (b) the corresponding reward gained

5.3 Performance of BO-RL with Multiple Destinations and Varying Number of Feedback Sources

This evaluation item examines whether the performance of BO-RL improves if more receivers provide feedback to the RL agent. RL agent receives the observation parameters from R1 with BO-RL and more than one destinations, for dynamically optimizing the queue size at R1 as shown in Fig. 8.

Tables 3 and Table 4 compare the results of average end to end delay and throughput of flows between the R1 configuration with or without BO-RL. 64 flows to transfer 1 MB Data each were generated with randomized sources and destinations so that each destination receives up to 16 flows.

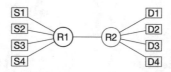

Fig. 8. The network configuration with 4 Sources and 4 Destinations

In this configuration, the network congestion on the link between R1 and R2 is severer than the previous experiment due to the increased number of flows. The end-to-end delay has been significantly reduced regardless of the number of receivers to provide feedback to the RL agent. However, this experience did not clarify the relationship between the number of feedback receivers and the ratio of delay reduction. On the other hand, average throughput slightly degraded in this experiment.

Table 3. Delay analysis with and without BO-RL

	R1 without BO-RL	The number of feedback sources in addition to R1 with BO-RL			
		1	2	3	4
StdDev	3.558	3.667	1.912	2.543	1.395
Max	111.442	84.135	85.074	78.392	83.127
Min	97.966	69.436	77.735	69.089	78.247
Avg	105.084	74.055	81.641	74.584	81.354
Med	105.613	73.265	81.466	74.846	81.729

Table 4. Throughput analysis with and without BO-RL

	R1 without BO-RL	The number of feedback sources in addition to R1 with BO-RL			
		1	2	3	4
StdDev	0.459	0.281	0.364	0.374	0.328
Max	1.937	1.048	1.454	1.657	1.430
Min	0.092	0.120	0.107	0.145	0.125
Avg	0.394	0.380	0.363	0.344	0.363
Med	0.230	0.262	0.230	0.254	0.253

5.4 Limitations and Potential Improvement

5.4.1 Network Topology and Traffic Pattern

The experiment was performed in a very small and simple network, and only TCP traffic was generated. While BO-RL shown the benefit of dynamic buffer optimization, it's benefit should also be explored in larger network typologies with variety of traffic patterns and applications. The number of BO-RL router can also be increased so that the buffer size optimization can happen in many parts in the network.

5.4.2 Number of Observation Parameters and Choice of RL Agent Algorithm

Our system used the following 4 parameters: the average queue size, the number of packets dropped from queue, the average throughput, and the average latency were used as observation parameters. Timestamp, which was used to measure the end-to-end per-packet delay, was obtained using the feature of NS3. In the real world networking scenario, precisely measuring one-way delay is not easy. However, there are some techniques [22–24] can be used to measure the one-way delay in the real world deployment. The observation parameters can be increased to feed more feedback to the RL agent, such as packet size, inter-packet arrival time, 5-tuple information (source/destination IP addresses and port numbers, protocol), etc., with DQN extension for the RL agent algorithm like Double DQN [25], Prioritized Experience Replay [26], Dueling DQN [27], Noisy Nets for Exploration [28].

5.4.3 Dependency on End User Feedback

In the current BO-RL implementation, the RL agent collects feedback from destination hosts in the network. While this mechanism works in NS-3 environment, it's not feasible in the real network. In order to complete the BO-RL operation without depending on the interaction with destination hosts, the network needs to support measurement, feedback collection and reward calculation. Software Defined Networking (SDN) [29], where SDN Controller can collect the observation parameters directly or indirectly from the switches, is a promising approach.

6   Conclusion

This paper proposed BO-RL, a use case for Reinforcement Learning based technique on dynamic buffer optimization in data plane aiming for improving both end-to-end delay and throughput in the network. We conducted the preliminary evaluation of the PoC implementation of BO-RL, and observed that the use of BO-RL introduced the significant improvement to both end-to-end delay and throughput by 17% and 12% respectively compared with the router without BO-RL when the RL agent collects the feedback from the router and only one destination of the traffic. However, if the RL agent collects feedback from multiple destinations, while a minimum of 22.30% improvement was observed in the end to end delay, the throughput improvement was not so significant. Considering the size of network and traffic patterns used in the experiment setup, as well as the observation parameters using that the RL agent was developed, BO-RL still has room for the further improvement as discussed in Sect. 5.4 for future work.

References

1. Pham, T.A.Q., Hadjadj-Aoul, Y., Outtagarts, A.: Deep reinforcement learning based qos-aware routing in knowledge-defined networking. In: Qshine 2018 - 14th EAI International Conference on Heterogeneous Networking for Quality, Reliability, Security and Robustness, Ho Chi Minh City, Vietnam (2018)

2. Chiariotti, F., Kucera, S., Zanella, A., Claussen, H.: Analysis and design of a latency control protocol for multi-path data delivery with pre-defined QoS guarantees. IEEE/ACM Trans. Netw. **27**(3), 1165–1178 (2019)
3. Yao, H., Yuan, X., Zhang, P., Wang, J., Jiang, C., Guizani, M.: Machine learning aided load balance routing scheme considering queue utilization. IEEE Trans. Veh. Technol **68**(8), 7987–7999 (2019)
4. Kong, Y., Zang, H., Ma, X.: Improving TCP congestion control with machine intelligence. In: NetAI 2018: ACM SIGCOMM 2018 Workshop on Network Meets AI & ML, Budapest, Hungary, 24 August 2018 (2018)
5. Alizadeh, M., et al.: Data center TCP (DCTCP). ACM SIGCOMM Comput. Commun. Rev. **40**(4), 63–74 (2010)
6. Vamanan, B., Hasan, J., Vijaykumar, T.: Deadline-aware datacenter TCP (D 2 TCP). ACM SIGCOMM Comput. Commun. Rev. **42**(4), 115–126 (2012)
7. Wu, H., Feng, Z., Guo, C., Zhang, Y.: ICTCP: incast congestion control for TCP. In: Proceedings ACM CoNEXT (2010)
8. Wilson, C., Ballani, H., Karagiannis, T., Rowtron, A.: Better never than late: Meeting deadlines in datacenter networks. In: Proceedings ACM SIGCOMM Conference (SIGCOMM), pp. 50–61 (2011)
9. Bouacida, N., Shihada, B.: Practical and dynamic buffer sizing using learn queue. IEEE Trans. Mob. Comput. **18**(8), 1885–1897 (2019)
10. Kim, M.: Deep Reinforcement Learning based Active Queue Management for IoT Networks (2019). https://doi.org/10.13140/RG.2.2.24361.65126
11. Radwan, A., To, H.T., Hwang, W.J.: Optimal control for bufferbloat queue management using indirect method with parametric optimization. Sci. Programm. **2016**, p. 10 (2016). Article ID 4180817
12. Handley, M., Raiciu, C., Agache, A., Voinescu, A., Moore, A.W., Antichi, G., Wójcik, M.: Re-architecting datacenter networks and stacks for low latency and high performance. In: Proceedings of SIGCOMM 2017, Los Angeles, CA, USA, 21–25 August 2017, P. 14 (2017)
13. Elsayed, M., Erol-Kantarci, M.: Deep reinforcement learning for reducing latency in mission critical services. In: IEEE Global Communications Conference (GLOBECOM), Abu Dhabi, United Arab Emirates, vol. 2018, pp. 1–6 (2018)
14. Krishnan, N.N., Torkildson, E., Mandayam, N.B., Raychaudhuri, D., Rantala, E., Doppler, K.: Optimizing throughput performance in distributed MIMO Wi-Fi networks using deep reinforcement learning. In: IEEE Transactions on Cognitive Communications and Networking (2019)
15. Nsnam. (n.d.). 3. Accessed 27 February 2021, https://www.nsnam.org/
16. François-Lavet, V., Henderson, P., Islam, R., Bellemare, M.G., Pineau, J.: An introduction to deep reinforcement learning. In: CoRR (2018)
17. Watkins, C.J.C.H., Dayan, P.: Q-learning. Mach. Learn. **8**, 279–292 (1992)
18. Mnih, V., Kavukcuoglu, K., Silver, D., et al.: Human-level control through deep reinforcement learning. Nature **518**, 529–533 (2015)
19. Nwankpa, C.E., Ijomah, W., Gachagan, A., Marshall, S.: Activation Functions: Comparison of Trends in Practice and Research for Deep Learning (2018). arXiv:1811.03378 [cs.LG]
20. Kingma, D.P., Jimmy, B.: Adam: A Method for Stochastic Optimization (2014). arXiv:1412.6980 [cs.LG]
21. Tensorflow. (n.d.). Acceessed 27 Feb 2021, https://www.tensorflow.org/
22. Hernandez, A., Magana, E.: One-way delay measurement and characterization. In: Third International Conference on Networking and Services, ICNS 2007, Athens, pp. 114–114 (2007)

23. Kumar, R., et al.: End-to-End network delay guarantees, for real-time systems using SDN. In: IEEE Real-Time Systems Symposium (RTSS), Paris, pp. 231–242 (2017). https://doi.org/10.1109/RTSS.2017.00029
24. Zhang, T., Liu, B.: Exposing end to ond delay in software-defined networking. Int. J. Reconfigurable Comput **2019**, p. 12 (2019). Article ID 7363901, https://doi.org/10.1155/2019/7363901
25. van Hasselt, H., Guez, A., Silver, D.: Deep reinforcement learning with double Q-learning (2016)arXiv:1509.06461 [cs.LG]
26. Schaul, T., Quan, J., Antonoglou, I., Silver, D.: Prioritized experience replay (2016). arXiv:1511.05952v4 [cs.LG], Accessed 25 Feb 2016
27. Wang, Z., Schaul, T., Hessel, M., van Hasselt, H., Lanctot, M., de Freitas, N.: Dueling Network Architectures for Deep Reinforcement Learning (2016). arXiv:1511.06581v3 [cs.LG], Accessed 5 Apr 2016
28. Fortunato, M., Azar, M.G., Piot, B., Menick, J., Osband, I., Graves, A., Mnih, V., Munos, R., Hassabis, D., Pietquin, O., Blundell, C., Legg, S.: Noisy networks for exploration (2019). arXiv:1706.10295v3 [cs.LG], Accesssed 9 July 2019
29. Farhady, H., Lee, H.Y., Nakao, A.: Software-defined networking: a survey. Comput. Netw. **81**, pp. 79–95 (2015). ISSN 1389-1286, https://doi.org/10.1016/j.comnet.2015.02.014

Formal Modeling, Verification, and Analysis of a Distributed Task Execution Algorithm

Amar Nath[1]([✉]) and Rajdeep Niyogi[2]

[1] Sant Longowal Institute of Engineering and Technology, Deemed University, Longowal, India
amarnath@sliet.ac.in
[2] Indian Institute of Technology Roorkee, Roorkee, India
rajdpfec@iitr.ac.in

Abstract. Task execution in a dynamic environment is a challenging and important problem in multi-robot systems. In such an environment, the arrival time, location of a task, states and availability of robots are not known in advance. Recently a distributed algorithm for task execution in such an environment has been suggested. The distributed algorithm is supposed to be run on different components (robots) of the distributed system (multi-robot system) in an asynchronous model. It needs to be formally proved or verified that the algorithm behaves as expected. In this paper, formal modeling, verification, and analysis of the distributed algorithm are carried out. SPIN model checker is used for modeling and verification of the algorithm. We identify some interesting safety and liveness properties of the algorithm and perform extensive simulations of the algorithm. The verification results ascertain that the behavior of the algorithm is as expected.

1 Introduction

In an automated environment (e.g., office, workshop, and warehouse), a team of mobile robots may be used to perform routine tasks like moving (carrying) a heavy box from one location to another and lifting a heavy object. If a robot r detects a task at some location, it attends the task and determines the size of the team needed for its execution. Since r has only the local view of the environment and does not know the current states and locations of other robots, it cannot determine a team of robots available for the task. With such insufficient information, a team cannot be formed by r, and subsequently, the task cannot be executed. Thus, the necessary information has to be acquired by r by communicating with other robots. This necessitates the design of a distributed algorithm for task execution in such a dynamic environment.

A distributed approach for task execution in a dynamic environment is presented in our previous work [1], where a communicating automata based model

© The Author(s), under exclusive license to Springer Nature Switzerland AG 2021
L. Barolli et al. (Eds.): AINA 2021, LNNS 225, pp. 370–382, 2021.
https://doi.org/10.1007/978-3-030-75100-5_32

is used to capture the behavior of robots. We use a distributed algorithm corresponding to the automata based model in this paper. A distributed algorithm is supposed to be run on different components (robots) of the distributed system (multi-robot system) in an asynchronous model. It needs to be formally proved or verified that the algorithm behaves as expected.

Several model checking tools have been developed (e.g., SPIN [2], UPPAAL [3], MCMAS [4], and PRISM [5]), to formally verify the system under study. The tools verify certain properties and accept properties specification and design models constructed using only specific languages. SPIN accepts properties specified in linear temporal logic (LTL), and the system design is modeled usingPromela. MCMAS accepts properties specified using CTL (Computation Tree Logic) and ATL (Alternating Time Logic) and the model is constructed using ISPL (Interpreted Systems Programming Language).

There exists other distributed approaches [7–9] for team formation. However, these works have not considered formal verification of the algorithms. SPIN model checker has been suggested to find multi-agent plans in [10].

In this paper, formal modeling, verification, and analysis of the distributed algorithm [1] are carried out using SPIN model checker [2]. Some important safety and liveness properties of the algorithm [1] are identified and verified to ensure the correctness of the algorithm.

The rest of the paper is structured as follows. The distributed algorithm for task execution is given in Sect. 2. The formal modeling is carried out in Sect. 3. The formal verification results and analysis are given in Sect. 4. Section 5 gives the conclusion.

2   Distributed Algorithm for Task Execution

The distributed algorithm corresponding to the communicating automata based model [1] is given below.

The algorithm consists of two functions *send* and *receive* shown in Algorithm 1 and 2 respectively. The agents are assumed to be autonomous and collaborative, and whenever possible they express their willingness to be part of a team. The states of an agent are IDLE, ANALYZE, BUSY, and PROMISE. IDLE state means that an agent is not engaged in any kind of activity; BUSY state means that an agent is engaged in an activity; ANALYZE state means that an agent has initiated the team formation process; and PROMISE state means that an agent has expressed its willingness to be the part of a team. The agent who initiates the team formation process is called initiator and the other agents available in the environment are referred to as non-initiator. The messages are of the following types: Request, Engaged, Willing, Not-required, and Confirm. Whenever an agent finds a complex task (that needs multiple agents) at some location l and time t, it becomes the initiator and executes the send function.

In the *send* function, an initiator i broadcasts a Request message, and waits for some time, say Δ. It is assumed that broadcast would be within a certain communication range which need not necessarily cover the entire working environment and Request would be delivered only to the agents present in the environment at that moment in time. A non-initiator j, who has the necessary skills, will send either a Willing or an Engaged message if its state is IDLE or PROMISE respectively. Otherwise, it will ignore the Request message. The initiator increases its counter c when it receives a Willing message. After Δ time has elapsed, i checks if there are enough agents available to form a team ($c \geq k - 1$). If yes, i selects the team with minimum cost and sends *Confirm* message to the members of the team and sends a Not-required message to ($c - (k - 1)$) agents, if any (line 5–9, Algorithm 1). The selection of a team is done using a cost function [1]. If no, i sends a Not-required message to all c agents who expressed their willingness to help. Also, i will change its state from ANALYZE to IDLE or PROMISE depending on its queue status.

Algorithm 1 Send function of i

Send function of initiator i for a task:
1 *state* := ANALYZE; broadcast Request$_i$; // Request
   start *timer* and wait for Δ unit of time;
2 **if** $(c \geq (k-1))$ **then**
3   for each possible team, calculate minimum cost as per [1];
        select the team that has minimum cost, say Γ;
        send Confirm$_i$ to these $(k-1)$ members of Γ;
        send Not-required$_i$ to $c - (k-1)$ non-initiators;
        ⟨*state* := BUSY; make Q_i empty⟩;
        initiate execution of τ when $(k-1)$ members of Γ arrive at l;
4 **else**
5   send Not-required$_i$ to c non-initiators;
        if $Q_i = \emptyset$ **then**
6 | *state* := IDLE;
7   **else**
8 | *state* := PROMISE;
            | send Willing$_i$ to the agent who's *id* is at front element of Q_i;
9   **end**
10 **end**
11 **if** Willing *message is received in a state* $\neq$ ANALYZE *from an agent j* **then**
12   send Not-required$_i$ to j;
13 **end**

Send function of Non-initiator j for a task:
14 **case** $((Q_j \neq \emptyset)$ *and* $(flag = true))$ **do**
15   send Willing$_j$ to the agent who's *id* is at front element of Q_j;
       $flag := false$;
16 **end**
17 **case** $((Q_j \neq \emptyset)$ *and* $(flag' = true))$ **do**
18   send Engaged$_j$ to the rear element of Q_j;
       $flag' := false$;
19 **end**

Fig. 1. Send function of i

The *receive* function of an agent is given in Algorithm 2. The computations are done based on the current state that may be IDLE (line no. 17–24), PROMISE (line no. 25–42), BUSY (line no. 12–16 and 43–47), and ANALYZE (line no. 2–10). Within a state, the type of message is checked and appropriate actions are taken. For example, when state is IDLE, if a Request message is received, it becomes PROMISE, the identifier of the sender is enqueued, and $flag$ is set

Fig. 2. Receive function of j

to true; all these actions are done atomically (denoted by $\langle\ldots\rangle$). Now the agent sends a Willing message to the sender (initiator) and $flag$ is set to false (line 17–19 Algorithm 1).

3   Formal Verification of the Distributed Algorithm

In this section, we present a formal verification process of the distributed algorithm given in Sect. 2. The key concern of the algorithm for task execution in the dynamic environment is the communication. Inter-agent communication is required to ensure the proper coordination between agents/robots for efficient mission execution in a dynamic environment. Linear temporal logic is used for modeling and verifying the validity of the algorithm.

3.1   Formal Modeling and Verification of the Algorithm Using SPIN Model Checker

We need to ascertain whether the distributed algorithm satisfies some properties; for example, the initiator may send *Confirm* messages only when it has received a sufficient number of *Willing* messages; if an agent gets a *Confirm* message in the *Promise* state, then it goes to the *Busy* state; if an agent broadcasts request message then, it must eventually receive either *Willing* or *Engaged* messages; if an agent has sent *Willing* message, then it must eventually receive either *Confirm* or *Not-Required* message. We have used the SPIN model checker for the verification of the algorithm.

3.1.1   LTL Model Checking

Model-checking is a technique for verifying that finite-state systems satisfy temporal logic specifications [11, 12]. The process of algorithm verification via model checking has three steps: (i) Modeling of the system under consideration. (ii) Specification of desirable properties of the system. (iii) Verifying whether the system satisfies the specifications.

First, the system's design is encoded in some formal modeling language (e.g., Promela) to obtain a model of the system. Second, the properties that the model should satisfy must also be stated in an unambiguous, formal notation (e.g., LTL). Third, to verify the system, a tool (e.g., SPIN model checker) will take as input the model and the specification (e.g., properties written in LTL) and determines whether or not the system meets the specification.

Linear temporal logic (LTL)

For a set of atomic propositions (AP), an LTL formula consists of Boolean operators $\wedge$ (and), $\neg$ (negation), and temporal operators, $\bigcirc$ (next), $\bigcup$ (until). Other temporal operators include $\square$ (always) - a property is satisfied now and forever into the future and $\Diamond$ (eventually) - a property is satisfied at some time in the future.

Syntax: LTL formulas over the set AP are formed according to the following grammar: $\varphi ::= true \mid p \mid \varphi_1 \wedge \varphi_2 \mid \neg\varphi \mid \bigcirc\varphi \mid \varphi_1 \bigcup \varphi_2$ where:

- p is an atomic proposition ($p \in$ AP).
- $\bigcirc\varphi$ means φ is true in the next moment in time.
- $\Diamond\varphi$ means φ is true eventually in future moments, $\Diamond\varphi = true \bigcup \varphi$
- $\square\varphi$ means φ is true in all future moments, $\square\varphi = \neg\Diamond\neg\varphi$
- $\varphi_1 \bigcup \varphi_2$ means that φ_1 is true until φ_2 is true.

By combining the temporal modalities $\Diamond$ and $\square$, new temporal modalities are obtained as follows: $\square\Diamond\varphi$ means "*infinitely often φ*" and $\Diamond\square\varphi$ means "*eventually forever φ*".

Semantics of LTL: The semantics of an LTL formula φ is defined by a satisfaction relation, $\models \subseteq (\mathbb{N} \to 2^p) \times \mathbb{N} \times \Phi$, where Φ is set of LTL formulas. For $\sigma : \mathbb{N} \to 2^p, j \in \mathbb{N}$ and $\varphi \in \Phi$, we write $(\sigma, j) \models \varphi$ to mean $(\sigma, j, \varphi) \in \models$. The terminology is as follows: φ holds at position j of σ; σ satisfy φ at j; and σ satisfies φ at $\sigma(j)$. For more details we refer to [12].

LTL Model Checking: In order to check if an LTL formula ψ holds in a transition system S, first a nondeterministic Büchi automata (NBA) is constructed for the negation of φ. Now construct a product transition system of S and the NBA, and then check the emptiness of the product automaton. For the details of this construction, we refer to [12].

3.1.2   Modeling of the Algorithm in Promela

The modeling of the distributed algorithm is done in Promela. Promela programs consist of processes, channels, and variables. In Promela *mtype* is used for defining symbolic names of numeric constants. As mentioned in the Algorithm 1, we have abstracted five types of messages, declared as follows:

```
mtype = {REQUEST, WILLING, ENGAGED, CONFORM, NOTREQ};
```
which is functionally equivalent to the sequence of macro definitions:
```
#define REQUEST 5; #define WILLING 4; #define ENGAGED 3;
#define CONFORM 2; #define NOTREQ 1;
```
Similarly the states of the agents are defined:
```
#define IDLE 1; #define READY 2; #define PROMISE 3; #define BUSY 4;
```

To transfer the messages among active process Promela uses the channel. The keyword *chan* is used to declare the channel. The channel declaration is of two types: local and global. In the channel declaration, its capacity can be initialized either by zero or some constant value. The messages are stored in first-in-first-out (FIFO) manner. In this paper, agents (processes) communicate in an asynchronous manner, and the channel capacity is defined by MESSAGE_BUFFER. This means that a channel, *chan*, can store at most MESSAGE_BUFFER messages. Each message is defined to have two fields of type *mtype* and *int*.

```
chan to_agent [N+1] = [MESSAGE_BUFFER] of {mtype, int};
```
where, #define MESSAGE_BUFFER 4

The total number of agents in the environment is N and the number of other agents k required to execute a joint action are defined as:
```
#define N 6;  #define k 4
```

The processes in Promela are created by using keyword **proctype** and then **run** statement is used to initialize the processes. It is, therefore, possible to create multiple processes simultaneously. All processes run asynchronously, and each process may take the finite number of parameters in SPIN. In this paper, the to_agent channel provides the communication means between different agents. Each agent (process) is modeled in Promela as given below. A portion of the Promela model (code) for the algorithms discussed in Sect. 2, is given below.

```
1    mtype = {REQUEST, WILLING, ENGAGED, CONFORM, NOTREQ, SELF1, SELF2};
2    /* TODO: Define states here */
3    #define IDLE  1;  #define READY 2
4    #define PROMISE 3; #define BUSY 4
5    #define MESSAGE_BUFFER 4; #define N 6
6    #define K 4
7     chan to_agent [N+1] = [MESSAGE_BUFFER] of
8    {mtype, int};
9    hidden int idx; /* used in for loops */
10sent_will, sent_eng, sent_conf, sent_nreq;
11   bool received_req, received_will, received_eng, received_conf,
     received_nreq;
12te_ready, state_promise, state_busy; /
13   bool at_Loc_Init;
14   #define sufficient_willing (c>=K)
15   int c;
16   ltl p0{[] (sent_conf -> sufficient_willing)}
17   ltl p4{[] (received_conf -> <> at_Loc_Init)}            /*L3*/
18      proctype agent(int i) {          /* to define the agent */
19      chan to_me = to_agent[i];
20      xr to_me;
21      byte state = IDLE;
22      int from, to;
23      bool is_willing[N + 1];
24      chan Q = [MESSAGE_BUFFER] of {int};
        do
25   ::to_me?REQUEST(from)->received_req=true;
26      if
27         :: state == IDLE ->state = PROMISE; state_promise = true;
```

The do loop used here is an infinite loop in which a nondeterministic choice is made between all instruction sequences starting with ::. Promela's do ... od provides a form of guarded commands in a CSP-like manner [13], where guards restrict the set of sequences that can be chosen.

In the above code, there are 5 outer choices, since there are 5 different messages. We now explain the working of the first outer choice. When an (non-initiator) agent i receives a *Request* message from an initiator agent (whose identifier is *from*) via channel *to_me* (the guard to_me[i]?REQUEST(from)), if the state is *Idle*, it changes the state to *Promise* and sends a *Willing* message via channel *to_me* to the initiator; if the state is not *Idle* it sends an *Engaged* message via channel *to_me* to the initiator.

4   Experimental Results and Analysis

The verification results with identified safety and liveness properties are discussed here. Safety and liveness are two important kinds of properties provided by all distributed systems. Safety property asserts that *"nothing bad happens"* in the system. We identified some safety properties of the algorithm given in Algorithm 1 and 2. The default safety property check of SPIN ensures the absence of deadlock, and unreachable code. SPIN model checker checks a safety property by trying to find a trace leading to the "bad" thing and if there is no such trace found, it means that the property is satisfied. Liveness property asserts that *"something good will eventually happen"*.

4.1   Safety Properties

We have identified the following safety properties:

Safety 1: Initiator may send *Conform* messages only when it has received sufficient number of *Willing* messages. This property is expressed in Promela (SPIN) as: [ ](sendConform ->   c$\geq$(k-1))

Safety 2: If an agent is in Promise state and it receives Conform message, then its next state will be Busy.

[ ]((promise && receiveConform) -> X busy)

We have performed several experiments with the varying number of agents N and the number of agents required k for different safety properties (Safety1, Safety2). We have performed several experiments and the results of the runs (like the one shown above) are summarized in Table 1 and Table 2 for $k = 2$ and $k = 3$ respectively with varying number of agents N, for the default property.

Table 1. Results with varying N and $k = 2$ with default safety property of the SPIN

# agents (N)	No. of states	Memory (MB)	Time (sec.)	Verified?
3	22,127	71.422	0.073	Yes
4	310,312	184.313	1.31	Yes
5	3,217,309	2011.656	24.3	Yes
6	3,443,944	2011.266	26.6	Yes
7	2,999,766	2010.875	27.9	Yes
8	2,640,990	2010.875	29.6	Yes
9	2,381,396	2010.484	30.2	Yes
10	2,073,344	2011.266	30.3	No
11	1,893,540	2010.875	29.8	Yes
...	...	...	...	...
20	986,432	2010.484	27.9	Yes

Table 2. Results with varying N and $k = 3$ with default safety property of SPIN

# agents (N)	No. of states	Memory (MB)	Time (sec.)	Verified?
4	5,048,791	2011.266	18.6	Yes
5	4,213,500	2010.047	20.3	Yes
6	3,444,635	2011.656	26.9	Yes
7	2,999,164	2010.484	28.9	Yes
8	2,640,990	2010.875	29.4	Yes
9	2,382,352	2011.266	29.8	Yes
10	2,073,760	2011.656	29.8	Yes
11	1,893,160	2010.484	29.3	Yes
...	...	...	...	...
20	98,7027	2011.656	32.5	Yes

- The time, memory and the total number of states explored are low with the small value of N and k.
- If the value of N is gradually increased and k is kept fixed ($k = 2$), the time, memory and states explored also increases with a rapid rate up to some extent, i.e. $N = 6$. After $N = 6$, we found that the value of time is still increasing. Now, after $N = 6$, the states explored has started decreasing slowly and memory used becomes almost constant.
- The value of time start decreasing after $N = 10$.

The evaluation of results indicates that the time, memory, and the number of states explored during verification depend on the total number of agents present in the environment. After some number of agents ($N = 6$) available in the environment time, memory and states explored start decreasing and become constant because of the high availability of idle agents in the environment, which makes ($c \geq (k - 1)$) easily satisfied, and no further work is required. For a fixed value of N, when we vary k, the running time and states explored almost remain the same.

The verification results corresponding to the safety properties (Safety1, Safety2), given above, for $k = 2$ are given in Tables 3 and 4 respectively. We have also performed several experiments by varying k for a fixed value of N for these properties. The pattern is similar to that obtained for the default property.

Table 3. Verification results for safety property (**Safety 1**) with $k = 2$

# agents (N)	No. of states	Memory (MB)	Time (sec.)	Verified?
3	22,127	71.179	0.031	Yes
4	310,312	182.898	0.437	Yes
5	2,073,075	1023.914	3.760	Yes
6	1,689,900	1023.914	4.070	Yes
7	1473750	1023.914	4.470	Yes
8	1306725	1023.914	4.530	Yes
9	1169175	1023.914	4.510	Yes
10	1021797	1023.914	4.370	Yes

Table 4. Verification results for safety property (**Safety 2**) $k = 2$

# agents (N)	No. of states	Memory (MB)	Time (sec.)	Verified?
3	22,127	71.179	0.031	Yes
4	310,312	182.898	0.421	Yes
5	2,073,075	1023.914	3.76	Yes
6	1,689,900	1023.914	4.41	Yes
7	1,473,750	1023.914	4.70	Yes
8	1,473,750	1023.944	4.56	Yes
9	1,169,175	1023.914	4.59	Yes
10	1,021,797	1023.914	4.38	Yes

4.2   Liveness Properties

We also have identified the following liveness properties:

Liveness 1: If an agent broadcasts *Request* message then, it must eventually receive either *Willing* or *Engaged* messages. This property is expressed in Promela (SPIN) as:

    [ ](sendRequest -> <> (receiveWilling || receiveEngaged))

Liveness 2: If an agent has sent *Willing* message then it must eventually receive either *Conform* or *Not-Required* message. This property is expressed in Promela (SPIN) as:

    [ ](sendWilling -> <> (receiveConform || receiveNotrequired))

Liveness 3: If an agent has received a *Conform* message, then it will be eventually at the location of the initiator. This property is expressed in Promela (SPIN) as:

    [ ](receiveConform -> <> (at_loc_Init))

The verification results corresponding to the liveness properties (Liveness 1, Liveness 2, Liveness 3), for $k = 2$ are given in Tables 5, 6 and 7 respectively.

Table 5. Verification results for liveness property (**Liveness 1**) with $k = 2$

# agents (N)	No. of states	Memory (MB)	Time (sec.)	Verified?
3	25,781	72.594	0.11	Yes
4	407,268	221.813	2.25	Yes
5	4,180,734	2010.875	29.60	Yes
6	3,424,694	2011.656	26.40	Yes
7	2,979,835	2010.875	28.50	Yes
8	2,641,520	2011.266	25.10	Yes
9	2,371,908	2010.875	22.00	Yes
10	2,062,962	2010.875	21.70	Yes

Table 6. Verification results for liveness property (**Liveness 2**) with $k = 2$

# agents (N)	No. of states	Memory (MB)	Time (sec.)	Verified?
3	23,361	71.813	0.078	Yes
4	352,623	200.719	1.58	Yes
5	4,180,736	2010.875	26.90	Yes
6	3,424,007	2011.266	26.00	Yes
7	2,979,834	2010.875	28.10	Yes
8	2,640,460	2010.484	29.40	Yes
9	2,371,432	2010.484	29.00	Yes
10	2,063,376	2011.266	27.90	Yes

Table 7. Verification results for liveness property (**Liveness 3**) with $k = 2$

# agents (N)	No. of states	Memory (MB)	Time (sec.)	Verified?
3	22,685	71.422	0.086	Yes
4	320,782	188.219	1.40	Yes
5	4,180,736	2010.875	25.20	Yes
6	3,424,008	2011.266	26.70	Yes
7	2,980,432	2011.266	29.60	Yes
8	2,640,990	2010.875	30.00	Yes
9	2,371,908	2010.875	29.80	Yes
10	2,062,962	2010.875	30.19	Yes

5   Conclusion

In this paper, we considered the problem of formal modeling, verification, and analysis of a distributed algorithm for task execution in a dynamic multi-robot environment. We modeled the distributed algorithm model in Promela, and subsequently, some important safety and liveness properties were identified and verified using SPIN model checker. The correctness of the algorithm is thus formally verified. Extensive experiments with a varying number of agents are performed. The results are quite encouraging, and it confirms the expected execution of the algorithm.

The evaluation of results indicates that the time, memory, and the number of states explored during verification depend on the total number of agents present in the environment. After some number of agents $(N = 6)$ available in the environment time, memory and states explored start decreasing and become constant because of the high availability of idle agents in the environment, which makes $(c \geq (k - 1))$ easily satisfied, and no further work is required.

The memory used and the number of states explored is almost constant with all values of k and for fixed $N = 12$. The time taken has increased up to $k = 4$, and after $k = 4$, it starts decreasing and becomes constant. The time, memory, and the number of states explored depend on the number of agents (N) available in the environment, not k. Hence, the number of agents present in the environment is the deciding factor for verification time, memory and states explored.

Acknowledgment. The authors thank the anonymous referees for their valuable comments that were helpful for improving the paper. The second author was in part supported by a research grant from Google.

References

1. Nath, A., Arun, A.R., Niyogi, R.: An approach for task execution in dynamic multirobot environment. In: Australasian Joint Conference on Artificial Intelligence, pp. 71–76 (2018)
2. Holzmann, G.J.: The model checker SPIN. IEEE Trans. Softw. Eng. **23**(5), 279–295 (1997)
3. Larsen, K.G., Pettersson, P., Yi, W.: UPPAAL in a nutshell. Int. J. Softw. Tools Technol Transf. **1**(1–2), 134–152 (1997)
4. Lomuscio, A., Qu, H., and Raimondi, F. MCMAS: a model checker for the verification of multi-agent systems. In: International Conference on Computer Aided Verification, 2009, pp. 682–688. Springer, Heidelberg. https://doi.org/10.1007/978-3-642-02658-4_55
5. Kwiatkowska, M., Norman, G., Parker, D.: PRISM: probabilistic symbolic model checker. In: International Conference on Modelling Techniques and Tools for Computer Performance Evaluation, 2002, pp. 200–204. Springer, Heidelberg. https://doi.org/10.1007/3-540-46029-2_13
6. Tkachuk, O., Dwyer, M.B.: Environment generation for validating event-driven software using model checking. IET Softw. **4**(3), 194–209 (2002)
7. Abdallah, S., Lesser, V.: Organization-based cooperative coalition formation. In: Proceedings IEEE/WIC/ACM International Conference on Intelligent Agent Technology, pp. 162–168 (2004)
8. Coviello, L., Franceschetti, M.: Distributed team formation in multi-agent systems: stability and approximation. In: 51st IEEE Conference on Decision and Control (CDC), pp. 2755–2760 (2012)

9. Tośić, P.T., Agha, G.A.: Maximal clique based distributed coalition formation for task allocation in large-scale multi-agent systems. In: International Workshop on Massively Multiagent Systems, pp. 104–120. Springer, Heidelberg (2004). https://doi.org/10.1007/11512073_8

10. Wooldridge, M.: An automata-theoretic approach to multiagent planning. In: Proceedings of the First European Workshop on Multiagent Systems, Oxford University, pp. 1–15 (2003)

11. Clarke Jr, E.M., Grumberg, O., Kroening, D., Peled, D., Veith, H.: Model Checking. MIT press, Cambridge (2018)

12. Baier, C., Katoen, J.P.: Principles of Model Checking. MIT press, Cambridge (2008)

13. Hoare, C.A.R.: Communicating sequential processes. The origin of concurrent programming, pp. 413–443 (1978)

Deep Learning Library Performance Analysis on Raspberry (IoT Device)

Luciana P. Oliveira[1]([✉]), José Humberto da S. S. Santos[2], Edson L. V. de Almeida[1], José R. Barbosa[1], Arley W. N. da Silva[3], Lucas P. de Azevedo[3], and Maria V. L. da Silva[3]

[1] IFPB Campus João Pessoa, Av. Primeiro de Maio, 720 - Jaguaribe, João Pessoa, PB, Brazil
luciana.oliveira@ifpb.edu.br, edson.vieira@academico.ifpb.edu.br
[2] UFCG Campus Campina Grande, R. Aprígio Veloso, 882 - Universitário, Campina Grande, PB 58428-830, Brazil
humberto.soares@ufcg.edu.br
[3] IFPB Campus Itaporanga, R. Projetada, SN - Vila Moco, Itaporanga, PB, Brazil
arley.silva@ifpb.edu.br,
{lucas.azevedo,maria.silva.7}@academico.ifpb.edu.br

Abstract. The use of machine learning techniques in Internet of Things (IoT) applications is increasing by the diversity of the IoT scenario. In addition, machine learning libraries contribute to this growth by promoting stable and easy-to-manipulate implementations, reducing the need to develop complex algorithms and bringing researchers' attention to other end-application activities. However, library diversity can make it difficult to choose from, since factors such as computational resource use are critical in an IoT scenario where such elements are limited. This work analyzes the performance, energy consumption and resource usage of two of the major machine learning libraries on the market, through standardized applications. It was found that the Pytorch 1.1.0 can be used with lower consumption.

1 Introduction

The Machine Learning (ML) concept proposes to solve complex and repeating activities through the use of mathematical models and computational resources, making it possible to develop computer programs capable of learning a certain task or predicting a certain result [1].

Artificial Neural Network (ANN) is an ML algorithm widely used to solve many problems, due to its ability to adapt in many situations and satisfactory results when compared to other algorithms [2]. This algorithm is inspired by the functioning of the human brain, using sets of neurons to formulate models capable of associating unknown inputs to previously known classes (outputs). Despite the complexities, several IoT applications reported in the literature present ANN as a viable alternative for solving various problems [3,4]. However, in applications involving IoT, factors such as manipulation of considerable data volumes, limited resource and low cost components are considered [2,5]. Such features require in-depth analysis of which modules to use in IoT and ANN applications.

Today, several machine learning libraries offer optimized and standardized implementations of the algorithms [6–9]. Avoiding the need to redo the code and allowing attention to be directed to the end applications. However, choosing a suitable library for a given situation can be challenging when considering scenarios as limited resources, as is often the case in IoT applications [14].

In this context, the main objective of this research is to compare two ML libraries in Raspberry (IoT device) considering three training scenarios (one linear regression from real estate data and two classifications with image processing: one in character recognition and one to identify model images). Therefore, this article presents the measurement results using answers to the following questions:

- Do any of the libraries have the least resource use in all scenarios?
- What is the relationship between runtime and energy for each library?

In order to answer such questions, this article presents in Sect. 2 concepts of Artificial Intelligence and IoT to offer support for the understanding of the scenarios that will be evaluated and the constraints of IoT devices. In Sect. 3, the methodology is described in term of the components and phases to collect and analyze data that will be used to compare libraries. In Sect. 4, the following items are presented: results of the processing consumption, memory, runtime, and power consumption of each library. Finally, Sect. 5 presents conclusions.

2 Background

2.1 Artificial Intelligence, Machine Learning and Neural Networks

Artificial Intelligence (AI) is a concept approached in 1956 at the Dartmouth office, considering artifacts might be able to learn any way to accomplish a specific task that would be performed by humans. Despite the seeming fiction theme initially, the concept received attention of the scientific community, and, in recent decades, with the increase of computational resources such as processing, volatile memory and storage, it has been possible to develop new approaches and applications using computers. Today, machine learning is one from several concepts in Artificial Intelligence [10].

The concept of machine learning currently encompasses a large set of algorithms and methods as illustrated in Fig. 1(a). The main techniques are based on statistical models, conditioned repetition algorithms and natural systems modeling.

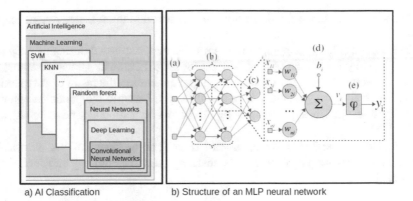

a) AI Classification   b) Structure of an MLP neural network

Fig. 1. AI.

Today the large volume of data, produced and made available by IoT scenarios, drives the development of applications involving ML and Internet of the Things. In this context, ANN algorithms have been gaining prominence due to the ability to adapt the most diverse approaches and problems explored [2].

ANN are used in the functioning of the human brain in clusters of connected neurons that form densely connected graphs, used by model generators by adjusting connections during learning. Figure 1(b) illustrates an example of ANN structure and an artificial neuron. The arrangement of layers in an ANN may vary by network type, but the most common topology is an input layer, N intermediate layers, also calls hidden layers and an output layer.

The artificial neuron (Fig. 1b) consists of a node of a graph, in which each edge that connects the vertices (nodes) has a weight that is adjusted during learning. For model adjustment, each i-neuronium sum the products between the weights (w_{ni}) and the inputs (x_{ni}) plus an adjustment value (b_i), according to Eq. 1.

$$v_j = \sum_{i=1}^{m} w_{ni}x_{ni} + b_i \qquad (1)$$

Finally, the signal is directed to a normally nonlinear function called the activation function (φ), which limits the output value of the neuron (v_i) to a specific range, resulting in a value of (y_i).

During the presentation of a training set for the multilayer perceptron (MLP), the classification signal runs through the network from the input to the output generating the result of the classification, then an error signal is generated in the reverse direction. Each weight is initialized with a random value that is adjusted during the step training. In this work, the weight adjustment is performed through the Backpropagation algorithm [12]. This algorithm performs the adjustment based on the error signals.

The Backpropagation algorithm uses the errors generated from the difference between the expected output and the output obtained by the network to optimize the MLP model [12]. For this, its applied the rule of the derivatives chain is applied to perform the computation in the reverse direction. Finally, the values of the weights are updated with the result of the derivation plus a constant denominated rate of learning (or bias).

Each weight is initialized with a random value that is adjusted during the step training. During the presentation of a training set for the MLP, the classification signal runs through the network from the input to the output generating the result of the classification. An error signal is generated from the difference between the predicted result and the obtained result, this signal is used by the Backpropagation algorithm to adjust the weights [12].

This algorithm and more complex variations such as Convolutional Neural Networks and deep learning algorithms are constantly used in cardiology and clinical analysis such as [13].

2.2   Internet of the Things - IoT

The Internet of Things (IoT) is constituted of smart objects that help automate services, improve everyday life, and aid decision making. They are characterized by the ability to collect and transmit data. Some of theses devices processing data to extract information or knowledge before to transmit data and such functionalities can be executed by artificial intelligence techniques (neural network, convulsion neural networks, decision tree and others).

In addition, they have Internet connectivity but have low computing power and some have a limited (non-rechargeable) battery life. For example, the ESP32 is a widely used IoT microcontroller with low cost, WiFi and Bluetooth as transmission capabilities, a coprocessor and advanced power modes.

On the other hand, Raspberry Pi is considered a minicomputer that has a modular structure. Its main features are portability and programmability, which provides its use in different types of applications. This IoT device can operate on Linux operating systems such as Raspibian or ubuntu. This minicomputer has a variety of interfaces allowing the use of different modules as needed and also to execute machine learning libraries to execute the processing of the data [14].

In this context, the challenge in an IoT environment is to execute artificial intelligence mechanisms, considering the resource constraints of smart objects. This is because the machine learning and other mechanisms implemented on smart objects can influence the demand for processing, and hence the power consumption that will be investigates in this study.

3   Methodology

This work is classified as its approach as a quantitative research, considering that its results and its samples can be quantified numerically [15]. Regarding its nature, this work is classified as applied research, since it aims to generate knowledge for practical application.

During the experiments in Raspberry (in Fig. 2(c)), some codes with the python language (sourceforge.net/projects/aina2020/files/deep_learning/) were run to analyze the deep learning libraries. These codes allow the automation to extract the execution time (real), the memory utilization and energy consumption. The collect procedure was executed ten times for each scenario and library combination, and then the runtime, resident memory and energy consumption were averaged.

In terms of energy consumption, a Digital Minipa ET-2231 multimeter was used. This multimeter is capable of measuring DC and AC voltage across buses and power lines. In electronic circuits, it can measure current, frequency, duty cycle, capacitance, test diodes, cable and wire continuity, and measure component resistance. A feature of this multimeter is its connection via USB enabling communication with the computer, facilitating the collection of information for the experiment. The multimeter used in the experiment can be observed in Fig. 2(a).

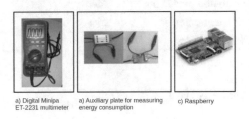

a) Digital Minipa   a) Auxiliary plate for measuring   c) Raspberry
ET-2231 multimeter   energy consumption

Fig. 2. Energy measurement

Moreover, a plate in Fig. 2(b) was designed to maintain greater accuracy in measurements. Such a plate makes it possible to interrupt the current, facilitating the measurement process. The plate was made of fiberglass, using thermal transfer engraving, through the corrosion process with iron perchloride. It consists of 4 connectors: Two-way KRE post and two alligator clips.

3.1 Deep Learning Library

Several optimized implementations of deep learning algorithms can now be found, reducing the need for development and allowing attention to be directed to the final application. Among the most used in the literature is possible to highlight Tensorflow and PyTorch, according to their official descriptions:

- **Tensorflow**: It is an open source machine learning platform with a comprehensive and flexible set of machine learning tools, libraries and resources.
- **PyTorch**: It is a library optimized for deep learning using GPUs and CPUs.

Among the main criteria for choosing these libraries are the amount of materials and quality of documentation available on the Internet, active community, stability of implementations and diversity of applications reported in the literature. Both libraries support application development involving features such as deep learning techniques, several neural networks, benchmarking metrics, and so on. But details are presented in Table 1.

Table 1. Other features: tensorflow e pytorch

Feature	Tensorflow	PyTorch
Developed by	Google	Facebook
Creation date	Feb. 2017	Oct. 2018
Open source	Yes	Yes
Main language for use	Python	Python
Multiple network topologies	Yes	Yes
Community	Bigger	Smaller
Graphic views	Static	Dynamic
Distributed processing	Yes	Yes
GPU processing	Yes	Yes
CPU processing	Yes	Yes

4   Experiment Description

Several applications are proposed in the literature involving concepts of IoT and Machine Learning [2]. To reduce the scope, this work analyzed two types of applications: (i) image classification that is usually used in industrial automation processes, allowing adjustments of mechanical arms, quality control of fruits, in addition to applications such as identification of fire spots, analysis of animal behavior and etc. [16, 17, 26]; and (ii) linear regression that is normally used when it is desired to predict a future value from past and present entries, such as the next stock exchange variation or the expectation of real estate values in some region.

These applications will be evaluated considering Pytorch 1.1.0, Pytorch 1.4.0 and TensorFlow 2.3.1. They will be presented in more detail in the following sections.

4.1   Image Classification

Usually a Convolutional type network (CNN) is used to perform the image classification. This type of network has the following information as input: a tensor formed by the number of images x image width x image height x image depth. Then, a mathematical operation called convolution is applied, which performs the multiplication of the matrix block (image) by a kernel mask, generating a map of shapes of equal or smaller size that is directed to the next layer. Similar to a neural network explained previously, each set of information tied to a pixel corresponds to an input neuron, so an image of size 100×100 pixels generates 10,000 weights between two densely connected layers. However, through convolution, this value can be reduced according to the modified kernel, so the same example can result in 400 pesos when using a 5×5 size kernel.

In this context, two scenarios were used: classification of clothing, with Fashion Mnist dataset [23], with database available at https://github.com/zalandoresearch/fashion-mnist, and classification of handwritten digits, with MNIST Dataset [24] available at http://yann.lecun.com/exdb/mnist. Each scenario has a dataset comprised of

60,000 small square 28 × 28 pixel grayscale images of items and the total number of the test dataset was 10,000 images. In first scenario, the dataset has 10 types of clothing (0: T-shirt/top, 1: Trouser, 2: Pullover, 3: Dress, 4: Coat, 5: Sandal, 6: Shirt, 7: Sneaker, 8: Bag, 9: Ankle boot). In second scenario, images of handwritten single digits between 0 and 9.

4.2   Linear Regression

The application for classification using numerical data can be performed with a conventional implementation of a neural network, using each column of the database as a network input. This type of application is one of the most common, as the data and description of them can be obtained from several sources normally present in IoT scenarios. For application, Boston Housing Dataset [25], a database available at http://lib.stat.cmu.edu/datasets/boston, was used, where it has 14 variables related to the prices of each residence and the characteristics of its location, relative to the city of Boston, Massachusetts, USA. Each sample is made up of statistical data on the number of residences within 25000ft2, industrial areas, local crime rates, among others. The implemented application adds the PRICE field, present in target, so that the algorithm can work on the prices properly.

4.3   Scenarios IoT

A Raspberry Pi 3B computer (Fig. 2(c)) was used to perform the experiment. This choice was motivated due to the amount of memory and processing present in the device, sufficient to execute a complex algorithm such as an ANN. This type of computer is constantly used for IoT project design, home automation, media centers and others [19], due to the low cost, ease of handling and increased computing resources compared to commonly used microcontrollers [22].

The version has a Quad Core 1.2 GHz Broadcom BCM2837 64bit processor, 1 GB RAM, attached wirelles chip, 100 Gbits support Ethernet connection, 4 USB 2.0 ports, HDMI connection, micro SD memory card support, display ports Toutch screen. Depending on an external power supply of 800 mA [20].

Raspbian Buster Lite 2019 operating system was used for installation, access and execution of the evaluated libraries. This system offers the main tools present in a conventional operating system, besides the possibility of installing other additional resources that may be needed [21].

5   Results

The tests were performed with three applications (two image classification and one linear regression), respectively involving image processing using two CNN network for image recognition and a MLP network for price prediction using numerical data.

It is important to consider that the tests were performed only for the network training stage, as this is the stage that usually requires the most time and computational resources.

Table 2 shows the results of the CNN network execution, an application was used using the dataset fashion, this is present in the Tensorflow and PyTorch/TorchVision documentation, the networks in both libraries were configured as similar as possible, being used 6 hidden layers of 32 neuron, when possible.

Table 2. Scenario 1: main fashion – classification with image processing and using CNN

Metric	TensorFlow 2.3.1	Pytorch 1.1.0	Pytorch 1.4.0
Runtime (h:mm:ss)	31:31.75	36:46.98	01:37:35
CPU%	71.39654071	97.81109595	99.06338704
Memory%	72.14316126	23.22992269	22.14530339
Power (W)	4380.996956	5245.388272	12858.9161

Because it involves image processing, this type of network demands a high rate of CPU consumption. It is possible to observe that both libraries had similar behavior, but the Pytorch library performed the task with shorter memory consumption.

Table 3 presents the results of other CNN network with 6 hidden layers of 16 neurons, using the dataset MNIST, composed to handwritten digits in image format.

Table 3. Scenario 2: handwritten digits - classification without image processing and using MLP

Metric	TensorFlow 2.3.1	Pytorch 1.1.0	Pytorch 1.4.0
Runtime (h:mm:ss)	36:25:84	01:15:32	01:41:29
CPU%	79.84820607	64.09488372	93.0056865
Memory%	69.16402944	31.00126246	29.29963738
Power (W)	5398.766018	9295.956035	13872.23185

Results from the third scenario in Table 4 used fewer resources than scenario 1 and scenario 2, for both libraries, because it work with numeric datas, more accessible to processing in major processors, inclusive in Raspberry PI ARM. Also, for every scenario, the tensor flow used the fewest resources.

Table 4. Scenario 3: linear regression using MLP

Metric	TensorFlow 2.3.1	Pytorch 1.1.0	Pytorch 1.4.0
Runtime (h:mm:ss)	29:02.19	4:38.21	4:40.64
CPU%	36.12668588	25.4923913	25.46129032
Memory%	31.14899135	29.77681159	30.28853047
Power (W)	3002.291787	435.768093	440.269105

6 Conclusion

The presented results of this paper aimed at investigating the characteristics of measurements in context of IoT, energy and artificial intelligence. This results has helped to understand the resource consumption of the real experiments in raspberry. After analyzing each measurement, the linear regression can be used with lower energy consumption.

The results of this paper suggest that Pytorch 1.1.0 can be used with lower consumption (memory and CPU) in all scenarios, lower energy consumption and quickly in some scenarios. This articles encourages researches to continually execute real experiment in new IoT devices, because experiments with real devices contain important information to be used by energy consumption model and others.

In this direction, the results achieved in this study can be used as a guide for researches to identify which parameters and metrics best fit for new articles can be more broadly compared to existing publications.

Acknowledgments. The authors would like to thank the Federal Institute of Paraíba (IFPB)/Campus João Pessoa for financially supporting the presentation of this research and, especially thank you, to the IFPB Interconnect Notice - No. 01/2020.

References

1. Angra, S., Ahuja, S.: Machine learning and its applications: a review. In: International Conference on Big Data Analytics and Computational Intelligence (ICBDAC), pp. 57–60, March 2017
2. Zantalis, F., Koulouras, G., Karabetsos, S., Kandris, D.: A review of machine learning and IoT in smart transportation. Future Internet **11**(4), 94 (2019)
3. Niranjan, A., Mishra, G.: Etas - efficient traffic signal assistance system using deep learning. In: ICCCNT, pp. 1–5, July 2018
4. Jan, T., Sajeev, A.S.M.: Boosted probabilistic neural network for IoT data classification. In: DASC/PiCom/DataCom/CyberSciTech, pp. 408–411, August 2018
5. Ta-Shma, P., Akbar, A., Gerson-Golan, G., Hadash, G., Carrez, F., Moessner, K.: An ingestion and analytics architecture for IoT applied to smart city use cases. IEEE Internet Things J. **5**(2), 765–774 (2018)
6. Chollet, F., et al.: Keras (2015). https://keras.io
7. Abadi, M., et al.: TensorFlow: large-scale machine learning on heterogeneous systems 2015, software available from tensorflow.org (2015). https://www.tensorflow.org/
8. Paszke, A., et al.: Automatic differentiation in PyTorch. In: NIPS Autodiff Workshop (2017)
9. Chen, T., et al.: MXNet: a flexible and efficient machine learning library for heterogeneous distributed systems. In: Neural Information Processing Systems,Workshop on Machine Learning Systems (2016). http://arxiv.org/abs/1512.01274
10. Kononenko, I.: Machine learning for medical diagnosis: history, state of the art and perspective. Artif. Intell. Med. **23**(1), 89–109 (2001)
11. McCulloch, W.S., Pitts, W.: A logical calculus of the ideas immanent in nervous activity. Bull. Math. Biophys. **5**(4), 115–133 (1943)
12. Werbos, P.: Generalization of backpropagation with application to are current gas market model. Neural Netw. **1**, 339–356 (1988)

13. Zhang, J., et al.: Automated detection and localization of myocardial infarction with staked sparse autoencoder and treebagger. IEEE Access **7**, 70 634–70 642 (2019)
14. Hentschel, K., Jacob, D., Singer, J., Chalmers, M.: Supersensors: Raspberry Pi devices for smart campus infrastructure, pp. 58–62, August 2016
15. Gerhardt, T.E., Silveira, D.T.: Métodos de pesquisa. EditoraUFRGS (2009)
16. Valletta, J.J., Torney, C., Kings, M., Thornton, A., Madden, J.: Applications of machine learning in animal behaviour studies. Anim. Behav. **124**, 203 – 220 (2017). http://www.sciencedirect.com/science/article/pii/S0003347216303360
17. Wang, G.: Machine learning for inferring animal behavior from location and movement data. Ecol. Inform. **49**, pp. 69–76 (2019). http://www.sciencedirect.com/science/article/pii/S1574954118302036
18. Netzer, Y., Wang, T., Coates, A., Bissacco, A., Wu, B., Ng, A.Y.: Reading digits in natural images with unsupervised feature learning (2011)
19. Salih, F., Omer, S.A.M.: Raspberry pi as a video server. In: International Conference on Computer, Control, Electrical, and Electronics Engineering (ICCCEEE), August 2018
20. Sandeep, V., et al.: Globally accessible machine automation using raspberry Pi based on internet of things. In: ICACCI, pp.1144–1147, August 2015
21. Fundation, R.P.: Raspberry Pi 3 model b (2016). https://www.raspberrypi.org/products/raspberry-pi-3-model-b
22. Sandeep, V., et al.: Globally accessible machine automation using raspberry Pi based on internet of things. In: ICACCI, pp. 1144–1147, August 2015
23. Xiao, H., Rasul, K., Vollgraf, R.: Fashion-MNIST: a Novel Image Dataset for Benchmarking Machine Learning Algorithms. arXiv:1708.07747, August 2017
24. LeCun, Y., Bottou, L., Bengio, Y., Haffner, P.: Gradient-based learning applied to document recognition. In: Proceedings of the IEEE, vol. 86(11), pp. 2278–2324, November 1998
25. Harrison, D., Rubinfeld, D.L.: Hedonic housing prices and the demand for clean air. J. Environ. Econ. Manage. **5**, 81–102 (1978). http://www.sciencedirect.com/science/article/pii/0095069678900062
26. Menezes, A.H., Kelvin, R.D.O., Oliveira, L.P., Oliveira, P.J.D.S.: IoT Environment to Train Service Dogs, S3C (2017). https://doi.org/10.1109/S3C.2017.8501386

Network Topology-Traceable Fault Recovery Framework with Reinforcement Learning

Tatsuji Miyamoto[1,2]([✉]), Genichi Mori[1,2], Yusuke Suzuki[1,2], and Tomohiro Otani[1,2]

[1] KDDI Corporation, Tokyo, Japan
{tt-miyamoto,ge-mouri,uu-suzuki,tm-otani}@kddi.com
[2] KDDI Research, Inc., Saitama, Japan
{tt-miyamoto,ge-mouri,uu-suzuki,otani}@kddi-research.jp

Abstract. Network Function Virtualization (NFV) is the focus of much attention especially in the field of network operation automation because of its efficient usage of resources and lower capital expenditures. However, NFV introduces additional complexity into network monitoring and management due to the network function being independent of the hardware. Moreover, maintaining the workflow still requires tremendous effort in order to sustain the automation function. In fact, a vast amount of workflows or program codes for automated deployment and the failure recovery operation have to be created and maintained per type of service and failure case. To address the above problems, previously we proposed an artificial intelligence-assisted workflow management framework. This work demonstrated that a scheme for fault-recovery operation automation. However, the conventional issue is to fix the dimensions of the input vector for machine learning (ML) algorithms to prevent the relearning repetition even if the network configuration and topology is changed. To address the above issue, this paper proposes a reinforcement learning-based fault recovery framework by applying the reinforcement learning (RL) algorithm which can adapt to changes of network topology and configuration. We demonstrate the effectiveness of the proposed framework with testbed results.

1 Introduction

Network management in telecom networks is becoming ever more complex because of 5G, and novel advances in software technology will become crucial especially in procedures for network lifecycle management. At the same time, demands for service quality such as in dealing with service outages and latency will still remain or grow even higher. Automating network management is an essential role of management, and thus ETSI ISG NFV has defined the management and orchestration (MANO) [1]. However, implementing intelligence into the orchestrator, and interworking with Operation Support Systems (OSSs) are

© The Author(s), under exclusive license to Springer Nature Switzerland AG 2021
L. Barolli et al. (Eds.): AINA 2021, LNNS 225, pp. 393–402, 2021.
https://doi.org/10.1007/978-3-030-75100-5_34

issues that still must be resolved. Linux foundation networking launched the Open Network Automation Platform (ONAP) project aimed at achieving automated management especially for NFV [2]. ONAP provides data analytics for troubleshooting and closed-loop functioning for interworking with design tools in addition to providing the orchestration function. These activities provide operators with standard APIs that simplify the operational procedures and accelerate the achievement of automation. However, from a maintenance perspective, problems still remain. The types of software errors will likely increase in Virtual Network Function (VNF) applications. To maintain automation, operators need to update the automation mechanisms per type of error. In addition, the implementation of automation programming (e.g. shell script) strongly depends on individual specialist knowledge and skills. The maintainability of the scripts consequently becomes complicated which leads to an increase in maintenance costs despite the se of virtualization technology for OPEX saving. To address this problem, a policy-based approach has been proposed that can simplify the complex operational process for implementing the automation mechanism. The policy-based approach presented in [3,4] essentially defines the design of actions and operations to control resources based on monitored information through sophisticated scripting (e.g., OpenStack Congress [5]). In [6], S. Calo et al. proposed the Watson Policy Management Library (WPML) [7] as a generic policy management library and authoring tool that can be used to define easy- to-use policy definitions and templates for creating services and service assurance activities. Although the policy-based approach can be used to provide automated responses to defined actions and rules, it is difficult to apply when operators are exposed to unexpected behavior.

Taking the above background into account, we have conducted two different works. The first is designing a sustainable business process. By referring to the specification for ETSI MANO defined by the Interfaces and Architecture (IFA) working group in ETSI NFV as an enhanced Telecom Operation Map (eTOM) process for fault recovery operation, we designed a standard for the auto-healing process in TMForum [8]. However, maintaining the deepest level of business processing is an issue that machine learning (ML) should address due to vendor-specific implementation. We therefore proposed an Artificial Intelligence (AI)-assisted closed-loop operation framework [9]. This work demonstrated a scheme for fault-recovery operation cases. However, the conventional issue is to fix the dimensions of the input vector for ML algorithms in order to prevent the relearning repetition even if the network configuration or topology is changed. In order to use ML techniques, network data must first be transformed so that it can be utilized by ML techniques.

Various data structures exist to represent attributes of graphs such as the matrix [10]. Graph classification is an important data mining task, and various graph kernel methods have been proposed recently for this task. Once the kernel matrix has been constructed, it is possible to classify the graphs with a Support Vector Machine (SVM) [11,12], using the supplied kernel matrix. These methods have proven to be effective, but they tend to have high computational overheads especially for large graphs [13]. To prevent the computation complexity, an alternative approach to constructing a feature-vector for graph classification is proposed [14]. However, the representation has shortcomings such as whether the network may change over time in the case of failures when adding or removing the nodes in an overlay network. In such cases, an adjacency matrix would have to change its dimensions, namely, the input of the ML algorithm would no longer be scale-free. To tackle these issues, we propose a network topology-traceable fault recovery framework by applying RL. In the framework, we designed a data representation procedure for preparing a set of input data for RL.

RL generally employs a simulator for training itself due to training which requires huge number of trials. On the other hand, we should assume that a 5G network on NFV is a complex structure which makes it difficult to simulate for all patterns of network infrastructure behavior. Thus, we validate the effectiveness of the proposed framework on a testbed by considering the following perspectives.

1. How many trials are required for convergence of learning?
2. How accurate is the network simulator required to be?

This paper addresses the above challenges and proposes a fault recovery framework by applying RL. Section 2 describes the proposed workflow generation framework. Section 3 describes the results of the experiment to validate the effectiveness of the proposed method. Finally, Sect. 4 provides the conclusions.

2   Proposal of Fault Recovery Framework

To deal with the various patterns of network faults, we implemented a systems of fault recovery automation as shown in Fig. 1. First, the NW Data Analyzer collects via the monitoring protocol (e.g., SNMP) and information such as performance monitoring (PM) data is collected from network infrastructure. Second, the NW Data Analyzer estimates the root cause of the node and labels its performance indexes. These data and vector values of the network topology graph are transformed by a NW State Converter into understandable information for machine learning. Finally, operation API of the workflow engine that suits the network conditions is triggered by a reinforcement learning engine. Through repetition of the above steps, the generation accuracy of the optimum workflow can be improved dynamically.

The scope of the proposal has two parts: the topology representation method of the NW State Convertor and Reinforcement Learning Engine which are depicted as the red boxes in Fig. 1.

2.1   Data Representation

The operator makes a decision to select the fault recovery workflow based on the status of the network such as topology, failure location and root cause. In order to use this information as ML input, we represent a topology graph and performance metric of the failure node as a multiple-layer matrix. Figure 2 shows

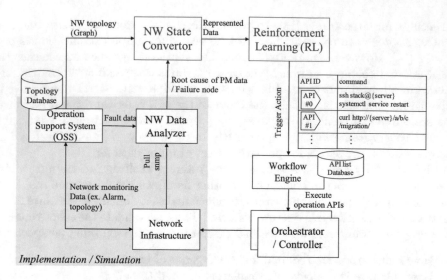

Fig. 1. Proposed fault recovery framework

the process of data representation. Regarding to network topology, it can be represented as a graph structure. On the other hand, when dealing with failures, it is required to identify the location of each interface separately. Therefore in Step 1, we add interface connectivity to the graph structure and represent the graph of the network and interface topology as a matrix. In this case, the topology can be represented by using the adjacency matrix. In Step 2, an additional matrix which indicates the failure location and root cause is lined up as the failure matrix. When failure occurs, the element of the faulre matrix which corresponds to the failure location and properties of the root cause are flagged as '1'. The other elements are '0' which means the normal state. Through the process of representation, the NW state convertor outputs the multiple-layer matrix and inputs to the RL engine.

2.2   Reinforcement Learning

Figure 3 shows the overview of the learning process for fault recovery operation. Since we consider RL algorithms in this paper, determination of environment, policy function and action value become crucial in this context. We consider tasks in which the agent interacts with an environment through a sequence of observations, actions and rewards. The goal of the agent is to select actions in a way that maximizes cumulative future rewards. Thus, we use a deep convolutional neural network to approximate the optimal action value function [15].

$$Q^*(s,a) = \max_{\pi} E\left[r_t + \gamma r_{t+1} + \gamma^2 r_{t+2} + ... | s_t = s, a_t = a, \pi\right] \quad (1)$$

which is the maximum sum of rewards r_t representing the change in network infrastructure discounted by γ at each time step t. We define the optimal action-

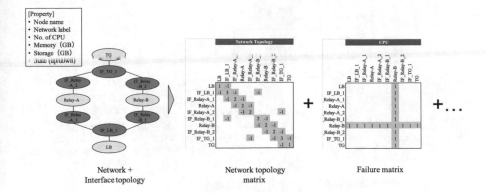

Fig. 2. Process of network topology representation

value function $Q^*(s, a)$ as the maximum expected return achievable by following any policy, after seeing some sequence s and then taking some action a, $\pi = P(a|s)$ is behavior policy mapping sequences to actions (or distributions over actions). The action corresponds to executing recovery operation, and the environment corresponds to a network state including network topology and failure matrix, respectively. The reward is calculated based on whether the action affects the state positively or not. At each time-step the agent selects an action a_t from the set of APIs of the workflow engine. The action is passed to the network infrastructure and modifies its internal state and the failure matrix.

By using the proposed framework, it is assumed that we can obtain the effect of autonomously improving the selection accuracy of tasks. In the next chapter, we will validate this assumption and describe the results that verify the effectiveness of the proposed method.

3 Experimental Results

We validate the effectiveness of the proposed framework on our testbed with the following three verification cases.

- CASE I: CPU usage in a particular single process is 100%. Recover by restarting the process.
- CASE II: Memory usage in two or more processes is 100%. Recover by restarting the VM.
- CASE III: Similar to case II, CPU usage in two or more processes is 100%, but it is not recovered by restarting the VM. Recover by rebuilding the VM.

In order to reproduce the trial of the above cases, we have developed a network simulator using OpenAI gym [16]. In this simulator, the faulty node can be fixed with Relay-B, and the CASE I and CASE II situations can be randomly generated. By preventing the complexity of the network simulator perspective, we define the scope of observation of the agent in RL as corresponding to 1 hop

398   T. Miyamoto et al.

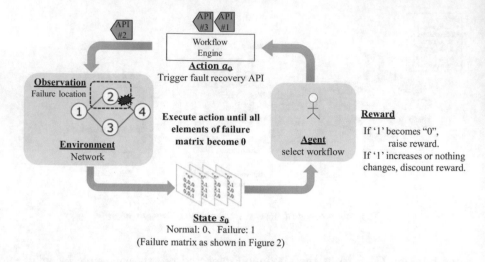

Fig. 3. Reinforcement learning process for fault recovery operation

from the failure node. Regarding the action in RL, we implement 20 APIs of fault recovery action as shown in Table 1. The reward is calculated as shown in Listing 1. We consider that the attribute "time to recovery" and "high or low risk" affect the operator's decision making. Therefore, we implement the following coefficients as a reward design.

1. Steps to recovery
2. Time to recovery
3. Risk level

Since 1) has been incorporated into the code shown in Listing 1, we implement the coefficients 2) and 3) for each API as shown in Table 2.

Listing 1. Reward Design

```
1 Def step():
2     # define initial reward from −0.1
3     reward = −0.1
4
5     action = random or DNN selection
6
7     if NW state dont change:
8         reward += −1.0
9
10    if NW is recovered:
11        reward += +1.0
12
13    return reward
```

In the following subsection, we validate the effectiveness of the proposed framework on a testbed from the same perspectives mentioned in Sect. 1.

Table 1. List of fault recovery actions

API#	Node	Action	API#	Node	Action
0	LB	Restart failed process	10	Relay-B	Restart failed process
1	LB	Restart failed server	11	Relay-B	Restart failed server
2	LB	Isolate failed server	12	Relay-B	Isolate failed server
3	LB	Instantiate standby server	13	Relay-B	Instantiate standby server
4	LB	Provisioning configuration	14	Relay-B	Provisioning configuration
5	Relay-A	Restart failed process	15	TG	Restart failed process
6	Relay-A	Restart failed server	16	TG	Restart failed server
7	Relay-A	Isolate failed server	17	TG	Isolate failed server
8	Relay-A	Instantiate standby server	18	TG	Instantiate standby server
9	Relay-A	Provisioning configuration	19	TG	Provisioning configuration

Table 2. Validated reward design of recovery

	API#0, 5,10,15	API#1, 6,11,16	API#2, 7,12,17	API#3, 8,13,18	API#4, 9,15,19
Time to recovery	−0.1	−0.3	−0.2	−0.2	0
Risk level	0	−0.3	−0.5	0	0

3.1 How Many Trials Are Required for Convergence of Learning?

Figure 4 shows the F1-score of RL in the proposed fault recovery framework when changing the learning rate. The result indicates the proposed framework can retain its accuracy of selecting fault recovery action when the parameter of the learning rate of RL is set lower than 0.01. Figure 5 shows the learning results of RL by using a network simulator in CASE I & III, CASE II & III, respectively. The average-loss value calculation result indicates that the learning process can be converged. On the other hand, the average-q calculation satisfies the requirement of higher than 0.8 in about 4,000 steps (=total number of actions) for each scenario. From the result the proposed framework requires a tremendous amount of time for trials to prepare enough training data of RL. Thus, we should consider that the proposed framework needs to prepare a network simulator from a time efficiency perspective for application to a real network operation.

3.2 How Accurate Is the Network Simulator Required to Be?

In order to validate the requirement of the network simulator, we evaluate the tolerance of the proposed framework against unexpected behavior of the network when fault recovery action is executed. In this experiment, we randomly injected a '1' value into the element of the failure matrix which is generated after the execution of fault recovery action. Figure 6 shows the F1-score when changing a noise ratio of the failure matrix in the proposed framework. The result indicates that the proposed framework can achieve higher performance until 13% of the

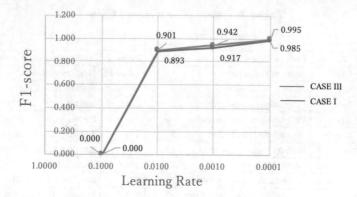

Fig. 4. F1-score of RL in proposed fault recovery framework when changing the learning rate

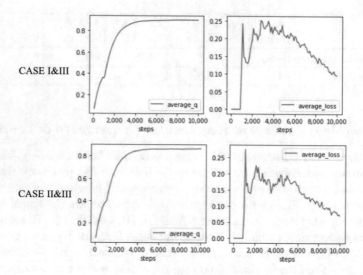

Fig. 5. Learning result of RL of proposed fault recovery framework

noise ratio. It shows that the behavior between the network simulator and real network against the execution of fault recovery action should be coincident 87% for applying the proposed framework. Actually, it depends on the acceptability criterion of the assurance functionalities.

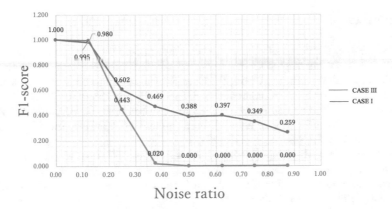

Fig. 6. F1-score when changing the noise ratio of the failure matrix in the proposed framework

4   Conclusions

This paper proposes a reinforcement learning-based fault recovery framework by applying an RL algorithm which can adapt to changes of network topology and configuration. In the framework, we designed a data representation procedure for preparing a set of data for RL which is formed as a matrix of the network-topology and failure state. We validated the effectiveness of the proposed framework by implementing and simulating network infrastructure. The result indicated that the learning process can be converged. On the other hand, the proposed framework requires a tremendous amount of failure injection/recovery operation trials for preparing enough training data of RL. Thus, we should consider that the proposed framework needs to prepare a network simulator from the time efficiency perspective for application to a real network operation. In order to validate the network simulator requirement, we developed a simulator by using OpenAI Gym and evaluated the tolerance of the proposed framework against unexpected behavior of the network when fault recovery action is executed. The result showed that the behavior between the network simulator and implemented network infrastructure should be 87% coincident for application to the proposed framework. The future work is how to shorten the time for trials in the training process or the development of a network simulator by satisfying the acceptability criterion of assurance functionalities.

Acknowledgements. This work was conducted as part of the project entitled "Research and development for innovative AI network integrated infrastructure technologies" (JPMI00316) supported by the Ministry of Internal Affairs and Communications, Japan.

References

1. ETSI TR,GS NFV-MAN001,v1.1.1, December 2014
2. https://www.onap.org/ ONAP
3. Makaya, C., Freimuth, D., Wood, D., Calo, S.: Policy-based NFV management and orchestration. In: IEEE Conference on NFV-SDN, pp. 128–134 (2015)
4. Romeikat, R., Bauer, B., Bandh, T., Carle, G., Schmelz, L., Sanneck, H.: Policy-driven workflows for mobile network management automation. In: Proceedings of the IWCMC 2010, pp. 1111–1115 (2010)
5. OpenStack Congress. https://wiki.openstack.org/wiki/Congress
6. Calo, S., Wood, D., Zerfos, P.: Technologies for federation and interoperation of coalition networks. In: 12th International Conference on Information Fusion (Fusion), pp. 1385–1392, July 2009
7. https://www.ibm.com/developerworks/community/groups/service/html/
8. https://inform.tmforum.org/features-and-analysis/2015/06/catalyst-recover-first-resolve-next-closed-loop-control-for-managing-hybrid-networks/
9. Miyamoto, T., Kuroki, K., Miyazawa, M., Hayashi, M.: AI-assisted workflow management framework for automated closed-loop operation. In: Network Operations and Management Symposium (NOMS), 2018 IEEE/IFIP, pp. 1–6, May 2018
10. Ramon, J., Gartner, T.: Expressivity versus efficiency of graphs kernels. In: 1st International Workshop on Mining Graphs, Trees and Sequences (2003)
11. Scholkopf, B., Smola, A.J.: Learning with Kernels: Support Vector Machines, Regularization, Optimization, and Beyond. MIT Press, Cambridge, MA, USA (2001)
12. Vapnik, V.N.: The Nature of Statistical Learning Theory. Springer-Verlag, New York (1995)
13. Gärtner, T., Flach, P., Wrobel, S.: On graph kernels: hardness results and efficient alternatives. In: Schölkopf, B., Warmuth, M.K. (eds.) Learning Theory and Kernel Machines, pp. 129–143. Springer, Berlin, Germany (2003)
14. Li, G., Semerci, M., Yener, B., Zaki, M.J.: Graph classification via topological and label attributes. In: International Workshop on Mining and Learning with Graphs (MLG), vol. 2, pp. 1–9 (2011)
15. Mnih, V., Kavukcuoglu, K., Silver, D., Rusu, A., Veness, J., Bellemare, M., Graves, A., Riedmiller, M., Fidjeland, A., Ostrovski, G., Petersen, S., Beattie, C., Sadik, A., Antonoglou, I., King, H., Kumaran, D., Wierstra, D., Legg, S., Hassabis, D.: Human-level control through deep reinforcement learning. Nature **518**(7540), 529–533 (2015)
16. https://gym.openai.com/

Mobile Sensing Data Analysis for Road State Decision in Snow Country

Yoshitaka Shibata[1]([✉]), Akira Sakuraba[1], Yoshikazu Arai[2], Yoshiya Saito[2],
and Jun Hakura[2]

[1] Regional Coorporate Research Center, Iwate Prefectural University, Sugo, Takizawa 152-89,
Iwate, Japan
{shibata,a_saku}@iwate-pu.ac.jp
[2] Faculty of Software and Information, Iwate Prefectural University, Sugo, Takizawa 152-52,
Iwate, Japan
{arai,y-saito,hakura}@iwate-pu.ac.jp

Abstract. In order to keep safety driving on the road in snow country, it is important to know the road state and predict the dangerous locations in advance. In this paper, the data from various in-vehicle sensor state of a snow country using various in-vehicle sensors are analyzed to identify the road state using AI technology. The various sensor data and the decided road state are organized as road state GIS platform system to widely show the road states on map. The accuracy of decision of road state is evaluated based on the prototype of the proposed road state information platform.

1 Introduction

Mobility and transportation is the most important means to maintain safe and reliable daily life and economic activity in the world. Autonomous driving is limitedly starting in some developed countries on highway and exclusive roads where the driving lanes are clear, the driving direction is the same and opposite driving car lanes are completely separated [1]. On the other hand, autonomous driving does not become popular in snow countries because of bad road surface conditions in winter. Even people are getting old and driving ability are getting worse, they have to drive their cars to continue their daily lives. In fact, more than 90% of car accidents in snow area in winter in Japan is due to slipping car on snowy or iced road [2]. More advanced road state sensing and decision system to identify the dangerous locations are indispensable for driving in winter.

In order to resolve those problems, we introduce a new generation wide area road surface state information platform based on crowd sensing. In road sensing, the sensor data from various environmental sensors including 9 axis dynamic sensor, far infrared road surface temperature sensor, air humid/temperature sensor, laser sensor, and GPS attached on vehicle are integrated to precisely determine the various road states and identify the dangerous locations. This road information is transmitted to the neighbor vehicles and road side servers using V2X communication network. The received road information is displayed on the monitor while indicating the dangerous location and

L. Barolli et al. (Eds.): AINA 2021, LNNS 225, pp. 403–413, 2021.
https://doi.org/10.1007/978-3-030-75100-5_35

alerting by voice annotation. This road information is also collected in wide area as big-data to the cloud computing server through the roadside servers and analyzed to predict the future road states and unobserved local roads by combining with open weather data and 3D terrain data along roads. Thus, wide area road state GIS information platform can be attained and safer and more reliable driving even in winter season in snow countries and can extend to autonomous driving.

In the following, the related works with road state information by sensor and V2X technologies are explained in section two. A general system and architecture of the proposed road surface state information platform are explained in section three. Next, crowed sensing system to collect various sensor data and to identify road states is introduced in section four. Preliminary experiment by prototyped platform to evaluate our proposed system is precisely explained in section five. In final, conclusion and future works are summarized in section seven.

2   Related Works

There are several related works with the road state sensing method using environmental sensors. Particularly road surface temperature is essentially important to know snowy or icy road state in winter season by correctly observing whether the road surface temperature is under minus 4 °C or over.

In the paper [3], the road surface temperature model by taking account of the effects of surrounding road environment to facilitate proper snow and ice control operations is introduced. In this research, the fixed sensor system along road is used to observe the precise temperature using the monitoring system with long-wave radiation. They build the road surface temperature model based on heat balance method.

In the paper [4–6], cost effective and simple road surface temperature prediction method in wide area while maintaining the prediction accuracy is developed. Using the relation between the air temperature and the meshed road surface temperature, statistical thermal map data are calculated to improve the accuracy of the road surface temperature model. Although the predicted accuracy is high, the difference between the ice and snow states was not clearly resolved.

In the paper [7], a road state data collection system of roughness of urban area roads is introduced. In this system, mobile profilometer using the conventional accelerometers to measure realtime roughness and road state GIS is introduced. This system provides general and wide area road state monitoring facility in urban area, but snow and icy states are note considered.

In the paper [8], a measuring method of road surface longitudinal profile using build-in accelerometer and GPS of smartphone is introduced to easily calculate road flatness and International Road Index (IRI) in offline mode. Although this method provides easy installation and quantitative calculation results of road flatness for dry or wet states, it does not consider the snow or icy road states.

In the paper [9], a statistical model for estimating road surface state based on the values related to the slide friction coefficient is introduced. Based on the estimated the slide friction coefficient calculated from vehicle motion data and meteorological data is predicted for several hours in advance. However, this system does not consider the other factors such as road surface temperature and humidity.

In the paper [10], road surface temperature forecasting model based on heat balance, so called SAFF model is introduced to forecast the surface temperature distribution on dry road. Using the SAFF model, the calculation time is very short and its accuracy is higher than the conventional forecasting method. However, the cases of snow and icy road in winter are not considered.

In the paper [11], blizzard state in winter road is detected using on-board camera and AI technology. The consecutive ten images captured by commercial based camera with 1280×720 are averaged by pre-filtering function and then decided whether the averaged images are blizzard or not by Convolutional Neural Network. The accuracy of precision and F-score are high because the video images of objective road are captured only in the daytime and the contrast of those images are almost stable. It is required to test this method in all the time.

In the paper [12], road surface state analysis method is introduced for automobile tire sensing by using quasi-electric field technology. Using this quasi-electric field sensor, the changes of road state are precisely observed as the change of the electrostatic voltage between the tire and earth. In this experiment, dry and wet states can be identified.

In the paper [13], a road surface state decision system is introduced based on near infrared (NIR) sensor. In this system, three different wavelengths of NIR laser sensors is used to determine the qualitative paved road states such as dry, wet, icy, and snowy states as well as the quantitative friction coefficient. Although this system can provides realtime decision capability among those road states, decision between wet and icy states sometimes makes mistakes due to only use of NIR laser wavelength.

With all of the systems mentioned above, since only single sensor is used, the number of the road states are limited and cannot be shared with other vehicles in realtime. For those reasons, construction of communication infrastructure is essential to work out at challenged network environment in at inter-mountain areas. In the followings, a new road state information platform is proposed to overcome those problems.

3 Road Surface State Information Platform

In order to resolve those problems in previous session, we introduce a new generation wide area road surface state information platform based on crowd sensing and V2X technologies as shown in Fig. 1. The wide area road surface state information platform is organized mainly by mobile wireless nodes, so called Smart Mobile Box (SMB) and roadside wireless nodes, so called Smart Relay Shelters (SRS). Each SMB is furthermore organized by a sensor server unit and wireless communication unit, and installed in a vehicle. The sensor server unit includes various sensor devices such as semi-electrostatic field sensor, an acceleration sensor, gyro sensor, temperature sensor, humidity sensor, infrared sensor and sensor server. Using those sensor devices, various road surface states such as dry, rough, wet, snowy and icy roads can be qualitatively and quantitatively decided.

On the other hand, the wireless communication unit in SMB and SRS includes multiple wireless network devices with different N-wavelength (different frequency bands) wireless networks such as IEEE802.11b/g/n (2.4 GHz), 11ac (5.6 GHz), 11ad (28 GHz), ah (920 MHz) and organizes a cognitive wireless node. The network node can selects the best link among the cognitive wireless network depending on the observed network quality, such as RSSI, bitrate, error rate by Software Defined Network (SDN). If none of link connection is possible, those sensor data are locally and temporally stored in the database unit as internal storage until the vehicle approaches to the region where a link connection to another mobile node or roadside node is possible. When the vehicle enters the region, it starts to connect a link to other vehicle or roadside node and transmit sensor data by DTN Protocol. Thus, data communication can be attained even though the network infrastructure is not existed in challenged network environment such as mountain areas or just after large scale disaster areas.

Thus, in our system, SRS and SMB organize a large scale information infrastructure without conventional wired network such as Internet or wireless network such as cellular network. The collected sensor data and road state information on a SMB are directly exchanged to near SMB by V2V communication. Those received data from near SMB are displayed on the digital map of the viewer system while alerting the dangerous locations such as icy road location. Thus, the vehicle can know and pay attention to the ahead road state before passing through.

On the other hand, when the SMB on own vehicle approaches to the SRS, the SMB make a connection to the SRS, delivers the sensor data and the decided road state information in the other SMBs through the N-wavelength wireless network. At the same time, those collected sensor data are sent to the cloud computing in data center through the cloudlet in SRS. By combining those sensor data form SRSs on various locations with public mesh weather data such as temperature, pressure, amount of rain and snow as open data, a wide area road states can be geographically and temporally predicted and provided as road state Geographic Information System (GIS) information service through public cellular network not only for current running drivers but for even ordinal users.

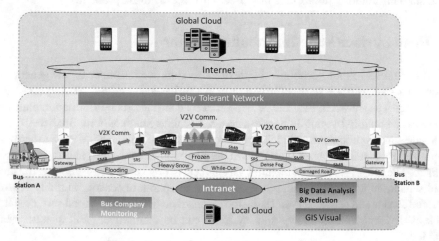

Fig. 1. Road surface state information platform

Thus, the proposed system can provide safer and more reliable driving environment even though winter season in snow countries. This network not only performs various road sensor data collection and transmission functions, but also provide Internet access network functions to transmit the various data, such as sightseeing information, disaster prevention information and shopping and so on as ordinal public wide area network for residents. Therefore, many applications and services can be realized.

4   Road State Sensing System

In order to detect the precise road surface conditions, such as dry, wet, dump, showy, frozen roads, various sensing devices including 9 axis dynamic sensor, far infrared road surface temperature sensor, air humid/temperature sensor, laser sensor, and GPS are integrated to precisely and quantitatively detect the various road surface states and determine the dangerous locations on GIS in sensor server as shown in Fig. 2.

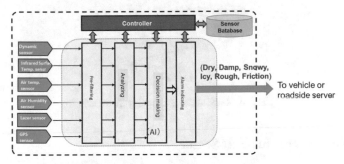

Fig. 2. Road state sensing system

4.1   9-Axis Dynamic Sensor

9-axis dynamic sensor including 3-axis accelerator, 3-axis gyro sensor and 3-axis electromagnetic sensor is used to measure vertical amplitude of roughness and horizontal along the road as well as horizontal drift due to snow ruts. The signal from 9-axis dynamic sensor is sub-sampled and converted by A/D convertor with several mseconds, and then transformed from time series to frequency spectra density. By reducing high frequency noise by low pass filter, the residual essential signal is input to the decision maker to qualitatively determine the road surface state. The quantitative values of the 9-axis dynamic sensor can be also used to control the velocity and steering angle of autonomous driving car (Fig. 3).

4.2   Far-Infrared Temperature Sensor

The far-infrared temperature sensor observes the road surface temperature to determine whether the road surface is snowy or icy without touching the road surface as shown in Fig. 4. The dynamic range of the observed temperature is very wide from −40–500 °C. Since its response time and sampling rate are 0.1 and 10 Hz, realtime decision of road surface state can be realized. However, since the emissivity of infrared amount depends on the objective material of road, calibration due to the difference of materials of road must be made to maintain the measurement accuracy.

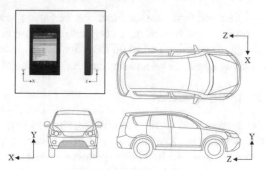

Fig. 3. Axis dynamic sensor

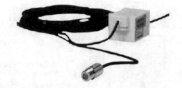

Fig. 4. Far-infrared temperature sensor

4.3 Air Temperature/Humidity Sensor

The air temperature sensor is used to measure the air temperature outside of vehicle and know the relation to the air and road surface temperatures. If the relation between both temperatures can be clearly understood, the road surface temperature can estimate by this relation. Then, since the road surface temperature is related to the road state, the future road state can be predicted using the future road surface temperature if other whether condition is the same. Since the element to detect air temperature part is platinum resistance thermometer sensor, the measurement accuracy is high within plus minus 0.3° although the response time of air temperature is long around 1–2 min.

On the other hand, humidity sensor is used to know the air is dry or wet and snowy or icy. If air is wet, the probability of snowy road state is high, whereas, if air is dry, the probability of icy road state is high in winter. The element of humidity sensor part is made by polymer capacity typed sensor, the measurement accuracy is low with plus minus 3%.

Fig. 5. Air temperature/humidity sensor

Thus, more detailed road states in both winter and summer seasons can be determined by combination of those sensors among possible road states (Fig. 5).

4.4 Near Infrared Laser Sensor

The far-infrared laser sensor precisely measures the friction rate of snowy and icy road states. Using three laser beams with different wave lengths ($\lambda 1$, $\lambda 2$, $\lambda 3$) and considering their reflection and absorption rates on the road surface, the road states such as dry, wet, snowy and icy states in addition to the friction rate, can be determined (Fig. 6).

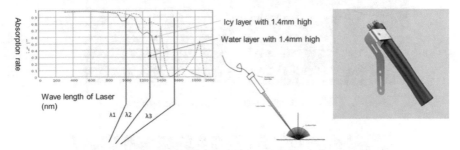

Fig. 6. Near-infrared laser sensor

4.5 Global Positioning System

GPS is required to obtain the exact current time and the locations of the vehicle. The all of the sampled sensor data are linked with the sampled time and the location by longitude and latitude indexes and managed as database such as the relation database in sensor server. When the vehicle goes through tunnel, the longitude and latitude indexes cannot be detected. In this case, the locations of the vehicle can be calculated using 9-axis dynamic sensor. As results, all of the time and location data can be always extracted.

5 Prototype and Field Experiment

In order to evaluate the sensing function and accuracy of sensing road surface states, both sensor and communication servers, and various sensors are set to the vehicle as shown in Fig. 7. The experimental vehicle running trials was made in the Morioka city which is a snow country in Northern Jan during January to February at both daytime and night about 4 h to evaluate each sensor data and decision accuracy in realtime on the winter road with various road states such as dry, wet, snowy, damp and icy states shown in Fig. 8. In order to evaluate the road surface decision function, the video camera is also used to compere the decision state and the actual road surface state.

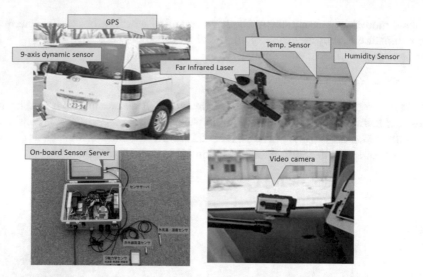

Fig. 7. Crowd road surface sensing system

Fig. 8. Experimental field at daytime and night

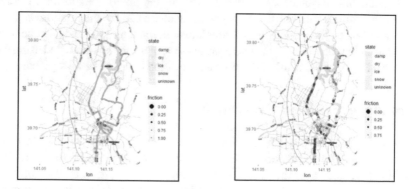

Fig. 9. Result 1: road state distribution for daytime and night

5.1 Evaluation and Analysis of Various Sensor Data

Figure 9 shows the road state distribution on both daytime in dry and damp states and night in snowy and icy road states. The gray, green, yellow and red mean dry, damp, snowy and icy, respectively. At daytime, since it is cloudy, the road surface state is dry and damp. Whereas at night, the road surface sate is snowy and icy. By evaluating and comparing those idenfied road states by the sensors and the video camera, our road surface state sensing system can identify the real winter states more than about 80% accuracy.

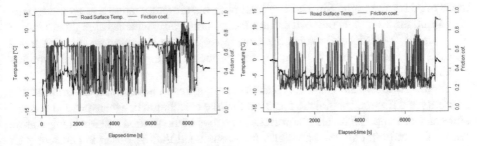

Fig. 10. Result 2: road surface temp. & friction coefficient

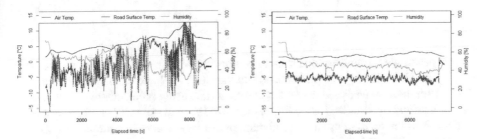

Fig. 11. Result 3: air, road surface temps and humidity

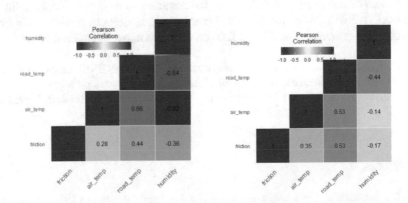

Fig. 12. Result: pearson correlation

Figure 10 shows the road surface temperature and the friction coefficient at both the same daytime and night. At daytime, the road surface temperature is higher and the friction coefficient is larger in average, whereas at night, the surface temperature is lower and almost constant and the friction coefficient is lower in average. It is concluded that night cold road surface is more dangerous than daytime and the dangerous locations can be precisely identified on the road map.

Figure 11 shows air temperature, road surface temperature and humidity at both the same daytime and night. Since at daytime, both the air temperature and humidity vary as elapsed time and then the road surface temperature vary, whereas at night, all the air temperature, humidity and road surface are lower and almost constant. It is concluded that the night time is easy to be snowy or icy due to lower temperature and humidity compared with daytime.

Figure 12 shows the Pearson correlation among road temperature, air temperature, humidity and friction coefficient. It is clear that the correlation between the road temperature and air temperature, and the correlation between the friction and road temperature are is high at both daytime and night. It is concluded that by observing correctly air temperature and road surface temperature, the average road state in the same area can be detected whereas by measuring road temperature and the friction coefficient, precise dangerous locations, such as snowy or icy state can be identified.

6 Conclusions

In this paper, we introduced road state information platform to realize safe and reliable driving in winter in snow countries based on various in-vehicle environmental sensors. The sensor data are analyzed to identify the road state using AI technology. The various sensor data and the decided road state are organized as road state GIS platform system to widely show the road states on the road map. In order to evaluate the proposed system, a prototype is designed and developed. The accuracy of decision of road state is evaluated on the actual public road. From this field experiment, our proposed system could provide reasonable accuracy to identify the correct road states although more experimental measurement data are required to improve the accuracy of road state identification. As future work, we will deploy the EV based connected autonomous driving based on our road state information platform in challenged road and network environment.

Acknowledgement. The research was supported by JSPS KAKENHI Grant Numbers JP 20K11773, Strategic Information and Communications R&D Promotion Program Grant Number 181502003 by Ministry of Affairs and Communication and Strategic Research Project Grant by Iwate Prefectural University in 2020.

References

1. SAE International: Taxonomy and Definitions for Terms Related to Driving Automation Systems, for On-Road Motor Vehicles. J3016_201806, June 2018
2. Police department in Hokkaido: The Actual State of Winter Typed Traffic Accidents, November 2018. https://www.police.pref.hokkaido.lg.jp/info/koutuu/fuyumichi/blizzard.pdf

3. Takahashi, N., Tokunaga, R.A., Sato, T., Ishikawa, N.: Road surface temperature model accounting for the effects of surrounding environment. J. Japan. Soc. Snow Ice **72**(6), 377–390 (2010)
4. Fujimoto, A., Nakajima, T., Sato, K., Tokunaga, R., Takahashi, N., Ishida, T.: Route-based forecasting of road surface temperature by using meshed air temperature data. In: JSSI&JSSE Joint Conference, pp. 1–34, September 2018
5. Saida, A., Sato, K., Nakajima, T., Tokunaga, R., Sato: A study of route based forecast of road surface condition by melting and feezing mass estamation method using weather mesh data. In: JSSI&JSSE Joint Conference, pp. 2–57, September 2018
6. Hoshi, T., Saida, A., Nakajima, T., Tokunaga, R., Sato, M., Sato, K.: Basic consideration of wide-scale road surface snowy and icy conditions using weather mesh data. Monthly report of Civil Engineering Research Institute for Cold Region, No. 800, pp. 28–34, January 2020
7. Fujita, S., Tomiyama, K., Abliz, N., Kawamura, A.: Development of a roughness data collection system for urban roads by use of a mobile profilometer and GIS. J. Japan Soc. Civ. Eng. **69**(2), I_90–I_97 (2013)
8. Yagi, K.: A measuring method of road surface longitudinal profile from sprung acceleration and verification with road profiler. J. Japan Soc. Civ. Eng. **69**(3), I_1–I_7 (2013)
9. Mizuno, H., Nakatsuji, T., Shirakawa, T., Kawamura, A.: A statistical model for estimating road surface conditions in winter. In: Proceedings of Infrastructure Planning (CD-ROM). The Society of Civil Engineers, December 2006
10. Saida, A., Fujimoto, A., Fukuhara, T.: Forecasting model of road surface temperature along a road network by heat balance method. J. Civ. Eng. Japan **69**(1), 1–11 (2013)
11. Okubo, K., Takahashi, J., Takechi, H., Sakurai, T., Kokubu, T.: Possibility of blizzard detection by on-board camera and AI technology. Monthly report of Civil Engineering Research Institute for Cold Region, No. 798, pp. 32–37, November 2019
12. Takiguchi, K., Suda, Y., Kono, K., Mizuno, S., Yamabe, S., Masaki, N., Hayashi, T.: Trial of quasi-electrical field technology to automobile tire sensing. In: Annual Conference on Automobile Technology Association, 417-20145406, May 2014
13. Casselgren, J., Rosendahl, S., Eliasson, J.: Road surface information system. In: Proceedings of the 16th SIRWEC Conference, Helsinki, 23–25 May 2012. https://sirwec.org/wp-content/uploads/Papers/2012-Helsinki/66.pdf
14. Ito, K., Shibata, Y.: V2X communication system for sharing road alert information using cognitive network. In: Proceedings of 8th International Conference on Awareness Science and Technology, Taichung, Taiwan, pp. 533–538, September 2017
15. Ito, K., Shibata, Y.: Experimentation of V2X communication in real environment for road alert information sharing system. In: Proceedings on IEEE International Conference on Advanced Information Networking and Applications (AINA 2015), Gwangju, Korea, pp. 711–716, March 2015
16. Ito, K., Shibata, Y.: Estimation of communication range using Wi-Fi for V2X communication environment. In: The 10th International Conference on Complex, Intelligent, and Software Intensive Systems (CISIS 2016), Fukuoka, Japan, pp. 278–283, July 2016

Towards Industrial 5G: An LTE+ Case Study Report

Alexander Gogolev[✉] and Markus Aleksy

ABB Corporate Research Center, Ladenburg, Germany
{Alexander.Gogolev,Markus.Aleksy}@de.abb.com

Abstract. This manuscript evaluates the robustness of the small LTE+ campus setup prototype for industrial automation, addressing architecture bottlenecks and the communication latencies. It shows that a channel overload or a simple Denial-of-Service attack can have a crucial impact on the communication latencies in such setups. The findings suggest that future LTE+ small campus setups could benefit from certain enhancements in their wired segment.

1 Introduction and Motivation

The ongoing progress in the development of wireless communication technologies, such as Long-Term Evolution (LTE), Narrowband Internet of Things (NB-IoT), and 5G enable new types of industrial applications. Traditional industrial systems usually rely on wired communication technologies to connect sensors and actuators. However, these type of communication needs to be enhanced by wireless technologies in future to address new developments, such as mixed reality applications, automated guided vehicles, moving robots and drones or achieving higher flexibility required by increasing demand for highly customized products and adaptable production facilities.

Additionally, the future of industrial automation promises to flatten the hierarchy in automation systems, unify the hardware and software, and bring the Industrial Internet of Things (IIoT) to the field level [1].

Enabling devices with IIoT on the application level is often suggested with Open Platform Communications and Unified Architecture (OPC UA) [2], and its early evaluations already show promising results [3–7]. 5G is another technology that could foster IIoT development for industrial automation on the communication plane. With its ultra-reliable-low-latency communication (URLLC) 5G vendors promise to unlock the use cases previously inaccessible due to limitations of the wired communication. Already there is evidence that indeed a low-latency communication can be achieved with 5G [8,9]. But there are still aspects that need careful assessment. For instance, Industry 4.0 concept suggests that new automation technologies should offer open standard and non-vendor binding interfaces and tools (as does, for instance, another next-gen communication platform, IEEE TSN [10]), and the 5G is currently developing such capabilities in 5G ACIA Alliance [11]. Another aspect is the communication reliability and

L. Barolli et al. (Eds.): AINA 2021, LNNS 225, pp. 414–423, 2021.
https://doi.org/10.1007/978-3-030-75100-5_36

robustness—it is known that mobile networks can handle high throughput, but in automation it is more important to deliver "little data on time", rather than "a lot of data delivered a bit later". Deterministic wireless communication studies show promising approaches [12–16], but if 5G aims to compete with wired industrial communication in terms of determinism, its robustness needs to be revisited.

Below we extend the assessment of the new communication technologies for industrial automation [6,7,17,18] with an LTE+ campus network. The test setup is rated as "LTE+" since at the moment of the experiment there is no finalized standard to directly attribute the capabilities of the studied campus cell to the 5G. This paper reflects on the findings from the perspective of communication robustness, since it is one of the most vital aspects for industrial systems.

The paper is organized as follows. Next section details the 5G application in the scope of the industrial automation. Section 3 explains the test system setup. Then, Sect. 4 presents the use cases and assumptions that guided our research. Sections 5 and 6 present and discuss test results. Finally, Sect. 7 concludes our work.

2   Industrial 5G: Campus Networking

The today's business of 5G vendors is creation and operation of the end-user mobile networks and their cooperation. In such systems network management is typically handled by the network operator, or a service provider–a single entity that exclusively controls throughput allocation and connection guarantees. Supporting multiple simultaneous subscriptions which can be streaming video feeds at high data rates, entails handling sporadic data bursts, often combined with high user mobility, both in terms of distance and speed. Such networks need to ensure data exchange with high throughput, but not-so-high-reliability, due to low critical data exchange.

In industrial automation (IA), whether it is process automation, or mobile robots, the end systems is not typically moving at speeds of up to 300 km/h, for distances on hundreds of kilometers (as could a typical high-speed train passenger). Industrial end systems then could be relatively slow and their mobility range could be confined within a limited space of a factory territory, i.e., a few square kilometers. Industrial automation, however, requires uninterrupted service for its end systems, because of, e.g., safety reasons. To guarantee such uninterrupted service IA often chooses to control all of the automation pyramid vertically: from the cloud to the factory floor. This way an automation architecture can be traced, verified and validated to ensure the necessary safety and reliability. Or, an architecture could be modular, allowing division into segments, for maintenance, redundancy or other purposes.

To meet requirements from different IA use cases and still maintain some generics, 5G campus networks seem to be a good candidate for pilot evaluations. 5G campus is a cluster of 5G network devices bounded by licensed frequencies and deployed within a limited territory. Such an approach has little challenges when it comes to reducing of the users' speed and possible mobility range.

However, vertical/modular integration would require such a campus network to expose some interfaces to the existing systems (typically, running fieldbuses), or the systems with upcoming technologies (e.g., running IEEE TSN), and local engineers and their tools. This would require a significant re-work of the current mobile network centralized and proprietary management. 5G Alliance for Connected Industries and Automation [11] established a task work group to define a standard exposure interface for such network management, to foster the adoption of 5G in IA.

But the largest and the most important hitch is the reliability of the data exchange which can affect the safety and execution continuity of the automated process. To a certain extent the safe execution of the process control can be ensured by the higher-level control software that could, for instance, shut down the system in a fail-safe manner to guarantee safely operation in case of interrupted communication. But, this still leaves the process continuity at question, which can lead to disruptions in the functioning of the factory.

The aspect of communication reliability and robustness in 5G campus network needs assessment in at least two of its segments:

- Wireless communication
- Overall system architecture

The robustness of wireless segment has been addressed with physical enhancements such as multiple-input and multiple-output (MIMO), orthogonal frequency-division multiplexing (OFDM), etc., and we leave it outside this paper. However the latter can be evaluated with a network engineering tool-set on the given example of the small cost-effective LTE+ campus network.

2.1   Industry Performance Requirements

Communication performance and robustness are crucial for industrial automation systems, and are typically assessed with timing parameters of data delivery, such as latency and jitter. Industrial automation spans timing requirements from motion control that require those in μs-range to, for instance, process automation, that often deals with millisecond-granular control cycles. 5G claims to match the latter latencies thus paving its way into such industries.

Robustness of wireless communication is, due to limitations of shared communication media, typically worse than that of wired communication. Hence URLLC for Industrial IoT systems is often motivated with the assumption that it will not hinder the communication robustness. Rather it claims to add flexibility, opening up new use cases that were previously inaccessible. Since the purely wireless aspects are omitted in this paper, the main focus is on the campus network architecture.

2.2   5G and Latency Accumulation

The discussions of 5G URLLC for industrial applications place the latency in data exchange in a millisecond, or even sub-millisecond range [8,9]. Such claims, however, do not always clearly specify the jitter, packet loss rate, and overall determinism.

As mentioned above, IA spans over multiple fields with rather varying requirements. For instance, motion control's μs-granular control loops are clearly unachievable with 5G URLLC. It could satisfy, on the other hand, the latency requirements for process automation or mobile robots, given that communication is stable and robust. We skip the assessment of stability and robustness in the wireless segment, but rather focus on the architecture of the 5G campus network that needs to reflect these requirements (which differ from the requirements of the "regular" user mobile networks).

Therefore, the open questions are:

1. Can the suggested architecture of the 5G campus network cope with high-load traffic?
2. How would the channel overload, or a denial-of-service (DoS) attack affect the end-to-end latency?
3. What is the quantifiable impact of such overload?

We try to reflect on this below, using a real-life LTE+ setup.

3   Test Setup Description

The test setup is shown in Fig. 1 and consists of:

1. 1 LTE+ Base Station
2. 2 LTE User Equipment devices, attached to the Linux machines
3. 1 management PC (Linux)
4. 1 non-managed Gigabit Ethernet bridge
5. 4 Linux PCs, connected to the bridge and management PC

We refer to the setup as LTE+, since at the moment of the testing there is no finalized standard that could directly attribute the capabilities of the setup to "Industrial 5G". This means that the HW and SW is operational but cannot be strictly speaking referred to as "5G" because the feature sets first need to be defined and rated in the standard.

3.1   User Equipment

User Equipment is a term used in mobile networks to designate the end user connection point on the system. It is, typically, a mobile phone or a modem. The test setup uses two PCs with two commercial-off-the-shelf (COTS) LTE UEs (User Equipment – a modem), communicating to the base station (BS).

We have observed that for such UEs (available as COTS and thus often used for testing), the communication latency does not always correlate with data rate/speed. In our setup the goal is to minimize the former, since automation devices rarely produce much traffic, but do require low latency communication. For this reason, they need to be configured in specific modes of operations, which in some cases requires some low-level access and change of firmware to provide lower latency instead of higher speed. These modes are commonly referred as "stick" and "router" modes, and the latter generally offers higher throughput (and is typically set as default mode). The "stick" mode may reduce the effective throughput but does show better latency. Thus, in our test the UEs are configured to run in the "stick" mode.

3.2   5G Enablement

Base station is then connected to a management PC(PC_m) over a Gigabit Ethernet cable and a switch, as shown in Fig. 1.

PC_3 and PC_4 are connected to PC_m, and exchange data with the PC_1 and PC_2. They use independent physical network interfaces on the PC_m to avoid traffic interference. All machines run Ubuntu 18.04 LTS, except PC_m – it runs a vendor-chosen Linux version with hardware binding (required to secure the BS management).

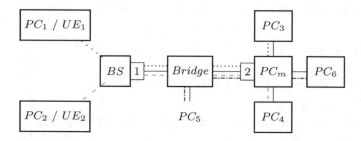

Fig. 1. Network setup with two LTE (UE) devices, and an LTE+ base station (BS). Solid gray line represents the Gigabit Ethernet connections, dashed lines – wired traffic, dotted lines – traffic over wireless link. Blue dotted line and the green dashed one – ICMP traffic between UEs and test PCs, loose dotting and dashing designates the wireless link. Red dash-dotted line designates disturbance traffic.

The network topology is schematically represented in Fig. 1, where bridge *Bridge* connects base station BS to the base-station management PC – PC_m. The physical network topology is organized in a way that all of the traffic to/from UEs is going through the ports $\boxed{1}$ and $\boxed{2}$ on the base station and PC_m, respectively. Four independent physical interfaces of PC_m are connected to the *Bridge*, PC_3, PC_4 and PC_6, respectively.

It can be seen that in the test setup PC_1 and PC_2 (to the left of the BS) communicate to their counterparts (to the right of the BS) – PC_3 and PC_4 using both wireless and wired links. They use physically independent UE devices (i.e., modems) with different QoS configuration: PC_1 has a wireless bearer configured in Qci 8, where PC_2 uses Qci 9 (higher priority). Qci could be explained as the broadband equivalent of the Ethernet QoS priority distinction, though not exactly right so, due to fundamental differences in defining the QoS in mobile networks [19]. What is important here is that PC_2 communicates on a "more important link" than PC_1. Once the data from PC_1 and PC_2 crosses over to the BS it is forwarded to the bridge and further towards PC_m. From the PC_m it is forwarded to PC_3 and PC_4 using physically separate interfaces. Thus, the data between PC_1 and PC_3 and PC_2 and PC_4 only shares the wired connection between the BS and the PC_m.

In a wireless segment (between UEs and BS) the Qci distinction ensures that in case of saturation (around 1 Gbps of combined throughput), the link with higher Qci receives better service. However, once the traffic leaves the BS towards PC_m and further, it should be served using Ethernet QoS mechanisms (e.g., out of TSN toolbox). TSN combines several such mechanisms of which some pre-date the TSN itself and therefore are available since a while. In the direction UEs $\rightarrow$ PC, even if the wireless media is saturated with high-density data-exchange, BS would ensure the delivery of the data from the high-importance link towards the "trunk" port on the PC_m. However, if the communication is bi-directional (and it typically is) the reverse direction (UEs $\leftarrow$ PCs) is also important. Here the determinism of the data exchange also needs to be ensured, but with Ethernet-based toolset, as mentioned above. But, even if the respective tool-set is in place and properly configured, it would be beneficial to have similar support on the BS side of the wired link. Moreover, the starvation of the PC_m hardware resources still poses a challenge: if PC_m is not the most capable one, processing gigabits of traffic can exhaust its computing capabilities.

To mitigate some of the most obvious bottlenecks the test setup connects both PC_3 and PC_4 with separate physical network interfaces. Respective forwarding rules are then added to the forwarding tables of the PC_m.

Generating high-throughput disturbance traffic incurs additional CPU load on the PCs, which may affect the performance measurements. To avoid this, the PCs that generate disturbance and the ones performing the actual communication are physically independent machines. PC_6 generates the disturbance traffic towards the PC_5 that has to go out of the same physical interface of the PC_m as the traffic towards the BS. Thus, the setup generates "non-important" disturbing traffic and "important" Internet Control Message Protocol (ICMP) traffic on physically independent machines, which then share the same physical interface on the PC_m. The latter, as one can notice, represents the single point of failure in this architecture. However, the software bundle on the PC_m is prepared by the broadband vendor, and implements a network configuration of a certain complexity. This software implements certain features, and utilizes the available hardware to ensure the necessary determinism and guarantees for the traffic on the broadband side of the test network.

4   Methodology and Performance Reference

To evaluate the data exchange latency and robustness we use the Linux ping command (IMCP echo request/reply) with an interval of 20 ms (empirically derived). The round trip times (RTT, referred further as request latency) are registered on the client side and averaged over 100000 cycles.

The unidirectional disturbance traffic of 800 Mbps is generated using an Iperf tool [20]. It can be shown that for instance, Ostinato packet generator [21] provides a steadier flow of packets/frames and therefore can be better suited for such tests [17]. However, the steady data flow is just a specific use case of a saturated link, while a typical data flow is rather sporadic. In any case the Iperf has already provided the saturation sufficient to distinguish a serious latency growth, as shown in the following section.

Let us briefly reference previous findings with regard to latency in new industrial network technologies. In [6,17], the latency of the ICMP echo request is studied in a multi-hop TSN network showing that RTTs can reach sub-ms latency with Time-Aware Shaping [22] or Frame Preemption [23]. It was also shown that achieved latencies stay stable even with disturbance traffic up to 100% of the channel capacity.

5   Latency Measurements

To evaluate the impact of traffic saturation, the testing is performed using ICMP packets of 64 bytes. The disturbance is consequently tested with a different throughput ($\in[400, 500, 800]$ Mbps), using 1518B packet size.

The Table 1 presents the baseline latencies achieved in a network with and without disturbance.

Table 1. Average RTTs over LTE+ setup

LTE+ Session	Disturbance			
	None	400 Mbps	600 Mbps	800 Mbps
High priority (Qci 9)	20 ms	20 ms	22 ms	35 ms
Low priority (Qci 8)	20 ms	20 ms	22 ms	35 ms

The test did not exceed the 800 Mbps of the disturbance since it would saturate the link even further, and suffocating Ethernet links is out of scope of this paper as it has been studied previously.

It shows that e2e communication latency noticeably grows with 800 Mbps of induced traffic, or with effectively 80% of the channel utilization. It also shows that both sessions (with Qci = 9 and Qci = 8) are equally disturbed, despite different levels of QoS guarantees. It is known that in LTE+ networks QoS is ensured by means of Qci bearer classes (now also known as "5ci" in 5G ACIA [11], with similar function as QoS priority levels of Ethernet) which shape traffic sessions in the wireless segment: between BS and UEs. However, it is evident that Qci distinction in the wireless segment is not sufficient to ensure full e2e determinism: there are guarantees missing elsewhere in the system. Since the system is built of wired and wireless segments, and wireless segment is presumed to be handled with Qci mechanisms, it leaves the wired segment to blame. The test sessions cross over both segments in reverse directions:

- $UE_1 \leftrightarrow PC_3$
- $UE_2 \leftrightarrow PC_4$

One could argue that in the direction $UE_1, UE_2 \rightarrow PC_3, PC_4$ the BS and Qci based QoS will ensure that high important traffic is not disturbed. However, this argument does not hold for direction $PC_3, PC_4 \rightarrow UE_1, UE_2$ the traffic needs guarantees before it gets to the BS where Qci mechanisms can take over. So, if the communication channel is congested in the direction $PC_3, PC_4 \rightarrow UE_1, UE_2$ the communication session latency may worsen. This explains the observations above.

6 Analysis

In the real-life mobile operator setups, the network management functions are not likely to run on a single PC that has a single point of failure, such as in the described setup, but rather a HW-based routers or multi-port-and-cpu-core server blades that are difficult to saturate with a simple DoS attack. Still, since a BS has just one port to which such router or a server is likely to be connected, the same observations hold. Moreover, "campus network" is marketed as an entry concept to industrial market, which must be reasonably cost-competitive, and therefore is likely to use the simplest setup available, similar to the described one. In such a setup the observations above are even more just. One important aspect of the adoption of the 5G is the compatibility of the QoS, or service guarantee levels, which can be difficult due to the different media capabilities and essential difference in requirements. Therefore, matching the 5G's QCIs (5CIs) to, e.g., Ethernet QoS needs to be organized in either flexible, or well-motivated manner. Furthermore, Qci guarantees must be then translated into the wired part of the setup, to ensure the QoS throughout the cell, campus and the whole mobile network.

In addition to this, integration with existing and future automation systems requires a thoroughly weighted choice of interfaces that is, perhaps, aligned with existing solutions for other "yet to come" communication technologies, such as IEEE TSN. Such solutions and their motivation can be found in the state of the art discussions in, e.g., FLC group [24].

Next, performance difference with wired technologies such as TSN or other deterministic Ethernet fieldbuses, must be accounted for in 5G and reflected properly. Statements such as "5G campus can be perceived from outside as a single TSN bridge", can be misleading and should be avoided, since TSN bridges guarantee $1,000$ times lower latencies than current 5G.

7 Summary

This paper extends earlier work [6,7,17] on new industrial communication technologies assessing the LTE+ campus network from a point of communication reliability and robustness.

We show that the studied campus network has a bottleneck in its design that can severely hinder the communication determinism. We illustrate this by channel overloading that simulates a simple denial-of-service attack that significantly increases the e2e communication latency. This indicates that studied 5G campus network concept is yet not mature enough for robustness testing, nor for industrial deployment. However, it can be used as a testbed for early tests in performance and integration. Last, but not least—the Ethernet QoS mechanisms can be suggested for enhancing the wired segment of the 5G campus network, where possible.

Further research can, therefore, address issue of increasing the efficiency and determinism on the 5G side, for instance, adding software and hardware on those bottleneck interfaces (if overall architecture stays unchanged).

Acknowledegements. This research was supported by the German Federal Ministry of Education and Research (BMBF) under grant number 16KIS0721. The responsibility for this publication lies with the authors.

References

1. Unlock the value of Industry 4.0. https://www.ericsson.com/en/internet-of-things/industry4-0
2. OPC Foundation. Open platform communications unified architecture. https://opcfoundation.org/about/opc-technologies/opc-ua/
3. Veichtlbauer, A., Ortmayer, M., Heistracher, T.: OPC UA integration for field devices. In: 2017 IEEE 15th International Conference on Industrial Informatics (INDIN), pp. 419–424, July 2017
4. Pfrommer, J., Ebner, A., Ravikumar, S., Karunakaran, B.: Open source OPC UA PubSub Over TSN for realtime industrial communication. In: 2018 IEEE 23rd International Conference on Emerging Technologies and Factory Automation (ETFA), vol. 1, pp. 1087–1090 (2018)
5. OPC UA TSN a new solution for industrial communication. https://www.br-automation.com/smc/953ce46647cb909f0cce603249fb229e29f0a30a.pdf
6. Gogolev, A., Braun, R., Bauer, P.: TSN traffic shaping for OPC UA field devices. In: 2019 IEEE 17th International Conference on Industrial Informatics (INDIN), vol. 1, pp. 951–956, July 2019
7. Gogolev, A., Mendoza, F., Braun, R.: TSN-Enabled OPC UA in field devices. In: 2018 IEEE 23rd International Conference on Emerging Technologies and Factory Automation (ETFA), vol. 1, pp. 297–303 (2018)
8. Pilz, J., Mehlhose, M., Wirth, T., Wieruch, D., Holfeld, B., Haustein, T.: A tactile internet demonstration: 1ms ultra low delay for wireless communications towards 5g. In: IEEE Conference on Computer Communications Workshops (INFOCOM WKSHPS), vol. 2016, pp. 862–863 (2016)
9. Soldani, D., Guo, Y.J., Barani, B., Mogensen, P., Chih-Lin, I., Das, S.K.: 5G for ultra-reliable low-latency communications. IEEE Netw. **32**(2), 6–7 (2018)
10. Institute of Electrical and Electronics Engineers. Time-sensitive networking task group. http://www.ieee802.org/1/pages/tsn.html
11. 5G Alliance for Connected Industries and Automation. https://www.5g-acia.org/about-5g-acia/

12. Zhang, S., Zhang, H.: A review of wireless sensor networks and its applications. In: IEEE International Conference on Automation and Logistics, vol. 2012, pp. 386–389 (2012)
13. Zheng, L., Lu, N., Cai, L.: Reliable wireless communication networks for demand response control. IEEE Trans. Smart Grid 4(1), 133–140 (2013)
14. Jonsson, M., Kunert, K.: Towards reliable wireless industrial communication with real-time guarantees. IEEE Trans. Ind. Inform. 5(4), 429–442 (2009)
15. Fan, B., Krishnan, H.: Reliability analysis of DSRC wireless communication for vehicle safety applications. In: 2006 IEEE Intelligent Transportation Systems Conference, pp. 355–362 (2006)
16. Berger, A., Pichler, M., Haselmayr, H., et al.: Energy-efficient and reliable wireless sensor networks-an extension to IEEE 802.15.4e. JEURASIP J. Wireless Commun. Netw. 126, 1–2 (2014)
17. Gogolev, A., Bauer, P.: A simpler TSN: traffic preemption vs. scheduling. In: 2020 IEEE Conference on Emerging Technologies and Factory Automation (ETFA), vol. 1, September 2020
18. Abukwaik, H., Gogolev, A., Groß, C., Aleksy, M.: Opc ua realization for simplified commissioning of adaptive sensing applications for the 5g iiot. Internet Things 11, 100221 (2020)
19. The 3rd Generation Partnership Project (3GPP). Policy and charging control architecture. https://portal.3gpp.org/desktopmodules/Specifications/SpecificationDetails.aspx?specificationId=810
20. iPerf – a tool for active measurements of the maximum achievable bandwidth on IP networks. https://iperf.fr/
21. Ostinato – packet generator. https://ostinato.org/
22. Institute of Electrical and Electronics Engineers. 802.1Qbv – Enhancements for Scheduled Traffic http://www.ieee802.org/1/pages/802.1bv.html
23. Institute of Electrical and Electronics Engineers. 802.1Qbu – Frame Preemption. https://www.ieee802.org/1/pages/802.1bu.html
24. OPC Foundation. Field Level Communications (FLC) Initiative. https://opcfoundation.org/flc/

Thermal Environment Simulator for Office Building with Multi-agent-Based Simulation Technique

Kodai Murakami[✉], Toshihiro Mega, and Noriyuki Kushiro

Faculty of Computer Science and Systems Engineering, Kyusyu Institute of Technology, Kitakyushu, Japan

Abstract. There is a growing interest to introduce split-ductless air conditioning systems into an office building both for realizing energy saving and comfort to office workers. Building energy management systems (BEMS) are expected to establish sophisticated controlling methods for the split-ductless air conditioning system. For creating the methods, simulators for thermal environment in a building are required. However, existing simulators necessitate a building information model (BIM) and huge amount of data for parameter tuning. Moreover, the existing simulators have difficulties to express spatial and temporal dynamics in each space in a building. The spatial and temporal dynamics in a space are indispensable for creating methods for controlling split-ductless air conditioning systems, because the systems independently manage devices dispersed in a room. A thermal environment simulator with multi-agent-based simulation (MAS) technique was proposed in the paper. The simulator can work without BIM and visualize spatial and temporal dynamics with limited numbers of parameters, which are easily obtained from BEMS. The simulator with MAS was implemented and applied to a field test in winter and summer seasons. As a result, high precision for reproducing spatial and temporal thermal dynamics was confirmed through the field test.

1 Introduction

Building Energy Management Systems (BEMS) has been introduced to reduce greenhouse gas emissions into office buildings [1]. Electricity energy consumption for air conditioning reaches above 30% in office buildings [2, 3]. The air conditioning systems are regarded as a major target of energy reduction to the BEMS. Energy consumption for air conditioning systems is affected by various of factors: outdoor temperature, humidity, density of office workers in a space. The BEMS ought to identify these factors and decide how to control the air conditioning systems efficiently. However, the operation for energy reduction in air conditioning systems often causes deterioration of comfortabilities for office workers, due to trade-off between saving energy and realizing comfort.

Various kinds of research have been conducted to relax incompatible features among saving energy and realizing comfort by the BEMS. For examples, [4] balanced energy efficiency and office workers' comfortability with machine learning techniques. [5] proposed algorithms by utilizing information elicited from WEB to improve comfortability.

Energy prediction based on huge historical data with machine learning techniques [6] is one of promising solutions for relaxing the trade-off, because the results of prediction help the BEMS plan optimal ways and control schedule for the air conditioning systems. [7] provided a prediction method, which combined accuracy and readability, with heterogeneous mixture modeling. Both accuracy and readability of the model contribute to create an optimal control and schedule method to BEMS.

Recently, split-ductless air conditioning systems are introduced into an office building. The split-ductless air conditioning systems realize energy saving and comfort to office workers by managing independent devices dispersed in a room. However, the existing simulators necessitate a building information model (BIM) and huge amount of data for parameter tuning. The existing simulators have difficulties to express spatial and temporal dynamics in each space in a building. The spatial and temporal dynamics are indispensable information for creating methods for controlling split-ductless air conditioning systems.

In this paper, a thermal environment simulator with multi-agent-based simulation (MAS) technique was proposed. The simulator works without BIM and visualize spatial and temporal dynamics with limited numbers of parameters, which are easily obtained from BEMS. The simulator was implemented and applied to a field test. As a result, high accuracy was confirmed through the experiment.

The rest of paper is organized as follows: Sect. 2 summarizes modeling approaches used in simulator for BEMS. Section 3 proposes concept of thermal environment simulator for office building with MAS technique, and Sect. 3.1 explains method for identifying parameters required for the proposed simulator. Section 4 summarizes the field test of the simulator and results, and Sect. 5 concludes the paper.

2 Thermal Environment Simulator for Office Building

This section surveys existing thermal simulation methods used in BEMS. Thermal environment simulator with MAS technique is proposed base on the surveys.

2.1 Categories of Modeling Approaches for Simulation

Modeling approaches in BEMS are classified into the following three categories [8, 9]. Features of each approach are summarized in Table 1.

2.1.1 Category1: Forward Approach

Forward approach is regarded as a white box approach. It contains descriptions of mathematical models on physics [9]. The mathematical models utilize the results of building information modeling (BIM) as a basis of simulation and requires many physical constants for calculation.

The forward approach is classified into the following three methods: Computational Fluid Dynamics (CFD), zone, and multi-zone approaches [10]. The CFD approach describes flow of field on a micro perspective. In this approach, heat transfer in a building is expressed as fluid in three dimensions [11]. [12] proposed volume element model for

Table 1. Features for each modelling approach

Categories	Features
Forward approach	Pros: High accuracy of simulation base on physics Cons: Requiring huge number of parameters for simulation and necessity of complex mathematical models for simulation
Data driven approach	Pros: Simple mathematical or statistical model Cons: Less parameters for simulation and huge training data
Grey box approach	Pros: Limited mathematical model Cons: Identification unknown parameters

3D dynamic building thermal simulation, and applies CFD approach in order to examine local thermal effect. Zone approach is a simplified version of the CFD approach, in which a room is divided into multiple zones and the physical equations are solved for each cell in each zone. [13] presents a zonal model to predict air temperature distribution in buildings with calculating mass air flows between different zones. Multi-zone approach is a method of modeling airflow and contaminant transfer between different rooms or between outside or wall-to-room by means of node-to-node transfer, and assumes that temperature and contaminant concentration are uniform within a single node. [14] evaluates assumption that air momentum effects can be neglected in a multizone airflow network models.

Several researches with forward approach for improving accuracies were challenged by solving complex mathematical equations explaining thermal environment in building through long period of tuning time [18, 19]. These challenges increase necessities for more minute BIM and more granular parameters, which are not easy to obtain from the existing BEMS.

2.1.2 Category2: Data-Driven Approach

The data-driven approach, known as a black box approach, constructs prediction model with huge training data with machine learning techniques [15]. The data-driven approach constructs the prediction model without any mathematical equations on physical, it requires just data for learning, and statistical models on machine learning techniques. For example, [16] predicted room temperature and relative humidity by autoregressive linear and nonlinear neural network models accurately.

Data driven approach has a merit to predict objective variables at specific time and at specific position in a building, however it has clear restrictions to explain temporal and spatial thermal dynamics in a space, due to nonexistence of physical rules. Therefore, it has difficulty to apply for determining control methods for the distributed air conditioning systems.

2.1.3 Category3: Grey Box Approach

Gray box approach is an intermediate modeling approach between the white box (forward) and the black box (data-driven) approaches. The gray box approach uses restricted number of mathematical equations on physics and identified parameters required for the equations on black box approach. In [17], plural of gray box models was estimated their accuracies for predicting thermal dynamics in multi-zones of office building. As a result, the gray model was confirmed that they realized high accuracies in comparison with the white box and black box approaches. However, the gray-box approach remains difficulties to select an appropriate mathematical model from several kinds of governing equations, and to identify parameters including the selected mathematical model with data elicited by BEMS.

2.2 Solutions for Issues in Existing Simulation Model

The following are requirements for thermal simulator in this study:

1. The simulator should predict spatial and temporal dynamics in each space in a building for controlling and scheduling split-ductless air conditioning systems.
2. The simulator does not require BIM for constructing prediction model, because the simulator should be applied to any scales of building, which has difficulty to obtain BIM.
3. The simulator should identify all the parameters required for simulation by data easily obtained from BEMS.

2.2.1 Concept of Thermal Environment Simulator for Office Building with Multi-agent-Based Simulation Technique

To solve the requirements, we propose a new model to describe thermal dynamics with fewer parameters. Multi-agent-based simulation modeling (MAS) is introduced as the modeling technique for thermal dynamics. MAS is a bottom-up approach to express an entire system through interactions among autonomous, rule-based actors called as agents, and simulate complex behaviors in the system by defining simple interaction rules among agents [20].

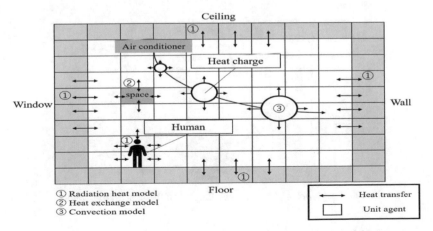

Fig. 1. Conceptual diagram of the proposed simulation model.

428   K. Murakami et al.

By introducing the MAS to thermal dynamics simulator, we are succeeded in reducing mathematical equations and parameters required for simulation. The simulator with MAS technique requires three basic heat conditions and three parameters. Figure 1 shows image of heat transfer in the room and overall diagram of the proposed simulation model. Heat transfer is determined by the following three types of heat transfer mechanism.

At the first, the heat transfer in which a space transfers heat obtained from a heat source to an adjacent space was defined as a heat exchange model as follows:

1. Heat exchange model - If the temperatures of the adjacent spaces are T_{space} and T'_{space}, respectively, the amount of heat transferred in unit time, Q_{space}, is

$$Q_{space} = \alpha \left| T_{space} - T'_{space} \right| \tag{1}$$

At the second, the effect of convective heat from the air that the air conditioning produces was modeled by assuming that particles called heat charge are emitted from the air conditioning. The heat charge is periodically emitted from the air conditioner and transfers heat to the surrounding space as it moves, as shown in Fig. 1. Heat transfer into space by heat charge was defined as follows:

2. Heat convection model - Assuming that the temperature of the emitted heat charge is T_{heat}, the temperature of the surrounding space is T_{space}, and the distance between the heat charge and the space is r, the amount of heat transferred from the heat charge to the space in unit time, $Q_{convection}$, is

$$Q_{convection} = \beta \frac{\left| T_{heat} - T_{space} \right|}{r^2} \tag{2}$$

The detailed conceptual diagram of the heat charge model is shown in Fig. 2. The heat charge is defined as properties: coordinates, energy, temperature, velocity, radius, and direction of movement.

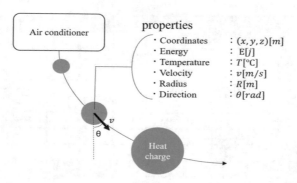

Fig. 2. Properties and conceptual diagram of the heat charge model.

Finally, radiant heat model is defined as follows, assuming that radiant heat is generated from people, equipment such as PCs, floors, ceilings, windows, and walls as heat sources, and that heat is transferred to adjacent spaces,

3. Radiation heat model - If the temperature of the heat source is $T_{barrier}$ and the temperature of the adjacent space is T_{space}, the amount of heat transferred $Q_{radiation}$ is

$$Q_{radiation} = \gamma \left| T_{barrier} - T_{space} \right| \tag{3}$$

Equations (1), (2) and (3) follow Fourier's law, which is a fundamental principle of heat transfer engineering.

Effect of forced convection from air conditioning ordinally needs to solve complex and numerous equations on fluid mechanics by using the CFD. In this study, the effects of pseudo forced convection come down to the Fourier's law, assuming that the heat charge interacts with surrounding spaces through agent simulation. This makes the simple. In addition, heat sources such as people and walls were also considered as agents to affect spatial and temporal dynamics of heat.

Equations (1), (2), and (3) are basic physical models that consolidate the constants necessary for fluid mechanism analysis, e.g., heat transfer coefficient, thermal conductivity, and viscous stress, respectively. The proportionality constants α, β, and γ are parameters, which represent amount of heat transfer at a unit temperature difference in unit time. Since the parameter are regarded as aggregation of many constants including various physical quantities, the proportionality constants α, β, and γ need to be identified anew. The parameters were determined experimentally in this study. Method to identify the parameters was also developed and details of the method are explained in Sect. 3.

3   Implementation of Thermal Environment Simulator with Multi-agent Modeling Technique

3.1   Implementation of Simulator with Multi-agent Modeling Technique

The simulator was developed by the MAS software "artisoc" [22] on the model proposed in Sect. 2. Figure 3 shows the overview of the simulator and office layout used for an example of simulation. Lists of functions of the simulator is shown in Table 2.

The simulator starts only with suction temperature of each air conditioners, which can be easily obtained by BEMS, and works on predetermined control schedule of the air conditioners. Users set conditions for simulations (date, time, outdoor temperature etc.) on the panel and directly observe spatial and temporal thermal dynamics as animation.

Measurement device for identifying the parameters required for the simulator was also developed in the study. The parameters are determined experimentally through the examination in an actual building prior to starting simulation. Various studies have been conducted to determine thermal properties in existing buildings. [21] proposed accurate mathematical algorithm for evaluating heat transfer by using infrared thermography. An overview of the device for identifying the parameters with multi-thermal cameras and the results of are presented in the next section.

Table 2. Functional requirements for the developed simulator

Functions	Details
Input (csv file)	BEMS data: initialization data control data: Air conditioning control file in minutes
Output (csv file)	Air conditioning information and observed temperature
Heat map	Red when the temperature is high, white when the temperature is low
Control panel	Setting of observation temperature position, air conditioning and input/output data
Graph visualization	Power consumption, observation temperature, indoor temperature

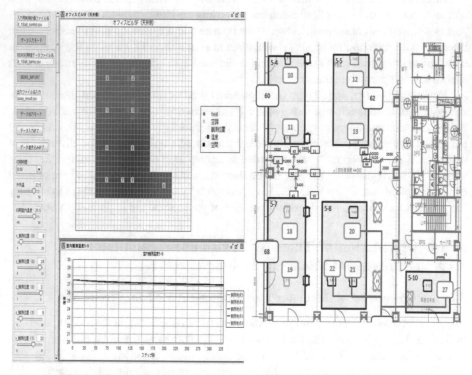

Fig. 3. The simulator we developed and the layout we used as a reference

3.2 Parameter Tuning in Lab and Methods for Tuning

As shown in the left image of Fig. 4, paper cylinders were hung at equal intervals from the ceiling in a space, and each paper cylinders were photographed with 6 thermal cameras (LIR-AX8) at every 10 s. The parameters were determined through calculating heat balance based on changes of temperature in paper cylinders for unit time. The image on right of Fig. 3 shows temperature of each paper cylinder.

 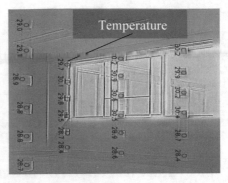

Fig. 4. Scene for the experiment with device (left), thermal image obtained from device (right).

Three proposed models (Eq. (1), (2) and (3)) are applied to describe respective heat exchange mechanism among agents. In the heat exchange model, heat is exchanged between spatial agents. In the convection model, heat is exchanged between heat charge and space. In the radiant heat model, heat is exchanged between objects such as floors and walls and space. Since the temperature of the space cannot be measured directly with thermal cameras, the parameters are determined by replacing paper cylinders as spatial agents. The corresponding positional relationships for each model are shown in Fig. 5.

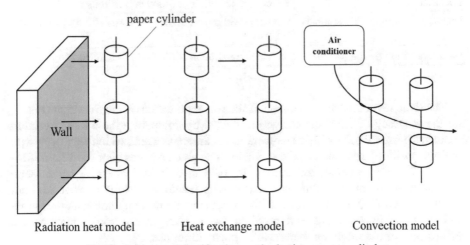

Fig. 5. Parameter identification method using a paper cylinder

Multi values should be determined in the radiation model, because effects from walls, windows, ceilings and floors are different. In this study, we assumed that walls, floors, and ceilings have almost the same thermal properties and give the same values. Each value of parameter was tuned through experiments on the simulator prior to field test and the results of parameter identification are shown in Table 3.

Table 3. Parameter identification results

Parameter value	Measured value [j/s · °C]
Heat exchange model (α)	0.520
Convection heat model (β)	0.610
Radiation heat model: window (γ)	0.039
Radiation heat model: wall, ceiling, floor, human (γ)	0.033

4   Field Test and Estimation of Simulation

4.1   Basic Setups for Field Test

Field test was conducted for evaluating accuracy of prediction with the simulator.

The simulator was installed into an office building (8000 m³) in Tokyo and evaluates by comparing the results of simulations with data obtained from the BEMS. Table 4 shows the details of the field test.

Table 4. Details of field test

Items	Details
Location	8-story office buildings (8000 m³) in Tokyo
Data acquisition period (summer)	8:00 am–7:30 pm on 11/Aug.–17/Aug., 20.Aug.–23/Aug. in 2020
Data acquisition period (winter)	8:00 am–7:30 pm on 14/Dec.–20/Dec. in 2020

The same layout for the experimental site was reproduced on the simulator (Fig. 3), and the accuracy of the simulator was evaluated by comparing the inlet temperature obtained from the BEMS with the temperature of the space agent, which were occupying the same position, calculated by the simulator. The green squares allotted to 10–13, 18–21, and 27 in Fig. 3 depict arrangement for devices of split-ductless air conditioning systems in the room. The simulator started with conditions of the floor at 8:00 am, which obtained from the BEMS, and simulates spatial and temporal dynamics in the floor 7:30 pm. Few minutes were required for simulate on notebook computer. The performance of the simulator seemed to be practical for use.

4.2   Parameter Tuning in Lab and Methods for Tuning

Figure 6 and 7 show comparison results from 8:00 am to 7:30 pm for each space depicted as 10–13 and 18–22 on 12th and 13th Aug., respectively. Differences between measured value and simulated value were small at every space and every time. We confirmed that the simulator can reproduce spatial and temporal thermal dynamics in the floor through the field test.

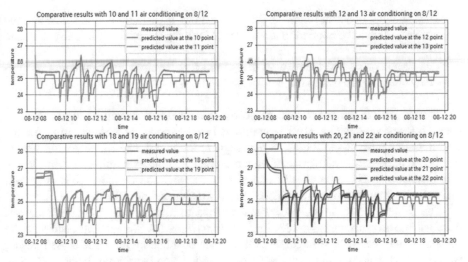

Fig. 6. Temperature comparison results for August 12

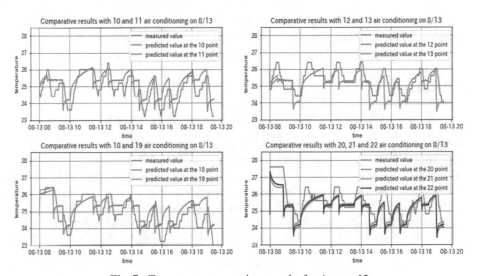

Fig. 7. Temperature comparison results for August 13

5 Conclusion

Thermal environment simulator for office building with multi-agent-based simulation (MAS) was proposed in this paper so as to create sophisticated controlling methods for the split-ductless air conditioning systems in BEMS.

The simulator satisfies the following requirements:

1. The simulator should predict spatial and temporal dynamics in each space in a building for controlling and scheduling split-ductless air conditioning systems.

2. The simulator does not require BIM for constructing prediction model, because the simulator should be applied to any scales of building, which has difficulty to obtain BIM.
3. The simulator should identify all the parameters required for simulation by data easily obtained from BEMS.

To solve the requirements, we propose a new model to describe thermal dynamics with fewer parameters. MAS is introduced as the modeling technique for thermal dynamics. By introducing the MAS technique to thermal dynamics simulator, we are succeeded in reducing mathematical equations and parameters required for simulation. The simulator works without BIM and visualize spatial and temporal dynamics with limited numbers of parameters, which are easily obtained from BEMS.

The simulator was implemented and applied to a field test. The simulator was installed into an office building (8000 m^3) in Tokyo area and evaluated its accuracies by comparing the results of simulation with the data obtained by the BEMS. As a result, we confirmed that the simulator can reproduce spatial and temporal dynamic thermal environment accurately.

Acknowledgements. We would like to express our special thanks to Mr. Fuyuki Sato in Mitsubishi Electric Building Techno Service. Discussions with him help progress of the study.

References

1. U.S. Energy Information Administration (EIA): Annual Energy Outlook 2020 (2020)
2. International Organization Standardization: ISO50001:2018 Energy management (2018)
3. The Energy Conservation Center, Japan (ECCJ): Energy Conservation for Office Buildings 2010 (2010)
4. Carreira, P., Costa, A., Mansur, V., Arsénio, A.: Can HVAC really learn from users? A simulation-based study on the effectiveness of voting for comfort and energy use optimization. Sustain. Cities Soc. **41**, 275–285 (2018). https://doi.org/10.1016/j.scs.2018.05.043
5. Papantoniou, S., Kolokotsa, D., Kalaizakis, K.: Building optimization and control algorithms implemented in existing BEMS using a web based energy management and control system. Energy Build. **98**, 45–55 (2015)
6. Yang, C., Létourneau, S., Guo, H.: Developing data-driven models to predict BEMS energy consumption for demand response systems. In: Ali, M., Pan, J.S., Chen, S.-M., Horng, M. (eds.) Modern Advances in Applied Intelligence. Lecture Notes in Computer Science (Lecture Notes in Artificial Intelligence), vol. 8481, pp. 188–197. Springer, Cham (2014)
7. Kushiro, N., Fukuda, A., Kawatsu, M., Mega, T.: Predict electric power demand with extended goal graph and heterogeneous mixture modeling. Information **10**(4), 134 (2019)
8. Coakley, D., Raftery, P., Keane, M.: A review of methods to match building energy simulation models to measured data. Renew. Sustain. Energy Rev. **37**, 123–141 (2014)
9. Kumar, A.: A review on modeling and simulation of building energy systems. Renew. Sustain. Energy Rev. **56**, 1272–1292 (2016)
10. Foucquier, A., Robert, S., Suard, F., Stephan, L., Jay, A.: State of the art in building modeling and energy performances prediction: a review. Renew. Sustain. Energy Rev. **23**, 272–288 (2013)
11. Wendt, John F. (ed.): Computational Fluid Dynamics, 3rd edn. Springer, Heidelberg (2009)

12. Yang, S., Pilet, T.J., Ordonez, J.C.: Volume element model for 3D dynamic building thermal modeling and simulation. Energy **148**, 642–661 (2018)
13. Inard, C., Bouia, H., Dalicieux, P.: Prediction of air temperature distribution in buildings with a zonal model. Energy Build. **24**(2), 125–132 (1996)
14. Wang, L., Chen, Q.: Evaluation of some assumptions used in multizone airflow network models. Build. Environ. **43**(10), 1671–1677 (2008)
15. Amasyali, K., El-Gohary, N.M.: A review of data-driven building energy consumption prediction studies. Renew. Sustain. Energy Rev. **81**(1), 1192–1205 (2018)
16. Mustafaraj, G., Lowry, G., Chen, J.: Prediction of room temperature and relative humidity by autoregressive linear and nonlinear neural network models for an open office. Energy Build **43**, 1452–1460 (2011)
17. Berthou, T., Stabat, P., Salvazet, R., Marchio, D.: Development and validation of a gray box model to predict thermal behavior of occupied office buildings. Energy Build. **74**, 91–100 (2014)
18. Khaddaj, M., Srour, I.: Using BIM to retrofit existing buildings. Procedia Eng. **145**, 1526–1533 (2016)
19. Volk, R., Stengel, J., Schultmann, F.: Building information modeling (BIM) for existing buildings – literature review and future needs. Autom. Constr. **38**, 109–127 (2014)
20. Macal, C.M., North, M.J.: Tutorial on agent-based modelling and simulation. J. Simul. **4**, 151–162 (2010)
21. Baldinelli, G., Bianchi, F., Rotili, A., Costarelli, D., Seracini, M., Vinti, G., Asdrubali, F., Evangelisti, L.: A model for the improvement of thermal bridges quantitative assessment by infrared thermography. Appl. Energy **211**, 854–864 (2018)
22. artisoc4.0: Specialized platform for MAS. https://mas.kke.co.jp/en/artisoc4/

Message Ferry Routing Based on Nomadic Lévy Walk in Wireless Sensor Networks

Koichiro Sugihara and Naohiro Hayashibara$^{(\boxtimes)}$

Kyoto Sangyo University, Kyoto, Japan
{i1986089,naohaya}@cc.kyoto-su.ac.jp

Abstract. Message ferry is a way of communication in Intermittently Connected Delay-Tolerant Wireless Sensor Networks. It is for collecting data from stationary sensor nodes. The efficiency of message delivery using message ferries depends on the routing scheme. Nomadic Lévy Walk that is a variant of Lévy Walk is an eligible candidate for a message ferry routing scheme. It includes homing behavior in addition to the behavior similar to Lévy walk with strategic moving the home position. In this paper, we conducted several simulation runs on a WSN to measure the message delivery ratio of message ferries with several routing schemes. We also evaluate that Nomadic Lévy Walk is suitable for message ferry routing. Our simulation result shows Nomadic Lévy Walk is practically useful as a message ferry routing for collecting data in WSNs.

1 Introduction

Various messaging protocols in Intermittently Connected Delay-Tolerant Wireless Sensor Networks (ICDT-WSNs) have been proposed so far [10]. This is because WSNs are attracted attention in various research fields. In this type of network, stationary sensor nodes which have wireless communication capability with limited communication range are located in a field, and messages and sensor data sent by a node are delivered to the destination in a Store-Carry-Forward manner because end-to-end paths do not always exist. It means that each message or data is carried to the destination by some mobile entities which repeat message passing to each other.

Now, we focus on the Message Ferrying based approach for data collection in ICDT-WSNs. Message ferries (or MF for short) are mobile nodes that move around the field to collect sensor data from wireless sensor nodes. Each sensor node sends data to an MF with short-range communication capability (e.g., Bluetooth) when it gets close.

Therefore, the efficiency of data collection depends on the routing scheme of MFs. Most of the message ferry routing schemes assume a fixed route; for instance, MFs move on a circular path [9,21,22]. However, the fixed route based routing scheme does not work if a part of the route is not available.

L. Barolli et al. (Eds.): AINA 2021, LNNS 225, pp. 436–447, 2021.
https://doi.org/10.1007/978-3-030-75100-5_38

We focus on a random walk based routing scheme for message ferry routing. In particular, Lévy walk has recently attracted attention due to the optimal search in animal foraging [7,20] and the statistical similarity of human mobility [11]. It is a mathematical fractal that is characterized by long segments followed by shorter hops in random directions. Lévy walk has also been used to improve the spatial coverage of mobile sensor networks [18], to analyze the properties of an evolving network topology of a mobile network [5], and to enhance grayscale images combined with other bio-inspired algorithms (i.e., bat and firefly algorithms) [6]. It is also considered to be particularly useful for data collection and finding outliers of sensor values to detect incidents in a large-scale delay-tolerant sensor network.

Nomadic Lévy Walk (NLW) is a variant of Lévy walk, which has been proposed by Sugihara and Hayashibara [16]. An agent starts from a sink node and returns to the sink node with the given probability α. Each sink node changes its position according to the given strategy with the probability γ.

We suppose to collect sensor data by using battery-driven autonomous electric vehicles as MFs. Moreover, they have an on-board camera to drive along the road and avoid obstacles. Each of them basically moves with a mobile power source vehicle. Then, it departs from the sink to explore sensor nodes for collecting data. It requires to go back to the sink to charge its battery.

In this paper, we apply NLW as a message ferry routing and measure the cover ratio in a WSN. In this context, agents and sink nodes are considered to be MFs and mobile power source vehicles. MFs are require to collect data from sensors efficiently. So, we measure the cover ratio that indicates the ratio of sensor nodes from which MFs collect data in our simulation. First, we analyze the simulation result with the parameter α and γ of NLW to clarify the impact of these parameters on the cover ratio. Then, we compare two message ferry routings, NLW and Random Way Point (RWP), regarding the cover ratio.

2   Related Work

We now introduce several research works related to message ferry in WSNs and WSNs. Then, we explain existing movement models that can be used for message ferry routing.

2.1   Message Ferry in WSNs and DTNs

Tariq et al. proposed Optimized Way-points (OPWP) ferry routing method to facilitate connectivity on sparse ad-hoc networks [4]. According to the simulation result, OPWP-based MFs outperforms other ferry routing method based on Random way point (RWP).

Shin et al. apply the Lévy walk movement pattern to the routing of message ferries in DTN [12]. They demonstrated message diffusion using message ferries based on the various configuration of Lévy walk. According to the simulation result, the ballistic movement of message ferries (i.e., smaller scaling parameter of Lévy walk) is efficient regarding the message delay.

Basagni et al. proposed the notion of a mobile sink node and its heuristic routing scheme called Greedy Maximum Residual Energy (GMRE) [3]. The motivation of the work is to prolong wireless sensor nodes that are deployed in a large field and send data to the sink periodically. The mobile sink as a data collection and processing point moves around the field to save the energy consumption of sensor nodes. The simulation result showed that the mobile sink with the proposed routing scheme improved the lifetime of sensor nodes.

Alnuaimi et al. proposed a ferry-based approach to collect sensor data in WSNs [1,2]. It divides a field into virtual grids and calculates an optimal path for a mobile ferry to collect data with a minimum round trip time. It utilizes a genetic algorithm and the node ranking clustering algorithm for determining the path.

2.2   Movement Models

Birand et al. proposed the Truncated Levy Walk (TLW) model based on real human traces [5]. The model gives heavy-tailed characteristics of human motion. Authors analyzed the properties of the graph evolution under the TLW mobility model.

Valler et al. analyzed the impact of mobility models including Lévy walk on epidemic spreading in MANET [19]. They adopted the scaling parameter $\lambda = 2.0$ in the Lévy walk mobility model. From the simulation result, they found that the impact of velocity of mobile nodes does not affect the spread of virus infection.

Thejaswini et al. proposed the sampling algorithm for mobile phone sensing based on Lévy walk mobility model [18]. Authors showed that proposed algorithm gives significantly better performance compared to the existing method in terms of energy consumption and spatial coverage.

Fujihara et al. proposed a variant of Lévy flight which is called Homesick Lévy Walk (HLW) [8]. In this mobility model, agents return to the starting point with a homesick probability after arriving at the destination determined by the power-law step length. As their result, the frequency of agent encounter obeys the power-law distribution though random walks and Lévy walk do not obey it.

Most of the works related to Lévy walk assume a continuous plane and hardly any results on graphs are available. Shinki et al. defined the algorithm of Lévy walk on unit disk graphs [14]. They also found that the search capability of Lévy walk emphasizes according to increasing the distance between the target and the initial position of the searcher. It is also efficient if the average degree of a graph is small [13].

Sugihara et al. proposed the novel mobility model called Nomadic Lévy Walk (NLW) as a variant of HLW and the sink relocation strategy based on a hierarchical clustering method [16,17]. They conducted simulations to measure the cover ratio of unit disk graphs. The simulation result showed that the mobility model cover a wide area with preserving homing behavior.

3   System Model

We assume a wireless sensor network so that each stationary sensor node is deployed in a field. So, we model the field by a unit disk graph $UDG = (V, E)$ with a constant radius r. Each node $v \in V$ is located in the Euclidean plane, and an undirected edge (link) $\{v_i, v_j\} \in E$ between two nodes v_i and v_j exists if and only if the distance of these nodes is less than $2r$. Nodes in a graph represent wireless sensor nodes. Note that r is the Euclidean distance in the plane. We assume that any pair of nodes in the graph has a path (i.e., connected graph). It does not mean network connectivity. It just means that there is a geometric route to any sensor node in the field. UDGs are often used for modeling road networks. We assume the field restricts the movement of mobile entities by links.

We also assume a mobile entity called a *message ferry* (MF), which collects messages and data from sensor nodes, and forwards them to the destination. This assumption is similar to the existing work [1,2,12,22]. Practically speaking, both MFs and mobile nodes have short-range communication capability such as Bluetooth, ad hoc mode of IEEE 802.11, Near Field Communication (NFC), infrared transmission, and so on. Each MF starts moving from its home position, called *sink*, and it goes back to the sink position occasionally.

There are two types of ferry interaction.

- *Ferry-to-Ferry interaction*. Message ferries exchange messages each other if they are at the same node (i.e., they are in their communication range).
- *Ferry-to-Node interaction*. Each sensor node can send data piggybacked on a message to a MF when it is physically close to the node (i.e., it is at the node in the graph).

We suppose that MFs are autonomic electric vehicles and sinks are mobile power source vehicles in practice. Thus, MFs are required to come back to their sink to charge their battery.

We also assume that each MF identifies its position (e.g., obtained by GPS), which is accessible from an MF, and it has a compass to get the direction of a walk. Each MF has a set of neighbor nodes, and such information (i.e., positions of neighbors) is also accessible. Moreover, an MF has an on-board camera to drive along the road and avoid obstacles. On the other hand, it has no prior knowledge of the environment.

4   Nomadic Lévy Walk

We proposed a variant of Lévy walk called Nomadic Lévy Walk in our previous work [16,17] to improve the ability of the broad area search while preserving the homing behavior. NLW is an extension of HLW [8], and holds the following properties in addition to HLW.

Nomadicity: Each sink moves its position with the given probability γ.
Sink relocation strategy: The next position of the sink is decided by a particular strategy.

The movement of each MF obeys HLW. Thus, their trajectory is radially from their sink. Moreover, they move their sink at the probability γ in Nomadic Lévy walk. In fact, the fixed sink restricts the area that each MF explores. The property of Nomadicity is expected to improve coverage by each MF.

We now explain the detail of the algorithm of Nomadic Lévy walk on unit disk graphs in Algorithm 4.

1: **Initialize:**
    $s \leftarrow$ the position of the sink
    $c \leftarrow$ the current position
    $o \leftarrow 0$ ▷ orientation for a walk.
    $PN(c) \leftarrow$ the possible neighbors to move.
2: **if** Probability: α **then**
3:     $d \leftarrow$ the distance to s
4:     $o \leftarrow$ the orientation of s
5: **else**
6:     d is determined by the power-law distribution
7:     o is randomly chosen from $[0, 2\pi)$
8: **end if**
9: **if** Probability: γ **then**
10:     $s \leftarrow P(c)$ ▷ update the position of the sink according to the given strategy.
11: **end if**
12: **while** $d > 0$ **do**
13:     $PN(c) \leftarrow \{x | abs(\theta_{ox}) < \delta, x \in N(c)\}$
14:     **if** $PN(c) \neq \emptyset$ **then**
15:         $d \leftarrow d - 1$
16:         move to $v \in PN(c)$ where v has the minimum abs(θ_{ov})
17:         $c \leftarrow v$
18:     **else**
19:         **break**   ▷ no possible node to move.
20:     **end if**
21: **end while**

At the begging of the algorithm, each MF holds the initial position as the sink position s. In every walk, each MF determines the step length d by the power-law distribution and selects the orientation o of a walk randomly from $[0, 2\pi)$. It can obtain a set of neighbors $N(c)$ and a set of possible neighbors $PN(c) \subseteq N(c)$, to which MFs can move, from the current node c. In other words, a node $x \in PN(c)$ has a link with c that the angle θ_{ox} between o and the link is smaller than $\frac{\pi}{2}$.

In unit disk graphs, it is not always true that there exist links to the designated orientation. We introduce δ that is a permissible error against the orientation. In this paper, we set $\delta = 90$. It means that MFs can select links to move in the range ± 90 with the orientation o as a center.

In a given probability α, the MF goes back to the sink node (line 2 in Algorithm 4). In this case, it sets the orientation to the sink node as o and the distance to the sink as d.

Each MF changes its sink position with the given probability γ (line 10 in Algorithm 4) and then starts a journey from the sink.

4.1   Sink Relocation Strategy

Each sink replaces its position according to the given strategy in NLW. Obviously, the sink position is one of the important factors for covering a wide area of a graph. Each sink manages the history H of the sink positions. We have proposed several sink relocation strategies for NLW [16, 17]. We use a strategy for preventing biased positions by using hierarchical clustering [17].

In this strategy, each sink assumes an ordered set of the history of the sink positions B_{hist} with the size h to store the coordinates of the locations at which the sink has been located in the past. It calculates the reverse orientation of each position in B_{hist}.

The sink positions could be biased at a particular area of the graph while relocating. As a result, the cover ratio could not be improved with MFs. For this reason, we use the unweighted pair group method with arithmetic mean (UPGMA) clustering method [15] to detect a bias of sink positions. UPGMA method is an agglomerative hierarchical clustering method based on the average distance of members in two clusters. It calculates a pairwise distance $pdist(A, B)$ between two clusters A and B. $pdist(A, B)$ is computed as follows.

$$pdist(A, B) - \frac{1}{|A| \cdot |R|} \sum_{a \in A} \sum_{b \in B} d(a, b) \tag{1}$$

$d(a, b)$ is the distance between the element a and b. In each step, a cluster A merges another cluster B so that $pdist(A, B)$ is the minimum value. Finally, all elements are in one cluster, and the merging process can be represented as a binary tree.

First, each sink constructs a matrix of $pdist(\{x\}, \{y\})$, where $x, y \in H$. $pdist(\{x\}, \{y\})$ is calculated by the Eq. 1. Note that the distance $d(x, y)$ is Euclidean distance. Then, two clusters $\{x\}$ and $\{y\}$ merge into one if $pdist(\{x\}, \{y\})$ is the minimum value and $pdist(\{x\}, \{y\}) \leq T$. This merging phase is repeated until there is no candidate to merge. As a result, the sink determines the past sink positions are biased if the size of the biggest cluster is more significant than $\lceil \frac{|H|}{2} \rceil$.

In the case of detecting a bias of sink positions, the next sink s will be relocated with the following condition, where $C \subseteq H$ is the biggest cluster.

$$pdist(C, \{s\}) > T$$

5   Performance Evaluation

We measure the average cover ratio, which indicates the ratio of sensor nodes from which MFs have collected data. It means that the efficiency of data collection by MFs. First of all, we show the cover ratio by MFs using Nomadic Lévy

Walk (NLW) as a message ferry routing with the parameter α and γ. Then, we compare the cover ratio with two message ferry routings; NLW and Random way point (RWP).

5.1   Environment

In our simulation, we distribute 1,000 nodes at random in the $1,000 \times 1,000$ Euclidean plane. Each node has a fixed-sized circle r as its communication radius. A link between two nodes exists if and only if the Euclidean distance of these nodes is less than $2r$. Thus, the parameter r is a crucial factor to form a unit disk graph. We automatically generated unit disk graphs such that two nodes have an undirected link if these circles based on r have an intersection.

5.1.1   Radius r of Unit Disk Graphs

We set $r \in \{35, 50\}$ as a parameter of the environments in our simulations. r is a compatible parameter with the diameter of a unit disk graph. Since nodes obtain more links to other nodes that are not connected by gaining r, it tends to become small. The average degrees of nodes $\overline{deg}(UDG)$ are 14.6 with $r = 35$ and 28.9 with $r = 50$, respectively. The diameters of unit disk graphs are 24 with $r = 35$ and 16 with $r = 50$.

5.2   Parameters

The scaling parameter λ and the homesick probability α are common to NLW and HLW. We set $\lambda = 1.2$ for all simulations because it is known as the parameter that realizes the efficient search on unit disk graphs [13].

The number of MFs is set as $k \in \{2, 5, 8\}$. Each MF has a particular sink node (e.g., mobile power source vehicle).

5.3   Performance Criteria

We use the following criteria for measuring the performance of MFs.

Cover ratio: This is a ratio of V, and a vertex covers $V' \subseteq V$ where $\forall v \in V'$ is visited by MFs at least once. Thus, the ratio is computed by $\frac{V'}{V}$. In this paper, we calculate the average cover ratio of ten simulation results for each configuration.

The cover ratio indicates how much MFs cover the field in terms of the ratio of sensor nodes from which MFs have collected data.

5.4   Simulation Result

We show the simulation results of the cover ratio of NLW based MFs on unit disk graphs. Then we compare the message delivery ratio with two routing schemes; NLW and RWP.

5.4.1   On the Parameter α of Nomadic Lévy Walk

Figure 1, 2 and 3 show the cover ratio of k MFs using NLW as a message ferry routing on unit disk graphs (UDGs) with $r = 50$ where $k = 2,\ 5,\ 8$, respectively. The parameter α is a probability with which each MF goes back to the sink. On the simulations, we configured the parameter of α in $\{0.2, 0.5, 0.8\}$ and $\gamma = 0.5$ to observe the impact of the parameter α on the cover ratio.

According to the results, the average cover ratio is inverse proportional to α. The cover ratio of $\alpha = 0.2$ with $k = 2$ improves 23.5% at 2000 simulation steps compared to that of $\alpha = 0.8$. The gap between them is getting smaller by increasing k. Then, it is only 3.8% with $k = 8$.

It means that the cover ratio improves with a lower probability that each MF goes back to the sink. Moreover, the impact of α on the cover ratio is emphasized with a small number of MFs. On the other hand, the impact could be negligible by increasing the number of MFs k.

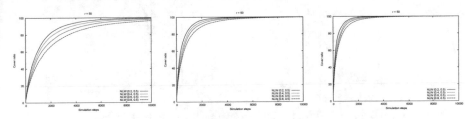

Fig. 1. Cover ratio of NLW with $\lambda = 1.2$, $\alpha \in \{0.2, 0.4, 0.6, 0.8\}$ and $\gamma = 0.5$ on UDG of $r = 50$ with $k = 2$.

Fig. 2. Cover ratio of NLW with $\lambda = 1.2$, $\alpha \in \{0.2, 0.4, 0.6, 0.8\}$ and $\gamma = 0.5$ on UDG of $r = 50$ with $k = 5$.

Fig. 3. Cover ratio of NLW with $\lambda = 1.2$, $\alpha \in \{0.2, 0.4, 0.6, 0.8\}$ and $\gamma = 0.5$ on UDG of $r = 50$ with $k = 8$.

Figure 4, 5 and 6 show the cover ratio of an MF with NLW on unit disk graphs with $r = 35$.

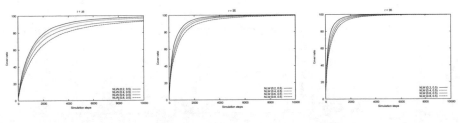

Fig. 4. Cover ratio of NLW with $\lambda = 1.2$, $\alpha \in \{0.2, 0.4, 0.6, 0.8\}$ and $\gamma = 0.5$ on UDG of $r = 35$ with $k = 2$.

Fig. 5. Cover ratio of NLW with $\lambda = 1.2$, $\alpha \in \{0.2, 0.4, 0.6, 0.8\}$ and $\gamma = 0.5$ on UDG of $r = 35$ with $k = 5$.

Fig. 6. Cover ratio of NLW with $\lambda = 1.2$, $\alpha \in \{0.2, 0.4, 0.6, 0.8\}$ and $\gamma = 0.5$ on UDG of $r = 35$ with $k = 8$.

The results have a similar tendency to the ones with $r = 50$. The average cover ratio of $\alpha = 0.2$ with $k = 2$ improves 20.3% at 2000 simulation steps compared to that of $\alpha = 0.8$. The gap between them is also getting smaller by increasing k.

5.4.2 On the Parameter γ of Nomadic Lévy Walk

Figure 7, 8 and 9 show the cover ratio of k MFs using NLW as a message ferry routing on unit disk graphs (UDGs) with $r = 50$ where $k = 2, 5, 8$, respectively. On their simulations, we configured the parameter of γ in $\{0.2, 0.5, 0.8\}$ and $\alpha = 0.2$ to observe the impact of the parameter γ on the cover ratio.

According to the results, The influence of the parameter γ is negligible on UDGs of $r = 50$. The difference between them is at most 3% regarding the cover ratio.

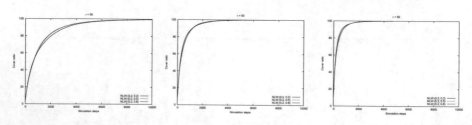

Fig. 7. Cover ratio of NLW with $\lambda = 1.2$, $\alpha = 0.2$ and $\gamma \in \{0.2, 0.5, 0.8\}$ on UDG of $r = 50$ with $k = 2$.

Fig. 8. Cover ratio of NLW with $\lambda = 1.2$, $\alpha = 0.2$ and $\gamma \in \{0.2, 0.5, 0.8\}$ on UDG of $r = 50$ with $k = 5$.

Fig. 9. Cover ratio of NWL with $\lambda = 1.2$, $\alpha = 0.2$ and $\gamma \in \{0.2, 0.5, 0.8\}$ on UDG of $r = 50$ with $k = 8$.

Figure 10, 11 and 12 show the cover ratio of an MF with NLW on unit disk graphs with $r = 35$. On UDGs of $r = 35$, the gap between the configurations of γ widens compared to the one on UDGs of $r = 50$. However, the gap is still at most 6% regarding the cover ratio.

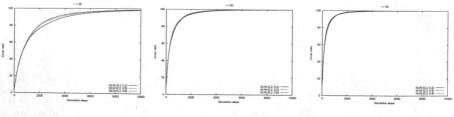

Fig. 10. Cover ratio of NLW with $\lambda = 1.2$, $\alpha = 0.2$ and $\gamma \in \{0.2, 0.8\}$ on UDG of $r = 35$ with $k = 2$.

Fig. 11. Cover ratio of NLW with $\lambda = 1.2$, $\alpha = 0.4$ and $\gamma \in \{0.2, 0.8\}$ on UDG of $r = 35$ with $k = 5$.

Fig. 12. Cover ratio of NLW with $\lambda = 1.2$, $\alpha = 0.6$ and $\gamma \in \{0.2, 0.8\}$ on UDG of $r = 35$ with $k = 8$.

5.4.3   Comparison of Nomadic Lévy Walk and Random Way Point

We compare the two message ferry routing schemes, NLW and RWP, regarding the average cover ratio.

MFs with RWP are exploring a UDG heading to way-points randomly located on the graph. On the other hand, MFs with NLW have to go back to the sink with a given probability α.

We clarify the impact of this homing behavior on the cover ratio through the comparison.

Figure 13, 14 and 15 show the average cover ratio of NLW and RWP on unit disk graphs with $r = 50$ and Fig. 16, 17 and 18 show that with $r = 35$.

On the parameter of UDG r, the difference is negligible regarding the average cover ratio.

RWP improves 12.6% and 15.4% with $k = 2$ at 1000 simulation steps regarding the average cover ratio compared to NLW in $r = 50$ and 35, respectively. However, the gap between them is less than 5% with $k = 5$, and it is almost the same with $k = 8$.

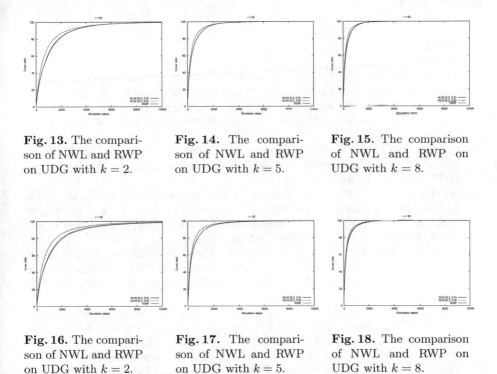

Fig. 13. The comparison of NWL and RWP on UDG with $k = 2$.

Fig. 14. The comparison of NWL and RWP on UDG with $k = 5$.

Fig. 15. The comparison of NWL and RWP on UDG with $k = 8$.

Fig. 16. The comparison of NWL and RWP on UDG with $k = 2$.

Fig. 17. The comparison of NWL and RWP on UDG with $k = 5$.

Fig. 18. The comparison of NWL and RWP on UDG with $k = 8$.

The cover ratio of NLW is worse than that of RWP with a small number of MFs. The gap could be negligible if we deploy a few additional MFs. According to the results, MFs with NLW have to go back to their sink; nevertheless, the cover ratio with NLW could be almost the same as that with RWP by adding a few MFs.

6   Conclusion

In this paper, we evaluated the cover ratio of NLW-based MFs in intermittently connected delay-tolerant wireless sensor networks (ICDT-WSNs). We also compared NLW-based MFs with RWP-based MFs regarding the cover ratio.

NLW-based MFs improve the cover ratio with a lower probability α that each MF goes back to its sink. On the other hand, the impact of γ on the cover ratio is negligible.

On the comparison of different routing schemes, RWP-based MFs is better than NLW-based MFs regarding the cover ratio, especially with a small number of MFs (i.e., $k = 2$). However, the difference between them is negligible by increasing k.

In this paper, we assume that each MF has a limited battery, and it should be charged periodically. MFs using NLW have to go back to their sink; nevertheless, the cover ratio with NLW could be almost the same as that using RWP by adding a few MFs. NLW-based MFs are practically useful in this usage scenario in ICDT-WSNs.

References

1. Alnuaimi, M., Shuaib, K., Abdel-Hafez, K.A.M.: Data gathering in delay tolerant wireless sensor networks using a ferry. Sensors 15(10), 25809–25830 (2015)
2. Alnuaimi, M., Shuaib, K., Abdel-Hafez, K.A.M.: Ferry-based data gathering in wireless sensor networks with path selection. Procedia Comput. Sci. 52, 286–293 (2015)
3. Basagni, S., Carosi, A., Melachrinoudis, E., Petrioli, C., Wang, Z.M.: Controlled sink mobility for prolonging wireless sensor networks lifetime. Wireless Netw. 14(6), 831–858 (2008)
4. Bin Tariq, M.M., Ammar, M., Zegura, E.: Message ferry route design for sparse ad hoc networks with mobile nodes. In: Proceedings of the 7th ACM International Symposium on Mobile Ad Hoc Networking and Computing, MobiHoc 2006, pp. 37–48. Association for Computing Machinery (2006)
5. Birand, B., Zafer, M., Zussman, G., Lee, K.W.: Dynamic graph properties of mobile networks under levy walk mobility. In: Proceedings of the 2011 IEEE Eighth International Conference on Mobile Ad-Hoc and Sensor Systems, MASS 2011, pp. 292–301. IEEE Computer Society (2011)
6. Dhal, K.G., Quraishi, M.I., Das, S.: A chaotic lévy flight approach in bat and firefly algorithm for gray level image. Int. J. Image Graph. Sign. Process. 7, 69–76 (2015)
7. Edwards, A.M., Phillips, R.A., Watkins, N.W., Freeman, M.P., Murphy, E.J., Afanasyev, V., Buldyrev, S.V., da Luz, M.G.E., Raposo, E.P., Stanley, H.E., Viswanathan, G.M.: Revisiting lévy flight search patterns of wandering albatrosses, bumblebees and deer. Nature 449, 1044–1048 (2007)
8. Fujihara, A., Miwa, H.: Homesick lévy walk and optimal forwarding criterion of utility-based routing under sequential encounters. In: Proceedings of the Internet of Things and Inter-cooperative Computational Technologies for Collective Intelligence 2013, pp. 207–231 (2013)

9. Kavitha, V., Altman, E.: Analysis and design of message ferry routes in sensor networks using polling models. In: 8th International Symposium on Modeling and Optimization in Mobile, Ad Hoc, and Wireless Networks, pp. 247–255 (2010)
10. Li, Y., Bartos, R..: A survey of protocols for intermittently connected delay-tolerant wireless sensor networks. J. Netw. Comput. Appl. **41**, 411–423 (2014)
11. Rhee, I., Shin, M., Hong, S., Lee, K., Kim, S.J., Chong, S.: On the levy-walk nature of human mobility. IEEE/ACM Trans. Netw. **19**(3), 630–643 (2011)
12. Shin, M., Hong, S., Rhee, I.: DTN routing strategies using optimal search patterns. In: Proceedings of the Third ACM Workshop on Challenged Networks, CHANTS 2008, pp. 27–32. ACM (2008)
13. Shinki, K., Hayashibara, N.: Resource exploration using lévy walk on unit disk graphs. In: The 32nd IEEE International Conference on Advanced Information Networking and Applications (AINA-2018) (2018)
14. Shinki, K., Nishida, M., Hayashibara, N.: Message dissemination using lévy flight on unit disk graphs. In: The 31st IEEE International Conference on Advanced Information Networking and Applications (AINA 2017) (2017)
15. Sokal, R.R., Michener, C.D.: A statistical method for evaluating systematic relationships. Univ. Kansas Sci. Bull. **38**, 1409–1438 (1958)
16. Sugihara, K., Hayashibara, N.: Message dissemination using nomadic lévy walk on unit disk graphs. In: Barolli, L., Hussain, F.K., Ikeda, M. (eds.) Proceedings of CISIS 2020, pp. 136–147. Springer (2020)
17. Sugihara, K., Hayashibara, N.: Performance evaluation of nomadic lévy walk on unit disk graphs using hierarchical clustering. In: Proceedings of the 34th Int'l Conferences on Advanced Information Networking and Applications (AINA-2020), pp. 512–522. Springer (2020)
18. Thejaswini, M., Rajalakshmi, P., Desai, U.B.: Novel sampling algorithm for human mobility-based mobile phone sensing. IEEE Internet Things J. **2**(3), 210–220 (2015)
19. Valler, N.C., Prakash, B.A., Tong, H., Faloutsos, M., Faloutsos, C.: Epidemic spread in mobile ad hoc networks: determining the tipping point. In: Proceedings of the 10th International IFIP TC 6 Conference on Networking - Volume Part I, NETWORKING 2011, pp. 266–280. Springer-Verlag (2011)
20. Viswanathan, G.M., Afanasyev, V., Buldyrev, S.V., Murphy, E.J., Prince, P.A., Stanley, H.E.: Lévy flight search patterns of wandering albatrosses. Nature **381**, 413–415 (1996)
21. Zhao, W., Ammar, M., Zegura, E.: A message ferrying approach for data delivery in sparse mobile ad hoc networks. In: Proceedings of the 5th ACM International Symposium on Mobile Ad Hoc Networking and Computing, pp. 187–198. Association for Computing Machinery (2004)
22. Zhao, W., Ammar, M., Zegura, E.: Controlling the mobility of multiple data transport ferries in a delay-tolerant network. In: IEEE INFOCOM 2005. Maiami, FL, USA (2005)

Image Similarity Based Data Reduction Technique in Wireless Video Sensor Networks for Smart Agriculture

Christian Salim[(✉)] and Nathalie Mitton

Inria, Villeneuve d'Ascq, France
{christian.salim,nathalie.mitton}@inria.fr

Abstract. Nowadays, to improve animal well being in livestock farming or beekeeping application, a wireless video sensor network (WVSN) can be deployed to early detect injury or Asiatic hornets attacks. WVSN represents a low-cost monitoring solution compared to other technologies such as the closed circuit television technology (CCTV). WVSNs are composed of low-power resource-constrained video sensor nodes (motes). These nodes capture frames from videos at a given frequency (frame rate) and wirelessly send them to the sink. The big amount of data transferred from the nodes to the sink consumes a lot of energy on the sensor node, which represents a major challenge for energy-limited nodes. In this paper, we introduce two complementary mechanisms to reduce the overall number of frames sent to the sink. First, the Transmission Data Reduction algorithm (TDR) run on the sensor node leverages the similarity degree of consecutive images. Second, the Inter-Nodes Similarity algorithm (INS) exploits the spatio-temporal correlation between neighbouring nodes in order reduce the number of captured frames. The results show a 95% data reduction, surpassing other techniques in the literature by 30% at least.

Keywords: Wireless video sensor networks · Data reduction · Geometric correlation · Inter-nodes similarity

1 Introduction

In recent years, the need for smart agriculture has increased [14]. Surveillance systems are deployed to support it, especially in remote sites in order for instance to monitor the animals behaviour, understand and protect the livestock and vegetables farming. To achieve it, a wireless video sensor network (WVSN) can serve as an energy-efficient and low cost solution. In WVSN, the system normally operates periodically, sending all the captured frames to the sink [12]. However, the system can be an event driven system sending only the frames that show a change in the area of interest. The wireless video sensor nodes are energy-limited nodes that monitor an area of interest according to their Field of Views (FOVs) and send the videos/frames to the sink. Shooting a video and capturing

L. Barolli et al. (Eds.): AINA 2021, LNNS 225, pp. 448–459, 2021.
https://doi.org/10.1007/978-3-030-75100-5_39

its frames consume a lot of energy but sending this great number of frames to the sink is even worse. In order to cover all the area of interest and prevent black holes, in general, WVSN are such that several FoVs overlap with each other. To reduce the energy consumption at the sensor node level, we aim to reduce the amount of data sent to the sink leveraging the overlap between FoVs and by applying a node embedded lightweight algorithm in order to cope with the nodes limited processing and energy resources.

In this paper, we propose a two-level data reduction approach. The first one, Transmission Data Reduction algorithm (TDR), runs on each node and selects only the critical frames to be sent to the sink. The second one, the Inter-Nodes Similarity algorithm (INS), leverages the spatio-temporal correlation between neighbour nodes to reduce the number of sent and captured frames. The TDR algorithm compares consecutive captured frames on the node based on a norm simple euclidean distance similarity method [11]. According to the degree of similarity with the last frame sent to the sink, the node decides whether to send this frame or not [12]. If the decision is to send the image, the node creates a new image img_{diff} that only includes the difference between the two compared images. img_{diff} is sent to the sink instead of the whole image. In this case, data reduction is attained when several consecutive frames represent a high similarity percentage (stable situation). To furthermore increase the data reduction, the INS algorithm computes the correlated degree of neighbouring nodes and compares their frames using the same norm simple euclidean distance metric. If the result detects a similarity for n consecutive frames, the node with the highest energy level sends the frames to the sink, the other one goes to idle mode and its frame rate is set to the minimum $FR = 1$. We compare this new method to another algorithm from the literature based on the colour and edge similarities and on the number of consecutive frames sent per period [10]. The results show that our approach reduces the amount of energy by 95% outperforming the other approach, while capturing all the events.

The remainder of this paper is organised as follows. Section 2, browses the related work. Section 3 describes the transmission data reduction algorithm while Sect. 4 details the INS algorithm The experimental results and the comparison with other methods are given in Sect. 5. Finally, Sect. 6 concludes and provides perspectives and future work.

2   Related Work

In this section, we discuss some previous works regarding data reduction in WSN and in WVSN in particular. A lot of techniques based on scheduling [3,6], intrusion location and clustering [4,13], compressive sensing [16], data similarity [10,11], and frame rate adaptation [7,15] exist for data reduction and for energy-efficiency. The authors in [13] used the kinematics functions to predict the next location of the intrusion in the area of interest in order to increase the frame rate adaptation of the targeted nodes. This approach reduced the energy dissipation of the overall network, since only the targeted nodes are capturing and not all the nodes in the zone of interest. This approach can come as a complementary solution to our method which focuses on the nodes detecting the intrusion and not on the projected positions of the intrusion. In [11] and [15], the authors worked on data reduction by adapting the frame rate of the nodes. This type of approaches is helpful and could be studied later on as a complementary work of the approach presented in this paper. The geometric correlation between several sensor nodes has been exploited by several authors such as [1,8,9]. In [1], the authors transform the coordinates of every point from 3D to a 2D coordinate system according to some reference points. Computing the disparity and the lengths of the reference vectors can determine whether for example two cameras are geometrically correlated according to a predefined threshold. In [10], the authors proposed a similarity function to detect the percentage of similarity between consecutive frames to attain data reduction by sending the critical frames to the sink. They merge it with a correlation overlapping technique at the coordinator level. The proposed colour-edge similarity function is time and energy consuming especially in the colour part where every channel is taken into consideration. They compared the frames on the coordinator level to further reduce the amount of frames. In this paper, the proposed method reduces the number of frames sent from each node to the coordinator by using the L2-norm distance to measure the similarity. It is a way less heavy to be implemented at the sensor node unlike other techniques that use the colour-edge and pixel similarities as in [7,10,15]. As explained in Sect. 3 it is a less complex technique regarding the execution time comparison with other approaches for the same purpose of data reduction. Unlike other approaches that use a 3D to 2D transformation to implement a similarity algorithm as in [1], we complement the approach by proposing a 2-D spatial correlation between neighbour nodes based on some geometrical conditions using simple equations. If 2 nodes are said "correlated", the Inter-Nodes Similarity algorithm (INS) runs on both nodes, the one with the most residual energy sends the data, the other goes into idle transmission mode and lowers its frame rate to its minimum $FR = 1$.

3 Data Reduction

In WSN, sensor-nodes operate periodically, and send a huge amount of data to the coordinator simultaneously. This scenario consumes a lot of energy on each sensor node representing the most important challenge in WSN in general. In WVSN, packets sent are multimedia, which emphasises even more the challenge of energy consumption. To overcome this challenge and to lower its importance on the feasibility of the network deployment, we proposed a data reduction method at the sensor node to reduce the number of raw images sent from each node to the sink. This technique combines two mechanisms presented by the Transmission Data Reduction algorithm (TDR) and the Inter-Nodes Similarity algorithm (INS).

3.1   Transmission Data Reduction

The Transmission Data Reduction algorithm (TDR) is described in Algorithm 1. TDR algorithm leverages the consecutive frames similarity as in [11], using the norm simple euclidean distance similarity method to compare images. If two consecutive sensed frames are estimated as similar (no new information is represented in the second frame), the node does not send the second frame to the coordinator. The node counts the number of consecutive similar frames, if it surpasses nb (the required number of consecutive similar frames), the state of the area of interest monitored by the node is considered as stable (situation = 1). This technique uses the norm L2 relative error function tested in C++ in OpenCV and compared to other techniques in [11] where the time to execute this technique is way less important than other techniques (edge, colour, ...) for the same purpose of data reduction. To apply this function, each image is transformed into a matrix. If the similarity $norm_{sim} = dist(fr_i, fr_{last})$ between those images does not surpass a certain threshold of similarity ($norm_{sim} < th_{sim}$), it means the images are not similar. In this case, the sensor-node creates and sends to the sink a new image called img_{diff} by adopting the absolute images difference function as shown in the equation below:

$$img_{diff} = absdiff(MATA, MATB) \tag{1}$$

This function computes the absolute value of the difference for each pixel, for each channel of the two matrices. The img_{diff} is an image that represents the difference between both images using less bits. In this case, the size in Bytes of the transmitted packets decreases because only the difference between images is sent rather than the full new image.

An example of the img_{diff} is shown in Fig. 2c representing the difference image img_{diff} of the frames presented in Figs. 2a and 2b taken from the dataset in [5][1]. As we can see, the shadow of the moving person and the person are the only differences taken into account. The last sent image to the coordinator is stored on the sensor node. Every new frame is compared to this last frame sent to the sink, it can replace this last frame if it represents a critical event or a change in the monitored zone as shown in Algorithm 1. In the remainder of this paper, the sent image img_{diff} is called critical frame. We called it critical, because the node sends only an image to the coordinator if an event or a change is happening in the area of interest. This whole process is explained in Fig. 1. Note that, the reconstruction algorithm at the sink level is out of the scope of this paper.

---

[1] http://changedetection.net.

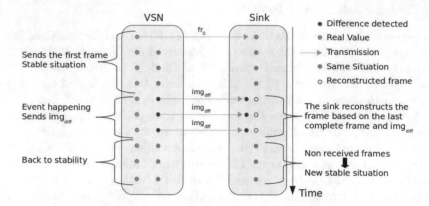

Fig. 1. Video sensor node and sink behaviours

(a) The First Frame   (b) The Second Frame   (c) The Difference Frame

Fig. 2. The difference scenario

Algorithm 1. Transmission Data Reduction Algorithm TDR on Sensor Node S_u

---
    Set fr_{last} as the last sent frame to the sink
2: Set $Situation_u = 0$
    Set $nb_{sim} = 0$ the number of consecutive similar frames
4: **while** $Energy > 0$ **do**
      for each frame i **do**
6:      **if** $norm_{sim} < th_{sim}$ **then**
          $nb_{sim} = 0$; $fr_{last} = fr_i$
8:         Sends img_{diff} to the sink
      else
10:        $nb_{sim} = nb_{sim} + 1$
          if $nb_{sim} > nb$ **then**
12:         $Situation_u = 1$ # Stable Situation
          end if
14:      **end if**
      end for
16: **end while**

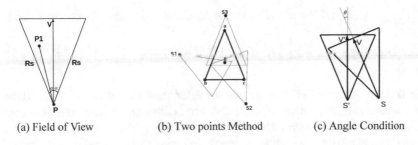

(a) Field of View   (b) Two points Method   (c) Angle Condition

Fig. 3. Overlapping model

4 Inter-Nodes Correlation: The Overlapping Method

In the above sections, we were interested in reducing the amount of sent frames from each node to the sink by applying a similarity function at the sensor node itself. To furthermore reduce the data transmission, we now focus on the spatial correlation between neighbouring sensor nodes. In our approach, two nodes are considered as neighbours if they are in the communication range of each other. Neighbour nodes exchange their vector of metrics defined in the video sensing model below to check whether their FoVs overlap enough to apply the learning process and the similarity function. We apply the same norm distance similarity function on frames captured by two overlapping neighbour sensor nodes. This technique reduces the number of the overall captured and sent frames from the nodes to the sink, by only sending one out of the two frames, depending on the residual energy of the two nodes as shown in Algorithm 2. Before describing the Inter-Nodes Correlation (INS) approach, we introduce the video sensing model, the characteristics of every video sensor node and the geometric condition to determine what nodes overlap and can thus proceed with the INS algorithm.

4.1 Video Sensing Model

We consider a 2-D model of a video sensor node where $z = 0$ (XOY plane) and all the captured frames are compared as 2-D images not taking account of the third dimension. A video sensor node S is represented by the Field of View (FoV) of its camera. Unlike a wireless sensor which range is omnidirectional, a FoV covers only a part of the surrounding area of a video sensor as shown in Fig. 3a. A FoV is a vector of 4-tuple $S(P, R_s, \overrightarrow{V}, \alpha)$ where P is the position of S, R_s is its sensing range, $\overrightarrow{V}$ is the vector representing the line of sight of the camera's FoV e.g. its sensing direction, and α is the offset angle of the FoV.

In [10] the authors presented the FoV with 4 points a,b,c and the gravity centre g as shown in Fig. 3b. The overlapping area is detected based on those points as explained in the upcoming subsections. A point P_1 is said to lay in the FoV of a video sensor node S at position P if and only if the two following conditions are satisfied:

1. $d(P, P_1) \leq R_s$, where $d(P, P_1)$ is the Euclidean distance between P and P_1 :
   $\|\overrightarrow{PP_1}\| \leq R_s$
2. The angle between $\overrightarrow{PP_1}$ and $\overrightarrow{V}$ is within $[-\alpha, +\alpha]$: $\overrightarrow{PP_1}.\overrightarrow{V} \geq \|\overrightarrow{PP_1}\| \times \|\overrightarrow{V}\| \times cos(\alpha)$.

In the remainder of this paper, we consider that all video nodes are identical, with the same sensing range R_s and the same offset angle α. In the next section, we discuss the spatial correlation of overlapping neighbour nodes based on some geometrical conditions, and how we use it in order to increase the data reduction at the sensor node level.

4.2   Overlapping Sensor Nodes

In this part, we describe the geometric condition which determines if two neighbours are overlapping nodes. This condition is a combination of the angle condition between the FoV's of the nodes and the ratio of the overlapping area between them.

4.2.1   The Angle Condition

The angle between two neighbouring sensor nodes is defined as the angle between the vectors of their FoVs. If the angle between the FoVs of two neighbouring nodes is wide, these two nodes can not take part in the similarity comparison function to decide which one among them sends the data to the sink. Indeed, that means that they are not sensing the same area of interest. A frame from two different perspectives can be widely different. To be able to define two neighbour sensor nodes as overlapping sensor nodes, the angle between their FoVs must not surpass a certain angle threshold. In order to determine the angle between vectors ($\mathbf{V}$ and $\mathbf{V'}$) of the sensor nodes S and S' respectively as shown in Fig. 3c, we use the scalar product between those sensor nodes. Since all sensor nodes are identical, both vectors $\mathbf{V}$ and $\mathbf{V'}$ have the same length l. The scalar product can be calculated in two formats. The first one according to their coordinates (x and y) where $\mathbf{V} = (X_V, Y_V)$ and $\mathbf{V'} = (X_{V'}, Y_{V'})$:

$$\mathbf{V}.\mathbf{V'} = X_V \times X_{V'} + Y_V \times Y_{V'} \tag{2}$$

The second format is given according to the length of each vector and to the angle between both, as follows:

$$\mathbf{V}.\mathbf{V'} = l^2 \times \cos(\mathbf{V}, \mathbf{V'}) \tag{3}$$

Where $l = \|\overrightarrow{V'}\| = \|\overrightarrow{V}\|$.
The below equation defines and computes the angle θ between the two vectors according to both formats of the scalar product:

$$\theta = \arccos((X_V \times X_{V'} + Y_V \times Y_{V'})/l^2) \tag{4}$$

If an angle threshold th_{angle} is defined as $30°$, the angle between $\mathbf{V}$ and $\mathbf{V}'$ must remain less than $30°$ so the two sensor nodes S and S' can proceed to the next step (the two points strategy), to be able at the end to take part in the similarity function process.

4.2.2   The Two Points Condition

Inspired from [10] we present below the two points condition as the second part of the geometric condition. A node S' satisfies the two points condition with another node S if g (the gravity centre of abc) belongs to the FOV of S alongside any other point between a, b and c from S's FOV as shown in Fig. 3b. In this figure, S_1, S_2 and S_3 satisfy this condition separately with S. In this scenario, each sensor node can be a candidate alongside S to apply the learning process and the similarity function between them on the one having the most residual energy. However, only S_1 and S_2 satisfy at the same time the angle and the two points conditions. For this purpose, S_1 and S_2 will be the nodes selected as two overlapping sensor nodes to test the inter nodes similarity algorithm INS as explained in the next section.

4.3   Inter-Nodes Similarity

As mentioned earlier, the comparison between the frames captured by overlapping sensor nodes is done at the start of each stable situation in the zone of interest. We used the same norm simple euclidean distance function to measure the similarity If the similarity computed ratio IN_{sim} is greater than a predefined threshold IN_{th} for the first frame as shown in Algorithm 2, it means that the two correlated nodes present the same information when shooting the zone of interest. In this case, only one out of the two sensor nodes is selected to send the data, depending on the residual energy RN of both nodes. If S_u has more energy left than S_v ($RN_{S_u} > RN_{S_v}$), S_u continues to send the data while S_v goes to transmission idle mode and its frame rate is reduced to $FR = 1$ frame per second on S_v. Once an event occurs in the FoV of S_v the frame rate is back to its maximum of 30 frames per second. Figure 4 shows the behaviour of two neighbour nodes while applying this overlapping technique and INS algorithm.

Algorithm 2. INS run on S_u

    Set $n_{sim} = 0$ the number of similar frames
    Get $situation_v$ and fr_{v_0} from S_v
3: **if** $situation_u = 1$ AND $situation_v = 1$ AND $IN_{sim} > IN_{th}$ **then**
    if $RN_{S_u} > RN_{S_v}$ **then**
        S_u continues sending the frames running TDR
6:   **else**
        $FR_u = 1$ (and stop sending)
    end if
9: **end if**

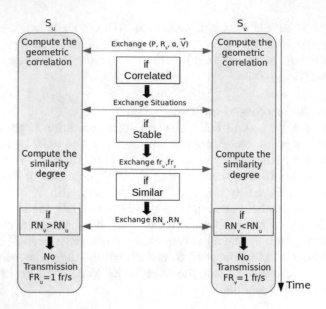

Fig. 4. Two overlapping sensor nodes behaviour running INS

5   Experimental Results

Several simulations have been conducted to validate our approach at the sensor node, aiming to minimise the number of frames sent from the node to the sink. We compare our approach to MASRA and SSA algorithms in [10] and PPSS algorithm in [2]. We have used a Matlab simulator. A scenario is introduced in Fig. 3c where 2 video-sensor nodes S1, S2 are deployed as overlapped neighbour nodes to monitor the same area of interest from different perspectives as shown in Fig. 3c. The angle threshold is set to 15°, however in this case $\alpha = 10$ and S_1 covers the gravity centre and one other point of the FoV of S_2, which makes them two overlapping neighbour nodes in order to apply the INS algorithm on both. The main purpose in our work is to send to the sink the frames that represent the critical situations while maintaining the information by applying TDR and INS algorithms. We have used 2 Microsoft LifeCam VX-800 cameras to shoot a short video of 120 s (3600 frames captured per node), each camera is connected to a laptop to do the processing via a Matlab simulator. Each camera captures 30 frames/s. In our scenario, an intrusion is detected at the following time-intervals:

S1: 40 s from 15 to 55 ; S2: 40 s from 20 to 60.

In the rest of the time, both sensor nodes had a stable situation. After conducting several simulations, the number of high similarities needed to detect a stable situation is set to $n = 60$ which means 2 s of stability. The similarity threshold th_{sim} is set to 85% in TDR algorithm and 75% in INS algorithm. Note that a full compressed image size is equal to $30KB$ and the difference image img_{diff}

Table 1. TDR algorithm over 120 s

Node	Situation	Captured frames	Sent frames
S_1	Stable	2300	2
S_1	Event	1200	240
S_2	Stable	2300	2
S_2	Event	1200	260
ALL	All	7200	504

Table 2. Overall reduction

Total frames	Type frames	Size frames
7200	Captured all	216000 KB
504	Critical all	15120 KB
504	img_{diff}	10080 KB

size is $20\,KB$. The thresholds and all the parameters can be adapted according to the application and the QoS required.

5.1   Transmission Data Reduction

Table 1 shows the number of frames sent by each node to the sink over 120 s using the TDR algorithm. This algorithm allows the reduction of the number and size of frames transmitted by the sensor-node as shown in Tables 1 and 2.

A transmission data reduction of more than 90% is reached by sending only the critical frames. However the size reduction reached 95% due to the image difference img_{diff} applied on every selected/critical frame. Once in a stable situation, the first frame is sent and due to similarity all the other frames are not transmitted to the sink.

5.2   Inter-Nodes Similarity

The INS algorithm runs on both nodes at the same time with TDR algorithm. In this section we consider that S_1 has more residual energy than S_2, which puts S_2 into the idle mode and reduces its frame rate to $FR_2 = 1$ frame per second. In this case the data reduction is emphasised also on the sensing process for S_2 as shown in Table 3. In this case, the number of frames sent to the sink remains the same as in TDR alone, but the number of captured frames reduces by more than 50% on S_2 and 33% overall.

Table 3. TDR and INS-120 s

Node	Situation	Captured frames	Sent frames
S_1	Stable	2300	2
S_1	Event	1200	240
S_2	Stable	133	2
S_2	Event	1200	260
ALL	All	4833	504

Table 4. S1 and S2 data reduction

Technique	Captured	Sent	Size in KB
All data	7200	7200	216000
MASRA+ SSA [10]	2540	1310	39300
PPSS [2]	3600	3600	108000
TDR + INS	4833	504	10080

5.3   Comparison

We compare to the scheduling algorithm $PPSS$ [2] and to the $MASRA$ algorithm [10] where the colour and edge similarities were both present to detect the difference between images. The difference between those approaches in terms of data reduction is shown in Table 4. As we can notice from Table 4, TDR and INS algorithms outperform MASRA algorithm [10] and PPSS algorithm [2] in terms of transmission data reduction by 70% and 90% respectively and in terms of the size of the data transmitted. This technique reduced the transmitted data by 95% and the sensed data by 33% of the overall data captured and transmitted. However, the sensing data reduction is better on MASRA algorithm by 45% which implements the frame rate adaptation of each node according to the scene but computes the geometric correlation on the sink not on the node, in this purpose they transmit 3 times more than our current approach. The overall gain remains positive by more than 30%.

6   Conclusion

In this paper, we introduced a distributed data reduction approach for wireless video sensor network. This approach is composed of two complementary algorithms, the transmission reduction algorithm (TDR) and the inter-nodes similarity (INS) algorithm. TDR detects the similarities between consecutive captured frames and decides to send only the difference image of each critical frame to the sink. The results show that the proposed algorithms did not miss any event in the recorded video sequence. The second part of the approach consists in a geometrical condition computed by neighbour nodes to detect whether they overlap and reduce accordingly their frame rate or transmission rate. Comparing our approach with MASRA and PPSS algorithms in terms of data reduction shows that our algorithm outperforms both algorithms by 30% and 90% respectively. As future works, first of all we aim to compare the relevance of our mathematical model by comparing it to other approaches. In order to increase the data reduction, the frame rate adaptation technique at the sensor node level can be taken into account.

Acknowledgements. This work was partially supported by a grant from CPER DATA, by LIRIMA Agrinet project and the Inria/CEFIPRA Associate team DC4SCM.

References

1. Dai, R., Akyildiz, I.F.: A spatial correlation model for visual information in wireless multimedia sensor networks. IEEE Trans. Multimedia **11**(6), 1148–1159 (2009)
2. Elson, J., Girod, L., Estrin, D.: Fine-grained network time synchronization using reference broadcasts. SIGOPS Oper. Syst. Rev. **36**, 147–163 (2002)
3. Feng, J., Zhao, H.: Energy-balanced multisensory scheduling for target tracking in wireless sensor networks. Sensors **18**(10), 3585 (2018)

4. Ghosal, A., Halder, S.: A survey on energy efficient intrusion detection in wireless sensor networks. J. Ambient Intell. Smart Environ. **9**, 239–261 (2017)
5. Goyette, N., Jodoin, P.M., Porikli, F., Konrad, J., Ishwar, P.: Changedetection.net: a new change detection benchmark dataset. In: IEEE Workshop on Change Detection (CDW-2012) (2012)
6. Jiang, B., Ravindran, B., Cho, H.: Probability-based prediction and sleep scheduling for energy-efficient target tracking in sensor networks. IEEE Trans. Mob. Comput. **12**(4), 735–747 (2013)
7. Monika, R., Hemalatha, R., Radha, S.: Energy efficient surveillance system using WVSN with reweighted sampling in modified fast haar wavelet transform domain. Multimed. Tools Appl. **77**(23), 30187–30203 (2018)
8. Mowafi, M.Y., Awad, F.H., Aljoby, W.A.: A novel approach for extracting spatial correlation of visual information in heterogeneous wireless multimedia sensor networks. Comput. Netw. **71**, 31–47 (2014)
9. Nu, T.T., Fujihashi, T., Watanabe, T.: Power-efficient video uploading for crowdsourced multi-view video streaming. In: IEEE Global Communications Conference (GLOBECOM) (2018)
10. Salim, C., Makhoul, A., Couturier, R.: Similarity based image selection with frame rate adaptation and local event detection in wireless video sensor networks. Multimed. Tools Appl. **78**, 5941–5967 (2019)
11. Salim, C., Makhoul, A., Couturier, R.: Energy-efficient secured data reduction technique using image difference function in wireless video sensor networks. Multimed. Tools Appl. **79**, 1801–1819 (2020)
12. Salim, C., Makhoul, A., Darazi, R., Couturier, R.: Combining frame rate adaptation and similarity detection for video sensor nodes in wireless multimedia sensor networks. In: IWCMC (2016)
13. Salim, C., Makhoul, A., Darazi, R., Couturier, R.: Kinematics based approach for data reduction in wireless video sensor networks. In: International Conference on Wireless and Mobile Computing, Networking and Communications (WiMob) (2018)
14. Salim, C., Mitton, N.: Machine learning based data reduction in WSN for smart agriculture. In: International Conference on Advanced Information Networking and Applications, pp. 127–138. Springer (2020)
15. Tayeh, G.B., Makhoul, A., Demerjian, J., Laiymani, D.: A new autonomous data transmission reduction method for wireless sensors networks. In: IEEE Middle East and North Africa Communications Conference (MENACOMM) (2018)
16. Veeraputhiran, A., Sankararajan, R.: Feature based fall detection system for elders using compressed sensing in WVSN. Wireless Netw. **25**, 287–301 (2019)

Using Clusters for Approximate Continuous k-Nearest Neighbour Queries

Wendy Osborn$^{(\boxtimes)}$ (ID)

Department of Mathematics and Computer Science, University of Lethbridge,
Lethbridge, AB T1K 3M4, Canada
wendy.osborn@uleth.ca

Abstract. This paper proposes several strategies for continuous k-nearest neighbour query processing in location-based services. All utilize a clustered dataset to create safe regions. The only previously proposed strategy to do so has limited validation of the safe region on the client, which leads to only approximate k-nearest neighbour results being obtained. This work improves upon both the creation of the safe region and the validation of the safe region on the client. An evaluation of our strategy, including a comparison versus the existing strategy, show significant improvements in accuracy, and up to over 90% accuracy. When compared to repeated k-nearest neighbour search, our strategy is computationally significantly faster for larger datasets. Obtaining significantly high accuracy at low computational costs is significant for our strategies.

Keywords: Continuous query processing · k-nearest neighbour · Approximate · Clustering · Location based services

1 Introduction

Location Based Services (LBS) [16] obtain and display information for a user on their mobile device (e.g., smartphone) from a remote server, given the type of information being sought and their current location. An example of such a query is the locations of the nearest k restaurants, given the user's current location. If the user stays in the same location, a one-time result containing the nearest k restaurants can be obtained from the server and transmitted to them. However, if the moving by foot or by car, the initial result will become invalid. Therefore, the location of the nearest k restaurants must be repeatedly updated for the user who is on the move. A continuous k-nearest neighbour query [12] takes a sequence of a user's locations on their trajectory and transmits them to a remove server to retrieve and/or update the *actual* k nearest points of interest (POIs) for to the user. However, an approximate continuous k-nearest neighbour query considers the trade-off between accuracy and efficiency by retrieving k POIs that are not guaranteed to be the closest to the user, but are considered "close enough" to meet their needs. Referring back to the example, the exact k nearest restaurants may be sought by the user as they are moving. However, the user may be satisfied with restaurants which are not be the actual k closest to them but still within walking or driving distance to them.

Strategies for handling continuous k-nearest neighbour queries have been proposed previously [1, 3–14]. These can be classified as one of the following:

- Repeated application of a standalone k-nearest neighbour strategy when the location of the user changes.
- Storing a subset of data on the mobile device, so that calls to the server can be reduced or minimized.
- Storing a safe region (and corresponding set of POIs that it contains) on the mobile device, so that calls to the server can be reduced or minimized. In a safe region, the set of points remains a valid result (e.g., a set of k-nearest neighbours) while the query remains inside of it.
- A hybrid approach that combines safe region and caching.

In addition, previously proposed continuous k-nearest neighbour strategies have one or more of the following limitations:

- Searching data or data structures (e.g., spatial access methods) repeatedly,
- Caching too much data on the user's mobile device,
- Complex strategies required to generate new safe regions when needed,
- Upfront knowledge of the user's travel trajectory, and
- Lack of use of a clustered dataset, which can help improve performance.

Specifically, most existing strategies do not consider the last limitation, which is the use of clustering in order to improve the performance of continuous query processing. A clustered dataset can assist in improving the performance of creating a safe region that resides on the user's mobile device. The only previously proposed strategy that considers the use of a clustered dataset for processing continuous k-nearest neighbour queries [14] is an approximate strategy. It only selects the clusters that overlap the query point, and only considers whether the query point is still inside the selected clusters. This alone does not guarantee an accurate k-nearest neighbour result, as the distance from the query point to the kth nearest neighbour may be greater than the distance to one or more sides of the safe region [14]. Therefore, the accuracy was found to be approximately 60% at worst, with at worst 80% of queries producing inaccurate results [14].

Therefore, in this paper several new strategies for continuous k-nearest neighbour query processing are proposed that improve upon the previously proposed strategy in [14]. Two aspects are the focus of this paper: the creation of the safe region using the clustered point set on the server, and additional validity testing of a safe region on the mobile device (i.e., client). An experimental evaluation show some significant improvements in accuracy over the strategy in [14] - up to almost 100% in some cases. In addition, our strategies are computationally, significantly faster for larger datasets over repeated searching.

2   Background

In this section, some background is summarized that is utilized for the work to be proposed here. First, the concepts clusterMBRs and superMBR are presented.

118.32 ,	4.83 ,	141.42 ,	77.61 ,	C0
55.62 ,	218.54 ,	86.18 ,	228.24 ,	C1
1.75 ,	38.69 ,	56.01 ,	85.18 ,	C2
166.11 ,	115.14 ,	204.77 ,	166.11 ,	C3
212.15 ,	55.62 ,	228.24 ,	100.89 ,	C4
60.77 ,	85.50 ,	112.75 ,	127.78 ,	C4
100.89 ,	177.06 ,	154.39 ,	212.15 ,	C5
4.83 ,	131.63 ,	51.91 ,	141.42 ,	C6

Fig. 1. Cluster MBRs

118.32 ,	77.61 ,	118.32 ,	77.61 ,	1
123.34 ,	69.35 ,	123.34 ,	69.35 ,	2
127.79 ,	60.78 ,	127.79 ,	60.78 ,	3
131.64 ,	51.95 ,	131.64 ,	51.95 ,	4
134.87 ,	42.81 ,	134.87 ,	42.81 ,	5
137.48 ,	33.51 ,	137.48 ,	33.51 ,	6
139.45 ,	24.05 ,	139.45 ,	24.05 ,	7
140.76 ,	14.48 ,	140.76 ,	14.48 ,	8

Fig. 2. Point set of cluster C0

Then, the strategy proposed in [14] is summarized, as this is the strategy used for comparison purposes in the evaluation (Sect. 4). Finally, the clustering strategies which are used in this work are summarized.

2.1 Safe Regions, ClusterMBRs and the SuperMBR

A safe region is a subregion of space (and the points or objects that reside in it) in which the result of the query remains valid while the query continues residing within the region. This approach is utilized by many strategies that process continuous spatial queries in order to minimize the number of requests to the server that must be made. One such type of safe region is a superMBR, which is formed by utilizing one or more clusterMBRs. A clusterMBR is a minimum bounding rectangle (MBR) that contains a set of points that were formed by a clustering strategy. Therefore, a clusterMBR is created for each cluster in a set, and will represent that cluster. First proposed in [15] for continuous region query processing, the superMBR is also utilized for approximate continuous k-nearest neighbour query processing in [14].

Figures 1 and 2 depict a set of clusterMBRs, and the points that are contained by cluster 0, The lower left-hand corner of a clusterMBR (lx, ly) is represented by the first two values, while the upper right-hand corner (hx, hy) is represented by the next two values. The last value in each record is the identifier for each cluster or point. For each record in Fig. 2, the first and third value are the same, as are the second and fourth values, since each is representing a point.

2.2 Continuous k-Nearest Neighbour Using the SuperMBR

The approximate continuous k-nearest neighbour strategy proposed in [14] utilizes a clustered dataset on the server to create a superMBR. The strategy takes a query point q and transmits it to a remote server, which creates an initial superMBR and corresponding point set to send back to the client. The client displays the k nearest neighbours from this set. This superMBR is considered valid until q moves outside of it. Whenever this happens, another request for a new superMBR and point set is made to the server.

On the server, one or more clusterMBRs are selected for creating the superMBR. In [14], two scenarios are considered: 1) q overlaps one or more

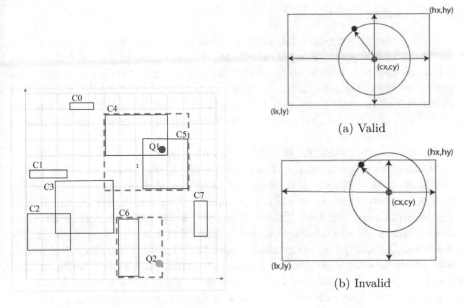

(a) Valid

(b) Invalid

Fig. 3. SuperMBR creation   **Fig. 4.** Valid and invalid cases

clusterMBRs, and 2) q overlaps no clusterMBRs. Figure 3 (from [14]) depicts both scenarios. As we can see, query Q1 overlaps clusterMBRs C4 and C5. Both are selected for creating the superMBR, which is represented in the figure by the orange rectangle that encloses both clusterMBRs. Query Q2, however, does not overlap any clusterMBR. In this scenario, the closest clusterMBR - C6 - is chosen, and the superMBR encloses both C6 and Q2. This is done in an attempt to reduce the number of requests made to the server.

One significant limitation of this strategy is with the validation approach for keeping an existing superMBR on the client, which is a simple test to ensure that q is still within the bounds of the superMBR. Figures 4a and 4b (from [14]) depicts some scenarios that can arise from this simple verification. In Fig. 4a we see the scenario in which the kth nearest neighbour is safely inside the superMBR. However, in Fig. 4b, we see that the distance between q and the kth nearest neighbour is greater than the distance between q and one side of the superMBR. As we can see from the circle formed from the kth on the outer edge, some of the circle resides outside of the superMBR. It is possible that one or more points in this area are closer to q, which makes the result in the superMBR invalid. In addition, Fig. 3 depicts the scenario in which a clusterMBR - C3 - overlaps the superMBR created for Q1. However, since C3 did not overlap Q1, it was not chosen to be part of the superMBR. The problem here is that C3 contains points that are part of a valid result, but will not be transmitted to the server.

2.3   Clustering Strategies

The work to be proposed in the next section can utilize a set of points that is clustered with any clustering strategy. For the evaluation, the following clustering strategies are utilized to see how they affect the strategies [2]: KMeans, Expectation-Maximization (EM), Farthest-First (FF), Density and Canopy. The implementation of these strategies in the WEKA[1] data mining software was utilized. WEKA was chosen due to the author's familiarity with it, and also due to its stable implementation.

KMeans uses partitioning to split a set of points into k clusters. First, k seed points are chosen randomly from the set of points, before all other points are assigned to the closest seed point. Then, a centroid point is calculated for each new cluster. Then, each point is re-assigned to a different cluster if it is closer to a different centroid. This is repeated while points continue to move between clusters. The partitioning-based method EM performs almost identically to KMeans. However, instead of using a distance function for assigning points to a cluster, EM uses a probability distribution function. FF also performs very similarly to KMeans. The main difference is that only the first seed point is chosen randomly. The other $k - 1$ seeds are the furthest away from each other and the first point seed. For the Density strategy, a base clustering algorithm (such as KMeans) is utilized to fit the data to a specified distribution. Finally, the Canopy strategy partitions the data into proximity regions, using two distances – one that dictates the distance a point can be from the centre of a cluster, and another that dictates the number of clusters formed.

3   Strategies for SuperMBR Creation and Validation

In this section, four strategies for approximate k-nearest neighbour query processing are proposed. All utilize a clustered set of points on the server to create a superMBR and corresponding point set for the client. In addition to utilizing the strategy for superMBR creation from [14], the strategies utilize an additional superMBR creation strategy, and a new client-side validation strategy. Both are proposed below.

3.1   Creation of SuperMBR on Server

The new strategy for creating a superMBR attempts to address the issue of missing points due to overlap with the superMBR but not with the query point. In addition, this strategy attempts to reduce the number of requests to the server that the client must make by incorporating additional clusterMBRs into the superMBR (i.e., which also leads to storing extra points on the client). However, a trade-off does exist since memory is limited on most mobile devices. To control this, the value β specifies the number of clusterMBRs that the superMBR can be composed of altogether.

---

[1] http://www.cs.waikato.ac.nz/ml/weka.

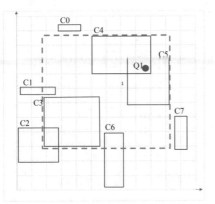 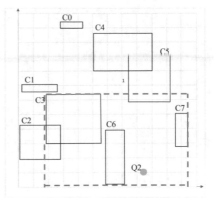

Fig. 5. Overlap   **Fig. 6.** No overlap

Both scenarios—initial overlap, and no initial overlap—are addressed in this strategy. First, the query q is tested for overlap against the set of clusterMBRs. If one or more clusterMBRs overlap q, then a superMBR is created that encompasses the initial chosen clusterMBRs. Then (if necessary), additional clusterMBRs are incorporated into the superMBR. This next step is performed by testing the remaining clusterMBRs for overlap with the existing superMBR. If any clusterMBRs are found to overlap the superMBR, the superMBR is expanded to enclose these clusterMBRs. This process repeats until β clusterMBRs have been incorporated. Figure 3 depicts an example for $\beta = 3$. Earlier, we saw that C3 overlapped the initial superMBR. It is now incorporated into the superMBR.

If no clusterMBRs are found to overlap q, then the β closest clusterMBRs to q are chosen for creating the superMBR. Figure 6 depicts an example for $\beta = 3$. The clusterMBRs C6, C7 and C3 are the closest to Q2.

3.2 Validity Test of SuperMBR on Client

The strategy proposed for testing the validity of a superMBR extends upon the strategy from [14] by also considering the boundaries of the superMBR with respect to q. As mentioned earlier, the distance of the kth nearest neighbour from q (i.e., $dist(q, k^{th}nn)$) may be greater than the distance from q to one or more sides of the superMBR (i.e., $dist(q, side)$, where $side = lx, hx, ly,$ or hy). This may lead to points just outside of the boundary actually being closer than the kth nearest neighbour. However, it may be possible that the kth nearest neighbour since the superMBR is valid, and no points outside of the boundary exist. To allow for some flexibility in this situation, the value α is proposed that specifies the difference between $dist(q, k^{th}nn)$ and $dist(q, side)$ that is permitted:

$$\alpha <= dist(q, k^{th}nn) - dist(q, side), \text{for all } side = lx, ly, hx, hy \qquad (1)$$

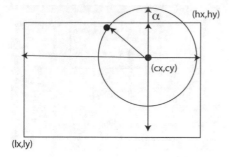

#Point	#*Queries*
500	20
1,000	28
5,000	63
10,000	89
50,000	198
100,000	283

Fig. 7. Validation strategy   **Fig. 8.** #Queries per data size

Figure 7 depicts this for $\alpha <= dist(q, k^{th}nn) - dist(q, hy)$. For no flexibility, $\alpha = 0$ (i.e., the kth nearest neighbour must be properly within the boundaries of the superMBR. Setting α to anything greater than zero will permit some flexibility. Our strategies incorporate both the strict and the flexible use of α.

3.3   Four Strategies

The strategies above for superMBR creation and validity testing, and the strategy for superMBR creation from [14] are applied in various way in the following four strategies:

- Strategy 1: Validity Test ($\alpha = 0$), superMBR creation strategy from [14],
- Strategy 2: ($\alpha = 1$), superMBR creation strategy from [14],
- Strategy 3: ($\alpha = 0$), new superMBR creation strategy,
- Strategy 4: ($\alpha = 1$), new superMBR creation strategy.

4   Evaluation

In this section, all four cluster-based continuous k-ne arest neighbour query processing strategies that are proposed above are evaluated. All strategies are evaluated for accuracy. In addition, all strategies are compared again the cluster-based approach proposed in [14] that is summarized above in Sect. 2. Finally, some observations concerning the running times of the strategies is presented.

All experiments utilize a simulated client-server application that was implemented on an IntelR CoreTM2 Duo CPU E8757 with 2.66 GHz CPU and 1835 MB RAM, running the CentOs Linux 7 OS.

Sets of n points were utilized for all experiments. Each set of n points is drawn from a region of size $\sqrt{n} * 10$ x $\sqrt{n} * 10$ units. Altogether, twelve sets of points were generated, with six sets assuming a random distribution and containing 500, 1,000, 5,000, 10,000, 50,000, and 100,000 points respectively, and the remaining six sets assuming an exponential (i.e., skewed) distribution and containing the same number of points as the uniform data sets.

For clustering our sets of points, we utilized the KMeans, Density (with a KMeans core), EM, FF, and Canopy implementations from WEKA. Other clustering algorithms existed in WEKA. However, they either did not work for larger sets of points, or they did not have the option to state the number of clusters, which was required for one set of tests below.

For all tests, a trajectory of query points that proceeds diagonally through the same space occupied by the points is utilized to evaluate all strategies. The number of query points for each trajectory depends on the number of points in the dataset, which is depicted in Fig. 8. The same trajectory is utilized for both the random and exponential sets of points of the same size.

The four tests that are conducted vary:

- The number and distribution of points,
- The number of nearest neighbours,
- The number of clusters formed in the set of points, and
- The clustering algorithm utilized to create the clusters.

In addition, the following percentage accuracies are calculated:

- **Perfect:** The percentage of completely accurate results (i.e. all k-nearest neighbours were correct),
- **Overall:** The percentage of overall accurate results, which includes both the completely accurate results and the partially correct results for queries that did not produce a set of completely accurate k nearest neighbours, and
- **Imperfect:** The percentage of accurate results in queries that did not result in completely true k nearest neighbours.

Due to space limitations, only the first two percentages are reported for a subset of the results that were generated during the evaluation. For each subset of results presented, there are five pairs of accuracy values. The first pair belongs to the strategy originally proposed in [14] while pairs 2 to 5 belong to the strategies Strategy 1 to Strategy 4 (see end of Sect. 3), respectively.

4.1 Results-Varying Data Size and Distribution

For these tests, the size of the point set varied between 500 and 100000 points. The number of nearest neighbours was set at 3, and the number of clusters set at 50 and created using KMeans clustering. Figures 9 and 10 depict the results of the 10000, 50000 and 100000 point sets for both uniform and exponential distributions, respectively. Overall, most strategies are achieving at least 80% accuracy, with some achieving 90+% accuracy.

For the larger data sets, results show modest to significant improvements for most of the proposed strategies over the original strategy. The accuracy for perfect results in the uniform point sets (Fig. 9) is approximately 11–12% higher than the original strategy for 10000 and 100000 points, while it is significantly higher for 50000 points, at almost 55% higher in most cases. The only exception – Strategy 2 for 100000 points—is significantly lower. As Strategy 2 permits the

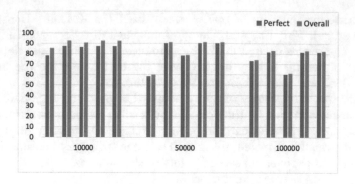

Fig. 9. Number of objects vs accuracy (uniform)

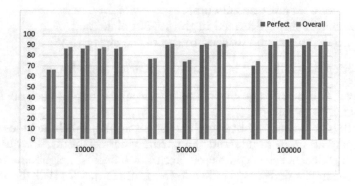

Fig. 10. Number of objects vs accuracy (exponential)

"sphere" of the kth result to extend outside of the superMBR, valid points may exist in this small region of space that are not being retrieved when needed. With a stricter requirement of the "sphere" being inside the superMBR (as with Strategy 1), these situations would be avoided. In addition, overall we see similar trends with the overall percentage of accuracy values.

For the exponential point sets (Fig. 10), we see more significant improvements across all of the strategies for both perfect and overall accuracies. The 50000 point set sees the most modest improvement at 17%, while the 10000 and 100000 point sets achieving more significant improvements at between 27–35%. For the tests on point set below 10000 points, the improvements are negligible for the most part—the only exception being the exponential 5000 point set, which achieves approximately a 50% improvement across all strategies. Therefore, given the results the strategies benefit the larger point set sizes the most.

4.2   Results-Varying Number of Clusters

For this set of tests, the number of clusters varied between 10 and 100. The uniformly distributed 10000-point set was used, as well as 3 near neighbours

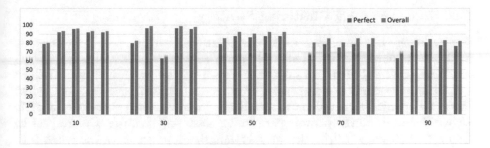

Fig. 11. Number of clusters vs. accuracy

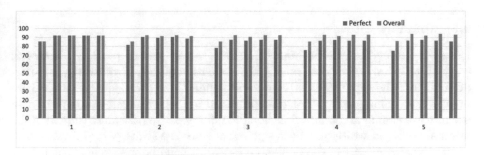

Fig. 12. Number of nearest neighbours vs. accuracy

and KMeans clustering. Figure 11 depicts the results that used 10, 30, 50, 70 and 90 clusters. From the results, we see that the more significant improvements happen in the smaller and larger number of clusters. For 10 and 30 clusters, improvements of between 17–21% are achieved, while those for 90 clusters are between 20–30%. Notice again that the one instance of poor performance occurs with Strategy 2. This also occurs in the results for some of the other clusters not shown here. Again, this is probably due to the nature of the data, and having some too close to other clusters that were fetched while the cluster those points resided in was overlooked due to the non-negative α value.

4.3 Results-Varying Number of Nearest Neighbours

For this set of tests, the number of nearest neighbours varied between1 and 5. The uniformly distributed 10000-point set was used, as well as 50 clusters created using KMeans clustering. Figure 12 depicts the results. It appears that the value of k does not significantly affect the improvement in accuracy. Although there is improvement as the value of k increases, the improvement is approximately 8% at the lower end and at most 16% at the upper end of k. However, the accuracies are 90% in many cases.

4.4   Results-Varying Clustering Strategy

Finally, for this last set of tests varied the strategies used for creating the clusters. Again, the uniformly distributed 10000-point set was used, as well as 50 clusters and 3 nearest neighbours. Results (not shown due to space limitations) show modest improvements of between 5–10% across all strategies, except for Fastest First (FF), which achieved more significant improvements—between 26–33%. FF was found to be the poorest performing clustering strategy with respect to the original strategy [14], so this may explain the improvement. The issues that exist in an FF clustering of data may have been addressed by the new strategies. Finally, overall, most accuracies were at least 80%, with several over 90%.

4.5   Running Time

Data concerning the running time of our strategies was not recorded. However, some observations were noted during the evaluation of the above strategies. First, even though the superMBR creation and validity test strategies proposed here involve more computation that that proposed in [14], all strategies appeared to execute in a similar amount of time and did not appear to take more time than the original proposed strategy. This is significant since the strategy in [14] was found to run significantly faster than repeated k-nearest neighbour searching, especially for the larger point sets. Repeated searching took up to 5–6 h to execute, where the original strategy, and the new ones proposed here, took a matter of a few seconds to execute. Therefore, even though the new strategies require more work, they are not increasing the running time in a noticeable way at all.

5   Conclusion

In this paper, several new strategies for approximate continuous k-nearest neighbour query processing are proposed. All utilize a clustered set of points on the server to create a superMBR and corresponding point set for the client. In addition to utilizing the original strategy for superMBR creation from [14], the strategies utilize an additional superMBR creation strategy, and a new client-side validation strategy. Both new superMBR strategies attempt to overcome limitations of the original strategy by creating a larger save superMBR, as well as additional and flexible validity testing of the superMBR on the client.

The new strategies were evaluated and compared again the original strategy. Results show that: 1) in many instances, at least 80% (and in some case, over 90%) accuracy in both the number of perfect results, as well as the overall accuracy, was achieved, 2) up to 50% improvement over the original strategy was achieved, and 3) the new strategies work very well for larger point sets. One limitation that was discovered was for Strategy 2, which achieved most of the worst performance among the strategies.

Some new research directions emerged from this work. The first is to explore additional extensions to superMBR creation so that results can be 100% accurate more often than what is being accomplished now. The second is to explore the use of other types of clustering strategies, as all of those we applied here are partitioning methods. In particular, it would be interesting to see how density-based clustering strategies would apply to our work. Finally, further evaluation of our work versus other safe-region based approaches is required. In addition, further evaluation is required to determine the running time, the amount of data transmission to and from the server, how many times a new safe region is required, etc., of all strategies.

References

1. Cheng, R., Lam, K.Y., Prabhakar, S., Liang, B.: An efficient location update mechanism for continuous queries over moving objects. Inf. Syst. **32**(4), 593–620 (2007)
2. Frank, E., Hall, M., Witten, I.: The weka workbench (2016). https://www.cs. waikato.ac.nz/ml/weka/Witten_et_al_2016_appendix.pdf. Accessed 14 Feb 2021
3. Gao, Y., Zheng, B.: Continuous obstructed nearest neighbor queries in spatial databases. In: Proceedings of the 2009 ACM SIGMOD International Conference on Management of Data, pp. 577–590 (2009)
4. Gupta, M., Tu, M., Khan, L., Bastani, F., Yen, I.L.: A study of the model and algorithms for handling location-dependent continuous queries. Knowl. Info. Syst. **8**(4), 414–437 (2005)
5. Huang, Y.K., Chen, C.C., Lee, C.: Continuous k-nearest neighbor query for moving objects with uncertain velocity. GeoInformatica **13**(1), 1–25 (2009)
6. Huang, Y.K., Chen, Z.W., Lee, C.: Continuous k-nearest neighbor query over moving objects in road networks. In: Proceedings of the International Conference on APWeb/WAIM, pp. 27–38 (2009)
7. Ilarri, S., Bobed, C., Mena, E.: An approach to process continuous location-dependent queries on moving objects with support for location granules. Jnl. Syst. Soft. **84**(8), 1327–1350 (2011)
8. Iwerks, G.S., Samet, H., Smith, K.P.: Maintenance of k-nn and spatial join queries on continuously moving points. ACM Trans. Database Syst. **31**(2), 485–536 (2006)
9. Ku, W.S., Zimmermann, R., Wang, H.: Location-based spatial query processing with data sharing in wireless broadcast environments. IEEE Trans. Mob. Comput. **7**(6), 778–791 (2008)
10. Lam, K.Y., Ulusoy, Ö.: Adaptive schemes for location update generation in execution location-dependent continuous queries. Jnl. Syst. Soft. **79**(4), 441–453 (2006)
11. Liu, F., Hua, K.A.: Moving query monitoring in spatial network environments. Mob. Netw. Appl. **17**(2), 234–254 (2012)
12. Mouratidis, K., Papadias, D.: Continuous nearest neighbor queries over sliding windows. IEEE Trans. Knowl. Data Eng. **19**(6), 789–803 (2007)
13. Osborn, W.: Continuous k-nearest neighbour strategies using the MQR-tree. In: Proceedings of the 21st International Conference on Network-Based Information Systems (2018)
14. Osborn, W., Anderson, C.: Approximate continuous nearest neighbour query processing in clustered point sets. In: Proceedings of the 11th Annual IEEE IEMCON Conference (2020)
15. Osborn, W., Keykavoos, F.: Continuous region query processing in clustered point sets. In: Proceedings of the 16th International Conference on Mobile Systems and Pervasive Computing, pp. 282–288 (2019)
16. Schiller, J.H., Voisard, A. (eds.): Location-Based Services. Morgan Kaufmann (2004)

Enhancement of Medical Information Security Through Lightweight Cryptography and Steganography in the Platform of Cloud-IoT Ecosystem

Humayra Binte Arfan, Zinia Anzum Tonni[✉], Saria Jahin Taluckder, and Rashed Mazumder

Jahangirnagar University, Dhaka, Bangladesh
rmiit@juniv.edu

Abstract. In the 21st century, an innovative method is the online medical integrated system. It has significant applications, such as reducing the expense of the healthcare system, enhancing the efficacy of the hospital management system, and improving the effectiveness of human capital. The online medical system can undoubtedly play a critical role in the COVID-19 pandemic scenario, and it does as well. A road map of the secured acquisition of medical data from COVID-19 patients, encoding, and distribution to the cloud will be seen in this article. The major aim of this paper is to design a framework for protecting cloud medical data where resources are very limited. Blowfish and Elliptic-curve hybrid cryptographic techniques have been used for encryption. Using the 2D Discrete Wavelet Transform (DWT) method at the fog layer of the cloud, the encrypted data is then embedded into a stego-image format. For a stego-image, the distortion of the cover image must be left unnoticeable. The PSNR, MSE, and SSIM values are most important for maintaining this hiding capacity. The PSNR value differed from 53.08 to 60.36, while the MSE value varied from 0.06 to 0.31 with different data sizes with respect to different formats of the grayscale cover image. For all experimental conditions, the Structural Similarity Index Test (SSIM) values were 1. The proposed model with its stego image PSNR, MSE, and SSIM values has demonstrated that this proposed model can be a suitable substitute to secure medical data in cloud.

1 Introduction

Medical information security is an essential method for securing the sensitive data but most often computerized image and relevant data of patient are transmitted across public networks [1]. There is another tool of Internet of Things (IoT) that has extensive usage in healthcare system, such as it can provide benefits like simultaneous reporting, end-to-end connectivity, affordability, and remote medical assistance services [2]. Cloud-based Medical Care has various

© The Author(s), under exclusive license to Springer Nature Switzerland AG 2021
L. Barolli et al. (Eds.): AINA 2021, LNNS 225, pp. 472–482, 2021.
https://doi.org/10.1007/978-3-030-75100-5_41

issues related to its security such as illegal access policy issues, and data protection. In addition, the exchange of medical information over public networks is vulnerable as there are different types of security threats. When data or health information is in public network (non-trusted communication channel), there is a high chance to leak that information. The COVID-19 epidemic poses an enormous challenges to the global health system where maintaining physical distance is mandatory. If the system is based on offline, we do not need to worry about the privacy issues. However, the offline health database management system is not suitable in this 21st century. So, the virtual medication and privacy issues both are important and mandatory nowadays. Therefore, a secure framework is needed for transmitting, storing, and processing of medical information.

2   Background Studies

Hamid et al. proposed a security model for privacy preservation of healthcare system using pair based cryptography for authenticating key exchange procedure [3]. Though there is a double security level, fog nodes may be an entry point of the attacker to steal the important data [4]. Mohammad Elhoseny et al. proposed a model combination of hybrid encryption scheme and steganography technique [5]. It gives the higher security in terms of PNSR, MSE and SSIM value. It can be unsuitable for resource constraint areas as AES and RSA both need large storage capacity [6]. Rajesh et al. proposed a novel tiny symmetric encryption algorithm (NTSA) providing enhanced security for the transfer of text files through the IoT network by introducing additional key confusions dynamically for each round of encryption [7]. Tan et al. proposed a method to secure the image using reversible watermarking technique having tamper detection ability [8]. It assures the integrity and authenticity of the image and offers very less run time with good quality of image.

Usman et al. proposed a new image steganography approach for securing medical data using Swapped Huffman tree coding to apply lossless compression and manifold encryption to the payload, before embedding into the cover image [9]. Baby et al. proposed data securing technique using DWT that used for hiding multiple color images into a single color image. Secret images are embedded into R,G and B color planes [10]. For combined method of cryptography and steganogarpy, some papers are remarkable. Sharma et al. proposed a method using blowfish and LSB steganography was applied to secure images from attackers [11]. Haddad et al. proposed a new joint watermarking JPEG-LS scheme that extracts the message from the compressed image bit stream [12]. It is based on the full lossless JPEG-LS en-coding and does not decompress the compressed image bit stream to access the message. Al Haj et al. proposed a system that combines encryption standards with watermarking techniques [13].

Chinnasamy et al. proposed a design of secure cloud storage for healthcare data by using a Hybrid encryption of user information using Blowfish and enhanced RSA algorithms [14]. To avoid false requests, the information is encrypted through a symmetric algorithm and keys are encrypted using an asymmetric algorithm. Ardiansyah et al. proposed a combination of two Steganography domains LSB and DWT coupled with Cryptography 3-DES method [15]. This model provides higher PSNR and less MSE value. Tseng et al. proposed a novel algorithm named Swapped Huffman code Table (SHT algorithm) joining compression and encryption based on the Huffman Coding [16]. This model shows that the security could be achieved without sacrificing the compression ratio and performance.

3   System Design

The proposed model composes of three continuous processes:

1. The confidential patient's data is compressed using Huffman encrypted compression technique to pass through the transmission line. After being encrypted from the IoT device, medical data will be transmitted to the fog layer.
2. Then, the medical data will be encrypted using a hybrid encryption scheme that is Blowfish and ECC encryption algorithms in fog layer.
3. Finally, the encrypted data is being concealed in a random medical cover image using 2D-DWT steganography technique producing a stego-image and uploaded to the cloud.

Figure 1 is describing our proposed model where we have used compression, cryptography, and steganography method.

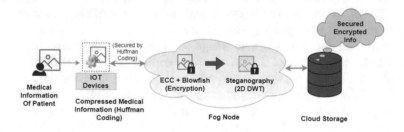

Fig. 1. System model

3.1   Terminal Devices

Data Compression Technique: Medical information contains very personal and vital details of individuals. For this reason, losing very little data can lead to many problems. Based on this, we have used Huffman coding for compressing and securing the medical information through the transmission line as it is a lossless compression technique.

3.2   Encryption in Fog Node

3.2.1   Incorporation Process into Fog Node

We have integrated the encryption technique into the fog node as IoT devices have such limited capacity. It is assumed that this fog node is situated very close to the cloud. So that with less time, output can be accomplished.

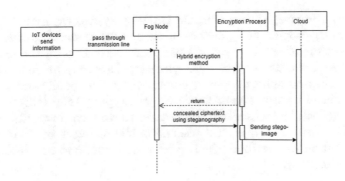

Fig. 2. Sequence diagram of data in fog node platform

3.2.2   Hybrid Cryptography and Steganography

Hybrid cryptography using Blowfish and ECC are considered to be used here for encryption. To generate ciphertext, Blowfish is used to encrypt data. Using ECC asymmetrical cryptography, Blowfish keys are distributed. Cipher text is embedded in a cover image using 2D-DWT steganography. So that the undesired individual may presume that the stego-image is the original data. If the intruder can assume the existence of stego-image authentication without understanding the key, he has to crack the blowfish encryption again, which has unpredictable key length and is difficult to unlock.

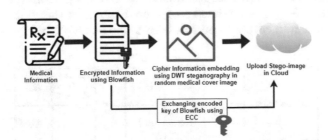

Fig. 3. Visualization of encryption process using cryptography and steganography.

3.2.3   Cryptography Using Blowfish

There are several cipher modes that can be used in Blowfish. We selected the Cipher Feedback (CFB) method for this project. Since it can be used for any size of data. Inputs in the Blowfish algorithm are sender-defined data and key. The key is encrypted by calling the cipher method, and the cipher text is eventually accessed using the CFB mode.

3.2.4   Key Sharing Using Elliptic Curve Cryptography

Symmetrical cryptographic algorithms exchange the sender-recipient key. However, at the time of sharing, a third party can eavesdrop on the transmission lines to access data. To prevent this, symmetric keys may be distributed using asymmetric cryptography, i.e. public cryptography. The public and private keys can be produced by using the Elliptic Curve Cryptography (ECC) algorithm. Blowfish's key can be transmitted from sender to recipient in cryptographic format with the assistance of the private and public recipient keys. As ECC is in our consideration to distribute the key between sender and recipient, it will be implemented further.

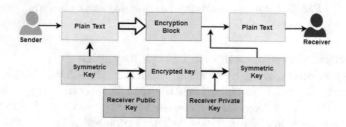

Fig. 4. Hybrid cryptography utilization

3.2.5   Steganography Scheme

We have used "Haar" filters in the 2-level 2D DWT algorithm in our project. In 2-Level 2D DWT, the sub band LL1 is split into a further four sub bands is seen in Fig. 5. Our encrypted text has been hidden in HH2 and HL2 sub bands.

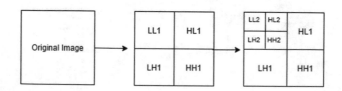

Fig. 5. Original image to 2-level 2D DWT transformation [5]

In the DWT algorithm, the cover image and encrypted text have been inserted in the input section. Finally we get the embedded image as output. Besides using the HH2 of LL1 sub band, HL2 can also be used if the information is too large to hide in single area.

4   Result Analysis

The implementation of proposed model is carried out using Python 3 in Jupyter Notebook and Matlab R2017a running on a personal computer with a 2.27 GHz Intel (R) Core (TM) I3 CPU, 16 GB RAM and Windows 10 as the operating system.

As a cover image, grayscale images from various formats are utilized here to legitimize our proposed technique. The text size in 15, 30, 45, 60, 100, 128, 256 bytes have been picked as plain content. The details of the experimented images of DICOM dataset [17, 18] that represents the Image(1), Image(2), Image(3) and DME dataset [19] that represents the Image(4) are specified in Table 1.

Table 1. Details of experimented images

No	Name	Resolution	Format
Image (1)	MRIBrain	400 × 400	PNG
Image (2)	Handskel	450 × 600	BMP
Image (3)	Kidney	630 × 429	JPG
Image (4)	Dieye3	768 × 576	Tiff

4.1   Visual Comparison

Figure 6 to Fig. 9 displays the embedded images where the encrypted text of 256 bytes has been used. If we analyze these images, it is almost impossible to make a distinction between the original and the embedded image through naked eye. At that point, we can say these embedded pictures are same as original one.

4.2   Compression Ratio

The data compression ratio is also expressed as the compression power, which determines the reduction in the size of the data. The values are expressed in Table 2 having the compressed and uncompressed size of different text size.

Fig. 6. Original and embedded image of image (1)

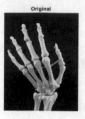

Fig. 7. Original and embedded image of image (2)

Fig. 8. Original and embedded image of image (3)

Fig. 9. Original and embedded image of image (4)

Table 2. Original size and compressed size after Huffman coding

Original size	Compressed size	Compression ratio
15 byte	8byte	1.875
30 byte	19 byte	1.579
45 byte	29 byte	1.552
60 byte	38 byte	1.579
100 byte	64 byte	1.563
128 byte	82 byte	1.561
256 byte	164 byte	1.561

4.3   Statistical Result

Table 3 shows the after effect of three measurable parameters PSNR, MSE, SSIM. The outcome is acquired from embedded grayscale pictures with different inserted text size in different image formats. From this table, it tends to be assessed that PSNR esteem practically reachable to 60 to 53. The MSE varied from .06 to .32. SSIM is consistently 1 which demonstrates no differences between original and embedded picture.

Table 3. Result of the statistical parameters for gray scale image with different text size.

Image (1)	Text size (byte)	PSNR	MSE	SSIM
	15	58.49	0.09	1
	30	58.42	0.09	1
	45	58.38	0.09	1
	60	58.27	0.10	1
	128	58.04	0.10	1
	256	57.57	0.1138	1
Image (2)	15	53.31	0.30	1
	30	53.28	0.30	1
	45	53.28	0.30	1
	60	53.24	0.31	1
	128	53.12	0.31	1
	256	53.08	0.32	1
Image (3)	15	60.36	0.06	1
	30	60.23	0.06	1
	45	60.17	0.06	1
	60	59.98	0.07	1
	128	59.59	0.07	1
	256	58.91	0.08	1
Image (4)	15	57.57	0.11	1
	30	57.56	0.11	1
	45	57.53	0.11	1
	60	57.54	0.11	1
	128	57.51	0.12	1
	256	57.44	0.12	1

4.4 Histogram Analysis

Figure 10 illustrates the histogram of both original and embedded image. Encrypted text of 256 byte data has been used in embedded image to have the histogram.

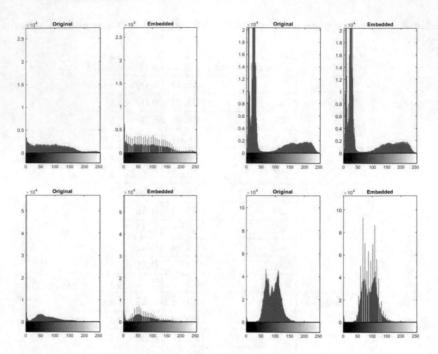

Fig. 10. Original and embedded histogram of Image (1), Image (2), Image (3), Image (4)

4.5 Lightweight Justification

Another thought is, how much size has been expanded of the embedded image contrasted with the original image. As we have chipped away at the asset requirement zone, it is additionally a major worry to direct the consumed space of the embedded picture. Table 4 shows that embedded image is not too large, so less space from memory will be required in the cloud.

Table 4. Size details of embedded image

Image name	Original image size (kb)	Embedded image size (kb)
Image (1)	96.2	110
Image (2)	29.2	242
Image (3)	23.3	117
Image (4)	78.6	200

Table 5 shows the memory usage and run session of proposed technique which have been performed creating a virtual environment for this project. This outcome demonstrated that proposed model takes less assets to execute its activity.

Table 5. Time and memory utilization

Text size (byte)	Encryption and Decryption time (ms)	Memory (MB)
100	14	100.2
128	17	100.3
256	24	101.7

4.6 Summary

Our proposed framework performs very well for JPG image having 58.91 PSNR worth and MSE esteem is .08 for embedding 256-byte text data . It likewise gives a decent outcomes to other picture format like tiff, png and bmp. This framework produces 56.75 PSNR esteem in average, SSIM is consistently 1 and MSE is .16 considering the tested picture formats.

5 Conclusion

The proposed architecture has developed using 2D DWT steganography technique and hybrid cryptographic algorithm, where Blowfish and ECC are considered to use. This architecture is suitable to process and transfer the medical information in a secure manner through the cloud storage using resource-constrained devices. It performs very well for jpg image format. However, the result of our proposed model is not satisfactory in bmp image format comparing to other formats. This result is obtained from different gray scale image format those having different text size. Though ECC is in our system design, we will try to implement this key exchanging technique in future. In our research model, we use only text as medical information. In near future, we will try to consider the medical images as input for secured transmission.

References

1. Avudaiappan, T., Balasubramanian, R., Pandiyan, S.S., Saravanan, M., Lakshmanaprabu, S.K., Shankar, K.: Medical image security using dual encryption with oppositional based optimization algorithm. J. Med. Syst. **42**(11), 1–11 (2018)
2. Patel, N.: Internet of things in healthcare: applications, benefits, and challenges. Internet: https://www.peerbits.com/blog/internet-of-things-healthcare-applications-benefits-andchallenges. html. Accessed 21 March 2019
3. Al Hamid, H.A., Rahman, S.M.M., Hossain, M.S., Almogren, A., Alamri, A.: A security model for preserving the privacy of medical big data in a healthcare cloud using a fog computing facility with pairing-based cryptography. IEEE Access **5**, 22313–22328 (2017)
4. Atlam, H.F., Walters, R.J., Wills, G.B.: Fog computing and the internet of things: a review. Big Data Cogn. Comput. **2**(2), 10 (2018)
5. Elhoseny, M., Ramírez-González, G., Abu-Elnasr, O.M., Shawkat, S.A., Arunkumar, N., Farouk, A.: Secure medical data transmission model for IoT-based healthcare systems. IEEE Access **6**, 20596–20608 (2018)

6. Bhardwaj, I., Kumar, A., Bansal, M.: A review on lightweight cryptography algorithms for data security and authentication in IoTs. In: 2017 4th International Conference on Signal Processing, Computing and Control (ISPCC), pp. 504–509. IEEE, September 2017

7. Rajesh, S., Paul, V., Menon, V.G., Khosravi, M.R.: A secure and efficient lightweight symmetric encryption scheme for transfer of text files between embedded IoT devices. Symmetry 11(2), 293 (2019)

8. Tan, C.K., Ng, J.C., Xu, X., Poh, C.L., Guan, Y.L., Sheah, K.: Security protection of DICOM medical images using dual-layer reversible watermarking with tamper detection capability. J. Digit Imaging 24(3), 528–540 (2011)

9. Usman, M.A., Usman, M.R.: Using image steganography for providing enhanced medical data security. In: 2018 15th IEEE Annual Consumer Communications & Networking Conference (CCNC), pp. 1–4. IEEE, January 2018

10. Baby, D., Thomas, J., Augustine, G., George, E., Michael, N.R.: A novel DWT based image securing method using steganography. Procedia Comput. Sci. 46, 612–618 (2015)

11. Sharma, M.H., MithleshArya, M., Goyal, M.D.: Secure image hiding algorithm using cryptography and steganography. IOSR J. Comput. Eng. (IOSR-JCE) (2013). e-ISSN, 2278-0661

12. Haddad, S., Coatrieux, G., Cozic, M., Bouslimi, D.: Joint watermarking and lossless JPEG-LS compression for medical image security. Irbm 38(4), 198–206 (2017)

13. Al-Haj, A., Abdel-Nabi, H.: Digital image security based on data hiding and cryptography. In 2017 3rd International Conference on Information Management (ICIM), pp. 437-440. IEEE, April 2017

14. Chinnasamy, P., Deepalakshmi, P.: Design of secure storage for health-care cloud using hybrid cryptography. In: 2018 Second International Conference on Inventive Communication and Computational Technologies (ICICCT), pp. 1717–1720. IEEE, April 2018

15. Ardiansyah, G., Sari, C.A., Rachmawanto, E.H.: Hybrid method using 3-DES, DWT and LSB for secure image steganography algorithm. In: 2017 2nd International conferences on Information Technology, Information Systems and Electrical Engineering (ICITISEE), pp. 249-254. IEEE, November 2017

16. Tseng, K.K., Jiang, J.M., Pan, J.S., Tang, L.L., Hsu, C.Y., Chen, C.C.: Enhanced Huffman coding with encryption for wireless data broadcasting system. In: 2012 International Symposium on Computer, Consumer and Control, pp. 622–625. IEEE, June 2012

17. McEvoy, F.J., Svalastoga, E.: Security of patient and study data associated with DICOM images when transferred using compact disc media. J. Digit Imaging 22(1), 65–70 (2009)

18. Hacklander, T., Martin, J., Kleber, K.: An open source framework for modification and communication of DICOM objects. Radiographics 25(6), 1709–1721 (2005)

19. Rabbani, H., Allingham, M.J., Mettu, P.S., Cousins, S.W., Farsiu, S.: Fully automatic segmentation of fluorescein leakage in subjects with diabetic macular edema. Invest. Ophthalmol. Vis. Sci. 56(3), 1482–1492 (2015)

Blockchain-Based Trust Management Approach for IoV

Amal Hbaieb[1,2,3](✉), Samiha Ayed[2], and Lamia Chaari[3]

[1] National School of Electronics and Telecommunications of Sfax, Sfax, Tunisia
[2] LIST3N Laboratory, University of Technology of Troyes, Troyes, France
`amal.hbaieb@utt.fr`
[3] CRNS-SM@RTS (Laboratory of Signals, systeMs, aRtificial Intelligence and neTworkS), Sfax, Tunisia

Abstract. Internet of Vehicles (IoV) brings new specific vehicular use cases further from Vehicular Ad-hoc Networks (VANETs) and offers sophisticated applications for more efficient Intelligent Transport Systems (ITS) in which autonomous driving technology is the quantum leap in smart connected vehicles. However, this innovation brings a security concern and remains cautious and suspicious as vehicular networks become more vulnerable. Traditional security techniques may not tackle all the security issues and provide the comprehensive protection in the current IoV system scenarios. It is therefore of paramount importance to organize trust in such environment and distinguish dishonest entity from trustworthy. Accordingly, trust management-based approaches are further needed to assist for effective vehicular networks deployment. In this paper, a trust management-based model is proposed to secure the vehicular communication system. The introduced trust approach executes a serie of security activities to ensure the trustworthiness of the exchanged data. In our scheme, vehicles also validate the trustworthiness of received data relying on location-trust metric. In this way, our approach will be efficient in calculating and maintaining trust scores in vehicular networks while accommodating related imperatives as Quality of Service (QoS) satisfaction.

1 Introduction

Internet of Vehicles (IoV) networks implement several autonomous vehicles capable of communicating with each other. This type of network is essentially characterized by the high mobility and the sensitive nature of transmitted data. Therefore, the field of use of IoV is confronted with very restrictive requirements in terms of Quality of Service (QoS) and security. Recently, there are several related works that have been interested in the study of IoV by looking at the network technical performance aspects. However, these works have not take into consideration the security aspect. From a security point of view, several works have focused on the study of security mechanisms in IoV without considering the impact of the proposed solutions on the performance of such network. Some of these solutions are based on cryptographic techniques. These techniques need to meet more stringent QoS requirements (e.g., low time complexity). Besides, their use requires and consumes a lot of memory and energy, which makes their consideration in

L. Barolli et al. (Eds.): AINA 2021, LNNS 225, pp. 483–493, 2021.
https://doi.org/10.1007/978-3-030-75100-5_42

the IoV unprofitable regarding network performances. In this paper, we aim to propose a secure management of IoV that allows later to maintain its performance. Our solution is based on the trust management concept. Trust management consists of a set of steps where a node assigns a trust value to another node during their interaction, in order to mitigate the maleffects of malicious and selfish nodes actions. The major of existing approaches for the trust management in IoV do not take into account the impact of the use of trust management techniques on the main QoS parameters. Thus, we propose to design a trust management scheme based on the Blockchain technology that will enable to deal with security and essential QoS requirements regarding its features. Our introduced trust approach executes a serie of security activities to ensure the trustworthiness of the exchanged data. In our scheme, vehicles validate the trustworthiness of received data relying on the reputation and the location metrics. In this way, our approach will be efficient in calculating and maintaining trust scores in IoV while accommodating related imperatives as QoS satisfaction. Our paper is organized into five sections as follows: Sect. 2 is dedicated to the related works and recent researches focused on Blockchain-based trust management in IoV. Section 3 provides our proposed system communication model where we identify the components of the envisaged communication system architecture, as well as our network model. Section 4 validates our proposal through a use case simulation. Finally, Sect. 5 concludes the paper.

2   Related Works

Blockchain technology [1] is having broad potential for the establishment of trust management in vehicular networks context. Indeed, the merits of Blockchain can be exploited to deal with the faced key issues related to centralization, security, and privacy, when maintaining, managing, tracking, and sharing data related to vehicles, traffic conditions, traffic violations, surveillance, and weather information. We refer here some ideas that were elaborated based on Blockchain to mainly address security [2–6], and privacy preservation aspect [7, 8] for vehicular networks. In [2], the authors proposed a Blockchain-based trust management scheme where the main process consists of trust rates generation and uploading step, trust value offsets calculation step (based on reputation and knowledge metrics), and miner election as well as consensus mechanism application for new blocks generation. The receiver entity relies on Bayesian inference rule to assess message-credibility and later generate trust scores for these received contents. Next, the computed trust values are uploaded to the RSUs nodes. Trust values offsets of vehicles nodes are obtained using weighted aggregation. The scheme applies Proof-of-work and proof-of-stake consensus. The Blockchain-based trust scheme in [7] is combined with a privacy preservation approach that is managed by identity-based group signatures for an efficient conditional privacy. The work uses Proof-of-work and Practical Byzantine Fault Tolerates for security reinforcement. Similarly, the work [8] exploited Blockchain features to build an anonymous reputation management scheme that addresses trust and privacy issues in VANET. The reputation management is based on experience and recommendations metrics. The scheme uses further reward and punishment models to improve the trust establishment. We refer also that the Named Data Networking (NDN) technology was proposed to assist in providing trustworthy communications within Blockchain-based networks models, such in [9] where authors explored this combination to enhance the trust in VANET. We note that the major of these works do not consider effectively the impact of their used schemes on the QoS parameters (e.g., by using less intensive computationally Blockchain). Therefore, we aim to tackle this limitation in our proposal.

3   Proposed System Communication Model

3.1   Architecture Overview

In this subsection, we introduce our proposed Blockchain-based trust management app-roach for IoV network. Figure 1 illustrates the overall architecture of the envisioned communication system. By keeping in mind the major QoS issues when deploying trust management solution in vehicular networks, we propose the elaboration of two-layered trust database by means of Blockchain: (1) a general Blockchain which will serve to exhibit the global trustworthiness of the communication system, and (2) a local Blockchain which will be assigned for the different zone areas of the network to pro-ceed a local trust management. The main use of local Blockchains is to meet lower delays while processing time-sensitive data and handling their security. The core of our proposed trust management model consists of ordinary,lightweight nodes, edge nodes, RSUs nodes, and trusted authority nodes. The ordinary nodes represent the main users of the communication system, which refer generally to the smart vehicles (as we will settle on V2V and V2I communications for our later simulation). The ordinary nodes act as smart vehicles and they are resource constraint nodes (e.g., storage, computation, and battery life). They consist of sensing, communication, and basic local processing layer. Each vehicle is installed with sensors, actuators, embedded gateway [10], and on-board units which has wireless communication capability to permit the vehicle to communi-cate with other nodes to share messages. The trusted authority nodes are assumed to be trusted (i.e., deployed for privacy issue dealing). For example, trusted authority node undertakes the work to register (further revoke) the new nodes that want to participate in the communication system. Trusted authority nodes also maintain the database that contains correlation between nodes' public address and real identities. The edge nodes handle trust-service requests/responses (e.g., trust values computation and trust values exhibition). We consider that the edge nodes are semi-trusted nodes. The edge nodes have more computational power than ordinary nodes, and have further cache storage. They host the local Blockchains which will be transferred later to RSUs; along with smart contracts execution and mining consensus performing. As usual, the RSUs nodes act as static nodes to provide required services (e.g., traffic information, road event, etc.) and undertake the computational work to add the local Blockchains into the global chain as we aforementioned.

Hence, the scalability and the availability of the network can be achieved as edge and RSUs nodes are placed in a distributed way. Regarding privacy, a cryptography-based mechanism can be integrated to ensure more reliable communication. The trusted authority nodes are deployed to help in privacy-preservation (e.g., certificates issuance for participating nodes). The proposal use Interplanetary File System (IPFS) along with the Blockchain technology to allow the storage of the large amount of generated data (particularly within an autonomous driving context) in a decentralized and distributed

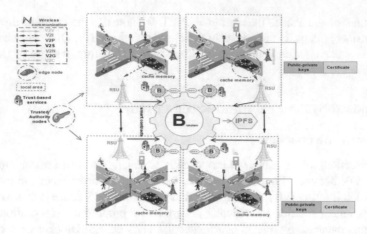

Fig. 1. Proposed network model: overall architecture.

way. The smart contracts can also be implemented to allow the trustworthy transaction and automate the network processes. IPFS hashes will be used in order to link and track the data in the Blockchain smart contracts. Moreover, to optimize the execution time of nodes queries, cache memory is introduced (e.g., to store frequently asked data). The procedure of nodes requests requires a large execution time that involves propagation delay and transmission delay. As IoV is a time-sensitive system, cache memory is applied at the edge nodes to meet this time complexity enhancement.

3.2 Main Procedures

The main phases of our Blockchain-based approach consist of (1) trust values computation by edge nodes, (2) local mining mechanism and blocks generation, (3) trust values uploading to the local Blockchains, (4) global mining mechanism and blocks generation, and (5) trust values uploading to the global Blockchain. The edge vehicles are responsible, as aforementioned, for determining the local trustworthiness of the IoV network and for hosting the internal Blockchains. In this subsection, we highlight the procedure of local trust management. The process of local trustworthiness assessment is preceded by an authentication procedure, and it stands also to apply an incentive algorithm to give reward or penalty to engaged nodes (i.e., messages senders, queried nodes). We term $V = [V_1, V_2, ..., V_i]$ the set of i lightweight vehicles in the IoV, $E = [E_1, E_2,..., E_k]$ the representation of k edge vehicles; $1 \leq k$ and $TA = [TA_1, TA_2..., TA_l]$ the set of l trusted authority nodes; $1 \leq l$ which are responsible to validate the identities of the nodes in the IoV communication system. The basic idea to gather the local trust information is to rely on the reputation of the sender node and to suggest to the recipient to evaluate the credibility of the received message. That is to say, the trust score refers to the combination of defined reputation value and message-credibility value. Basically, each sender will have an initialized reputation value. Figure 2 depicts the process applied in this local trust management phase. We assume that every lightweight vehicle can quickly query the trust score of the sender node (i.e., inquires about the trustworthiness of a node from which it received messages) from the nearest edge vehicle E_k. The trust values that k edge vehicles keep for j sender nodes are denoted as $R = [R_1, R_2, ..., R_j]$. In fact, each queried E_k calculates the sender trust value based on reputation and location metrics. We refer that the trust values are maintained in the E_k corresponding IPFS, while the most frequent ones are stored in the E_k respective cache. For example, once a lightweight vehicle obtains a message from another vehicle, it identifies the later' trust, describing how to react on the received content. Regarding authentication, a vehicle node V_i ($V_i \in$ V) that wants to join the IoV network uses its assigned pair of public-private keys (termed as $(Pu^m{}_i, Pv^m{}_i)$; $1 \leq m$, where m is the cardinality of generated keys) and receives a digital certificate from a trusted authority node TA_u.

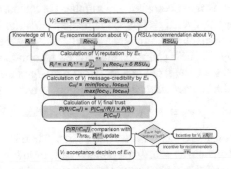

Fig. 2. Process of local trust management.

Obviously, the vehicle V_i generates the pair of keys $(Pu^m{}_i, Pv^m{}_i)$ to reach a classical encrypted communications, and the V_i associated certificate issued by the trusted authority node TA_u is used to correlate V_i public key $(Pu^m{}_i)$ and real identity. The delivered certificate by TA_u to V_i contains the V_i public key, the generated signature of the entire certificate (Sig_u), the information about the certificate issuer TA_u (IF_u), the expiration date of the given certificate (Exp_i), as well as the V_i trust value (R_i) (i.e., V_i initial reputation value, or V_i trust value given by an edge vehicle E_k):

$$Cert_{i,u}^m = Pu_i^m,\ Sisg_u,\ IF_u,\ Exp_i,\ R_i \tag{1}$$

Step 1: Calculation of Reputation Value
The reputation value of a sender node will be calculated by an edge vehicle based on the knowledge metric and available recommendations provided from other surrounding edge vehicles. The knowledge metric refers either to the sender' initialized reputation, or its trust established by the previous queried edge vehicle and retrieved from the IPFS (from the cache memory of the edge vehicle; if available). The number of queried recommenders as the related number of hops will be defined taken into account their impact on the calculation accuracy. The edge vehicle E_k can consider further the recommendation available from the nearest RSU to establish the reputation value. Moreover, we define three weightiness levels for exchanged events: (1) high level (concerning crucial and severe events, e.g., accident or information about road work), (2) ordinary level (e.g. beacon messages comprising driving status), and (3) soft level (concerning less important events, e.g., normal traffic information). For instance, let the vehicle V_i requests the

nearest edge vehicle E_k to exhibit the trust value of another vehicle V_j from which it receives a message about an event $Ev_h(Evh \in [Ev_1, Ev_2, ..., Ev_h]; 1 \leq h)$, and let the event Ev_h an accident event. Suppose that the edge vehicle E_k communicates with its direct neighbor E_q in order to proceed the required value calculation. The queried edge vehicle E_k computes the V_j reputation value as follows:

$$R_j^t = \alpha R_j^{t-1} + \beta \sum_{q=1}^{n^t} \gamma_q Rec_{q,j} + \delta RSU_{a,j} \tag{2}$$

where $R_j{}^t$ denotes the reputation value of the sender V_j given by the edge vehicle E_k at the current time t, n^t is the recommenders number (in the current scenario, we have only one recommender; E_q), $Rec_{q,j}$ refers to the recommendation of the edge vehicle E_q about V_j, $RSU_{a,j}$ represents the recommendation about V_j by the nearest placed RSU_a ($RSU_a \in [RSU_1, RSU_2, ..., RSU_b]$), γ_q stands for E_q recommendation credibility, and the variables α, β, and δ are the fixed weights of knowledge factor ($R_j{}^{t-1}$), edge nodes recommendations, and RSU_a recommendation, respectively. We point that $\alpha, \beta, \delta \in [0, 1]$, $\alpha + \beta + \delta = 1$, and $\sum_{q=1}^{n} \gamma_q = 1$. Next, E_k will pass to judge V_j message-credibility.

The edge nodes are considered as semi-trusted nodes. Therefore, after making the V_j message acceptance decision, a reward or a punishment for involved recommenders will be carried out with respect to their associated weights (regardless Ev_h weightiness level): e.g., there are two kinds of actions that will make recommender weight (γ_q) increasing. First, recommender gave a good recommendation and V_j message is admitted. Second, recommender did not support V_j when V_j message is dropped. Also, recommender weight will be decreased if it supported V_j whereas V_j message is denied, or if it gave bad recommendation whereas V_j message is accepted.

Step 2: Calculation of Message-Credibility Value
As indicated above, the second phase in the process of local trust establishment consists of evaluating the E_k degree of belief on the V_j message-credibility about the event Ev_h. This task is relied on the location metric, as the location proximity is an important attribute that can assess the data credibility (i.e., data are likely more accurate as the sender node V_j is close to the location of the event Ev_h). Let $C_m = [C_{m1}{}^t, C_{m2}{}^t, ..., C_{mj}{}^t]$ the set of messages-credibility values estimated by k edge vehicles. A $C_{mj}{}^t$ value corresponds to the credibility of Ev_h sent from V_j to V_i and defined by E_k. The calculation of the location proximity of V_j to Ev_h allows E_k to determine the $C_{mj}{}^t$ value.

$$C_{mj}^t = \frac{minimum(loc_{Vj}, loc_{Evh})}{maximum(loc_{Vj}, loc_{Evh})} \tag{3}$$

Where $|loc_{Vj} - loc_{Evh}| < rho_{loc}$, loc_{Vj} and loc_{Evh} are the locations of the sender node V_j and the event Ev_h, respectively. The rho_{loc} indicates the admissible threshold difference. If the marked condition is not satisfied, we assign value 0.1 to the credibility value $C_{mj}{}^t$. Now the reputation score and the message-credibility value of the sender node V_j are available, we can get the related final trust value using the Bayes rule (as we aim to predict the final trust value of the sender node by taking into account its message-credibility value that is derived after its reputation score establishment):

$$P\left(R_j^t/C_{mj}^t\right) = \frac{P\left(C_{mj}^t/R_j^t\right) \times P\left(R_j^t\right)}{P\left(C_{mj}^t\right)} \tag{4}$$

where $P(R_j^t)$ is the reputation of the sender node V_j, and $P(C_{mj}^t)$ is the V_j associated message-credibility. When $P(R_j^t/C_{mj}^t)$ exceeds the preset threshold for the trustworthiness value *(Thrs_{tr})*, the edge node E_k regards V_j message as true. The edge node E_k reports the ultimate trust score of V_j to the vehicle V_i which will execute the acceptance task of Ev_h. Once the E_k decision has been made on the validity of the event Ev_h, it performs a payoff-based algorithm for the sender node V_j in order to guide the update of its trustworthiness. This incentive algorithm counts on the weightiness level of the event E_k to adjust the old trust of the sender (see the following algorithm). The sender V_j is rewarded with a positive reinforcement of its old trust value if the event Ev_h is confirmed, otherwise, the V_j is penalized by diminishing its trustworthiness. For example, in the above scenario the event Ev_h corresponds to an accident, hence the old trust of V_j will be decreased in half once the accident is denied.

GENERAL PROCESS OF THE SENDER NODE PAYOFF

(Output){updated R_j^{t-1}}(Input){R_j^{t-1}:V_j old trust(IPFS); P(R_j^t/C_{mj}^t):V_j trust; Thrs$_{tr}$:trust threshold; w_h:E_k weightiness level; θ,σ,η:payoff coefficients;
begin
  if $w_h \in$ 'High' then if P(R_j^t/C_{mj}^t)≥ Thrs$_{tr}$ then
          R_j^{t-1}=P(R_j^t/C_{mj}^t)×θ; else R_j^{t-1}=P(R_j^t/C_{mj}^t)÷θ;
  else if $w_h \in$ 'ordinary' then if P(R_j^t/C_{mj}^t) ≥ Thrs$_{tr}$ then
          R_j^{t-1}=P(R_j^t/C_{mj}^t)×σ; else R_j^{t-1}=P(R_j^t/C_{mj}^t)÷σ;
  else if $w_h \in$ 'soft' then if P(R_j^t/C_{mj}^t) ≥ Thrs$_{tr}$ then
          R_j^{t-1}=P(R_j^t/C_{mj}^t)×η; else R_j^{t-1}=P(R_j^t/C_{mj}^t)÷η;
end.

Furthermore, we can suppose that many sender nodes have send, quasi simultaneously, the same message about a particular event Ev_h to vehicle V_i. The later node will query Ev_h validity as usual from the nearest edge vehicle E_k. E_k will execute the calculation steps of senders' reputation values and message-credibility values. The vehicle V_i accepts the event Ev_h from the sender V_j of which E_k has confirmed its reliability first. Finally, all senders will be rewarded in case of Ev_h confirmation. We can also extend to the scenario where the node V_j sends to V_i, quasi-simultaneously, many messages related to different events. In this case, E_k has to determine the coherent trust value of the sender V_j. Assume that V_i receives two different messages (Ev_h and Ev_g) from V_j. The queried edge vehicle E_k starts to estimate its belief on them (using the above corresponding equations). Thereafter, E_k compares the generated results ($P(R_j^t/C_{mj}^t)$ according to Ev_h and $P(R_j^t/C_{mj}^t)$ according to Ev_g) with the preset threshold $Thrs_{tr}$, to verify the validity of both Ev_h and Ev_g. E_k will get conflicting about V_j trust (e.g., Ev_h is validated and Ev_g is decayed, and note that V_j has now two associated trust values). At this end, E_k must estimate a new proper trust value for V_j to preserve the trust dataset consistency. This value can be deduced by calculating the geometric average of the obtained trust results (e.g., $P(R_j{}^t/C_{mj}{}^t)$-Ev_h and $P(R_j{}^t/C_{mj}{}^t)$-Ev_g), with the following manner:

$$P\left(R_j^t/C_{mj}^t\right) = \sqrt[nb_j]{\prod_{j=1}^{nb_j} P\left(R_j^t/C_{mj}^t\right)} \tag{5}$$

where nb_j is the number of sent messages from V_j, and $\prod_{j=1}^{nb_j} P(R_j{}^t/C_{mj}{}^t)$ represents the product over the generated trust values of V_j. Then, E_k compares one more time this ultimate value with the trust threshold $Thrs_{tr}$ to run the payoff algorithm for V_j. So, to recap the local trust management, each new node that want to participate the network has to register within the trusted authority node. Next, this participating entity generates private/public keys, and submits to trusted authority node its initial public key to prove its legal identity. The trusted authority node will issue an initial certificate to this node (will be used for tracking the real identity of node in case of disputes). Whenever receiving a message, the node can request the trust value of the sender from the nearest edge node. The edge node checks first requester identity. Then, it undertakes the work of required trust value computation. This procedure includes (1) reputation value calculation step and (2) message-credibility value calculation step. So, the edge node verifies its cache memory. If the old trust value of the sender is present at the cache, it will be used directly for the calculation, otherwise, the edge node gets access to the IPFS to retrieve this trust data; if available. Next, the edge node exhibits the new determined trust value to the requester node which will make the decision of sender message acceptance. Edge nodes begin the mining for adding the new calculated trust value into the local Blockchain. Also, the miner can be rewarded (e.g., valid transaction number-based). Edge nodes store the frequent required trust values in their caches. Upon being created, edge nodes upload the local Blockchain into the RSUs, which will check in turn the validity of this Blockchain and add it into the global chain. Likewise, the miner can be rewarded after a valid transaction. Similarly, the following steps can be executed, for example, while an event witness overall scenario. Whenever an event is occurred and ordinary node is a witness, it sends a report to the nearest edge node. The later will check the sender identity. Then, the edge node verifies the report validity, and accordingly it can update the trust value of the sender. Mining consensus process is triggered once the sender trust data is not on the local Blockchain. As aforementioned, the metrics used for trust value computation are reputation-based (for sender reliability assessment), and location-based (for message-credibility assessment). However, we can also take into account the time of the received event when estimating message-credibility as this factor helps in verifying message freshness. The problem of selfishness or low node participation can be tackled by the involvement of an incentive mechanism. For example, whenever an ordinary node produces a valid event report, it gets incentive (in term of trust adjusting, or may certificate revocation). Besides, an announcement protocol, such in work [7], may be involved to assist in nodes authentication (within an event witness context).

4 Simulation Results

In this section, we conduct the performance evaluation of our trust management scheme to validate its effectiveness. We implement the approach in the Python environment, where we apply the Ethereum Blockchain. To this end, we define the parameters' values used in our evaluation. We set the type of vehicles nodes before the beginning of the evaluation. We suppose that these nodes are authenticated (i.e., each node has already generated its pair of public-private keys, and has received a certificate from a trusted authority node). As aforementioned, the vehicles nodes in our model are grouped into areas, based on the location factor. Hence, in each area, we select 2%, 3%, 10% and 85% as maximum percentages of the members of RSUs, trusted authority, edge, and lightweight nodes, respectively. Also, we consider 40% as the maximum percentage of malicious lightweight nodes that provide false messages. The major nodes of IoV are assumed to be likely honest. Therefore, the selected proportions seem to be realistic. Our scenario is considered for a highway road, where the speed of vehicles varies between 90 km/h and 150 km/h, and the maximum nodes density is 150 nodes. In order to begin the calculation of the trust value, we initialize the reputation value for each sender at 0.4, also, we define the threshold of the trust at 0.7. Regarding the values of associated weights, we set 0.5, 0.3, 0.2 to α, β, and δ, respectively. We consider that the knowledge factor (where α is assigned to), is the most important metric that will be taken into account in the calculation of the trust value. Finally, the maximum number of hops between queried recommenders is defined by 2 hops. We evaluate our proposal in V2V and V2I environment with up to three different traffic density scenarios: 10, 50, and 150 vehicle nodes, where 2%, 5%, 10%, 20%, 25%, 30%, and 40% of lightweight vehicles are malicious, and where we suppose that the received events are critical (accident event). We assume that the recommenders are trusted. Under these parameters, we estimate our communication system trustworthiness. The proposal proves its effectiveness in term of detection rate of malicious nodes. The sender vehicles that reported invalid events have yielded a weak value of trust. The scheme proves also that it supports the scalability requirement. Figure 3 illustrates the outcomes of the detection rate of malicious vehicles. For instance compared to the approach proposed in [7], our scheme presents better performance concerning detection rate. Nevertheless, it is notable that the higher the percentage of dishonest recommenders in the system is, the easier for them to damage the communication system. We assess also the time complexity factor. Figure 4 depicts that the proposal reaches a good tradeoff between latency and trust management process accomplishment. The latency corresponds to the total spending time for establishing the trust in the communication system (i.e., trust values calculation, and scores upload and update in local and global Blockchains). Finally, Fig. 5 illustrates the communication overhead ratio obtained from carrying out our test. In this paper, the communication overhead is stated as the number of valid packets that are transferred between nodes. We evaluate this parameter over the percentage of malicious vehicles. As shown in Fig. 5, the ascending percentage of malicious vehicles in the communication system reduces the communication overhead. Thence, the communication overhead requirement is satisfied as the number of detected malicious nodes has a significant impact on such criteria.

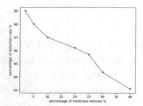
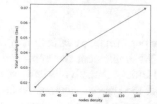
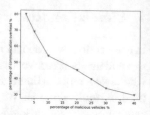

Fig. 3. Detection rate of malicious vehicles in the communication system.

Fig. 4. Total spending time for establishing trust in the communication system

Fig. 5. Communication overhead over malicious vehicles in the communication system

5 Conclusion

In this paper we present a Blockchain-based trust management approach for IoV communication system. By using two layers of Blockchain, each node is able to asses the trustworthiness of the entities that communicate with. The trust estimation is based on reputation and location metrics. By determining a few parameters, each participating node can perceive a reasonably accurate view of the IoV network. The implementation results demonstrate the efficiency of our work. The use of Blockchain solution allowed us to find the good trade off between security and QoS in IoV environment. The proposed network model shows that is can be dynamic, scalable, light and reliable. We plan to further take into account the privacy preserving and define a detailed adversary model in the future. Besides, the unavailable discussion regarding the implementation with real simulation tools in this paper will be the focus of our future work.

Acknowledgments. This work was supported by the Ministère de L'Enseignement supérieur, de la Recherche et de l'Innovation–France, and the FEDER; TBAN project.

References

1. Puri, V., Kumar, R., Van Le, C., Sharma, R., Priyadarshini, I.: A vital role of blockchain technology toward internet of vehicles. In: Handbook of Research on Blockchain Technology, pp. 407–416. Elsevier (2020)
2. Yang, Z., Yang, K., Lei, L., Zheng, K., Leung, V.: Blockchain-based decentralized trust management in vehicular networks. IEEE IoT J. **6**(2), 1495–1505 (2019)
3. Yang, Y.T., Chou, L.D., Tseng, C.W., Tseng, F.H., Liu, C.C.: Blockchain-based traffic event validation and trust verification for VANETs. IEEE Access **7**, 30868–30877 (2019)
4. Javaid, U., Aman, M.N., Sikdar, B.: DrivMan: driving trust management and data sharing in VANETS with blockchain and smart contracts. In: 2019 IEEE 89th Vehicular Technology Conference (VTC2019-Spring), pp. 1–5. IEEE (April 2019)
5. Dong, W., Li, Y., Hou, R., Lv, X., Li, H., Sun, B.: A blockchain-based hierarchical reputation management scheme in vehicular network. In: 2019 IEEE Global Communications Conference (GLOBECOM), pp. 1–6. IEEE (December 2019)
6. Chai, H., Leng, S., Zhang, K., Mao, S.: Proof-of-reputation based-consortium blockchain for trust resource sharing in internet of vehicles. IEEE Access **7**, 175744–175757 (2019)

7. Liu, X., Huang, H., Xiao, F., Ma, Z.: A blockchain-based trust management with conditional privacy-preserving announcement scheme for VANETs. IEEE IoT J. 7(5), 4101–4112 (2019)
8. Lu, Z., Wang, Q., Qu, G., Liu, Z.: Bars: a blockchain-based anonymous reputation system for trust management in vanets. In 2018 17th IEEE International Conference on Trust, Security and Privacy in Computing and Communications/12th IEEE International Conference on Big Data Science and Engineering (TrustCom/BigDataSE), pp. 98–103. IEEE (August 2018)
9. Khelifi, H., Luo, S., Nour, B., Moungla, H., Ahmed, S.H.: Reputation-based blockchain for secure NDN caching in vehicular networks. In: 2018 IEEE Conference on Standards for Communications and Networking (CSCN), pp. 1–6. IEEE (October 2018)
10. Hbaieb, A., Rhaiem, O.B., Chaari, L.: In-car gateway architecture for intra and inter-vehicular networks. In 2018 14th International Wireless Communications Mobile Computing Conference (IWCMC), pp. 1489–1494. IEEE (June 2018)

Improved Visual Analytic Process Under Cognitive Aspects

Samar Bouazizi[1]([✉]) and Hela Ltifi[2]

[1] Faculty of Sciences and Techniques of Sidi Bouzid, Computer Science and Mathematics Department, University of Kairouan, Kairouan, Tunisia
[2] REGIM-Lab.: Research Groups in Intelligent Machines, LR11ES48, National Engineering School of Sfax, University of Sfax, 3038 Sfax, Tunisia
hela.ltifi@ieee.org

Abstract. The Visual Analytics (VA) systems aids to ameliorate the decision-making activity, since these systems relates visual to automatic analysis to extract knowledge and explore data. In this paper, we present an adaptation of a cognitive Visual Analytics with a well-known Visual Analytics process under cognitive aspects: intelligent data collection, fusion and synchronization and their analysis using the adapted cognitive data mining technique. To prove the effectiveness of our approach, we have applied it during a formative assessment. We have used the Eye Tracking as well as the Electroencephalogram (EEG) as an intelligent data collection source and the Fuzzy Logic as data mining technique. We performed an evaluation for proving the utility and usability of this developed system.

Keywords: Cognitive Visual Analytics · Eye-tracking · EEG · Fuzzy logic · Formative assessment

1 Introduction

The importance of user interfaces in the decision support systems research field has been emphasized [1] as the main purpose is to more integrate the decision-maker into the decision-making process. Consequently, a new interactive system combining the capacity of automatic processing and the capacity of human reasoning has taken place. This system is called Visual Analytics (VA) systems [2].

Actually, visual analysis combines interactive visualization with automated methods of data analysis to provide interactive and scalable decision support. Given the evolution of the field of VA, researchers introduced several interactive and automatic methods and measures for sense-making and insight generation [3]. However, the data collection process is generally done using traditional tools and methods such as database management systems. In this context, we propose to integrate automatic cognitive data collection in the VA process applied in education field, specifically in a formative assessment. The collected data must be computationally analyzed. The uncertain nature of the collected data requires the use of specific technique: The Fuzzy Logic that is based on "degrees of truth" allowing representing the human behavior with the system [4].

L. Barolli et al. (Eds.): AINA 2021, LNNS 225, pp. 494–506, 2021.
https://doi.org/10.1007/978-3-030-75100-5_43

The rest of this paper is presented as follows: in Sect. 2, we provide a short study background concerning decision support systems and VA. Subsequently, we formally define our cognitive VA framework in Sect. 3. Within Sect. 4, we present our proposal application. Finally, Sect. 5 summarizes our work and paves the way for future research.

2 Background of the Study

To manage large quantities of complex data, to generate knowledge and to make appropriate decisions, Decision Support Systems (DSS) are intelligent and visual [1, 19]. Such DSS are named VA systems. The term "VA" related Visual Analysis to Automatic Analysis [2]. It is a multidisciplinary field which mainly integrates: (1) data mining, (2) information visualization and (3) perception and cognition. The VA couples the automated analysis techniques with interactive visualizations to understand, interpret and effectively decide from massive and complex data sets [2]. In the literature, several VA researches works proposed pipelines for designing VA systems. We begin with Wjik process [9].

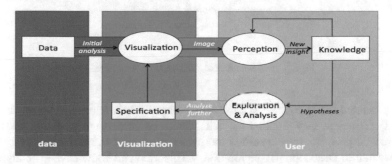

Fig. 1. The Wijk Visual Analytics process

As shown in Fig. 1, the analyst gains new ideas by perceiving the image. These ideas will be translated into knowledge, hence the emergence of new hypotheses. Finally, after exploration and analysis, the analyst provides suitable specifications which will be considered in the visualization to improve his/ her knowledge. Later, Kcim et al. [2] proposed VA process that is based on the combination of the human and the automatic data processing (cf. Fig. 2).

In the sub-process of automatic data analysis based on the data mining methods (cf. Fig. 2), the central phase makes it possible to automatically and intelligently extract the precious information called models or "patterns" from data by means of automatic analysis algorithms. Visualization is important in the sub-process of visual analysis of data and patterns. On the one hand, it provides interactive visual representations of the data and patterns generated by the mining method. On the other hand, it ensures efficient Analyst-Image interaction. Hence the improvement in the quality of the knowledge extracted by perceiving the visualization adequate to the decision-making situation. Sacha et al. [10], basing on the Keim et al. VA process [2], proposed an adaptation of the Keim et al.'s process.

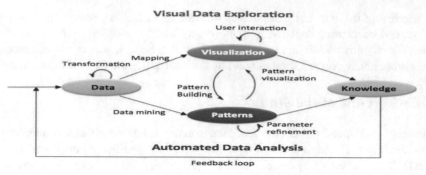

Fig. 2. The Keim et al. Visual Analytics process

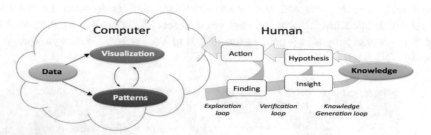

Fig. 3. The Sacha et al. Visual Analytics process

As presented in Fig. 3, the first part (computer part) is the same as proposed by Keim et al. [2]. However, they detailed the activities to be carried out to obtain knowledge in the second part. So, they specified the phases necessary to reach knowledge which are grouped into three loops of exploration, verification and knowledge generation. The first is used to explore, allowing to choose the appropriate actions on the visualization according to the results. The second loop consists of verifying the information making it possible to automatically formulate hypotheses on the data analysis from the results. The last loop aims to generate knowledge.

The review of these processes shows that the VA is still an area of research that needs further improvements. Our work takes place in this context.

3 Proposed Cognitive Visual Process

In recent years, a remarkable research development in the field of VA, the main purpose of which is to help users easily explore, analyze and interpret data. This visual analysis approach transforms information into knowledge that represents the science of analytical reasoning facilitated by interactive visual interfaces. Our contribution concerns the two parts of the visual analysis process of Keim et al. [2], which is illustrated in Fig. 4.

As presented by Fig. 4, we have adapted the process of Keim et al. [2] (visible in Fig. 4A) by integrating cognitive aspects (visible in Fig. 4B). Our adaptation is marked in green in Fig. 4B.

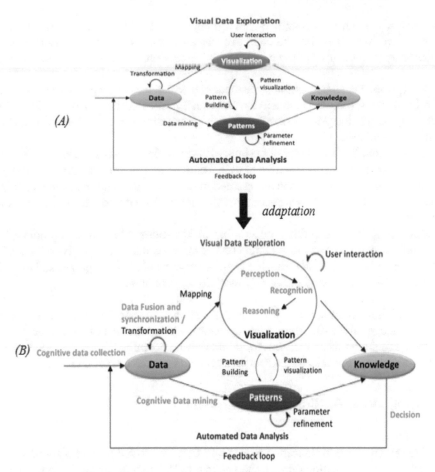

Fig. 4. The proposed cognitive Visual Analytics process

The process begins by a cognitive data collection. Contrary to traditional data collection, we propose to use artificial intelligence to mimic the way the human mind reads data. This method learns to automatically recognize a lot of data rather than traditional configuration by experts. It is more adapted to collect data from visual interfaces. A set of artificial intelligence tools can be applied such as the eye tracking [11], etc. The need for this method comes from the interest presented by different fields of application such as those of monitoring and assessments. In the case where different data collection tools have been used, it is a question of merging and synchronizing them. Otherwise, an only transformation is made on the collected data to prepare them for data mining. The data fusion and synchronization consist of searching connection between the collected data from the used cognitive tools for finally creating a single dataset before automatic and visual analysis.

The treated data will be used in the data mining algorithm and mapped for creating visualization at the same time. Concerning the automatic analysis, the applied data mining algorithm must be adapted to the cognitive collected and synchronized data. For this reason, we named it cognitive data mining in the Fig. 4B. Concerning the visual analysis, we have specified the cognitive visualization process of the user/decision-maker/analyst. It consists of three main steps: perception, recognition and reasoning.

a) Perception. Visual perception is the means by which users interpret visual interfaces on a computer screen. it is an activity that usually relies on information delivered by one's senses. It is the process of collecting and processing sensory information and the resulting awareness.

b) Recognition. It is the ability of the human brain to identify what has been previously perceived. Recognition is a cognitive skill that allows users to retrieve information stored in memory and compare it to the represented information. Recognition is, in fact, a type of recovery. It consists of having access to information from the past stored in the human memory.

c) Reasoning. It is a cognitive process that makes it possible to pose a problem in a thoughtful way with a view to obtaining one or more results. The objective of reasoning is to better define/understand what has been perceived and then recognized and to verify its reality, using alternately different "laws" and experiences.

Combining visual and automatic analysis allows users to better generate insights and knowledge for making appropriate decisions. This proposal, as they will be demonstrated throughout this article, aims to add interesting potential for VA to support human decision-making. For this, we present its application on the Education field.

4   Application for Formative Assessment

A.   Context

Visual Analytics in Education field proposes methods and platforms for delivering education information through the Internet [13], trying to reach the improvement of the learning performance of the students (groups and individuals) and forecast their academic performance level in each subject. Many conditions are considering like the students' environment, social competences and emotional maturities in our context, we propose to design and develop a Visual Analytics system, based on the proposed process, which allows formative assessments in university courses. Following we present the process application.

B.   Process Application

Cognitive Data Collection
This research came from the difficulties that teachers face in recognizing whether their students are benefiting from formative assessment, or not. For this, we combined the eye tracking data and EEG signals taken from students while editing, compiling and executing a JAVA program. The data collection was performed while making the students in front of the computer screen doing their tasks. Hence, 55 students in the same class were equipped with: The Electroencephalogram (EEG) device and Eye tracker:

1. The Electroencephalogram (EEG). Which is a test made to the brain for measuring the electrical activity. The EEG signals are voltage signals emerged in the neural activities. These signals can be used either to classify the state of mind or find a disorder in neural activity [13]. The EEG also allows evaluating the mental workload in many situations (mental fatigue, memory load while performing tasks) because the signals are sensitized to variations in mental workload.

The EEG device used is Biosemi Active amplifiers at rate of 512 Hz. The data was recorded from 72 channels. This device sets new standards for multi-channel, high resolution bio potential measurement systems for research applications. The EEG signals was band-pass filtered from 0.1 to 100 Hz. Each participant is instrumented with biosemi active amplifiers.

2. Eye tracker. Eye Tracking is a powerful tool that has been implemented for analyzing the user attention while during information processing tasks [11]. It is considered as an attention index to evaluate cognitive conflicts in human-machine interaction [16]. Typically, the recorded data have temporal information on the area of interest that participants are interested in or the fixation of the eyes on the screen.

An Eyelink 1000 tracker is used to collect the eye movement data. This device has high speed camera. It has been designed to interchangeably fit into different mounting options and assures the unprecedented low noise, fast sampling rates and stability [15]. Every participant putted this device, while the formative assessment in progress, and keep doing its tasks as usual.

Data Fusion and Synchronization

After recording data from the eye tracker and EEG, we have combined these data with EEGLab MATLAB toolbox. The original raw EEG data obtained was preprocessed in EEGLab. The parameters of each recorded technique (eye tracking and EEG signals) were set as input in the fuzzy logic technique with MAMDANI fuzzy inference system and the output was the students' concentration level. We have eight input parameters which are: area of interest, heat map, fixation, saccade, artifact and frequency, brightness, noise. The first four parameters are the parameters of the eye tracking technique, the following two parameters are the parameters of the EEG signals and the last two parameters made to improve the student's concentration level since the brightness degree as well as the noise level strongly influences on the concentration of the student.

Automatic Analysis

The cognitive data mining in our context was done using the fuzzy logic technique. The objective is to perform an automatic analysis for the prepared data in order to predict whether a participant is distracted or concentrated.

The fuzzy logic is a computing approach based on "degrees of truth" rather than the usual "true or false" Boolean logic (1 or 0). It deals with the '0' and the '1' as extreme cases of the truth, but also the various states of this truth located between the two. It seems closer to the functioning of our brain.

Motivation for Using the Fuzzy Logic. In this work we deal with human being which means that we deal with the uncertainty because, in our case, a student could have eyes fixed on screen and high EEG signal frequency but in reality, he/she is distracted.

The choice of the fuzzy logic is based on the following reasons: the fuzzy logic technique supports ambiguity and uncertainty that can be issued from the EEG and eye tracker devices. Also, it offers flexibility as well as simplicity of use. Fuzzy algorithms are often robust, in the sense that they are not very sensitive to changing environments and erroneous or forgotten rules. In addition, the fuzzy methods usually have a shorter development time than the conventional methods.

To apply the fuzzy logic, we have associated linguistic variables to the parameters of the prepared data as presented in Table 1. To treat uncertainty in recorded data is to express them in fuzzy numbers form related to the linguistic variables of the membership functions. A fuzzy number is equal to a quantity whose value is imprecise. In Table 2, are visible the fuzzy numbers of the linguistic terms of the fuzzy logic output variable named Concentration_level. The result will imitate more truthfully the behavior of the student.

Table 1. Linguistic variables of the input parameters.

Parameters	Linguistic variables	Description
Area of interest	Poor focus, average focus, high focus	The area that the participant focuses in the most
Heat map	Poor n_o_g_f, average n_o_g_f, high n_o_g_f	Indicate the rate of participants' number of gaze fixation
Fixation	Low fixation, average fixation, high fixation	Whether the participant fixed his eyes on screen for longtime
Saccade	Short, average, long	The rate of participants' saccade
Artifact	Low, medium, high	The rate of signals that recorded on the EEG and not generated by the brain
Frequency	Low, average, high	The frequency of the EEG signal
Brightness	Low, medium, high	The brightness degree on the screen in front of the student
Noise	Low, medium, high	The noise level in the class

The fuzzy logic application is supposed to treat the recorded results and reasons using linguistic variables. It applies a set of rules and takes place in three steps:

1. Fuzzification. The numeric variables are converted to linguistic variables with a membership function defined in the interval [0, 1].

μ_A: X --> [0,1] : *quantifies the grade of membership of the element in X to the fuzzy set A Where A is the fuzzy set and each element of X is mapped to a value between 0 and 1.*

2. Motor Inference. To apply the fuzzy logic rules in order to tolerate imprecision and uncertainty. We have developed 52 rules. A fuzzy rule is a linguistic IF–THEN construction,

Table 2. Linguistic terms of the output variable 'concentration_level'.

Linguistic terms	Fuzzy numbers: From–To	Mean
Distracted	0–10	5
Motivated not concentrated	5–15	10
Not motivated and concentrated	10–20	15
Motivated and concentrated	15–25	20

If <input variable is A> then <output variable is B >
Where: A and B are descriptors of knowledge pieces containing linguistic variables and the fuzzy rule expresses a relationship between inputs and outputs variables.

We used the Mamdani's fuzzy inference method [16]. Mamdani's work was founded on fuzzy algorithms for complex systems and decision processes [17].

3. Defuzzification. To produce single output and to transform this linguistic output variable into numerical data.

Visual Analysis

Visual Mapping. The generated visualization is presented in Fig. 5. It is subdivided into three main parts: Fig. 5A for students' information, Fig. 5B for visual Analysis and Fig. 5C for automatic analysis.

The goal of this visualization is to interactively analyze the data of one participant to decide whether this participant is benefiting of this formative assessment. It allows teachers to evaluate the concentration of their students (participants). The interface presents the formative assessment of the student colored in red (cf. Fig. 5A).

The representation of the collected data is presented in Fig. 5B, which is subdivided into two parts: the eye tracker part that indicates the data collected by the EyeLink (participants' fixation and saccade) and the part of EEG signal that indicates the curve generated by the EEGLab with the indication of the participants' fixation and saccades time during the record phase. The legend in the right helps to specify the fixation (in blue) and saccade (in green). The third part is the fuzzy logic (cf. Fig. 5C) contains three parts: the rule (generated in the Mamdani fuzzy inference system) that correspond to the current student data, the parameters (the inputs) values of the recorded data from the current student and finally the result which indicate the concentration level of this student.

Cognitive Activities of the Visualization

The visual analysis of the developed interface involves three main cognitive activities, as presented in Sect. 3.

1. Perception. The visual interface integrates several visual elements (i.e. Histograms, Heat map, Angular histogram, Main Sequence and curves) related to the eye tracker and

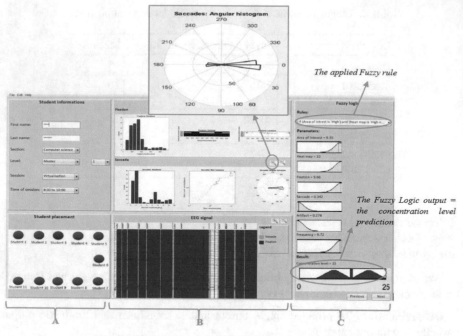

Fig. 5. The proposed cognitive Visual Analytics process

EEG data/parameters (cf. Fig. 5B). They are graphic representation that structure and explore causal relationships between data parts. The interactivity mechanisms (zoom in, zoom out and Legend) support visualization flexibility. In fact, the teacher can click over nodes (cf. Fig. 5A) to display monitoring information of other. In addition, the teacher can visualize computerized results generated by Fuzzy Logic technique.

2. Recognition. The teacher interacts with the visual representations to gain knowledge associated to the students' concentration level, to retrieve data associated to the displayed information from the dataset using the interaction techniques, and to visually interpret the differences and/or similarities between the displayed data. To understand the generated result, the teacher can apply the interaction mechanisms by zooming or filtering, which can help him /her to concentrate and pay attention. He/she enthusiastically interplays between global and local views of the displayed data. Teacher can take different viewing positions while representations can reveal other important information.

3. Reasoning. The teacher looking at the interactive visual histograms, heatmaps, main sequence and curves can determine why the students has such concentration level and then analyzes this information using his/her visual reasoning skills.

After acquiring knowledge on the monitored students' concentration level generated by the system and his/her tacit knowledge, the teacher (i.e., decision-maker) can react to improve the quality of its formative course.

After acquiring knowledge on the monitored students' concentration level generated by the system and his/her tacit knowledge, the teacher (i.e., decision-maker) can react to improve the quality of its formative course.

5   Evaluation

A. Utility Evaluation

To evaluate the utility of our adapted VA system, we have applied the confusion matrix [18] (Fig. 6) to assess the prediction ability of our fuzzy logic. Figure 6 presents our confusion matrix concerned by a dataset composed of 55 students where all instances are classified according to the output linguistic variables.

```
=== Confusion Matrix ===

a  b  c  d    <-- classified as
8  0  0  0  |  a = Distracted
0  9  1  0  |  b = Motivated_not_concentrated
0  0  6  0  |  c = Concentrated_not_motivated
0  0  0  26 |  d = Concentrated_and_motivated
```

Fig. 6. The confusion matrix

Figure 7 presents the predictive measures generated by the confusion matrix. These measures are calculated by output linguistic variable (Distracted, Motivated_not_concentrated, Concentrated_not_motivated, and Concentrated_and_ motivated). For example, for the "Concentrated_ and_motivated" class the True Positive (TP) Rate (TPR), which corresponds to cases where the instances are correctly classified, is equal to 1. The False Predictive (FP) Rate, corresponding to the instances classified as "Concentrated_and_motivated" but actually belong to another class, is equal to 0. The accuracy rates (precision) are interesting, the lowest of which is 85,7%.

=== Detailed Accuracy By Class ===

	TP Rate	FP Rate	Precision	Recall	F-Measure	MCC	ROC Area	PRC Area	Class
	1,000	0,000	1,000	1,000	1,000	1,000	1,000	1,000	Distracted
	0,900	0,000	1,000	0,900	0,947	0,937	1,000	1,000	Motivated_not_concentrated
	1,000	0,023	0,857	1,000	0,923	0,915	1,000	1,000	Concentrated_not_motivated
	1,000	0,000	1,000	1,000	1,000	1,000	1,000	1,000	Concentrated_and_motivated
Weighted Avg.	0,980	0,003	0,983	0,980	0,980	0,977	1,000	1,000	

Fig. 7. Prediction measures

B. Usability Evaluation

The aim is to assess our visual mapping usability; we have made a controlled experiment where 30 persons have participated. In this study, we have designated 3 categories of academic participants: novice, knowledge-intermittent and expert. Table 3 presents the user profiles that have been involved for the evaluation task.

Table 3. Participants profile.

Types of participants	Number	Age	Skill
Novice	10 students	18–20	Modest experience in using computer
Knowledge-intermittent	10 scientific students	21–34	Good experience in using computer: 15–17 years
Expert	10 computer science teachers	35–45	Expert experience in using computer

Evaluation Protocol. We divided the participants randomly into groups of five. We asked them each individually to remember strategy implementation problems that they had already experienced. We collected that individual experience and then let the groups work for 45 min. Half of the groups using non-visual user interface support and half of the groups using our he visual system. After that, we issued a post experiment survey, where we collected subjective metrics such a satisfaction with the outcome and satisfaction with the process. We also captured all the results. 45 min later, after a distraction task, we issued a recall test to see how much these participants remember from the meeting they had.

Evaluation Results. We found that the results are very intriguing. Not only we did find that the productivity of the teams using our visual tool was significantly higher but also the quality of their work was higher even more. So, on individual basis the recall and knowledge gains that we tested were much higher in the visual tool support groups than in the non-visual groups. Through this controlled experiment, we have shown that our tool is very relevant for knowledge acquiring and decision-making.

6   Conclusion

The proposed process integrates cognitive aspects from data collection [21] to the automatic and visual analysis. It allows studying and analyzing cognitively collected data to gain related knowledge [20]. The proposed process was applied to support the formative assessments. We have developed several visualization techniques to visually analyze cognitive data sets issued from Eye Tracker and EEG devices. These data are displayed using interactive graph representations (histogram, heatmap, etc.) to be analyzed and visually interpreted (perceived, recognized, and reasoned) for generating related knowledge. We have developed the Fuzzy Logic technique as a data mining technique able to deal with and analyze the collected cognitive data. We have evaluated the system utility by verifying the prediction ability of the fuzzy logic using the matrix confusion, which shows interesting results. In addition, we have conducted a controlled experiment-based usability evaluation for this visual tool. It showed that the knowledge gains were significant. Such evaluation reflected the feasibility of the suggested process.

Future outlook of this work concerns the application of the proposed process for acquiring knowledge other kinds of patterns such as the association rules and probabilities. We plan also to propose a cognitive VA process that generates decisions from knowledge.

References

1. Ltifi, H., Kolski, C., Ben, A.: Combination of cognitive and HCI modeling for the design of KDD-based DSS used in dynamic situations. Decis. Support Syst. **78**, 51–64 (2015)
2. Keim, D., Andrienko, G., Fekete, J.-D., Görg, C., Kohlhammer, J., Melançon, G.: Visual analytics: definition, process, and challenges. In: Kerren, A., Stasko, J.T., Fekete, J.-D., North, C. (eds.) Information Visualization. Lecture Notes in Computer Science, vol. 4950, pp. 154–175. Springer, Heidelberg (2008)
3. Blascheck, T., John, M., Kurzhals, K., Koch, S., Ertl, T.: VA2: a visual analytics approach for evaluating visual analytics applications. IEEE Trans. Vis. Comput. Graph. **22**(1), 61–70 (2016)
4. Ltifi, H., Amri, S., Ben Ayed, M.: Fuzzy logic–based evaluation of visualizations generated by intelligent decision support systems. Inf. Vis. **17**(1), 3–21 (2016)
5. Borra, S., Dey, N., Bhattacharyya, S., et al. (eds.): Intelligent Decision Support Systems. Applications in Signal Processing. De Gruyter, Berlin, Boston (2019)
6. Li, G., Kou, G., Peng, Y.: A group decision making model for integrating heterogeneous information. IEEE Trans. SMC Syst. **48**(6), 982–992 (2018)
7. Fayyad, U.M., Djorgovski, S.G., Weir, N.: Automating the analysis and cataloging of sky surveys. In: Fayyad, U.M., Piatetsky-Shapiro, G., Smyth, P., et al. (eds.) Advances in Knowledge Discovery and Data Mining, pp. 471–494. The MIT Press, Menlo Park, CA (1996)
8. Ltifi, H., Benmohamed, E., Kolski, C., Ben Ayed, M.: Enhanced visual data mining process for dynamic decision-making. Knowl. Based Syst. **112**, 166–181 (2016)
9. Wijk, V.J.J.: The value of visualization. In: IEEE Visualization, pp. 79–86 (2005)
10. Sacha, D., Stoffel, A., Stoffel, F., Kwon, B.C., GKeim, D.A.E.: Knowledge generation model for visual analytics. IEEE Trans. Vis. Comput. Graph. **20**(12), 1604–1613 (2014)
11. Rayner, K.: Eye Movements and Visual Cognition: Scene Perception and Reading. Springer, New York (2012)
12. Wang, S.-L., Hwang, G.-J.: The role of collective efficacy, cognitive quality, and task cohesion in computer-supported collaborative learning (CSCL). Comput. Educ. **58**(2), 679–687 (2012)
13. Pradhan, C., Shariar Rahaman, M., Sheikh, A., Kole, A., Maity, T.: EEG signal analysis using different clustering techniques. In: Abraham, A., Dutta, P., Mandal, J.K., Bhattacharya, A., Dutta, S. (eds.) Emerging Technologies in Data Mining and Information Security: Proceedings of IEMIS 2018, Volume 2, pp. 99–105. Springer, Singapore (2019)
14. Dehais, F., Causse, M., Vachon, F., Tremblay, Sébastien.: Cognitive conflict in human–automation interactions: a psychophysiological study. Appl. Ergon. **43**(3), 588–595 (2012)
15. Research SR: Eyelink 1000 users' manual, version 1.5. 2 (2010)

16. Mamdani, E.H., Assilian, S.: An experiment in linguistic synthesis with a fuzzy logic controller. Int. J. Man Mach. Stud. **7**, 1–13 (1975)
17. Zadah, L.A.: Outline of a new approach to the analysis of complex systems and decision processes. IEEE Trans. Syst. Man Cybern. **3**(1), 28–44 (1973)
18. Mohana Shankar, P.: Pedagogy of Bayes' rule, confusion matrix, transition matrix, and receiver operating characteristics. Comput. Appl. Eng. Educ. **27**(2), 510–518 (2019)
19. Jemmaa, A., Ltifi, H., Ayed, M.: Multi-agent architecture for visual intelligent remote healthcare monitoring system. In: Abraham, A., Han, S.Y., Al-Sharhan, S.A., Liu, H. (eds.) Hybrid Intelligent Systems. Advances in Intelligent Systems and Computing, vol. 420, pp. 211–221. Springer, Cham (2016)
20. Ogiela, L.: Syntactic approach to cognitive interpretation of medical patterns. In: Xiong, C., Huang, Y., Xiong, Y., Liu, H. (eds.) Intelligent Robotics and Applications, pp. 456–462. Springer, Heidelberg (2008). https://doi.org/10.1007/978-3-540-88513-9_49
21. Ogiela, L.: Cognitive informatics in automatic pattern understanding and cognitive information systems. In: Yingxu Wang, D., Zhang, W.K. (eds.) Advances in Cognitive Informatics and Cognitive Computing, pp. 209–226. Springer, Heidelberg (2010)

Secure Distribution of Factor Analysis of Mixed Data (FAMD) and Its Application to Personalized Medicine of Transplanted Patients

Sirine Sayadi[1,2(✉)], Estelle Geffard[2], Mario Südholt[1], Nicolas Vince[2], and Pierre-Antoine Gourraud[2]

[1] STACK Team, IMT Atlantique, Inria, LS2N, Nantes, France
{Sirine.Sayadi,Mario.Sudholt}@imt-atlantique.fr
[2] University of Nantes, Nantes University Hospital, INSERM, Research Center in Transplantation and Immunology, UMR 1064, ATIP-Avenir, Nantes, France
Estelle.Geffard@etu.univ-nantes.fr,
{Nicolas.Vince,Pierre-Antoine.Gourraud}@univ-nantes.fr

Abstract. Factor analysis of mixed data (FAMD) is an important statistical technique that not only enables the visualization of large data but also helps to select subgroups of relevant information for a given patient. While such analyses are well-known in the medical domain, they have to satisfy new data governance constraints if reference data is distributed, notably in the context of large consortia developing the coming generation of personalised medicine analyses.

In this paper we motivate the use of distributed implementations for FAMD analyses in the context of the development of a personalised medicine application called KITAPP. We present a new distribution method for FAMD and evaluate its implementation in a multi-site setting based on real data. Finally we study how individual reference data is used to substantiate decision making, while enforcing a high level of usage control and data privacy for patients.

1 Introduction

Big data analysis techniques are increasingly popular to extract new information from massive amounts of data to improve decision making, notably in the medical sector. A major challenge for clinicians consists in safely making correct treatment decisions based on ever growing amounts of patient data. This problem calls for new analysis techniques and algorithms, in particular, for precision medicine.

Precision Medicine, that is using genetic or molecular profiling for optimizing care of small patient groups, will probably become the standard of care in the next decades. It represents a deep revolution in health care also because not only patients will be covered but also healthy individuals. For this (r)evolution of medicine to become reality, it is necessary to deliver the evidence of the efficiency and cost-effectiveness of precision medicine. This, in turn, requires decisions concerning patients to be substantiated by an analysis of large-scale reference data that is relevant for their personal situation compared to others [2].

Dimension reduction is a major technique for transforming large multi-dimensional data spaces into a lower dimensional subspaces. This is while preserving significant characteristics of the original data. Among dimension reduction methods, the most common method is Principal Component Analysis (PCA) [3], which enables dimension reduction for quantitative data variables. Other methods are Factor Analysis of Mixed Data (FAMD) [5], which performs dimension reduction for mixed (quantitative and qualitative) data variables, and dictionary learning (DL) [4] one of the most powerful methods of extracting features from data.

FAMD analysis provides simplified representations of multi-dimensional data spaces in the form of a point cloud within a vector subspace of principal components. If two points are close to each other in this cloud, a strong global similarity exists between them with respect to the selected principal components. In the biomedical field, this kind of analysis is frequently used to present patients groups in a simplified and visual way for a large range of complex clinical data encompassing quantitative data (for example obtained from biological exams) and qualitative data (for example gender information). The result are actionable representations of each patient's individual characteristics compared to those of others.

In the context of a French public-private partnership KTD-innov and a H2020 EU project EU-train, FAMD has been used for dimension reduction as part of the clinical decision support system KITAPP (the kidney transplant application) [6]. This precision medicine web-application computes predictive scores and represents distributions of patients' variables in a subgroup of reference patients after kidney transplantation. The application is conceived to relay the intuition and experience of clinicians by means of on-demand computations and graphical representations. One of KITAPP's key functionalities consists in the "contextualization" of patients relative of a population of reference (POR). To this end, it first uses FAMD for dimension reduction, then applies a percentile statistical modelling [21] algorithm, and visualizes the relations of patients to the POR.

Medical studies often involve large national or international collaborations (such as our KTD-innov and EU-Train projects). Simple centralization schemes for the placement of data and computations are frequently not applicable in this context because data and computations may not be shared due to legal reasons, security/privacy concerns and performance issues). To deploy this kind of medical services in larger contexts distributed systems and algorithms for precision medicine have to be provided. One of the main lines of research around the KTD-innov and EU-TRAIN projects is the implementation of a reference database integrated into a distributed computing infrastructure allowing a secure access to data while respecting the European GPDR data protection regulation [12]. However, very few distributed algorithms have been developed for and applied to the domain of precision medicine.

Data sharing and analysis placement are generally difficult due to governance, regulatory, scientific and technical reasons. Analyses are often only possible "on premise." Furthermore, researchers and institutions may be averse to lose control over both data usage. In addition, huge volumes of data are intrinsically difficult to share or transfer, notably because of cost of the use of computational, storage and network resources.

On the other hand distributed architectures enable more flexible data governance strategies and analysis processes by freeing them from centralization constraints [9]. Decentralized databases enable performing local calculations on patient data, without any individual data circulating outside the clinical centers generating the data. To this end, one may strive for distributed statistical calculations. Fully-distributed analyses have been proposed, see Sect. 2, for contextualizing the state of a patient relative to POR data stored in a distributed database. Such algorithms have to meet requirements of scalability, security and confidentiality [10], as well as availability properties and right to privacy properties [11]. These criteria are difficult to satisfy, however, because the statistical significance and accuracy of analyses often directly depend on the number of cases or individuals included in the database.

A solution to these problems can be based on harnessing distributed analyses that manipulate sensitive data on the premises of their respective owners and harness distributed computations if non sensitive, aggregated, summarized or anonymized data is involved.

In this paper, we present two main contributions:

- We motivate and define requirements for distributed algorithms for dimension reduction in the context of the KITAPP project.
- We present a novel distributed FAMD algorithm for dimension reduction in the presence of sensitive data in precision medicine and apply it to a contextualization problem.

The rest of this article is organized as follows. Section 2 presents related works. Section 3 presents the kidney transplantation application (KITAPP) and its use of FAMD dimension reduction. Section 4 presents our distributed algorithm FAMD and a corresponding implementation. Section 5 provides experimental results, notably concerning privacy requirements, and a performance evaluation. Finally, Sect. 6 summarizes our findings and proposes some future work.

2 Related Work

Parallel versions of PCA dimension reduction algorithms have already been proposed. Liang et al. [14] propose a client-server system and send singular vectors and singular values $U\Sigma V$ from the client to the server. Feldman et al. [15] have shown how to compute PCAs by sending smaller matrices $U\Sigma$ instead of sending all matrices of a singular value decomposition, thus improving on the communication cost. Wu et al. [16] have introduced an algorithm that improves on the storage and data processing requirements and harnesses Cloud computing for PCA dimension reduction. These proposals send matrices of synthesized data of the original data and not real data. This is very interesting for biomedical analyses in order to ensure privacy of patient data. Imtiaz et al. [17] have improved Feldman et al.s' proposal by adding privacy guarantees using differential privacy.

To the best of our knowledge, no distributed FAMD algorithm has already been proposed. In this paper, we propose a distributed FAMD on the basis of a distributed PCA algorithm. Our algorithm is structured into two parts (similar to Pagès [5]):

1. Transform qualitative data into quantitative one using complete disjunctive tables [13], thus transforming the original FAMD dimension reduction problem into a PCA one.
2. Perform the dimension reduction of the distributed PCA based on Feldman *et al. s'* proposal [15].

3   The Kidney Transplantation Application (KITAPP)

Chronic kidney failure affects approximately 10% of the world population and can progressively lead to end-stage kidney disease requiring replacement therapy (dialysis or transplantation). Kidney transplantation is the best treatment for end-stage kidney disease [19].

3.1   KITAPP Overview

Data from approximately 1500 renal transplantation, including clinical and immunological items, were collected since 2008 as part of a French national project.

KITAPP enables personalized contextualization algorithm to be harnessed to compare data trajectories of a given patient (POI) to a sub-population with similar characteristics (POR) selected by filters or distance measures. The information relative to a graft is selected from similar cases at the time of the graft. With the help of clinicians and knowledge of the existing body of research, we defined a set of variables to select the sub-population of reference.

We propose three population contextualization algorithms: compare a given patient's data to PORs with

1. similar characteristics selected by filters or approaches based on statistical analysis;
2. the nearest neighbor method or
3. the cluster method.

With our filter approach (1), the POR is defined according to selected filters made available to the clinician, such as age, gender and Body Mass Index (BMI). Methods (2) and (3) are based on the results of an FAMD. Following this analysis, we can then select a POR by close neighbor method (2): by selecting the N individuals most similar to a POI or we can select a POR by clustering (3): by selecting the individuals in the same cluster as our POI.

The visualization of contextualized information is done by comparing a POI's biological data (creatinemia) and its evolution over time post-transplantation (clinical visits) to a POR that is represented by their median and percentile values.

3.2 Motivation for Distributed Analyses for KITAPP

We intend to harness the KITAPP application as part of large-scale cooperations with many (national and international) partners. To meet the challenge of harnessing medical data while keeping sensitive data on premise or ensure strong data protection if data is moved, computations are often performed today over distributed databases that are linked to a computation integrator that enables a center to interact with and access some data from remote sites. Each clinical center collects, stores and controls their own patients' data. The founding principle of the architecture is that no data of individuals circulates outside the centers. However, this sharing paradigm is very restrictive and inhibits a large range of potential analyses to be performed - either because sensitive data cannot be appropriately protected or the analysis cannot be performed sufficiently efficient.

The need for local storage and distribution of reference data is motivated by the actionable value and publication value. It provides the possibility of controlling locally who has accessed to data, what are the usages of the local data and how to limit then should it be needed. The use of distributed infrastructure is a central element of multi-stakeholders data governance.

We are therefore working on more general distributed analysis architectures and implementations that ease collaboration as part of multi-centric research projects, where each center can control and account for their own patients' data usage even if located remotely. Contextualization then has to be performed with respect to large-scale distributed medical databases that are maintained at different sites.

4 Distributed FAMD

In the following we first provide an overview over the architecture and properties of our algorithm before defining it in detail.œ

4.1 Overall Architecture and Properties

Factor analysis of mixed data (FAMD) [5] is a method of dimension reduction of variables including mixed quantitative and qualitative data into fewer components for information synthesis reasons. This analysis can defined, for instance using matrix operations, as follows:

$$FAMD = PCA + MCA \tag{1}$$

where PCA is a principal component analysis dimension reduction for quantitative variables and MCA is a multiple correspondence analysis dimension reduction for qualitative variables.

Overall, our algorithm works as follows. As a first step, we transform the qualitative variables into quantitative ones using complete disjunctive coding (CDC) [20] that is performed locally on each site. In a second step we perform dimension reduction by means of a distributed PCA in order to obtain a secure and distributed FAMD algorithm.

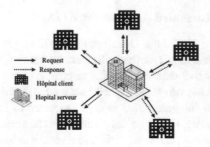

Fig. 1. Collaboration architecture

We harness the distributed cooperation architecture shown in Fig. 1. Data transformation and dimension reduction analysis are performed locally at multiple sites separately. The coordination between the sites is done by an aggregator site, which receives synthetic data and performs the overall dimension reduction. Imtiaz *et al.* [17] have proposed a secure and distributed algorithm for PCA dimension reduction. This algorithm uses differential privacy as a security technique and synthetic data for communication between nodes.

The resulting algorithm has two important properties :

- *Low communication cost:* The communication cost of parallel and distributed FAMD algorithms essentially depends on the size of the matrices transferred between sites. Many dimension reduction algorithms require matrices of size $D \times D$ to be sent, where D is the number of data items to be analyzed (that is, transplantation data in our case). In contrast, Imtiaz *et al.*' [17] algorithm requires matrices to be sent of type $D \times R$ where R is the number of variables and (typically) $R \ll D$.
- *Security/privacy awareness:* Our algorithm satisfies two interesting characteristics: (1) differential privacy is used for data protection and (2) communication between the sites and the aggregator involves only synthesized data P_S and not the original data, which minimizes possibilities of data theft and also supports data protection.

We harness the same principle and properties while providing two new contributions: (1) a transformation of qualitative variables into quantitative variables to obtain a secure and distributed algorithm FAMD dimension reduction and (2) a scalable distributed implementation and evaluated it by analyzing real-world biomedical data on a realistic grid environment.

4.2 Algorithm Definition

Algorithm 1 presents our secure and distributed FAMD algorithm. Lines 1–16 implement the first step, the transformation of a full FAMD problem into a (qualitative) PCA problem. Each site begins by calculating, for each quantitative variable, the corresponding mean μ_k, standard deviation σ_k and Centering and Reduction Function $X_{i,k} = \frac{1}{\sigma_k}(x_{i,k} - \mu_k)$. For each qualitative variable, Complete Disjunctive Coding using

---

Algorithm 1: Distributed FAMD Algorithm

Input : Data matrix $X_s \in \mathbb{R}^{D \times N_s}$ for $s \in [S]$ of N elements and P variables, with C quantitative variables and M qualitative variables; ε, δ. privacy parameters, j. reduced dimension;

Output: V_j: Matrix of eigenvectors on top j

1 **foreach** *site* $s \in S$ **do**

2 $\quad$ **foreach** *element* $i \in N$ **do**

3 $\quad\quad$ **foreach** *element* $k \in C$ **do**

4 $\quad\quad\quad$ Compute the mean of the variable μ_k;

5 $\quad\quad\quad$ Compute the standard deviation of the variable σ_k;

6 $\quad\quad\quad$ Compute the Centering and Reduction Function $X_{i,k} = \frac{1}{\sigma_k}(x_{i,k} - \mu_k)$;

7 $\quad\quad\quad$ **return** $X_{i,k}$;

8 $\quad\quad$ **end**

9 $\quad\quad$ **foreach** *element* $k \in M$ **do**

10 $\quad\quad\quad$ Apply the Complete Disjunctive Coding using (ade4 package on R);

11 $\quad\quad\quad$ Compute the effective of the modality N_k;

12 $\quad\quad\quad$ Compute the proportion $p_k = N_k/N$;

13 $\quad\quad\quad$ Compute the Indicator Weighting Function $X_{i,k} = \frac{x_{i,k}}{\sqrt{p_{,k}}}$;

14 $\quad\quad\quad$ **return** $X_{i,k}$;

15 $\quad\quad$ **end**

16 $\quad$ **end**

17 $\quad$ Compute $A_s = \frac{1}{N_s} X_s X_s^T$;

18 $\quad$ Generate $D \times D$ symmetric Matrix E where $E_{i,j} : i \in [D], j \leq i$ drawn i.i.d. from $N(0, \Delta_{\varepsilon,\delta}^2)$ where $\Delta_{\varepsilon,\delta} = \frac{1}{N s_\varepsilon} \sqrt{2 log(\frac{1.25}{\delta})}$, $E_{i,j} = E_{j,i}$;

19 $\quad$ Compute $A_s = A_s + E$;

20 $\quad$ Perform $SVD(A_s) = U\Sigma U^T$;

21 $\quad$ Compute $P_s = U\Sigma^{1/2}$;

22 $\quad$ Send P_s to the aggregator;

23 **end**

24 Compute $A = \frac{1}{s} \Sigma_{s-1}^s P_s P_s^T$;

25 Perform $SVD(A) = V\Lambda V^T$;

26 Send V_j to all sites ;

27 **return** V_j;

---

the ade4 package from the R language is then applied in order to transform the qualitative variables into a quantitative variable, followed by the computation of the modalities N_k and proportions $p_k = N_k/N$ to compute the indicator weighting function $X_{i,k}$.

The second step, the dimension reduction proper, is implemented on lines 17–23. Each site calculates the (second moment) matrix $A_s = \frac{1}{NS} X_s X_s^T$. The application of the scheme of differential privacy (following Dwork *et al.*' proposal [18]) is performed by generating the noise matrix E of size $D \times D$ and the estimated differential privacy matrix $A_s = A_s + E$ on line 19. Each site then performs the Singular Value Decomposition $SVD(A_s)$) of matrix A_s to compute the matrix $(P_s = U\Sigma^{1/2})$ and broadcast it to the aggregator.

At the aggregator site, the server computes, see lines 24–26, the matrix $A = \frac{1}{s}\sum_{s=1}^{s} P_s P_s^T$ of all sites. It performs next the global Singular Value Decomposition $\tilde{SVD}(A) = V\Lambda V^T$.

5   Experimentation

In this section we report on experiments involving analyses over real medical data that we have carried out on a real heterogeneous large-scale grid infrastructure. We report on our setup, and evaluate our implementation w.r.t. three criteria.

5.1   Setup

Our experiments have been carried out on renal transplantation data available in the European database Divat [22]. In order to compare with results from the KITAPP project, we have applied our distributed algorithm to its analyses on 11,163 transplantation data. We started by divided the data file before transfer and analysis on the different sites.

We have implemented our distributed algorithm and executed it in a grid-based environment featuring different distributed architectures, ranging from placing all clients on different (geo-distributed) machines to placing them as one cluster on only one machine. This distributed environment constitutes a realistic architecture of a medical collaboration involving the research and clinical centers, the partners of the KITAPP project. We have implemented our distributed algorithm using the Python and R programming language using 860 lines of code. The whole distributed system can be deployed and executed on an arbitrary number of sites of the Grid'5000 infrastructure using a small script of only eight commands.

The Grid'5000 platform is a platform, built from eight clusters in two European countries for research in the field of large-scale distributed systems and high performance computing. For our experiment, we have reserved a machine as a server (aggregator) executing a Python program to manage the analysis, client interactions and generation of the final result. To create a number of client sites we have reserved machines distributed over five different sites in France.

5.2   Results and Performance Evaluation

The KITAPP-motivated FAMD dimension reduction analysis we employed as a test case has been executed on the basis of 11,163 transplantation operations characterized using 27 qualitative and quantitative variables distributed over five sites. We have set the dimension reduction parameter j on the server to two. Figure 2 shows the resulting two-dimensional subspace after application of our distributed FAMD analysis.

In order to distribute the POR selection for POI contextualization between sites, we have applied the K-means unsupervised clustering technique to the result of the distributed FAMD dimension reduction analysis. FAMD and clustering enables grouping of patient data according to their similarity and proximity relative to principal components. Figure 3 presents the result of k-means clustering, three independent data clusters

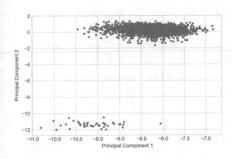

Fig. 2. Distributed FAMD for 5 sites.   **Fig. 3.** Clusterning of distributed FAMD.

that correspond exactly to the result of the (centralized) sequential algorithm that is used as part of the KITAPP project.

Each cluster is characterized by specific variables combinations. The green cluster corresponds to living donors. The yellow cluster corresponds to deceased donors and the purple one to deceased donors with expanded criteria, such as aged *we* > 50 years or subject to hypertension or creatinine levels $\geq 133\,\mu\mathrm{mol/L}$. Note that we always obtain the same clusters independent from the number of sites that participate in the distributed FAMD analysis if we operate it with the same data, which shows a strong scalability potential of our proposed algorithm.

In the following we evaluate three properties of our implementation. (i) the quality of the reduction technique in the presence of noise introduced by the differential privacy technique using a notion of captured energy, (2) execution time and (3) communication cost. For evaluation purposes we consider three architectures: our distributed FAMD reduction technique (denoted"DPdis" below), a more centralized version where W where all the second moment matrix A_s of each clients are aggregated at the server (denoted "fulldis"), and a fully-centralized FAMD version (denoted "pooled").

Captured Energy/Utility. Following Imtiaz *et al.* the captured energy q is used to evaluate the quality of V_j principal directions based on the difference in information utility between the case where all data is centralized q_{pooled}, all second moment matrix A_s of each site are distributed $q_{fulldis}$ and secure distributed proposed FAMD algorithm q_{DPdis} by data size and number of sites.

The captured energy is defined as the matrix multiplication $q = tr(V_j(A)^T A V_j(A))$ measuring the amount of optimal eigenvalues captured in the subspace FAMD. For any other sub-optimal sub-spaces, the value would be less than the optimal value.

– *Energy per site.* We have varied the number of S sites that participate in this analysis by keeping the total number of samples $N = 11163$ (i.e. we decreased the size N_s of each site). Figure 4 shows a deterioration in the performance of $q_{fulldis}$ and q_{DPdis} for an increasing number of sites. This decrease in performance is explained by the decrease in the number of elements per site. In addition, the presence of high variance noise degrades the number of eigen-directions stronger than the noise which are detected by the PCA instead of capturing all the j directions which present the data.

- *Energy by data size.* Figure 5 shows an increase in performance of captured energy q as a function of the elements number per site. $q_{fulldis}$ and q_{DPdis} have almost the same performance in captured energy for the two variation of sites number and data size, q_{pooled} always keeps better performance.

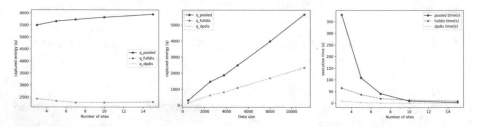

Fig. 4. Captured energy (q) by sites number.

Fig. 5. Captured energy (q) by data size.

Fig. 6. Execution time by sites number.

Execution Time. We have varied the number of S sites by keeping the global number of samples $N = 11163$ (i.e. we decreased the size N_s of each site) and we have measured the execution time. Figure 6 shows that execution time decreases with increasing sites number. Our proposed approach $dpdis$ always keeps the least execution time. This is due to the Lower communication cost explained in 4.1 section.

Communication Cost/Data Sharing: The lower cost of communication of our proposed algorithm introduced in the previous section allows for a minimum sharing of data (matrices Ps are shared and not A_s). In the case of our experiments with distribution on 5 sites with global samples equal to $11,163$ and 27 features, the quantity of data shared by all clients is equal to $11,163 \times 27 = 301.401$ values instead of sending $sqr(11,163) = 124.612.569$ values.

6   Conclusion and Future Work

FAMD dimension reduction is an important tool for transforming complex data into lower-dimensional sub-spaces while preserving important characteristics of the original data. This technique is generally useful to reduce complexity and support decision making. In this paper, we have motivated the use of dimension reduction for geo-distributed biomedical collaborations that require distributed models and implementations of biomedical algorithms with distributed implementation. Its evaluation on a real geo-distributed grid infrastructure using real data has valided its efficiency, scalability, privacy protection properties.

As future work, we will focus on extensions of federated learning as a more general method for the definition of secure and distributed biomedical analyses.

References

1. Hood, L.: Systems biology and p4 medicine: past, present, and future. Rambam Maimonides Med. J. (2013)
2. Gourraud, P., Henry, R., et al.: Precision medicine in chronic disease management: the multiple sclerosis bioscreen. Ann. Neurol. **76**(5), 633–642 (2014)
3. Jolliffe, I.T.: Principal Component Analysis. Springer Series in Stat. Springer (2002)
4. Shakeri, Z., Sarwate, A.D., Bajwa, W.U.: Sample complexity bounds for dictionary learning from vector and tensor valued data. In: Rodrigues, M., Eldar, Y. (eds.) Information Theoretic Methods in Data Science, Chapter 5. Cambridge University Press (2019)
5. Pagès, J.: Multiple Factor Analysis by Example using R. Chapter 3 (2014)
6. Herve, C., Vince, N., et al.: P218 The kidney transplantation application (KITAPP): a visualization and contextualization tool in a kidney graft patients' cohort. Hum. Immunol. **78**, 216 (2017)
7. KTD-Innov. www.ktdinnov.fr. Accessed 07 Jan 2021
8. EU-TRAIN. eu-train-project.eu . Accessed 07 Jan 2021
9. Brous, P., Janssen, M., et al.: Coordinating decision-making in data management activities: a systematic review of data governance principles. In: International Conference on Electronic Government. Springer (2016)
10. Boujdad, F., Gaignard, A., et al.: On Distributed Collaboration for Biomedical Analyses WS CCGrid-Life (2019)
11. Scheel, H., Dathe, H., Franke, T., Scharfe, T., Rottmann, T.: A privacy preserving approach to feasibility analyses on distributed data sources in biomedical research. Stud. Health Technol. Inform. **267**, 254–261 (2019)
12. Regulation (EU) 2016/679 of the European Parliament and of the Council of 27 April 2016 on the protection of natural persons with regard to the processing of personal data and on the free movement of such data, and repealing Directive 95/46/EC (General Data Protection Regulation). http://data.europa.eu/eli/reg/2016/679/oj. Accessed 07 Jan 2021
13. Greenacre, M., Blasius, J.: Multiple Correspondence Analysis and Related Methods. Chapman and Hall/CRC Press, London (2006)
14. Liang, Y., Balcan, M., Kanchanapally, Y.: Distributed PCA and k-Means Clustering (2013)
15. Feldman D., Schmidt, M., Sohler, C.: Turning big data into tiny data: constant-size coresets for K-means, PCA and projective clustering. In: Proceedings of the Twenty-fourth Annual ACM-SIAM Symposium on Discrete Algorithms, SODA 2013, pp. 1434–1453 (2013)
16. Wu, Z., Member, Li, Y., Plaza, A., Li, J., Xiao, F., Wei, Z.: Parallel and distributed dimensionality reduction of hyperspectral data on cloud computing architectures. IEEE J. Select. Top. Appl. Earth Obser. Remote Sens. **9**, 1–9 (2016)
17. Imtiaz, H., Sarwate, A.: Differentially private distributed principal component analysis. In: 2018 IEEE International Conference on Acoustics, Speech and Signal Processing (ICASSP), pp. 2206–2210 (2018)
18. Dwork, C., McSherry, F., Nissim, K., Smith, A.: Calibrating noise to sensitivity in private data analysis. In: Halevi, S., Rabin, T. (eds) Theory of Cryptography. TCC 2006. Lecture Notes in Computer Science, vol. 3876. Springer, Berlin, Heidelberg (2006)
19. Hill, N., Fatoba, S., et al.: Global prevalence of chronic kidney disease - a systematic review and meta-analysis. PLoS One (2016)
20. Mellinger, M.: Correspondence analysis in the study of lithogeochemical data: general strategy and the usefulness of various data-coding schemes. J. Geochem. Explor. **21**(1–3), 455–469 (1984). ISSN 0375-6742
21. Sayadi, S., Geffard, E., Südholt, M., Vince, N., Gourraud, P.: Distributed contextualization of biomedical data: a case study in precision medicine. In: AICCSA 2020 - 17th IEEE/ACS International Conference on Computer Systems and Applications, pp. 1–6, November 2020

22. Divatfrance. www.divat.fr. Accessed 07 Jan 2021
23. Balouek, D., Carpen Amarie, A., Charrier, G., et al.: Adding virtualization capabilities to the grid'5000 testbed. In: Cloud Computing and Services Science. Springer (2013)

Comparative Study of Traditional Techniques for Unsupervised Autonomous Intruder Detection

Anik Alvi[✉], Tarem Ahmed, and Mohammad Faisal Uddin

Independent University, Bangladesh (IUB), Dhaka, Bangladesh
{alvi,tarem,faisal}@iub.edu.bd

Abstract. In this paper we investigate five traditional techniques to extract features within a face image, and we evaluate them by applying the Kernel-based Online Anomaly Detection (KOAD) algorithm. The main objective of this work is to explore the various fundamental feature extraction techniques that can be used to identify whether a person's face is covered by a mask or not. Although face covering or wearing a mask is recommended during this global COVID-19 pandemic, deliberate face occlusion is considered to be a suspicious activity in a normal scenario. Even during this pandemic, it may be considered suspicious if someone is covering his/her face inside an ATM booth or an apartment complex during odd hours, for instance. Our proposed framework detects such intrusion activity by combining a traditional face detection algorithm with KOAD. Comparative analysis is performed for each filter used and we show that our proposed system achieves high detection accuracy with low computational complexity, while also providing the added benefits of being adaptive, portable, and involving low infrastructural costs.

1 Introduction

In the modern world, visual surveillance is present in almost every place. From malls to banks, offices to homes, every place is monitored with many cameras for security reasons. It becomes tedious and monotonous for a security officer to stare at multiple screens for a long periods of time. This leads to many suspecting activities going unnoticed. To avoid such incidence, there is a need for automated visual surveillance algorithms which can detect any intrusion activity removing the burden of human security officers to constantly monitor. Before the COVID-19 pandemic, wearing a mask was an unusual activity and was regarded as suspicious, especially in sensitive places like ATM booths and offices. Now, not wearing a mask in these places can be considered an anomalous activity. For both scenarios, detecting and informing such activity to security officers has become an important issue. We hope that the present pandemic scenario is a temporary one, upon the end of which camera operators would go back to signaling the wearing of a mask as the anomalous activity to be alerted to.

We investigate a framework consisting of Face Detection, Feature Extraction and Unsupervised Pattern Matching steps to develop a system of autonomously detecting anomalous images from an image sequence. The objective is to develop a framework

L. Barolli et al. (Eds.): AINA 2021, LNNS 225, pp. 519–530, 2021.
https://doi.org/10.1007/978-3-030-75100-5_45

which will signal the absence of a mask in the present pandemic state-of-affairs, and conversely the presence of any face covering in the old normal future. To this end, we incorporate the legacy Viola-Jones Face Detector [2], perform a comparative study of the fundamental feature extraction techniques of Wavelet Decomposition [3], the Gabor Filter [4], and Canny [5], Laplacian [6], and Sobel Edge Detectors [7], and the Kernel-based Online Anomaly Detection (KOAD) [8,9] algorithm to learn in unsupervised manner.

Whilst a plethora of commercial tools for mask detection have recently sprung up during the pandemic [10], a vast majority of these software remain proprietary. Their algorithmic foundations are not available in the public domain, which make it difficult to customize the software or to further build on it. We believe that developing image processing algorithms based on established theoretical foundations will lead to robust schemes which stand the test of time and offer a wider scope of applications.

1.1 Related Work

Face mask detection has been a new area of research due to the pandemic. With the government wanting to monitor whether the citizens are properly wearing masks or not, many software companies [10] are linking existing face detection and occluded face detection algorithms to detect whether a person is wearing a mask or not. Recently, a deep learning model like Inception V3 [11] was trained in a masked face data set and achieved almost full accuracy. For the case of visual intruder detection, research is being performed to recognize posture [12] and detect intruders. Bouchrika et al. [13] used gait analysis to identify and track a person. Even occluded face detection can be considered as intruder detection from the assumption that the person is concealing his/her face. Algorithms like Face Attention Network [14] have used anchor-level attention to highlight the features from an occluded face to detect it. One of the disadvantages of all these algorithms is that they are not hardware independent, and so require expensive sophisticated hardware to be installed.

1.2 Our Contribution

In this paper, we investigate traditional algorithms that can be used to detect whether a person is wearing a mask or not. Our proposed method use a face detector [2] to extract faces from images, use various filters to extract features from the face, and uses a pattern matching algorithm [15–17] to identify whether a person may be flagged as an intruder or not. Our paper is limited to detecting whether a face is covered by mask or not. In the real-world environment, different objects may appear which is outside the scope of this paper. The following primary contributions are presented, with each experimentally demonstrated:

1. To the best of our knowledge, this work constitutes the first thorough exploration of the legacy techniques to the problem at hand.
2. Results of our proposed approach are compared for two different environment settings. While the first setting is appropriate for public areas (e.g. ATM booths, Governmental facilities, Airports, etc.) under surveillance, the second setting is more

appropriate for home or small office settings where more familiar faces appear regularly. Accuracy comparisons between the filters is presented for two different environmental settings with two different data sets.

3. To make our algorithm more efficient in comparison to other algorithms in terms of space and time, we have successfully utilized a compression algorithm with the system.

1.3   Outline of Paper

The rest of the paper is organized as follows. Section 2 describes our proposed framework of the traditional techniques. Section 3 presents our experiments and compares performance within each traditional technique. We conclude in Sect. 4 with a layout of our future work.

2   Proposed Framework

2.1   Face Detector

We used the fundamental Viola Jones face detection algorithm [2] to extracting faces from an image. It uses Haar-like features which consists of dark regions and light regions. To apply these Haar-like features in face images, it first computes an integral image which is an intermediate representation of an image which basically adds side-by-side pixels. So, when Haar-like features are applied and a darker region needs to be subtracted from a lighter region, it can easily be done using integral image.

The AdaBoost algorithm is then used to find a strong classifier which is a linear combination of weak classifiers. To find the weak classifier, at each iteration, the algorithm finds the error rate of each feature and selects the feature with the lowest error rate for that iteration.

A cascade classifier is then used which is made up of multi-stage classifiers. Each stage is a strong classifier which is produced by AdaBoost algorithm. If an input image receives a negative result at a specific stage of the classifier, then it is immediately discarded; the image needs to pass through all the stages with positive result to be considered a face, and hence proved to be very quick in detecting a face from an image.

Although, it is not effective to detect masked faces since it cannot find the difference in pixels that is required by Haar-like features. So, instead we extract the whole image for masked faces to get a larger variation within the images.

2.2   Feature Extraction

2.2.1   Wavelet Decomposition

After extracting images from the face detector, the images are converted to grayscale and standard two-dimensional Haar wavelet decompositions [3] are performed. The reason for doing so is to work with the images in the frequency domain rather than space domain. This is because in the space domain, with minor camera movements, differences between specific pixels in different images may arise, which is not desirable. Moreover, it is better to consider and compare each image as a whole. The wavelet decomposition represents an image in a way that reflects variation in neighboring pixel intensities, and to perform image compression. This makes it suitable to be fed into algorithms and to look for patterns between higher order statistics of the pixels.

2.2.2   Gabor Filter

Gabor filters have been previously used in face recognition [4]. It also enhances step edges in an image but does not filter out information such as illumination and hue from the original image.

2.2.3   Canny Edge Detector

An elegant way of eliminating noise is provided by the Canny edge detector [18]. The Canny edge detector filters out information such as illumination and hue from the original image, while preserving the important structural properties [5]. This results in an "edge" image where the step edges are enhanced.

2.2.4   Laplacian Edge Detector

The Laplacian edge detector is based on the Laplace operator [6]. It is similar to the Canny edge detector but is more sensitive to noise within the image. The step edges enhancement in Laplacian is also less compared to Canny step edges.

2.2.5   Sobel Edge Detector

The Sobel edge detector uses the Sobel operator, which is a gradient based method on the first order derivatives [7]. Since Sobel uses two kernels, one on the horizontal axis and one on vertical axis, so better step edge enhancement is obtained than Canny or Laplacian edge detectors.

2.3   Kernel-Based Online Anomaly Detection Algorithm

The Kernel-based Online Anomaly Detection (KOAD) algorithm is a learning algorithm based on the so-called "kernel trick" which uses a kernel function to map the input image onto a feature space of much higher dimension [19]. The idea is to use a suitable kernel function that computes the inner product of the input vectors in the feature space without explicit knowledge of the features vectors themselves. The kernel function used is:

$$k(x_i, x_j) = \phi(x_i), \phi(x_j) \tag{1}$$

where x_i, x_j denote the input vectors and ϕ represents mapping onto feature space.

Considering a set of multivariate measurements $\{x_t\}_{t=1}^{T}$. When these measurements are associated with mapping ϕ, the feature vectors that depict normal behaviour are expected to cluster. It is then able to explain the region of normality using a relatively small dictionary of approximately linearly independent elements $\{\phi(\tilde{x}_j)\}_{j=1}^{m}$ [20]. Comparing this with approximation threshold v, the equation of projection error is:

$$\delta_t = min_a \| \sum_{j=1}^{m} a_j \phi(\tilde{x}_j) - \phi(x_t) \|^2 < v \tag{2}$$

where $a = \{a_j\}_{j=1}^{m}$ is the optimal coefficient vector.

Fig. 1. Sample images from data set 1.

Fig. 2. Sample images from data set 2.

This projection error is calculated for each timestep in KOAD. This error measure δ_t is then compared with two thresholds v_1 and v_2, where $v_1 < v_2$, and if $\delta_t < v_1$, KOAD infers that x_t is sufficiently linearly dependent on the dictionary and represents normal behaviour, otherwise if $\delta_t > v_2$, it concludes that x_t is far away from the realm of normality and immediately raise a "Red1" alarm to immediately signal an anomaly.

In the case if $v_1 < \delta_t < v2$, KOAD infers that x_t is sufficiently linearly independent from the dictionary to be considered an unusual event. It may indeed be an anomaly, or it may represent an expansion or migration of the space of normality itself. So, KOAD raises an "Orange" alarm, keeps track of the contribution of the relevant input vector x_t in explaining subsequent arrivals for l timesteps, and then takes a firm decision on it.

3 Experiments

3.1 Data Sets

Our experiments were performed on two data different sets. The first data set has been created by us, where we have collected 100 different people images from the internet. Within this data set, five intruder images were inserted at random time intervals. Sample images from Data Set 1 are presented in Fig. 1. For the second data set, we use a Georgia Institute of Technology face database [21]. This data set contains 750 images of 50 different people with each person having 15 images of different pose, expression and illumination. From this data set [21], we have taken 190 images and included 10 masked images at random time steps. Figure 2 represents the sample images of this data set. From these two data sets, we have considered them as representative of two different environmental scenarios. Data Set 1 is relevant for ATM surveillance and Data Set 2 is appropriate for small office or home surveillance.

Fig. 3. Gabor edge images corresponding to sample raw images from data set 1.

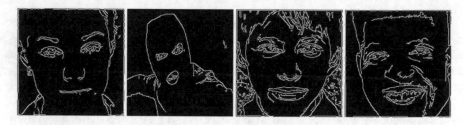

Fig. 4. Gabor edge images corresponding to sample raw images from data set 2.

Fig. 5. Canny edge images corresponding to sample raw images from data set 1.

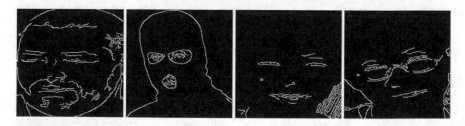

Fig. 6. Canny edge images corresponding to sample raw images from data set 2.

Face detector was not required for Data Set 2 since the images were cropped to 150 × 150 pixels with background also removed. The images from both these data sets were pre-processed using the filters mentioned earlier. Each filter represented the same input image in different ways. The results obtained are present using the figures that follow.

Fig. 7. Laplacian edge images corresponding to sample raw images from data set 1.

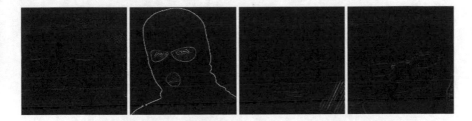

Fig. 8. Laplacian edge images corresponding to sample raw images from data set 2.

Fig. 9. Sobel edge images corresponding to sample raw images from data set 1.

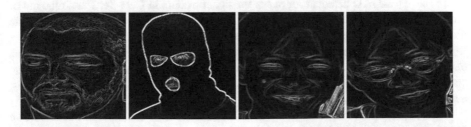

Fig. 10. Sobel edge images corresponding to sample raw images from data set 2.

3.2 Results

Data Set 1 contained five intruders and Data Set 2 contained 10 intruders, inserted at random time intervals. Figures 3, 4, 5, 6, 7, 8, 9, 10 shows the different feature vectors for each technique. So, for each method, we ran KOAD over a range of values for the thresholds v_1 and v_2 and found suitable thresholds that gave a high detection rate for both data sets. In order to check the detection rate, we needed to compare them with the time steps in which anomalies were identified in the data set. For Data Set 1, anomalous time steps were: $t = 27, 50, 66, 70, 80$ and for Data Set 2, anomalous timesteps were: $t = 21, 32, 53, 74, 104, 116, 127, 148, 169, 185$.

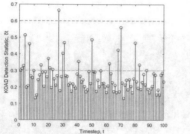 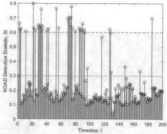

Fig. 11. Progression of the KOAD detection statistic δ_t with time when the Gabor Filter used as the feature extractor. Each time step represents each image. KOAD efficiently detects both correct and falsely claimed anomalies. The left sub-figure corresponds to Data Set 1 and the right sub-figure corresponds to Data Set 2.

Table 1 summarizes the results found from each method.

Gabor and wavelet decomposition yield similar results for same threshold parameters. Sobel also has similar threshold values and achieves a higher detection rate with lowest false alarm rate for Dataset 2. As a result, Sobel with KOAD performs well if the same images of different orientations exist within a data set. Laplacian and Canny require further investigation for better results.

Figures 11, 12, 13, 14 present the progression of the KOAD detection statistic δ_t versus timestep t for each filter and the values found in each of the two datasets. The green dotted line in each figure denotes the lower threshold v_1 and the red dotted line denotes the upper threshold v_2.

In Fig. 11, KOAD detects four out of the five intruders in Data Set 1 and nine out of the 10 intruders in Dataset 2. Although the false alarm rate increases slightly from Data Set 1 to Data Set 2, high accuracy is nonetheless achieved.

In Fig. 12, KOAD also detects four out of the five intruders in Data Set 1 and nine out of the 10 intruders in Data Set 2 for almost the same parameters for KOAD. False alarm rate also rises slightly from Data Set 1 to Data Set 2 due to increase in images, but high accuracy is still achieved.

Just like for the previous methods, KOAD in Fig. 13 also detects four out of the five intruders in Data Set 1 and nine out of the 10 intruders in Data Set 2 for almost the same parameters for KOAD. False alarm rate decreases significantly from Data Set 1 to Data Set 2, showing that it performs much better in data sets that contain repeated images of different orientations.

Table 1. Detection rate and false alarm rate of different methods for two data sets.

Method for feature extraction	Threshold, v_1	Threshold, v_2	Data set 1: Detection rate	Data set 1: False alarm rate	Data set 2: Detection rate	Data set 2: False alarm rate
Gabor	0.3	0.6	80	12.632	90	14.737
Wavelet decomposition	0.3	0.6	80	21.053	90	36.842
Sobel	0.4	0.6	80	34.737	90	3.684
Laplacian	0.2	0.4	80	56.842	50	27.895
Canny	0.3	0.6	Invalid	100	Invalid	100

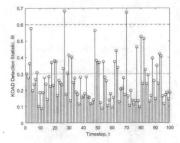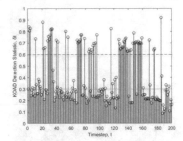

Fig. 12. Progression of the KOAD detection statistic δ_t with time when wavelet decomposition used as the feature extractor. Each time step corresponds to each image. KOAD also efficiently detects both correct and falsely claimed anomalies. Left sub-figure corresponds to Data Set 1 while right sub-figure corresponds to Data Set 2.

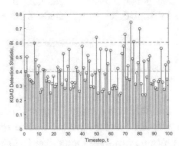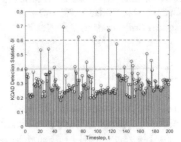

Fig. 13. Progression of the KOAD detection statistic δ_t with time with Sobel edge detector used as the feature extractor. Each time step corresponds to each image. KOAD also efficiently detects both correct and falsely claimed anomalies. Left sub-figure corresponds to Data Set 1 and right sub-figure corresponds to Data Set 2.

In Fig. 14, KOAD achieves moderate performance. In Data Set 1, it detects four out of the five intruder images, with high false alarms. In Data Set 2, it detected five out the 10 intruders with a lower false alarm. Further research is needed to achieve high accuracy with using the Laplacian edge detector.

The representation from the Canny edge detector, when applied to KOAD, could not differentiate between normal face image and intruder images. Further research is needed here. The results of using the Canny edge detector are thus not presented here in the interest of space.

Since every filter gives a different feature representation of the same image, the parameters of KOAD are changed for better detection rate with each filter. Optimum settings were identified by trial and error. However, the KOAD algorithm parameters were kept constant for each filter across the two different data sets, to conclude that the results are not sensitive to parameter settings after an initial training period. Detailed complexity and parameter sensitivity analysis is not presented here in the interest of space.

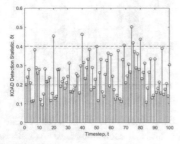 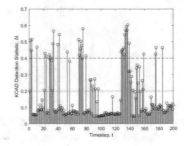

Fig. 14. Progression of the KOAD detection statistic δ_t with time with the Laplacian edge detector used as the feature extractor. Each time step corresponds to each image. KOAD moderately detects both correct and falsely claimed anomalies. Left sub-figure corresponds to Data Set 1 and right sub-figure corresponds to Data Set 2.

4   Conclusion and Future Work

In this paper, we investigated traditional methods that can extract features from a face. Five traditional techniques have been used to extract features within a face image and we evaluated them by applying Kernel-based Online Anomaly Detection (KOAD) algorithm. We created two frontal face image datasets with two different environment settings. We also included a traditional face detector to this application to extract faces. KOAD processed each face and signalled an anomaly if the face was covered, suspecting it to be an intruder. Being a learning algorithm, the system became recursive and easily signalled if any intruder was present. Its computational complexity is independent of time. KOAD can achieve almost accurate detection with low false alarm rates and have proven to be adaptive to changes in the environment. Furthermore, feature extraction techniques like Gabor, Sobel and Haar wavelet decomposition also proved to be effective methods for KOAD to use.

The future work will focus on analysing the parameter sensitivity for this application. We would like to compare this work with benchmark work performed by other people similar to this application. We would also like to search for different image enhancement techniques that can be used in Laplacian and Canny edge detectors to achieve better detection performance.

References

1. Valera, M., Velastin, S.: A review of the state-of-the-art in distributed surveillance systems. In: Velastin, S., Remagnino, P. (eds.) Intell. Distrib. Video Surveill. Syst., pp. 1–30. Institution of Electrical Engineers, London, UK (2008)
2. Viola, P., Jones, M.: Rapid object detection using a boosted cascade of simple features. In: Proceedings of the 2001 IEEE Computer Society Conference on Computer Vision and Pattern Recognition. CVPR 2001, Kauai, HI, USA, pp. I (2001). https://doi.org/10.1109/CVPR.2001.990517
3. Zhang, D.: Wavelet ransform. In: Fundamentals of Image Data Mining. Texts in Computer Science. Springer, Cham (2019)
4. Bui, L., Tran, D., Huang, X., Chetty, G.: Face recognition based on Gabor features. In: 3rd European Workshop on Visual Information Processing, Paris. France, pp. 264–268 (2011). https://doi.org/10.1109/EuVIP.2011.6045542
5. Canny, J.: A computational approach to edge detection. In: IEEE Trans Pattern Anal. Mach. Intell. PAMI-8(6), 679–698 (1986). https://doi.org/10.1109/TPAMI.1986.4767851.
6. Wang, X.: Laplacian operator-based edge detectors. IEEE Trans Pattern Analy. Mach. Intell. **29**(5), 886–890 (2007). https://doi.org/10.1109/TPAMI.2007.1027
7. Kanopoulos, N., Vasanthavada, N., Baker, R.L.: Design of an image edge detection filter using the Sobel operator. IEEE J Solid-State Circ. **23**(2), 358–367 (1988). https://doi.org/10.1109/4.996
8. Ahmed, T., Coates, M., Lakhina, A.: Multivariate online anomaly detection using kernel recursive least squares. In: Proceedings IEEE INFOCOM, Anchorage, AK, USA (2007)
9. Ahmed, T., Oreshkin, B., Coates, M.: Machine learning approaches to network anomaly detection. In: Proceedings of ACM/USENIX Workshop on Tackling Computer Systems Problems with Machine Learning Techniques (SysML). Boston MA, USA (2007)
10. Datakalab. Company Webpage (2016).https://www.datakalab.com/
11. Chowdary, G.J., Punn, N.S., Sonbhadra, S.K., Agarwal, S.: Face mask detection using transfer learning of inceptionV3. In: Bellatreche, L., Goyal, V., Fujita, H., Mondal, A., Reddy, P.K. (eds) Big Data Analytics. BDA, : Lecture Notes in Computer Science, vol. 12581. Springer, Cham (2020)
12. Nar, R., Singal, A., Kumar, P.: Abnormal activity detection for bank ATM surveillance. In: 2016 International Conference on Advances in Computing, Communications, and Informatics (ICACCI), Jaipur, India, (2016), pp. 2042–2046.https://doi.org/10.1109/ICACCI.2016.7732351.
13. Bouchrika, I., Carter, J.N., Nixon, M.S.: Towards automated visual surveillance using gait for identity recognition and tracking across multiple non-intersecting cameras. Multimed. Tools Appl. **75**, 1201–1221 (2016). https://doi.org/10.1007/s11042-014-2364-9
14. Wang, J., Yuan, Y., Gang, Y.: Face Attention Network: An Effective Face Detector for the Occluded Faces. ArXiv abs/1711.07246: n. pag (2017)
15. Anika, A., Karim, K.L.D., Muntaha, R., Shahrear, F., Ahmed, S., Ahmed, T.: Multi image retrieval for kernel-based automated intruder detection. In: IEEE Region 10 Symposium (TENSYMP). Cochin, India, pp. 1–5 (2017)

16. Ahmed, T., Pathan, A.K., Ahmed, S: Adaptive algorithms for automated intruder detection in surveillance networks. In: 2014 International Conference on Advances in Computing, Communications and Informatics (ICACCI), New Delhi, India, pp. 2775–2780 (2014). https://doi.org/10.1109/ICACCI.2014.6968617

17. Ahmed, T., Ahmed, S., Ahmed, S., Motiwala, M.: Real-time intruder detection in surveillance networks using adaptive kernel methods. In: 2010 IEEE International Conference on Communications, Cape Town, South Africa, pp. 1–5 (2010)

18. Au, C., Skaff, S., Clark, J.: Anomaly detection for video surveillance applications. In: Proceedings of IEEE ICPR, Hong Kong, China (2006)

19. Scholkopf, B., Smola, A.: Learning with Kernels. MIT Press, Cambridge, MA, USA (2001)

20. Engel, Y., Mannor, S., Meir, R.: The kernel recursive least squares algorithm. IEEE Trans. Sign. Proc. **52**(8), 2275–2285 (2004)

21. Georgia Tech Face Database. ftp://ftp.ee.gatech. edu/pub/users/hayes/facedb/

A Novel Dynamic Authentication Method for Sybil Attack Detection and Prevention in Wireless Sensor Networks

Wadii Jlassi[1]([✉]), Rim Haddad[2], Ridha Bouallegue[3], and Raed Shubair[4]

[1] Innov'COM Lab/Sup'Com, National Engineering School of Tunis, University Tunis El Manar, Tunis, Tunisia
[2] Laval University, Quebec, Canada
Rim.haddad@eti.ulaval.ca
[3] Innov'COM Lab/Sup'Com, University of Carthage, Tunis, Tunisia
Ridha.bouallegue@supcom.tn
[4] New York University, Abu Dhabi, UAE
raed.shubair@nyu.edu

Abstract. One of the most concerns in wireless sensor networks is security. Wireless sensor networks are vulnerable to attacks in every layer of the network. They are extremely necessitous for securing network protection. Sybil attack is one of those attacks in which a legal node is converted into a Sybil one which is a replica node with a different personality using a similar ID. This kind of identity theft attack can be classified as Sybil attack. This attack mainly influences routing protocols and operations such as voting, data aggregation and reputation evaluation. This paper proposes a dynamic and accurate method for Sybil attack detection and prevention. The proposed method generates a routing table that holds information about deployed nodes. The base station verifies the keys of nodes that will communicate and transmit data between them. Data transmission occurs between nodes once they get the signal from the base station. The message authentication method will progress data transmission in the network and enhance the throughput compared to RPC.

Keywords: Wireless sensor networks · Sybil attack · Random password comparison · Message authentication method

1 Introduction

Wireless Sensor Networks (WSNs) consist of small sensor nodes co-working to monitor and obtain information about the area of interest [1]. They are termed ad hoc networks as they need no infrastructure. Therefore, WSNs are used in many areas such as environmental conservation, domestic applications, Military surveillances, and so on. The basic task of sensor networks is to sense the events, gather efficient data and send it to their requested destination [2]. The parameters of the WSN such as the sensing data, total nodes, the life span of sensors, the energy consumption, geographical location of the sensor placement, the environment, and the context are critical in the design of WSN.

Clustering is used to divide the network into several sections called clusters with the objective to prolong the lifetime of the wireless sensor network. Each cluster has one node which is called cluster head. This node collects data from the other member nodes of the cluster and then sends it to base station. Now the main focus area of research in clustered sensor networks is energy and security. Security of such networks is a challenge especially when they are employed in critical applications [3]. The important goal of security implementation is to reduce energy consumption and data security.

Wireless sensor networks are susceptible to attack starting from the physical layer and going all the way up the stack to the application layer. Based on the WSN layers, various types of attacks are classified as follows. Malicious node, Desynchronization belongs to the Application layer. Desynchronization, Flooding belongs to Transport Layer. Sybil, Selective forwarding, Sinkhole, Wormhole belongs to Network layer. Collision, Exhaustion belongs to the Datalink layer. Jamming, Interference, Node Replication Attack belongs to the physical layer.

Sybil attack is a type of security threat. A threat is a set of circumstances that has the potential to cause loss or harm [4]. It can be easily launched in WSN due to its nature of unstructured and distributed topology and wireless communication. It is one of the primary attacks that would facilitate the onset of many different attacks in the network. The Sybil attacker also called the adversary captures a legal node or inserts an illegal node in the network. This malicious node forges fake identifiers or duplicates existing nodes identifiers in different areas of the network. This attack decreases the network lifetime, creates suspicion in the genuineness of the data, reduces the effectiveness of fault-tolerant schemes and poses a threat to geographic routing protocols. Also, malicious node attracts a heavy and considerably disrupts routing protocols. It disturbs operations such as data aggregation, voting and reputation evaluation. A malicious node which enters the network with multiple IDs is shown in Fig. 1.

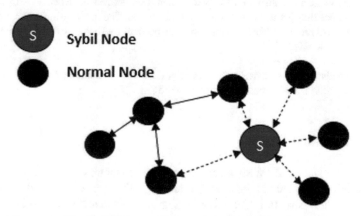

Fig. 1. Wireless sensor network with Sybil node

Various algorithms are proposed to detect and confront this attack in WSN, used different techniques such as radio resource testing technique [5], Neighboring Information [6], node location based on RSSI [7], random key pre-distribution, Random password comparison [8], RSSI of cluster head, Leach routing protocol [9], Neighboring Information and broadcast two-hop packets [10], observer nodes and neighboring information [11]. These techniques require heavy computational complexity which leads to reduction in energy, since energy consumption in WSN is limited, and have a limitation in their detection range.

This paper concentrates on the security in wireless sensor networks and proposes a dynamic and accurate algorithm to detect and prevent Sybil attack. The rest of the paper is organized as follows: Sect. 2 reviews related works (existing works). Section 3 describes the proposed algorithms. Results of simulations and discussions are presented in Sect. 4. Finally, Sect. 5 concludes the paper and reference papers are included.

2 Related Works

Wireless sensor nodes perform the operation of transmitting the data from the source to the destination which should be made in an efficient way so that the data transmission between the sender and the receiver will be in an effective manner. The sensor nodes have limitations in storage, power, latency, constraint bandwidth and reduced corporal size [12]. Since WSN nodes have limitations, safekeeping is the most important role needed to detect and prevent malicious activities in the network [13].

Sybil attack was introduced for the first time by Douceur [14] in peer-to-peer networks. It is a matter of critical importance and consternation in network security leading to many fake identities that can cause disruption in the network [15]. The detection techniques of sybil attack are classified as localized, distributed and centralized. In other words, they could be classified based on various characteristics such as key distribution, broadcast, multicast, location, random or group deployment.

In [16], a solution based on a centralized base station (BS) has been proposed. Each node in WSN sends the list of neighbors and their locations to BS. The same node ID in two lists with inconsistent locations results in a clone detection. BS has the mission to remove the clones.

In [17], a position-based Sybil attack detection approach has been proposed verifying the physical position of each node. Sybil nodes can be detected using this approach because they will appear to be at exactly the same position as the malicious node that generates them.

An RSSI-based locating scheme has been introduced in [18]; it uses proportion of RSSIs from different receivers in order to calculate nodes locations in the network. This is sufficient to detect Sybil nodes since all of them are positioned in nearby locations.

A protocol has been proposed in [19] which uses identifier-based encryption. This is used to avoid Sybil attacks by not allowing malicious nodes to acquire various identifiers.

In [20], an algorithm has been proposed to find Sybil nodes in WSNs. It consists of three phases. In the first phase, they send ID and power value to the head node, the head node checks for nodes with power value under the threshold value. In the second phase, if the nodes are very close then the distance between the receiver and the sender is zero so the node suffers from Sybil attack. In the third phase, the routing procedure in the cluster is verified if there was a hop between the Sybil identities.

Like [21], has provided a method to avoid Sybil attacks. The information on routes is collected by an intelligence algorithm during network activity. The Sybil nodes are detected according to their energy changes when the network is active. Another algorithm [22] has been provided to avoid Sybil attacks in wireless sensor networks, based on clients' puzzles and learning Automata (LA). A Sybil attack detection based on a trust-based model [23] has been provided which incorporates the concept of fuzzy expert system and neural network.

A lightweight trust system [24] has been proposed which uses energy as a metric parameter for a hierarchical WSN for dealing with Sybil attack in WSNs. This system can reduce the communication overhead in the network. The evaluation of this system views efficiency and scalability for detecting Sybil attacks in terms of true and false positive detection in a heterogeneous WSN.

A mobile agent-based Sybil attack detection method has been proposed in [25] which uses three parameters: random key pre-distribution, random password and threshold value. The network performance evaluation before and after the detection of Sybil nodes based on: the throughput and the packet delivery ratio. Another algorithm called Sybil attack Detection Algorithm (SDA) [26] based on mobile agent has been proposed to detect and prevent the sybil attack. It uses three parameters namely random key pre-distribution, random password generation and threshold value. The suspect node is informed to the BS and the BS alerts the network nodes.

Like [27], has provided a lightweight, dynamic algorithm for detecting Sybil nodes in mobile wireless sensor networks. This algorithm uses watchdog nodes to monitor the traffic passively and assign bit wise tags to mobile nodes based on their movement behaviors, and then detects Sybil nodes according to the labels and node id, during detection phase.

Another algorithm [11] has been proposed which uses some observer nodes to detect Sybil attacks in mobile wireless sensor networks. Two phases proposed in this algorithm: monitoring phase which observer nodes record the number of meeting occurrences of other nodes in a vector called history, for R time periods. Then, detection phase in which observer nodes cooperate and identify Sybil nodes based on the content of history vectors.

A novel Sybil attack detection protocol (NoSad) [28] is proposed. This protocol is a localized method using intra-cluster communication and RSSI value to identify and isolate the Sybil attack in WSN.

In general, the main disadvantages of the existing algorithms for detection Sybil attack are: high rate of false detection, high cost (need for nodes locations), no scalability and complexity of detection algorithms.

3 Proposed Work

The main objective of this paper is to propose a method for detecting and preventing Sybil attacks in wireless sensor networks. The network G is composed of N number of nodes deployed randomly and a base station BS.

The proposed algorithm generates a routing table (rtable RPC) which holds the information of each node's id, time and password. The intermediate node in the route is identified between source and destination. The intermediate node's information is compared with the RPC database during the communication in the middle of nodes. Based on the comparison results, it decides whether it is a Sybil or a normal node. A Sybil node will not be able to submit the dynamic password which is allowed to all the nodes in the network. The system model goes through the following steps:

- Step 1: The network G is composed of N number of nodes deployed randomly.
- Step 2: Generate table ti for ni, ti = {c1, c2, c3} where ci = NID, time, pwd.
- Step 3: Discover the route among the elected nodes S, D are source and destintion nodes.
- Step 4: Check the route and verify the Sybil activity then pass the data.

Random password creates a new password every few second for each node and sends it to all nodes in the network. After communication between source and destination, the destination node's id, time delay and the random password corresponding to the time delay will be compared with RPC database. If the information matches, the node is considered to be a normal node and the source node will send data otherwise the node is considered as a Sybil node. The nearest neighbor is found by calculating the minimum distance from the source node as given in (1).

$$dij = \sqrt{(xj - xi)^2 + (yj - yi)^2} \, \forall i, \, j \text{ denotes node index} \tag{1}$$

The Random password comparison method algorithm is given below:

- Network G, which has n number of nodes, G = {n1, n2, n3….ni, nn}
- Table T = {t1, t2, t3….ti,….tn} where ti is table for ith node.
- Table T is commonly available
- in BS where it can be authenticate all nodes in G.
- c1, c2, c3 are the three columns in the table ti where c1 is the node id, c2 is the time, c3 is the password generated for node
- Elect any two nodes S, D as Source and Destination Node in G.
- Node D sends a data request message to node S.
- A route discovered between S and D using the sub-procedure DISROUT(S, D) is obtained.
- Now S sends data in the same route verifying for Sybil activity by CHKROUT (S, D) in all the nodes found in the route. Then node S sends a data to D.

The algorithm gives the values c1, c2 and c3 for each node and they are stored in the rtable. After discovering the route between source and destination, every nearest node should submit their information (id, time password) which is compared with the rtable values. If the information matches, the node is chosen as the nearest neighbor and added to the route. Otherwise, it means that the current node is discarded and another node is chosen for the same process. This process stopped if the destination node is discovered.

RPC generates the route by adding the genuine node in its path from source to destination node using several sub procedures DISROUT(S, D) and CHKROUT(S, D).The sub procedures detect the route from source to destination before transmitting the data. The Sybil node is detected during the data transmission.

Figure 2 represents respectively sub procedures DISROUT (S, D) and CHKROUT (S, D).

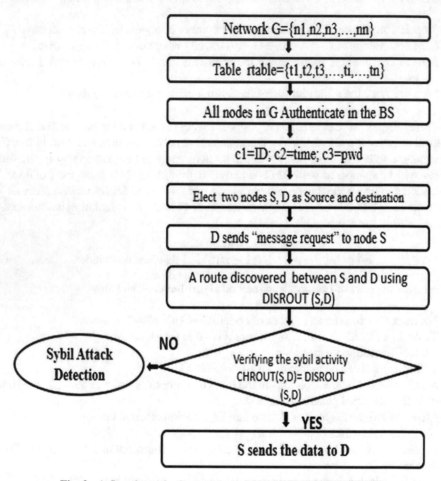

Fig. 2. A flowchart of sub procedures DISROUT AND CHKROUT

Even though the detection ratio is high, this technique has its own drawback such as high energy consumption on the nodes near the BS or sink due to forwarding packets.

Application of RPC needs a lot of time for road research and distance calculation. We propose to use RPC method with message authentication method to improve the data transmission between nodes. In the network G, a node Ni sends a request messages to node Nj with its key, as msg(Ni).Node Nj submits its key message with msg(Nj),and later, both keys are verified by the base station and an ok signal produced for sending the data. Data transmission occurs between Ni and Nj once they get the signal from the base station.

The message authentication and passing method are given in Fig. 3.

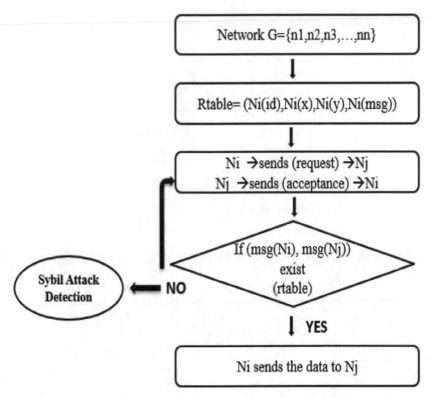

Fig. 3. A flowchart of the message authentication method

4   Simulation Results

In this section, to prove the performance of our proposed method, NS-2 software is used for simulation. Table 1 shows the initial parameters of Wireless Sensor Network for simulating 100 nodes with a network size of 100 * 100.

Figure 4 presents the network deployment with 100 nodes and one Sybil node detection. All nodes are constructed under a single BS. The number of created nodes is 100 and 0 is the base station. We analyze the performance by calculating the average delay of data packet transfer and the throughput.

As shown in Fig. 5, our proposed method's average delay of the data packet transfer of the network is very less than RPC. The Average delay extends when the number of nodes increase respectively for the different methods. For message authentication, the delay is less compared to RPC due to the time of detecting of Sybil node by the sub procedures DISROUT and CHECKROUT.

Figure 6 shows a Comparison of Throughput between the existing method RPC with the proposed method. In the case of message authentication method, the throughput will

Table 1. Simulation parameters

Parameter	Level
Area	100 m × 100 m
Nodes	100
Packet size	50 bytes
Transmission protocol	UDP
Simulation time	100 s
Propagation	Two ray ground
Placement	Random
Application Traffic	CBR
Routing Protocol	AODV
Initial energy	100 J

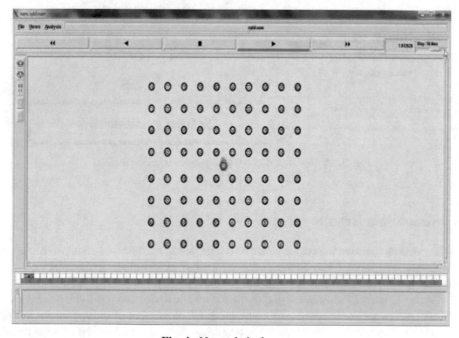

Fig. 4. Network deployment

reach 98% while in RPC, the throughput will be 78%. The proposed method improves security and increases the communication in the wireless sensor networks compared to RPC.

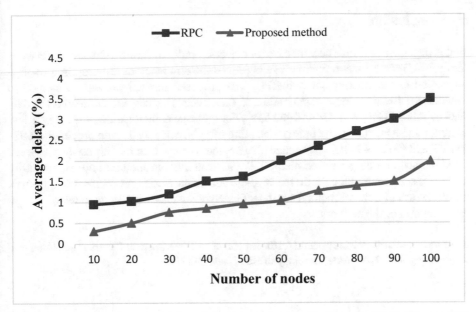

Fig. 5. Comparison of average delay of the data packet transfer between RPC with proposed method.

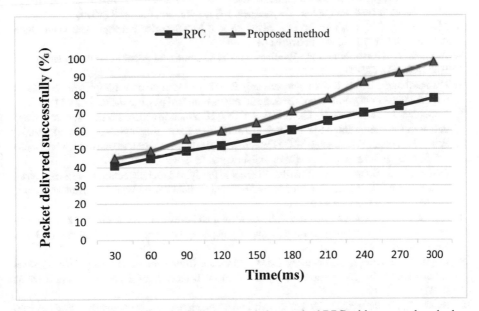

Fig. 6. Comparaison of Throughput between existing method RPC with proposed method.

5 Conclusion

In this paper, we present a new method for detecting and preventing Sybil attacks in wireless sensor networks. Simulation results proved the efficiency of the proposed method compared to the existing method RPC. The proposed method presents a method to validate the route by the authentication of each node. Besides, we compared the data packet transfer using our method and RPC. For message authentication, the delay is less compared to RPC due to the time of detecting of Sybil node by the sub procedures DIS-ROUT and CHECKROUT. We made also another comparison of Throughput between RPC method and the proposed method. We noticed that our method improves security and increases communication in the wireless sensor networks. As future work, we will study the use of a novel method for detecting Sybil attacks with a guarantee of energy consumption of the network.

Acknowledgments. We are grateful for the support of the Department of Electrical and Computer Engineering at the New York University of Abu Dhabi (NYU).

References

1. Rawat, P., Singh, K.D., Chaouchi, H., Bonnin, J.M.: Wireless sensor networks: a survey on recent developments and potential synergies. J. Supercomput. **68**, 1–48 (2014)
2. Manjunatha, T.N, Sushma, M.D., Shivakumar, K.M.: Security concepts and sybil attack detection in wireless sensor networks (2016)
3. Akyildiz, I.F., Kasimoglu, I.H.: Wireless sensor and actor networks: research challenges. Ad Hoc Networks (2014)
4. Dhamodharan, U.S.R.K., Vayanaperumal, R.: Detecting and preventing sybil attacks in wireless sensor networks using message authentication and passing method (2015)
5. Newsome, J., Shi, E., Song, D., Perrig, A.: Te sybil attack in sensor networks: analysis and defenses. In: International Symposium on Information Processing in Sensor Networks (2004)
6. Ssu, K.F., Wang, W.T., Chang, W.C.: Detecting sybil attacks in wireless sensor networks using neighboring information. Proc. Comput. Networks **53**, 3042–3056 (2009)
7. Demirbas, M., Song, Y.: An RSSI-based scheme for Sybil attack detection in wireless sensor networks. In: IEEE Computer Society International Symposium on World of Wireless, Mobile and Multimedia Networks (2006)
8. Amuthavalli, R., Bhuvaneswaran, R.S.: Detection and prevention of sybil attack in wireless sensor network employing random password comparison method. J. Theoricaland Appl. Inf. Technol. **67**(1), 236–246 (2014)
9. Jangra, A., Priyanka, S.: Securing LEACH protocol from Sybil attack using jakes channel scheme (JCS). In: International Conferences on Advances in ICT for Emerging Regions (2011)
10. Rafeh, R., Khodadadi, M.: Detecting Sybil nodes in wireless sensor networks using two-hop messages. Indian J. Sci. Technol. **7**(9), 1359–1368 (2014)
11. Jamshidi, M., Ranjbari, M., Esnaashari, M., Qader, N.N., Meybodi, M.R.: Sybil node detection in mobile wireless sensor networks using observer nodes. Int. J. Inf. Visualization **2**(3), 159–165 (2018)
12. Birami, K.S.: Sybil attack in Wireless Sensor Network (2016)

13. Sharmila, S., Umamaheswari, G.: Detection of Sybil attack in Mobile Wireless Sensor networks. Int. J. Eng. Sci. Adv. Technol. **2**(2), 256–262 (2012)
14. Douceur, J.R.: The sybil attack. In: International Workshop on Peer-to-Peer Systems, pp. 251–260. Springer, Heidelberg (2002)
15. Sharmila, S., Umamaheswari, G.: Detection of sybil attack in mobile wireless sensor networks. Int. J. Eng. Sci. Adv. Technol., 256–262 (2017)
16. Eschenauer, L., Gligor, V.D.: A key-management scheme for distributed sensor networks. In: Proceedings of the 9th ACM Conference on Computer and Communications Security, pp. 41–47 (2002)
17. Manjunatha, T.N., Sushma, M.D., Shivakumar, K.M.: Sybil Attack Detection Through on Demand Distance Vector Based Algorithm in Wireless Sensor Networks (2015)
18. Triki, B., Rekhis, S., Boudriga, N.: An RFID based system for the detection of sybil attack in military wireless sensor networks. In: Computer Applications and Information Systems (wccais), World Congress on, pp. 1–2. IEEE (2014)
19. Sujatha, V., Anita, E.M.: Detection of sybil attack in wireless sensor network (2015)
20. Sharmila, S., Umamaheswari, G.: Detection of Sybil Attack in Mobile Wireless Sensor Networks (2016)
21. Muraleedharan, R., Ye, X., Osadciw, L.A.: Prediction of Sybil Attack on WSN using Bayesian Network and Swarm Intelligence. Wireless Sensing and Processing (2008)
22. Jamshidi, M., Esnaashari, M., Meybodi, M.R.: An algorithm for defending sybil attacks based on client puzzles and learning automata for wireless sensor networks. In: 18th National Conference of Computer Society of Iran (2017)
23. Sarigiannidis, P., Karapistoli, E., Economides, A.A.: Detecting Sybil attacks in wireless sensor networks using UWB ranging-based information. Expert Systems with Applications, pp. 7560–7572 (2015)
24. de Sales, T.B.M., Perkusich, A., de Sales, L.M., de Almeida, H.O., Soares, G., de Sales, M.: ASAP-V: a privacy-preserving authentication and sybil detection protocol for VANETs. Inf. Sci. **23**(4), 208–224 (2016)
25. Pecori, R.: S-Kademlia: A trust and reputation method to mitigate a Sybil attack in Kademlia. Comput. Netw. **94**, 205–218 (2016)
26. Prasad, K., Mallick, C.: Mobile agent based sybil attack detection algorithm for wireless sensor network. In: International Conference on Emergent Trends in Computing and Communication (2015)
27. Jamshidi, M., Zangeneh, E., Esnaashari, M., Meybodi, M.R.: A lightweight algorithm for detecting mobile Sybil nodes in mobile wireless sensor networks. Comput. Electr. Eng. **64**, 220–232 (2017)
28. Angappan, A., Saravanabava, T.P.: Novel Sybil attack detection using RSSI and neighbour information to ensure secure communication in WSN. J. Ambient Intell. Humanized Comput. (2020)

An Effective Location-Based Alert Messages Dissemination Scheme for Software Defined Vehicular Networks

Raoua Chakroun$^{(\boxtimes)}$, Slim Abdellatif, and Thierry Villemur

LAAS-CNRS, 7 Avenue du Colonel Roche, Toulouse, France
{rchakrou,slim,villemur}@laas.fr

Abstract. Software Defined Networking (SDN) is considered as a key paradigm for future vehicular networks and its adoption may pave the way to novel approaches to support emerging and legacy ITS services, such as Alert Message (AM) dissemination. Historically, to meet the low dissemination delay requirements and full vehicle coverage, Vehicular Ad-hoc NeTworks (VANETs) were considered as the underlying assumption of most proposed AM dissemination schemes. With SDN, vehicular-to-infrastructure (V2I) links are a viable alternative that adds to Vehicle-to-vehicle (V2V) links. Moreover, a global view of the network at a centralized controller complemented with information on vehicles and their route helps to adjust the dissemination procedure to improve its performance and efficiency. In this paper, we propose a location-based AMs dissemination scheme that combines V2I broadcasts with V2V rebroadcasts in order to provide a low delivery delay, very limited collisions, high information coverage with insignificant signaling and network overhead. The originality of our scheme stands in the selection process of V2V rebroadcasts which is based on vehicles' location with respect to predefined rebroadcast points selected by the controller. Our proposal is evaluated and compared to legacy techniques. It shows significant improvements in delivery delays and network resource utilization.

1 Introduction

As an obvious evolution of legacy ITS (Intelligent Transportation Systems), Cooperative ITS (C-ITS) have been pushed in the front of the stage with the idea of enabling communications between vehicles (V2V: Vehicle-to-Vehicle), vehicles and infrastructure (V2I: Vehicle-to-Infrastructure), and more generally vehicles and all their surroundings (V2X: Vehicles-to-everything). The rationale is that thanks to these new communication alternatives, the perception of vehicles on their surrounding and their environment is enriched, its accuracy is improved and the perception horizon is enlarged. As a consequence, novel C-ITS services that further improve road safety, traffic efficiency, comfort and convenience to drivers and passengers are envisioned, e.g. cooperative maneuver, bird's eye view, cooperative awareness (e.g. Emergency vehicle warning, Alert for an accident), etc. [11, 15].

One of the challenges facing the development of C-ITS is the provision of network connectivity services that meet the various Quality of Service (QoS) expectations of C-ITS services. One direction that has adhesion from academia and Industry is to consider

L. Barolli et al. (Eds.): AINA 2021, LNNS 225, pp. 542–552, 2021.
https://doi.org/10.1007/978-3-030-75100-5_47

a hybrid multi-access vehicular network that combines, amongst: DSRC (Dedicated Short Range Communication) which is foreseen for rapid dissemination and delivery of critical safety messages, and LTE (Long Term Evolution) which is foreseen for heavier and less stringent message exchanges [15,20]. One promising approach towards this hybridization is to apply the Software Defined Networking (SDN) paradigm as leverage to unify the network control of the above-cited technologies with lots of promising opportunities and outcomes in terms of improved network resource usage and performance [7,9,16]. For these reasons, this work considers SDN-based vehicular networks.

Point-to-multi-point communications and more specifically Geo-Broadcasting are at the heart of many C-ITS services. Alert Message AM dissemination is such a service, which upon an emergency or a risky situation (accident or vehicle breakdown), alert messages are generated for some time and spread over a geographical area of interest as fast as possible to allow other vehicles to react properly very quickly [3]. Historically, message alert dissemination was carried out by a DSRC based VANET (Vehicular Ad-hoc NeTworks) operating on a dedicated communication channel by inviting vehicles to blindly rebroadcast the alerts. This method, known as simple flooding, causes a huge amount of unnecessary re-transmissions and collisions (known as the broadcast storm problem), thus wasting bandwidth, increasing dissemination delay, and lowering packet delivery ratio [17]. Lots of proposals from the literature tried to mitigate the broadcast storm problem by controlling the vehicles that are allowed to rebroadcast (using some form of random selection or clustering techniques to group vehicles) or reducing the contention. But, most were designed for VANETs and neither assume the presence of V2I links nor take advantage of the benefits that a global view of the network at a centralized controller can bring to the dissemination scheme.

In this paper, we propose an alert messages dissemination scheme that primarily exploits V2I transmissions and complements with V2V transmissions in order to reach vehicles located in areas that are not covered (white zones) or poorly covered (grey zones) by transmissions from the infrastructure. The originality of our scheme stands in the selection process of relay vehicles which is based on a local decision at each vehicle based on its location with respect to predefined rebroadcast points that are computed by the controller based on its knowledge of the radio performance and coverage as well as road traffic conditions. In addition, our proposal induces marginal extra signaling and network overhead.

The rest of the paper is organized as follows: Sect. 2 presents an overview of existing work on AM dissemination. Section 3 describes the system model as well as the proposed location-based dissemination algorithm. Section 4 evaluates and discusses our simulation results. Finally, Sect. 5 concludes the paper.

2 Related Work

Alert message dissemination has been widely studied in a VANET/V2V context. All aim at mitigating the effect of the "broadcast storm" while ensuring high information coverage. These works can be broadly classified as follows.

Lots of the proposed schemes try to control the flooding procedure of AMs either by reducing the number of rebroadcasters or by reducing the contention between transmitting vehicles. In the former case, some vehicles are selected to relay the AMs [2,14,19].

This selection is based on vehicles' characteristics, distance from the sender, local density, interests, transmit power, etc., and helps to limit re-transmissions and contention at the cost of reduced information coverage. In the latter case, MAC level protocol parameters are adjusted on a per vehicle basis in order to statistically assign different back-off periods to vehicles [6]. By so doing, collisions are reduced with no impact on information coverage. However, useless transmissions are not avoided.

Another class of proposals set aside the flooding logic and try to guide AMs dissemination by organizing vehicles in clusters (groups of vehicles) and defining how AM are propagated between and within clusters [8,10,18]. Many criteria are used for cluster formation as well as cluster-head selection, e.g. neighborhood, direction/destination, relative velocity, etc. A last class of proposals is those based on routing protocols that proactively or reactively compute point-to-multi-point routes to all known vehicles [4]. They clearly exhibit the most predictable information coverage and efficient resource usage, but this comes at the cost of increased network overhead and complexity embedded in vehicles.

3   Location-Based Alert Messages Dissemination

3.1   System Model

An SDN controller is assigned to each region, it manages all the RSUs that fall inside. Each controller identifies whites and grey zones in its region and selects K rebroadcast zones where vehicles can rebroadcast an AM sent by an RSU in order to enhance the overall coverage and reach all the vehicles located in grey or white zones. These rebroadcast zones are defined by a rebroadcast point $P_i(x_i, y_i)$, $i \in [1, K]$, with x_i and y_i are the GPS coordinates of P_i, and a radius d_{max} in the order of a few meters from the rebroadcast point. The contribution of this paper concerns the dissemination procedure, as a consequence, a basic broadcast point placement is adopted as explained hereafter.

Real experiments in [20] show that the packet loss rate and delay increase when the distance between vehicles and RSU is greater than a threshold R_{th} which also depends on mobility, due to the wireless propagation channels. According to [15], an RSU will be deployed every 2 km in Europe with a theoretical coverage defined by $R = 850$ m.

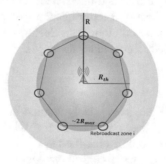

Fig. 1. Rebroadcast zones

We assume that the controller is able to define a threshold distance R_{th} for each RSU from which the quality degrades and the probability of reception decreases. From this threshold radius (Fig. 1), it can build a regular polygon with $K, K \in [5, 17]$ equal sides of $2R_{max}(\pm 50\,\text{m})$ (where R_{max} is the maximum radius coverage of a vehicle), to avoid

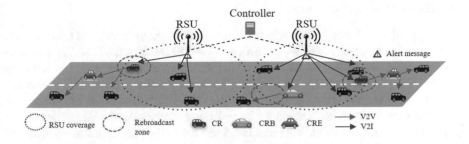

Fig. 2. Location-based alert messages dissemination

Algorithm 1: Alert message dissemination

Input:
AM : Alert Message
$P_i, i \in [1, k], d_{max}$: rebroadcast zones
V_j : current node, $Pos(V_j)$: its position, $Pos(V_j)$: its direction
S : Sender

1 AM.Reception()   /* AM received */
2 **if** $S = RSU$ **then**
3      $P_i \leftarrow F(Pos(V_j), dir(V_j))$ /* find the nearest rebroadcast zone */
4   **if** $d(P_i, V_j) \leq d_{max}$ **then**
5         $V_j \leftarrow CR$ /* V_j is a Candidate Relay */
6         $T \leftarrow T_1$ /* Calculate time to wait T */
7   **else if** $d_{max} < d(P_i, V_j) \leq R_{max}$ **then**
8         $V_j \leftarrow CRB$ /* V_j is a Candidate Relay Backup */
9         $T \leftarrow T_2$
10 **else**
11   **if** *(AlreadyReceive (AM) = True) and (V_j = CR or V_j = CRB)* **then**
12   **if** $d(P_i, S) < d(P_i, V_j)$ **then**
13   Stop (t) and Discard()
14   **else if** *(AlreadyReceive (AM) = True) and (V_j = CRE)* **then**
15   Stop (t) and Discard()
16   **else if** *$V_j \notin \{RSU\}$ and $TTL \neq 0$* **then**
17        $V_j \leftarrow CRE$ /* V_j is a Candidate Relay Exceptional */
18        $T \leftarrow T_3$
19   **else**
20   Discard ()
21 LaunchTimer (T)   /* start timer */

---

collisions between two vehicles in two adjacent rebroadcast zones, whose polygon vertices represent the rebroadcast points. In this case, a rebroadcast zone i is represented by the vertex P_i with a margin d_{max} of a few meters. If a rebroadcast zone does not serve any road area, it will be omitted.

3.2 Dissemination Procedure

The proposed AM dissemination scheme is described in Algorithm 1. At handover, on the re-association request with a new RSU, a vehicle V_j receives the list of the rebroadcast zones that are under the coverage of the RSU. An AM is broadcasted with the position of the sender and a predefined Time To Live TTL (used to control the rebroadcasts of vehicles located in white zones as shown in Fig. 2). Vehicles that receive an AM check their location: inside, around, or away from the rebroadcast zones. The vehicle closest to the broadcast point is the best candidate for the rebroadcast. This is achieved by assigning, on an AM reception, a personalized arbitration waiting time to each vehicle. The closer a vehicle is from the rebroadcast point, the lower is its waiting time.

Let $P_i = F(pos(V_j), dir(V_j))$ a function which takes the GPS position of vehicle $pos(V_j) = (x_j, y_j)$ and its direction as parameters, and returns the closest rebroadcast zone in its direction.

Referring to Algorithm 1, if a vehicle V_j receives an AM, first, it checks the sender:

Step 1 [Lines 2–3] If the sender is an RSU, the node finds the closest rebroadcast point P_i to its location, and it checks its position according to the rebroadcast zone, otherwise, the message is received from a relay vehicle, go to step 4

Step 2 [Lines 4–6] If V_j is in the rebroadcast zone, the node is declared as Candidate Relay CR and the vehicle in the rebroadcast zone closest to the rebroadcast point has the highest priority to rebroadcast, so it has the minimum waiting time T_1. Another factor that we consider is the distance between the RSU (sender) and V_j (receiver) to avoid the simultaneous transmissions of multiple vehicles sitting at the same distance from the rebroadcast point. Hence, a relaying metric G_1 that is used to calculate the time to wait T_1, is given by :

$$G_1 = (1 - \alpha)\frac{d(V_j, S)}{R} + \alpha(1 - \frac{d(V_j, P_i)}{d_{max}}) \qquad (1)$$

where : $d(V_j, S)$ indicates the distance between vehicle V_j and RSU, R is maximum transmission coverage of the RSU, $d(V_j, P_i)$ is the distance between vehicle V_j and rebroadcast point P_i, d_{max} is the radius of a rebroadcast zone. α determines the influence of each component (distance from the rebroadcast point and distance from the RSU) to compute G_1. Hence, if we consider a DSRC enabled vehicle, the waiting time T_1 is derived as follows:

$$T_1 = (CW_{min} - G_1 CW_{min})T_s \qquad (2)$$

where CW_{min} and T_s are respectively the minimum contention window and the time slot duration of DSRC's medium access technique, namely CSMA/CA (Carrier Sense Multiple Access/Collision Avoidance) [1]. Go to step 7, otherwise, go to step 3

Step 3 [Lines 7–9] If V_j is in the radius coverage R_{max} of rebroadcast point, then, the node is declared as Candidate Relay Backup CRB and the relaying metric G_2 is given by :

$$G_2 = (1 - \alpha)\frac{d(V_j,S)}{R} + \alpha(1 - \frac{d(V_j,P_i)}{R_{max}}) \tag{3}$$

Hence for a DSRC enabled vehicle, the waiting time T_2 is as follows:

$$T_2 = [CW_{min} + [(CW_{max} - CW_{min}) - G_2(CW_{max} - CW_{min})]]T_s \tag{4}$$

where CW_{max} is the maximum contention window related to unicast transmissions. T_2 is set between CW_{min} and CW_{max} to avoid collisions with nearby CR vehicles. This is particularly useful when the road traffic is low or sparse since, in case of no vehicle in a broadcast zone, a close-by vehicle is given the opportunity to rebroadcast the AM. Go to step 8, Otherwise discard the message

Step 4 [Lines 10–11] If V_j receives the same message another time, and V_j is a CR or a CRB go to step 5, otherwise, go to step 6

Step 5 [Lines 12–13] If V_j and the sender node are in the same rebroadcast zone, i.e. the distance between the rebroadcast point and the sender is less than the distance between the rebroadcast point and V_j, stop the timer, the procedure exits, otherwise, go to Step 9

This is to allow only the closest vehicle to the rebroadcast point to rebroadcast (if both vehicles are in the same rebroadcast zone). It also avoids that a CR or CRB related to a rebroadcast point to be prevented from rebroadcasting when it receives a rebroadcast AM related to another rebroadcast point.

Step 6 [Lines 14–15] If V_j receives the same message another time, and V_j is a Candidate Relay Exceptional CRE, stop the timer, the procedure exits, otherwise, go to Step 7

Step 7 [Lines 16–18] If V_j is not attached to any RSU and the TTL is not expired, V_j is declared as CRE. In this case, the farthest vehicle from the sender, but still in its transmission range, has the highest priority to rebroadcast, so the relaying metric G_3 is given by:

$$G_3 = \frac{d(V_j,S)}{R_{max}} \tag{5}$$

Hence the waiting time T_3 is calculated as follows:

$$T_3 = (CW_{min} - G_3CW_{min})T_s \tag{6}$$

Otherwise, Discard message

Step 8 [Line 21] Start back-off Timer T

Step 9 Having waited for the time T, V_j decrements TTL and forwards the message.

4   Performance Evaluation

The objective is to assess the performance of our method: (1) to reach all the vehicles sitting in the area of interest in a very short time, (2) to avoid collisions, and (3) to effectively use network resources. Also, our proposal is compared to the flooding technique [5] and V2I broadcasts (with no V2V rebroadcasts). The following performance metrics are used:

Collision Ratio: The collision ratio is the percentage of MAC collisions divided by the number of sent packets.

Information Coverage: computed as the total number of vehicles that successfully receive the AM divided by the number of all the vehicles.

Packet Delivery Ratio: is the ratio between the number of successfully received packets at the vehicles of the area of interest and all transmitted packets

Dissemination Delay: The dissemination delay is the total time required to deliver the AM to **all** the vehicles in the area of interest.

4.1 Simulation Setup

In the experiments, an event based network simulator Netsim was used. In order to make the most realistic simulations, SUMO (Simulation of Urban Mobility) bidirectionally coupled with Netsim as described in [12] was used. Netsim includes implementations of IEEE 1609.4 and IEEE 802.11p communication standards. It further includes Basic Safety Message (BSM) handling and beaconing for cooperative awareness messages (CAMs). In order to create realistic road traffic scenarios on a real map with SUMO, we used OpenStreetMap as the geographic data source and selected an area of 2 km × 2 km located in the city center of Toulouse, France. The traffic of vehicles was generated randomly. We inserted vehicles into the network topology at a constant rate (30, 75, 100, 200, 300, 400, and 500 vehicles). The speed of vehicles was varied from 0 to 20 m/s, i.e. vehicles had different accelerations at different timestamps. The maximum transmission range of each vehicle was set to $R_{max} = 250$ m. In all simulated scenarios, there was two RSUs separated by a distance of 2 km. Each one was responsible for the transmission of alert messages to vehicles located in its transmission range. The other simulation parameters are shown in Table 1.

4.2 Rebroadcast Zone Selection

For the performance evaluation of our dissemination procedure, we considered a basic and empirical rebroadcast points/zones selection method described hereafter. The goal is to compute a threshold distance R_{th} associated to the two considered RSUs and build our polygon from which we derive the locations of rebroadcast points. To that end, each RSU broadcasts a control message every 100 ms for 500 s. After each packet, we logged the distance of each vehicle from the RSUs, the vehicles that received the message, and the packet status (success or error). From the performance analysis, we observed that the packet loss rate dramatically increases when the distance between the vehicle and RSU is strictly greater than $R_{th} \simeq 600$ m. From this threshold, we built a regular octagon for each RSU with 8 sides of 459 m($0.765R_{th}$ [13]) length and 8 vertices which represent our rebroadcast points. We did not place all the rebroadcast points as the RSUs are in the corners; we had a total of 9 rebroadcasts zones shown in Fig. 3.

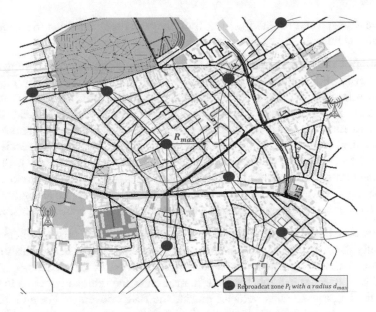

Fig. 3. Toulouse's city scenario and rebroadcast zones selection

Table 1. Simulation configuration parameters

Parameter	Value
Simulation time	500 s
Alert start	10 s
Packet generation rate	10 packets/s
AM packet size	1460 bytes
Propagation model	Nakagami m = 3
Slot time	16 μs
CW_{min}	15
CW_{max}	1023
d_{max}	16 m
α	0.8

4.3 Performance Analysis

Collision Ratio. Figure 4 shows the collision ratio as a function of vehicle density. Even for high traffic density, we observe that, compared to flooding, our method significantly reduces packet collisions (around 0.001%). This is due to the fact that only vehicles around rebroadcast zones are allowed to relay the message with different waiting times for each vehicle. The occurrence of some packet collisions is due to the behavior of the CSMA/CA mechanism. If the channel is busy, relay nodes must wait for a random back-off time before re-transmission. So when two vehicles are very close to each other, they have almost a similar waiting time, thus, they may rebroadcast at the same time if the channel is busy.

Information Coverage. Figure 5 shows the percentage of information coverage versus vehicle density. For low traffic density, the percentage of information coverage of the proposed technique is close to other techniques, but it is still better. This is due to the fact that, in such a scenario, the presence of vehicles around all rebroadcast zones at each AM transmission is not guaranteed. Indeed, vehicles can join and leave the network quickly according to the random mobility of SUMO without receiving any message. When increasing traffic density, flooding is a bad option as shown in the figure. Our method outperforms flooding and V2I approaches, and in contrast to these latter, the information coverage strictly increases as more vehicles are covering the area of interest. This means that rebroadcasts are taking place without causing collisions.

Dissemination Delay. Figure 6 shows the dissemination delay versus vehicle density. The results show that the dissemination delay of our technique is significantly lower than those of the other techniques, especially in high traffic density. Indeed, at each message period, there is a relay vehicle in or around the rebroadcast zone to transfer quickly the message to other vehicles outside the RSUs coverage. For an urban area, the frequent collisions inherent to the flooding technique cause an additional delay, and the mono-broadcast V2I technique takes a longer time to reach the maximum number of vehicles in the area, as the vehicle must be in the coverage of an RSU to receive the message. In low density, fewer vehicles are available in the network, so the presence of vehicles around each rebroadcast zone from the first AM message is not guaranteed, thus, our technique takes more time to reach all vehicles, e.g. for vehicles density of 30 vehicles, the total time required is 518.8 ms. Indeed, the average speed of vehicles is between 7 and 13 m/s, the vehicle can only move approximately 1 m in 100 ms, therefore vehicles take more time to draw near the rebroadcast zones. It is worth noting that, in this low density scenario, more than 75% of vehicles have the message delivered within 100*ms*.

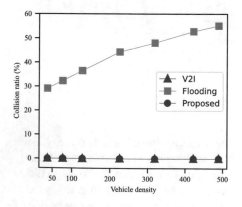

Fig. 4. Collision ratio vs. vehicle density

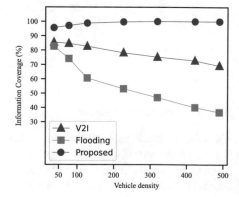

Fig. 5. Information coverage vs. vehicle density

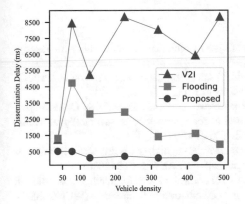

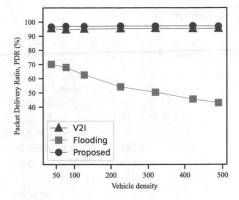

Fig. 6. Dissemination delay vs. vehicle density

Fig. 7. Packet delivery ratio vs. vehicle density

Packet Delivery Ratio. Figure 7 shows the packet delivery ratio versus vehicle density. The results show that the packet delivery ratio of our proposal is slightly better than the V2I technique. This is due to the fact that relay vehicles rebroadcast only reach their neighbors. Because of the blind re-transmission which may cause a collision or a useless transmission, flooding suffers from poor packet delivery ratios.

5 Conclusion and Future Works

In this paper, we propose an alert message dissemination scheme based on vehicle locations to provide high coverage with fast delivery. The main peculiarities of our solution are: First, the combination of V2I transmissions, where RSUs broadcast alert messages and V2V re-transmissions, where some selected relay vehicles rebroadcast the messages after a personalized waiting time, Second, the definition of broadcast zones by the SDN controller where relay vehicles are allowed to rebroadcast the message to reach all vehicles in the area of interest and notably those located in white or poorly covered areas. In this paper, we have proposed a static and simple method to define rebroadcast zones. The results obtained from our study prove the effectiveness of our scheme by maintaining its high delivery ratio, avoiding packet collisions, and ensuring rapid dissemination and high information coverage.

The main perspective of this work is to propose a machine-learning-based dynamic placement method of rebroadcast zones, which adapts to traffic conditions, weather, signal quality, etc.

Acknowledgements. This work is funded by Continental Digital Service France (CDSF) in the framework of the eHorizon project.

References

1. Ahizoune, A.: A protocol for broadcasting messages in vehicular networks. thesis manuscript (2011)
2. Liu, B., Jia, D., Wang, J., Lu, K.: Cloud-assisted safety message dissemination in VANET-cellular heterogeneous wireless network. IEEE Syst. J. **11**, 128–139 (2017)
3. Salim, F.D., Loke, S.W., Rakotonirainy, A., Srinivasan, B., Krishnaswamy, S.: Collision pattern modeling and real-time collision detection at road intersections. In: Intelligent Transportation Systems Conference (ITSC), pp.161–166 (2007)
4. Aldabbagh, G., Rehan, M., Hasbullah, H., Rehan, W., Chughtai, O.: A driver safety information broadcast protocol for VANET. Appl. Math. Inf. Sci. **10**, 451–468 (2016)
5. Ciccarese, G., Blasi, M.D., Marra, P.: On the use of control packets for intelligent flooding in VANETs. In: IEEE Wireless Communications and Networking Conference (WCNC 2009), Budapest, pp. 1–6 (2009)
6. Virdaus, K., Kang, M., Shin, S., Lee, C.G., Pyun, J.: A counting-based broadcast model of emergency message dissemination in VANETs. In: International Conference on Ubiquitous and Future Networks (ICUFN), Milan, pp. 927–930 (2017)
7. Zheng, K., Kan, L., HOU, H., Lu, Q., Meng, N., Hanlin, L.: Soft-defined heterogeneous vehicular network: Architecture and challenges. IEEE Network **30**, 72–80 (2016)
8. Liu, L., Chen, C., Qiu, T., Zhang, M., Li, S., Zhou, B.: A data dissemination scheme based on clustering and probabilistic broadcasting in VANETs. Veh. Commun. **13**, 78–88 (2018)
9. Salahuddin, M.A., Al-Fuqaha, A., Guizani, M.: Software-defined networking for rsu clouds in support of the internet of vehicles. IEEE Int. Things J. **2**, 133–144 (2014)
10. Ali, M., Malik, A.W., Rahman, A.U., Iqbal, S., Hamayun, M.M.: Position-based emergency message dissemination for Internet of vehicles. Distrib. Sens. Netw. **15**, 1550147719861585 (2019)
11. Boban, M., Kousaridac, A., Manlakis, K., Eichinger, J., Xu, W.: Use Cases, Requirements, and Design Considerations for 5G V2X, IEEE Vehicular Technology Magazine (2017)
12. Netsim simulator. https://www.tetcos.com/vanets.html
13. Polygon construction. https://debart.pagesperso-orange.fr/geoplan/polygone-regulier.mobile.html
14. Oliveira, R., Montez, C., Boukerche, A., Wangham, M.S.: Reliable data dissemination protocol for VANET traffic safety applications. Ad Hoc Netw. **63**, 30–44 (2017)
15. Report of the working group: communication technologies for cooperative ITS. https://www.ecologie.gouv.fr/sites/default/files/Rapport
16. Toufga, S., Abdellatif, S., Owezarski, P., Villemur, T., Relizani, D.: Effective prediction of V2I link lifetime and vehicle's next cell for software defined vehicular networks: a machine learning approach. In: IEEE Vehicular Networking Conference (VNC), Los Angeles, CA, United States (2019)
17. Tseng, Y.C., Ni, S.Y., Chen, Y.S., Sheu, J.P.: The broadcast storm problem in a mobile ad hoc network. In: The Fifth Annual ACM/IEEE International Conference on Mobile Computing and Networking, pp. 152–162 (1999)
18. Velmurugan, V., Manickan, J.M.L.: A efficient and reliable communication to reduce broadcast storms in VANET protocol. Cluster Computing (2019)
19. Wu, L., Nie, L., Fan, J., He, Y., Liu, Q., Wu, D.: An efficient multi-hop broadcast protocol for emergency messages dissemination in VANETs. Chinese J. Electron. **26**, 614-623 (2017)
20. Xu, Z., Li, X., Zhao, X., Zhang, M.H., Wang, Z.: DSRC versus 4G-LTE for connected vehicle applications: a study on field experiments of vehicular communication performance. J. Adv. Transp. **2017**, (2017)

Energy-Efficient Fuzzy Geocast Routing Protocol for Opportunistic Networks

Khuram Khalid[1], Isaac Woungang[1(✉)], Sanjay Kumar Dhurandher[2], and Jagdeep Singh[3]

[1] Department of Computer Science, Ryerson University, Toronto M5B 2K3, Canada
{khuram.khalid,iwoungan}@ryerson.ca
[2] Department of Information Technology, Netaji Subhas University of Technology, New Delhi, India
[3] Department of Computer Science and Engineering, Sant Longowal Institute of Engineering and Technology, Longowal, Punjab, India
jagdeep@sliet.ac.in

Abstract. This paper proposes an energy-efficient version of our recently proposed Fuzzy-based Check-and-Spray Geocast (FCSG) routing protocol (called EFCSG) for opportunistic networks (OppNets). The proposed EFCSG protocol utilizes the Mamdani fuzzy econtroller along with fuzzy attributes such as node's residual energy, buffer space, and movement to determine the relay nodes that are eligible to participate in the routing process. Through simulations using real mobility traces, EFCSG is shown to outperform the Energy Efficient Check-and Spray Geocasting (EECSG), the Fuzzy Geocast Routing Protocol for OppNets (FCSG), and the Fuzzy Geocasting mechanism in OppNets (F-GSAF), in terms of delivery ratio, average latency, number of hop count, and overhead ratio, under varying Time-to-Live (TTL) and buffer size.

1 Introduction

OppNets are a type of challenged networks where high node mobility and inter-mittent connectivity are frequent [1]. In this context, there is no guarantee of an end-to-end path between the sender and receiver of a message like in traditional network to perform the message routing. Instead, a store-carry-and-forward mechanism [2] is utilized, where every node receiving the data packets is expected to first store the message in its local buffer (if space permits), then eventually pass the message opportunistically to an encounter node in its neighbourhood (assuming that this node is eligible to carry the data packets), and so, until the message eventually gets to its intended destination.

Several routing protocols for OppNets implementing this paradigm have been discussed in the literature [3], each of which have used its own opportunistic route decision selection mechanism. These mechanisms involve the use of various different parameters including the node's attributes or a combination of them, the hope being that the selected nodes become the best possible forwarders of the message toward its destination. Examples of such mechanisms include

L. Barolli et al. (Eds.): AINA 2021, LNNS 225, pp. 553–565, 2021.
https://doi.org/10.1007/978-3-030-75100-5_48

geocasting techniques [4], fuzzy logic-based techniques [5], a combination of these techniques, and much more [3].

In a geocast technique [4], the message is first delivered from the source node to the destination region (also called destination cast) instead of a selected node, then it is disseminated within that destination cast using some kind of control flooding mechanism. Doing so requires the consideration of temporal and spatial domains, along with some constraints imposed in the availability and accessibility of the nodes' resources [4]. In such technique, every node that resides within the destination cast is a potential receiver of the message if the message destination address corresponds with its current location. A node that does not reside to the destination cast cannot receive the message. It should also be noted that frequently visited locations (or Point of Interests) are those locations that are often visited during the control flooding process. From an implementation standpoint, a cast is defined as a point (pair of two dimensional coordinates) of polygon; and to identify the position of a point outside or inside the polygon, some methods such as the Angle Summation test, Crossing test, and Triangle test algorithms can be utilized [4,6].

Another way to achieve an effective message routing in OppNets is by using fuzzy-logic [7] in the design of the utility function that decides on the selection of suitable forwarders of a message from source to its destination. This is motivated by the fact that one of the unique features of fuzzy logic is that diverse (and sometimes conflicting parameters) of a given problem can be integrated and evaluated as a single parameter to be used for the purpose of decision making. In a nutshell, fuzzy logic [7] is a process that involves performing an analysis of some information using the so-called fuzzy sets, where each of these sets is used to describe the output and input variables, noting that every fuzzy set is characterized by a membership function. The role of this function is to map each input value to a membership value that belongs in the range [0,1]. It should be noted that a membership value closer to 1 reveals that the corresponding input belongs to the fuzzy set with a high degree. On the other hand, a small membership value mean that the considered fuzzy set is not suitable for that input. Indeed, the relationship between the inputs and corresponding outputs are governed by the so-called fuzzy rules, which ultimate goal is to dynamically characterize the overall behavior of the system. Therefore, a fuzzy system is typically composed of fuzzification and rule base, inference engine and de-fuzzification components.

The fuzzification component considers a crisp numeric value as input and uses the rule base activation to map it to a fuzzy output (known as a suitable ensemble of fuzzy sets), then assigns it a degree of membership to any of the fuzzy sets. The inference engine component is the one that determines in which way the obtained fuzzy sets can be aggregated to form the expected fuzzy solution space. Considering the fuzzy output as entry, the de-fuzzification component converts it into a crisp value that can be used for decision making or action control. This paper integrates fuzzy and geocast techniques in designing a routing protocol for OppNets (called Energy-efficient Fuzzy Check-and-Spray Geocast (EFCSG)). Two phases are involved in the design, namely: (1) Message forwarding to the destination cast using a fuzzy logic technique; and (2) Intelligent control flooding of the message to all the nodes within the geocast region using the so-called Check-and-Spray mechanism.

The paper is organized as follows. Section 2 discusses some representative work on fuzzy geocast routing protocols for OppNets. Section 3 describes the proposed EFCSG protocol. Section 4 presents the simulation results on the performance of the EFCSG protocol. Section 5 concludes the paper.

2   Related Work

The use of fuzzy logic for designing routing protocols for challenged networks such as OppNets is recent. Few representative works in this direction are described in [5–13].

Dhurandher et al. [5] proposed a geocast routing protocol for OppNets, in which some fuzzy attributes are used in a fuzzy controller to decide on the best forwarders nodes to carry the message toward the destination node. In their proposed approach, a multi-copying spray mechanism is utilized for relaying the message up to the destination cast. Next, within this cast, the message is flooded in an intelligent manner, in such a way the undelivered copies of the messages that are already delivered to the destination are removed from the buffers of nodes.

Rahimi and Jamali [10] proposed a DTN routing protocol for vehicular networks based on fuzzy logic. In their scheme, a fuzzy controller is constructed using parameters like node's velocity, number of nodes, node's direction, and node's distance from destination, to determine the suitability of a node to be selected as best forwarder to carry it toward the destination. This suitability is quantified by the calculation of the so-called node' chance value; the highest that value, the better chance the node has to be selected.

Banerjee and Dutta [11] presented an adaptive geocasting scheme for mobile ad hoc networks, in which a node that belongs to the destination cast is expected to receive a notification as being member of a geocast group and is intimated to stay within a prescribed logically bounded network region for a limited time period, until it is ready to accept the multicast data packet to be forwarded toward the destination. In this process, the location information is used to reduce the multicast delivery overhead.

Jain et al. [12] proposed a routing scheme for DTNs, which is also based on fuzzy logic. In their scheme, the message size, the number of message replicas, and message remaining battery, are used as inputs to a fuzzy controller to make the decision on choosing the messages to be scheduled when a node is encountered. This involves the design of a mechanism to determine the number of message replicas in the network.

Mottaghinia and Ghaffari [13] proposed a distance and energy-aware routing protocol for delay tolerant mobile sensor networks, in which a fuzzy controller uses the energy of sensor nodes and the distance from the sink as parameters to determine the next hop selection of a source node. Another fuzzy controller is designed for buffer management purpose, which uses the number of message replicas, the survival time, and message size as input parameters.

Recently, Khuram et al. [8] proposed a fuzzy-based geocast routing protocol for OppNets (called FCSG), in which the output of the considered fuzzy controller is used as pattern to determine the direction of the node's movement and the node's speed when a source node meets an encounter. Based on this, the direction movement of any relay node selected to forward the message toward the destination is obtained as the angle between the line that joins the sender node to the center of the destination cast and the line along which the node is currently moving. As such, the smaller this angle is, the better is the chance of successful message delivery to destination. This paper proposes an energy efficient version of the FCSG scheme (called EFCSG) and evaluate its performance against that of the FCSG [8], EECSG [9] and F-GSAF [5] schemes, in terms of delivery ratio, average latency, number of hop count, and overhead ratio, under varying Time-to-Live (TTL) and buffer size.

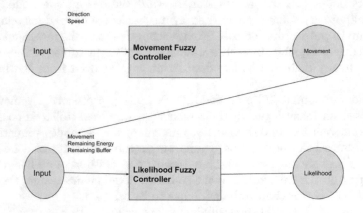

Fig. 1. Fuzzy controllers used in phase 1 of EFCSG design [8].

3   EFCSG Design

Assuming that a source node, say SN, wants to send a message m to a destination node, say D, the design of the proposed EFCSG routing scheme consists of two phases as shown in Algorithm 1.

Notations used in Algorithm 1:

- C: number of remaining copies of a message
- VH: very high
- H: high
- M: medium
- L: low
- VL: very low

Table 1. Movement controller [8].

Movement controller		
Direction	Speed 2	Movement
VL	L	L
VL	M	L
VL	H	L
L	L	L
L	M	L
L	H	M
M	L	L
M	M	M
M	H	H
H	L	M
H	M	H
H	H	H

- SN: sender node
- RN: receiver node
- m: message
- L(): likelihood function

- **Phase 1: Forwarding the Message Towards the Destination Cast**

  In this phase, we apply a multi-copying spray technique involving the use of two fuzzy controllers. More precisely, the message m generated by SN is assigned a payload data, a cast definition (set of two dimensional points forming the geographic cast), a pair of epoch times that define the lifetime of the message, and an integer C representing the maximum number of remaining copies of m that is allowed to generate and send forward to any potential encountered node during the forwarding process of m.

  When node SN opportunistically encounters another node, say RN, the likelihood of RN is calculated by means of a sequence of fuzzy controllers (as depicted in Fig. 1).

  In Fig. 1, the so-called Movement Controller uses as inputs the RN speed and direction (angle between the line that joins SN to the center of the geocast region and the line on which RN is moving at the moment), along with parameters given in Table 1 to determine the movement of RN toward the destination cast. It should be noted that the variable representing the direction varies from 0 to 180 (in degree) whereas the variable representing the speed varies from 0 to 15 milliseconds (ms).

  Next, the so-called Likelihood Controller considers the RN movement obtained from the first controller, along with the buffer space (in bytes)

Table 2. Likelihood controller [8].

Likelihood controller			
Energy	Buffer 2	Movement	Likelihood
L	L	L	VL
L	L	M	L
L	L	H	M
L	M	L	VL
L	M	M	L
L	M	H	M
L	H	L	VL
L	H	M	M
L	H	H	H
M	L	L	L
M	L	M	M
M	L	H	H
M	M	L	L
M	M	M	M
M	M	H	H
M	H	L	L
M	H	M	M
M	H	H	H
H	L	L	L
H	L	M	M
H	L	H	M
H	M	L	L
H	M	M	H
H	M	H	VH
H	H	L	L
H	H	M	VH
H	H	H	VH

and remaining energy of RN, as inputs to determine the likelihood of RN (denoted $L(RN, m)$) to receive m, based on parameters that are given in Table 2. If $L(RN, m) \geq L(SN, m)$, then m is forwarded to RN and its C value is decreased accordingly based on the current likelihood value of RN as per Table 2; otherwise, RN is ignored, its C value is kept unchanged, and a new search for another suitable RN is launched. In this process, if at some point in time, the value C of RN reach 0 while message m has not yet reach its destination D, then no additional message copies will be generated and eventually, message m will be loss in the network.

Algorithm 1. EFCSG Algorithm

Initialization during message generation:
m is created
Sender node and destination geocast region are defined
C is initialized

Phase 1: Moving a message towards its destination cast

1: **for** each SN that meets the next RN **do**
2:     drop expired messages from buffer
3:     **for** each m in the SN buffer **do**
4:         **if** m already exists in RN **then**
5:             skip this m and go to the next m in the for loop
6:         **end if**
7:         **if** C > 0 **then**
8:             **if** RN is located in geocast area **then**
9:                 forward message to RN
10:                 C = 0
11:                 // In case RN is not located within the
12:                 // geocast region, use Tables I and II to
13:                 // calculate the likelihood of RN to receive
14:                 // the message.
15:             **else if** L(RN,m) > L(SN,m) **then**
16:                 forward a copy of message to RN
17:                 **switch** L(RN,m) **do**
18:                     **case** VH: C = C-5, break;
19:                     **case** H: C = C-4, break;
20:                     **case** M : C = C-3, break;
21:                     **case** L : C = C-2, break;
22:                     **case** VL : C = C-1, break;
23:             **end if**

Phase 2: Distributing message copy to all nodes inside the geocast

24:         **else if** C = 0 and m is in geocast region **then**
25:             **if** RN is inside the destination geocast **then**
26:                 **if** SN energy is greater than a preset threshold value **then**
27:                     **if** RN does not have message m already **then**
28:                         Forward m to RN
29:                     **end if**
30:                 **end if**
31:             **end if**
32:         **end if**
33:     **end for**
34: **end for**

- **Phase 2: Forwarding the Copies of Message m to all Nodes Within the Geocast Region**

  In this phase, an energy-efficient controlled flooding technique is implemented, which consists of two steps: (1) Checking the energy level of node RN (inside the geocast region) against an energy level threshold T; and checking the presence of the already existing message m in the RN cast as well as the current location of RN; and (2) Flooding the message m to only those nodes RN for which the checking in (1) was successful. Afterwards, RN becomes SN and Phase 1 is invoked again. This process is continued until its termination. It should be noted that for the calculation of T, the residual energy of the source/intermediate node ($CurrentEnergy(node)$) and the residual energy of the neighbouring node of SN ($neighbourEnergy(node)$) are calculated as follows:

$$CurrentEnergy(node) = getCurrentEnergy(srce/inter, node)$$

$$neighbourEnergy(node) = getCurrentEnergy(neighbour)$$

where $getCurrentEnergy()$ is a function returning the residual energy of the corresponding node. Based on these values, we get

$$T = neighbourEnergy(node)/count$$

where $count$ is the total number of neighbouring nodes of the current node.

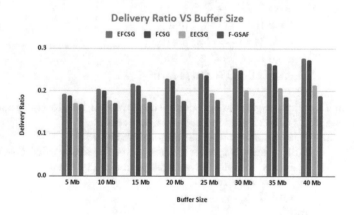

Fig. 2. Delivery ratio versus buffer size.

4 Performance Evaluation

The performance of the proposed EFCSG routing protocol is evaluated using the ONE simulator [14], then compared against the performances of the FCSG [8], EECSG [9] and F-GSAF [5] schemes, in terms of delivery ratio, average latency, number of hop count, and overhead ratio, under varying Time-to-Live (TTL) and buffer size, using the INFOCOM 2006 real mobility traces [15].

For these simulations, the delivery ratio is determined as the ratio of messages successfully delivered to the destination at the end of simulations vs. the total number of messages generated in the network. Besides, the overhead ratio is considered a measure of the bandwidth efficiency, calculated as:

$$\frac{\text{Number of relayed messages} - \text{Number of delivered messages}}{\text{Number of delivered messages}} \quad (1)$$

4.1 Simulation Results

In Fig. 2, it is observed that the delivery ratio increases when the buffer size is increased. Also, in terms of delivery ratio performance, E-FCSG is about 1.73% better than FCSG, 21.76% better than EECSG, and 31.65% better than F-GSAF. This is due to the fact that when the buffer size is increased, more messages are stored in the node's buffer, therefore the higher the chances of message delivery.

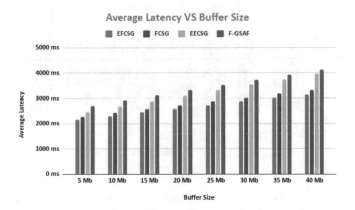

Fig. 3. Average latency vs. buffer size

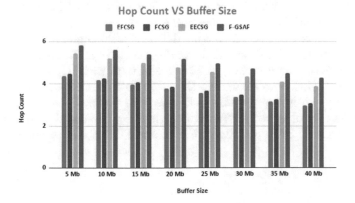

Fig. 4. Hop count vs. buffer size

In Fig. 3, it is observed that the average latency increases as the buffer size is increased. The reason is that when the size of the buffer increases, more messages stay in the node's buffer and less messages are dropped during buffer management, causing the average latency to increase. In terms of latency performance, it is observed that E-FCSG is about 5.22% better than FCSG, 17.31% better than EECSG, and 22.37% better than F-GSAF.

In Fig. 4, it is found that the hop count decreases as the buffer size is increased. This is due to the fact that, the more space the buffer size has, the more time the messages can reside in a relay node and in this case, the hop count decreases. In terms of hop count performance, it is E-FCSG is about 2.65% better than FCSG, 21.24% better than EECSG, and 27.47% better than F-GSAF.

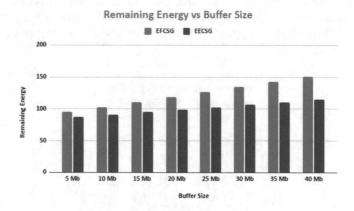

Fig. 5. Remaining energy vs. buffer size

In Fig. 5, it is observed that the remaining energy increases as the buffer size is increased. This is due to the fact that, the energy threshold check within the geocast area saves the nodes from further spending energy in scanning, receiving, and sending packets. From the result, it is found that E-FCSG is about 21.78% better than EECSG in terms of remaining energy.

In Fig. 6, it is observed that the delivery ratio increases as the TTL is increased. In terms of delivery ratio performance, it is found that E-FCSG is about 1.09% better than FCSG, 11.92% better than EECSG, and 31.05% better than F-GSAF. This is attributed to the fact that in the EECSG design, a node gets more time to find a suitable relay node to forward the message. Similar type of results are observed for the average latency, hop count, and average ratio metrics as depicted in Fig. 7, Fig. 8, Fig. 9, and Fig. 10, respectively.

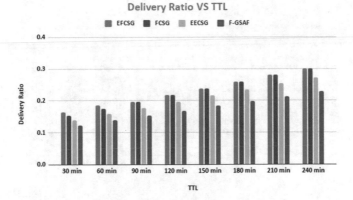

Fig. 6. Delivery ratio vs. TTL

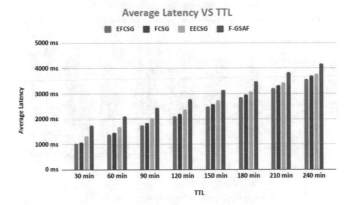

Fig. 7. Average latency vs. TTL

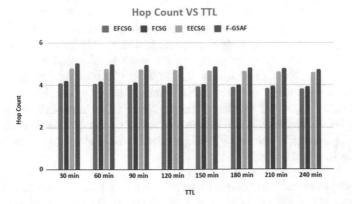

Fig. 8. Hop count vs. TTL

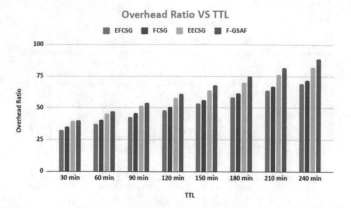

Fig. 9. Overhead ratio vs. TTL

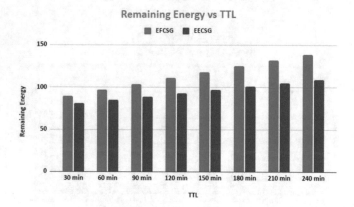

Fig. 10. Remaining energy vs. TTL

5   Conclusion

In this paper, we have proposed an energy-efficient version of the fuzzy-based check-and-spray geocast routing scheme (called EFCSG) for OppNets. Using the INFOCOM 2006 real mobility traces under, we have shown by simulations that the proposed EFCSG outperforms the FCSG, EECSG, and F-GSAF routing schemes with respect to delivery ratio, average latency, number of hop count, and overhead ratio, under varying Time-to-Live (TTL) and buffer size. As future work, we plan to investigate the message security in the propose EFCSG scheme, by designing a security-aware EFCSG protocol to defend against network attacks such as blackhole attacks.

Acknowledgements. This work is partially sponsored by a grant held by the 2nd author, from the National Science and Engineering Research Council of Canada (NSERC), Reference number: RGPIN-2017-04423.

References

1. Lilien, L., Kamal, Z.H., Bhuse, V., Gupta, A.: The concept of opportunistic networks and their research challenges in privacy and security. In: Mobile and Wireless Network Security and Privacy, pp. 85–117. Springer, Boston (2007)
2. Huang, C.M., Lan, K.C., Tsai, C.Z.: A survey of opportunistic networks. In: 22nd International Conference on Advanced Information Networking and Applications-Workshops (AINA Workshops 2008), pp. 1672–1677. IEEE (2008)
3. Kuppusamy, V., Thanthrige, U.M., Udugama, A., Förster, A.: Evaluating forwarding protocols in opportunistic networks: trends, advances challenges and best practices. Fut. Internet **11**(5), 113 (2019)
4. Rajaei, A., Chalmers, D., Wakeman, I., Parisis, G.: GSAF: efficient and flexible geocasting for opportunistic networks. In: 2016 IEEE 17th International Symposium on a World of Wireless, Mobile and Multimedia Networks (WoWMoM), pp. 1–9. IEEE (2016)
5. Dhurandher, S.K., Singh, J., Woungang, I., Takizawa, M., Gupta, G., Kumar, R.: Fuzzy geocasting in opportunistic networks. In: International Conference on Broadband and Wireless Computing, Communication and Applications, pp. 279–292. Springer, Cham (November 2019)
6. Ma, Y., Jamalipour, A.: Opportunistic geocast in disruption-tolerant networks. In: 2011 IEEE Global Telecommunications Conference-GLOBECOM 2011, pp. 1–5. IEEE (December 2011)
7. Zadeh, L.A.: Soft computing and fuzzy logic. In: Fuzzy Sets, Fuzzy Logic, and Fuzzy Systems: Selected Papers by Lotfi a Zadeh, pp. 796–804 (1996)
8. Khalid, K., Woungang, I., Dhurandher, S.K., Singh, J.: A fuzzy based check-and-spray geocast routing scheme for opportunistic networks. IEEE Syst. J. (2020, submitted). ManuscriptID ISJ-RE-20-10831
9. Khalid, K., Woungang, I., Dhurandher, S.K., Singh, J., Rodrigues, J.J.P.C.: Energy-efficient check-and-spray geocast routing protocol for opportunistic networks. Information **11**(11), 504 (2020)
10. Rahimi, S., Jamali, M.A.J.: A hybrid geographic-DTN routing protocol based on fuzzy logic in vehicular ad hoc networks. Peer-to-Peer Netw. Appl **12**(1), 88–101 (2019)
11. Banerjee, A., Dutta, P.: Fuzzy controlled adaptive geocasting in mobile ad hoc networks. Int. J. Inf. Technol. **14**(2), 109–131 (2008)
12. Jain, S., Chawla, M., Soares, V.N., Rodrigues, J.J.: Enhanced fuzzy logic-based spray and wait routing protocol for delay tolerant networks. Int. J. Commun. Syst. **29**(12), 1820–1843 (2016)
13. Mottaghinia, Z., Ghaffari, A.: Fuzzy logic based distance and energy-aware routing protocol in delay-tolerant mobile sensor networks. Wireless Pers. Commun. **100**(3), 957–976 (2018)
14. Keränen, A., Ott, J., Kärkkäinen, T.: The ONE simulator for DTN protocol evaluation. In: Proceedings of the 2nd International Conference on Simulation Tools and Techniques, pp. 1–10 (March 2009)
15. Scott, J., Gass, R., Crowcroft, J., Hui, P., Diot, C., Chaintreau, A.: CRAWDAD dataset cambridge/haggle (v. 2009-05-29). CRAWDAD Wireless Network Data Archive (2009)

Time Synchronization for Automotive Prototyping and Testing Platforms

Esraa Elelimy[(✉)] and Andrew Azmy

Valeo, Cairo, Egypt
{esraa.elelimy,andrew.azmy}@valeo.com

Abstract. In typical automotive prototyping and testing platforms, data streams between several heterogeneous components. Each of these components will usually have its own internal clock. However, these clocks will differ in their resolution, drift rate, and other properties. Furthermore, some components will provide unstamped data, especially when integrating automotive and non-automotive components such as digital applications and simulation tools. As a result, having a time synchronization and data stamping mechanism is indispensable to these platforms to accurately report when an event has occurred or when a sensor or an application has sent their data. In this paper, we present an abstracted time synchronization approach that facilitates synchronizing and time stamping data from diverse automotive and non-automotive elements, regardless of the component's underlying communication protocol. Moreover, this approach can achieve high synchronization accuracy even when individual components are distributed over multiple networks.

1 Introduction

Prototyping is an integral part of developing complex products. In the automotive industry, prototyping platforms are regularly used in the early development stages to test and validate software applications to ensure their safety and functionality before the production phase. Moreover, prototyping platforms speed up the process of exploring new innovative ideas and novel concepts [4]. For such platforms, the ability to test a wide range of applications is dependent on the components integrated within the platform. Consequently, a prototyping platform usually consists of various elements such as Electronic Control Units (ECUs), Front Cameras, Radars, and many other sensors and controllers.

In some cases, building a physical prototyping platform might be infeasible; one approach in such situations is to develop virtual prototyping platforms. However, these platforms do not eliminate the need for physical ones [1,11]. Other approaches involve a combination of physical and virtual components. For example, in cases where a specific sensor is not available, a model could be developed using simulation tools such as CarMaker [8], SCANeR studio [21] or unity [22] and data from that model could be passed to the other platform components.

L. Barolli et al. (Eds.): AINA 2021, LNNS 225, pp. 566–576, 2021.
https://doi.org/10.1007/978-3-030-75100-5_49

Whether physical, virtual, or a combination of both, these platforms require a robust time synchronization mechanism as many automotive applications are time-sensitive. One example of such applications is Advanced Driving Assistance Systems (ADAS), which include several Cameras, radar, and ECUs; all of these components need to have a standard reference time and be synchronized together. Another example would be data logging from different sensors, which also requires precise time stamping.

In addition to the previous examples, and with the recent growing appeals for connecting the vehicles with smart-gadgets and IoT applications [5], the prototyping platforms need to encompass all these new tools and applications. This integration adds more diversity to the communication protocols and the data types flowing in the vehicle. Subsequently, exposing several limitations to the previous synchronization methods used in prototyping platforms.

This paper presents a time synchronization mechanism that facilitates synchronization between automotive and non-automotive components, either digital applications or simulation tools, as it is built upon an abstracted platform for interoperability called Let's DO [3] that allows integrating heterogeneous components in a seamless manner. Additionally, the presented approach has the ability to synchronize such components even when they are distributed across different networks.

2 Background and Related Work

A Vehicle network contains various physical components; each uses a different underlying communication protocol. The standard protocols found in a vehicle network are FlexRay [15], Ethernet [2], and Controller Area Network(CAN). However, each of these protocols has different properties and capabilities, making time synchronization a complicated task [9]. For example, FlexRay runs on a predefined time pattern. However, CAN supports priority-based transmission, meaning that when multiple messages are transmitted, the one with higher priority dominates the bus. Ethernet has inherent latencies, so when using Ethernet-based communication for time-sensitive tasks there will be a need for timing information for the transmitted data [9] that is why various standards are available for time synchronization. As mentioned before, connecting the vehicle with smart-gadgets adds more protocols to the vehicle network and presents new challenges as most of these gadgets and applications do not provide stamped data.

A plentiful of literature have discussed time synchronization, and several methods and platforms were developed to address this problem. However, these platforms were mostly addressing specific sub-problems in the automotive industry and did not generalize or perform well when the problem is expanded. In addition to that, the ability to communicate and synchronize with either digital applications or simulation tools was not discussed enough in the literature.

In the following sections, we will mention several platforms and methods that were meant to solve the time synchronization problem, we will discuss their shortcomings and limitations that led to the development of the synchronization mechanism which will be introduced in this paper.

2.1   Time Sensitive Networking

AUTOSAR 4.2.2 and IEEE 802.1AS, also known as Time-Sensitive Networking (TSN), are widely used in the automotive industry for time synchronization. AUTOSAR harmonizes the time synchronization between FlexRay, CAN, and Ethernet using a modified version of the IEEE 802.1AS standard [9]. However, this approach still has some limitations. Mainly, it requires a statically predefined network topology, and the synchronization mechanism strictly depends on that topology. Besides, integrating new components to such topology can not be easily configured, especially when they have different characteristics, such as digital gadgets.

2.2   Synchronization in RTmaps

RTmaps [18] is a platform for prototyping automotive applications. It provides an internal synchronization module to adjust the clocks across multiple distributed machines [13]. However, it does not provide a smooth way to integrate and synchronize with non-automotive digital components, which provide unstamped data or simulation tools that usually use its own reference time.

2.3   Synchronization in ROS

While Robot Operating System (ROS) [20] allows integration with heterogeneous components and applications, the framework itself does not support time-synchronization across multiple machines [16], making it not ideal for automotive prototyping.

To integrate components that do not have reference time in ROS, the developers will have to implement their synchronization mechanism as done by Fregin *et al.* in [7] where they implement Precision Time Protocol (PTP) for synchronizing four cameras for an ADAS system; as ROS does not provide such capabilities. The lack of time synchronization in ROS adds more complexity and overhead for the developers when implementing time-sensitive applications.

2.4   Precision Time Protocol

IEEE 1588 standard, also known as Precision Time Protocol (PTP), [17] is broadly used for time synchronization. PTP was developed to overcome the limitations in previous protocols such as the Network Time Protocol (NTP) [19] and Global Positioning System (GPS), as PTP provides higher accuracy than NTP and allows for hardware time stamping. PTP is also suitable for indoor applications as it does not depend on satellite signals, which makes it more robust than GPS in such situations [14].

According to the IEEE 1588 standard, clocks are arranged in a master-slave hierarchy where each slave synchronizes with its master based on information from several messages exchanged between them [6]. At the base of this hierarchy is the Grand-Master clock, which is usually selected by the best master clock algorithm and is considered as the time source to the network. This clock's choice depends on its properties and accuracy relative to the other clocks in the network. The master clock sends cyclic multicast sync messages to all the other clocks (slave clocks) in the network, while the slaves respond using a unicast message to the master.

The messages exchanged between the master and each slave starts with a sync message fired from the master, followed by a follow-up message containing the timestamp at which the sync message was sent. The sync message arrives at the slave end at time t2, and shortly after that, the follow-up message will arrive carrying the timestamp t1. After that, the round-trip communication time needs to be calculated to accurately get the time offset between the master and the slave clock. As a result, the slave clock will send a delay request message at time t3, reaching the master at time t4, and the master will respond with a delay response message containing the actual time it received the request [14]. Figure 2 illustrates the sequence of messages exchanged between the master and each slave. Using t1, t2, t3, t4, the slave clock can now calculate the time offset and update its internal clock according to the following equations (Fig. 1):

$$D1 = t2 - t1 \tag{1}$$

$$D2 = t4 - t3 \tag{2}$$

$$Offset = \frac{D1 - D2}{2} \tag{3}$$

$$MeanPathDelay = \frac{D1 + D2}{2} \tag{4}$$

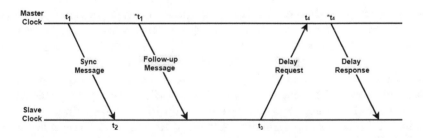

Fig. 1. PTP timing diagram

3   Time Synchronization in Let's DO

Previous platforms and synchronization methods had several pitfalls and limitations. This paper targets two of these limitations. Firstly, the lack of the ability to synchronize with digital applications and IoT gadgets. Secondly, the complexities involved when synchronizing hardware components and applications that use different communication protocols.

In the following section, we will introduce Let's DO [3], a platform for interoperability between different software components. After that, we will present an abstracted synchronization mechanism built on Let's DO and addresses the previously mentioned shortcomings and is the main contribution of this paper. Finally, we will then present the results obtained using that module in different configurations.

3.1   Let's DO

Let's DO, introduced in [3], is a platform for interoperability and data exchange between heterogeneous components. It provides standardized and abstract communication interfaces allowing different elements and applications to communicate with each other, including digital applications, simulation tools, and other automotive and non-automotive components.

In Let's DO, each computing element is called a node. Nodes communicate with each other using the message-oriented-middleware libraries that have abstracted and standardized APIs for communication. These standardized APIs allow nodes to send and receive messages to each other over the Let's DO network regardless of their original underlying communication protocols, as Let's DO offer a variety of pre-defined messaging adapters and utility libraries.

Let's DO has two messaging standards, Message Queuing Telemetry Transport (MQTT) and Data Distribution Service (DDS). While these standards allow interoperability and efficient communication among non-homogenous components, they lack having synchronization capabilities.

3.2   Method

The time synchronization mechanism introduced in this paper is built as a utility library within the Let's DO framework [3], employing its messaging libraries. It is also based on the IEEE 1588 standard [17] with the addition of several modifications to be suitable for prototyping platforms.

Firstly, we created a timestamp message that could be easily imported and used in all Let's DO messages. We also added utilities to fill the timestamp for any node whether it supports software or hardware time stamping.

The PTP standard is originally built upon IP and User Datagram Protocol (UDP) [14], which complicates synchronizing components that use other protocols. To address this issue, the synchronization mechanism presented here is built upon the standardized and abstracted Let's DO APIs for sending and receiving messages, making it independent of the underlying protocols used for communication by each node.

In a Let's DO network, each distributed node will either be a slave node or a master node. Using Let's DO standard APIs for publishing and subscribing, the master node will publish a sync message and a follow-up message periodically every 10 ms. These messages will be published to a pre-specified topic to which all the other nodes, slave nodes, are subscribing. Each slave node will publish an independent delay request to the master node, in addition to the timing information in the delay request, we added an additional ID attribute to identify the correspondence between each slave and its message request. The master will then publish a delay response to the node with the same ID as in the delay request. Based on these messages exchanged, each slave node can now calculate its offset and delay according to Eqs. (1) to (4).

Our module supports having a dynamically changing network topology, where new nodes can enter an existing Let's DO network at any time and still able to synchronize with the master. This feature was not available in the previous synchronization platforms used in automotive applications. Besides, this synchronization module can synchronize Let's DO nodes running either on the same network or different networks.

3.3   Experimental Results

We performed several experiments to characterize the accuracy of the introduced synchronization module in Let's DO. As previously mentioned, Let's DO supports both MQTT and DDS as messaging standards. The developer can use the desired standard during run-time through simple configurations. The nodes used in the following experiments were configured to use MQTT.

In the first experiment, all the nodes were connected to the same WiFi network with WPA2 security and a link speed of 65 Mb/s. Figure 2 shows the network architecture for this experiment. In the second experiment, each node existed in a different network and they were communicating through a common broker on a server as indicated in Fig. 3.

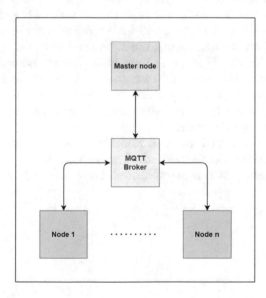

Fig. 2. Distributed nodes communicating over a local network

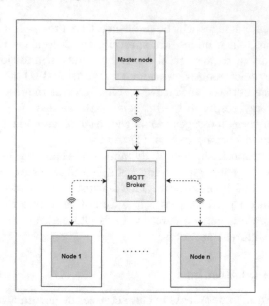

Fig. 3. Distributed nodes communicating over multiple networks

Both experiments were repeated several times to ensure the reproducibility of the results. In each trial, the offset between the nodes and the master node was recorded.

Figure 4 shows the average offset between the nodes and the master when all the nodes were in the same network. As indicated, the average offset was 3.7 ms with a standard deviation of 7.2 ms, and Root Mean Square Error (RMSE) of 8.4 ms. For automotive prototyping, having a synchronization accuracy of approximately four milliseconds is enough to test various algorithms and applications as most of the sensors used will have a refresh rate in sub-milliseconds.

For the second set of experiments, the average offset was six milliseconds with a standard deviation of 10.6 ms, and Root Mean Square Error (RMSE) 12.2 ms as indicated in Fig. 5. In the second case, the synchronization accuracy dropped compared to the first experiments which were expected due to the lower quality of service and long delays in the messages transmitted when the computing nodes do not exist on the same network.

In [12], Mani *et al.* tested the synchronization accuracy over MQTT when no additional mechanisms were employed. It was shown that the synchronization RMSE was 162.5 ms, which is significantly higher than what was presented in our results.

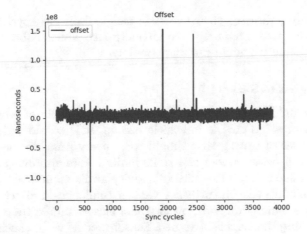

Fig. 4. Synchronization accuracy on a Wifi Network

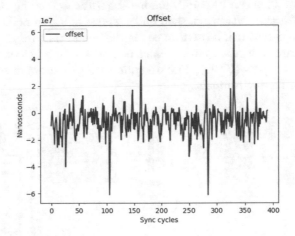

Fig. 5. Synchronization accuracy over different networks

4   Applications

Digitization is currently transforming every aspect of day-to-day life. Every industry is racing to evolve to meet digital lifestyle demands and expectations. The automotive industry is not an exception. Manufacturers are modifying their product development strategies to meet consumer-driven requirements, which has started by integrating digital services into vehicles' infotainment systems and are targeting novel solutions to further enable mobility as a service.

The need for such transformation goes deeper, reaching the development cycle of almost all vehicle systems. This need has been greatly emphasized with the increasing demand of remote access to all business aspects since the start of the recent pandemic (COVID-19) [10].

Manufacturers directed their efforts to make all physical systems needed by their engineering teams remotely and virtually accessible, enabling them to continue being productive while working remotely.

4.1   Testing ADAS System

One of the significant activities in the development cycle of advanced driver-assistance systems (ADAS) is in-vehicle testing. It is considered as one of the conclusive phases of testing that should cover every possible use case and scenario, taking the system under test to its limits before approving its release.

Converting such a step to a virtual activity avails numerous advantages and savings, but simultaneously presents various challenges and required adaptations. Firstly, a simulation software that can closely mimic the whole environment surrounding the vehicle is needed. Sensor models within such environment would send relevant information to another computing node running the software under test. This can be a PC running the software, or the actual hardware control unit with the software flashed on it. The simulation node also sends out reference (ground-truth) information to another node that responsible for judging the performance of the system under test based on a set of predefined key performance indicators (KPIs). The diagram in Fig. 6 shows a sample for such distributed testing environment.

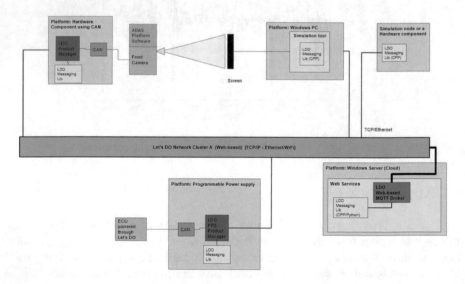

Fig. 6. Sample of a distributed testing environment

For such a loop, temporal synchronization of the exchanged data is crucial for formulating a useful judgement of the system performance. Synchronization issues in the network connecting all the aforementioned nodes might hinder all testing results useless.

5   Conclusion

In this paper, we presented the importance of having a synchronization mechanism for automotive prototyping platforms. We also discussed the limitations of existing platforms for prototyping as there is either the limitation of connecting heterogeneous components or the lack of synchronization support. We then introduced the synchronization module in the Let's Do framework addressing the previously mentioned gaps in the literature; having a mechanism embedded in an abstracted framework like Let's DO solves synchronizing heterogeneous components and digital applications since it is independent of any particular protocol. It was also shown that the presented module achieves a synchronization accuracy that is sufficient for time-sensitive applications in the automotive industry.

References

1. Cioroaica, E., et al.: Prototyping automotive smart ecosystems. In: 2018 48th Annual IEEE/IFIP International Conference on Dependable Systems and Networks Workshops (DSN-W), pp. 255–262. IEEE (2018). https://doi.org/10.1109/DSN-W.2018.00072
2. Kozierok, C.M., Correa, C., Boatright, R.B., Quesnelle, J.: Automotive Ethernet: The defnitive guide (2005)
3. ElHakim, R., et al.: Let's DO - Automotive Platform for Interoperability
4. Elverum, C.W., Welo, T.: The Role of early prototypes in concept development: insights from the automotive industry. Procedia CIRP **21**, 491–496 (2014). https://doi.org/10.1016/j.procir.2014.03.127
5. Fehling, C., et al.: Your coffee shop uses cloud computing. IEEE Internet Comput. **18**(5), 52–59 (2014). https://doi.org/10.1109/MIC.2014.101
6. Flammini, A., Ferrari, P.: Clock synchronization of distributed, real-time, industrial data acquisition systems. In: Sciyo, MV (ed.) Data Acquisition (2010). https://doi.org/10.5772/10458
7. Andreas, F.: ROSCon 2017: building a computer vision research vehicle with ROS. https://roscon.ros.org/2017/presentations/ROSCon%202017%20Computer%20Vision%20Research%20Vehicle.pdf
8. GmbH, IPG Automotive. CarMaker. Accessed 25 Aug 2020, https://ipg-automotive.com/products-services/simulation-software/carmaker/
9. Jesse, B.: Time Synchronization in Automotive Ethernet Networks Balancing Act Between AUTOSAR, IEEE, and TSN. Vector (2016). https://assets.vector.com/cms/content/know-how/_technical-articles/Ethernet_Timesync_Automobil-Elektronik_201608_PressArticle_EN.pdf
10. Kude,T.: Agile Software Development during COVID19. http://knowledge.essec.edu/en/innovation/agile-software-development-during-after-COVID19.html
11. Kulkarni, A., et al.: Virtual prototyping used as validation tool in automotive design. Semant. Scholar (2011). https://doi.org/10.36334/modsim.2011.a5.kulkarni
12. Mani, S.K., et al.: A System for Clock Synchronization in an Internet of Things (2018). ArXiv:1806.02474 [Cs]

13. Nashashibi, F.: /Sup RT/M@ps: a framework for prototyping automotive multi-sensor applications. In: Proceedings of the IEEE Intelligent Vehicles Symposium 2000 (Cat. No. 00TH8511), pp. 99–103. IEEE Xplore (2000). https://doi.org/10.1109/IVS.2000.898325

14. Neagoe, T., et al.: NTP versus PTP in computer networks clock synchronization. In: 2006 IEEE International Symposium on Industrial Electronics, vol. 1, pp. 317–362. IEEE Xplore (2006). https://doi.org/10.1109/ISIE.2006.295613

15. Corporation, N.I.: Flexray automotive communication bus overview (2015)

16. Clock - ROS Wiki. http://wiki.ros.org/Clock

17. IEEE Standard for a Precision Clock Synchronization Protocol for Networked Measurement and Control Systems. In: IEEE Std 1588–2008 (Revision of IEEE Std 1588-2002), pp.1–300 (2008). https://doi.org/10.1109/IEEESTD.2008.4579760

18. Intempora RTMaps. https://intempora.com/products/rtmaps/

19. Ntp.Org: Home of the Network Time Protocol. http://www.ntp.org/

20. Robot Operating Sytem (ROS). https://www.ros.org

21. SCANeR Studio. https://www.avsimulation.com/scanerstudio/#

22. Unity. https://unity.com/products/unity-simulation

Localization of Wireless Sensor Networks Using Negative Knowledge Bilateration Techniques

Saroja Kanchi[(✉)]

Department of Computer Science, Kettering University, Flint, MI 48504, USA
skanchi@kettering.edu

Abstract. In this paper, we propose an anchor based localization technique for localization of nodes in wireless sensor networks given range measurements. While much of the research has focused on gathering location information based on existing range measurements, this paper focuses on obtaining location using the knowledge of lack of range information between two nodes termed as *negative knowledge*. Three variations of the negative knowledge technique are used to develop localization algorithm which include negative knowledge bilateration, joint negative knowledge bilateration and double negative knowledge bilateration. The simulation results show that the proposed algorithm localizes significantly more nodes than traditional trilateration. This improvement holds in cases of sparse and dense random networks, connected random networks, and, H-shaped random networks. The number of nodes localized by the proposed algorithm is significantly higher for varying percentage of anchors and varying network radii for each of these types of networks.

1 Introduction

With the explosion of mobile devices incorporated with sensors and the rapid growth of IoT technology, algorithms and protocols for wireless sensor networks have received considerable attention. Sensors have limited capability in the realms of computation, energy, storage and mobility. When deployed, they transmit sensed data to a centralized server or to nearest sink node, and, the sensed data is used for various applications such as surveillance, monitoring etc. Often, when deployed in an environment that is subject to mobility such as underwater missions, vehicles involved in combat etc. their locations change over time. Since equipping sensors with GPS consumes tremendous energy, the locations of deployed sensors becomes unknown over time. Localization of sensors is defined as the problem of finding geo-location of sensor nodes under different models of communication and different deployment architectures. Localization algorithms have received tremendous attention in last three decades as described in the survey papers [3–5] in the literature.

Techniques for localization of nodes in wireless sensor networks can be categorized using various assumptions about available information for localization: whether sensor nodes remain stationary or mobile during execution; whether anchor nodes (anchor based) present or no anchor nodes (anchor free) are present; whether distance information between the nodes is available (range-based) or is not available (range-free). In addition, 3D localization, indoor localization, underwater localization and localization under noisy information require vastly different techniques [5].

Radio range of sensor node is the radius within which a node can transmit and receive signals to and from other sensor nodes. The range information between nodes can be obtained using RSSI (Received Signal Strength Indicator), ToA (time of arrival of a signal) or TDoA (Time difference between arrival of signals), AoA (Angle of Arrival of signal). RSSI is the cheapest of the four, however, it is more error prone. A range based distributed localization algorithm using Multi Dimensional Scaling (MDS) was proposed in [6]. The researchers in [9] present a range based source localization in which distance measurement of the path loss model uses ratio approach and also least squares approach. A received signal strength indicator (RSSI) based localization technique called cognitive maximum likelihood (C-ML) was developed in [8] to localize nodes with the positioning scheme that relies on the statistical path loss model. Oğuz-Ekim et al. propose a range based maximum-likelihood (ML) algorithm in [7]. Bayesian approach was used in [14] in which position is achieved by using iterative least square (ILS) rather than lateration. In [10], the authors provide an anchor free localization technique using the range measurements and triangulation. An extended range and redundant association of intermediate nodes (TERRAIN), is presented in [10]. TERRAIN uses (assumption based coordinates) ABC and transmits data to the other nodes to construct final mapping. Moore et al. [12] propose a range measurement based anchor free localization algorithm, using trilateration technique in 'Robust Distributed network Localization with noisy range measurements' (RODL). It adopts a cluster based localization, which uses quadrilaterals to avoid flip ambiguities.

Range-free localization with beacons using centroid approach was presented in [1], in which localization is accomplished by each node computing the centroid of the locations of the anchors from which it gets information. The algorithm reduces the accumulation of error due to low anchor density by finding the location using approximate point in triangle test (APIT). In the range-free localization presented in [13], connectivity information is used by finding number of hops between nodes. DV-Hop technique combines the radio range information with the proximity to anchors to deal with the issue of low anchor density within the proximity of the node. MDS (Multidimensional Scaling) [2] is an anchor free technique used in both range based and range free localization and hop distance is used in the absence of range measurements to compute distances between nodes.

It is well known that a node's location can be uniquely computed in 2D if the Euclidean distances to three other nodes whose locations are known are available. This technique of location determination is termed as trilateration. Multilateration is the extension of trilateration technique used to consolidate the distances received from several nodes. When range measurements and anchors are available, multilateration is used to propagate location information from one node to another thus increasing the number of anchors over time in the network. Collaborative multilateration is presented in [15]. In [11], the limitation of trilateration-based approach is removed by proposing localization of wheels. Applying the rigidity theory, researchers in [16] propose the concept of verifiable edges and derive the conditions for an edge to be verifiable. On this basis, they design a localization approach with outlier detection, which explicitly eliminates ranges with large errors before location computation. In [17], authors propose a distributed localization protocol, GROLO, that is able to perform efficient distributed localization through an adaptive global rigidity mechanism especially designed for the resource limited IoT nodes. In order to increase the positioning accuracy of the multi-channel location algorithm in a multipath environment, [18] proposes an optimal multi-channel trilateration positioning algorithm (OMCT) by establishing a novel multi-objective evolutionary model. The authors in [19] provide a localization algorithm based on shadow edges as a necessary and sufficient localizability condition.

In this paper, we propose a localization algorithm based on negative knowledge bilateration. The algorithm is compared to trilateration technique and we show by simulation that when localization includes negative knowledge bilateration techniques, the algorithm localizes significantly more nodes in a variety of network or varying density and with various number of anchors.

Section 2 provides the background of the WSN models as a network graph, the model used and the trilateration technique. Section 3 describes the various negative knowledge bilateration techniques used in the algorithm. Section 4 presents the algorithm and discusses implementation details. Section 5 presents the results of simulation and Sect. 6 provides conclusion.

2 Background

The $2D$ localization problem of wireless sensor network considered here is the problem of finding mapping of the sensors in a WSN on to the $2D$ space such that given range measurements between pairs of sensor nodes equals with the Euclidean distance between the sensor nodes specified by the mapping. If we represent the underlying network graph $G = (V,E)$ with vertices representing sensor nodes and edge weights $e(u,v)$ representing the distance between vertices u and v when range measurement between u and v is given, the problem of localization is to determine the unique mapping $m()$ of vertices V to points in $2D$, such that $dist(m(u),m(v)) = e(u,v)$ of the graph G for every pair of vertices $u,v \in V$ where $dist$ denotes the Euclidean distance between corresponding points on the plane. It is known that the underlying graph of a network has to be globally rigid [20] for the nodes in the network to be localizable. Here we consider networks that are random networks, connected networks and networks of H-shape and find large number of localizable nodes in all these types of networks.

Given a WSN, the range information between sensor nodes is available if and only if they are within the radio range of each other. It is assumed all nodes in the network have the same radio range and it is called the *radius* of the network. It is assumed that all nodes remain stationary for the duration of the algorithm. There are some special nodes marked *anchors* which are equipped with GPS and they know their locations a-priori. The localization problem therefore is to find the geo-locations of the unlocalized (non-anchor) nodes.

When a node has distance to three nodes whose locations are known in 2D, the node's location can be uniquely found using *trilateration*. The technique of using trilateration is described in Sect. 3. The process of iterative trilateration begins by identifying an unlocalized node in the network and localizing it if it has available, the range measurements to three neighbors whose locations are known. Inversely, nodes that are adjacent to three anchors can be localized initially as it is done in the algorithm in this paper. These localized nodes now act as anchors to help localize other unlocalized nodes. The process of iterative trilateration cannot progress when unlocalized nodes do not have distances to three neighbors that are localized. The success of trilateration is therefore sensitive to the location of the anchors within the network and order of localization of the unlocalized nodes. Trilateration stops progressing, even if there are several nodes with known locations in the network. Our algorithm progresses beyond trilateration by using negative knowledge bilateration techniques.

Trilateration: Here, we first describe how an unlocalized node can find its unique location using the geometric technique of trilateration. That is, given the distances d_1, d_2 and d_3 from the unlocalized node A to three localized nodes P_1, P_2, and P_3 respectively, the location of the node A can be found using geometric properties on 2D plane. See Fig. 1.

Given, $P_1 = (x_1, y_1)$, $P_2 = (x_2, y_2)$, and $P_3 = (x_3, y_3)$, then the coordinates of $A = (x, y)$ can be found by solving the following three equations:

$$\left.\begin{array}{l}(x - x_1)^2 + (y - y_1)^2 = d_1^2 \\ (x - x_2)^2 + (y - y_2)^2 = d_2^2 \\ (x - x_3)^2 + (y - y_3)^2 = d_3^2\end{array}\right\} \tag{1}$$

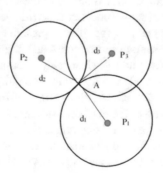

Fig. 1. Trilateration

which essentially is a solution to the equation

$$2MX = b \tag{2}$$

where

$$M = \begin{bmatrix} 2(x_3-x_1) & 2(y_3-y_1) \\ 2(x_3-x_2) & 2(y_3-y_2) \end{bmatrix} \tag{3}$$

with

$$b = \begin{bmatrix} d_1^2 - d_3^2 - x_1^2 - y_1^2 + x_3^2 + y_3^2 \\ d_1^2 - d_2^2 - x_2^2 - y_2^2 + x_3^2 + y_3^2 \end{bmatrix} \tag{4}$$

and

$$X = \begin{bmatrix} x \\ y \end{bmatrix} \tag{5}$$

3 Negative Knowledge Bilateration Techniques

Bilateration refers to a node having distance information to two neighbor nodes as opposed to three that is required for trilateration. this information provides two possible positions for the unlocalized node. clearly, additional information is needed in order use bilateration for determining the location of the unlocalized node. Given two possible positions of a node, lack of distance information to a localized node is used determine which of the two positions of the unlocalized node is indeed the correct geo-location. In the next section, we describe the details of three supplemental information used with bilaterable nodes, namely, negative knowledge bilateration, joint negative knowledge bilateration, and, double negative knowledge bilatcration. To maximize the number of localized nodes, we run the trilateration technique after each of the negative knowledge technique that are applied.

Negative Knowledge Bilateration: Without having distances to three localized nodes, if an unlocalized node has distances available to two localized nodes, we call the node *bilaterable.* As a first step, the novel technique of negative knowledge, that is, lack of distance to a localized node that is within the sensor radius of one of the two possible locations reveals that one of locations would not be correct location, due to the missing distance information. Thus the negative knowledge of missing distance is used to confirm the location of the unlocalized node A.

Assume that the node A is adjacent to only two other localized nodes in the graph, say P_1, and P_2, with distances d_1 and d_2, then clearly, A must be at the intersection of two circles with centers P_1, and P_2 and radii d_1 and d_2 respectively. Let us call the two possible locations of node A, P_A and $P_{A'}$. See Fig. 2. We propose a novel approach where missing distance information from a localized node is used to determine the location of A as either P_A or $P_{A'}$. Let us assume that there is a localized node B within the sensor radius s of the possible location P_A of node A, and since there is no edge between B and A then the location of A would be at $P_{A'}$. Similarly, if there is a node B, within the sensor radius s of the possible location $P_{A'}$ of node A, and since there is no edge between B and A then the location of A would be at P_A. This is proved in Theorem 1 below:

582   S. Kanchi

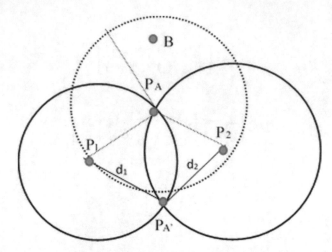

Fig. 2. Negative knowledge bilateration

Theorem 1: Assume that the possibility of trilateration is examined for all unlocalized nodes in a network graph. Given distances two localized nodes P_1 and P_2 from A, and a localized node B within sensor radius of one of the two possible locations, then the node A is localizable.

Proof: We examine the possibility of the point A being at location P_A and $P_{A'}$. Assume that there a sensor node that located within the sensor radius around the center of P_A. Assume that there is a localized node B within this radius. Since there is no edge between the node A and the node B, clearly node A cannot be at location P_A. The reason why there is no edge between A and B is because if there was an edge, we would have localized the node A using trilateration technique as it is also adjacent to P_1 and P_2. A must be located at location $P_{A'}$. Thus the absence of edge between a localized node B and the node A tells us the location of node A.

To use the negative knowledge bilateration, we identify an unlocalized node A with two edges to localized nodes and find the two intersection points of the circles formed by centers P_1 and P_2 and radii d_1 and d_2 as shown below.

$$\left. \begin{aligned} (x-x_1)^2 + (y-y_1)^2 &= d_1^2 \\ (x-x_2)^2 + (y-y_2)^2 &= d_2^2 \end{aligned} \right\} \tag{6}$$

However, it is possible that there is no localized node within the sensor radius of P_A in which case, we try to find a localized node around radius s of $P_{A'}$. If neither circles have a localized node, then we move to localize other nodes in the network and return to localizing A using Joint Bilateration or Double Negative Bilateration as described below.

Joint Bilateration: Assume that the two possible positions P_A and $P_{A'}$ of bilaterable node A. Also assume that Negative Knowledge Bilateration above has been attempted to resolve the position of A and the position of A was not resolved. For each pair of bilaterables A and B with $P_A, P_{A'}$ as possible positions for A, and $P_B, P_{B'}$ for B in the WSN, where $e(A,B)$ is given, we compute $dist(P_A,P_B)$, $dist(P_{A'},P_B)$, $dist(P_A,P_{B'})$, and $dist(P_{A'},P_{B'})$. One of these four must equal $d(A,B)$ since the all possible positions of bilaterables A and B have been determined. We can thus jointly localize both A and B to the corresponding locations that match $e(A,B)$. When $e(A,B)$ is not available as edge in the graph, we then attempt to localize using double negative knowledge bilateration as described below.

Double Negative Knowledge Bilateration: We now explore the case where there is no localized node B within the sensor radius of two possible locations of the unlocalized node P_A and $P_{A'}$. In this case, we would not be able to confirm the correct location of A. Assume that there is a localized node B that is within the radius s (adjacent to P_1 or P_2) of the graph but is not adjacent to A. If the Euclidean distance $dist(A,B) < s$, then the correct position of A is $P_{A'}$, since graph would have an edge between A and B and clearly A has only two edges to localized nodes P_1 and P_2. Conversely, if $dist(A',B) < s$, the A is localized to position of P_A (Fig. 3).

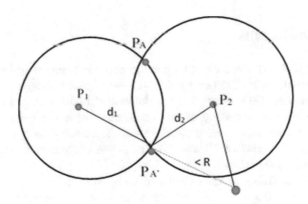

Fig. 3. Double negative knowledge bilateration

4 Algorithm

The vertices of the WSN graph that are not anchors are initially set to *unlocalizd* and all anchor nodes are set to *localized*. The goal of the localization algorithm is to find location of unlocalized nodes using techniques described in Sect. 2 and Sect. 3. Once the location is determined, the unlocalized node becomes *localized*.

Negative Knowledge Bilateration Algorithm:

```
1. Mark each anchor in the WSN as a localized node.

2. Repeat:
     a. Repeat:
               Perform trilateration by finding
               trilaterables among all adjacencies of
               localized nodes.
        Until (no nodes can be marked localized)
     b. Mark a node "bilaterable" if it is adjacent to
        precisely two localized nodes during the process
        above.
     c. For each node that is bilaterable, apply negative
        knowledge localization described in Section 3.
     d. Perform joint bilateration as described in Section
        3 for each pair of bilaterables which are adjacent
        to each other in the WSN.
     e. For each remaining bilaterable, use double
        negative knowledge bilateration, if applicable

3. Until (no additional localized node is found)
```

In the centralized version of the code above, each node sends its neighborhood information i,e its adjacencies to the centralized server and the server computes the location of each node.

5   Simulation Results

The simulation was performed using by placing nodes in a 100×100 square for random and connected graphs. Each data point is collected over average of 50 networks of the given parameters. Node density is the number of nodes that is within the radio range of nodes. For node density of nd, the number of nodes generated would be $(nd*10,000/\pi*s^2)$, where s is the network radius. The node density of 8 for random networks and node density of 10 was used to generate connected graphs and these represent highly sparse graphs.

Given a specified radius, nodes were placed using uniform distribution with real value co-ordinates in the 100×100^2. For connected networks, 1-connectivity was checked using breadth first search on the underlying graph. The percentage of anchors used comparing the performance was 15%, which is the lowest we have seen in the literature. The anchors were uniformly distributed in the network. The edges of the network graph connect nodes whose Euclidean distance is less than or equal to the radius of the network.

Figures 4 and 5 show random network of radius 10 and connected network of radius 10 respectively as a sample graph. Figure 6 shows the results of localization on connected networks of various percentage of anchors on a sparse graph of radius 12. Note that the number of nodes localized using the negative knowledge localization techniques is significantly higher for all anchor percentages. Similar results can be seen for random networks in Fig. 8. Figure 7 shows the number of nodes localized for various radii from sparse to dense graph. Again it is noted that the new algorithm performs significantly better for various radii. The percentage of anchors is set to 15% in this case. Similar results can be found for random networks as shown in Fig. 9. Figure 10 shows a H-shaped graph of radius 10 and Fig. 11 shows the performance of the algorithm for H-shaped graph.

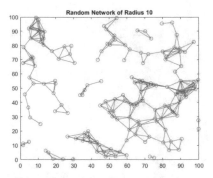

Fig. 4. Random network

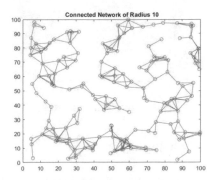

Fig. 5. Connected network

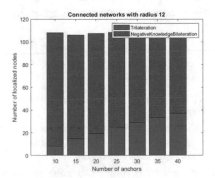

Fig. 6. Connected graphs: anchors varied

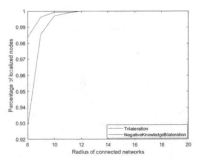

Fig. 7. Connected graphs: radius varied

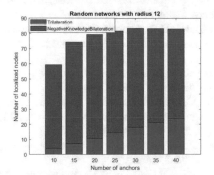

Fig. 8. Random graphs: anchors varied

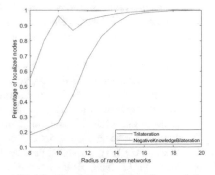

Fig. 9. Random graphs: radius varied

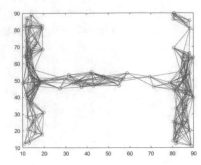

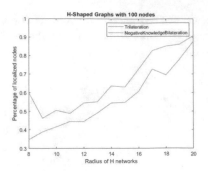

Fig. 10. H-Graph with radius 10

Fig. 11. H-Graphs: radii varied

6 Conclusion

In this paper, we have presented an improved localization algorithm which used lack of distance information in three different ways to localize more nodes than traditional trilateration. Simulation results show that significantly more nodes are localized using proposed algorithm for random sparse and dense networks, connected networks and H-shaped networks.

References

1. Bulusu, N., Heidemann, J., Estrin, D.: GPS-less low-cost outdoor localization for very small devices. IEEE Pers. Commun. **7**(5), 28–34 (2000)
2. Shang, Y., Ruml, W., Zhang, Y., Fromherz, M.P.: Localization from mere connectivity. In: Proceedings of the 4th ACM International Symposium on Mobile Ad Hoc Networking & Computing, pp. 201–212, June 2003
3. Mao, G., Fidan, B., Anderson, B.D.: Wireless sensor network localization techniques. Comput. Networks **51**(10), 2529–2553 (2007)
4. Amundson, I., Koutsoukos, X.D.: A survey on localization for mobile wireless sensor networks. In: International Workshop on Mobile Entity Localization and Tracking in GPS-less Environments, pp. 235–254. Springer, Heidelberg, September 2009
5. Chowdhury, T.J., Elkin, C., Devabhaktuni, V., Rawat, D.B., Oluoch, J.: Advances on localization techniques for wireless sensor networks: a survey. Comput. Networks **110**, 284–305 (2016)
6. Ji, X., Zha, H.: Sensor positioning in wireless ad-hoc sensor networks using multidimensional scaling. In: IEEE INFOCOM 2004, vol. 4, pp. 2652–2661. IEEE, March 2004
7. Oğuz-Ekim, P., Gomes, J.P., Xavier, J., Stošić, M., Oliveira, P.: An angular approach for range-based approximate maximum likelihood source localization through convex relaxation. IEEE Transactions on Wireless Communications **13**(7), 3951–3964 (2014)
8. Bandiera, F., Coluccia, A., Ricci, G.: A cognitive algorithm for received signal strength based localization. IEEE Trans. Signal Process. **63**(7), 1726–1736 (2015)
9. Xu, Y., Zhou, J., Zhang, P.: RSS-based source localization when path-loss model parameters are unknown. IEEE Commun. Lett. **18**(6), 1055–1058 (2014)
10. Savarese, C., Rabaey, J.M., Beutel, J.: Location in distributed ad-hoc wireless sensor networks. In: 2001 IEEE International Conference on Acoustics, Speech, and Signal Processing. Proceedings (Cat. No. 01CH37221), vol. 4, pp. 2037–2040. IEEE, May 2001

11. Yang, Z., Liu, Y., Li, X.Y.: Beyond trilateration: On the localizability of wireless ad-hoc networks. In: IEEE INFOCOM 2009, pp. 2392–2400. IEEE, April 2009
12. Moore, D., Leonard, J., Rus, D., Teller, S.: Robust distributed network localization with noisy range measurements. In: Proceedings of the 2nd International Conference on Embedded Networked Sensor Systems, pp. 50–61, November 2004
13. Singh, S.P., Sharma, S.C.: Range free localization techniques in wireless sensor networks: a review. Procedia Comput. Sci. **57**(7–16), (2015). 3rd.
14. Coluccia, A., Ricciato, F.: RSS-based localization via Bayesian ranging and iterative least squares positioning. IEEE Commun. Lett. **18**(5), 873–876 (2014)
15. Savvides, A., Park, H., Srivastava, M.B.: The bits and flops of the n-hop multilateration primitive for node localization problems. In: Proceedings of the 1st ACM International Workshop on Wireless Sensor Networks and Applications, pp. 112–121, September 2002
16. Yang, Z., Wu, C., Chen, T., Zhao, Y., Gong, W., Liu, Y.: Detecting outlier measurements based on graph rigidity for wireless sensor network localization. IEEE Trans. Vehicular Technol. **62**(1), 374–383 (2012)
17. Wu, H., Ding, Z., Cao, J.: GROLO: realistic range-based localization for mobile IoTs through global rigidity. IEEE Internet Things J. **6**(3), 5048–5057 (2019)
18. Fang, X., Chen, L.: An optimal multi-channel trilateration localization algorithm by radio-multipath multi-objective evolution in RSS-ranging-based wireless sensor networks. Sensors **20**(6), 1798 (2020)
19. Oliva, G., Panzieri, S., Pascucci, F., Setola, R.: Sensor networks localization: extending trilateration via shadow edges. IEEE Trans. Automat. Contr. **60**(10), 2752–2755 (2015)
20. Eren, T., Goldenberg, O.K., Whiteley, W., Yang, Y.R., Morse, A.S., Anderson, B.D., Belhumeur, P.N.: Rigidity, computation, and randomization in network localization. In: IEEE INFOCOM 2004, vol. 4, pp. 2673–2684. IEEE, March 2004

UAVs Route Planning in Sea Emergencies

Nicholas Formica[1], Leonardo Mostarda[2], and Alfredo Navarra[1(✉)]

[1] Mathematics and Computer Science Department,
University of Perugia, Perugia, Italy
`nicholas.formica@studenti.unipg.it`, `alfredo.navarra@unipg.it`
[2] Computer Science Division, University of Camerino, Camerino, Italy
`leonardo.mostarda@unicam.it`

Abstract. We consider the *First Boat Rescue* (FBR) problem, a new challenging variant of the well-known *Electric Vehicle Routing* problem. It comes from practical scenarios of rescue activities in the sea. In particular, usually boats relatively close to the coast that require some medical intervention are served by rescue lifeboats, even though most of the times the intervention is unnecessary. As alternative and cheaper approach, here we propose the usage of drones, equipped with basic medical tools that might suffice to solve the request. The main issue is that drones are battery powered and their lifespan is rather short. Hence, recharging activities are required in order to accomplish their missions. To this respect, we exploit a recent development about a charging station composed of a buoy that recharges itself by means of the sea movement and provides a charging pad for a drone. It comes out that by suitably dispose a set of such buoys in the area of interest to be monitored, it is possible to serve many boats alarms by means of drones, without wasting fuel and time for rescue lifeboats. We present a resolution algorithm for FBR along with various simulations concerning random and ad-hoc scenarios.

1 Introduction

Boating is the leisurely activity of travelling by boat, or the enjoyment useof a boat, such as powerboats, sailboats, or man-powered vessels, for travelling as well as sports activities, such as fishing or water-skiing. It is a popular activity, and there are millions of boaters worldwide. The statistics that are reported in 2013 by the Italian Ministry of Infrastructure and Transport show that boating activity is usually performed by lots of boats with small and medium size. More precisely, 41% of boats have an overall length that is less than 10 m, 58.8% have a length between 10 and 24 m while only 0.3% of the boats have a length that exceeds 24 m.

The work has been supported in part by the Italian National Group for Scientific Computation GNCS-INdAM, project "Travelling Salesman Problem 2.0" – Progetti di Ricerca 2020, and by the Italian project "HALYomorpha halys IDentification" (HALY-ID), funded by ICT AGRIFOOD MIPAAF, CUP J99C20000820003.

Every year, boating activities require to deal with a high number of rescuing calls. In Italy in 2013, 6166 boats called for help. During the summer period (35 days) about 1152 people called for aid in Marche region. The help procedure requires a rescue lifeboat to reach the boat asking for help and assist it. The main problem is that most of the help calls are false alarms. More precisely, only 2.4% of the calls requires rescuing, thus lifeboat assistance. False alarms cause waste of resources (like fuel) but most importantly they prevent the lifeboat to quickly assist other boats that are really in need. In order to solve this problem we propose the use of telemedicine via Unmanned Aerial Vehicles (UAVs). A UAV that is equipped with audio/video devices and few medical sensors (e.g., temperature, pulse oximeter, heartbeat and glucometer) is sent to the boat. A doctor from a remote base station can connect to the UAV, evaluate the emergency level and decide whether or not a lifeboat is needed. Our solution also requires the use of an energy sea buoy [12] that can be used to recharge the UAV when long distance needs to be covered.

1.1   Related Work

Our problem is closely related to the Electric Vehicle Routing Problem (eVRP) [4,7] which extends the classical Vehicle Routing Problem (VRP) in order to consider the limited autonomy range of electric vehicles. This limitation is solved by planning battery charging/replacing stations. eVRP models are usually based on the following assumptions: (i) the vehicles can fully or partially charge their batteries at every charging station, and (ii) the battery charging is linear. The author in [8] presents a survey on the electric vehicle routing problem. They focus on battery electric vehicles for logistic purposes by considering two main problems that are limited driving range and the need for an additional recharging infrastructure. They incorporate the maximum-route distance into the vehicle-scheduling model and consider battery swapping or full-fast charging. They model the single-depot vehicle-scheduling problem and propose a heuristic. The work presented in [2] is quite close to ours. The routing problem is related to electric vehicles that can use different recharge technologies and can perform partial recharges. A set of feasible routes is a feasible solution if all customers are visited once and no more than a fixed number of vehicles is used. The goal is to minimise the total recharging cost. This is estimated by considering the number of recharge cycles after which a battery is replaced. In [10], the authors consider partial nonlinear charging. They propose a hybrid metaheuristic that is applied to a set of realistic instances. A set of customers are served by using an unlimited and homogeneous fleet of electric vehicles and charging stations can handle an unlimited number of vehicles simultaneously. Each customer is visited exactly once; each route starts and ends at the depot and the objective function minimises the total route time. In [13], the authors study a Multi-Depot Green Vehicle Routing problem where vehicles start from different depots, serve customers, and, at the end, return to the original depots. The limited capacity of vehicles forces them to visit fuel stations. The objective is to minimise the total carbon emissions. A Two-stage Ant Colony System is proposed to find a feasible and acceptable solution for this NP-hard problem. In [1], the authors present a green vehicle routing problem. They assume a graph that is composed of vertices representing customer locations, alternative fuelling stations (AFSs), and a depot. A solution is a set of vehicle tours with minimum distance. Each vehicle starts and returns from a depot, visits a set of customers without exceeding the autonomy driving range. Each tour may include a stop at one or more AFSs to allow the vehicle to refuel en route. In [9], the authors present a general eVRP where electric vehicles can recharge at a charging station during their delivery operations. Their formulation consider the costs associated with time and electricity consumption. All vehicles start and end at the same depot. The proposed eVRP finds the minimum travel time cost and energy cost. Our problem, and in general the eVRP, are closely related to the classical VRP [5,11] and to the Multi Depot Vehicular Routing Problem (MVR) [6] which in turn generalises the Travelling Salesman Problem. Such problems are known to be NP-hard. They study how goods can be delivered from various depots by using a set of home vehicles that can serve a set of customers via a given road network. MVR finds a route for each vehicle that starts and finishes at its depot and the global transportation cost is minimised. In [3], the authors present the Steiner Travelling Salesman Problem which is closely related to ours. In fact, each salesman is assigned to visit a subset of nodes, called terminals, possibly passing through some further nodes not required to be visited. The goal is to find minimum-weight closed walks.

1.2   The Marche Region Case Study

We have considered the Italian Marche region case study. In particular, the area of the cost selected in the sea of Marche concerns 175 km × 40 km mission region. A possible configuration of *base stations* and *charging stations* is shown in Fig. 1A. Sea buoys act as charging stations for drones. The charging stations can charge drones with "contact based" charging systems[1] (see Fig. 2). The pad provides up to 500W loss-free charging system for the drones. Oil platforms can also be used to host charging stations for charging drones. Figure 1B shows the deployment of oil platforms in the sea of Marche region.

We considered a consumer grade drone[2] which can be seen in Fig. 2a. It has 4280 mAh battery which can allow the drone a 27 min flight. The drone can fly at maximum speed of 94 km/h. The battery can be charged in about 27 min with linear quick charging. Audio, video and some sensors (such as oxygen, temperature) will be used during the rescuing. These are powered by a different battery thus rescuing do not affect the drone energy consumption. The rescue time when serving a boat is 15 min in average.

---

[1] https://skycharge.de/charging-pad-outdoor.
[2] The DJI Inspire 2.

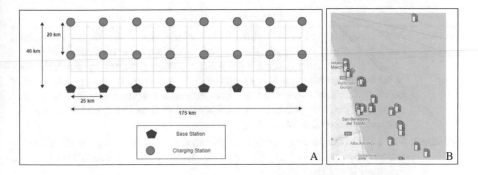

Fig. 1. A) Charging buoy grid, B) Oil platforms

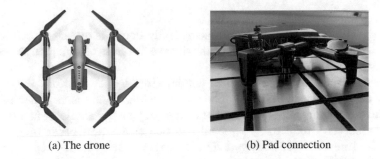

(a) The drone   (b) Pad connection

Fig. 2. The contact-based charging pad for the drones.

Our problem can be seen as a novel variation of the eVRP. A set of drones need to save a set of boats by going via one or more charging buoys. Each buoy can host at most on drone at the time. The goal is to minimise the time needed to rescue all the boats. This may require a drone to partially charge its battery. In our problem, drones are not required to return at the departing base station. The organisation of buoys into grid can be a peculiarity that can be further analysed. The Marche region scenario suggests similar case studies such as drone patrolling and woods monitoring.

2   Problem Formalisation

In this section, we provide the formalisation of our problem. We consider a fixed infrastructure of buoys given by a rectangular grid G of n rows and m columns. Each vertex of the grid represents a buoy, i.e. a recharging station. The bottom row of the grid represents base stations from where drones can start their journeys. The number of drones that can be exploited for a mission is denoted by k, whereas the number of boats that may require intervention is denoted by B. Boats can appear in any point within the area delimited by G.

First Boat Rescue (FBR)	
Input:	A rectangular grid G embedded in the plane of size $n \times m$ buoys where the bottom row represents the possible starting base stations; k drones; B boats. Each buoy is associated with a charging power e_b. Each drone is associated with a battery of power e_d; Each boat is represented by a point in the area covered by G. Furthermore, the time required to fully or partially recharge a drone on a buoy, as well as the speed of flying, depend on the used technology
Solution:	For each drone, a trajectory that starts and ends in a base station (not necessarily the same one) that can be traced by the drone without draining its battery
Goal:	Minimise the total time required to serve all the B boats

Clearly, the resolution of the problem highly depends on the chosen technology. In particular, the capabilities of the drones in terms of flying autonomy and speed are crucial.

From the point of view of pure combinatorial optimisation, the problem becomes easily unfeasible. In fact, one may consider an instance where plenty of boats surround a single buoy so that the charging energy available is not enough to guarantee the drone to fly over all the boats, whereas other buoys cannot be exploited because too far apart. On the other hand, the depicted scenario is rather unrealistic. This suggests to consider "reasonable" instances dictated by practical observations. Moreover, depending on the traffic, i.e., the number of boats one may expect to serve and the number of available drones, one parameter that can be suitably tuned to increase the percentage of successful missions concerns the density of the underlying grid. Our investigation is then intended to provide a suitable heuristic for the resolution of the problem in practical scenarios. Moreover, our study might be useful also to understand what should be a satisfying size for the underlying grid in order to be able to serve all the expected boats.

3   Resolution Algorithm

In this section, we describe our heuristic for the resolution of the FBR problem. Algorithm 3 shows a portion of the corresponding pseudo-code which concerns the computation of the trajectories that the selected drones must trace.

In order to provide the correct input to Algorithm 3, we need to solve an instance of the *Multi-Depot Vehicle Routing* (MVR) problem [4,6,7]. As observed in Sect. 1.1, this is a well-known variant of the standard VPR where there are many depots (one for each vehicle) from where vehicles can start and end their journeys. In practice, by ideally considering the drones with infinite battery, the resolution of MVR provides a preliminary solution for FBR. The solution of MVR is provided by a set of trajectories, one for each vehicle (drone) involved so as to minimise the time required to serve the requests (boats). Each trajectory τ

Algorithm: FBR
Input: Solution S of MVR as a set of trajectories

1 **for** *each trajectory* $\tau = (v_0, v_1, \ldots, v_c) \in$ S **do**
2 $\quad$ **for** $i = 1$ *to* c **do**
3 $\quad\quad$ **if** e_d *is not enough to cover* (v_{i-1}, v_i) **then**
4 $\quad\quad\quad$ Let v_b be the buoy that minimises $T = t(v_{i-1}, v_b) + t(v_b, v_i)$;
5 $\quad\quad\quad$ **if** $e_d + charge - T > 0$ OR $e_b > \frac{max_b}{2}$ **then**
6 $\quad\quad\quad\quad$ $e_d = e_d + charge - T$;
7 $\quad\quad\quad\quad$ $c = c + 1$;
8 $\quad\quad\quad\quad$ **for** $j = c$ **down to** $i + 1$ **do**
9 $\quad\quad\quad\quad\quad$ $v_j = v_{j-1}$;
10 $\quad\quad\quad\quad$ $v_i = v_b$;
11 $\quad\quad\quad\quad$ $Cost_\tau = Cost_\tau + T$;
12 $\quad\quad\quad$ **else**
13 $\quad\quad\quad\quad$ Backtrack with a new buoy not tested before, if any;

Fig. 3. Algorithm for FBR starting from a solution of MVR.

is represented by a sequence $(v_0, v_1, \ldots, v_c)$ of $c + 1$ points, for some $c > 0$, that must be reached by a vehicle. Moreover, each trajectory τ is associated with a $Cost_\tau$ which represents the time needed by a vehicle to trace τ. Clearly, MVR does not take into account the battery constraint for the vehicles, hence no intermediate stops (buoys) are necessary. It turns out that an optimal solution to MVR represents a lower bound to our FBR. Again, as MVR is NP-hard [4,6,7], of course the same holds for FBR. In our implementation, we refer to the code provided in https://github.com/brandhaug/mdvrp-ga to solve MVR, which makes use of genetic techniques.

The obtained solution is then processed by means of Algorithm 3 in order to check (see Line 3) whether the computed trajectories can be traced as they are, or (Lines 4–6) they need some recharging activities according to the capabilities of the drones we consider. If a buoy v_b, with charging power e_b and maximum charging power max_b, is inserted between two consecutive points v_{i-1} and v_i of a computed trajectory τ (see Lines 7–10), then $Cost_\tau$ must be updated of quantity $t(v_{i+1}, v_b) + t(v_b, v_1) - t(v_{i+1}, v_i)$ (Line 11), where function $t(x, y)$ returns the time needed by the drone to move from x to y, whereas x and y range among the set of buoys and boats.

The algorithm also applies some backtrack method (up to four hops backward, see Line 13) when the current trajectory cannot be traced. This might happen for instance if the buoys around the current tested position have been already assigned to other drones and the remaining charge is not sufficient to reach the desired target. To this respect, it is also worth noting that the order in which the various trajectories returned by solving MVR are considered might imply different solutions or even lead to impossibility for FBR. For this reason, in our implementation we consider all (or a subset of) the permutations to increase the chances to find a (good) solution.

Our heuristic also optimises the times required for charging activities. In fact, drones are assumed to start with the full charge but, when buoys are exploited, only the percentage of battery required to reach the subsequent buoy is considered as charging requirement. This allows the use of the same buoy to serve different drones, with the only constraint that only one drone per time lands on such a buoy.

4   Experiments

In this section, we first remind all the settings under which our simulations have been conducted. In particular, the chosen parameters directly come from the practical scenario described in Introduction. Then, we show the results we have obtained for ad-hoc and random instances.

4.1   Settings

Based on the practical scenario described in Introduction, as field of action we consider a rectangular grid G of size 3×8. As shown in Fig. 1A, this allows the coverage of an area of 175×40 km^2. In fact, according to the characteristics of the buoys and those of the drones, the chosen setting allows a good compromise in terms of density and feasibility in real scenarios.

- Subsequent columns are distanced 25 km one from the other; subsequent rows are distanced 20 km one from the other.
- The bottom row of the grid represents the base stations from which drones can start to fly; we assume at most one drone per base station;
- Initially, every buoy b (and every drone d, resp.) is assumed to be fully charged with a power of max_b (max_d, resp.) corresponding to 27 min of charging (flying, resp.);
- As observed, the considered drone can fly for 27 min at a maximum speed of 94 km/h; We assume a constant average speed of 80 km/h;
- Once a boat is reached, no battery consumption is assumed as the drone provides further instrumentation self-supplied;
- The time required for serving a boat is assumed to be of 15 min on average;
- A drained battery of the drone requires 27 min to get fully charged;
- The time required by a drone d to accomplish all its planned trajectory τ_d is given by the difference between the arrival time and the starting time;
- The time required by a drone d to accomplish its mission (i.e., for serving all the boats it is assigned to) is given by the difference between the time in which it finishes to serve the last boat of its trajectory and the starting time.

Boats are randomly distributed in the area covered by G. We vary the number of boats from 5 to 20. Concerning the number of drones, it is assumed that up to 8 drones (one per base station) can be used. However, according to the input instance, it is not always necessary nor recommended to make use of the whole swarm.

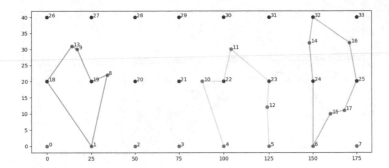

Fig. 4. Solution for $B = 10$.

We also conduct some experiments with specific locations for the boats with the intent to stress the system in what we intuitively consider extreme cases. These would come out with very low probability from the uniform distribution.

4.2   Simulations

The first experiments reported in Figs. 4, 5 and 6 concern the uniform distribution of boats from 5 to 20. The case with 5 boats is not reported because is very similar to the case of 10. When the density of the boats is up to 10, the resolution of the problem is rather natural and straightforward from the solutions obtained for MVR.

In Fig. 4, we find a solution for FBR when 10 boats are considered. Numbers associated to buoys and boats are just meant as ids to univocally refer to such elements. The obtained solution makes use of 3 drones since the usage of further drones would increase the total time of intervention. The solution has been obtained as a natural extension of the one obtained for MVR by using the considered genetic algorithm. The first trajectory from the left requires a time mission of 1 h and 42.25 min which includes 45 min required to serve three boats and 22 min for the charging activity at buoy 19. The total time to trace the whole trajectory is instead of 2 h and 43 min. This is given by the mission time plus the charging activity at buoy 18 (of 24 min) and the time required to trace the segment from 18 to 1. It is worth mentioning that the time required to trace the whole trajectory and to serve the three boats, produced by the resolution of the underlying MVR problem, was of 1 h and 42 min.

The longest trajectory is the third one. It requires a total of 3 h and 14 min, against 2 h and 15 min obtained for MVR. The second trajectory shows a case where the drone starts and ends its journey in two different base stations.

In Fig. 5, we find a solution for FBR when 15 boats are considered. The longest trajectory is the second one which requires a total of 4 h and 54.75 min, against the 3 h and 39.75 obtained for MVR. Actually the length of this trajectory seems to suggest that it would be better to split it between two drones. However, the usage of another drone may possibly make the whole solution more costly,

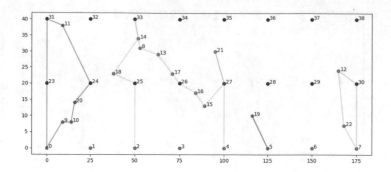

Fig. 5. Solution for $B = 15$.

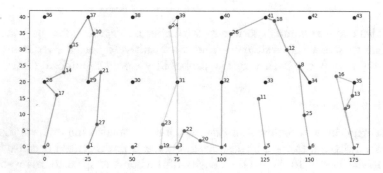

Fig. 6. Solution for $B = 20$.

because it could require to trace longer trajectories in order to ensure the two drones can find enough energy on the buoys for recharging activities. To this respect, it is interesting to see how the buoy 33 has been used for allowing the drone to keep on tracing its trajectory.

In Fig. 5, we find a solution for FBR when 20 boats must be served. Considering that the field of intervention contains 24 buoys, then 20 boats constitute a good threshold above which it might happen that no solution can be found, at least within 'reasonable time', i.e. without waiting for buoys to get back their charge. In fact, we have to keep in mind that we are dealing with rescue activities. By our experiments, we observed that $B = 20$ is a good limit that, under the studied scenario and with the uniform distribution, guarantees to find effective solutions. Of course in a context where more boats are expected, the density of the buoys must be reconsidered.

The longest trajectory is the first one which requires a total of 3 h and 53.25 min, against the 2 h and 31.5 min obtained for MVR. As it can be seen, just a few of the available buoys remain unused and also the number of drones involved is 6 out of the 8 that are available.

Concerning ad-hoc examples, in Fig. 7, we report two cases where 10 boats are thicken in a very restricted area. In Fig. 7a, by means of two drones it has been possible to find a solution. Contrarily, with the density shown in Fig. 7b,

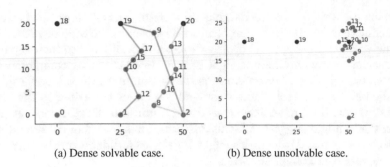

(a) Dense solvable case.   (b) Dense unsolvable case.

Fig. 7. Dense cases with $B = 10$.

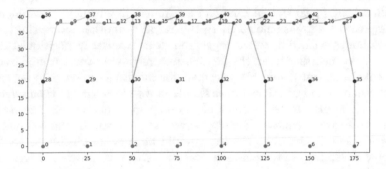

Fig. 8. Solution for $B = 20$ in the case where boats are faraway from the depots.

no reasonable solution exists. Clearly, in both cases it would be better to use a rescue lifeboat rather than drones as some non-false alarms might be easily faced. Another ad-hoc positioning of the boats is shown in Fig. 8. We consider 20 boats disposed faraway from the depots. The obtained solution is indeed rather satisfactory. It makes use of just three drones and the longest trajectory requires 5 h and 4 min against the 3 h and 23.25 min obtained from the resolution of the underlying MVR problem. Moreover, the time of mission, is just of 3 h and 56.5 min, i.e., the drone requires 1 h and 7.5 min to come back to the depot.

5   Conclusions and Future Works

We introduced the *First Boat Rescue* (FBR) problem, a new challenging variant of the well-known *Electrical Vehicular Routing* Problem. Starting from the official data provided by the Italian Ministry of Infrastructure and Transport, we have envisioned a possible solution that exploits advanced technologies based on drones. Hence, we presented a resolution algorithm along with an extensive case study in order to show various peculiarities.

Apart for its practical interest, FBR represents an interesting combinatorial problem that deserves further investigation. We know it is NP-hard in general and that sometimes even feasibility cannot be reached, provided that it is not reasonable to wait for the recharging time of the buoys. However, it would be challenging to characterise the problem with respect to the topology of the underlying network, to the relation between drones and buoys and to their capabilities. In particular, many parameters might be considered that may completely change the resolution strategy. For instance, is the rectangular grid the best way to dispose the buoys? It is reasonable to spread the buoys according to the expected traffic of boats. So far, we have considered a uniform distribution but perhaps different data might suggest different infrastructures. Another consideration may refer to the weather. Trajectories might take into consideration the current wind or the possibility of thunderstorms.

Furthermore, the practical environment suggests that an online version of FBR should be considered. How trajectories might be modified on-the-fly if boats are assumed to appear in real time and not known in advance? Dynamicity certainly represents a great challenge. To this respect, another interesting scenario considers buoys with different charges and drones suitably positioned in various buoys, not necessarily on the depots. Actually, one question might concern what would be the most suitable placement for the drones in order to better serve possible requests and whether drones should move to re-arrange themselves once one or more drones started a mission. In doing so, drone might become better prepared to further requests. Finally, as we have shown, all the technology for real experiments is available. Hence, it would be very interesting to verify how the proposed strategy may behave in practice. This may provide suggestions for modifications in case the parameters mentioned above, and not considered so far in our solutions, reveal to be decisive to accomplish required missions.

References

1. Erdoğan, S., Miller-Hooks, E.: A green vehicle routing problem. Transp. Res. Part E: Logistics Transp. Rev. **48**(1), 100–114 (2012). https://doi.org/10.1016/j.tre.2011.08.001
2. Felipe, Á., Ortuño, M.T., Righini, G., Tirado, G.: A heuristic approach for the green vehicle routing problem with multiple technologies and partial recharges. Transp. Res. Part E: Logistics Transp. Rev. **71**, 111–128 (2014). https://doi.org/10.1016/j.tre.2014.09.003
3. Gabrel, V., Mahjoub, A.R., Taktak, R., Uchoa, E.: The multiple Steiner TSP with order constraints: complexity and optimization algorithms. Soft Comput. **24**(23), 17957–17968 (2020). https://doi.org/10.1007/s00500-020-05043-y
4. Ghorbani, E., Alinaghian, M., Gharehpetian, G.B., Mohammadi, S., Perboli, G.: A survey on environmentally friendly vehicle routing problem and a proposal of its classification. Sustainability **12**(21), 9079 (2020). https://doi.org/10.3390/su12219079
5. Han, M., Wang, Y.: A survey for vehicle routing problems and its derivatives. In: IOP Conference Series: Materials Science and Engineering **452**, 042,024 (2018). https://doi.org/10.1088/1757-899x/452/4/042024
6. Karakatic, S., Podgorelec, V.: A survey of genetic algorithms for solving multi depot vehicle routing problem. Appl. Soft Comput. **27**, 519–532 (2015). https://doi.org/10.1016/j.asoc.2014.11.005

7. Lalla-Ruiz, E., Erdelić, T., Carić, T.: A survey on the electric vehicle routing problem: Variants and solution approaches. J. Adv. Trans. **2019**, 5075,671 (2019). https://doi.org/10.1155/2019/5075671

8. Li, J.Q.: Transit bus scheduling with limited energy. Transp. Sci. **48**(4), 521–539 (2014). https://doi.org/10.1287/trsc.2013.0468

9. Lin, J., Zhou, W., Wolfson, O.: Electric vehicle routing problem. Transp. Res. Procedia **12**, 508–521 (2016). In: Tenth International Conference on City Logistics, 17—19 June 2015, Tenerife, Spain. https://doi.org/10.1016/j.trpro.2016.02.007. http://www.sciencedirect.com/science/article/pii/S2352146516000089

10. Montoya, A., Guéret, C., Mendoza, J.E., Villegas, J.G.: The electric vehicle routing problem with nonlinear charging function. Transp. Res. Part B: Methodol. **103**, 87–110 (2017). https://doi.org/10.1016/j.trb.2017.02.004

11. Mor, A., Speranza, M.G.: Vehicle routing problems over time: a survey. 4OR **18**(2), 129–149 (2020). https://doi.org/10.1007/s10288-020-00433-2

12. Viet, N.V., Wu, N., Wang, Q.: A review on energy harvesting from ocean waves by piezoelectric technology. J. Model. Mech. Mater. **1**(2) (2017). https://doi.org/10.1515/jmmm-2016-0161

13. Zhang, W., Gajpal, Y., Appadoo, S.S., Wei, Q.: Multi-depot green vehicle routing problem to minimize carbon emissions. Sustainability **12**(8), 3500 (2020). https://doi.org/10.3390/su12083500

SQUAB: A Virtualized Infrastructure for Experiments on BGP and its Extensions

Naoki Umeda[1]([✉]), Naoto Yanai[1], Tatsuya Takemura[1], Masayuki Okada[2], Jason Paul Cruz[1], and Shingo Okamura[3]

[1] Osaka University, Suita, Japan
n-umeda@ist.osaka-u.ac.jp
[2] University of Nagasaki, Sasebo, Japan
[3] National Institute of Technology, Nara College, Nara, Japan

Abstract. The border gateway protocol (BGP), which is known as a backbone protocol of the Internet, is constantly the target of many hijack attacks. To combat such attacks, many extensions of BGP have been developed to make BGP more secure. However, to perform experiments to evaluate their performance, most BGP extensions require the utilization of platforms, such as testbeds, with high operating costs. In this paper, we propose *SQUAB (Scalable QUagga-based Automated Configuration on Bgp)*, a lightweight evaluation tool for protocols under development and for protocols that will be developed by a user with actual devices locally. SQUAB can configure BGP networks automatically, and thus it can significantly reduce the overhead of experiments on BGP and its extensions. Unlike conventional testbeds, SQUAB can set up BGP networks locally and its execution requires only a computational resource of a typical laptop computer. We used SQUAB in experiments to check the validity of functions based on network topologies in the real world. Our results show that SQUAB can configure a network composed of 50 routers within 52.9 s and consumes only 354.7 MB of memory. Furthermore, as a case study, we confirm that SQUAB can also provide experimental networks, even including other protocols, e.g., BGPsec.

1 Introduction

Border gateway protocol (BGP) [22] is a protocol designed to exchange route information among networks called *autonomous systems (ASes)* by assigning unique numbers called AS numbers. BGP is utilized as a backbone protocol of the Internet, and many vulnerabilities have been reported [20]. In fact, BGP was found to be targeted by an average of 4.82 hijacks per day [28], and all hijacks inflicted serious damage. Therefore, guaranteeing the validity of route information is an important and urgent issue in Internet security.

Many countermeasures against BGP hijacking have been proposed in the past years [20]. However, evaluations of the effectiveness and risks of these countermeasures are insufficient. This lack of evaluation is mainly due to BGP itself, i.e., it cannot lend itself well to supporting experiments [24].

© The Author(s), under exclusive license to Springer Nature Switzerland AG 2021
L. Barolli et al. (Eds.): AINA 2021, LNNS 225, pp. 600–613, 2021.
https://doi.org/10.1007/978-3-030-75100-5_52

Therefore, many efforts have been devoted to developing tools that can effectively evaluate BGP and its extensions. In general, an evaluation tool should be able to support network configurations and manage interconnections between BGP routers in an experiment [24]. However, existing evaluation tools require users to configure networks manually, which is a challenge even for experts, when designing BGP configurations that operate as intended [12, 16]. This means that the existing tools have limitations in scaling experiments. Notably, when users want to evaluate their designed protocol locally, they need to manually set up a network environment for their experiments. Such manual setup, therefore, requires additional effort from users during their experiments.

In this paper, we present *SQUAB (Scalable QUagga-based Automated Configuration on Bgp)*, a novel tool that configures BGP environments automatically for local experiments. It is a user-friendly tool for networks of BGP routers, i.e., BGP networks. SQUAB can configure BGP networks automatically based on only an easy interface written by a user. SQUAB makes use of virtual containers, making it lightweight, i.e., it can be executed by a typical laptop computer. SQUAB enables users to set up their experimental environment locally, and thus users will not need to register and pay for a service, which are necessary in conventional testbeds.

As a preferable use case, SQUAB can be used as the first step in the evaluation of protocols under development or that will be developed by a user as a proof-of-concept. Evaluating the design of a new protocol in a testbed often requires extra overhead due to additional processes, e.g., user registration and setting up a server, as described above. In contrast, after developing a software router of their designed protocol, users can use SQUAB to evaluate their protocol for any network configuration scale by introducing a container image of the router. We demonstrate that SQUAB can configure networks mixing BGP and BGPsec [15], which provides the validity of route information by using digital signatures for BGP, through a reference implementation for BGPsec. We also discuss further insights about use cases and the limitations of SQUAB. We released SQUAB and the reference implementation described above publicly[1].

In summary, the contributions of this paper are as follows:

- Design and implementation of SQUAB, a tool that allows users to perform experiments on BGP easily.
- Public release of SQUAB and BGPsec reference implementation.

2   Research Motivation

In this section, we first describe BGP and define the requirements of SQUAB. We then describe the technical difficulty in satisfying the requirements.

---

[1] https://github.com/han9umeda/SQUAB.

2.1 Border Gateway Protocol

In BGP, ASes exchange the route information, which includes IP prefixes and routes to a destination, i.e., *AS_PATH*, with their neighbors. BGP routers then decide the best routes to each IP prefix in accordance with the received route information and the static policy defined locally by the operators if there are multiple routes for the IP prefix. In doing so, the ASes append their own AS numbers to *AS_PATH* and advertise the IP prefix and the *AS_PATH* to the neighbors as the best route. As a result, chains of *AS_PATH* to reach each IP prefix are configured. BGP requires operators to prepare various configuration files, e.g., information on the neighbors, policies in transmission, and reception of route information between ASes. Such a complicated setting may lead even experts to misconfigure BGP settings. Meanwhile, operators of each AS just prepare for its configuration file, and thus BGP has no function to manage the whole Internet from a higher perspective. BGP cannot guarantee the validity of the route information nor detect incorrect route advertisement, i.e., misconfiguration or route hijacking. Consequently, despite the development of many BGP extensions, it remains unclear if these extensions are operable on network topologies in the real world.

2.2 Goal

Our goal is to develop a tool that allows users to flexibly evaluate the validity of their protocol design. Such a tool should have the following properties:

- Automated configuration: Users only need to use one command to set up an experimental environment for BGP. A BGP configuration is difficult to set up, and thus an automated configuration is useful for all users.
- Lightweight: Execution time is short enough, and memory requirement is low enough for a common laptop computer.
- Local deployment: Experimental environments are deployed locally. In doing so, a tool should be stand-alone, and any other resources, such as a server, for experiments, are unnecessary. Consequently, users will not need to register and pay for a service.

When designing an experimental tool, its safety and extendability should be taken into account. Hereafter, safety means never disrupting Internet borders, and extendability means that a new protocol can be quickly introduced into the tool as long as an implementation of the protocol is available.

Meanwhile, we assume the following conditions for designing a tool that satisfies the three properties described above. First, routers are required to be software routers that can be deployed on the tool to be developed. In particular, we focus on Quagga Routing Suite [7] and its extensions in the current specification (see Sect. 3.3 for details). Second, software routers are implemented in advance, i.e., the development of the software routers themselves is out of the scope of this paper.

2.3   Technical Difficulty

Achieving the goals described in the previous subsection is *non-trivial*. In general, a configuration file in the conventional BGP defines a setting for some ASes. To configure experimental environments at the network-level, a user often needs to prepare all configuration files to make them consistent. This is different from the technology of designing a single router of observation point [24]. As described in Sect. 1, operating BGP precisely is often challenging even for experts [12,16].

To avoid the challenge above, existing works have utilized testbeds that can support network configurations in their experiments. However, the use of testbeds inherits additional overhead. Experimental environments may be set up locally using numerous virtual machines, but such an approach often requires a user to have knowledge about virtual machines and network topologies. Thus, the use of testbeds does not achieve the lightweight and local deployment properties.

3   Design of SQUAB

In this section, we present SQUAB (Scalable QUagga-based Automated Configuration on Bgp). We first describe the design concept of SQUAB and then explain the modules used as building blocks of SQUAB. Finally, we present the interface of SQUAB, which corresponds to our central idea, and the workflow.

3.1   Design Concept

Our design concept is to virtually configure experimental networks consisting of virtualized containers for BGP software routers through a user-friendly interface.

Loosely speaking, a user specifies a configuration from three points of view: the existence of an AS, the function for each BGP router, and the connections between ASes. After a user specifies the three information through the user-friendly interface, SQUAB will automatically create networks including virtual containers of routers. SQUAB automatically completes settings, e.g., such that BGP routers connect with each other, and thus the user can begin experiments immediately. Therefore, SQUAB achieves the automated configuration property. Moreover, the user can execute the automated configuration described above on a computer locally, and thus SQUAB achieves the local deployment property.

SQUAB achieves the lightweight property by combining virtual containers to construct an experimental environment. In a conventional experimental environment, virtual machines that virtualize not only applications but also hardware, such as CPU and memory, are utilized. However, a virtual machine also virtualizes functions that are not necessary for the experiment, thus increasing the load. On the other hand, by utilizing a virtual container, hardware virtualization becomes unnecessary, and then only the application level, i.e., the routers are virtualized. Consequently, SQUAB drastically reduces the overhead caused by the experimental environment.

3.2   Modules and Architecture

SQUAB utilizes virtual containers and software routers as elemental technologies. A virtual container is a virtual environment that packages the application execution environment and isolates resources, such as processes and networks, from the host OS. As mentioned above, virtual containers virtualize only at the application layer in comparison with a conventional virtual machine which virtualizes from the hardware layer. Virtual containers make applications run faster, and they use fewer resources than virtual machines. Moreover, by virtualizing the application itself, it is possible to provide an execution environment without depending on the host environment. Docker[2] is a known platform for virtual containers [19].

A software router is a software that can provide the functions of a router on the host OS by packaging the network router. Quagga and BIRD [1] are known open-source BGP software routers used in network operations. There are many extensions of these software routers that support BGP extension protocol, e.g., BGP-SRx [5] for BGPsec.

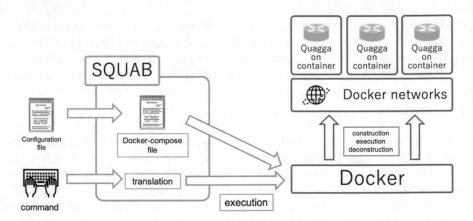

Fig. 1. The architecture of SQUAB. SQUAB executes Docker to create a network consisting of Docker containers, which include Quagga routers.

Figure 1 shows the architecture of SQUAB and its position. SQUAB receives configuration files and commands, and executions Docker. SQUAB runs each module in a coordinated manner.

We provide the behavior of each module in detail in the next section.

3.3   Methodology

We describe the design of SQUAB in detail. In the following, we describe the methodology using Quagga as a software router and Docker as a virtual container.

---

[2] https://www.docker.com/.

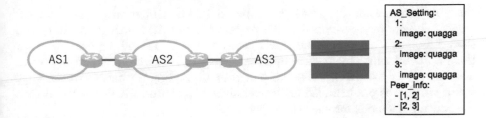

Fig. 2. Example of a network topology to be evaluated and the SQUAB configuration file: The left-side shows the topology to be evaluated and the right-side shows an example of the SQUAB configuration file to set up the topology.

3.3.1   Setup of Network Configuration

Figure 2 shows the topology and *SQUAB configuration file* which the user gives as input to SQUAB. The users manually create the file corresponding to the topology they want to configure. The SQUAB configuration file is written in the YAML format and can be rewritten to change the experimental patterns. In the hash nest of the AS_Setting key, the user inputs the AS numbers and docker container image names each AS uses as a router, e.g., Quagga. In the hash nest of the Peer_info key, the user inputs the connection information for each AS.

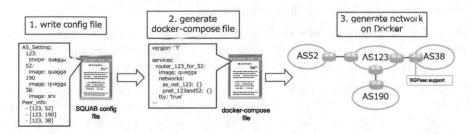

Fig. 3. The workflow of SQUAB for creating experimental environments: The workflow consists of three processes. By writing the SQUAB configuration file shown in Fig. 2, SQUAB will create a corresponding docker-compose file and its resulting networks on Docker.

3.3.2   Workflow

Figure 3 shows the SQUAB environment creation workflow. First, SQUAB reads the SQUAB configuration file provided as input by a user, and then it generates the corresponding docker-compose file. The docker-compose file is described so that the docker engines can interpret the contents similar to the input files. As the first step of generating a docker-compose file, SQUAB reads all AS information from the user-input SQUAB configuration file. SQUAB assigns a router-specific IP address to each AS to avoid duplication, and then it writes to the docker-compose file to generate the corresponding docker network. Next, SQUAB reads the peer information from the user-input SQUAB configuration file. SQUAB creates the docker networks and AS routers for the AS-to-AS connection, and then it assigns a network IP address to each network to avoid duplication. After that, SQUAB runs the docker-compose to set up all containers and networks, and then it creates *BGP configuration files* by giving arguments to the shell script which was pre-installed in the container of a router. Finally, SQUAB runs the routing daemons for each container.

The various experimental data are collected by calling standard UNIX commands or the shell scripts pre-installed in the container of a router. We utilized Python 3.8.5, Quagga 0.99.22.4, and Docker Desktop 3.0.4 to implement SQUAB.

4   Experiments

We developed SQUAB according to the design described above and used it in experiments to validate its functions and evaluate its performance. As a case study, we used SQUAB to configure networks mixing BGP routers and BGPsec routers.

4.1   Purpose Experiments

We conducted the following three experiments.

Function Validation: We confirm that SQUAB sets up the topology as defined in the SQUAB configuration file, and then confirm that routers correctly exchange the route information in the experimental environment.

Performance Evaluation: We measure the running time of SQUAB to set up and remove the experimental environment. We also measure the memory consumption of the containers by using SQUAB. Hence, we confirm that SQUAB is a lightweight tool that requires only machine resources, e.g., a typical laptop computer.

Case Study: We configure networks such that BGP and BGPsec [15], which is a security extension of BGP, are mixed, and then check if their routers correctly exchange route information. Hence, we confirm that it is easy to embed a software router that supports a BGP extension protocol into SQUAB.

4.2   Experimental Settings

We describe the experimental settings. The experiments were conducted on MacBook Pro 2017(macOS 11.1) with 2.3 GHz Dual-Core Intel Core i5, 8 GB memory.

Function Validation: We validate the functions of SQUAB using an actual topology centered on NAITWAYS(AS57119). Concretely, we obtain the topology information from BGPlay[3], an Internet topology visualization tool. Based on the

---

[3] https://stat.ripe.net/widget/bgplay.

information, we manually prepare the SQUAB configuration file and then input it into SQUAB. After that, we observe the behavior of SQUAB.

Performance Evaluation: We measure the running time for setting up or removing the experimental environment using SQUAB by changing the number of routers placed in a linear topology from 2 to 50. Then, we measure the memory consumption of all the containers by using the `docker stats` command. We measured the memory consumption three times for each pattern and then take their average.

Case Study: We modify SQUAB to handle BGP-SRx, a software router running BGPsec. At that time, we checked how much work the modification required. In addition, we configure networks mixing BGP routers and BGPsec routers, and then confirm that the routers correctly exchange the route information and that the function is working correctly as BGPsec.

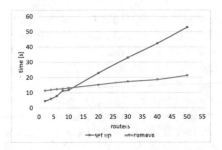

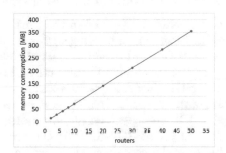

Fig. 4. The running time of SQUAB: The blue line represents the time for setting up an environment and the orange line represents the time for removing the environment.

Fig. 5. The memory consumption of SQUAB: The memory consumption of all containers to set up the environment is measured.

4.3 Results

We show the results of the three experiments describe above.

Functional Validation: We confirm that SQUAB correctly sets up the topology centered on NAITWAYS(AS57119) and then records all the details on a video[4] as a demonstration. At the beginning of the video, the topology centered on NAITWAYS output by BGPlay is displayed. Next, the video shows the SQUAB configuration file corresponding to the topology at 0:05. SQUAB in Python script format is run at 0:11. SQUAB then reads the SQUAB configuration file and creates a network and containers of the router. After waiting for all the routers on the topology exchange route information, the script that collects the setting information and routing tables of all the routers is run at 2:00, and

---

[4] http://www-infosec.ist.osaka-u.ac.jp/media/aina2021_squab_demo.mp4.

then its results are displayed from 2:10 to 3:50. The results show the IP prefixes and the corresponding AS_PATHs, which are consistent with the results of the exchange of route information in the topology displayed at the beginning of the video. Thus, we confirm that SQUAB can set up the environment correctly. In addition, the script that removes the experimental environments is run at 4:00 and the results confirm that SQUAB can remove the environment.

Performance Evaluation: Figure 4 shows the the running time of SQUAB for setting up and removing each topology. The setting up took 11.9 s for 10 routers and 52.9 s for 50 routers, showing that the time tends to increase by about one second per router. On the other hand, the removing took 13.1 s for 10 routes and 21.4 s for 50 routes, showing that the time tends to increase by about 0.12 s per router.

Figure 5 shows the memory consumption of all the containers to set up the environment. The memory requirement of SQUAB is 71.0 MB for 10 routers and 354.7 MB for 50 routers, showing that each router tends to consume 7 MB.

Case Study: We checked the workload for enabling SQUAB to handle BGP-SRx. First, for embedding in the function to set up the experimental environment, we needed to add 44 lines of code to the original 326 lines of code, i.e., 370 lines of code totally. We also created a Dockerfile that builds a container image of a BGP-SRx router and a script to generate both a key and a certificate for signing and verification. Besides, we modified the script for Quagga, which writes the BGP configuration file in a router, to correspond with the BGPsec configuration file of BGP-SRx. As the script to install BGP-SRx into the image for building the container image, we adopt the original code of BPG-SRx[5]. Meanwhile, certificates into the `rpkirtr_svr`, the RPKI emulator by the BGP-SRx suite, cannot be registered due to a bug. Thus, we need to operate the container of `rpkirtr_svr` and then register the certificate manually after setting up the experimental environment.

We write the SQUAB configuration file corresponding to the topology of NAITWAYS utilized in the experiments on the functional validation and then run SQUAB. Figure 6 shows the setting information and routing tables of the BGP-SRx router. As shown in the figure, the routers correctly exchange the route information. In addition, we confirm that routers correctly conduct the verification by BGPsec through the existence of v(v,-), which indicates the issuance of a certificate.

---

[5] https://github.com/usnistgov/NIST-BGP-SRx.

```
bgpd> naitway_router_1103_for_30781_1
spawn telnet localhost 2605
Trying 127.0.0.1...
Connected to localhost.
Escape character is '^]'.

Hello, this is QuaggaSRx (version 0.4.2.9).
Copyright 1996-2005 Kunihiro Ishiguro, et al.


User Access Verification

Password:
bgpd> show ip bgp summary
BGP router identifier 10.10.10.18, local AS number 1103
RIB entries 15, using 1680 bytes of memory
SRx host localhost, port 17900
Peers 1, using 4568 bytes of memory

Neighbor        V   AS MsgRcvd MsgSent  TblVer  InQ OutQ Up/Down  State/PfxRcd
172.16.72.2     4 30781     20      15       0    0    0 00:11:09           7

Total number of neighbors 1
bgpd> show ip bgp
BGP table version is 0, local router ID is 10.10.10.18
Status codes: s suppressed, d damped, h history, * valid, > best, i - internal,
              r RIB-failure, S Stale, R Removed
Validation:   v - valid, n - notfound, i - invalid, ? - undefined
SRx Status:   I - route ignored, D - SRx evaluation deactivated
SRxVal Format: validation result (origin validation, path validation)
Origin codes: i - IGP, e - EGP, ? - incomplete

   Ident    SRxVal SRxLP Status Network       Next Hop       Metric  LocPrf Weight Path
*> 01A95B4D n(n,-)        172.16.1.0/24   172.16.72.2                     0 30781 206610 i
*> CCD263D6 n(n,-)        172.16.81.0/24  172.16.72.2                     0 30781 57119 51405 i
*> 1CA3E636 n(n,-)        172.16.163.0/24 172.16.72.2                     0 30781 57119 i
*> 4F8E26AF n(n,-)        172.22.192.0/24 172.16.72.2                     0 30781 39351 i
*> 6EE9BB1F v(v,-)        172.26.73.0/24  172.16.72.2                     0 30781 29075 3333 i
*> 34F0A415 n(n,-)        172.26.137.0/24 172.16.72.2         0           0 30781 i
*> 97C4E4C1 n(n,-)        172.28.120.0/24 172.16.72.2                     0 30781 29075 i
*> -------- v(v,-)        172.30.116.0/24 0.0.0.0            0       32768 i

Total number of prefixes 8
```

Fig. 6. The setting information and routing tables of the BGP-SRx router: From the routing table of the BGP-SRx router, it can be seen that there is the character of v(v,-) in the line of the router that issued the certificate.

5   Discussions

In this section, we discuss the performance of SQUAB based on the experimental results, use cases, and limitations of the current specification.

5.1   Performance

We first discuss the running time of SQUAB. The running time is different between setting up and removing an environment according to Fig. 4. For setting up an environment, each container with a router needs to generate the BGP configuration file and run the daemon. Therefore, the running time for setting up an environment is longer than that for removing an environment.

Next, the memory consumption of SQUAB is 7 MB per router according to Fig. 5. This result indicates that, for example, the memory consumption is about 700 MB even when a network topology consisting of 100 routers with 50 inter-AS connections is set up. We thus believe that SQUAB enables a typical laptop computer to run experiments even with medium-scale topologies.

5.2   Use Cases

SQUAB is applicable to the existing protocols for BGP extensions. We describe D-BGP [23] and APVAS [14] as use cases of SQUAB.

First, we describe the application to D-BGP, a general-purpose protocol for BGP extensions that supports them on the same network. D-BGP has a function to tunnel routers that have no support for the extensions. Although the function of D-BGP has not been evaluated on the network composed of only D-BGP routers, SQUAB can provide networks composed of various routers, which support only BGP extensions except for D-BGP. A prototype implementation of D-BGP is publicly available[6]. It is implemented as an extension of Quagga, and hence, by making a container image composed of the prototype of D-BGP and existing configure file generation script, SQUAB can deal with D-BGP.

Next, we describe the application to APVAS, a protocol for BGPsec that reduces the memory consumption of routers more than state-of-the-art protocols by introducing aggregate signatures [10], which can aggregate individually generated signatures of BGPsec into a single short signature. Although the performance of APVAS has not been evaluated on the actual topology, SQUAB can potentially provide the evaluation. The latest APVAS prototype is implemented as an extension of BIRD, and a user can use SQUAB to evaluate the performance of APVAS by modifying the syntax specification of the BGP configuration file from Quagga to BIRD. In this way, SQUAB can be applicable for software routers other than Quagga.

5.3   Limitations

We describe several limitations of the current specification of SQUAB. First, the current specification does not support configuration of AS policy, e.g., the local-preference [13] to override the AS path length metric of the base BGP operation and the multi exit discriminator (MED) attributes [18] to provides a mechanism for BGP speakers to convey to an adjacent AS the optimal entry point into the local AS.

Second, network topologies generated by SQUAB are limited. Concretely, the networks are composed of only AS border routers. In other words, SQUAB cannot represent components of an intra-AS to configure and maintain routing tables within a single AS, e.g., the network latency due to complicated networks or random errors inside the AS. To introduce the functions described above,

---

[6] https://github.com/RS1999ent/quagga_evolvabilty.

implementation to simulate intra-AS is necessary. Likewise, the latency on inter-AS communication should be considered to represent networks in the real world further.

Finally, if users want to conduct an experimental evaluation of new protocols, they need to develop their software routers in advance. Moreover, the scripts of SQUAB to set up routers may need updates. However, the automation of developing software routers and updating the scripts is out of the scope of the current specification. These topics are reserved for future work.

6   Related Works

In this section, we briefly review related works in terms of experimental tools for BGP, testbed developments, and BGP security issues in the past decades.

Experimental Tools for BGP: Although there are many tools for measuring empirical values on BGP [11,17,21,26], they cannot interact with the existing ecosystem. This means that evaluation, including the current policy and connectivity of actual ASes, is limited. In contrast, PEERING [24] is the first tool that takes into account ecosystems in the real world. However, PEERING specializes in experiments concerning borders and is thus unsuitable for experiments to measure each AS's connectivity on the Internet.

Testbed Developments: Many existing testbed environments, such as Cloud-Lab [2], XSEDE [9], and Emulab [4], provide hardware as a computer resource in contrast to SQUAB. More specifically, Emulab [4] provides the testbed environments specialized for network experiments, and also there are many Emulab-based tools by customizing to each application such as EPIC [25] and DETER [3]. Emulab provides not only computer resources but also network connections representing network delays as experimental environments. However, the use of Emulab is restricted to a responsible project leader, e.g., a professor or a person who has permission from the leader, and this approval may unexpectedly consume time. PlanetLab [6] and WIDE [8] are projects where participants bring machine resources and then configure networks to share an experimental environment. However, the participants need to meet several conditions, such as providing two Linux machines with global IP addresses. Compared with the testbeds mentioned above, SQUAB is based on containers whose cost is lightweight and thus it allows experiments on only a typical laptop computer without any third-party resources. Therefore, anyone can use SQUAB to quickly start experiments. The closest tool to SQUAB is DOCKEMU [27], which uses Docker to set up experimental network environments. However, DOCKEMU is based on NS-3, a network simulation tool, while SQUAB provides network emulation by virtue of Docker networks.

7   Conclusion

In this paper, we proposed SQUAB, a lightweight evaluation tool for experiments on BGP and its extensions. We publicly released our implementation of

SQUAB, including its reference implementation for BGPsec. In the design of a new protocol, the effectiveness of the protocol should be evaluated through various experiments, and SQUAB makes such evaluation easy. We thus believe that SQUAB will assist researchers in developing future BGP extensions and their social implementation. We are in the process of introducing AS policy support such as the local-preference attributes and the multi exit discriminator (MED) attributes.

Acknowledgments. This research was supported in part by the Japan Society for the Promotion of Science KAKENHI Numbers 18K18049, and Innovation Platform for Society 5.0 at MEXT.

References

1. The bird internet routing daemon. https://bird.network.cz/
2. Cloudlab https://www.cloudlab.us/
3. The deter project. https://deter-project.org/
4. emulab. https://www.emulab.net/portal/frontpage.php
5. Nist bgp secure routing extension (bgp / srx) prototype https://www.nist.gov/services-resources/software/bgp-secure-routing-extension-bgp-srx-prototype
6. Planetlab https://planetlab.cs.princeton.edu/
7. Quagga routing suite. https://www.nongnu.org/quagga/
8. Wide. https://www.wide.ad.jp/
9. Xsede. https://www.xsede.org/
10. Boneh, D., Gentry, C., Lynn, B., Shacham, H.: Aggregate and verifiably encrypted signatures from bilinear maps. In: Proceedings of EUROCRYPT 2003, pp. 416–432 (2003)
11. Chen, K., Choffnes, D.R., Potharaju, R., Chen, Y., Bustamante, F.E., Pei, D., Zhao, Y.: Where the sidewalk ends: extending the internet as graph using traceroutes from p2p users. In: Proceedings of the 5th International Conference on Emerging Networking Experiments and Technologies, pp. 217–228 (2009)
12. Fogel, A., Fung, S., Pedrosa, L., Walraed-Sullivan, M., Govindan, R., Mahajan, R., Millstein, T.: A general approach to network configuration analysis. In: Proceedings of NSDI 2015, pp. 469–483. USENIX Association (2015)
13. Griffin, T., Huston, G.: Bgp wedgies. RFC 4264 (2005). https://rfc-editor.org/rfc/rfc4264.txt, published: RFC 4264
14. Junjie, O., Yanai, N., Takemura, T., Okada, M., Okamura, S., Cruz, J.P.: APVAS: reducing memory size of as_path validation by using aggregate signatures (2020). CoRR **abs/2008.13346**, https://arxiv.org/abs/2008.13346
15. Lepinski, M., Sriram, K.: BGPsec Protocol Specification (2017). https://doi.org/10.17487/RFC8205, https://rfc-editor.org/rfc/rfc8205.txt, published: RFC 8205
16. Mahajan, R., Wetherall, D., Anderson, T.: Understanding BGP misconfiguration. In: Proceedings of SIGCOMM 2002, pp. 3–16. ACM (2002)
17. Mao, Z.M., Rexford, J., Wang, J., Katz, R.H.: Towards an accurate as-level traceroute tool. In: Proceedings of the 2003 conference on Applications, Technologies, Architectures, and Protocols for Computer Communications, pp. 365–378 (2003)
18. McPherson, D.R., Gill, V.: BGP MULTI_EXIT_DISC (MED) Considerations. RFC 4451 (2006). https://www.rfc-editor.org/rfc/rfc4451.txt, published: RFC 4451

19. Merkel, D.: Docker: lightweight linux containers for consistent development and deployment. Linux J. **2014**(239), 2 (2014)
20. Mitseva, A., Panchenko, A., Engel, T.: The state of affairs in BGP security: a survey of attacks and defenses. Comput. Commun. **124**(2018), 45–60 (2018)
21. Peterson, L., Bavier, A., Fiuczynski, M.E., Muir, S.: Experiences building planetlab. In: Proceedings of the 7th Symposium on Operating Systems Design and Implementation, pp. 351–366 (2006)
22. Rekhter, Y., Hares, S., Li, T.: A Border Gateway Protocol 4 (BGP-4) (2006). https://doi.org/10.17487/RFC4271, https://rfc-editor.org/rfc/rfc4271.txt, published: RFC 4271
23. Sambasivan, R.R., Tran-Lam, D., Akella, A., Steenkiste, P.: Bootstrapping evolvability for inter-domain routing with d-bgp. In: Proceedings of SIGCOMM 2017, pp. 474–487. ACM (2017)
24. Schlinker, B., Arnold, T., Cunha, I., Katz-Bassett, E.: Peering: virtualizing BGP at the edge for research. In: Proceedings of CoNEXT 2019, pp. 51–67. ACM (2019)
25. Siaterlis, C., Genge, B., Hohenadel, M.: Epic: a testbed for scientifically rigorous cyber-physical security experimentation. IEEE Trans. Emerg. Topics Comput. **1**(2), 319–330 (2013)
26. Staff, R.N.: Ripe atlas: a global internet measurement network. Internet Protoc. J. **18**(3), 1–4 (2015)
27. To, M.A., Cano, M., Biba, P.: Dockemu–a network emulation tool. In: 2015 IEEE 29th International Conference on Advanced Information Networking and Applications Workshops, pp. 593–598. IEEE (2015)
28. Vervier, P., Thonnard, O., Dacier, M.: Mind your blocks: on the steal thiness of malicious BGP hijacks. In: Proceedings of NDSS 2015. Internet Society (2015)

Scalability Support with Future Internet in Mobile Crowdsourcing Systems

Peron Sousa and Antonio A. de A. Rocha[✉]

Institute of Computing, Fluminense Federal University, Rio de Janeiro, Brazil
peron_rezende@id.uff.br, arocha@ic.uff.br

Abstract. The popularization of mobile devices with multiple resources, the real-time location, and the availability of "human computing", provided by a global crowd of users, motivated the emergence of Mobile Crowdsourcing Systems. This new paradigm presents several requirements such as assertiveness in task assignment, security, mobility support and scalability. In this paper, scalability is the ability to meet increasing demand without performance loss, a point ignored in many proposals that make use of the traditional model (client-server). The solution of this problem is also the main focus of this work. To address the issue we review the state of the art and analyze where each proposal failed. We designed a new architecture and worked on its implementation to carry out a proof of concept that confirmed its feasibility. Experiments evaluated the performance of the components that make up the solution and the results showed that the architecture offers a reduction in latency between 26.31% and 32.03%. Another interesting point is that the new architecture, unlike the state of the art, has no overhead and no need to manage dozens of machines around the world. It was also the first with mobility support for Mobile Crowdsourcing Systems.

1 Introduction

Mobile Crowdsourcing System (MCS) is a new paradigm that seeks to take advantage of an already installed base of devices with multiple features to locate workers and assign them tasks that have restrictions of time and space. Moreover, these tasks require human cognition to be performed quickly, cheaply, and with greater efficiency.

An example of an MCS request is the location of parking spaces. People can easily identify available spots and report with images and/or text messages. If we were to use an autonomous system based on ultrasound to find it, not only would we need special hardware, but also sophisticated algorithms to ensure data reliability [1].

New challenges have arisen with the advent of these systems and scalability is one of them, should MCS intend to support a multitude of users scattered around the globe, who are in constant movement. In our proposal, scalability is the ability to meet growing demand without loss of performance; a point ignored in many proposals which make use of the traditional model (client-server). In this model, the backend is responsible for registering users, distributing tasks, receiving responses and, in some cases, consolidating information. However, this type of system does not scale, losing performance, and may stop when a large number of users and tasks arrive in a short time. In the state-of-the-art, we find solutions involving load balancing with active replication of servers, proxy, Software Defined Networking (SDN), elastic computing, edge computing, and Device to Device (D2D). However, they have limitations [2].

Our goal is to demonstrate how our architecture, called Snow Flurry (inspired by the eponymous mobiles of sculptor Alexander Calder), solves the issue of scalability more efficiently than state-of-the-art and with mobility support. In this work, the efficiency is measured by latency and proved by experiment comparative with another solution. To achieve the objectives, we used the Global Name Service (GNS) of the MobilityFirst architecture. This GNS (also known as Auspice) enables a Globally Unique Identifier (GUID), mobility, and contextual search [3,4]. Because it is able to address the issue of mobility and most of the scalability issues involved in MCS architectures.

With the realization of the proof of concept (PoC) we verified the viability of our proposal. With performance tests, volume, and infrastructure analyzes we show: that the architecture offers a reduction in latency between 26.31% and 32.03% when compared to the state-of-the-art; dispenses the monitoring and maintenance of a complex infrastructure with approximately 200 hosts. These contributions occur along with meeting the requirements for a scalable architecture. In addition, it is also the first, to the best of our knowledge, to support mobility.

2   Snow Flurry's Proposal and Implementation

In this section we will introduce the proposed Snow Flurry architecture through each of its elements; this project aims to scale and support mobility for crowdsourcing systems. Then, we will introduce the implementation and explain how the Snow Flurry architecture works. We will also show the role of each component throughout the process and discuss commonalities with other MCS.

2.1   Proposed Architecture

Figure 1 presents an overview of the Snow Flurry architecture. In it, the optional elements are represented by dashed edges surrounding the Content Distribution Network (CDN). Therefore, the mandatory components are: the notification service, the backend and the mobile device.

Notification Service: We designed Snow Flurry to work with a notification service that uses the push method. These services allow you to send messages to the user without the need to take the initiative to check for new messages. Also, the push method increases the number of tasks completed and reduces overhead [5].

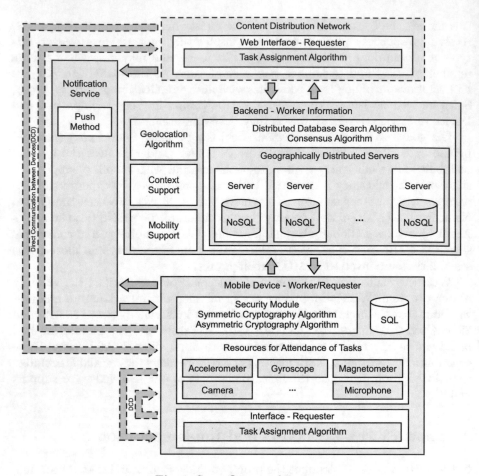

Fig. 1. Snow flurry architecture

Backend: The backend is the component responsible for keeping the workers' information, from the register (with name, email, geolocation, etc.) to the result of their work in some scenarios. This item becomes critical when we think of a global reach crowdsourcing system that should support millions of users around the world. Using a client/server architecture provided by a supercomputer would result in a solution with a critical point of failure and varying levels of quality of service, as users' perceived latency would vary with each other's geographic location, that is, the further away from the server, the worse would be the application response time. One solution to these points is to distribute servers geographically.

Geographically Distributed Servers solves two issues: (1) prevents the system from having a critical point of failure and (2) reduces latency, improving the quality of service offered to the user. *Consensus Algorithm* is needed to be used for maintaining replicated data consistent. Whenever a change occurs on one server, it is necessary to write that change on others. Yet, simultaneous changes can occur on different servers and a decision is needed to define which change will be considered.

Distributed Database Search algorithm is necessary to locate data across multiple servers. It has some challenges. In our specific cases, one of them is join operations between distributed tables. An alternative is the use of document-oriented databases (a type of NoSQL). *NoSQL Servers* (with document-oriented databases) solve the problem of interactions by avoiding them through denormalization, yet they are possible by using key-value, column family, and graphs. It means a more dynamic structure (such as JSon) that can be used for context definition, such as saying that a crowdsourcing worker is a taxi driver, photographer, or even both. This brings us to another need for Snow Flurry architecture, *Context Support*, which jointly with NoSQL databases can provide a new structure aligned with system needs whenever changes prove necessary. Although most modern NoSQL databases already offers, specific *Geolocation Algorithms* still need to be produced for each emerging scenario.

Finally, *Mobility Support* here is the ability of the device to change identifiers without losing established connections, i.e., being able to change IP and, consequently, network without having to restart an operation (such as a download in progress). Therefore, we understand that this support must allow both worker and requester to be in transit while performing tasks and transmitting responses in mobile crowdsourcing systems.

Mobile Device: While in crowd sensing systems smartphones are more requested due to low user interaction and the multiple features offered by these devices (such as GPS, accelerometer and cameras), in mobile crowdsourcing systems we may consider other types of devices, (such as, tablets and notebooks) because the greater interest is in the cognitive power of the user. But, *Resources for Attendance of Tasks* may also be useful. Ideally, the worker's device has GPS, accelerometer, cameras, among other features to perform tasks (which directs, but does not limit the system for smartphone users).

In addition to enabling campaign launching, the in-app requester interface also brings with it a strategy that enables scalability in mobile crowdsourcing systems, which is the distribution of the task assignment algorithm. By embedding this algorithm in the app, we remove from the backend the burden of centralizing the task assignment process and all the computational effort involved in this event, diluting this burden between requesters' devices. The app needs to have a lightweight local relational database (*SQL local DB*) to enable the tracking of tasks, both from the point of view of workers and the management of requesters. This information only exists on the end devices, i.e., it will not be kept at the backend of the system.

From a security perspective, one of the safest communication models that exists is the combined use of symmetric with asymmetric cryptography algorithms. Thus, the *security module* in the app must enable the creation of asymmetric cryptographic keys and user registration at the backend with the consequent submission of the public key created for it. Afterwards, updates to the base must occur through commands signed by some symmetric cryptography algorithm, this signature must be encrypted with the private key generated by the asymmetric cryptographic algorithm before sending with the command to the backend.

Finally, at bottom left part of Fig. 1, we can see represented the *Direct Communication between Devices (DCD)*, when the worker's mobile device communicates directly with the requester's mobile device, so the arrows exit and return to the mobile device component. And, at the left side of Fig. 1, the dotted arrows connecting the mobile device to the content distribution network component represent a direct communication between a requester using a web interface and a worker using the app.

2.2 Implementation

The Snow Flurry implementation consists of 3 main components: an Android App, GNS Auspice and Firebase Cloud Messaging. The interface with MCS can be made exclusively by the App we developed for Android devices, i.e., serve both workers and requesters. We will call workers the users who respond to the tasks of the requesters. Therefore, the requesters are the users who create the tasks in the MCS. We will call campaign the creation of a task (or set of related tasks) and its distribution to more than one worker. Optionally, the architecture can rely on an Internet page composed of HTML/JavaScript and distributed by a global CDN service.

As mentioned, the second element is a GNS. For this work we chose the Auspice [4], because its project considered the popularization of mobile devices and prepared for such event. It is able to keep billions of records (user accounts) and support millions of queries/updates per day, sent from all over the world [3]. The Android application enables the registration in Auspice and the data update. Among these data we highlight the geolocation, the context information and the token of the notification service. This communication is illustrated by the number one in Fig. 2. Unlike Peng et al. [6], in which the backend is updated every 2 min, our proposal only forwards an update to the GNS if a change occurs.

In Fig. 2, the number two illustrates the search for workers by the requester. At this point, we send the geolocation of the point of interest (PoI) within a radius of 1 km (by default). The PoI is obtained when the map is opened and changes as we move it, see an example in Fig. 3. The frontend receives the data and presents them, redoing this process every 5 s (by default).

Figure 3 illustrates the user interfaces of the worker. It allows to draw vector shapes on the map. On the screen we see five workers, three with the blue background, one green and one red. Each one defines its own radius of action and informs this to Auspice. The requester views the derived perimeter by clicking on the marker. In the example of Fig. 3, compared to the green worker, the blue collaborator serves a larger area.

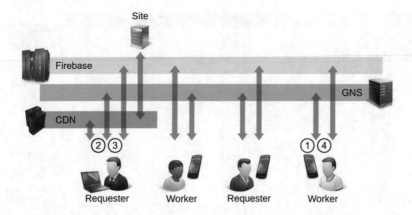

Fig. 2. Snow Flurry operation, from task assignment to results reception

In the process, once the location has been chosen, it is defined the task with a given context and deadline. In this example we indicate the context (color) and add the necessary information to the task. In the sequence, the frontend identifies the workers who serve the context (color) and assigns them the task through Firebase Cloud Message, circle number three from Fig. 2. Then, the event indicated by circle number four of Fig. 2 occurs. In it, Firebase delivers the notification to the device. This delivery uses the token that the machine received from Firebase. This token is persisted in the GNS by the app and collected by the requester, circle one and two of Fig. 2.

The GNS can store tokens from other notification services, such as the Apple Push Notification Service (APNS) and the Microsoft Push Notification Service (MPNS). We opted for Firebase, due to its integration with the Android Studio development environment, but there are other ways the worker can access the tasks. Some solutions use the pull method, in which the worker searches for their existence. Firebase works with push, in this method the task is sent to the worker. According to Kandappu et al. [5], push increases the number of completed tasks and reduces the overhead.

Tasks are routed preferably by notification services. In these services, when the worker is disconnected, the tasks/notifications are stored for a period. After this wait, without successful delivery, the notification is discarded. If the worker informed the token that identifies him in these services to Auspice, the application will send notifications through them. That way, if the user leaves their device unplugged for a long time (vacation, for example), they will not receive old notifications that have missed the requester's deadline. We can also notify workers through the Android app. In Fig. 3, we see how the solution looks in the application during the worker search step and view of their radius, as seen on page to the Internet. Similarly, we can view the opening of a task through the application.

Fig. 3. User interfaces of the worker

Circle number one in Fig. 2 also represents the completion of the task, at this time the worker sends the data and/or files to the Auspice. In its structure, Auspice uses a document-oriented database (MongoDB). Due to the way it was implemented, we can use the same function used to create a context for storing data in JSon format. After the deadline, the requester collects the responses, as shown by circle number two in Fig. 2. In this way, when using the Auspice, we are (1) able to focus on the development of the functionalities of the MCS, (2) save lines of code, (3) reduce environment settings, and (4) we do not need to monitor and maintain replicas. Our proposal uses the Spark Plan (free), its has no limits for Cloud Messaging.

For proof of concept we have developed the two types of features that consume GNS services, the web page and the app for Android devices. Requesters can use both the web page and the application, while workers only the application. The Android application allows registration in Auspice and data update, among which we highlight the geolocation, context information and notification service token. This communication is illustrated by the number one circle in Fig. 2. Unlike Peng et al. [6], which updates the rear every 2 min, Snow Flurry only forwards an update to GNS if a change occurs.

The DCD was not implemented in PoC. In direct communication, due to the mobility of the devices involved, there is the possibility that one of the tips or both of them change network, exchanging IP. To avoid loss of communication we can use MSocket. It is a Java library (analogous to the conventional Socket) that allows the mobility of both ends without loss of connection. This solution is part of the research on the MobilityFirst architecture and works together with Auspice, using cryptographic keys in communication. Apart from some additional code adaptations, we can simply exchange Socket and ServerSocket for MSocket and MServerSocket, respectively. This way, it is possible to provide mobility to an application that works with a TCP communication without relying on a DNS [7]. We don't use this library in implementing Snow Flurry, because it still needs some adaptations to run perfectly on the Android environment. Mobility support was provided by an Auspice API that is consumed by HTTP requests.

3   Comparing Snow Flurry to the State-of-the-Art

In this section we briefly present the state-of-the-art in scalability for crowdsourcing and compare it to Snow Flurry proposal. In the proposal of Yan et al. [8], mCrowd is presented, which is a mobile crowdsourcing platform acting as a proxy between users (requesters and workers) and the platforms Amazon Mechanical Turk (MTurk) and ChaCha. By analyzing the proposal, we can conclude that mCrowd must face a similar problem to that found in the client-server model. Even if partner platforms are able to ensure scalability, mCrowd may represent a bottleneck between them and users. To avoid this problem, Hara et al. [9] proposed a decentralized system, where special servers, known as Server Access Network Entities (SANE), are distributed geographically to dilute the load. Auspice can perform this entire operation for our proposal in a similar way, having several servers and positioning the information geographically close to those interested. SANE architecture uses Distributed Hash Table (DHT) to organize its components and the work presents an implementation with a relational database, MySQL. As the architecture has a very specific use, it acts as a proxy between crowdsourcing clients and crowdfunding servers, the relational database model is sufficient. In addition to not being tied to a specific use, the Snow Flurry architecture also has algorithms for the proper positioning of information. Auspice uses an algorithm called Gigapaxos [10] that has the ability to quickly manage and reconfigure a very large number of replica groups. Li et al. [11] conceptualized a decentralized blockchain-based framework for crowdsourcing called CrowdBC. The idea is not to depend on any trusted third party, to guarantee user privacy and low transaction fees.

In the survey of Xiao et al. [2], we find a study on scalability for mobile crowd sensing (application similar to crowdsourcing, but with little or no human interaction, and it focuses on the collection of sensor data). One of those systems is proposed by Marjanović et al. [12] and uses edge computing for crowd sensing application. Basically, this work advocates a cloud architecture with intermediate servers (edge) closer to users to reduce latency. Our proposal already does this, since Auspice has servers spread around the world that organize the data in a way that leaves them geographically close to users. Marjanović's et al. research was the one that focused on scalability the most, but the work concentrated efforts on finding out how many intermediate servers would be needed to make their proposal viable and confined themselves to saying that edge computing can provide ultra-low latency (less than 1 ms), without presenting evaluations.

Medusa, by Ra et al. [13], is another a crowd sensing solution that offers services with a high level of abstraction in specifying the steps required to complete a detection task and which employs a distributed run-time system to coordinate the execution of these tasks between smartphones and a cluster in the cloud. It works similarly to mCrowd of Yan et al. [8], as it also intermediates with MTurk. However, Ra et al. [13] states that Medusa needs not to be restricted to a single machine and that it can work with as many logically independent simultaneous servers as needed. The work presents an implementation with MySQL without giving further details on how the application handles database growth.

Table 1 summarises a comparison among state of the art architectures, including the Snow Flurry proposal. From the table, we can note that Snow Flurry covers most of the aspects that other architectures do not, except for the item "Anonymity" which will be the subject of future work.

Table 1. Comparison between architectures

	Centered Element	Content Distribution Network	Task Assignment Distributed Algorithm	Push Method Notification	Cloud and/or Elastic Computing	Geographically Distributed Servers	Database Full Replication	Search in Distributed Database	Context Search	Geolocation	Information Distribution Method	Dynamic Structure (NoSQL)	Mobility Support	Encryption	Anonymity	Application and/or Source Code Available
Geocrowd [14]	✓	-	✗	✗	✗	-	-	-	-	✓	-	-	-	-	-	✗
gMission [15]	✓	-	✗	✗	✗	-	-	-	-	✓	-	-	-	-	-	✗
CrowdBC [11]	✗	-	✗	-	✗	✓	✓	✓	-	-	✗	-	-	✓	✓	✗
mCrowd [8]	✓	-	✗	-	✓	✗	-	-	-	-	-	-	-	-	-	✗
SANE [9]	-	-	-	-	-	✓	✓	✗	-	-	-	-	-	-	-	✗
Edge [12]	✗	✗	-	-	✓	✓	✓	✗	-	✓	✗	-	-	-	-	✗
Medusa [13]	✓	✗	✗	-	✓	-	✓	-	-	✓	✗	-	✗	-	-	✓
Snow flurry	✗	✓	✓	✓	✗	✓	✗	✓	✓	✓	✓	✓	✓	✓	✗	✓

4   Experiment

In this section we will describe the scenario, the metric used in the comparison and how the values were obtained. We will present the results obtained in the tests, discuss implementation points that affect the performance and to make considerations about the complexity and costs involved in a global infrastructure.

Due to the unavailability of MCS, we will present a comparison of Snow Flurry implementation against the Medusa by Ra et al. [13], a mobile crowd sensing system.

To evaluate the performance of a single instance of GNS, we used a Linode machine, consisting of 1 CPU core with 64-bit, running Ubuntu 16.04 operating system, 2 GB RAM, 30 GB SSD storage, 2 TB transfer, 40 Gbps input, 1000 Mbps output and located in Newark, NJ, USA. We installed Java version 1.8.0_151, the GNS and we carried a load of 5000 workers in the system.

To evaluate the performance of our solution we will use as latency metric on each type of communication. To rule out the delay due to the network, the requests were triggered from the machine itself where the implementation was. We performed 30 repetitions in each case to obtain the boxplots shown in Fig. 4.

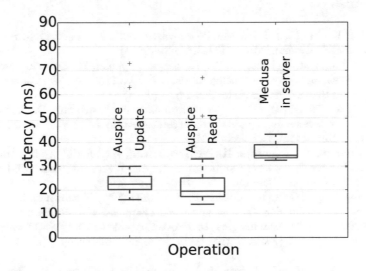

Fig. 4. Performance of requests without network delay.

We can see that actual overhead costs are only 34.47 ms on average for Medusa, which allows for a yield of 1740 instances of tasks per minute on a single server. The creators of Medusa claim that this component is highly parallelizable since each task instance is relatively independent and can be assigned a separate Task Controller. Thus, if the task transfer rate becomes a problem, it would be possible to leverage the cloud computing elastic resources to scale the system [13].

Following the same principle, we can see that our solution is able to serve more requests per machine, since the average times of 25.4 and 23.43 ms, enable, respectively, the attendance of 2362 and 2560 requests per minute. Values representing a capacity increase between 35.75% and 47.13%, with a decrease in latency between 26.31% and 32.03%. If our proposal worked with elastic computing, these data would already indicate a possible savings compared to the reduction in the provisioning of resources.

5 Conclusion

In this paper we discuss the problem of scalability in mobile crowdsourcing systems and the techniques used in state of the art. We presented a new architecture, called Snow Flurry, which use a global name service and cloud services for worker notification. We established a comparative table with the other solutions, which was possible to realize that the only point of attention of Snow Flurry is the absence of anonymity. We conducted a proof of concept with the Snow Flurry implementation and performed comparative experiments with a mobile crowd sensing system, due to unavailability of mobile crowdsourcing system solutions. The results presented a reduction in latency between 26.31% and 32.03%.

References

1. Ganti, R.K., Ye, F., Lei, H.: Mobile crowdsensing: current state and future challenges. IEEE Commun. Mag. **49**(11), 32–39 (2011)
2. Xiao, Y., Simoens, P., Pillai, P., Ha, K., Satyanarayanan, M.: Lowering the barriers to large-scale mobile crowdsensing. In: Proceedings of the 14th Workshop on Mobile Computing Systems and Applications, p. 9 (2013)
3. Sharma, A., Tie, X., Uppal, H., Venkataramani, A., Westbrook, D., Yadav, A.: A global name service for a highly mobile internetwork. ACM SIGCOMM Comput. Commun. Rev. **44**(4), 247–258 (2015)
4. Raychaudhuri, D., Nagaraja, K., Venkataramani, A.: Mobilityfirst: a robust and trustworthy mobility-centric architecture for the future internet. ACM SIGMOBILE Mob. Comput. Commun. Rev. **16**(3), 2–13 (2012)
5. Kandappu, T., Misra, A., Cheng, S.-F., Jaiman, N., Tandriansyah, R., Chen, C., Lau, H.C., Chander, D., Dasgupta, K.: Campus-Scale Mobile crowd-tasking: deployment & behavioral insights. In: Proceedings of the 19th ACM Conference on Computer-Supported Cooperative Work & Social Computing, pp. 800–812. New York (2016)
6. Peng, X., Gu, J., Tan, T.H., Sun, J., Yu, Y., Nuseibeh, B., Zhao, W.: Crowdservice: serving the individuals through mobile crowdsourcing and service composition. In: 31st IEEE/ACM International Conference on Automated Software Engineering, pp. 214–219 (2016)
7. Bronzino, F., Nagaraja, K., Seskar, I., Raychaudhuri, D.: Network service abstractions for a mobility-centric future internet architecture. In: Proceedings of the Eighth ACM International Workshop on Mobility in the Evolving Internet Architecture, pp. 5–10. ACM (2013)
8. Yan, T., Marzilli, M., Holmes, R., Ganesan, D., Corner, M.: mCrowd: a platform for mobile crowdsourcing. In: Proceedings of the 7th ACM Conference on Embedded Networked Sensor Systems, pp. 347–348 (2009)
9. Hara, T., Springer, T., Bombach, G., Schill, A.: Decentralised approach for a reusable crowdsourcing platform utilising standard web servers. In: Proceedings of the ACM Conference on Pervasive and Ubiquitous Computing Adjunct Publication, pp. 1063–1074 (2013)
10. Venkataramni, A., Gao, Z., Gu, T., Anantharamu, K.: Scalable fine-grained reconfigurable replica coordination. University of Massachusetts (UMass), Technical Report (2018)

11. Li, M., Weng, J., Yang, A., Lu, W., Zhang, Y., Hou, L., Liu, J., Xiang, Y., Deng, R.H.: Crowdbc: a blockchain-based decentralized framework for crowdsourcing. IEEE Trans. Parallel Distrib. Syst. **30**(6), 1251–1266 (2019)
12. Marjanović, M., Antonić, A., Žarko, I.P.: Edge computing architecture for mobile crowdsensing. IEEE Access **6**, 10:662–10:674 (2018)
13. Ra, M.R., Liu, B., La Porta, T.F., Govindan, R.: Medusa: A programming framework for crowd-sensing applications. In: Proceedings of the 10th International Conference on Mobile Systems, Applications, and Services, pp. 337–350 (2012)
14. Kazemi, L., Shahabi, C.: Geocrowd: enabling query answering with spatial crowdsourcing. In: Proceedings of the 20th International Conference on Advances in Geographic Information Systems, pp. 189–198 (2012)
15. Chen, Z., Fu, R., Zhao, Z., Liu, Z., Xia, L., Chen, L., Cheng, P., Cao, C.C., Tong, Y., Zhang, C.J.: gmission: a general spatial crowdsourcing platform. Proc. VLDB Endowment **7**(13), 1629–1632 (2014)

HyDiLLEch: A WSN-Based Distributed Leak Detection and Localisation in Crude Oil Pipelines

Safuriyawu Ahmed[1]([✉]), Frédéric Le Mouël[1], Nicolas Stouls[1],
and Gislain Lipeme Kouyi[2]

[1] Univ Lyon, INSA Lyon, Inria, CITI, EA3720, 69621 Villeurbanne, France
{Safuriyawu.Ahmed,frederic.le-mouel,nicolas.stouls}@insa-lyon.fr
[2] Univ Lyon, INSA Lyon, DEEP, EA7429, 69621 Villeurbanne, France
gislain.lipeme-kouyi@insa-lyon.fr

Abstract. One of the major failures attributed to pipeline transportation of crude oil is oil leakages and spills. Hence, it is monitored via several classical leak detection techniques (LDTs), which are more recently implemented on centralised wireless sensor networks (WSN)-based leak detection and monitoring systems (LDMS). However, the LDTs are sometimes prone to high false alarms, and the LDMS are sensitive to single points of failure. Thus, we propose HyDiLLEch, a distributed leakage detection and localisation technique based on a fusion of several LDTs. In this work, we implemented HyDiLLEch and compared it to the individual LDTs in terms of communication efficiency and leakage detection and localisation accuracy. With HyDiLLEch, the number of nodes detecting and localising leakages increases by a maximum of four to six times, thereby eliminating single points of failures. In addition, we improve the accuracy of localisation in nodes physically-close to the leak and maintain an average of 96% accuracy with little to no communication overhead.

1 Introduction

Oil leakages and spills (OLS) in pipelines occur due to ageing infrastructures, corrosion, and most commonly third-party interference. OLS has both economic and environmental effects. Some of them include the annual loss of up to 10 billion USD in the United States [1], pollution of water and land resources, fatal accidents caused by resulting explosions, etc. Therefore, several measures are put in place to help monitor, detect and predict such failures in a timely manner.

These measures can be broadly categorised into *human-based* and *non-human-based* leakage detection and monitoring systems (LDMS). The human-based LDMS comprises the use of community-based surveillance, security personnel, specialised helicopters, etc., for detecting and localising leakages. Although they work, their drawbacks include long delays, high cost, inefficiency, etc. This is ascertained with the recorded loss of crude oil from the pipeline network of Shell Nigeria to the tune of 11000 barrels per day in 2018 due to failures, an increment of nearly 550% compared to the previous year [2]. The non-human-based (sensorised) LDMS, on the other hand, detect leakages via the supervisory control and data acquisition (SCADA), fibre optic or copper installed along the length of the pipeline. More advanced sensorised LDMS include the use of multisensory network, i.e. Wireless Sensor Networks (WSN), which employ various detection techniques and are considered to be more efficient than others [3,4].

However, existing works such as [5] proposed WSN-based LDMS that are centralised, making them susceptible to Single Point of Failures (SPOF). This is in addition to other constraints (energy consumption, robustness, etc.) associated with WSNs accompanied by sensitivity, accuracy, high false alarm rates, etc. [6] related to the used leak detection techniques (LDTs). Together, these challenges might explain the limited adoption of sensorised LDMS for pipelines, as shown in [7].

Thus, in this article, we present HyDiLLEch: a fusion of several detection techniques suitable for resource-constrained networks. HyDiLLEch is aimed at improving the resilience to failures (communication, node, third party interference, etc.) by removing the SPOF associated with centralised systems in a distributed manner. We examine its efficiency in terms of *accuracy* of localisation, *energy consumption*, and *communication overhead* compared to the classical LDTs. The remainder of this article is structured as follows. In Sect. 2, related work is discussed. In Sect. 3, we detail some required hydraulics background. Our contribution is presented in Sect. 4 and discussed in Sect. 5. We conclude and present future works in Sect. 6.

2   Related Work

Sela *et al.* [8] worked on a robust placement of sensors in a pipeline network using robust mixed integer optimisation (RMIO) and robust greedy approximation (RGA) as enhancement of the nominal versions. In most cases, RMIO and MIO outperformed the other versions in the conducted test. [9] worked on fault detection in water pipelines using an approximate solution of the minimum set cover problem by adopting the minimum test cover approach. Both works are based on the assumption that a single sensor can detect failures in multiple pipelines, making them robust to node failures.

Rashid *et al.* [6] used machine learning for leakage detection and classification in pipelines. They compared several machine learning algorithms like the support vector machine (SVM), K-nearest neighbour (KNN) and Gaussian mixture model (GMM), and SVM outperformed the rest in terms of sensitivity, specificity, and accuracy for leak size estimation. They, however, did not consider leakage localisation. Another work based on artificial neural networks (ANN) is that of Roy [10]. He proposed the use of ANN, consisting of input, hidden and output layers as an optimisation tool for accurate leakage detection in pipeline networks. While results from conducted experiments showed improved accuracy in leak detections, ANN may not be suitable for resource-constrained networks like WSNs.

Beushausen *et al.* [11] worked on detecting transient leaks using a statistical pipeline leak detection method through the analysis of the pressure, flow, and modified volume balance of crude oil. Although the system successfully detected some of the transient leaks in the pipeline, localisation error was up to 10 km, amongst other challenges. Authors in [12] proposed a non-intrusive leak detection method for fluid pipelines. They estimated the effects of air bubbles in the proposed system's efficiency and evaluated its accuracy and suitability in detection. Also demonstrated in [5] are two leak detection techniques, i.e. the negative pressure wave method and the gradient method. Experimental results showed that both methods could be used to detect and locate leaks. The gradient-based is more energy efficient due to its low sample rate but has slightly lower accuracy compared to the other method.

Whereas these works have considered various aspects of LDMS, the industrial implementation in the midstream is still in the early stages [7]. Research [13] shows that majority of failures in pipelines are caused by vandalism and third-party interference. We have therefore taken into account such situations to propose a decentralised detection and localisation of leakages to address these challenges. To the best of our knowledge, no work has considered distributed leakage detection based on their causal factors.

3   Leakages in Pipelines

Software LDTs such as the pressure point analysis (PPA), the gradient-based method (GM) and the negative pressure wave method (NPWM), amongst others, have proven to be more efficient than others. They are non-invasive LDTs and easy to implement and deploy on existing infrastructure [5]. Therefore, HyDiLLEch is a combination of these techniques. In this section, we give a short reminder about leakages in pipelines and how these techniques work.

Due to frictional resistance, in a transmission pipeline with a steady state, i.e. presenting no leakage, the pressure decreases with distance. Figure 1 illustrates the Bernoulli's principle, which shows the pressure gradient changes of the crude oil as it travels along a pipeline. P_0 and P_L are the pressures at the inlet and the outlet of the pipeline, respectively in a steady state. P_0 decreases with distance along the pipeline resulting in a relatively constant pressure gradient (PG) for every measurement points. In a leak case, a negative pressure wave (NPW) - travelling in opposite directions from the point of leak - is generated by the occurrence of a leak [5,14]. This results in different pressures P_0^{leak} at the inlet and P_L^{leak} at the outlet of the pipeline compared to P_0 and P_L.

Methods to detect and localise a leakage include PPA, GM and NPWM, amongst others. PPA is a detection technique used in determining dissipation of flow in pipeline fluid implemented by measuring the pressure at various points along the pipeline using the well known Bernoulli equation (Eq. 1).

$$z_a + \frac{P_a}{\rho g} + \frac{V_a^2}{2g} = z_b + \frac{P_b}{\rho g} + \frac{V_b^2}{2g} + E_{ab} \qquad (1)$$

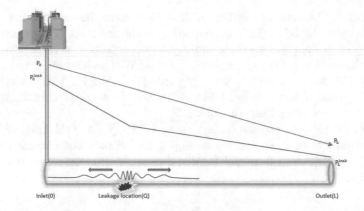

Fig. 1. The pressure gradient distribution in a pipeline before and after leakage

where E_{ab} is the energy head loss, V = velocity, d = Pipeline's inner diameter, L = the distance from point a to point b,

P_a = pressure at point a, ρ = fluid mass density , g = gravitational force and z_a = the elevation at point a.

The GM makes use of the two steady states, i.e. the states before a leak (PG_{0-Q}^{leak}) and after a leak (PG_{Q-L}^{leak}). Given the difference in these pressure gradients as shown in Fig. 1, leakage can be localised with Eq. (2).

$$\overline{Q} = \frac{L \times dPG_{Q-L}^{leak} + (dp_0 - dp_L)}{dPG_{Q-L}^{leak} - dPG_{0-Q}^{leak}} \tag{2}$$

where $\overline{Q}$ = the estimated leak location, L = the pipeline length, dp_0 = average increment in the pipeline's initial cross-section, dp_L = average increment in pipeline final cross-section, $dPG_{0-Q}^{leak}/dPG_{Q-L}^{leak}$ are the average increments in the pressure gradient before/after the leak point.

The NPWM, on the other hand, makes use of the arrival of the NPW front at the sensor nodes. The NPW speed c can be calculated with (Eq. 3). But, the amplitude of the wave attenuates with distance. Thus, (Eq. 4) gives us the maximum detectable distance of the wave front from the point of leakage.

$$c = \frac{1}{\sqrt{\rho(\frac{1}{K} + \frac{d}{Y.w})}} \tag{3}$$

$$A_b = A_a * e^{-\alpha D} \tag{4}$$

where ρ = Fluid density, K = Fluid's modulus of elasticity, d = Pipeline's inside diameter, Y = The Young's modulus and w = Pipeline's wall thickness, A_b = the amplitude at sensing point b, A_a = the initial amplitude i.e. at sensing point a, α = attenuation coefficient and D = the distance between two sensing points.

Each of the techniques mentioned above has some limitations. For example, the accuracy of the GM is tightly coupled with the accuracy of the sensor nodes. The NPWM is dependent on the ability to accurately detect the wave front, which reduces the detecting distance. The NPWM is also less power-efficient than the GM due to its higher sampling rate [5]. Using PPA to detect leakages in transient states and localise leaks is impossible without combining it with other changes resulting from the leak [15].

Therefore, we propose combining these LDTs (PPA, GM and NPWM) by taking advantage of their various strengths to improve the accuracy and sensitivity of leakage detection and localisation while minimising their individual weaknesses.

4   Contributions

We propose a distributed architecture based on several sensors deployed on the pipeline with a mesh connection that enables pre-processing of information among geographically close sensors. This allows distributed detection and localisation of leakages at the sensor level. For the communication aspect, we propose the use of a LPWAN in short and long-range communication and cellular network as a backhaul as they are more widely deployed.

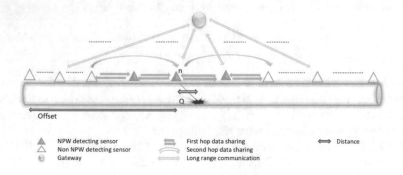

Fig. 2. Detection and localisation of leakages

Based on this architecture, we introduce *HyDiLLEch*, an LDT based on the fusion of PPA, GM, and NPWM. It aims to minimise false positives and improve accuracy through a distributed algorithm by defining appropriate distance between nodes and interconnecting them until 2-hops. This is achieved through the spatial correlation of data shared among geographically close sensor nodes. The data containing pressure and the NPW information is shared with neighbouring sensors in a single-hop or double-hop manner. Detection and localisation are implemented locally, i.e. on the sensors using this *partial information* shared amongst the nodes.

Propagation-Based Node Placement: Some of the factors that affect the performance and efficiency of an LDMS is the placement of sensors on a pipeline. Several approaches exist based on the maximum communication range [16], the definition of the shortest distance between the event and a sensor [8,9], or the placement of nodes only at the key junctions [1]. By our side, we assume that leakage events are stochastic in nature, cannot be pre-determined and may occur at any point in the pipeline. Also, researches [2,13] shows that over 90% of leakages are caused by third party interference and vandalism, where the mean value of the failure rate is 0.351 per km-year. Hence, we propose the deployment of several sensors along a pipeline segment with two constraints :

- The distance between sensors must be less than half the sensor's maximum communication range to ensure a two-hop interconnection.
- The distance between sensors and event source must be sufficiently small to make small leakage events detectable by at least three nodes considering the amplitude of the NPW (Eq. 4). This is to ensure detection and localisation in the event of node failure.

In this work, we consider a linear and uniform deployment of sensors on the pipeline. We limit the number of collaborations among the sensors to a maximum of 2-hops. This allows us to minimise the various interferences resulting from such collaborations and optimising the energy consumption for all the sensors. Therefore, 2-hops data sharing represents additional information from two upstream nodes and two downstream nodes (Fig. 2).

Leakage Detection: In HyDiLLEch, leakage detection is implemented by making use of PPA to pre-estimate the expected pressure at every sensor node location. According to Eq. (1), we can find the pressure gradient PG_{0-L} for a pipeline. Thus, the pressure gradient in a steady state at every point for a horizontal pipeline can be estimated as follows:

$$PG_{0-L} = (\frac{P_0}{\rho g} - \frac{P_L}{\rho g})\frac{1}{L} \tag{5}$$

A threshold is set to accommodate the difference between the real value and possible calibration errors from the sensor readings. The first step in detecting leakage is by comparing the sensed pressure and the pre-estimated pressure threshold. A difference between these values represents a possible leakage. Then we check the pressure values to determine the pressure gradient on both side of the current node i ($PG_{(i-1)-(i)}$, $PG_{(i)-(i-1)}$) and finally, we check for the arrival of the NPW front. Thus, we can correctly detect the presence of leakages and reduce false positives.

Leakage Localisation: Each node detecting a leakage does its localisation estimation by two methods. The first one is GM, using pressure gradients and Eq. (2), while the second is considering the time of arrival of the NPW front [17]. It can be calculated by:

$$\overline{Q} = node's\ offset + c \times \delta t \tag{6}$$

where c is the negative wave speed estimated by Eq. (3), δt the difference in arrival time of the signal at the upstream and downstream nodes and *node's offset* is the distance between the nodes multiplied by the detecting node's index.

Algorithm 1. HyDiLLEch (Double-Hop)

1: {Initialisation while no leakage}
2: Form Neighbourhood
3: Get PG in steady state
4: Set upper, and lower threshold values for PG
5: **for ever do**
6:     Get pressure data from neighbours
7:     Calculate local gradients $PG_{(i-1)-(i)}$, $PG_{(i-2)-(i)}$, $PG_{(i)-(i+1)}$ and $PG_{(i)-(i+2)}$

8:     Detect NPW
9:     **if** local PG is outside the threshold *and* NPW detected, share NPW data **then**
10:        Try to localise using GM data (Eq. 2)
11:        Try to localise using NPW data (Eq. 6)
12:     **else**
13:        No leak detected
14:     **end if**
15: **end for**

Finally, HyDiLLEch is defined with a neighbourhood of 1-hop (2 nodes) or 2-hops (4 nodes) to deal with a possible node failure - respectively referred to as HyDiLLEch-1 and HyDiLLEch-2. Algorithm 1 shows the main structure of the 2-hops version.

Both versions result in an increment in the number of sensor node detecting and localising leakages (NDL) with relatively high localisation accuracy. The results of the simulation are discussed next in the following section.

5   Simulations and Results

Metrics used to determine the efficiency of LDTs include *accuracy* of leakage localisation, *sensitivity* to leaks, amongst others. In this work, we focus on these two metrics to determine the efficiency of HyDiLLEch compared to the classical LDTs, including the energy consumption and communication overhead due to the distributed approach of our LDT.

Using NS3, we simulate crude oil propagation and leakages on a horizontal transmission pipeline for a single-phase laminar flow within industrially-defined operational criteria. The pipeline properties, such as the length, material, etc., are those of existing long haul transmission pipelines sourced from the Department of Petroleum Resources (DPR) of Nigeria. Tables 1 and 2 outlines the simulation parameters.

Table 1. Pipeline and oil characteristics

Material	Carbon steel
Pipeline Length (L)	20 km
Wall thickness (w)	0.323 m
Inside diameter (d)	0.61 m
Height/elevation (z)	0 m
Oil kinetic viscosity	2.90 mm^2/s
Temperature	50 °C
Oil density (ρ)	837 kg/m^3
Inlet pressure (P_0)	1000 psi
Reynolds no (R_e)	1950
Velocity (V)	2 m/s
Molecular Mass (m)	229
Oil elasticity (K)	$1.85 \times 10^5 psi$
Carbon steel elasticity (Y)	$3 \times 10^6 psi$
Gravitational force (g)	9.81 m/s^2
Constant (e)	2.718
Coefficient of friction (λ)	0.033
Wave speed (c)	14.1 m/s

Table 2. Network simulation parameters

Number of sensors	21
Number of gateways	1
PHY/MAC model	802.11ax Ad hoc
Transmit power	80 dBm
Transmit distance	20 km
Error model	YANS
Propagation Loss	Log-distance
Path Loss (L_0)	46.67 dB
Reference distance (d_0)	1 m
Path-Loss Exponent (σ)	3.0
Packet size	32 bytes
Data rate	1 Kbps
Distance between sensors	1 Km

These parameters enable us to calculate the values of the NPW speed using Eq. (3). Expected pressure (P_L) at the pipeline's outlet in a steady state is firstly estimated by the Bernoulli equation defined in Eq. (1). The resulting value is then used to determine the pressure gradient using Eq. (5). As a preliminary work, all tests are conducted in ideal conditions, i.e. no communication or node failures and results are discussed in the following subsections.

5.1 Localisation Accuracy

In our first simulations, we implemented two LDTs (NPWM and GM) in a centralised way [18]. While they both localise the leakage at an average of 99% and 98%, respectively, with different variance, the centralised detection and localisation make them susceptible to SPOF and less robust to other types of failures, i.e. communication failure and others outlined in Sect. 3. In this section, we explain how HyDiLLEch performs relative to this drawback.

Both single-hop (HyDiLLEch-1) and double-hop (HyDiLLEch-2) versions of HyDiLLEch were simulated, and the results are represented in Fig. 3 and 4 respectively. The number of NDL are *four* for HyDiLLEch-1 and *six* for HyDiLLEch-2, a significant increment compared to *one gateway* in the centralised version. These sensors, which are represented as N_{1-6} in Fig. 3 and Fig. 4, make use of the spatially correlated data (pressure gradient and negative pressure wave arrival) received from neighbouring sensors.

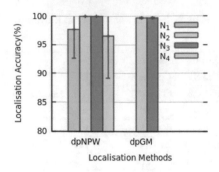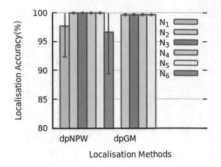

Fig. 3. Single-hop HyDiLLEch   **Fig. 4.** Double-hop HyDiLLEch

Thus, in HyDiLLEch-1, the average localisation accuracy on all four nodes is above 96%, with nodes N_2 and N_3 maintaining the highest accuracy of about 99%. HyDiLLEch-2 has similar results with two additional NDLs. Nodes physically-close to the leak -N_2, N_3, N_4, N_5- are also maintaining the highest accuracy of about 99%.

Note that from Figs. 3 and 4, we have several nodes localising using different principal information (dpGM or dpNPW). The hybrid method is kept loosely coupled to allow robustness of detection and to improve fault tolerance in cases where one or more performance influencing factors such as noise, communication failure, node failures, etc. is particularly detrimental to one of the localisation techniques. The choice of the most accurate estimated leak location will be made through a consensus algorithm to be implemented in the data and service layer in our future work.

5.2   Communication Efficiency

To compare the communication efficiency of the LDTs, we consider the cost of communication by the number of packets used by each LDT. We also evaluate their energy consumption based on the sampling rate and the radio energy consumption of the net devices. Obtained results are discussed next.

Communication Overhead: Figure 5 shows the number of packets exchanged between the sensors and the gateway (for NPWM and GM-based on the total number of packets needed to localise) and amongst the sensors for HyDiLLEch-1 and HyDiLLEch-2. The GM has the lowest number of exchanged packets compared to NPWM and the two versions of HyDiLLEch. As expected, data correlation among neighbours increased the overhead in the number of packets in HyDiLLEch. Thus, HyDiLLEch-1 relative to GM has an overhead of about 50% with a corresponding 200% increment in the number of NDL. HyDiLLEch-2 has approximately 60% overhead with a 400% increment in the number of NDL as well. On the contrary, both versions of HyDiLLEch requires a lower number of packets when compared to NPWM packets with an even higher increment of 400% and 600% in the number of NDL respectively.

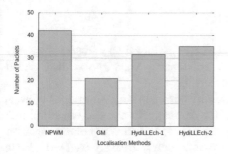

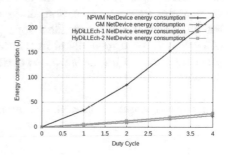

Fig. 5. Communication overhead by number of packets

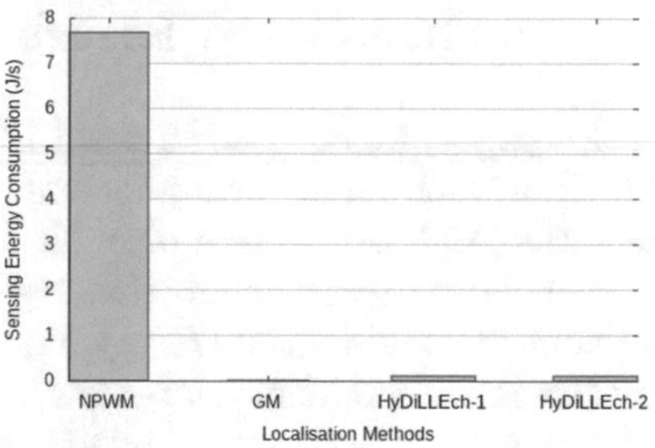

Fig. 6. Sampling energy consumption of the sensors

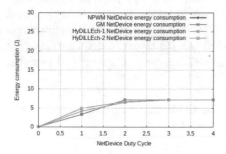

Fig. 7. Radio energy consumption

Fig. 8. Cumulative energy consumption

Energy Consumption: We evaluated the energy consumption of each LDT based on *the sampling rate* of the sensors and *radio energy consumption* by the network duty cycle. Results shown in Fig. 6 reveals the energy consumed per second by the sensors. The high sampling rate of NPWM results in more energy consumption compared to both GM and HyDiLLEch. Combining both techniques in HyDiLLEch allows first determining the offset region before enhancing the location accuracy by the expected time of arrival of the NPW. Thus, we need less than 2% of NPWM sampling rate, allowing us to take advantage of the low energy consumption of GM while maintaining the high localisation accuracy of NPWM.

In the radio energy consumption case, we analyse the rate of change of the energy consumed by the netdevice for every duty cycle. From Fig. 7, the most significant change can be seen during the first and second duty cycles. This is due to the difference in connections required among the participating nodes by each LDT. Once connections are established and as the cycle increases, the energy consumption converges and becomes stable for each of the methods. Figure 8, on the other hand, shows the cumulative energy consumption of all the methods. Energy consumption compared to NPWM is reduced by approximately 86% and 83% for HyDiLLEch version 1 and 2, respectively, in the first duty cycle, and increase by 6–7% compared to GM per NDL. Both versions show a more linear rate of change in consumption as the cycle increases, which is very similar to that of GM.

6   Conclusion and Future Works

In this paper, we proposed a new LDT –HyDiLLEch– based on a fusion of several detection techniques using a node placement strategy that allows distributed detection and localisation of small to big-sized leakages.

Simulation results show that HyDiLLEch in both the single and double-hops versions increased the NDLs by four to six times, respectively, thereby eliminating the SPOF problem related to the classical LDTs. In terms of communication efficiency, both versions of HyDiLLEch performed comparatively well with little to no communication overhead. We also showed that the global energy consumption of HyDiLLEch is relatively low compared to NPWM and similar to GM. In our future work, we will examine the fault tolerance and resilience of HyDiLLEch by introducing communication, node, etc. failures to the system. The choice of the most accurate estimated leak location will be made through an asynchronous consensus to be implemented in the data and service layers. Once the data and service layers are implemented, experiments will be carried out on real pipeline networks.

Acknowledgment. This work was supported by the Petroleum Technology Development Fund.

References

1. Slaughter, A., Bean, G., Mittal, A.: Connected barrels: Transforming oil and gas strategies with the Internet of Things. Technical Report Deloitte Center for Energy Solutions (2015)
2. SPDCN, "Security, theft, sabotage and spills," Shell Petroleum Development Company of Nigeria Limited, Technical Report 2017
3. Henry, N.F., Henry, O.N.: Wireless sensor networks based pipeline vandalisation and oil spillage monitoring and detection: Main benefits for Nigeria oil and gas sectors. In: The SIJ Transactions on Computer Science Engineering & its Applications (CSEA), vol. 3, no. 1 (January 2015)
4. Yu, H., Guo, M.: An efficient oil and gas pipeline monitoring systems based on wireless sensor networks. In: International Conference on Information Security and Intelligent Control, pp. 178–181 (August 2012)
5. Ostapkowicz, P.: Leak detection in liquid transmission pipelines using simplified pressure analysis techniques employing a minimum of standard and non-standard measuring devices. Eng. Struct. **113**, 194–205 (2016)
6. Rashid, S., Akram, U., Qaisar, S., Khan, S.A., Felemban, E.: Wireless sensor network for distributed event detection based on machine learning. In: 2014 IEEE International Conference on Internet of Things (iThings), and IEEE Green Computing and Communications (GreenCom) and IEEE Cyber, Physical and Social Computing (CPSCom), pp. 540–545 (September 2014)
7. Slaughter, A., Mittal, A., Bansal, V.: Bringing the digital revolution to midstream oil and gas. Technical Report (2018)
8. Sela, L., Amin, S.: Robust sensor placement for pipeline monitoring: mixed integer and greedy optimization. Adv. Eng. Inf. **36**, 55–63 (2018)

9. Perelman, L.S., Abbas, W., Koutsoukos, X., Amin, S.: Sensor placement for fault location identification in water networks: a minimum test approach. Automatica **72**, 166–176 (2016)
10. Roy, U.: Leak detection in pipe networks using hybrid ANN method. Water Conserv. Sci. Eng. **2**(4), 145–152 (2017)
11. Beushausen, R., Tornow, S., Borchers, H., Murphy, K., Zhang, J.: Transient leak detection in crude oil pipelines. In: International Pipeline Conference (2004)
12. Santos, A., Younis, M.: A sensor network for non-intrusive and efficient leak detection in long pipelines. In: 2011 IFIP Wireless Days (WD), pp. 1–6 (October 2011)
13. Ambituuni, A., Hopkins, P., Amezaga, J.M., Werner, D., Wood, J.M.: Risk Assessment Of A Petroleum Product Pipeline In Nigeria: The Realities Of Managing Problems Of Theft/sabotage (2015)
14. Sowinski, J., Dziubinski, M.: Analysis of the impact of pump system control on pressure gradients during emergency leaks in pipelines. In: E3S Web of Conferences, vol. 44, p. 00166 (2018)
15. Sheltami, T.R., Bala, A., Shakshuki, E.M.: Wireless sensor networks for leak detection in pipelines: a survey. J. Ambient Intell. Humanised Comput. **7**, 347–356 (2016)
16. Guo, Y., Kong, F., Zhu, D., Tosun, A.S., Deng, Q.: Sensor placement for lifetime maximization in monitoring oil pipelines. In: Proceedings of the 1st ACM/IEEE International Conference on Cyber-Physical Systems, ICCPS 2010 (2010)
17. Wan, J., Yu, Y., Wu, Y., Feng, R., Yu, N.: Hierarchical leak detection and localization method in natural gas pipeline monitoring sensor networks. Sensors (Basel) **12**(1), 189–214 (2012)
18. Ahmed, S., Mouël, F.L., Stouls, N.: Resilient IoT-based monitoring system for crude oil Pipelines. In: Proceedings of the 7th International Conference on Internet of Things: Systems, Management and Security (IOTSMS), IEEE (2020)

Tactful Opportunistic Forwarding: What Human Routines and Cooperation Can Improve?

Rafael Lima Costa[1,2,3]([✉]), Aline Carneiro Viana[3], Artur Ziviani[4],
and Leobino Nascimento Sampaio[2]

[1] École Polytechnique/IPP, Palaiseau, France
[2] Federal University of Bahia (UFBA), Salvador, Brazil
{rlimacosta,leobino}@ufba.br
[3] Inria, Palaiseau, France
aline.viana@inria.fr
[4] National Laboratory for Scientific Computing (LNCC), Petrópolis, Brazil
ziviani@lncc.br

Abstract. Opportunistic D2D forwarding algorithms have leveraged human mobility characteristics to improve cost-effective content delivery. Most previous proposals focused on traditional or simplistic human-centered metrics to improve performance in scenarios such as cellular data offloading. Still, there is a need to approximate algorithm's metrics to inherent in-depth aspects of human mobility hidden into real datasets while leveraging more realistic scenarios interesting to mobile operators. This work proposes TOOTS, a novel *human-aware opportunistic D2D forwarding strategy* for cost-effective content delivery on cellular networks. TOOTS features a dissemination policy and a forwarding algorithm that *leverages wireless encounter patterns, temporal, spatial, geographic, and direction awareness* to improve cost-effectiveness delivery. These characteristics are extracted from NCCU and GRM datasets. We compare TOOTS with the most popular state-of-art social-aware algorithm, Bubble Rap, combined with three dissemination policies. The TOOTS evaluation results show increased performance in terms of delivery rate, delivery latency, and overhead.

Keywords: Human-centered computing · Social computing · Network architectures · Device-to-Device (D2D) forwarding

1 Introduction

The advent of 5G wireless networks will drive new business models in nearly every vertical industry. Device-to-Device (D2D) communication and opportunistic forwarding are considered enablers of 5G [1] and 6G [19] network communication and, consequently, emerging applications. They represent a new paradigm

L. Barolli et al. (Eds.): AINA 2021, LNNS 225, pp. 638–652, 2021.
https://doi.org/10.1007/978-3-030-75100-5_55

to offload traffic from wireless communication networks, which have attracted multiple research initiatives involving user participation [13]. D2D opportunistic forwarding algorithms tackled cost-effective and timely delivery of data [5,9]: that is, delivering as much content as possible with less overhead and delay. In these scenarios, contents (or messages) are forwarded user-to-user, from source to destination in an opportunistic fashion (i.e., relying on user devices' intermittent connectivity).

Previous research discussed opportunistic network deployment to deal with networking performance challenges, such as traffic growth (e.g., through data offloading). Most initiatives focused on proposing algorithms and evaluating their metrics. However, they still lack more evaluations with real traces to show the benefits of using opportunistic communication in real-world applications (i.e., in scenarios with more realistic settings). These algorithms typically consider user encounters and individual mobility [5,9,10], points of interest (PoIs) [9], and time-evolving social ties between node pairs [8]. Apart from that, not much was done to approximate the evaluation metrics to broader inherent aspects of human mobility while targeting QoE and QoS [2]. There is thus a lack of initiatives beyond traditional techniques or limited human-mobility features to identify routines (spatiotemporal patterns), related consequences (e.g., wireless encounters), and movement decisions (e.g., motion direction) with more granularity and precision. This work focuses on metrics able to reflect human mobility features during periods of the day. Still, each population has particular habits that can change wireless contact dynamics [4]. Therefore, such metrics must be evaluated throughout different populations, while spatiotemporal aspects must fit each population's habits. In this work, we combine features extracted from a real-world and a synthetic dataset and apply them into a novel Tactful Opportunistic cOmmunicaTion Strategy (TOOTS). This strategy features a *dissemination policy and a forwarding algorithm*. In the former, we leverage users' spatiotemporal properties and induced wireless encounters to choose content-disseminating nodes that have shown previous encounter routines with destination nodes. The forwarding algorithm relies on nodes' popularity, direction and displacement, network-cell (as PoIs) visiting, and proximity.

In [2], we presented and detailed the features of TOOTS, including mathematical formulations and results from metrics characterizations using MACACO (a private European Dataset) [17]. In [2], we discussed mobility datasets handling and characterization. The same methodology is applied here with different datasets. Conversely, the word *tactful*, which is part of the strategy's name, means having or showing skill and sensitivity in handling with people. In [3], we discuss the Tactful Networking paradigm, whose goal is to add perceptive senses to wireless networks by assigning it with human-like capabilities of observation, interpretation, and reaction to daily-life features and associated entities. Our contributions with TOOTS are thus the following:

- The TOOTS design combining human-aware metrics in a more granular and precise way than state-of-art to improve networking and system performance;

- The comparison of the TOOTS effectiveness in each stage using a real-world and a synthetic dataset with state-of-art enhanced Epidemic and Store-wait-forward, and Bubble Rap alternatives;
- TOOTS reaches 100% delivery rate with respectively 28%, and 73% less delivery latency, and with 16%, and 27% less overhead in the real-world and synthetic datasets.
- Sharing the notion of going beyond traditional techniques when dealing with human-aware metrics to reach a superior system and networking performance. We also emphasize the need to evaluate such opportunistic strategies through more realistic scenarios and real datasets, incentivizing mobile network operators to deploy opportunistic networks more widely.

The results mentioned above cope with [3], which talks about the need to have a more in-depth look into human characteristics to build tactful networking solutions. The remainder of this work is organized as follows. Section 2 discusses related work. Section 3 presents the mobility models (datasets) used and motivates the strategy design. Section 4 discusses TOOTS, including its architecture, features, and algorithms for content-dissemination and forwarding. Section 5 presents the experimentation results, analysis, and discussions. Section 6 concludes this paper and points out future work.

2   Related Work

Over the years, different works extracted user mobility metrics to assist routing algorithms and opportunistic forwarding strategies. Due to the dynamic human mobility, choosing the nodes to forward the content is challenging.

In [5], the authors introduced Bubble Rap, a distributed forwarding algorithm for delay-tolerant networks. The algorithm exploits centrality and community detection, characteristics remarked as less volatile than user mobility. Bubble Rap outperformed other state-of-art algorithms and is used as a benchmark with recently published research [9,14]. Despite its broad contributions, the major shortcomings relate to small scenarios (available traces had less than 100 nodes) evaluations. Further, community detection algorithms are computationally expensive and require parameters calibration, which is not feasible in real-time D2D scenarios.

In [8], the authors propose dLife, a forwarding algorithm that relies on contact duration and degree during daily hour periods. They capture the social behavior and interactions dynamism, resulting in improved delivery rate, cost (i.e., overhead), and latency. Like Bubble Rap, dLife was simulated through small traces. The evaluation setup and scenario also had shortcomings (e.g., content size ranging from 1 to 100 kb).

GROUPS-NET, an opportunistic algorithm that relies on social group meetings, is introduced in [10]. This algorithm requires a centralized computation of group-to-group paths whenever a given device wishes to send content to a destination. This requirement is challenging in larger populations. Furthermore, GROUPS-NET is validated within two campus datasets, lacking other kinds of environment evaluations. In large-scale scenarios, they had a similar delivery ratio as Bubble Rap but with reduced overhead.

In [9], the algorithm SAMPLER combines individual mobility, PoIs, and social-awareness for opportunistic routing. Despite showing improved delivery ratio, reduced overhead, and delivery latency, this work has a few shortcomings. The algorithm requires static relay points deployment and social community computations, which require parameters calibration. Second, there is a lack of details about the simulation scenario and parameters to turn possible reproducibility for comparison reasons. Finally, the evaluations need more realistic settings. For example, the authors considered any contact as enough to forward a message. Other parameters, such as message time-to-live, node buffer size, and more realistic content-size, must be evaluated.

TOOTS Positioning: From the related initiatives described herein, we consider that understanding human-behavior and context from mobility is important for networking solutions. Different from such studies, we use metrics with a more granular link with time. We focused on identifying different human activities during periods of the day. Through analyzing the datasets and our metrics, we found that their coefficients vary according to the period [2]. Considered human-mobility features are routines (spatiotemporal patterns), related consequences (e.g., wireless encounters), and movement decisions (e.g., motion direction). Metrics and insights based on generic aspects might not apply to all populations, so datasets of different kinds (i.e., different environments) and more users need to be evaluated. Facets such as node density, proximity, transportation mode, and even cultural factors can change contact dynamics. Therefore, it is necessary to choose the temporal aspects and other characteristics accordingly.

3   Rationale

Datasets Description: We use the real-world NCCU trace [16], which features the displacements (given by GPS coordinates) of 115 users on a campus (3764 m × 3420 m) throughout two weeks. Additionally, we adopted the synthetic GRM dataset, which captures social regularity by human mobility [11]. The GRM dataset has 1000 users displacements in a 1500 m x 1500 m area throughout two weeks and applies here to simulate a larger network (i.e., with more nodes).

Overhead and Latency Evaluation: In Fig. 1, we evaluate the Epidemic forwarding with a consumer set size equal to 1% of each dataset amount of nodes. Note that 1% represents a limited number of consumers, which puts the Epidemic strategy in a challenging scenario. Depending on datasets' characteristics and the strategy's kind, this can result in lower bound results for dissemination strategies, which will also be used in the TOOTS evaluation hereafter. Although mostly evaluated in the literature, we illustrate the performance constraints related to the Epidemic forwarding algorithm using the two mentioned datasets. The Epidemic algorithm always tries forwarding a content to an encountered node that does not have it. Epidemic forwarding is known as having smaller delays and a higher delivery ratio; however, it brings a heavy network overhead burden.

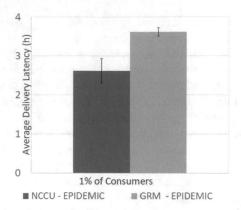

Fig. 1. Epidemic forwarding average delivery latency performance evaluation.

Source and destination nodes were chosen randomly with a content generation per hour during one full day. The experiments were carried 30 times, and so the confidence intervals are presented.

The Epidemic algorithm delivers the contents with an average latency of 2.6 and 3.6 h, respectively, for NCCU and GRM. The delivery-rate was 100% on those experiments, but with 98.44% and 99.62% average network "infection" (overhead). The latter means that most network nodes became forwarders. For reaching such a delivery rate and small delivery latency, the network was flooded, which is not feasible in the real-world and is highly costly. Furthermore, one of the Epidemic Forwarding characteristics is a higher average of hops (i.e., intermediary nodes) for delivering the message. For example, there are 6.62 and 8.85 average hops in these experiments for content delivery on NCCU and GRM. This characteristic brought an insight and motivation for our strategy: taking the content closer to the destination could reduce hops, delay, and overhead by combining rational dissemination with intelligent forwarding decisions. In the next section, we introduce TOOTS.

4   TOOTS Architecture and Strategy Description

TOOTS Architecture: Figure 2 describes the architecture for TOOTS. It features a mobile network owned by a cellular operator with an infrastructure of mobile edge computing (MEC) [7]. In this MEC site, the operator stores contents commonly requested by its subscribers. The operator's goal is to offload content through opportunistic D2D communication, assuming the users are willing to cooperate through incentives (e.g., data plan savings). We assume that the operator controls the offloading process while the forwarding decisions happen in the user devices based on locally calculated metrics. TOOTS learns from users' mobility for seven days. Upon having a set of user requests for a given content, the operator runs our Tactful Dissemination Policy (TDP) (Sect. 4.1) to choose a set of disseminator nodes. In Fig. 2, we illustrate the offloading of two different contents through the chosen disseminators. Subsequently, an algorithm (Sect. 4.2) runs locally (in the node) for taking a forwarding decision.

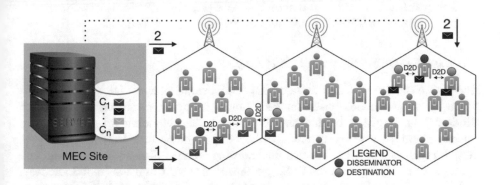

Fig. 2. TOOTS architecture.

TOOTS Metrics: TOOTS combines a novel temporal approach described below with other insights into a Tactful Dissemination Policy (TDP)(Sect. 4.1); the other metrics herein described are combined into a forwarding algorithm (Sect. 4.2). Following, we resume these metrics, which are detailed (including mathematical formulations) and evaluated in [2]. We discuss the intuitions behind the use of each metric in the descriptions of the algorithms.

- **Temporal Approach**: We divided the day into six non-uniform periods of different durations. This decision is justified by associating the periods to the instants the population travels and has longer confinements (e.g., the period from 06:00 to 09:59 accounts for most individuals' home-to-work displacement, while from 10:00 to 13:59 there is more confinement at work and smaller displacements on lunchtime). With this temporal division, the intention is to extract decision factors increasingly accurate and closer to the real human routines. In [2], we found metrics coefficients heterogeneity per period in MACACO Dataset. The same phenomenon was observed here with NCCU and GRM, which reinforces our intuition.
- **Social Awareness Centrality Degree** $(C_{D_p}(u))$: Measures the social bonds of a user u in a period of the day p, that is, his number of encounters given a specific communication range. Someone with a higher degree is more "popular" (i.e., has further encounters).
- **Coverage Area Radius of Gyration** $(R_{G_p}(u))$: Quantifies in meters a user (u) individual mobility related to a center of mass, calculated from his movements in a period (p).
- **Sojourn Time** $(ST_p^c(u))$: Quantifies in hours the stay of a user u in a network cell c in a period p. We divided [2] the limited geographic space from each dataset into cells (original local operators' cells were too large) and calculated users' stay.

- **Destination Proximity as Geographical Awareness** ($MP_p^c(u)$): We calculate the maximum proximity a node u reached towards a cell c in a period p through a geodesic formula.
- **Geographic Direction Awareness**: Analyzes the last 30 min of node mobility to determine if its displacement is towards a given cell.

4.1   1st TOOTS Phase: TDP Policy

Herein, we detail our Tactful-Based Dissemination Policy (TDP). TDP (Algorithm 1) chooses nodes based on their social behavior according to each time window. The intuition behind this is to start the offloading process with the content closer to their consumers. Following, we show how we apply the predictive regularity of user routines for choosing content disseminators. From the dynamic contact graph $G_t = (V, E_p)$, where V is the set of users (mobile nodes), and E_p is the set of edges found during the period p of the week k. We assume that there is a logically centralized operator entity that, at the end of a week k, receives the following information from each one of its nodes $u \in V$:

- **(a)**: $\sigma(u)_p$ - the set of users v encountered by u in each period p.
- **(b)**: $\Delta_{CLID_p}(u)$ - the average local improved centrality degree metric [15] for a user u in a period p. This metric accounts for the number of contacts (i.e., encounters), considering their duration and earliness. With that said, higher coefficients belong to nodes with more contacts that had longer durations and happened earlier in a time window (i.e., in a period p).

Algorithm 1: SelectDisseminators

> **input** : G_t, C, c, p
> **output**: D(c)
> 1 **begin**
> 2 $D(c) \longleftarrow \emptyset$
> 3   **while** $C \neq \emptyset$ **do**
> 4   let $u \in V$ maximizing $|\dfrac{|\sigma_p(u) \cap C|}{|C|} + \Delta_{CLID_p}(u)|$
> 5 $D(c) \longleftarrow u$
> 6 $C \longleftarrow C - \sigma(u)$
> 7   **end**
> 8   Return $D(c)$
> 9 **end**

During the week $k+1$, upon having a set of consumers $C(c) \in V$ interested in a content c, Algorithm 1 runs for choosing disseminator nodes. We justify the use of $\sigma(u)_p$ obtained at the week k by the fact that the policy tries to select nodes that had direct contact with the consumers, and due to their routines, they are most likely to repeat those interactions [12]. We prioritize the direct contacts, i.e., higher proximity, as they are most likely to repeat, possibly minimizing the spending of network and user device resources. Further, the C_{LID} measures neighborhood coverage capabilities through contacts and is used to identify the nodes' popularity in the network, giving importance to their contacts' duration and earliness. We use this metric because, in an opportunistic communication scenario, a contact that can transmit a message (i.e., the encounter lasts enough to send the data) earlier in the time window might decrease the total delivery latency. Furthermore, depending on the content size, choosing short contacts can waste node resources without proper message transmission.

Algorithm 1 selects a set of disseminators $D(c)$ for offloading a content c upon period p. $D(c)$ starts empty as the algorithm runs when there is a set of Consumers C requisitions for a content c. The policy selects (line 4) the user with the higher coefficient for the direct contacts with Consumers $(\sigma_p(u) \cap C)$, normalized by C's size, summed with the $\Delta_{C_{LID_p}}(u)$ (u's average C_{LID} in p). The algorithm then removes the Consumers covered by u (line 6). In this way, we guarantee a next user u with a similar coefficient for direct contacts do not get selected. Algorithm 1 loops (line 3) until the consumer set is not fully covered ($C \neq \emptyset$). The policy assumes that there are no isolated users in the traces. In Sect. 5, we evaluate the performance of the proposed dissemination policy with different sizes of consumer set (1%, 5%, and 10% of each dataset amount of nodes).

4.2 2nd TOOTS Phase: Human-Aware Forwarding

After choosing the disseminator set $D(c)$ for a content c, the operator sends the content to each node $u \in D(c)$, for storing in its local buffer. From this moment, upon an encounter, any node carrying c runs Algorithm 2 to decide on the next forwarding neighbor. The input is: the destination node d (that is, $d \in C$), the content c, the period p, the encountered node v, and l as the coordinates of d's cell. Each node stores the metrics coefficients locally by period p during the whole simulation, and a node u transmits c to v only if a certain algorithm condition is satisfied. First, if v already has c (line 2), u waits for the next encounter. Second, if $v = d$ (line 4), that is, v is the destination, c is transmitted. The content transmission is successful if the edge e between u and v is persistent enough (there is no topology change causing a disconnection) until the time t necessary to transfer c.

As explained in Sect. 4, the geographic space was divided into cells. If v is at l (line 6), the nodes avg. Centrality degree in the period p are compared. If v has a higher avg. CD, it means v met more nodes during p, so c is forwarded to v, given his higher dissemination capacity inside that cell. In the contrary case (line 9), if $\Delta_{RG_p}(v) > \Delta_{RG_p}(u)$ and $\Delta_{ST_p^{cell}}(v) > \Delta_{ST_p^{cell}}(u)$, v is less "popular" (lower $\Delta_{C_{D_p}}$) than u. Nevertheless, v can potentially cover a larger area inside the cell l (given his higher RG) and routinely stays longer in l (given his higher avg. ST). Hence, c is then forwarded to v (line 10). If v is not in l, his recent mobility towards l is checked. If the direction test is true (line 12), i.e., v moves

---

Algorithm 2: HumanAwareForwarding

> **input** : d, c, p, v, l
> **output**: A forwarding Decision
> 1 **begin**
> 2   **if** $c \in v$ **then**
> 3   exit(0)
> 4   **else if** $v = d$ **then**
> 5 $v \longleftarrow c$
> 6   **else if** $v.cell = l$ **then**
> 7   **if** $\Delta_{C_{D_p}}(v) > \Delta_{C_{D_p}}(u)$ **then**
> 8 $v \longleftarrow c$
> 9   **else if** $\Delta_{R_{G_p}}(v) > \Delta_{R_{G_p}}(u)$ *and* $\Delta_{ST_p^l}(v) > \Delta_{ST_p^l}(u)$ **then**
> 10 $v \longleftarrow c$
> 11   **else**
> 12   **if** $testDirection(v, l)$ **then**
> 13 $v \longleftarrow c$
> 14   **else if** $\Delta_{MP_p^l(v)} < \Delta_{MP_p^l(u)}$ **then**
> 15 $v \longleftarrow c$
> 16   **end**
> 17 **end**

---

towards the aim and has the potential to reach or to get closer to l, c is sent to v. Finally, if the direction test is false, the Alg. checks the avg. Destination proximity. If $\Delta_{MP_p^l(v)} < \Delta_{MP_p^l(u)}$, c is forwarded to v, meaning that node v got closer (or visited) l during p and it is more likely to repeat this behavior.

5 Experimentation Results and Analysis

As stated previously, the operator's goal is to offload content through opportunistic communication. The contents consist of 60 s advertising videos with the average size of a YouTube 720P HD 30fps video, ranging from 11–14 MB. TOOTS learns from users mobility for seven days; one day is for content generation, with an offloading task started by the operator every hour, and three days are the delivery deadline [15]. The consumers of each content are chosen randomly, with different consumer set sizes. The operator injects the contents on the network through their subscribers, who run a forwarding algorithm upon each contact. Every node has an 802.11/11 Mbps network interface. The communication range is 30 m (avg. for WiFi Direct). This scenario simulation is through the Opportunistic Network Environment (ONE) Simulator [6], with NCCU and GRM imported as mobility models. Each experiment was carried 30 times. The confidence intervals appear when necessary.

5.1   TDP Results and Analysis

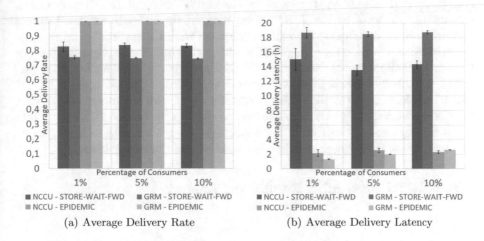

(a) Average Delivery Rate   (b) Average Delivery Latency

Fig. 3. Tactful dissemination policy average delivery rate and average delivery latency performance evaluation.

This section evaluates the TDP policy only. For this, we modify two literature forwarding strategies, named store-wait-forward (a.k.a., direct delivery) and epidemic forwarding. This modification consists of using the dissemination nodes given by the TDP policy as the starting content sources, executing then each of the literature strategies. Figure 3 plots the performance of the TDP-enhanced store-wait-forward and epidemic forwarding. The store-wait-forward transmits the message only if the encountered node is the destination (consumer). In both traditional and TDP-enhanced versions, the store-wait-forward has zero overhead. The traditional epidemic had 98.44%, and 99.62% overhead, as shown in Sect. 3. On the other hand, the TDP-enhanced epidemic might get a comparable delivery rate and decreased delay. TDP-enhanced store-wait-forwarding has a potentially higher delivery rate and decreased delivery latency, as its success, though still depending on direct encounters, is improved by the TDP's initial content dissemination. Improvements might still take time (or never happen) depending on the source and destination social bonds.

Figure 3(a) reveals that combined with store-wait-forward, TDP induces an average delivery rate of 82% on NCCU and 75% on GRM, regardless of the number of consumers evaluated (1%, 5%, and 10%). TDP enhancement assures choosing nodes with past direct encounters with the consumers, which favors the delivery rate increase. Compared with the traditional store-wait-forward, the TDP-enhanced average delivery rate with 1% of consumers was 18% higher on NCCU and 60% higher on GRM. In Fig. 3(a), as expected, the TDP-enhanced epidemic forwarding reaches a 100% average delivery rate in all scenarios. Besides, the TDP enhancement favors the decrease of the average hop count of the traditional Epidemic forwarding.

In Fig. 3(b), we evaluate the average delivery latency. We see that regardless of the size of consumers set, the contents take on average 13h–15h to be delivered with TDP-enhanced store-wait-forward on NCCU and 18h–19h on GRM, respectively. Thanks to the TDP policy, most of the delay-tolerant content is forwarded in an acceptable time [15]. Compared with the traditional store-wait-forward with 1% of consumers, TDP-enhanced reduced the average delivery latency by 23% on NCCU and 48% on GRM. As expected, when the forwarding is the enhanced epidemic, the average delay is smaller (18.6% less on NCCU and 64.5% less on GRM with 1% of consumers), but with comparable network infection (overhead), which is unfeasible or very costly in real scenarios. From these findings above, we look for a forwarding algorithm able to increase the delivery rate with as lower as possible overhead and delivery latency.

5.2   TOOTS Forwarding Results and Analysis

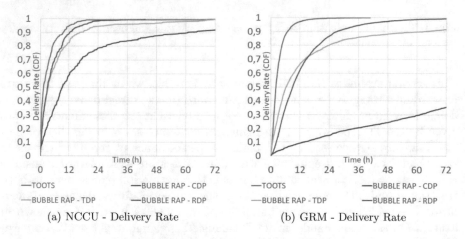

(a) NCCU - Delivery Rate   (b) GRM - Delivery Rate

Fig. 4. Delivery rate performance comparison of TOOTS, bubble rap-CDP, bubble rap-TDP, and bubble rap-RDP on NCCU and GRM datasets.

Finally, TOOTS (i.e., the full strategy) is evaluated in terms of delivery rate, delivery latency, and overhead within NCCU and GRM datasets. The chosen scenario is with random consumer sets with a size equal to 1% of dataset nodes (as used for the epidemic performance evaluation). TOOTS' performance is compared with Bubble Rap, the most popular social-based forwarding algorithm for delay-tolerant content. In Bubble Rap, each node gets a GlobalRank, and a LocalRank. The first measures the popularity in the whole network, while the second stores the popularity inside a community. Both centrality ranks are calculated with the C-Window technique, which measures nodes' popularity as the average number of unique nodes encountered throughout a time window of fixed length from the last 24 h. Furthermore, Bubble Rap identifies social communities, and each node must belong to at least one. Given these metrics, the forwarding strategy transmits a content c carried by u to v when: v is the destination node (d); v has higher GlobakRank or belongs to d's community; v belongs to d's community and has higher LocalRank.

We combine Bubble Rap with three dissemination policies responsible for injecting the contents into the network upon the start of the offloading process by the operator:

- **Random-Based Dissemination Policy (RDP)**: selects origin nodes randomly (i.e., working like the traditional Bubble Rap);
- **Centrality-Based Dissemination Policy (CDP)**: selects the higher centrality node as the origin;
- **Tactful-Based Dissemination Policy (TDP)**: proposed in Sect. 4.1. Selects the nodes with more direct encounters and higher C_{LID}.

In Fig. 4(a), we evaluate the delivery rate for both TOOTS and Bubble Rap on NCCU trace. TOOTS achieved a 100% delivery ratio and had the fastest delivery (90% of the contents delivered in up to 9 h). Bubble Rap-CDP failed in delivering all contents, even in the NCCU's smaller population, which has overall high centrality degree nodes. Furthermore, using a CDP in the real-world is not feasible, as it creates a bottleneck and tends to drain the most "popular" users' devices resources. Bubble Rap-RDP reaches a 100% delivery ratio but is expected to show higher overhead (analysis to follow). Finally, Bubble Rap-TDP reaches a 100% delivery ratio but takes slightly more time to forward all contents successfully than TOOTS.

In the GRM trace (Fig. 4(b)), TOOTS delivers 100% of the contents, with an average of 97% in up to 12 h. Bubble Rap-CDP shows an even worst delivery performance than in NCCU. That happens due to (i) communication bottlenecks, as there is a higher node density, and each node interface can transmit only one content at a time, and (ii) the fact that the GRM dataset has overall nodes with much lower centrality degrees. Bubble Rap-RDP delivers 100% but is also expected to show higher overhead (analysis to follow). Bubble Rap-TDP reaches close to 94% delivery ratio. This result can also be explained by GRM's lower centrality degrees (explanation to follow), calling for more time to reach 100% delivery ratio in Bubble Rap-TDP.

In Fig. 5(a), we evaluate the delivery latency. TOOTS has the lower average delivery latency on both datasets, followed by Bubble Rap-RDP on NCCU and Bubble Rap-TDP on GRM. We remind that on GRM, only TOOTS and Bubble Rap-RDP reached 100% delivery rate in up to 72h. On GRM, Bubble Rap-CDP shows a much higher average latency, justified by the bottlenecks on the fewer higher centrality nodes and by a dataset characteristic: from the content-generation time till the end of the simulation, over 90% of the nodes have low centrality degrees (up till 0.4). These dataset characterization results were not plotted as we focus on the full strategy proposal and its results.

Finally, in Fig. 5(b), we find the Bubble Rap-CDP with the smaller overhead on both datasets. On NCCU, this happens as the higher centrality degree node can find the destination directly for most of the contents generated. On

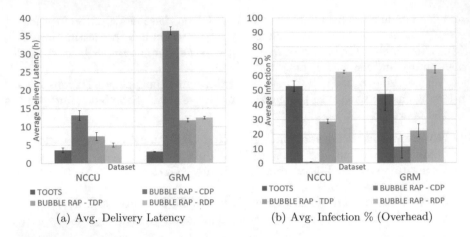

(a) Avg. Delivery Latency   (b) Avg. Infection % (Overhead)

Fig. 5. Delivery latency and overhead performance comparison of TOOTS, bubble Rap-CDP, bubble Rap-TDP, and bubble Rap-RDP on NCCU and GRM datasets.

the other hand, still, a bottleneck arises. Despite the lower overhead, Bubble Rap-CDP has a shallow delivery rate compared with the other proposals. The time windows used by Bubble Rap here also might be a reason for the slower delivery performance. Bubble Rap-RDP presents the worst-case in terms of overhead. Compared with Bubble Rap-TDP, there is respectively 53% and 65% less overhead on NCCU and GRM. On the other hand, we must remember this combination does not deliver 100% of the contents on GRM when the deadline is up to 72h. Using other real datasets with larger populations would be interesting to evaluate this strategy. TOOTS is the fastest strategy able to deliver 100% of the contents on both datasets. Its overhead is respectively 10% and 17% smaller than Bubble Rap-RDP.

6   Conclusion and Future Work

Despite remarked as an enabler of future networking generations since 4G/LTE, opportunistic communications have not been widely deployed by cellular operators. Recent research discusses the D2D-enabled 5G and 6G [18,19]. Novel initiatives in opportunistic communication will arise through a deeper understanding of the human behavior hidden into the datasets and the use of AI [19].

This work discussed a more realistic application scenario and introduced a human-aware opportunistic communication strategy named TOOTS. Compared to Bubble RAP and in both evaluated datasets (i.e., NCCU and GRM), TOOTS shows increased delivery performance and reduced delivery delay and overhead. Besides, TOOTS does not rely on complex communities calculation requiring parameter calibration. These characteristics make Bubble Rap challenging to be implemented in real-world scenarios.

Aiming for even better performance, future work related to TOOTS must evaluate the involved decision factors individually. In particular, in the real dataset NCCU, TOOTS showed a better correlation with the leveraged human characteristics. The time-window proposal also made it possible to calculate more precise metrics reflecting the routine-based human mobility behavior at different daily periods.

There is still a lack of datasets with larger populations to be evaluated for real-world data offloading applications. We found that a dissemination strategy to inject the contents into the network is valuable in a scenario such as the one presented. Further, research must also evaluate the presented algorithm and other similar proposals regarding different contact-range, available bandwidth, variable content-size, node buffer size, and node energy constraints.

Furthermore, when it comes to human-aware metrics in the big data era, there are also many possibilities. Thus, future research shall invest efforts in proposing different metrics to identify human routines and social aspects from mobility data, but from other sources (e.g., personality traits [3]).

Finally, it is necessary to discuss different application scenarios to have a clearer idea of what opportunistic D2D communication will be able to do or enhance. Showing opportunistic gains in more realistic scenarios and innovative applications can motivate cellular network operators to invest more efforts in the context of human-centered networks.

Acknowledgements. This work was supported in part by the CAPES (88881.135682/2016-01; 88887.468235/2019-00), CNPq (465.560/2014-8; 310.201/2019-5; 432064/2018-4), FAPERJ(E-26/203.046/2017), FAPESB (TIC0004/2015), INCT-CiD, EMBRACE associate team Inria project, and ANR MITIK project (PRC AAPG2019).

References

1. Akyildiz, I.F., Nie, S., Lin, S.C., Chandrasekaran, M.: 5g roadmap: 10 key enabling technologies. Comput. Netw. **106** 17–48 (2016) Supplement C
2. Costa, R.L., Carneiro Viana, A., Ziviani, A., Sampaio, L.N.: Towards human-aware d2d communication. In: IEEE DCOSS, pp. 173–180 (2020)
3. Costa, R.L., Viana, A.C., Ziviani, A., Sampaio, L.N.: Tactful networking: humans in the communication loop. IEEE TETCI, pp. 1–16 (2020)
4. Cuttone, A., Lehmann, S., González, M.C.: Understanding predictability and exploration in human mobility. EPJ Data Sci. **7**(1), 2 (2018)
5. Hui, P., Crowcroft, J., Yoneki, E.: Bubble rap: Social-based forwarding in delay-tolerant networks. IEEE TMC **10**(11), 1576–1589 (2011)
6. Keränen, A., Ott, J., Kärkkäinen, T.: The one simulator for dtn protocol evaluation. In: Proc. ICST, Simutools (2009)
7. Mao, Y., You, C., Zhang, J., Huang, K., Letaief, K.B.: A survey on mobile edge computing: The communication perspective. IEEE Commun. Surv. Tutor **19**(4), 2322–2358 (2017)
8. Moreira, W., Mendes, P., Sargento, S.: Opportunistic routing based on daily routines. In: IEEE WoWMoM, pp. 1–6 (2012)
9. Nunes, I.O., Celes, C., Nunes, I., Vaz de Melo, P.O.S., Loureiro, A.A.F.: Combining spatial and social awareness in d2d opportunistic routing. IEEE COMMAG **56**(1), 128–135 (2018)

10. Nunes, I.O., de Melo, P.O.S.V., Loureiro, A.A.F.: Leveraging d2d multihop communication through social group meeting awareness. IEEE Wirel. Commun. **23**(4), 12–19 (2016)

11. Nunes, I.O., Celes, C., Silva, M.D., Vaz de Melo, P.O., Loureiro, A.A.: GRM: group regularity mobility model. In: Proceedings ACM MSWiM, pp. 85–89 (2017)

12. Oliveira, E.M.R., Viana, A.C., Sarraute, C., Brea, J., Alvarez-Hamelin, I.: On the regularity of human mobility. Pervasive Mob. Comput. **33**, 73–90 (2016) (Supplement C)

13. Rebecchi, F., Dias de Amorim, M., Conan, V., Passarella, A., Bruno, R., Conti, M.: Data offloading techniques in cellular networks: a survey. IEEE Commun. Surv. Tutorials **17**(2), 580–603 (2015)

14. Tao, J., Wu, H., Shi, S., Hu, J., Gao, Y.: Contacts-aware opportunistic forwarding in mobile social networks: a community perspective. In: IEEE WCNC, pp. 1–6 (2018)

15. Thilakarathna, K., Viana, A.C., Seneviratne, A., Petander, H.: Design and analysis of an efficient friend-to-friend content dissemination system. IEEE TMC **16**(3), 702–715 (2017)

16. Tsai, T., Chan, H.: Nccu trace: social-network-aware mobility trace. IEEE COMMAG **53**(10), 144–149 (2015)

17. Viana, A., Jaffres-Runser, K., Musolesi, M., Giordano, S., Fiore, M., Olmo, P.: Mobile context-Adaptive CAching for COntent-centric networking (2013). http://macaco.inria.fr Accessed Jan 2021

18. Waqas, M., Niu, Y., Li, Y., Ahmed, M., Jin, D., Chen, S., Han, Z.: A comprehensive survey on mobility-aware d2d communications: principles, practice and challenges. IEEE Commun. Surv. Tutorials **22**(3), 1863–1886 (2020)

19. Zhang, S., Liu, J., Guo, H., Qi, M., Kato, N.: Envisioning device-to-device communications in 6g. IEEE Netw. **34**(3), 86–91 (2020)

Cooperative and Coordinated Mobile Femtocells Technology in High-Speed Vehicular Environments
Mobility and Interference Management

Rand Raheem$^{(\boxtimes)}$, Aboubaker Lasebae, Miltos Petridis, and Ali Raheem

Faculty of Science and Technology, Middlesex University, London NW4 4BT, UK
{R.H.Raheem,A.Lasebae,M.Petridis,A.Raheem}@mdx.ac.uk

Abstract. In future networks, most users who will be accessing wireless broadband will be vehicular. Serving those users cost-effectively and improving their signal quality has been the main concern of many studies. Thus, the deployment of Mobile Femtocell (Mobile-Femto) technology on public transportation is seen to be one of the promising solutions. Mobile-Femto comes with its mobility and interference challenges. Therefore, eliminating the Vehicular Penetration Loss (VPL) and interference while improving signal quality and mobility for train passengers is the main concern of this paper. The initial system-level evaluation showed that the dedicated Mobile-Femto deployment has great potential in improving users' experience inside public transportation. The Downlink (DL) results of the Proposed Interference Management Scheme (PIMS) showed significant improvement in Mobile-Femto User Equipment (UE) gains (up to 50%) without impacting the performance of macro UEs. In contrast, the Uplink (UL) results showed noticeable gains for both macro UEs and Mobile-Femto UEs.

1 Introduction

There are always new challenges on future mobile networks especially with the traffic explosion in today's networks. Previous studies predicted that from 2020 the number of connected devices to the internet will exceed 50 billion which in return will require a tremendous increase in the network capacity. Future networks will enable users to be connected to the internet from anywhere at any time and with anything. This is without neglecting that future transportation systems may play an integral role in wireless networks by providing additional communication capabilities and becoming part of the internet communication infrastructure. Thus, to achieve that sort of connectivity, it is highly desirable to reduce the environmental impact of mobile communication systems by intelligently deploying new wireless nodes to enhance UEs wireless connectivity. Mobile-Femto technology is seen to be a promising solution for future networks to improve passengers' wireless connectivity inside public transportation [1, 2].

Nowadays, one of the most used transportation services that save time and effort are High-Speed Trains (HSTs). HSTs are used all over the world e.g. the TGV Eurostar in Europe, the Shinkansen train in Japan and the HST in China with a speed exceeds the 350 km/h [3]. Due to the well-shield carriages, the HSTs are more exposed to Doppler

L. Barolli et al. (Eds.): AINA 2021, LNNS 225, pp. 653–666, 2021.
https://doi.org/10.1007/978-3-030-75100-5_56

frequency shift, high VPL (up to 40 dB), low Handover (HO) success rate, and the UL/DL interference issues [3]. The European "Shift2Rail" project [4] has made it clear that there is a persistent need for seamless wireless connectivity and high quality of services to be provided for passengers in future rail development. Although plenty of research has discussed GSM-R networks [5] and 5G networks [6]; not much researches dealt with the communications network regarding future railway services. On the other hand, Mobile Communications Enablers for Twenty-twenty (2020) Information Society (METIS) is one of the few projects that are concerned with the possibility of improving the wireless network connectivity by simply using low power small Base stations (BSs). This project has also introduced the proposal of placing mobile BSs in vehicles, such as cars, trucks, buses, and trains to improve vehicular UE performance.

Therefore, the main contribution of this paper is improving the performance and signal quality of train passengers via installing small-cells technology inside train carriages. That is without neglecting the importance of discussing the main two challenges here; mobility and interference that accompany the deployment of such technology.

2   Related Work

The European 5G project group (METIS) predicted that from 2020, most users who require reliable connection to the internet will be train passengers where the number will exceed 300 users' devices per train. Whereas, the E-train projects from the International Union of Railways (UIC) confirmed that the railway services will spread widely in the next few years and dominate people's lives. In addition, there will be more railway services that are focused on realizing the objective of smart, green, and integrated transport systems [7]. That has created significant motivation to improve the performance of passengers in public transportations, reduce signal outage, and mitigate interference in future networks. Providing indoor coverage on HSTs with outdoor BSs is not the most convenient solution due to high VPLs caused by the Faraday cage characteristics of railcars [8]. This fact has led to poor signal quality inside train carriages were offering broadband services is not always possible. Although, broadband access on trains could be achieved by installing BSs close to the railways; however, this solution is not convenient to train operators due to the high investment needed to deploy such equipment. Besides, such solutions increase the number of unnecessary HOs; as a result, this has focused the research community on offering solutions that take advantage of the existing wireless infrastructure to propose efficient methods to manage mobility in a seamless way for the end UEs. Hence, a cooperative moving Relay Node (RN) system deployed on HSTs has been introduced to enhance the coverage area for HSTs passengers and to overcome the VPL issues [9, 10]. In contrast, a novel HO scheme that utilises the Coordinated Multipoint Transmission (CoMP) in high-speed Femtocell networks has been proposed to ensure seamless, deployable, and efficient HO procedure for the train passengers when the train switches between different coverage areas [11, 12]. Thus, installing small cells inside HSTs to serve train passengers is the future of next-generation networks. However, this solution comes with its deployment challenges e.g. interference and mobility issues especially in the HST environment. Therefore, this paper pays more attention to the previous two challenges and provides optimal solutions to improve signal quality inside HST environments.

3 Mobile Femto Scenario in High Speed Train Environment

Mobility management is an essential component in mobile cellular communication systems because it offers clear benefits to the end UE in the HST environments i.e. low delay services to voice and real-time video connections. Thus, coming up with an effective solution to overcome the above challenges was the main motivation behind proposing the Cooperative and Coordinated Mobile-Femtos (CCMFs) system in HST environments. The CCMFs would be implemented on trains or other large-spatial-dimensions vehicles where they are connected to one another through the CrX2 coordination interface as shown in Fig. 1. This allows high optimisation for the group HO procedure and Backhaul (BH) link connection.

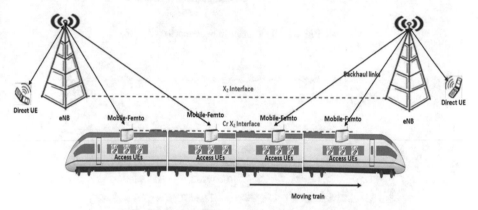

Fig. 1. CCMFs scenario in HST environments

However, one of the main challenges in deploying the CCMF technology is the excessive interference caused by the different access and BH links. That has been classified by the following DL and UL interference scenarios. The possible DL interference scenarios in HST CCMF are illustrated in Fig. 2. Interference (1) is caused by neighboring eNBs' signals to the macro UEs while the BH links interference (2) occurs because of the high Transmission Power (TP) of neighboring eNBs to neighboring Mobile-Femtos. As in interference (2); interference (3) is caused by the high TP of the serving eNB to the Mobile-Femto UEs within its coverage. In contrast, interference (4) occurs when the Mobile-Femto TP is very high to the extent that it causes interference to the macro UEs within the same eNB coverage. Finally, interference (5) is the interference between the Mobile-Femtos themselves.

On the other hand, the possible UL interference scenarios in HST CCMF are demonstrated in Fig 3.

The UL interference (1) is caused by Macro UEs to other neighboring eNBs, while interference (2) is caused by the BH UL transmission of the Mobile-Femto to neighboring eNBs. Interference (3) occurs when the Mobile-Femto UE causes interference to the Macrocell signals while interference (4) is caused by the macro UE to the access links. As for DL, interference (5) occurs between the access links of the many deployed CCMFs within the same Macrocell coverage.

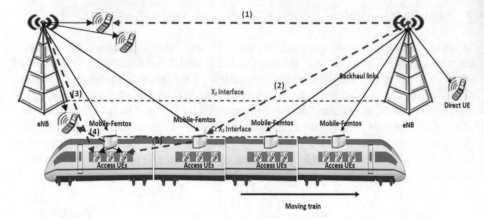

Fig. 2. DL interference scenarios

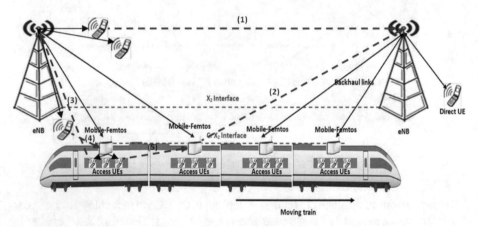

Fig. 3. UL interference scenarios

However, interference is not the only challenge in the HST environments because mobility plays another significant role in this environment. That is because; the forwarding of the data from the source eNB (SeNB) to the target eNB (TeNB) is the main goal of offering seamless HOs in future networks. Data forwarding typically takes place over the X2 interface between eNBs. In the presence of Mobile-Femto deployments, handing-over the Mobile-Femto itself between Macrocells and forwarding the data is a major challenge in today's networks. That HO procedure takes a place whenever the SINR between the Mobile-Femto and the source eNB gets weaker. In that case, the Mobile-Femto initiates a HO procedure to maintain its connectivity with the hosting network. The choice of target eNBs depends on the hosting train path and direction as discussed in our previous work [13]. However, the mobility of the installed CCMF in HSTs makes it even more challenging for the cellular operator to control the generated interference of these small cells. Therefore, there are several techniques that could be used by service providers to mitigate the interference between eNBs and CCMF and between the CCMFs themselves in future networks.

3.1 Coverage Optimisation

An important challenge for the CCMFs is optimising their radio coverage area dynamically. This has helped in achieving the desired level of performance for mobile transmission, avoiding undesired interference, and reducing power consumption [14, 15]. Consequently, optimising the coverage plan procedure is important to have good radio conditions everywhere within the vehicular environment. This can be achieved by mitigating the impact of path loss (PL) which in turn can be achieved by installing Mobile-Femtos in appropriate locations with appropriate TP [16]. The NLOS PL equation of the BH links is given by

$$PL(L) = 34.53 + 38\log_{10}(L) \tag{1}$$

Where L is the distance in meters and no less than 20 m.
On the other hand, the LOS PL equation of the access links is given by

$$PL(L) = 30.18 + 26\log_{10}(L) \tag{2}$$

The NLOS PL model is for the BH link between the Mobile-Femto and the eNB. In contrast, a constant PL has been used for the LOS access link between Mobile-Femtos and their served UEs within the same train carriage.

3.2 Transmission Power Control

Controlling the TP of eNBs and Mobile-Femtos is another efficient method to mitigate interference in current and future networks. However, insufficient TP with poor Femtocell deployment leaves the network with dead zones (holes) [17]. Whereas, if the Femtocell TP is high with dense Femtocells deployment, this can create excessive interference between the different parties within the network. In the CCMFs scenario, the channel between the Femtocell and its UEs is considered to be less affected by the interfered signals from other BSs. This is due to the short distance between the transmitter and the receiver. In addition, the high penetration loss of the train environment (well shielded carriages) prevents outside signals from passing through to train passengers. This makes it clear that train carriage chassis can work as a barrier to isolate train passengers from outside signals. Thus, train passengers can be served only by the installed Femtocell within their carriage; as a result, the interference can be mitigated.

Hence, the previous two possible solutions have been used for years to reduce the impact of interference in current and future networks. However, in HST environment there is always worst case scenario due to the high mobility challenge. Therefore, to overcome such scenarios, the following interference mitigation scheme is presented to demonstrate its positive impact on future networks and vehicular UEs performance.

4 Proposed Interference Management Scheme

The frequency scheme shown in Fig. 4 divides the Macrocell coverage into center and edge zones.

F_0 sub-band is allocated for center zones, whereas F_1, F_2 and F_3 are assigned to edge zones of Macrocell$_1$, Macrocell$_2$ and Macrocell$_3$ respectively. Thus, applying the frequency reuse scheme between the CCMF in HST environment has positive impact in mitigating the generated interference between transmission signals. That is because the Mobile-Femto inside the train carriage allocates frequency sub-bands that have not been used by Macrocells or other neighboring Mobile-Femtos. It can only reuse frequencies of further Mobile-Femtos that would not cause interference to its served UEs. However, in real life scenarios the frequency reuse scheme works by receiving Macrocell dimensions, no. of Mobile-Femtos and locations. More network characteristics are required like BS TPs, received power from serving and interfering cells and the white Gaussian noise. Based on the received Macrocell characteristics and used TPs, the inner cell radius is calculated to differentiate between centre and cell edge zones as illustrated in Fig. 5.

The following algorithm summaries the frequency reuse scheme discussed earlier.

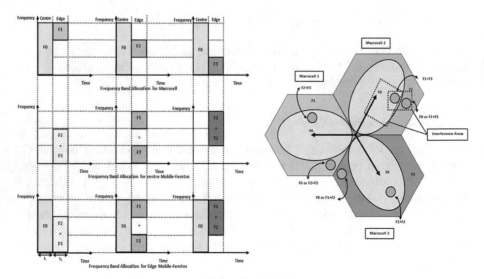

Fig. 4. The PIMS

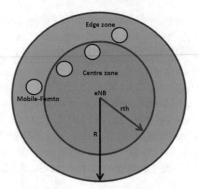

Fig. 5. Inner cell radius

Algorithm 1. PIMS for Macrocell/CCMF

```
Create Network topology
for inner cell radius equal to 0: R
for frequency band equal to 0: Total-subcarriers
Allocate frequency band for macrocells
    if inner cell radius >= distance mfemto
        Mobile-Femto belongs to center zone
        Allocate frequency band for mfemto
    else
        if inner cell radius < distance mfemto
        Mobile-Femto belongs to edge zone
        Allocate frequency band for mfemto;
    endif
        endif
for all UEs equal to 1 : U calculate
Signal to noise ratio
Capacity
Throughput
endfor
endfor
endfor
```

5   System Level

In order to study the impact of the PIMS on UEs; the SINR, throughput and capacity will be calculated. The Macro and Mobile-Femto UEs can be interfered by the DL/UL signals of neighboring eNBs and Mobile-Femtos. Therefore, on subcarrier n the received Macro UE SINR is given by

$$SINR_{m(D),n} = \frac{P_x^{eNB}\left|G_{1,eNB,n2}\right|^2 PL(x)\varepsilon}{\sum_{eNB'=1}^{neNB'} P_x^{eNB'}\left|G_{1,eNB',n}\right|^2 + \sum_{MFemto=1}^{nMFemto} P_x^{MFemto}\left|G_{2,MFemto,n}\right|^2 + P_{noise}} \tag{3}$$

The serving eNB and the neighbouring eNB' TPs are presented by P_x^{eNB} and $P_x^{eNB'}$ respectively. The channel gain between the Macro UE m and serving eNB is presented by $G_{1,eNB,n}$ while the channel gain from neighbouring eNBs is denoted by $G_{1,eNB',n}$. In contrast, the TP of neighbouring Mobile-Femtos is presented by P_x^{MFemto} whereas the channel gain between the Macro UE m and the neighbouring Mobile-Femto is presented by $G_{2,MFemto,n}$. The PL model has already been calculated and it is expressed by PL where x is the distance between the Macro UE and its serving eNB. The VPL and the white noise power are presented by ε and P_{noise} respectively. On the other hand, the Mobile-Femto UEs can be interfered by eNB signals or any other neighbouring Mobile-Femto signals, thus, on subcarrier n the received Mobile-Femto UE SINR is given by

$$SINR_{mf(A),n} = \frac{P_x^{MFemto} C_{loss}}{\sum_{eNB=1}^{neNB} P_x^{eNB} \left| G_{1,eNB,n} \right|^2 + \sum_{MFemto'=1}^{nMFemto'} P_x^{MFemto'} \left| G_{2,MFemto',n} \right|^2 + P_{noise}} \tag{4}$$

In Eq. (4), there is only a constant system loss since there is no channel gain over the LOS access link with the served Mobile-Femto where the only existing channel gain $(G_{2,MFemto',n})$ is between the Mobile-Femto UEs and other neighbouring Mobile-Femtos. In contrast, on subcarrier n, the Macro UEs capacity can be calculated as

$$C_{m,n} = BW. \log_2 \left(1 + \alpha SINR_{m(D),n} \right) \tag{5}$$

where the divided bandwidth for subcarrier n is represented by BW and α is a constant coding margin for target Bit Error Rate (BER) which is given by $\alpha = -1.5/\ln(5BER)$. As a result, the overall throughput of the serving eNB M is given by

$$Throughput_M = \sum_m \sum_n \beta_{m(D),n} C_{m(D),n} \tag{6}$$

$\beta_{m,n}$ represents the assigned subcarrier for Macro UEs. When $\beta_{m(D),n} = 1$, the subcarrier n is assigned to Macro UE m, otherwise $\beta_{m(D),n} = 0$. Thus, every Macro UE in every time slot is allocating different subcarrier within the same Macrocell. The $\sum_{m=1}^{N_m} \beta_{m(D),n} = 1$ for $\forall$ k as N_m is the no. of Macro UEs and k is the available Physical Resource Blocks (PRBs). On the other hand, the expected $\sum_{mf=1}^{N_{mf}} \beta_{mf(A),n} = 3$ for $k \in F_{Mobile-Femto}$ where the no. of Mobile-Femto UEs is given by N_{mf} and $F_{Mobile-Femto}$ represents the available sub-bands allocated to Mobile-Femtos. This makes it clear that the PIMS reuses the full frequency bands three times in every Macrocell. As a consequence, the capacity of the Mobile-Femto UE on subcarrier n is given by

$$C_{mf,n} = BW. \log_2 \left(1 + \alpha SINR_{mf(A),n} \right) \tag{7}$$

On the other hand, the overall throughput of the serving Mobile-Femto can be given by

$$Throughput_{MFemto} = \sum_{mf} \sum_n \beta_{mf(A),n} C_{mf(A),n} \tag{8}$$

Hence, the subsequent results will show that the PIMS is greatly capable of preventing the interference among the Macro and CCMFs UEs.

6   Results and Discussion

The system-level simulator written in Matlab [18] has been used to evaluate the network performance before and after implementing the PIMS in the HST environment. Figure 6 illustrates network topology and train path where the CCMFs will be installed inside train carriages. In our scenario the x-position of the tracks does not change and the train moves in a straight line throughout the cell edges of Macrocells at a constant speed of 100 km/h.

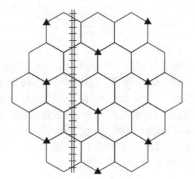

Fig. 6. Network layout

As it is shown above, the eNBs are uniformly deployed and have a 600 m cell radius thus, the inter-site distance is 1200 m. Two loads have been considered 10 and 25 mobile macro UEs. A full eNB buffer is considered where there are always buffered data ready for transmission for each node. The urban NLOS Microcell model has been considered which is more appropriate to describe the communication links and to eliminate the interference issue as much as it is possible in such High speed environment. The back-haul links use omni-directional antenna, instead of 7 dBi directional antenna. The access link log normal fading Standard Deviation (STD) is 0 dB. The reason for those changes is because the MFemto and the vehicular UE are inside. Table 1 represents some of the considered parameters for simulating the required scenarios while more detailed tables can be found in our previous work [19]. In our previous work we have discussed the situations where choosing different parameters have an impact on the achieved performance with an in depth comparison between different scenarios of being served by a Macrocell, fixed or mobile femtocells.

6.1   Downlink Scenario

Figure 7(a) illustrates the gained throughput of cell edge UEs who are being served by the CCMF when the PIMS is applied for the DL signals. The achieved results show that the gain is greater for small MFemto TPs and small VPLs compared to the case when there is no use of the PIMS. In that case, the interference from the Macro network is at its highest as the low VPL does not isolate the indoor UEs from outdoor interference signals. However, high VPL provides natural protection against outside interference.

662   R. Raheem et al.

Table 1. Simulation parameters

Parameter	Value	Parameter	Value
System bandwidth	10 MHz	No. of BH links	5
Traffic module	Full buffer	No. of Mobile-Femto UEs	1–5
Mobile-Femto type	In-band, full duplex	Train position	Cell edge
No. of cells	19	VPL	10–30 dB
No. of Macro UEs	10–25	Mobile-Femto TP	0–20 dBm (DL)
No. of carriages	5	Cell edge UE PL	PL > 120 dB

Thus, when the VPL and the TP are both high, the gain from the PIMS is low as the SINR is high in this case. Whereas Fig. 7(b) clearly shows that using the PIMS does not drop Macro UE's performance significantly as the loss is estimated to be about 3% at its most based on the used parameters. Please note that the different colors in the two charts represent the achieved gain for different TPs (0, 10 and 20 dBm) and VPLs (10, 20 and 30 dB). The changes in each parameter were needed for experimental purposes to study the impact of each on the achieve UE gain.

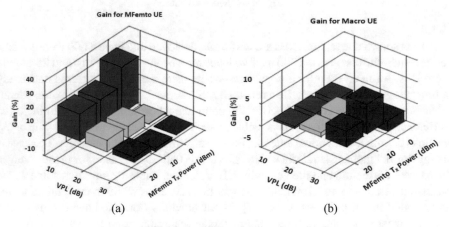

Fig. 7. (a) Vehicular UE's gain. (b) Macro UE's gain

6.2 Uplink Scenario

Figure 8 illustrates the UL Macro UE's performance in HST environment with and without the used PIMS. It also shows the performance of the network without the Macro UE Transmit Power Control (TPC). Consequently as expected, TPC improves the cell-edge Macro UE performance, and the PIMS improves this even further.

On the other hand, Fig. 9 illustrates the performance for train passengers (Mobile-Femto UEs) with and without the use of the PIMS.

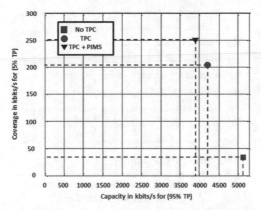

Fig. 8. UL performance of Macro UEs with the PIMS and 10 Macro UEs

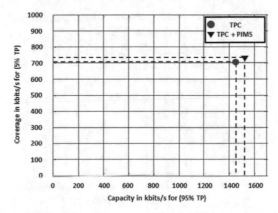

Fig. 9. Vehicular UL performance with the PIMS and 10 Macro UEs

Hence, it is obvious that the achieved results reflect the improvement in both Macro and Mobile-Femto UE performance after applying the PIMS. However, with the PIMS, the 95% throughput of macro UEs has slightly deteriorated. This is due to the fact that cell-edge UEs are given priority in the PRBs allocation while leaving fewer options for the Macro UE. This is in fact a characteristic of most Fractional Frequency Reuse (FFR) methods, which are more concerned with improving the cell-edge UEs performance. However, the simulation was repeated by increasing the no. of UEs in the cell to 25 UEs per cell. The results are shown in Fig. 10 and Fig. 11 for Macro UEs and train passengers respectively.

When the 95% performance of Macro UEs and Mobile-Femto UEs are being compared, it has clearly been noticed a large difference in the achieved performance between the two. This is due to the fact that the BH capacity is divided between the Mobile-Femto UEs inside the train carriages where each user gets an equal share. Furthermore, the PIMS does not degrade the 95% TP compared to the case without the PIMS. The reason is that the PIMS improves the BH performance; since the BH links are always being treated as cell edge UEs, giving them an advantage.

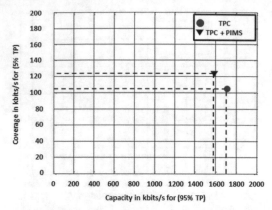

Fig. 10. UL performance (Macro UEs) with the PIMS and 25 Macro UEs

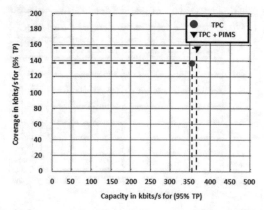

Fig. 11. UL performance (vehicular UEs) with the PIMS and 25 Macro UEs

7   Conclusion

The future of the railway industry is expected to rely upon smart transportation systems that provide train passengers with new services. These smart railways are more into improving users experience when they are using public transportation. The CCMFs technologies were the optimal solution to mitigate the impact of mobility, VPL, PL while boost users performance and internet services inside train carriages. However, every technology comes with its challenges and the main challenge in HST environment is the interference. Therefore, a frequency reuses scheme was introduced to mitigate the impact of the generated interference in HSTs environment and maintain signal services inside train carriages.

Acknowledgments. The authors would like to thank all those who contributed to the completion and success of this work. Also, the Faculty of Science and Technology at Middlesex University played significant role in backing this work at all stages of the study.

References

1. Raheem, R., Lasebae, A., Aiash, M., Loo, J., Colson, R.H.: Mobile Femtocells utilisation in LTE vehicular Environments: vehicular penetration loss elimination and performance enhancement. J. Veh. Commun. **9**, 31–42 (2017)
2. Raheem, R., Lasebae, A., Aiash, M., Loo, J.: Mobility management for vehicular user equipment in LTE/Mobile Femtocell Networks. Int. J. Inf. Syst. Serv. Sect. **9**(4), 60–87 (2017)
3. Ai, B., He, R., Zhong, Z., Guan, K., Chen, B., Liu, P., Li, Y.: Radio wave propagation scene partitioning for high-speed rails. Int. J. Antennas Propag. **2012**, 7 (2012)
4. Shift2Rail: Rails Proejcts. https://www.shift2rail.org. Accessed 03 Jan 2021
5. Zhong, Z.D.: Key Issues for GSM-R and LTE-R. In: Dedicated Mobile Communications for High-speed Railway. Advances in High-speed Rail Technology, pp. 19–55. Springer, Germany (2018). https://doi.org/10.1007/978-3-662-54860-8_2
6. METIS project, Deliv. D1.1: Scenarios, Requirements and KPIs for 5G Mobile and Wireless System, Technical report. https://cordis.europa.eu/docs/projects/cnect/9/317669/080/delive rables/001-METISD11v1pdf.pdf. Accessed Apr 2014
7. HORIZON 2020 Work Programme 2018–2020. Smart, Green and Integrated Transport Revised, EC Decision C(2020)6320. https://ec.europa.eu/programmes/horizon2020/en/h2020-section/smart-green-and-integrated-transport. Accessed Sept 2020
8. Vivas, P.N., Campo, C., Rubio, C.G., Carrion, A.R.: Communication technology for vehicles. In: 4th International Workshop, Nets4Cars/Nets4Trains, Vilnius, pp. 174–185 (2012). https://doi.org/10.1007/978-3-642-29667-3
9. Scott, S., Leinonen, J., Pirinen, P., Vihriala, J., Phan, V.V., Latva-aho, M.: A cooperative moving relay node system deployment in a high speed train. In: IEEE 77th Vehicular Technology Conference (VTC2013-Spring), Germany (2013). https://doi.org/10.1109/VTCSpring.2013.6691818
10. Small Cells and infrastructure Blog: Small Cells on the Train – A 2 hop solution. https://smallcells.3g4g.co.uk/2015/07/small-cells-on-train-2-hop-solution.html. Accessed 05 Jan 2021
11. Chae, S., Nguyen, T., Jang, M.Y.: A novel handover scheme in moving vehicular femtocell networks. In: 2013 5th International Conference on Ubiquitous and Future Networks (ICUFN), Vietnam (2013). https://doi.org/10.1109/ICUFN.2013.6614800
12. Pan, Meng-Shiuan., Lin, Tzu-Ming., Chen, Wen-Tsuen.: An enhanced handover scheme for mobile relays in LTE-a high-speed rail networks. IEEE Trans. Veh. Technol. **64**(2), 743–756 (2015)
13. Raheem, R., Lasebae, A., Loo, J.: Mobility management in LTE/Mobile Femtocell Networks: outage probability and drop/block calls probability. In: IEEE 3rd International Conference on Emerging Research Paradigms in Business and Social Sciences, ERPBSS-2015, Dubai, United Arab Emirates (2015)
14. Ma, T., Pietzuch, P.: Femtocell Coverage Optimisation Using Statistical Verification. https://lsds.doc.ic.ac.uk. Accessed 20 Jan 2021
15. Lu, Z., Bansal, T., Sinha, P.: Achieving user-level fairness in open-access femtocell-based architecture. IEEE Trans. Mob. Comput. **12**(10), 1943–1954 (2013). https://doi.org/10.1109/TMC.2012.157
16. Alqudah, Y.A.: On the performance of Cost 231 Walfish Ikegami model in deployed 3.5 GHz network. In: Proceedings of the International Conference on Technological Advances in Electrical, Electronics and Computer Engineering (TAEECE), Konya, Turkey (2013). https://doi.org/10.1109/TAEECE.2013.6557329
17. Raheem R., Lasebae A., Aiash M., Loo J., Colson R. H.: Interference management and system optimisation for femtocells technology in LTE and future 4G/5G networks. In: Advances in Communications: Reviews' Book Series, vol. 1. IFSA Publishing, Barcelona, Spain (2017)

18. Nt.tuwien.ac.at. Vienna LTE-A Simulators – nt.tuwien.ac.at (2021). https://www.nt.tuwien.ac.at/research/mobile-communications/vccs/vienna-lte-a-simulators/. Accessed 13 Feb 2021
19. Raheem, R.: Interference management and system optimisation for femtocells technology in LTE and future 4G/5G networks. Ph.D. Thesis submitted to Middesex Univerity. https://eprints.mdx.ac.uk/21255/1/RRaheemThesis.pdf. Accessed 14 Feb 2021

Resource Allocation for Millimeter Wave D2D Communications in Presence of Static Obstacles

Rathindra Nath Dutta[✉] and Sasthi C. Ghosh

Advanced Computing and Microelectronics Unit, Indian Statistical Institute,
203 B. T. Road, Kolkata 700108, India
{rathin_r,sasthi}@isical.ac.in

Abstract. Device-to-device communications using millimeter-wave frequencies has already been well acclaimed due to its improved spectral reuse, higher data rate and lower delay. Whereas, such high frequency communications with smaller wavelengths suffers from high propagation and penetration losses, thus requires short distance obstacle-free line-of-sight (LOS) communications. We allow relay aided communications in cases where direct LOS communication path is blocked by some obstacle. In this work we proposed an algorithm for efficient frequency channel allocation for the direct, as well as relay aided, blockage free LOS communications for real time application scenario. We also control the transmission powers allocated to the devices for better channel allocation and links activation, which in turn improves the overall system throughput. Simulation results show that our proposed algorithm produces better solutions in comparison to an existing work in terms of link activation, system throughput, fairness and energy efficiency.

1 Introduction

Modern smart user equiptments (UEs) with their bandwidth intensive applications like video calls, video streaming services and mobile gaming have already saturated the capacities of traditional cellular networks. To cope up with it, several cooperative communication strategies have been proposed among which communication in the millimeter-wave (mmWave) frequency has gained considerable interest for enabling device-to-device (D2D) communication in next generation cellular networks to satisfy such high data rate demand [10,13].

In D2D communications, two nearby devices are allowed to directly communicate with each other with limited or no involvement of the base station. The D2D communication in mmWave frequency can dramatically improve the spectral reuse, data rate, delay and energy efficiency. However, these mmWave signals suffer from extremely high propagation and penetration losses making it ideal for short distance obstacle-free line-of-sight (LOS) communications [9–11]. Using directional antennas and beamforming technique, one can establish LOS communication links between two devices with almost zero interference to other

L. Barolli et al. (Eds.): AINA 2021, LNNS 225, pp. 667–680, 2021.
https://doi.org/10.1007/978-3-030-75100-5_57

devices outside the beamwidth [1, 10]. Thus using beamforming and mmWave one can have high speed pseudo-wire like communications within the beamwidth, when LOS exists between the corresponding transmitter-receiver pair.

In cases where a direct communication suffers from blockage due to obstacles, relay assisted communications are being used. Optimal relay selection and resource allocation problems have been studied in [4, 6]. However, these works do not consider presence of obstacles in the communication paths. In case of static obstacles, prior knowledge of obstacle locations can be made available by satellite imaging or they can be learned [11]. The relay selection problem in the presence of static obstacles have also been studied by several authors [5, 11]. In [5] authors consider a 3D model for the obstacles and find a relay-assisted path giving maximum throughput among all candidate paths for backhaul communications. In [11] authors learn the obstacles without resorting to satellite imaging and provide relay assisted LOS path when direct communication is blocked.

Be it direct or relay assisted, all active links must be assigned with frequencies, making frequency allocation a challenging problem in D2D communications. With limited number of available frequency channels, reuse of frequencies is a common practice. In this setting, frequency allocation to the active links while keeping the interference level as low as possible, is a challenging task, in fact proven to be NP-complete. The problem is more complex for relay aided real-time application scenario, where data received at a relay device in a time slot must be sent out immediately in the next time slot. The frequency allocation problem for D2D communications has been explored in [4, 6, 7, 13, 15]. The work which is closest to ours, is the resource allocation strategy investigated in [7]. Here the authors have proposed a framework for efficient frequency channel allocation as well as controlling the transmission power of the devices. They begin by a preliminary grouping of the devices with respect to assigned frequencies and later reform these groups by exchanging and adding new links. However, they neither considered the presence of the obstacles nor allowed real time relay aided communications.

To the best of our knowledge none of the works including the ones mentioned above have considered the frequency assignment problem with relay selection in the presence of static obstacles under multiple frequency reuse scenario supporting real time relay aided application scenario. To this end, we developed a framework having the knowledge of static obstacle locations for a D2D overlay communication network in which we aim to optimally allocate channel resources and maximize the number of links activated to improve overall system throughput. We propose an algorithm which allocates frequency channels to the requesting D2D pairs having direct LOS communication paths. In cases where direct LOS path is blocked due to obstacles we allow one-hop relay assisted LOS communications whenever possible. We are considering a real-time application scenario where data received at a relay device in a time slot must be sent out immediately in the next time slot. Therefore our algorithm solves the problem for two consecutive time slots because a subset of links activated in the first slot must also be activated in the second slot. Moreover our algorithm incorporates power control mechanisms for efficient channel allocation by maximizing the number of links activated while maintaining energy efficiency and fairness of the resource allocation.

The rest of the paper is arranged as follows. The system model is given in Sect. 2. In Sect. 3, we mathematically formulate the problem under consideration. Our proposed algorithm is described in Sect. 4 and its simulation results is presented in Sect. 5. Finally we conclude this work with Sect. 6.

2   System Model

We consider a single cell D2D overlay communications scenario in the presence of static obstacles where the UEs can be classified into D2D pairs and relays depending upon their intention to communicate. There are M D2D pairs who are willing to communicate and N relays which are providing relaying services to them. Half duplex communication is assumed and hence a UE can not communicate and relay simultaneously. We assume time is discretized into time slots $t_0, t_1, \cdots$ with a small Δt time span for each slot. We denote a pair of successive time-slots as a *superslot*. We also discretize the entire region as a grid of small squared cells of size $a \times a$.

Mobility Consideration: We consider all D2D devices as pseudo-stationary, that is they do not change their position for the duration of a superslot. For a particular superslot, position of an UE can be determined with great accuracy [8]. Thus we can solve the problem independently for each superslot. Position of a device in some unit grid cell is approximated by the center of the small squared cell.

Obstacle Modelling: Each obstacle is approximated by its minimum enclosing rectangle. We assume that sizes and locations of these rectangles corresponding to respective static obstacles are known a priori through satellite imagery. The grid cells containing some portion of an obstacle are marked as *blocked*. In Fig. 1 all the grid cells containing the obstacle, denoted by a black rectangle, has been marked in grey as blocked.

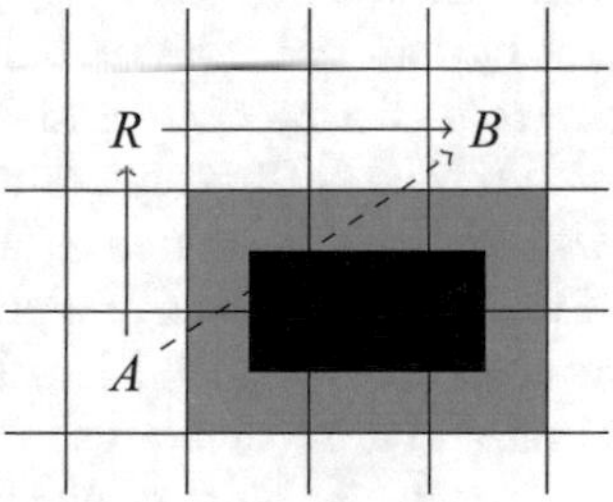

Fig. 1. Relay aided communication in the presence of an obstacle

Direct and Relay Aided Communication: We call an LOS link of a D2D pair to be blocked if there is a blocked cell on the LOS path between them. For a D2D pair if there is an obstacle-free LOS path they can communicate directly. Whereas, if their LOS path is blocked due to some obstacle in between, the D2D pair may communicate via an intermediate relay providing LOS path between them. We are only considering one hop relay assisted D2D communications in such cases. We further assume that an idle UE can serve as a relay for only one D2D pair. We denote $\mathscr{D}$ to be the set of D2D pairs having LOS links and $\mathscr{D}_R$ as the set of D2D pairs requiring a relay. As shown in Fig. 1 direct LOS path is blocked for the $A - B$ D2D pair, which may communicate via a relay R.

Candidate Relays: For a D2D pair, all idle UEs having LOS links within its vicinity form the set of candidate relays for that pair. An idle UE can be present in more than one candidate sets, but can only be selected for a single D2D pair for relaying. We denote $\mathscr{R}_k$ to be the set of candidate relays for the D2D pair k and $\mathscr{R} = \bigcup_{k \in \mathscr{D}_R} \mathscr{R}_k$.

Communication Channel: We assume mmWave directional antennas are being used for the D2D communications. For this we use *flat-top model* of beamforming with directional antennas [1,10] where mainlobe antenna gain $G_m = 2\pi/\theta$ and sidelobe gain $G_s = 0$. Here θ is the beamwidth. We assume the directional antennas are automatically aligned to each other based on the device positions and we ignore the alignment overhead being negligible compared to actual data transmission [2]. Following [1,12], the received power for the transmitted power P can be computed as $R_{T_x,R_x} = k\phi G_{T_x} G_{R_x} P d_{T_x,R_x}^{-\alpha}$, where d_{T_x,R_x} is the distance between the transmitter (T_x) and the receiver (R_x), α is the pathloss exponent, $k = (\frac{\lambda}{4\pi d_0})^2$ with d_0 and λ being the reference distance for the antenna farfield and the wavelength respectively, ϕ is the shadowing random variable, G_{T_x} and G_{R_x} are the antenna gains at the transmitter and the receiver respectively. If the receiver falls within beamwidth of the transmitter, we set $G_{T_x} = G_m$, and $G_{T_x} = G_s$ otherwise. Similarly, if the transmitter falls within beamwidth of the receiver, we set $G_{R_x} = G_m$, and $G_{R_x} = G_s$ otherwise.

Interference Consideration: Consider a D2D pair $k = (k_1, k_2)$, $k \in \mathscr{D}_R$ where device k_1 needs to send some data to device k_2 via relay r, $r \in \mathscr{R}_k$. In the first time slot of a superslot, k_1 plays the role of a transmitter, r becomes the receiver and k_2 does not take part. In the second slot, r becomes the transmitter, k_2 plays the role of a receiver and k_1 sits idle. While for a D2D pair $i = (i_1, i_2)$, $i \in \mathscr{D}$ with device i_1 directly transmitting to device i_2, i_1 plays the role of a transmitter while i_2 receives. Such a D2D pair can be opportunistically scheduled in either of the two time slots of a superslot. For a specific time slot, where a transmitter-receiver pair (T_x, R_x) is scheduled for transmission, T_X may cause interference to all active receivers other than R_x receiving in the same frequency channel as of R_x only if those receivers fall inside the *critical region* of (T_x, R_x) as defined below.

Critical Region: Consider a transmitter T_x transmitting with beamwidth θ to a receiver R_x where the line joining T_x and R_x makes an angle ψ with respect to the positive X axis. We define critical region CR_{T_x,R_x} of (T_x, R_x) as the $\psi \pm \theta/2$ region as depicted in Fig. 2.

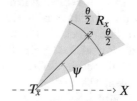

Fig. 2. Critical region

3   Problem Formulation

We define the following binary variables $\forall\, i \in \mathscr{D}, k \in \mathscr{D}_R, r \in \mathscr{R}_k, f \in \mathscr{F}, s \in \{1,2\}$.

$$X_{i,f,s} = \begin{cases} 1 & \text{if } i\text{-th D2D pair communicates directly using frequency } f \text{ in slot } s \\ 0 & \text{otherwise} \end{cases}$$

$$Y_{k,r,f,s} = \begin{cases} 1 & \text{if } k\text{-th D2D pair communicates via relay } r \text{ using frequency } f \text{ in slot } s \\ 0 & \text{otherwise} \end{cases}$$

For a D2D pair $i \in \mathscr{D}$ transmitting using frequency $f \in \mathscr{F}$, the interferences during the first slot and the second slot of a superslot are given in (1) and (2) respectively. Here $R_{a,b}$ denotes the received signal power at device b transmitted from device a.

$$I_{i,f,1} = \sum_{j\in\mathscr{D},i\neq j,i_2\in CR_{j_1,j_2}} R_{j_1,i_2} X_{j,f,1} + \sum_{k\in\mathscr{D}_R,i_2\in CR_{k_1,r}} \sum_{r\in\mathscr{R}_k} R_{k_1,i_2} Y_{k,r,f,1} \quad (1)$$

$$I_{i,f,2} = \sum_{j\in\mathscr{D},i\neq j,i_2\in CR_{j_1,j_2}} R_{j_1,i_2} X_{j,f,2} + \sum_{k\in\mathscr{D}_R} \sum_{r\in\mathscr{R}_k,i_2\in CR_{r,k_2}} R_{r,i_2} Y_{k,r,f,2} \quad (2)$$

The corresponding SINR value is given in (3), where η_0 denotes thermal noise.

$$SINR_{i,f,s} = \frac{R_{i_1,i_2}}{\eta_0 + I_{i,f,s}}, \quad \forall i \in \mathscr{D}, f \in \mathscr{F}, s \in \{1,2\} \quad (3)$$

Similarly, for a D2D pair $k \in \mathscr{D}_R$ transmitting via relay $r \in \mathscr{R}_k$ using frequency $f \in \mathscr{F}$, the interferences during the first and the second slots of a superslot are given in (4) and (5) respectively.

$$\hat{I}_{k,r,f,1} = \sum_{i\in\mathscr{D},r\in CR_{i_1,i_2}} R_{i_1,r} X_{i,f,1} + \sum_{k'\in\mathscr{D}_R,k'\neq k,r\in CR_{k'_1,r'}} \sum_{r'\in\mathscr{R}_{k'}} R_{k'_1,r} Y_{k',r',f,1} \quad (4)$$

$$\hat{I}_{k,r,f,2} = \sum_{i\in\mathscr{D},k_2\in CR_{i_1,i_2}} R_{i_1,k_2} X_{i,f,2} + \sum_{k'\in\mathscr{D}_R,k'\neq k} \sum_{r'\in\mathscr{R}_{k'},k_2\in CR_{r',k'_2}} R_{k'_1,k_2} Y_{k',r',f,2} \quad (5)$$

For each $k \in \mathscr{D}_R, r \in \mathscr{R}_k, f \in \mathscr{F}$ the SINR values for the two slots are given in (6).

$$\widehat{SINR}_{k,r,f,1} = \frac{R_{k_1,r}}{\eta_0 + \hat{I}_{k,r,f,1}}, \quad \widehat{SINR}_{k,r,f,2} = \frac{R_{r,k_2}}{\eta_0 + \hat{I}_{k,r,f,2}} \quad (6)$$

In order to satisfy the required data rate, the SINR value of each active link i must be above some threshold T_i. This is formalized as the following constraints.

$$\widehat{SINR}_{i,f,s} \geq T_i\, X_{i,f,s} \qquad \forall i \in \mathscr{D}, f \in \mathscr{F}, s \in \{1,2\} \quad (7)$$

$$\widehat{SINR}_{k,r,f,s} \geq T_k\, Y_{k,r,f,s}\ \forall k \in \mathscr{D}_R, r \in \mathscr{R}_k, f \in \mathscr{F}, s \in \{1,2\} \quad (8)$$

Furthermore, we must also have the following constraints on the binary variables.

$$\sum_{f \in \mathscr{F}} (X_{i,f,1} + X_{i,f,2}) \qquad \leq 1, \quad \forall i \in \mathscr{D} \tag{9}$$

$$\sum_{r \in \mathscr{R}_k} \sum_{f \in \mathscr{F}} Y_{k,r,f,s} \leq 1, \quad \forall k \in \mathscr{D}_R, s \in \{1,2\} \tag{10}$$

$$\sum_{k \in \mathscr{D}_R, r \in \mathscr{R}_k} \sum_{f \in \mathscr{F}} Y_{k,r,f,s} \leq 1, \quad \forall r \in \mathscr{R}, s \in \{1,2\} \tag{11}$$

$$\sum_{f \in \mathscr{F}} Y_{k,r,f,1} - \sum_{f \in \mathscr{F}} Y_{k,r,f,2} = 0, \quad \forall k \in \mathscr{D}_R, r \in \mathscr{R}_k \tag{12}$$

Constraint (9) ensures that a direct link can be scheduled at either of first or second slot. Constraints (10) and (11) ensure that for relay aided communication, a pair can choose only one relay and a particular relay can be chosen by at most one pair, respectively. Constraint (12) ensures that if a source to relay link is scheduled in the first slot, the relay to destination link must be scheduled in the second slot.

For each $k \in \mathscr{D}_R, r \in \mathscr{R}_k$ the *effective SINR* of the two slots is computed as:

$$\widehat{SINR}_{k,r} = \min(\sum_{f \in \mathscr{F}} \widehat{SINR}_{k,r,f,1} \, Y_{k,r,f,1}, \sum_{f \in \mathscr{F}} \widehat{SINR}_{k,r,f,2} \, Y_{k,r,f,2}) \tag{13}$$

We aim to maximize the total number of links activated as expressed bellow.

$$\max \left\{ \sum_{i \in \mathscr{D}} \sum_{f \in \mathscr{F}} \sum_{s \in \{1,2\}} X_{i,f,s} + \sum_{k \in \mathscr{D}_R} \sum_{r \in \mathscr{R}_k} \sum_{f \in \mathscr{F}} Y_{k,r,f,1} \right\} \tag{14}$$

Throughput obtained from the active links can be computed using Shannon's capacity formula *data rate* $= B \log(1 + SINR)$, where B is the channel bandwidth.

4   Scheduling and Resource Allocation Algorithm

We begin by preprocessing the blocked cells to compute a visibility graph which we later use to check for an obstacle-free LOS path between any two cells. With the candidate relay sets computed for each relay aided D2D pairs we proceed for the resource allocation. We first perform an initial frequency and power assignment for the direct D2D pairs which is later refined with exchanges and addition of new direct, as well as relay aided, D2D links. We also optimize the power allocation during the frequency assignment process itself for efficient resource utilization. Finally we compute the system throughput as per the final frequency and power assignment.

Precomputing Visibility Graph
Consider two grid cells with indices (i_1, j_1) and (i_2, j_2) respectively and a line l joining the centers of these two cells. We say two cells are *visible* to each other if and only if the LOS path from any point in one cell to any point in the other cell is not obstructed by any obstacle. Considering each cell is of size $a \times a$, the two cells are visible to each other if and only if there is no blocked cells within the $2a$ strip centered around the line l. We reduce the search space by considering only the blocked cells in the 2D range $[(i_1, i_2), (j_1, j_2)]$. The blocked cells in this range can efficiently be obtained by storing all blocked cells in a 2D *range tree* that uses fractional cascading [14].

For each cell we check visibility of all other cells within a circle of radius d. Here d is the *transmission range*, i.e., the distance up to which signal strength does not fall below the required threshold using the Friis free space propagation model [3]. We store this data in a binary array *visible* of size $O(hvd^2)$, where h and v are the number of horizontal and vertical grid cells respectively. Algorithm 1 formalizes this procedure.

Algorithm 1: Compute visibility graph

---

1 Store the blocked cells in a 2D range tree T_R
2 Create the *visible* array
3 **for** $i_1 = 1$ *to* h **do**
4   **for** $j_1 = 1$ *to* v **do**
5   **for** $i_2 = i_1 - d$ *to* $i_1 + d$ **do**
6   **for** $j_2 = j_1$ *to* $j_1 + d$ **do**
7   Run rectangular range query $[(i_1, i_2), (j_1, j_2)]$ on T_R and store the resultant cells in $\mathscr{B}$
8   Let l be the line joining the center of grid cells (i_1, j_1) and (i_2, j_2)
9   **if** $\exists b \in \mathscr{B}$ *such that* $b \in 2a$ *wide strip around* l **then**
10                     $visible[(i_1, j_1), (i_2, j_2)] = visible[(i_2, j_2), (i_1, j_1)] = false$
11   **else**
12                     $visible[(i_1, j_1), (i_2, j_2)] = visible[(i_2, j_2), (i_1, j_1)] = true$

Selecting Candidate Relays
An idle device r may act as a potential relay for a D2D pair $k = (k_1, k_2) \in \mathscr{D}_R$ if and only if r is within the transmission range of k_1 and k_2 is also in the transmission range of r. Therefore, in order to act as a potential relay for this D2D pair, r must reside in the intersection of the two d radius circles centered at the cells of k_1 and k_2. To reduce the search space for finding the candidate relay devices, we consider the minimum enclosing rectangle containing the intersection of these two circles and perform a 2D *range search* with respect to this rectangle.

Figure 3 depicts this idea. Such an axis parallel rectangle can uniquely be identified by any of its two opposite corners. In this figure P and Q are two such corners such that $P.i < Q.i$ and $P.j < Q.j$, where $A.i$ and $A.j$ denote row and column indices of the grid cell containing some point A. Let C_1 and C_2 be the points where the two circles intersect with each other. We can compute $P.i, P.j, Q.i$ and $Q.j$ as follows.

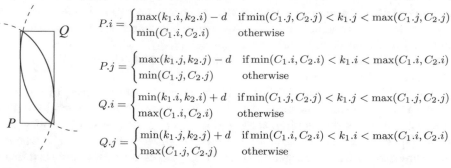

$$P.i = \begin{cases} \max(k_1.i, k_2.i) - d & \text{if } \min(C_1.j, C_2.j) < k_1.j < \max(C_1.j, C_2.j) \\ \min(C_1.i, C_2.i) & \text{otherwise} \end{cases}$$

$$P.j = \begin{cases} \max(k_1.j, k_2.j) - d & \text{if } \min(C_1.i, C_2.i) < k_1.i < \max(C_1.i, C_2.i) \\ \min(C_1.j, C_2.j) & \text{otherwise} \end{cases}$$

$$Q.i = \begin{cases} \min(k_1.i, k_2.i) + d & \text{if } \min(C_1.j, C_2.j) < k_1.j < \max(C_1.j, C_2.j) \\ \max(C_1.i, C_2.i) & \text{otherwise} \end{cases}$$

$$Q.j = \begin{cases} \min(k_1.j, k_2.j) + d & \text{if } \min(C_1.i, C_2.i) < k_1.i < \max(C_1.i, C_2.i) \\ \max(C_1.j, C_2.j) & \text{otherwise} \end{cases}$$

Fig. 3. Relays

The candidate relay sets $\mathscr{R}_k$, $\forall k \in \mathscr{D}_R$ can be obtained using the Algorithm 2, where $dist(k_1, r)$ is the distance between k_1 and r.

Algorithm 2: Candidate relay selection

---

1 Store all idle devices as potential relay nodes in a 2D *range tree* T_R
2 **foreach** $k \in \mathscr{D}_R$ **do**
3   Compute the two intersection points C_1 and C_2
4   Compute locations of P and Q as stated in Figure 3
5   Run a *rectangular range query* $[(P.i, Q.i), (P.j, Q.j)]$ on T_R and store the results in $\mathscr{R}_k$
6   **foreach** $r \in \mathscr{R}_k$ **do**
7   **if** $dist(k_1, r) > d$ *or* $dist(k_2, r) > d$ **then**
8              $\mathscr{R}_k = \mathscr{R}_k \setminus r$

The Scheduling and Resource Allocation Algorithm

With F number of orthogonal frequency channels available, we denote $\mathscr{F}_1 = \{1, 2, \ldots, F\}$ and $\mathscr{F}_2 = \{F + 1, F + 2, \ldots, 2F\}$ as the set of frequencies available for the first and the second time slot respectively. We define $\mathscr{F} = \mathscr{F}_1 \cup \mathscr{F}_2$ as the total frequency set for the two slots. We construct an *interference graph* G in which vertices are the D2D pairs in $\mathscr{D}$ and we put an edge from vertex i to vertex j with $i \neq j$ if and only if there exists an obstacle-free LOS path from i_1 to j_2 and $j_2 \in CR_{i_1, i_2}$. Let $\mathscr{F}$ be the set of available colors. We apply a simple greedy coloring algorithm on G to color as many vertices as possible with the given set of colors in $\mathscr{F}$. With this greedy coloring, we color the vertices of the graph by ordering the vertices by their degrees and allocate a conflict free color to each vertex whenever possible. More often than not it might be the case that the whole graph cannot be colored with only $|\mathscr{F}|$ colors. This coloring represents an initial frequency assignment. For each frequency $f \in \mathcal{F}$ we define a group G_f containing all those D2D pairs whose corresponding vertices are colored with color f. Since two D2D pairs belonging to two different frequency groups can not interfere with each other, we handle each of these frequency groups independently of each other. The D2D pairs having no frequency assigned to them are put into a set $\mathcal{C}$. We now allocate minimum possible power satisfying the required SINR threshold T_i to each D2D pair i in G_f by using the following linear programming (LP) based power control algorithm: minimize $\sum_{i \in G_f} P_i$ subject to constraints $P_{min} \le P_i \le P_{max}$

and $R_{i_1,i_2} \ge T_i(\eta_0 + \sum_{j \in G_f, i \neq j, i_2 \in CR_{j_1,j_2}} R_{j_1,i_2})$. We should note that the initial frequency assignment obtained by the graph coloring has already taken care of the interference, i.e., no device in a particular group G_f can interfere any other device in the same group. But at later stages of our algorithm we possibly add more D2D pairs into a particular group which might introduce interference in that group. This justifies the presence of the interference term (the $\sum$ term) in the second constraint in our power control LP. This initial frequency and power allocation process is formalized as Algorithm 3. Next we revise each frequency group $G_f, f \in \mathcal{F}$ by exchanging a D2D pair in G_f by a D2D pair in $\mathcal{C}$ such that maximum power allocated to a D2D pair in G_f reduces. This in turn improves the chance of adding more D2D pairs to G_f. We repeat this process until no such exchange is possible. At this point we run Algorithm 5 to add more direct D2D pairs from $\mathcal{C}$ into G_f. When there are no direct D2D pairs which can be added in this way we move our attention to relay aided D2D pairs and invoke Algorithm 6. This refinement process is formalized as Algorithm 4.

Algorithm 3: Initial frequency and power assignment algorithm

---

1 Create vertices corresponding to each pair $i \in \mathcal{D}$ for the interference graph G
2 **foreach** $i \in \mathcal{D}$ **do**
3   **foreach** $j \in \mathcal{D}$ **do**
4   **if** $i \neq j$ and $j_2 \in CR_{i_1,i_2}$ **then**
5   Add an edge between vertices i and j into G

6 Apply greedy graph coloring on G
7 Set $G_f = \{i|$ vertex i gets color $f\}$ $\forall f \in \mathcal{F}$
8 Set $\mathcal{C} = \{i|$ vertex i does not get any color $f \in \mathcal{F}\}$
9 **foreach** $f \in \mathcal{F}$ **do**
10   Run linear programming based power control algorithm on G_f to find P_i for $i \in G_f$

In Algorithm 5, to add more direct D2D pairs into a frequency group G_f we adopt a simple greedy strategy. We find the D2D pair $\hat{i}$ from $\mathcal{C}$ causing minimum interference into G_f while transmitting at P_{max}. We then apply the power control algorithm on $G_f \cup \{\hat{i}\}$. If there exists a feasible power allocation we permanently move $\hat{i}$ from $\mathscr{C}$ into G_f otherwise we discard $\hat{i}$ for this group. We repeat this process with rest of the D2D pairs in $\mathscr{C}$ until no new pair can be added into G_f. Here R_{T_X,R_X}^{max} denotes the received power at R_X from transmitter T_X transmitting at P_{max}.

Algorithm 4: Group refinement algorithm

1 **foreach** $f \in \mathscr{F}$ **do**
2 $\quad$ set $\mathscr{C}_{tmp} = \mathscr{C}$
3 $\quad$ **while** $\mathscr{C}_{tmp} \neq \Phi$ **do**
4 $\quad\quad$ let i be a D2D pair in $\mathscr{C}_{tmp}$
5 $\quad\quad$ set $\hat{i} = \underset{i \in G_f}{\operatorname{argmax}} \{P_i\}$
6 $\quad\quad$ set $G_{tmp} = G_f \cup \{i\} \setminus \{\hat{i}\}$
7 $\quad\quad$ run power control algorithm on G_{tmp}
8 $\quad\quad$ **if** *there is a feasible power allocation for G_{tmp} and* $\underset{i \in G_{tmp}}{\max} P_i < P_{\hat{i}}$ **then**
9 $\quad\quad\quad$ set $G_f = G_{tmp}$ and $\mathscr{C} = \mathscr{C} \cup \{\hat{i}\} \setminus \{i\}$
10 $\quad\quad$ set $\mathscr{C}_{tmp} = \mathscr{C}_{tmp} \setminus \{i\}$
11 $\quad$ run Algorithm 5 for G_f
12 run Algorithm 6

Algorithm 5: Adding direct D2D pairs

1 set $\mathscr{C}_{tmp} = \mathscr{C}$
2 **while** $\mathscr{C}_{tmp} \neq \Phi$ **do**
3 $\quad$ set $\hat{i} = \underset{i \in \mathscr{C}_{tmp}}{\operatorname{argmin}} \left\{ \sum_{j \in G_f, j_2 \in CR_{i_1,i_2}} R_{i_1,j_2}^{max} \right\}$
4 $\quad$ set $G_{tmp} = G_f \cup \{\hat{i}\}$
5 $\quad$ run power control algorithm on G_{tmp}
6 $\quad$ **if** *there is a feasible power allocation for G_{tmp}* **then**
7 $\quad\quad$ set $G_f = G_{tmp}$
8 $\quad$ set $\mathscr{C}_{tmp} = \mathscr{C}_{tmp} \setminus \{\hat{i}\}$
9 set $\mathscr{C} = \mathscr{C} \setminus G_f$

Algorithm 6 deals with relay aided D2D pairs from $\mathscr{D}_R$. To establish communication of a D2D pair $k \in \mathscr{D}_R$ via some relay $r \in \mathscr{R}_k$ we need to assign some frequency $f_1 \in \mathscr{F}_1$ for the D2D pair $k' = (k_1, r)$ in slot 1 and some frequency $f_2 \in \mathscr{F}_2$ for the pair $k'' = (r, k_2)$ in slot 2. This is equivalent to adding the D2D pair k' into group G_{f_1} and k'' into group G_{f_2}. We again adopt a greedy selection strategy to schedule the relay aided D2D pairs. We select a D2D pair $\hat{k} \in \mathscr{D}_R$ with relay $\hat{r} \in \mathscr{R}_{\hat{k}}$ along with two frequencies $\hat{f}_1 \in \mathscr{F}_1$ and $\hat{f}_2 \in \mathscr{F}_2$, such that k' and k'' cause minimum interferences to the corresponding groups among all other possible choices. We again run the power control algorithm for the groups $G_{\hat{f}_1} \cup \{k'\}$ and $G_{\hat{f}_2} \cup \{k''\}$. If there exists feasible power allocations for both cases, we permanently add k' and k'' to the respective groups and mark $\hat{r}$ as used. We mark $\hat{k}$ as done and do not consider it for further iterations. We repeat this process until no more choice remains.

Based on the final frequency and power assignment we obtain the system throughput as per the Shannon capacity formula. The worst-case time complexity of the proposed scheduling and resource allocation algorithm is $O(MF\beta)$ where M is the number of requesting D2D pairs, F is the number of available frequencies and β is the time complexity for solving LP with at most M number of variables and $2M$ constraints.

Algorithm 6: Adding relay aided D2D pairs

1 set $C_{tmp} = \mathscr{D}_R$ and $R_{used} = \Phi$
2 **while** $C_{tmp} \neq \Phi$ **do**
3     $\hat{k}, \hat{r}, \hat{f}_1, \hat{f}_2 =$

$$\underset{k \in C_{tmp}, r \in \mathscr{R}_k \setminus R_{used}, f_1 \in \mathscr{F}_1, f_2 \in \mathscr{F}_2}{\operatorname{argmin}} \left\{ \sum_{i \in G_{f_1}, i_2 \in CR_{k_1,r}} R^{max}_{k_1,i_2} + \sum_{j \in G_{f_2}, j_2 \in CR_{r,k_2}} R^{max}_{r,j_2} \right\}$$

4   set $k' = $ D2D pair $(\hat{k}_1, \hat{r})$ and $k'' = $ D2D pair $(\hat{r}, \hat{k}_2)$
5   run power control algorithm on both $G_{\hat{f}_1} \cup \{k'\}$ and $G_{\hat{f}_2} \cup \{k''\}$ separately
6   **if** *there is a feasible power allocation for both* $G_{\hat{f}_1} \cup \{k'\}$ *and* $G_{\hat{f}_2} \cup \{k''\}$ **then**
7   set $G_{\hat{f}_1} = G_{\hat{f}_1} \cup \{k'\}$ and $G_{\hat{f}_2} = G_{\hat{f}_2} \cup \{k''\}$
8   assign relay $\hat{r}$ to D2D pair $\hat{k}$
9   set $R_{used} = R_{used} \cup \{r\}$
10   set $C_{tmp} = C_{tmp} \setminus \{\hat{k}\}$

---

5 Simulation Results

We consider a simulation environment similar to [7]. We have taken a cell of size $500\,\text{m} \times 500\,\text{m}$ with total 200 UEs among which requesting D2D pairs varying from 30 to 95 and while the rest act as relays. The distance between a D2D pair varies from 1 m to 20 m. For the mmWave D2D communications in 60 GHz frequency we adapt the channel parameters from [10, 12]. More specifically, channel bandwidth is 20 MHz, beamwidth is 45°, the maximum transmission power is 24 dBm, SINR threshold is 15 dB, thermal noise is -174 dBm/Hz, pathloss exponent is 1.88, and the shadowing random variable follows log normal distribution with zero mean and 1.2 dB standard deviation. Here $\lambda = \frac{c}{f}$ where c is the speed of light and $f = 60$ GHz. For the visibility computation we consider the grid cells are of size $5\,\text{m} \times 5\,\text{m}$.

We compare our proposed algorithm with the D2D resource allocation and power control (DRAPC) algorithm presented in [7] in terms of number of D2D pair links successfully activated, overall system throughput achieved, energy efficiency and fairness of resource allocation. For the energy efficiency we consider number of bits transmitted per unit power measured in $bit/mJoule$, while the fairness of the throughput of the requesting pairs is computed following Jain's fairness index [7].

Since DRAPC does not consider presence of any obstacles, for a fair comparison we first consider a simulation environment with no obstacles. As evident in Fig. 4 our proposed algorithm outperforms the DRAPC algorithm in all of the four criteria. This improvement can be attributed to the fact that, given a channel assignment, DRAPC uses a naive greedy strategy for power optimization while our proposed algorithm optimally solves an LP for the same. By minimizing the power assignment to a frequency group our algorithm widens the scope for packing more D2D pairs in the same group while satisfying the required SINR.

Furthermore, we also consider two more simulation environments, one with 15 obstacles and another with 25 obstacles each of size $10\,m \times 10\,m$ placed in the area uniformly at random. In Fig. 5 we again observe the improvement of the results as the number of obstacles are increased from zero to 15 and 25. This improvement comes from the fact that, while the DRAPC algorithm can only activate D2D pairs having direct LOS communication paths, our algorithm can significantly improve the solution by considering relay aided communications.

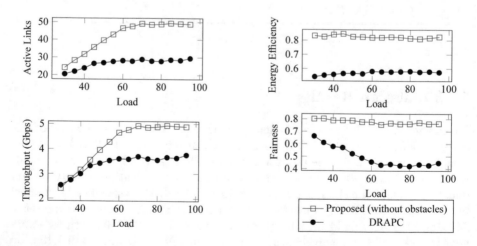

Fig. 4. Experimental results with no obstacles

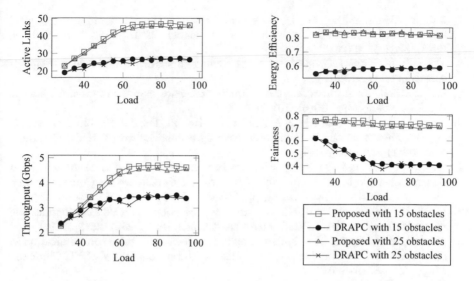

Fig. 5. Experimental results with varying number of obstacles

6 Conclusion

We have addressed the channel and power allocation problem in context of mmWave D2D overlay communications in the presence of static obstacles. To this end, we have proposed an algorithm which by extensive simulations is shown to outperform the DRAPC algorithm in terms of number of links successfully activated, system throughput achieved, energy efficiency and fairness of resource allocation.

References

1. Cai, L.X., Cai, L., Shen, X., Mark, J.W.: Rex: a randomized exclusive region based scheduling scheme for mmwave wpans with directional antenna. IEEE Trans. Wireless Commun. **9**(1), 113–121 (2010)
2. Congiu, R., Ghadikolaei, H.S., Fischione, C., Santucci, F.: On the relay-fallback tradeoff in millimeter wave wireless system. In: IEEE Conference on Computer Communications, INFOCOM Workshops, San Francisco, CA, USA, April 10-14, pp. 622–627 (2016)
3. Friis, H.T.: A note on a simple transmission formula. Proc. IRE **34**(5), 254–256 (1946)
4. Hoang, T.D., Le, L.B., Le-Ngoc, T.: Joint mode selection and resource allocation for relay-based d2d communications. IEEE Commun. Lett. **21**(2), 398–401 (2017)
5. Hu, Q., Blough, D.M.: Relay selection and scheduling for millimeter wave backhaul in urban environments. In: IEEE 14th International Conference on Mobile Ad Hoc and Sensor Systems (MASS), pp. 206–214 (2017)
6. Kim, T., Dong, M.: An iterative Hungarian method to joint relay selection and resource allocation for d2d communications. IEEE Wireless Commun. Lett. **3**(6), 625–628 (2014)

7. Lai, W., Wang, Y., Lin, H., Li, J.: Efficient resource allocation and power control for lte-a d2d communication with pure d2d model. IEEE Trans. Veh. Technol. **69**(3), 3202–3216 (2020)
8. Moore, S.K.: Superaccurate GPS chips coming to smartphones in 2018. https:// spectrum.ieee.org (2017)
9. Pi, Z., Khan, F.: An introduction to millimeter-wave mobile broadband systems. IEEE Commun. Mag. **49**(6), 101–107 (2011)
10. Qiao, J., Shen, X.S., Mark, J.W., Shen, Q., He, Y., Lei, L.: Enabling device-to-device communications in millimeter-wave 5g cellular networks. IEEE Commun. Mag. **53**(1), 209–215 (2015)
11. Sarkar, S., Ghosh, S.C.: Relay selection in millimeter wave d2d communications through obstacle learning. In: 2020 International Conference on Communication Systems Networks (COMSNETS), pp. 468–475 (2020)
12. Singh, D., Ghosh, S.C.: Mobility-aware relay selection in 5g d2d communication using stochastic model. IEEE Trans. Veh. Technol. **68**(3), 2837–2849 (2019)
13. Tehrani, M.N., Uysal, M., Yanikomeroglu, H.: Device-to-device communication in 5g cellular networks: challenges, solutions, and future directions. IEEE Commun. Mag. **52**(5), 86–92 (2014)
14. Willard, D.E.: New data structures for orthogonal range queries. SIAM J. Comput. **14**(1), 232–253 (1985)
15. Yu, G., Xu, L., Feng, D., Yin, R., Li, G.Y., Jiang, Y.: Joint mode selection and resource allocation for device-to-device communications. IEEE Trans. Commun. **62**(11), 3814–3824 (2014)

Bandwidth Slicing with Reservation Capability and Application Priority Awareness for Future Vehicular Networks

Abdullah A. Al-khatib[1]([✉]), Abdelmajid Khelil[1], and Mohammed Balfaqih[2,3]

[1] Institute for Data and Process Science, Department of Computer Science,
Landshut University of Applied Sciences, Landshut, Germany
{s-aalkha,khelil}@haw-landshut.de
[2] Department of Computer and Network Engineering,
University of Jeddah, Jeddah, Saudi Arabia
[3] Department of Automobile Transportation, South Ural State University, Chelyabinsk, Russia

Abstract. This paper addresses the bandwidth slicing for different types of connected vehicle applications with diverse demands in heterogeneous networks. We develop a joint network and device levels slicing scheme based on bandwidth reservation and allocation optimization. The scheme enables vehicles to reserve bandwidth in advance for their critical applications. It categorizes and prioritises vehicular applications based on their degree of criticality. The reserved bandwidth is first allocated. Then, the sub-optimal allocation of the remaining bandwidth is determined and allocated according to degree of criticality of vehicular applications to achieve maximal resource utilization. Moreover, we employ machine learning to accurately predict the time for bandwidth re-slicing, whereas exceeding this time causes high cost of latency and consequently degrades the application performance. Our proposed scheme outperforms the conventional slicing allocation schemes in terms of efficiency and reliability of bandwidth utilization.

1 Introduction

The emergence of the fifth generation (5G) and beyond networks are expected to provide Vehicle-to-everything (V2X) communication and facilitate various new vehicular applications such as cooperative, assisted, or autonomous driving. These applications have great potential to address traffic related issues, including reducing traffic accidents and congestion, and improving the comfort of passengers and the energy efficiency during travel. Vehicular applications are considered as a real time traffic due to that, both high network connectivity and high bandwidth are required. Besides, ultra-low latency is needed for safety applications such as obstacle detection and lane change [1]. These requirements complicate the problem of providing robust network resources on-time and fulfilling the desired Quality of Service (QoS) [2].

To meet such crucial requirements, the 5G and beyond networks are intended to adopt many emerging technologies while exploiting heterogeneous networks (HetNets). The exploration of the unlicensed bands and underutilized spectrum through different wireless access technologies has been proposed [3]. For instance, Long-Term Evolution

L. Barolli et al. (Eds.): AINA 2021, LNNS 225, pp. 681–691, 2021.
https://doi.org/10.1007/978-3-030-75100-5_58

(LTE) devices can utilize WiFi unlicensed bands to support bandwidth critical applications. However, vendors and service providers in a HetNet must coordinate for resource management, which is a very difficult process hindering innovations in network architecture. Thus, the concept of network slicing has been identified, in the first release of 5G normative specifications [4], to facilitate multi-tenancy, multi-service support, and provide on-demand network services for different application scenarios. Network slicing refers to creating multiple end-to-end virtual networks over a common infrastructure. Each network slice is logically isolated to serve different types of applications with diverse requirements.

Software Defined Network (SDN) and Network Function Virtualization (NFV) paradigms are seen as promising enabler for network slicing, in which NFV enables implementing network functions that are realized in costly hardware platforms as software appliances placed on low-cost commodity hardwares or running in the cloud computing environment [5]. This architecture separates control plane and data plane to realize the flexible control of network traffic. Since the centralised SDN controller has a global network view, it is deployed to dynamically make routing decisions that satisfy diverse service demands with maximum cost-efficiency. Within a slice, the required bandwidth and computation resources can be allocated on-demand depending on a set of QoS parameters and other indicators such as latency, reliability, and energy efficiency.

Bandwidth slicing solutions are classified according to slicing level into network-level and device-level solutions. The former refers to allocating the aggregated bandwidth of a HetNet among the BSs. The latter is the process of allocating BS resources to its associated vehicles which might occur based on reactive or proactive approaches. The reactive allocation occurs after application performance degradation or significant demand for more bandwidth. In addition, it is accomplished upon vehicle request and negotiation with SDN controller resulting in high cost of signaling and latency. The proactive allocation, on the other hand, is challenging in terms of the required vehicles movement tracking and future position forecasting. It is necessary to ensure that the overhead of employing proactive method is reasonable compared to its benefit. In addition, resource allocation must be timely and accurate especially for delay-critical applications. The early allocation could lead to a waste of resources while the late allocation might lead to service interruption, and consequently degrading the user experienced service quality. This paper is extents to our prior work (reservation-based slicing [6]), with focus on a joint network and device level slicing scheme based on reservation and bandwidth allocation optimization in HetNets. The solution enables vehicles to reserve bandwidth in advance for their critical applications. Moreover, it categorizes and prioritises vehicular applications based on their degree of criticality. With the softwarization function, SDN controller aggregates the total bandwidth from HetNet. The SDN controller specifies how the bandwidth is allocated among the BSs. The network-level slicing rate for each BS is determined according to calculated device-level slicing rates. The bandwidth ratio of each BS is determined proactively in two steps: (i) The reserved bandwidth is assigned to vehicles that reserved resources for their delay-critical applications, and (ii) the sub-optimal allocation of the remaining bandwidth is determined and allocated based on criticality degree of vehicular applications to achieve maximal resource utilization.

Moreover, a machine learning model is utilized to predict the accurate time of bandwidth re-slicing, which consequently reduces the latency.

The remainder of the paper is organized as follows. We present network slicing concept and types as well as the requirements of different vehicular applications. Next, the related work is discussed and compared. Then, we describe our proposed priority and reservation-based slicing scheme including the prediction model of re-slicing time. Consequently, we present the performance evaluation of the proposed scheme. Finally, we conclude the paper and give an overview of our future work.

2   Network Slicing Concept and Types

The network slicing utilizes available physical or virtual resource of wireless networks to enable service providers to efficiently manage and customize their virtual networks, and thereby guarantee the demanded QoS [7]. The network slicing can be classified based on the type of resource and network function sharing into Core Network (CN) slicing and Radio Access Network (RAN) slicing [8]. The CN refers to dividing networks into more fine-granular network functions [9]. The functions are related to control plane and user plane functionalities, such as mobility management, session management, and authentication. Each function becomes flexible, scalable, programmable, and auto configurable and can be managed by SDN and NFV [10]. On the other hand, the RAN refers to dividing the physical resources (bandwidth and power) of various Radio Access Technologies (RAT) between multiple virtual slices. The bandwidth slicing is more challenging due to the different nature and parameters of wireless resources, such as time/frequency resources, frame size, and Hybrid Automatic Repeat Request (HARQ) options.

With respect to slicing level, the bandwidth slicing solutions are categorized into network-level and device-level solutions. In the former, the bandwidth resources of a HetNet are allocated among the BSs. In a process known as slicing, BS resources are allocated to its associated users; this process may be based on either one of reactive or proactive approaches. In reactive approach, the user first submits its demand for bandwidth; only then does the process of negotiation between BS and SDN controller begin. This process may result in significant latency, leading to service degradation during the negotiation process. The latency period is dependent on the time taken during the period of request and initiation of response. In contrast, in proactive approach, prediction or reservation are used prior to the allocation of user demand of BS resources. The latency is reduced, in proactive approach, by assigning the bandwidth resource proactively to vehicles application based on requirements criteria to ensure optimized network performance.

Implementing network slicing in vehicular networks enhances network agility and reconfigurability. It enables a network operator to guarantee diverse QoS requirements of different vehicular applications by composing slices flexibly. Based on the minimum level required of reliability and latency, the applications can be categorized and prioritized accordingly into three classes:

(i)   Internet and infotainment applications (priority-III):

This class aims to facilitate more infotainment experience such as web browsing, and video streaming. It is expected that such applications will be more on-demand upon the widespread of automobile driving, where drivers need media consumption. The requirement of low latency is not critical for these applications and the reliability is not a concern [8]. The required data rate is 0.5 Mb/s for web browsing, and up to 15 Mb/s for HD video [11].

(ii)   Traffic safety and efficiency applications (priority-II):

The main objective of traffic safety and efficiency applications is to reduce the traffic accidents and improve traffic efficiency. For instance, to notify vehicles in surrounding area about emergency stop event to trigger safer behavior, a vehicle sends a message about this event to Roadside Unit (RSU), which in turn relays the message to its surrounding vehicles. Such applications require low latency (i.e., 100 ms maximum latency) and high reliability (i.e., about 99%). The required data rate is about 1 Mb/s.

(iii)   Autonomous driving related applications (priority-I):

They enable a vehicle to operate with little to no human assistance. For example, to create an image of their surroundings and navigate traffic, bird's eye view application is used. They stream data from sensors such as cameras or radar at intersections to identify eventual pedestrians or free places. Compared to traffic safety applications, autonomous driving applications require ultra-low latency (i.e., 10 ms maximum latency), ultra-high reliability (i.e., at least 99.999%), and high data rate (i.e., at least 10 Mb/s).

3   Related Work

Several efforts have been dedicated for bandwidth slicing in different use cases including vehicular networks. Allocating the network bandwidth statically between operators was the main slicing. Due to the variation of users and load distribution on idea of the earliest bandwidth slicing solutions for network-level each cell, dynamic bandwidth allocation solutions have been proposed based on the allocated device-level slices which are determined reactively or proactively.

A reactive bandwidth allocation scheme for two-tier HetNet Networks was proposed in [12]. Optimal bandwidth allocation ratios of each BS were determined based on the service type of their end users to maximize the network utility. Although two types of applications were considered, there is no prioritization of critical applications. Moreover, the scenario of having multi-association was not considered. An architecture of Mobile Edge Computing (MEC) and SDN was proposed in [13] for autonomous vehicles to address the issue of high data traffic volume and high number of computing/storing tasks. A dynamic multi-resource management scheme by migrating computing task and radio bandwidth slicing was introduced. An algorithm of dynamic bandwidth allocation among multiple virtual wireless operators and service providers was presented in [14].

A weighted proportionally fairness/protection criteria was developed and considered to achieve desirable fair bandwidth allocation. Pareto-optimality of user association to base stations was also considered. The fairness was considered in terms of utility gains and capacity savings across different operators and service providers and their associated users. The main limitation of the proposed algorithm is the high complexity.

Several network slicing solutions have been proposed to allocate bandwidth reactively or proactively upon forecasting the network load and bandwidth usage in the next time window. However, the main limitation of adopting forecasting models is the high cost and complexity due to the need of vehicle movement tracking and predicting their future locations. A new metric was defined based on the available bandwidth that can be allocated to a single very active bearer. In [15], an optimization problem with the objective of minimizing network overheads while satisfying QoS requirements of each virtual network was formulated. A regressive moving average prediction method was developed to obtain the optimal allocation ratios with the consideration of load forecasting.

The authors in [15] proposed a mechanism for 5G Network Slice Broker that involves network slicing traffic forecasting and admission control. The slice forecasting module was developed to analyse the network slices traffic patterns and provide forecasting information to the admission control module. Based on this information, admission control policies are applied to take a decision on tenants requests for the next time window. Another solution for 5G Network Slice Broker was proposed in [16] to measure, predict usage, and provision slice resources by deriving the throughput of an individual wireless link. A short-term prediction model was developed to predict resource usage by a slice based on historical collected data of the network.

4   Bandwidth Slicing Scheme with Reservation Capability and Application Priority Awareness

The vehicles in the proposed scheme can reserve the needed bandwidth proactively. The reservation process must occur within a suitable time before the actual need. At the Fig. 1, a slice manager in charge of designing a proper bandwidth slice for the vehicles and interacts with the SDN controller to deploy the bandwidth slicing.

Vehicle v_i (i = 1, 2,..., n) issues a request to slice manager to reserve bandwidth slicing for a certain period at a certain time t and location in which v_i will be located at t. We categorize, and prioritise vehicular applications based on the minimum level required of reliability and latency into three classes. Each v_i has a positive constant weightω, where ω indicates that the corresponding vehicular application has a first, second or third priority class. The bandwidth slicing scheme is initiated by aggregating the total network bandwidth from all BSs. The network-level slicing is determined upon obtaining the total reserved bandwidth βrr by vehicles and number of active associated vehicles at each BS. First, the slice manager obtains the reserved bandwidth rate at each BS. If the total reserved bandwidth did not exceed the total bandwidth of HetNet, the slice manager then determines the sub-optimal allocation rate of the remaining bandwidth β_{rmg} at each BS based on the applications of the associated vehicles and their priorities.

To obtain the sub-optimal allocation of β_{rmg}, an optimization problem is formulated. The optimization objective is the utility maximization (see Fig. 2) of β_{rmg} for various

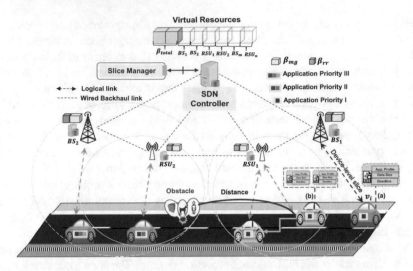

Fig. 1. Bandwidth slicing in two-tier HetNet

vehicle applications with different priorities, which is the summation of utilities achieved by all vehicles with various priorities. The network utility maximization depends on number of associated vehicles at each BS, bandwidth reservation ratio, and bandwidth device-level slicing ratio. The input of the formulated problem includes:

- β_t: The total aggregated network bandwidth from all BSs.
- βrr: The ratio of total reserved bandwidth.
- ω: The weight of vehicular application representing their prioritization level.
- X_{BS}: The association pattern (X_{BS}), where X_{BS} equal one if v_i is associated with the BS and X_{BS} equal zero if not.
- f_r: The sub-optimal device-level slicing ratio received from the BS for v_i, which is determined according to their X_{BS}, the priority level of their application, and the total number of associated vehicles with the BS.

The following constraints are considered in the problem formulation to maximize the utilization of β_{rmg}:

- Each vehicle can only associate with one BS during each bandwidth allocation period. It is assumed that there is no vehicle with multi association patterns.
- The association pattern must be binary in which a vehicle is either associated to a BS or not.
- The allocated bandwidth for each slice is the lowest acceptable bandwidth, which is not less than periodic packets arrival rate.
- The allocated bandwidth guarantees that the achievable rate is not less than the effective bandwidth.
- The summation of ratio βrr and ratio β_{rmg} is equal to one.

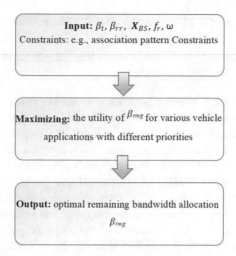

Fig. 2. Formulation components of the remaining bandwidth β_{rmg} optimization.

A case study of two-tier HetNet is shown in Fig. 1. The network consists of multiple Macro BSs and RSUs with an overlap coverage. The Macro BSs (B_m) are placed at the first tier. In the second tier, small RSUs (R_s) cells are placed. The B_m and R_s are directly connected to edge routers of the core network via wired backhaul links.

Alternative Concave Search (ACS) algorithm [17] is utilized to solve the optimization problem. It is a maximization method based on block-relaxation, where only the variables of an active block are optimized while the variables of the other blocks are fixed. In this case study, the bandwidth slicing ratios β_{Bm} and β_{Rs} are firstly fixed to obtain sub-optimal association patterns. Then, the sub-optimal bandwidth slicing ratios β_{Bm} and β_{Rs} (where β_{Bm} and β_{Rs} are β_{rmg} for B_m and R_s respectively) are obtained by fixing the association patterns. The ACS algorithm iteratively solves for sub-optimal bandwidth slicing ratios and association patterns.

5   Prediction of Re-slicing Time

To re-allocate the total bandwidth between BSs, the re-slicing process must be conducted in a certain time window. The βrr will be assigned to vehicles based on their reservation time and location. However, the re-slicing of the β_{rmg} must be timely accurate especially in the scenario of delay-critical applications. The early re-slicing of β_{rmg} could lead to a waste of resources while the late re-slicing might lead to service interruption, and consequently degrading the user experienced service quality [10].

An example of such scenario is shown in Fig. 1. Vehicle v_i acquires the allocated bandwidth slicing that satisfies QoS requirement of its current applications. For example, the slice manager during the slice reservation allocated 10 Mbps to Vehicle at position **(a)**. Vehicle was predicted to transmit an image of unknown obstacle ahead with 500 ms maximum latency requirements to its Edge computing to identify the obstacle. However, Vehicle at the position **(b)** decided to change lane and will use lane changing warning service with 100 ms maximum latency requirement. When Vehicle starts another application with lower latency requirement (i.e., higher QoS requirements), the slice manager in such scenario must be aware of maximum allowable latency of vehicle applications and increase the allocated bandwidth slice.

It is significant for the slice manager to predict the accurate time of bandwidth re-slicing. Exceeding this time causes higher latency and consequently degrades the QoS of applications. We employ a machine learning model for predicting the re-slicing time. We use a traffic simulation-based dataset as an input flow to train the machine learning model. Here, we apply our re-slicing time prediction model in the scenario of obstacle detection application. First step starts with the generation of the simulation dataset that consists of one million rows. Each row consists of five fields, Vehicle-ID, Section-ID for each road network link, a velocity, a distance, and the critical moment which is used in traffic management decision-making process. The critical moment is the time that delay-critical application performance will start degrading if the re-slicing process is not performed. The value of the critical moment depends on the type of vehicular application. The dataset considers Vehicle-ID, Section-ID for each road network link, velocities within the range (20–220) (km/h) and distances to an obstacle within the range (20–220) (meters). The dataset is divided into training dataset and testing dataset.

The training dataset is used in the second step to train different machine learning models and determine the critical moment of obstacle detection application. The examined models are Naive Bayes (NB), Decision Tree (DT), K-Nearest Neighbor (KNN), Support Vector Machine (SVM), and Random Forest (RF). These machine learning models learn to predict the suitable re-slicing time without resource wasting or degrading QoS of applications. The accuracy of these predictions depends on how well the model can be generalized from the training data. The first and second steps are time-consuming processes and thus they are performed offline in advance. They are performed once, then the trained machine learning models are saved and reused later. During the training phase, the KNN model outperformed the other model, thus it is adopted in our solution. In the third step, the re-slicing solution loads the trained machine learning model and decides the time of bandwidth re-slicing based on the data given in the first arrival flow.

6   Performance Evaluation

The case study presented in Fig. 1 is utilized to demonstrate the effectiveness of the proposed joint network and device levels bandwidth slicing scheme over two-tier HetNet. The transmit power is set to 40 dBm for each B_m and 30 dBm for each R_s. The distance between each B_m and its under-laid RSU is set as 400 m, as recommended by 3GPP [4]. The locations are fixed for B_m and R_s. The total number of vehicles is varied from 50 to 300 vehicles and uniformly distributed between B_m and R_s according to a Poisson Point Process (PPP). The total reserved bandwidth rate is set to 10%. The random data arrivals and current data queue backlogs of the Macro BS is 20 packet/s, and the average rate packet arrivals at each queue of either the Macro BS or an RSU is 5 packet/s. The default parameters values are obtained from 3GPP [4].

Figure 3 studies the effect of different reservation rates in each R_s. The number of vehicles in each R_s is increased gradually by 25 vehicles. The results show that the

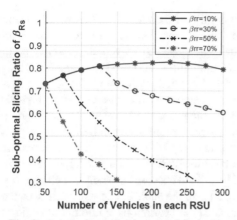

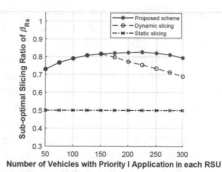

Fig. 3. Sub-optimal slicing ratio of β_{Rs} with different βrr ratio

Fig. 4. Sub-optimal slicing ratio of β_{Rs} achieved by different slicing solution with variations of number of vehicles with priority I application.

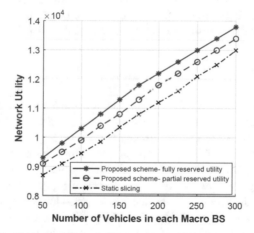

Fig. 5. Network utility of different bandwidth slicing solutions.

sub-optimal bandwidth ratio of the remaining bandwidth β_{Rs} for R_s increases gradually till the number of vehicles reaches the peak value (i.e., the bandwidth is completely allocated). After the peak value, the sub-optimal ratio is decreased gradually because the added vehicles share the slices with the existing vehicles in the network. The peak value is reached with a smaller number of vehicles when the reserved slicing rate βrr is higher. The peak values are 50, 75, 125, and 250 vehicles when the reserved slicing rates βrr are 10%, 30%, 50%, and 70%, respectively. The number of vehicles that reserve the bandwidth for their applications is a key decision to obtain a good system performance. Different scenarios have different suitable reserved slicing rate depending on the number of the vehicles and network parameters.

Figure 4 compares the sub-optimal slicing ratio of the remaining bandwidth β_{Rs} in our proposed scheme, dynamic slicing, and static slicing. The number of vehicles with priority I application in each RSU is gradually increased in 25 vehicle increments. In static slicing, fixed rate from the aggregated bandwidth is allocated to the RSUs. Thus, increasing the number of vehicles with priority I application has no effect on the sub-optimal slicing ratio. The bandwidth is totally allocated when the number of vehicles with priority-I reaches 250 vehicles in the proposed scheme and 150 vehicles in the dynamic slicing scheme. This is because our scheme allocates more bandwidth to vehicles with priority-I application instead of other applications, which consequently decreases the chance for applications with low priority to be associated with their preferred network.

The bandwidth resource might not be fully utilized because of the mismatch between the demand and actual bandwidth allocation. Hence, the achieved network utility of the macro BS by different slicing solution is evaluated with variations of number of vehicles as shown in Fig. 5. The network utility, and thus the network throughput of the bandwidth using the proposed scheme is significantly improved compared to static slicing. This can be explained as the static slicing allocates the bandwidth to BSs with a fixed rate without consideration of the demands of its associated vehicles applications. The network utility of the proposed scheme is higher if the reserved bandwidth is fully utilized because the vehicles applications requirements are considered. However, the network utility will be lower if the reserved bandwidth is not utilized or partially utilized.

7   Conclusion

In this article, we developed a joint network and device level slicing scheme based on reservation and bandwidth allocation optimization. The scheme categorizes and priori- tises vehicular applications based on their degree of criticality. Accordingly, the vehicles reserve bandwidth in advance for their critical applications. First, the slice manager allo- cates the reserved bandwidth, then obtains and allocates the sub-optimal of the remaining bandwidth according to degree of criticality of vehicle applications to achieve maximal resource utilization. Moreover, a machine learning model is employed to predict the accurate time of bandwidth re-slicing. Exceeding this time causes higher latency and consequently degrades the fidelity of applications. Finally, the numerical results con- firmed that the proposed scheme allocates the demanded bandwidth for delay-critical applications more efficiently than the current slicing solutions. Our future work will investigate the reservation pricing possibility and the scenario of reservation cancelation.

References

1. Chebaane, A., Khelil, A., Suri, N.: Time-critical fog computing for vehicular networks. Chapter in Wiley Book. Fog Comput. Theory Pract. (2020). ISBN: 978-1-119-55169-0. https://onlinelibrary.wiley.com/doi/abs/10.1002/9781119551713.ch17
2. Palattella, M.R., Soua, R., Khelil, A., Engel, T.: Fog computing as the key for seamless connectivity handover in future vehicular networks. In: Proceedings of the 34th ACM/SIGAPP Symposium On Applied Computing (SAC) (2019)

3. Huang, J., Qian, F., Gerber, A., Mao, Z.M., Sen, S., Spatscheck, O.: A close examination of performance and power characteristics of 4G LTE networks. In: Proceedings of the 10th International Conference on Mobile Systems, Applications, and Services (MobiSys), pp. 225–238 (2012)
4. Study on Management and Orchestration of Network Slicing for Next Generation Network (Release 15): Document TS 28.801, 3GPP (2018)
5. Kreutz, D., Ramos, F.M., Verissimo, P.E., Rothenberg, C.E., Azodolmolky, S., Uhlig, S.: Software-defined networking: a comprehensive survey. In: Proceedings of the IEEE, vol. 103(1), pp. 14–76 (2014)
6. Al-Khatib, A.A., Khelil, A.: Priority-and reservation-based slicing for future vehicular networks. In: Proceedings of the 3rd International Workshop on Advances in Slicing for Softwarized Infrastructures (S4SI@NetSoft), pp. 36–42 (2020)
7. Mei, J., Wang, X., Zheng, K.: Intelligent network slicing for V2X services toward 5G. IEEE Netw. 33(6), 196–204 (2019)
8. Zhou, G., Zhao, L., Liang, K., Zheng, G., Hanzo, L.: Utility analysis of radio access network slicing. IEEE Trans. Veh. Technol. 69(1), 1163–1167 (2019)
9. Ksentini, A., Frangoudis, P.A.: Toward slicing-enabled multi-access edge computing in 5G. IEEE Netw. 34(2), 99–10 (2020)
10. Wang, G., Feng, G., Quek, T.Q., Qin, S., Wen, R., Tan, W.: Reconfiguration in network slicing—optimizing the profit and performance. IEEE Trans. Netw. Serv. Manage. 16(2), 591–605 (2019)
11. Study on LTE-based V2X Services (Release 14): 3GPP TR 36.885 Std. (2016)
12. Ye, Q., Zhuang, W., Zhang, S., Jin, A.L., Shen, X., Li, X.: Dynamic radio resource slicing for a two-tier heterogeneous wireless network. IEEE Trans. Veh. Technol. 67(10), 9896–9910 (2018)
13. Peng, H., Ye, Q., Shen, X.S.: SDN-based resource management for autonomous vehicular networks: a multi-access edge computing approach. IEEE Wirel. Commun. 26(4), 156–162 (2019)
14. Caballero, P., Banchs, A., De Veciana, G., Costa-Pérez, X.: Multi-tenant radio access network slicing: statistical multiplexing of spatial loads. IEEE/ACM Trans. Netw. 25(5), 3044–3058 (2017)
15. Sciancalepore, V., Samdanis, K., Costa-Perez, X., Bega, D., Gramaglia, M., Banchs, A.: Mobile traffic forecasting for maximizing 5G network slicing resource utilization. In: Proceedings of the IEEE Conference on Computer Communications (INFOCOM), pp. 1–9 (2017)
16. Gutterman, C., Grinshpun, E., Sharma, S., Zussman, G.: RAN resource usage prediction for a 5G slice broker. In: Proceedings of the Twentieth ACM International Symposium on Mobile Ad Hoc Networking and Computing, pp. 231–240 (2019)
17. Gorski, J., Pfeuffer, F., Klamroth, K.: Biconvex sets and optimization with biconvex functions: a survey and extensions. Math. Meth. Oper. Res. 66(3), 373–407 (2007)

Proposal of a Disaster Response AI System

Tomoyuki Ishida[✉] and Tenma Hirose

Fukuoka Institute of Technology, Fukuoka, Fukuoka 811-0295, Japan
t-ishida@fit.ac.jp, s17b1042@bene.fit.ac.jp

Abstract. In the event of a large-scale natural disaster, the Disaster Countermeasures Headquarters refers to past disaster information and disaster response that are not stored in a database. It usually takes a lot of time to refer to this information. Therefore, there is a possibility that disaster response will be delayed. In this study, we have developed a method that automatically proposes disaster response such as shelter operation and transportation of relief supplies using AI technology from information such as past disaster response records and regional disaster prevention plans.

1 Introduction

In the event of a disaster, the disaster countermeasures headquarters is an organization established in the national, prefectural, and municipal governments to smoothly collect and transmit disaster information and make decisions on disaster response actions. The disaster countermeasures headquarters must function effectively to respond promptly and appropriately. When a disaster occurs, the disaster countermeasures headquarters will be busy with various disaster responses, such as evacuation shelter management, management and provision of relief supplies, and public relations activities. However, the Cabinet Office (Japan) has been developing an "Integrated Disaster Management Information System [1]" for sharing disaster prevention information among disaster prevention related organizations since 2005. It is not easy to transfer technology to municipalities since this system is a large-scale system. Besides, the Ministry of Internal Affairs and Communications started the operation of "L-alert (Public Information Commons) [2]" in 2011, which quickly and accurately conveys disaster information to residents. This system does not have a viewpoint of disaster response at the disaster site level. Currently, there are few approaches to support using information technology based on the disaster countermeasures headquarters of each municipality.

2 Previous Study

Takahagi et al. [3–5] developed comprehensive disaster prevention and mitigation information system consisting of a disaster information registration system, a disaster information sharing system, and a disaster information transmission common platform. The disaster information registration system digitizes various disaster information handled by the disaster countermeasures headquarters. The disaster information sharing system aggregates and analyzes the digitized disaster information. Besides, the disaster information transmission common platform quickly distributes disaster information to residents via various media and information transmission tools. However, Hirohara et al. [6] developed a decision-making support cloud system for disaster countermeasures using an interactive large-scale ultra-high-definition display. This system supports the organization of enormous amounts of information and decision-making by displaying digitized disaster information for each content on a large display installed at the disaster countermeasures headquarters. However, both studies did not consider the utilization of regional disaster prevention plans or past disaster response records. Thus, we implement a disaster response AI system that proposes a disaster response at the disaster countermeasures headquarters by AI using the disaster response job data of Chiba Prefecture in Typhoon Faxai and Hagibis in 2019.

3   System Architecture

Figure 1 shows the system's architecture. The system's architecture consists of the application server group and database server group. The application server group consists of a disaster response operation management server, disaster response operation evaluation server, and disaster response operation suggestion server. The database server group consists of disaster information, disaster response records, local government staff, and relief supply information storage.

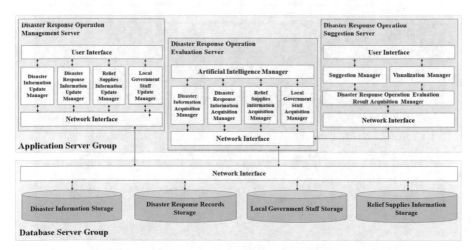

Fig. 1. System architecture of the disaster response AI system.

4 Prototype System

Figure 2 shows the menu screen of the disaster response AI system. The user can register disaster information, relief supply information, and staff information for disaster response by selecting the "Register Disaster Information," "Register Relief Supply Information," and "Register Staff Information" menus at the bottom of the screen. When the user selects "Register Disaster Information" from the menu screen of the disaster response AI system, the screen transitions to the disaster information list screen (Fig. 3). The user can register new disaster information by selecting the "Add" button on the disaster information list screen. Figure 4 shows the new disaster information registration screen. When the user edits the registered disaster information record, the disaster information edit screen is displayed by selecting the "Edit" button on the disaster information list screen. Furthermore, when the user deletes the registered disaster information record, the disaster information deletion screen is displayed by selecting the "Delete" button on the disaster information list screen.

When the user selects "Register Relief Supply Information" from the menu screen of the disaster response AI system, the screen transitions to the relief supply management screen (Fig. 5). The user can register new relief supply information by selecting the "Add" button on the relief supply management screen. Figure 6 shows the new relief supply information registration screen. On the relief supply information registration screen, the relief supplies are managed by registering the number of relief supplies stockpiled by the municipality.

Fig. 2. Menu screen of the disaster response AI system.

Fig. 3. Disaster information list screen for registering disaster information.

Fig. 4. New disaster information registration screen.

Fig. 5. Relief supply management screen.

Fig. 6. Relief supply information registration screen.

When the user selects "Register Staff Information" from the menu screen of the disaster response AI system, the screen transitions to the staff management screen (Fig. 7). The user can register new staff information by selecting the "Add" button on the staff management screen. Figure 8 shows the new staff information registration screen.

Fig. 7. Staff management screen.

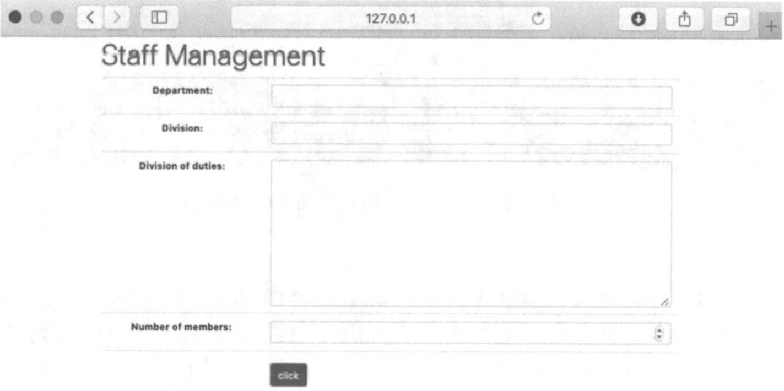

Fig. 8. Staff information registration screen

On the disaster response proposal screen (Fig. 9) of the disaster response AI system, the user can search the disaster response proposal derived from the past disaster response by AI. Based on the past disaster response evaluation, the number of relief supplies to be dealt with in the future is visualized as a disaster response proposal on this screen.

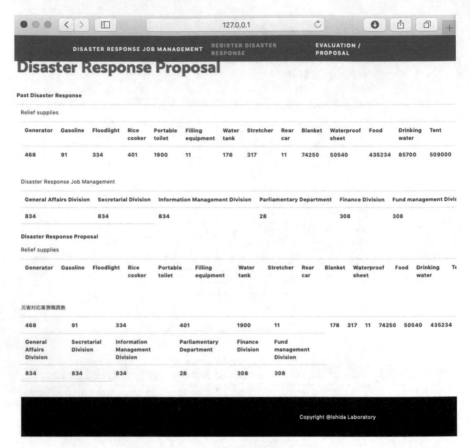

Fig. 9. Disaster response proposal screen.

Figure 10 shows the input/output sequence of the disaster response AI system. When the user registers disaster information in the disaster response AI system, the disaster information is stored in the database through the application server. Besides, when the user re-registers the registered disaster information, the disaster information in the database is updated through the application server. Furthermore, when the user deletes the registered disaster information, the disaster information stored in the database through the application server is deleted. The same flow applies to registration, editing, and deleting relief supply management information and staff management information.

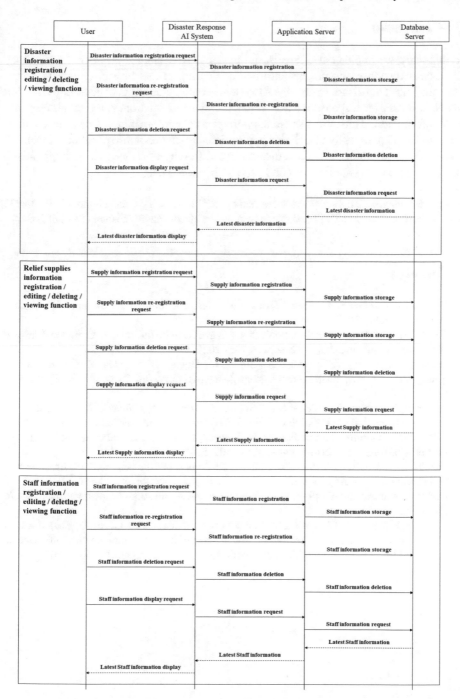

Fig. 10. Input/output sequence of the disaster response AI system.

5 Conclusion

In this study, we implemented the disaster response AI system to improve the efficiency of disaster response and prompt information sharing at the disaster countermeasures headquarters. This disaster response AI system accumulates regional disaster prevention plans, past disaster response records, relief supplies information, and disaster response staff information. Thus, this system supports the decision-making of disaster countermeasures headquarters by visualizing appropriate disaster response personnel allocation and relief supply information in a disaster event. In the future, we plan to implement a function for external applications and expand the stored data.

Acknowledgments. This study was supported by JSPS KAKENHI Grant Number JP19K04972 and the research grand of The Telecommunications Advancement Foundation (TAF) of Japan.

References

1. Cabinet Office (Japan), Integrated Disaster Management Information System Maintenance, Accessed January 2021
2. Foundation for Multi Media Communications, L-alert (Public Information Commons), https://www.fmmc.or.jp/commons/, Accessed January 2021.
3. Takahagi, K., Ishida, T., Sakuraba, A., Sugita, K., Uchida, N., Shibata, Y.: Construction of a mega disaster crisis management system. J. Internet Serv. Inf. Secur. (JISIS) 5(4), 20–40 (2015)
4. Takahagi, K., Ishida, T., Sakuraba, A., Sugita, K., Uchida, N., Shibata, Y.: Proposal of the disaster information transmission common infrastructure system intended to rapid sharing of information in a time of mega disaster. In: Proceedings of the 18th International Conference on Network-Based Information Systems, pp. 505–510 (2015)
5. Takahagi, K., Ishida, T., Uchida, N., Shibata, Y.: Proposal of the common infrastructure system for real-time disaster information transmission. In: Proceedings of the 30th International Conference on Advanced Information Networking and Applications Workshops, pp. 673–676 (2016)
6. Hirohara, Y,, Ishida, T., Uchida, N., Shibata, Y.: Proposal of a disaster information cloud system for disaster prevention and reduction. In: Proceedings of the 31st International Conference on Advanced Information Networking and Applications Workshops, pp. 664–667 (2017)

An Efficient Image Codec Using Spline-Based Directional Lifting Wavelet Transform and an Optimized EZW Algorithm

Rania Boujelbene[✉] and Yousra Ben Jemaa

L3S Laboratory, ENIT, University of Tunis-El Manar, Tunis, Tunisia
{rania.boujelbene,yousra.benjemaa}@enis.tn

Abstract. This paper proposes a new image compression system with an optimal spline-wavelet transform based on adaptive directional lifting and an improved Embedded zero-tree wavelet (EZW) coding algorithm. The performance of the proposed scheme is evaluated using objective and subjective criteria. Experimental results have shown promising performance compared with the results obtained by the JPEG2000 standard.

1 Introduction

Technology advancement has enabled people not only to capture a large amount of data but also to exchange these data. It has also greatly changed the speed and access to the internet, thus allowing a widespread use of Web applications such as games, bank services, streaming video platform (Netflix) and social networks which have been changing dynamically.

Thus, one should be able to compress these data in very efficient formats in order to store them or transmit them over given bandwidth channels without degrading the quality [1, 2].

As a result of this, image compression techniques have been developed with a lot of new algorithms and also with variations of the already existing ones.

In addition, it is always necessary to change these techniques in order to adapt them to the evolution of Web applications and media transmission or storage.

At present, the most advanced image compression systems are based on three processing steps namely decorrelation, quantization and entropy coding specially the arithmetic coding which is more commonly used. The first step consists in changing the representation space of the image. Indeed, a transform approach generally consists in applying a linear transformation to the intensity values representing the image in order to extract the main components. Among the most successful approaches in this field, wavelet decompositions consist in representing the image by a sum of small waves of varying intensity, size and position. A study on existing orthogonal wavelets [3] has been done and has proved that these

L. Barolli et al. (Eds.): AINA 2021, LNNS 225, pp. 701–712, 2021.
https://doi.org/10.1007/978-3-030-75100-5_60

types of wavelets are not symmetrical and introduce distortions during the base change. To fill this gap, other approaches are proposed and prove the need to study the bi-orthogonal wavelets [4] and more precisely the family of splines [5]. However, these approaches do not faithfully represent the detailed information of the image.

In the second step, in order to keep the progressive side of the wavelet transform, several methods of organization and quantization are required [6]. These include EZW, SPIHT, SPECK and EBCOT. These encoders are applied to the coefficients obtained by the wavelet transform [7].

Among the existing wavelet image compression techniques, EZW [8] is one of the first and powerful algorithms. It is shown effective, fast in execution and computationally simple method for the stream bit. Based on the conclusion which was presented in [9] and which states that the EZW is more promising than other existing coding schemes and can bring more improvements, many attempts to enhance its features and to reduce its limitations have been suggested in the literature [9].

Unlike existing approaches [10–13], we dealt with the two steps of image compression in order to construct a new efficient wavelet image compression system. Our contribution is emphasized as follows:

- Enhance the first step of decorrelation by employing a new spline wavelet transform based on directional lifting, which allows to further reduce the magnitude of the high-frequency wavelet coefficients.
- Improve the second step of quantification by using an enhanced EZW coding algorithm which aims at optimizing the binary coding by decreasing the total number of bits contained in the symbols.

The remaining sections of this paper are organized in the following way. In Sect. 2, we present the block diagram for the proposed image compression codec. We describe the principle of the spline-based directional lifting wavelet transform and the different steps of the proposed NE-EZW coding algorithm. The experimental results are presented in Sect. 3. We illustrate the robustness of the proposed compression system through a comparative study with the JPE2000 standard. Finally, Sect. 4 concludes the paper and presents the future work.

2 Proposed Codec Scheme for Wavelet Image Compression

We present here the block diagram of our optimal wavelet image compression scheme. Indeed, we are the first to present this scheme which is composed of three connected blocs as shown in Fig. 1.

A Spline wavelet transform based on adaptive directional lifting (ADL-SWT) represents the transform which combines the spline filter with the adaptive directional lifting (ADL).

After the wavelet transform step, we try to get a way to code the wavelet coefficients into an effective result by taking into account the storage space and

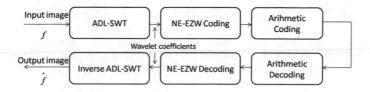

Fig. 1. Block diagram for optimal codec.

the redundancy. An enhanced EZW image coding (NE-EZW) is the best way which ensures a triple trade-off between the total number of symbols/the size of the compressed image/the quality of the reconstructed image. This algorithm consists in combining two adaptive approaches, which makes it possible to have fewer redundant symbols with fewer bits.

In addition, in order to further improve the performance of the proposed NE-EZW, we use arithmetic coding as an entropy coding.

Once the input image has been coded, it is saved or sent through the communication channel to the receiver who needs to use this code in order to reconstruct the input image. This is the decoding process which consists of the arithmetic decoding, the NE-EZW decoding and the inverse ADL-SWT transform.

2.1 ADL-SWT Transform

Instead of alternately using the lifting-based prediction in the horizontal or vertical direction, the ADL performs the prediction in windows of high pixel correlation. For lossy image compression, unlike conventional methods that use the ADL with the biorthogonal 9/7 filter [14], this technique mixes ADL with a spline wavelet filter. In fact, we have concentrated on the polynomial spline for the calculation of the filter taps. Lately, it was shown in [5] that the polynomial spline wavelet filter of fifth order provides the best performance as compared to the most efficient existing filters such as the biorthogonal 9/7.

Hence, to construct this performed scheme, the best spline filter of fifth order is combined with the ADL by incorporating the coefficients calculated by this filter into the ADL. Thus, the proposed ADL-SWT is employed as the representation of our image compression system. The proposed 2-D ADL-SWT involves two separable transforms. The schematic representation of this transform is shown in Fig. 2.

Let X[m,n] be a 2-D signal, where m and n represent the row and column indices, respectively. Firstly, carry out 1-D ADL-SWT on each image column, producing a vertical low-pass subband (L) and a vertical high-pass subband (H). Secondly, carry out 1-D ADL-SWT on each row of L and H.

After one-level decomposition, one low-pass subband (LL) and three high-pass subbands (LH,HL and HH) are generated. In other words, the subband decomposition structure of 2-D ADL-SWT is the same that of 2-D DWT.

For ADL-SWT, unlike DWT which does transform along the fixed direction, the selected filtering need to be encoded as side information. So, to reduce the

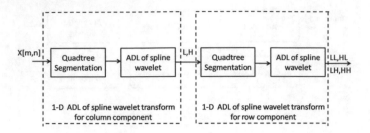

Fig. 2. 2-D Adaptive directional lifting spline wavelet transform.

overhead bits for the direction information, the image is divided into regions of approximately uniform edge orientations using quadtree segmentation. All the pixels in the local region are predicted and updated along the uniform direction which is chosen in a rate–distortion optimal sense. The main steps of the ADL-SWT transform are illustrated by Algorithm 1.

2.2 NE-EZW Coding Algorithm

To increase the compression performance, the NE-EZW is used which can be viewed as an extension of the EZW coding technique. Unlike the previous improved versions of EZW [15,16], the new algorithm ensures a triple trade-off between the total number of symbols/the size of the compressed image/the quality of the reconstructed image.

This algorithm suggests the use of two new added symbols in its dominant pass as compared to the original EZW. This increases the number of coefficients not to be encoded and decreases the total number of symbols by adding two steps namely remove the last T and de-bit encoding that aim at eliminating unnecessary symbols. In addition, the NE-EZW greatly reduces the amount of information by the use of the compressor cell operator in the binary coding step which aims at coding a consecutive number of symbols by a small number of bits.

The NE-EZW has six steps known as initialisation, dominant pass (or significance mapping pass), remove the last T, de-bit encoding, binary encoding and subordinate pass. The process of its coding algorithm is illustrated by Algorithm 2.

3 Experimental Results

3.1 Data and Evaluation Criteria

To assess the efficiency of the proposed approach, a number of numerical experiments have been performed on a number of natural and medical images. For this purpose, the standard Waterloo 8-bit greyscale image set [17] containing 12 images of different sizes (Lena, Barbara, Boat, Mandrill, Zelda, Goldhill, Peppers, House, Washsat, France, Montage, Library) and a two medical images [18] have been considered to perform the required tests.

Algorithm 1. ADL-SWT transform.

1: The image is adaptively divided into blocks of variable sizes based on hierarchical quadtree segmentation.
2: Each block of 16 × 16 can be divided into three modes (16 × 16, 8 × 8 and 4 × 4).
3: Perform the ADL of spline wavelet filter on these modes
4: **if** The enegry metric (sum of absolute coefficients in high frequency subband) related to one of these block modes is smaller than the other **then**
5:   Select this mode
6: **end if**
7: The output is divided to variable-size regions based on hierarchical quadtree segmentation.
8: Each block of 16 × 16 can be divided into three modes (16 × 16, 8 × 8 and 4 × 4).
9: Perform the ADL of spline wavelet filter on these modes
10: **if** The enegry metric related to one of these block modes is smaller than the other **then**
11:   Select this mode
12: **end if**
13: The inverse transform is performed to reconstruct the image by giving the optimal lifting direction as side information

---

For a fair comparison, the proposed codec has been compared with the JPEG2000 [19]. Peak Signal to Noise Ratio (PSNR), Structural Similarity Index Measure (SSIM) and Bjøntegaard Delta PSNR (BD-PSNR) were used in our experiments as a set of performance criteria. The PSNR is defined as follows:

$$PSNR(dB) = 10log_{10}[\frac{(Peak)^2}{MSE}], \tag{1}$$

where $Peak$ is equal to 255 for the images in 8 bits per pixel (bpp) and

$$MSE = \frac{1}{M \times N} \sum_{1}^{N} \sum_{1}^{M} (f(i,j) - \widehat{f}(i,j))^2, \tag{2}$$

where f, $\widehat{f}$ and $M \times N$ represent the original image, the reconstructed image and the total number of pixels in the image, respectively.

Besides PSNR, to further evaluate the quality and the distortion for the reconstructed image using the presented techniques, we use SSIM which is considered to be correlated with the quality perception of the human visual system (HVS). The SSIM is a decimal value between 0 (zero correlation with the original image) and 1 (exact same image). It is defined as follows:

$$SSIM = [l(f,\widehat{f})]^\alpha [c(f,\widehat{f})]^\beta [s(f,\widehat{f})]^\sigma \tag{3}$$

where: $(f,\widehat{f})$, $l(f,\widehat{f})$, $c(f,\widehat{f})$ and $s(f,\widehat{f})$ represent respectively two images, luminance comparison, contrast comparison and structural comparison between two images.

$\alpha > 0$, $\beta > 0$ and $\sigma > 0$ are used to adjust the importance of the three parameters.

To calculate the coding efficiency between different codecs based on PSNR measurements, a Bjøntegaard model was proposed by Gisle Bjntegaard. It is

---

Algorithm 2. NE-EZW.

1: $Th_0 = 2^{(log_2(max(abs(coeffs))))}$ /* Initialization   */
2: k=0
3: Dominant List=All coefficients
4: Character Encoding List=[ ]
5: Subordinate List=[ ]
6: **for** each coefficient x in the Dominant List /* Dominant Pass   */ **do**
7:    **if** $(abs(x) \geq Th_k)$ **then**
8:       **if** at least one descendant $\geq Th_k$ **then**
9:          **if** x>0 **then**
10:             Put P on Character Encoding List
11:          **else**
12:             Put N on Character Encoding List
13:          **end if**
14:       **else**
15:          **if** x>0 **then**
16:             Put P_t on Character Encoding List
17:             Don't code its descendants
18:          **else**
19:             Put N_t on Character Encoding List
20:             Don't code its descendants
21:          **end if**
22:       **end if**
23:    **else**
24:       **if** x is non-root of a zerotree (it is a descendant from a Zerotree Root) **then**
25:          Don't code this coefficient
26:       **end if**
27:       **if** x is a zerotree root **then**
28:          Put T on Character Encoding List
29:       **else**
30:          Put Z on Character Encoding List
31:       **end if**
32:    **end if**
33: **end for**
34: **for** each symbol(P, P_t, N and N_t) in the Character Encoding List **do**
35:    Put its magnitude on the Subordinate List
36:    Remove it from the Dominant List
37: **end for**
38: Scan the Character Encoding List in inversed order/* Remove last T   */
39: Delete all the T after (P, P_t, N or N_t) while encounter the first (P, P_t, N or N_t)
40: Encode the Character Encoding obtained from the previous step by making the use of
    decreasing bits coding/* De-Bit encoding Algorithm   */
41: Transform the output of the previous step into binary coding/* Binary encoding Algorithm   */
42: **for** each coefficient in the Subordinate List/* Subordinate Pass   */ **do**
43:    **if** magnitude of coefficient is in the Bottom Half of $[Th_k, 2Th_k]$ **then**
44:       output '0'
45:    **else**
46:       output '1'
47:    **end if**
48: **end for**
49: $Th_{k+1} = Th_k/2$ /* Halve threshold   */
50: k=k+1
51: Go to the Dominant Pass: Repeat until the reconstructed image is reached or until the
    number of transferable bits required is exceeded

---

used to calculate the average PSNR and bit rate differences between two rate-distortion (R-D) curves obtained from the PSNR measurement when encoding a content at different bit rates. The Bjøntegaard delta PSNR (BD-PSNR), which corresponds to the average PSNR difference in dB for the same bit rate is used in our experiments.

3.2 Performance Analysis of the Proposed Codec

We presented in this section the performance of our system in terms of reconstructed image quality.

The variations of the PSNR versus the bitrate for all test images is given by the curves represented in Figs. 3 and 4.

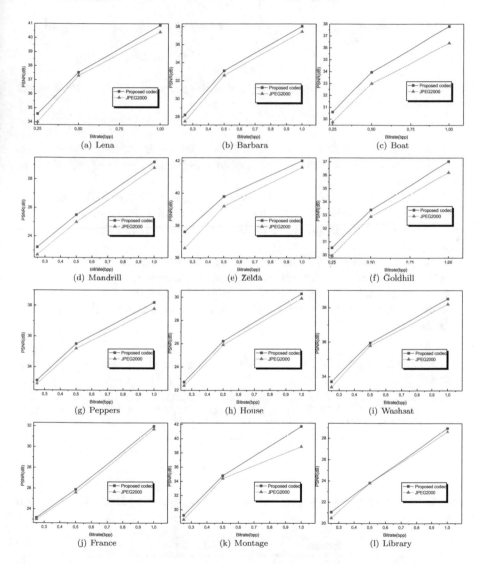

Fig. 3. PSNR (in dB) versus the bitrate (bpp) of the proposed and JPEG2000 codecs for all grayscale test images.

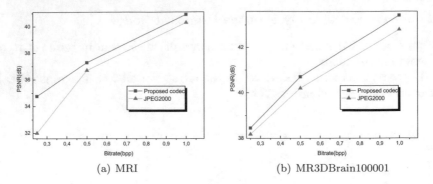

(a) MRI   (b) MR3DBrain100001

Fig. 4. PSNR (in dB) versus the bitrate (bpp) of the proposed and JPEG2000 codecs for all medical test images.

By comparing the different values of PSNR, we see clearly the effectiveness of our proposed codec as compared to the most-well-known codec JPEG2000 for all test images.

The coding efficiency was also evaluated using the average PSNR gain (i.e. BD-PSNR) measure. According to the simulation results shown in Tables 1 and 2, we conclude that the proposed codec outperforms the JPEG2000 standard for different test images.

Indeed, the average PSNR gain of the grayscale and medical test images reaches 0.48 dB and 0.72 dB, respectively. The performance, reported in Tables 3 and 4, are computed in terms of SSIM. It can be observed that for all test images the proposed codec produces high SSIM values when compared to the JPEG2000 for all bit-rates.

Table 1. BD-PSNR gain of the proposed algorithm against JPEG2000 for all grayscale test images

Image	Lena	Barbara	Boat	Mandrill	Zelda	Goldhill	Peppers	House	Washsat	France	Montage	Library
BD-PSNR(dB)	0.328	0.55	0.945	0.488	0.637	0.57	0.303	0.323	0.21	0.263	0.838	0.157
Average	0.477											

Table 2. BD-PSNR gain of the proposed algorithm against JPEG2000 for all medical test images

Image	BD-PSNR(dB)
MRI	0.948
MR3DBrain100001	0.483
Average	0.716

Table 3. SSIM versus the bitrate (bpp) of the proposed and JPEG2000 codecs for all medical test images.

Image	Bitrate (bpp)	Methods	
		Proposed codec	JPEG2000
MRI	0.25	0.9231	0.9187
	0.5	0.9412	0.938
	1	0.984	0.9832
MR3DBrain100001	0.25	0.9452	0.945
	0.5	0.9712	0.9636
	1	0.9834	0.9801

Figure 5 shows the reconstructed MRI medical image for the proposed and JPEG2000 codecs at 0.25 bpp. In this subjective test, we compare the perceptual quality between the two codecs. It can be observed that the visual quality of the image in Fig. 5(b) is better than the other.

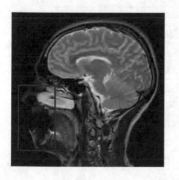 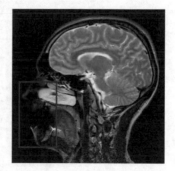

(a) JPEG2000   (b) Proposed codec

Fig. 5. Subjective assessment of the different codec for the MRI medical image at 0.25 bpp (a) PSNR = 32.1 dB (b) PSNR = 34.76 dB

Table 4. SSIM versus the bitrate (bpp) of the proposed and JPEG2000 codecs for all grayscale test images.

Image	Bitrate (bpp)	Methods	
		Proposed codec	JPEG2000
Lena	0.25	0.9771	0.9712
	0.5	0.9918	0.9859
	1	0.9981	0.9913
Barbara	0.25	0.9321	0.9268
	0.5	0.9759	0.9646
	1	0.9974	0.9968
Boat	0.25	0.955	0.9413
	0.5	0.9721	0.9696
	1	0.9987	0.9862
Mandrill	0.25	0.7932	0.7919
	0.5	0.885	0.8788
	1	0.976	0.97
Zelda	0.25	0.9611	0.9691
	0.5	0.9799	0.9775
	1	0.9941	0.9937
Goldhill	0.25	0.8315	0.8219
	0.5	0.9391	0.937
	1	0.9725	0.97
Peppers	0.25	0.9678	0.9602
	0.5	0.9744	0.973
	1	0.9891	0.9807
House	0.25	0.8123	0.8044
	0.5	0.9134	0.9091
	1	0.9628	0.961
Washsat	0.25	0.8966	0.8953
	0.5	0.9316	0.9307
	1	0.9867	0.9899
France	0.25	0.7188	0.7097
	0.5	0.84	0.833
	1	0.9356	0.9303
Montage	0.25	0.878	0.8737
	0.5	0.9536	0.94
	1	0.9897	0.9859
Library	0.25	0.5987	0.5981
	0.5	0.7482	0.748
	1	0.8746	0.8741

4   Conclusion

In this paper, we presented a novel wavelet image compression system which is constructed with a new spline wavelet transform based on adaptive directional lifting and an optimized coding algorithm based on the notion of zerotree. Experimental results have shown the superiority of the proposed system over the most known algorithm JPEG2000 in terms of PSNR, BD-PSNR and SSIM for different test images.

In the future, we plan to use our proposed image compression technique in the Wireless Multimedia Sensor Networks (WMSN) in order to check its efficiency in this field.

References

1. Doutsi, E.: Retina-inspired image and video coding. PhD thesis, Université Côte d'Azur (2017)
2. Bovik, A.C.: Handbook of Image and Video Processing. Academic press, Cambridge (2010)
3. Rasool, U., Mairaj, S., Nazeer, T., Ahmed, S.: Wavelet based image compression techniques: comparative analysis and performance evaluation. Int. J. Emerg. Technol. Eng. Res. (IJETER) **5**(9), 9–13 (2017)
4. Prasad, P.M.K., Umamadhuri, G.: Biorthogonal wavelet-based image compression. In: Artificial Intelligence and Evolutionary Computations in Engineering Systems, pp. 391–404. Springer (2018)
5. Boujelbene, R., Jemaa, Y.B., Zribi, M.: An efficient codec for image compression based on spline wavelet transform and improved SPIHT algorithm. In: 2017 International Conference on High Performance Computing & Simulation, HPCS 2017, 17–21 July 2017, Genoa, Italy, pp. 819–825. IEEE (2017)
6. Singh, A., Kirar, K.G.: Review of image compression techniques. In: 2017 International Conference on Recent Innovations in Signal Processing and Embedded Systems (RISE) (2017)
7. Hussain, A.J., Al-Fayadh, A., Radi, N.: Image compression techniques: a survey in lossless and lossy algorithms. Neurocomputing, **300**, 44–69 (2018)
8. Shapiro, J.M.: Embedded image coding using zerotrees of wavelet coefficients. IEEE Trans. Sign. Process. **41**(12), 3445–3462 (1993)
9. Boujelbene, R., Jemaa, Y.B., Zribi, M.: A comparative study of recent improvements in wavelet-based image coding schemes. Multim. Tools Appl. **78**(2), 1649–1683 (2019)
10. Boujelbene, R., Jemaa, Y.B.: A new structure of spline wavelet transform based on adaptive directional lifting for efficient image coding. Sign. Image Video Process. **14**(7), 1451–1459 (2020)
11. Boujelbene, R., Boubchir, L., Jemaa, Y.B.: Enhanced embedded zerotree wavelet algorithm for lossy image coding. IET Image Process. **13**(8), 1364–1374 (2019)
12. Jiang, X., Song, B., Zhuang, X.: An enhanced wavelet image codec: SLCCA plus. In: 2018 International Conference on Audio, Language and Image Processing (ICALIP), pp. 163–167. IEEE (2018)
13. Mander, K., Jindal, H.: An improved image compression-decompression technique using block truncation and wavelets. Int. J. Image Graph. Sign. Process. **9**(8), 17 (2017)

14. Antonini, M., Barlaud, M., Mathieu, P., Daubechies, I.: Image coding using wavelet transform. IEEE Trans. Image Process. **1**(2), 205–220 (1992)
15. Che, S., Che, Z., Wang, H., Huang, Q.: Image compression algorithm based on decreasing bits coding. In: 2009 Fifth International Conference on Information Assurance and Security, vol. 1, pp. 217–220. IEEE (2009)
16. Brahimi, T., Laouir, F., Boubchir, L., Ali-Chérif, A.: An improved wavelet-based image coder for embedded greyscale and colour image compression. AEU-Int. J. Electron. Commun. **73**, 183–192 (2017)
17. Greyscale and color databases for the evaluation of image and signal processing algorithms. http://links.uwaterloo.ca/Repository.html
18. Medical database for the evaluation of image and signal processing algorithms. http://www.international-education-biometrics.net/medeisa/
19. Xiaolin, W.: Lossless compression of continuous-tone images via context selection, quantization, and modeling. IEEE Trans. Image Process. **6**, 656–664 (1997)

An Innovative Monocular Mobile Object Self-localization Approach Based on Ceiling Vision

Alfredo Cuzzocrea[1,2](✉), Luca Camilotti[3], and Enzo Mumolo[3]

[1] IDEA Lab, University of Calabria, Rende, Italy
alfredo.cuzzocrea@unical.it
[2] LORIA, Nancy, France
[3] University of Trieste, Trieste, Italy
{camilotti,mumolo}@units.it

Abstract. This study deals with the estimation of the position of a mobile object using ceiling landmarks images acquired by a low resolution camera placed on a mobile object. The mobile object is moving in an indoor environment where light is given by electric lamps with circular holders. The images of the circular holders are projected on the image plane of the camera and are processed by means of computer vision algorithms. The pixels of the images of the light holders on the ceiling are mapped to the pixels of the images of the light holders on the image plane of the camera by means of a two dimensional dynamic programming algorithm (2D-DPA). The projection distortions are thus compensated and this reduces the estimation errors. The algorithm described in this paper estimates the distance from the camera lens to the center of the landmarks using only ceiling vision. Localization can be easily obtain from such distance estimations. The projections are geometrically described and the distance estimation is based on the pixels mapping information obtained by 2D-DPA.

Keywords: Localization · Ceiling landmarks · Mobile object · Two dimensional dynamic programming

1 Introduction

Self-localization of mobile objects is a fundamental requirement for autonomy. Mobile objects can be for example a mobile service robot, a motorized wheelchair, a mobile cart for transporting tasks or similar. Self-localization represents as well a necessary feature to develop systems able to perform autonomous movements such as navigation tasks. Self-localization is based upon reliable information coming from sensor devices situated on the mobile objects. There are many

This research has been made in the context of Excellence Chair in Computer Engineering – Big Data Management and Analytics at LORIA, Nancy, France.

sensors available for that purpose. The early devices for positioning are rotary encoders. If the encoders are connected to wheels or legs movement actuators, relative movements of the mobile object during its path [1] can be measured. Then, mobile object positioning can be obtained with dead-reckoning approaches. Dead reckoning [1] is still widely used for mobile robot positioning estimation. It is also true that dead-reckoning is quite unreliable for long navigation tasks, because of accumulated error problems. Other popular sensor devices for self-localization are laser or sonar based range finder devices and inertial measurement devices. In outside scenarios the most popular approaches are based on Global Positioning System (GPS). Due to the importance of self-localization, many other solutions for indoor environment have been proposed so far with different cost and accuracy characteristics. For example the Ultra Wide Band radio signal indoor localization systems [2], or the Bluetooth-based angle of arrival radio devices [3], or a combination of them. However these systems have serious limitations in cost and reliability, respectively. Another important type of sensors which may be used for cost effective self-localization are the CCD cameras, which require computer vision algorithms for localization such as for example visual odometry, [23]. Mobile objects vision self-localization is currently an open research field [5] and an increasing number of new methods are continuously proposed. As a matter of fact we have to consider that self-localization of mobile objects requires centimeter-level accuracy and Computer Vision is one of the most cost-effective techniques able to reach that accuracy. Consequently, some surveys of Computer Vision based self-localization techniques appeared recently in the literature, [6].

In this paper we describes a novel Computer Vision algorithm for estimating the distance from the camera lens to the center of ceiling landmarks with circular shape using a monocular low cost webcam. From the distance, mobile object localization approaches can be easily developed and a simple example is provided in this paper. The images of the ceiling landmarks are projected on the image plane of the camera. The projection is analytical described, but the projections distortions, which may arise especially when low cost devices are used, may affect the results. To take into account the projection distortions in order to obtain a better precision of the results, we use an approximation of the two-dimensional dynamic programming (2D-DPA) algorithm [4] which finds a sub-optimal mapping between the image pixels of the ceiling landmarks and the image plane pixels of the projected landmarks. Since optimum 2D-DPA is NP-complete, in fact, many approximations have been developed. For example, the 2D-DPA technique described by Levin and Pieraccini in [25] has an exponential complexity in the image size, while Uchida and Sakoe describe in [24] a Dynamic Planar Warping technique with a complexity equal to $O(N^3 9^N)$. Lei and Govindaraju propose in [26] a Dynamic Planar Warping approximation with a complexity of $O(N^6)$. However each approximation has some limitation in terms of continuity of the mapping. In this paper we use a approximation of the optimum 2D-DPA with a complexity of $O(N^4)$ [22] which is implemented on a GPU to obtain real-time performance. When the landmark is far from the camera or if the environments has low lighting conditions, an high quantization noise may arise in acquired images. However the algorithm we describe in this paper is particularly robust against noise due especially to the use of two-dimension DPA.

This paper is organized as follows. Section 2 reports on related work. Section 3 the localization problem is described, and in Sect. 4 the projection distortion is geometrically described, while in Sect. 5 the two-dimensional Dynamic Programming approximation is described. In Sect. 6 the proposed algorithm is sketched and in Sect. 7 the computer vision algorithms for the detection of landmarks on the image plane are reported. Section 8 sketches a possible global localization approach of the mobile object. In Sect. 9 we report some experimental comparison of the proposed algorithm with state of the art algorithm. Finally, Sect. 10 concludes the paper with concluding remarks and a suggestion of future works.

2 Related Work

Many papers on vision-based mobile robot self-localization appeared recently in the literature. For example Avgeris *et al.* describe in [18] a self-localization algorithm for mobile robots that uses cylindrical landmarks resting on the floor and a single pivotal camera with an horizontal angle of view of 30-degree. Each cylindrical landmark has a different color in order to be easily detected by the robot. However, frontal vision could be occluded by objects or people. Such interference can be avoided by placing the landmarks on the ceiling, so that the camera is tilted toward the ceiling. Ceiling vision has been used by many authors to perform mobile robot localization. One of the early proposals is described in [20] and is based upon a digital mark pattern and a CCD camera. The camera is tilted, so the horizontal distance from the ceiling mark pattern is obtained measuring the ratio between the length and the width of the pattern picture. Kim and Park, [7], acquire ceiling images in a small area with a fish-eye lens camera. Ceiling outlines are detected by means of adaptive binarization and segmentation. Robot pose is obtained after identification of the ceiling region and the determination of the center and the momentum of the region. Lan *et al.* describe in [8] a mobile robot positioning algorithm based on artificial passive landmarks placed on the ceiling and infrared sensors. The landmarks are made of reflective film 2D structures containing dots assigned to unique ID's. The infrared sensors consist of an infrared camera and an infrared LED array. A similar approach is described in [9] where artificial passive reflective landmarks are placed on the ceiling and an infrared camera plus an infrared LED source are used to capture the reflection the IR light on the landmark for estimating the robot pose. Wang *et al.* describe in [21] a vision control system which capture ceiling RGB images with a camera placed on the robot, convert the image to HSV color space and use V channel images to reduce the effect of illumination lamps. The common objects and the straight lines on the ceiling are detected by template matching and used to estimate the robot orientation. Other Computer Vision based approaches are based on the Free Space Density concept. For example A. Ribacki *et al.* use an upward facing camera to extract the ceiling boundaries for estimating the ceiling space density from the current image [11]. Other authors, for example [12,13] use the ceiling depth images for robot localization. In these approaches self-localization is obtained from Principal Component Analysis of ceiling depth images. Ceiling vision is used by many other authors to perform self-localization of mobile robots. For example Qing Lin *et al.* describe in [17] a visual odometry algorithm based on a monocular camera which points to the ceiling. The algorithm uses several local features detectors for matching the features between two sequential frames of the ceiling.

In addition, it should considered the emerging integration of these topics with the innovative *big data trend* (e.g., [14–16]). Here, the main research perspective is that to take into account the well-known *3V model* of big data, including *volume*, *velocity* and *variety*.

3 Problem Description

We show in Fig. 1 a mobile object in an indoor environment. The movable object is equipped with a camera set tilted towards the ceiling at an angle φ. We call h the distance between the camera and the ceiling. Moreover in Fig. 2 the horizontal and vertical angles of view of the camera, called θ_x, and θ_y respectively, are highlighted. The direction towards which the camera is oriented is shown with the 'Camera Direction' arrow. The ceiling landmark is shown in Fig. 1 with a segment with a greater thickness and the image plane of the camera is shown with a segment orthogonal to the camera direction. The ceiling landmark is projected to the landmark on the image plane. The visual landmarks positioned on the ceiling used in this approach are the lighting holders shown as that shown in Fig. 3. We choose landmarks with isotropic shapes on the plane because in this way the distortion components due to image rotation can be eliminated. The simpler isotropic shape is the circle. As shown in Fig. 3, the lines of pixels on the image plane are all parallel to the reference abscissa on the ceiling plane regardless of the angle of the camera with respect to the landmark. It is important to remark that each landmark must be distinguishable from the others and its

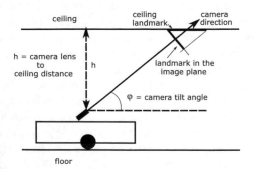

Fig. 1. A mobile object with a camera on it, tilted toward the ceiling.

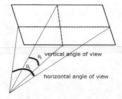

Fig. 2. The horizontal and vertical angles of view of the camera.

Fig. 3. An example of the circular lamp holder used in this paper

coordinate in the global reference system must be known. A schematic representation of a mobile object and some landmarks with the orthonormal reference
system centered on the camera lens is shown in Fig. 4. The reference abscissa
changes dynamically in relation to the direction of the focal axis. The reference
abscissa, in fact, is always normal to the focal axis and at the same time it is
parallel to the horizon.

The landmarks must be distinguishable from each other. There are many
possible solutions for making the landmarks distinct. A simple possibility is
to paint each holder with a different color. More recently, the characteristic
frequency of fluorescent lights has been used, for instance in [19]. In this paper
we used the simplest solution, namely we painted adjacent lamp holders with
different colors. For this reason the landmarks in Fig. 4 are represented with
different colors, where for simplicity the three circular landmarks positioned on
the ceiling are colored in red, blue and green. Figure 4 shows that the landmarks
which fall within the visual field of the camera are projected onto the image
plane of the camera. Of course we know in advance the physical position of
each landmark in the global reference system. On the other hand the landmark
colours can be detected using well known computer vision techniques.

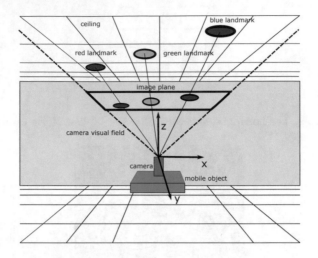

Fig. 4. Schematic representation of orthonormal reference system, landmarks and image plane.

4   Projective Transformations

The projective transformation is the linear transformation of coordinates reported in (1).

$$p' = Tp \tag{1}$$

where p represents a generic point in space expressed in homogeneous coordinates, relative to the orthonormal reference system S described by the quadruple $(O, \hat{i}, \hat{j}, \hat{k})$. The projected point p' is expressed in coordinates relative to the reference system S' described by the quadruple $(O', \hat{i}', \hat{j}', \hat{k}')$, where $\hat{i}' = \hat{i}$, $\hat{j}'$ has the direction of the segment $\overline{MQ}$ and $\hat{k}'$ has the direction of the normal to the segment $\overline{MQ}$.

Since p is expressed with the three components (x_p, y_p, z_p) and p' has the three components $(x_{p'}, y_{p'}, z_{p'})$, Eq. (1) can be also written as follows

$$\begin{pmatrix} x_{p'} \\ y_{p'} \\ z_{p'} \end{pmatrix} = T \begin{pmatrix} x_p \\ y_p \\ z_p \end{pmatrix} \tag{2}$$

Such a transformation maintains the properties of collinearity, that is, the points which in S belong to a line, are aligned in a line also in S'. However, projective transformation may not be defined for every point of S, in the sense that some points could be mapped in S' at infinity.

Let us view Fig. 4 from the left side, that is the $y - z$ plane of the orthonormal reference system which has its origin coinciding with the center of the camera lens. This plane is highlighted in Fig. 5, where the ceiling is at $z = h$, and the field of view of the camera is shown with points M and E. Let us assume that a landmark falls within the vertical angle of view. Then, the center of the landmark is the point C. On the other hand, if we view Fig. 4 from the front side, that is the $x - z$ plane, we obtain the system shown in Fig. 6. Of course the camera image plane, which is the plane normal to the focal axis in Fig. 4, is shown with the segment $M - Q$ in Fig. 5 and segment $G - I$ in Fig. 6.

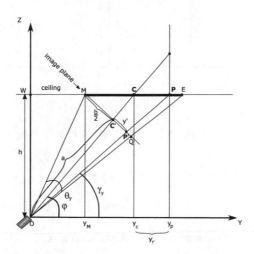

Fig. 5. Plane $y - z$ in orthonormal reference system

Suppose we fix a point P on the ceiling. If the point falls within the field of view of the camera it is shown as P in Fig. 5. Let (p_x, p_y, p_z), with $p_z = h$, be the coordinates of P. The point P is projected to the image plane of the camera to the point P', which has coordinates $(x_{p'}, y_{p'}, z_{p'})$. Also the center of the landmark in Fig. 5 is projected to the point C' and the segment $M - E$ is projected to the segment $M - Q$ in the image plane. In this model, the focal distance of the device or other characteristic parameters are not taken into account. It is in fact a purely ideal model, which has the only purpose of deriving the relations that define the projective transformation from the orthonormal system whose origin coincides with the center of the camera lens to the image plane system. The latter is chosen independently of the characteristics of the camera. With reference to the Figs. 5 and 6, we introduce the following geometric variables characteristic of the problem.

$$\Phi = \varphi + \frac{\theta_y}{2} - \frac{\pi}{2} \qquad (3)$$

– The distance a from the origin to the barycenter of the landmark projected on the image plane:

$$a = \overline{OC'} = \frac{h}{\sin(\varphi)} - h\left(\tan(\varphi) + \frac{1}{\tan(\varphi)}\right)\cos(\varphi) \qquad (4)$$

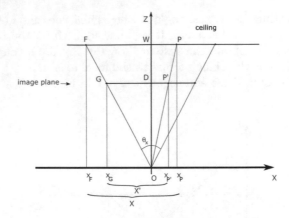

Fig. 6. Plane $x - z$ in orthonormal reference system

– The abscissa of the point P' on the image plane:

$$\frac{b}{2} = \overline{MC'} = \overline{C'Q} = h(\tan(\varphi) + \frac{1}{\tan(\varphi)} \sin(\varphi)) \tag{5}$$

Equations (4) and (5) are developed in **Appendix A**. Moreover, we define the following two variables:

$$G = -h\left(\tan\varPhi\tan\varphi + 1\right) \tag{6}$$

and

$$F = h\left(\tan\varphi - h\tan\varPhi\right) \tag{7}$$

We remark that the following considerations are based on three coordinate systems, namely an orthonormal reference system centered on the camera lens, shown in Fig. 4, an orthonormal reference system on the image plane and a system on the ceiling plane which is simply translated by h with respect to that centered on the camera lens. In general, points on the systems centered on the camera lens and on the ceiling are denoted with a capital letter, such as **P**, while that on the image plane of the camera are denoted with a capital letter plus an apex such as **P'**. In this case, **P'** is the **P** point projected on the image plane. If we look at the landmark seen from the orthonormal reference system centered on the camera lens, its barycenter is located at (x_c, y_c). A generic point on the ceiling has coordinate (x, y) and the same point projected on the image plane is (x', y'). The coordinates of a generic point on the landmark is given relative to its barycenter: $(x = x_c + x_r)$ and $(y = y_c + y_r)$. According to Figs. 6 and 5 the offsets x_r, y_r are projected to the image plane in x', y'.

Assume now we have an optimum mapping between images. In other words, assume that, having two images A and B, $A = \{a(i,j)|i,j = 1,...,N\}$ and $B = \{b(u,v)|u,v = 1,...,M\}$, we can estimate the mapping function

$$F(i,j) = \begin{bmatrix} u \\ v \end{bmatrix} = \begin{bmatrix} x(i,j) \\ y(i,j) \end{bmatrix} \tag{8}$$

which maps each pixel (i, j) of one image to the pixel (u, v) of the other image such that the difference between the two images is minimized, as shown in (9).

$$min \sum \sum \|a(i, i) - b(u, v)\| \tag{9}$$

where $u = x(i, j)$ and $v = y(i, j)$. Such mapping is performed through a two dimensional Dynamic Programming operation [24]. 2D-DPA is the base of image matching algorithms called Elastic Image Matching. Unfortunately, the Elastic Image Matching operation is NP-complete [29]. For this reason we devise an approximation which reduces the 2D-DPA operation complexity to $O(N^4)$, as described below.

The barycenter of the landmarks, (x_c, y_c), are estimated using the following Proposition.

Proposition 1. *By measuring the abscissa and ordinate (x', y') of a generic point on the landmark projected on the image plane we can estimate the coordinate (x_c, y_c) of the ceiling landmark using the following equations:*

$$x_c = \frac{h \cos(\varphi - \gamma_y)(x' - g)}{a \sin(\gamma_y)} + g - x_r \tag{10}$$

$$y_c = \frac{aG + ay_r \tan(\varphi) - (y' - \frac{b}{2})(y_r + F)}{y' - \frac{b}{2} - a \tan(\varphi)} \tag{11}$$

Proof. In **Appendix B** we give a sketch of the derivations.

A different estimation of the coordinates of the landmark barycenter is obtained for all the points P inside the landmarks. A sequence of barycenter coordinates x_c, y_c are thus obtained, of which we compute the expected value. The algorithm is thus sequentially divided into two parts: estimation of $\mathbf{E}(x_c)$ and $\mathbf{E}(y_c)$ by measuring the dimension y' and x' of the distorted image on the image plane.

The distance from the camera lens and the landmark in the ceiling reference system is thus the following:

$$d = \sqrt{\mathbf{E}(x_c)^2 + \mathbf{E}(y_c)^2)} \tag{12}$$

with reference to Figs. 6 and 5, where $C = (x_c, y_c, z_c)$ is the barycenter of the landmark in the reference system (O, i, j, k). We obtain he sub-optimal correspondence, pixel by pixel, between a reference image and a distorted image by means of approximated two dimensional dynamic programming, . Our algorithm therefore uses the deformation of the image to derive the distance of the landmark, i.e. it is intended to determine how the perspective has distorted the image.

The coordinates of the barycenter of the ceiling landmarks are obtained using the coordinate x' measured on the image plane and x_r using the mapping function, and in terms of y' and y_r. Clearly (x_r, y_r) and (x', y') are both known because they are derived from the coordinates of the pixels in the pattern and in the test images respectively. What associates the two pixels is the mapping relationship described in (8) obtained by 2D-DPA.

The characteristic that differentiates the algorithms present in the literature from the one developed in this paper is the statistical character of the obtained estimate. The algorithm based on dynamic programming is able to calculate a position estimate for each single pair of associated pixels from the mapping. The advantage is that a large number of points are used, which contribute to the calculation of the average distance value. This makes the estimate more truthful, especially when the landmark is very distant, which results in a smaller image and a greater quantization error.

5   2D Dynamic Programming Based Image Mapping Technique (2D-DPA)

For the sake of coherence with what we write below, we repeat now the mapping considerations summarized above about images A and B using instead images X and Y. Given the two images, $X = \{x(i,j)\}$ and $Y = \{y(u,v)\}$, the mapping of one image to the other is represented by the operation

$$D(X,Y) = \min \sum_{i=1}^{N} \sum_{j=1}^{N} \|x(i,j) - y(u,v)\|$$

where $u = x(i,j)$, $v = y(i,j)$ is the mapping function between the pixels of X and Y. The quantity $D(X,Y)$ gives a distance between the image X and the optimally deformed Y, the optimal warping function $x(i,j), y(i,j)$ gives an interpretation of the image X according to the generation model Y.

Given the i–th row of the X image and the j–th row of the Y images, namely $Y_j = (y_{j,1}, y_{j,2}, \ldots, y_{j,N})$, $X_i = (x_{i,1}, x_{i,2}, \ldots, x_{i,N})$ respectively, the distance between the two rows is obtained by applying a 1D-DPA [27] for finding a warping among the two rows as described in (13). Here the map M' is, say, over (n,m) coordinates, so that $M'_l = ((i_l, n_l), (j_l, m_l))$.

$$d(X_i, Y_j) = \frac{\min\limits_{M'} \sum_{l=1}^{M'} d(M'_l)}{M'} = \frac{\min\limits_{M'} \sum_{l=1}^{M'} \|x_{i_l,n_l} - y_{j_l,m_l}\|}{2N} \tag{13}$$

Finally, the distance between the two images is obtained by (14). In this case the map $\overline{M'}$ is between all the rows of X and Y. As before, $|\overline{M'}|$ is the length of the path.

$$D(X,Y) = \frac{\min\limits_{M'} \sum_{k} d(\overline{M'}_k)}{|\overline{M'}|}$$

$$= \frac{\min\limits_{M'} \sum_{k} d(X_{i_k}, Y j_k)}{|\overline{M'}|} = \frac{\min\limits_{M'} \sum_{k} \dfrac{\min\limits_{M'} \sum_{l=1}^{|M'|} d(M'_l)}{2N}}{2N}$$

$$= \frac{\min_{M'}\{\sum_k \min_{M'} \sum_{l=1}^{|M'|} \|x_{i,n_l} - y_{j,m_l}\|\}}{4N^2} \tag{14}$$

Let us assume that the images are of equal size, that is $N \times N$ pixels. Then the length of the optimum path between the two images is equal to $2N$. The local distances in each point of this path is obtained with other 1D-DPA with paths of length $2N$. The total length is the sum of $2N$ along the $2N$ long path, giving $4N^2$ at the denominator. The complexity of the described operation is $O(N^2 N^2) = O(N^4)$ where N is the image dimension.

6   Proposed Algorithm

The algorithm described in this paper is summarized in the following Algorithm. The inputs of the algorithm are the two gray-scale images img_A and img_B which are the landmark on the image plane and on the ceiling respectively. We perform the 2D-DPA algorithm on these two images to obtain the mapping function as result. the mapping function is represented with a linked list where each node is the map related to the two pixels. The function $get()$ give as result the value of the pixel on the image indicated as input and is used to get the values of the two pixels linked by the map on the two landmark images. To decide if the pixel is a landmark pixel or not, we consider their gray levels. The landmarks have a lower values with respect to the environment and thus if the pixel values is less then a thereshold, the pixel is a landmark pixel.

Input:img_A, img_B
Output: *distance*

img=Detect(img_A);   ▷ get the landmark in the image plane
id=identify(img);   ▷ identify the landmark
head=2D-DPA(img_A, img_B);
ptr=head;   ▷ head is the list of mapping function
repeat
    $pixA = get(img_A, ptr)$;   ▷ pixel of img_A
    $pixB = get(img_B, ptr)$;   ▷ pixcl of img_B
    if $(pixA \leq L)\&\&(pixB \leq L)$ **then**   ▷ the pixels are in the landmark
        Compute x_c, y_c with (10) and (11)
        $sum_y + = y_c$;
        $sum_x + = x_c$;
        counter++;
    $ptr = ptr \to next$;
until $ptr == NULL$
$y_c = y_s um/counter$;
$x_c = x_s um/counter$;
$distance = \sqrt{x_c^2 + y_c^2}$;
return *distance*

---

7   Computer Vision Approach for Extracting Landmark Images

We briefly summarize in this Section the computer vision operations we did on the image acquired from the ceiling. The problem is to detect from the image plane the isotropic images which represent the landmark. Another operation, which is not reported here, is the identification of the landmark. The simplest way is to draw the landmarks with different colors, since the computer vision operations to identify the colors are very simple. There are however many other ways which can be used for the identification, typically based on some type of code drawn inside the landmark. Of course the computer vision operations are slightly more complex than using different colors. More importantly, the computer vision operations to decode drawn codes could need greater camera resolution.

We report in Fig. 7 the Computer Vision algorithms we applied on the original image for extraction of isotropic images. The algorithms are described as follows:

- The acquired image is first transformed in gray-scale, and then its edges are obtained via the Canny's operator, obtaining the Edge image.
- From the Edge image, its contours are extracted, obtaining the Contour1 image.
- The Contour1 image is processed via morphological analysis. More precisely the opening operation with circular structuring element, is applied to Contour1 image in order to eliminate the little Side Dishes. The edges are then extracted again with the Canny operator, and then the contours are extracted again, finally obtaining the Contour2 image.
- Ellipse fitting is applied to Contours2 image. Based on the position and size of the found ellipses, square portions are cut out from original image. Most likely, the landmarks are contained in one of the extracted portions.

The results are shown in Fig. 8. These results refer to the input image shown in Fig. 3.

Fig. 7. Block diagram of the computer vision algorithms.

Fig. 8. Processed results, with reference to Fig. 3

8 Localization

The localization of the mobile object is an issue we leave open as starting from distance estimation several possible solution can be developed. However, just to point out a possible simple idea based on trilateration, we report Fig. 9. This figure shows a global reference system which is related to the indoor environment is shown. Another reference system which is rotated and translated with respect to the first one. The origin of second reference system is centered on the camera lens of the mobile object. Note that the x y planes shown in Fig. 9 correspond to the ceiling plane. The mobile object identify the landmarks and knows in advance their location coordinate in the global reference system. Our algorithm estimates the distance from the mobile object and the detected landmarks. Therefore, we can think to draw a circle with center on the landmark

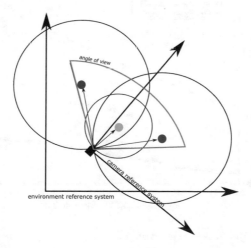

Fig. 9. Sketch of a possible localization by trilateration

and radius equal to the estimated distance. If at least three different landmarks are detected, the mobile object can be localized in global reference system.

9   Experimental Results

The experiments has been made using an Intel I7 CPU with **8**cores running at **3.07** GHz and a memory of **24** GB. Then, the two dimensional DPA algorithm has been written in the CUDA framework and executed on a NVidia Kepler TM GK110 device. A low cost **640 × 480** webcam is used for image acquisition. In Fig. 10 we report the average error of the estimation distance from the camera lens and the barycenter of the landmarks. As a general consideration regarding these results, if the camera tilt-angle is high (i.e. if the inclination of the optical axis is close to the perpendicular to the ceiling) the error is quite small, but the field of view turns out to be very limited. To take advantage of wider fields of view, higher tilt-angles must be used. In this case, however, the error is higher. Furthermore, if the light in the environment become worse, the average error increases. Our algorithm, however, is quite robust against noise. The curve drawn in Fig. 10 with solid line is obtained by the algorithm described in this paper. The curve in the middle is related to the approach developed in 2019 by Avgeris *et al.* and described in [18]. Finally, the higher curve is related to the work proposed by Ogawa *et al.* in [20]. Despite being quite old we include this result because its setting is very similar to this paper (the camera is directed towards the ceiling with a tilt angle equal to **30°**). The errors are in any cases well above that obtained by all the other algorithms.

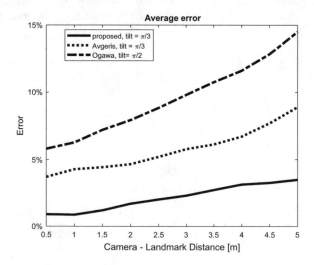

Fig. 10. Average errors of the estimated distance

10   Final Remarks and Future Work

In this paper we present an algorithm to measure the distance of a mobile object to the lightning lamps used as ceiling landmarks in indoor environment. The algorithm has many attractive features, mainly the accuracy, which is better than many other visual-based algorithms. Also, the distance measurements algorithm is robust against noise. Quantization noise can be high in low lighting condition of the environment and if the distance from landmarks and camera is high. The negative outcome of the algorithm is the high complexity of 2D-DPA which, even if polynomial, can lead to high computational times. In [22], however, we show how the 2D-DPA when implemented on a NVidia Kepler TM GK110 device leads to computation time less then 100 ms, for image size of **100 × 100** pixels.

This paper naturally opens to the development of localization algorithms based on our distance estimation algorithm. The global localization is in fact under development. Another open important issue is the landmark placement. Finally, the estimation of the orientation of the mobile object is another fundamental problem not addressed in this paper. The use of the characteristic frequencies of fluorescent lamp is an interesting method to identify the landmarks. In this case, adaptive and artificial intelligence metaphors, perhaps inherited by different scientific context (e.g., [30,31]), may be considered. Future works will be focused on these open points.

Acknowledgements. This research has been partially supported by the French PIA project "Lorraine Université d'Excellence", reference ANR-15-IDEX-04-LUE.

Appendix A

Referring to Fig. 5, we derive below the geometric variables reported in Sect. 4.

Consider first Eq. (4). $a = \overline{OC'} = \overline{OC} - \overline{C'C}$. From the right triangle $\triangle OCC_y$ we have $\overline{OC} = \frac{h}{\sin(\varphi)}$. Moreover, from $\triangle MC'C$ we have $\overline{C'C} = \overline{MC}\cos\varphi = (\overline{WC} - \overline{WM})\cos\varphi = (\frac{h}{\tan\varphi} + h\tan\Phi)\cos\varphi$. Therefore

$$a = \frac{h}{\sin(\varphi)} - h\left(\frac{1}{\tan\varphi} + \tan\Phi\right)\cos\varphi$$

Considering Eq.(5), we have $\frac{b}{2} = \overline{MC}\sin\varphi = h\left(\frac{1}{\tan\varphi} + \tan\Phi\right)\sin\varphi$.

Appendix B

We now report a sketch of the derivation of the two propositions reported in Sect. 4.

Let us start with 7. Regarding Fig. 5, the angle formed by segments $\overline{OR}$ and $\overline{OP}$ is equal to $(\Phi - \gamma_y)$, so $\tan(\Phi - \gamma_y) = \frac{\overline{RP}}{\overline{OR}} = \frac{[y\tan\varphi - h(\tan\Phi\tan\varphi + 1)]\cos\varphi}{h(\tan\varphi + y - h\tan\Phi)\cos\varphi}$.

In addition to simplifying the $\cos\varphi$, we use the definition of F and G reported above.

$$G = -h\left(\tan\Phi\tan\varphi + 1\right)$$

and

$$F = h\left(\tan\varphi - h\tan\Phi\right).$$

Then we have: $\tan\left(\Phi - \gamma_y\right) = \frac{y\tan\varphi + G}{y + F}$. We conclude that $y' = \frac{b}{2} + a\frac{y\tan\varphi + G}{y + F}$ By setting $y = y_r + y_c$ we obtain the landmark coordinate y_c reported in (11).

Going now back to 7, let us consider Fig. 6. For lack of space we only state that, according to considerations very similar to that just described, we can conclude that

$$x' = g + \frac{a\sin\gamma_y\left(x - g\right)}{h\cos\varphi - \gamma_y} \tag{15}$$

where $g = \overline{GD} = \overline{DI} = a\tan\frac{\theta_x}{2}$. As we did previously, we substitute $x = x_r + x_c$ in 15 and thus we can obtain x_c, described in (10).

References

1. Campbell, S., O' Mahony N., Carvalho, A., Krpalkova, L., Riordan, D., Walsh, J.: Where am I? localization techniques for mobile robots a Review. In: 6th International Conference on Mechatronics and Robotics Engineering, ICMRE, Barcelona, Spain, 12-15 February, pp. 43–47 (2020)
2. He, C., Xia, Y., Yu, C., Jiang, C.: A multi-hop distributed indoor localization algorithm for ultra-wide-band sensor network. In: 16th International Conference on Control, Automation, Robotics and Vision (ICARCV), 13–15 December, Shenzhen, China, pp.1335–1340 (2020)
3. Kumar, G., Gupta, V., Tank, R.: Phase-based Angle estimation approach in indoor localization system using bluetooth low energy. In: International Conference on Smart Electronics and Communication (ICOSEC), Trichy, India, pp. 904–912, 10–12 September 2020
4. Glasbey, C.A.: Two-dimensional generalisations of dynamic programming for image analysis. Stat. Comput. **19**(49), 49–56 (2009)
5. Wei, A.: Shang, survey of mobile robot vision self-localization. J. Autom. Control Eng. **7**(2), 98–101 (2019)
6. Morar, A., Moldoveanu, A., Mocanu, I., Moldoveanu, F., Radoi, I.E., Asavei, V., Gradinaru, A., Butean, A.: A comprehensive survey of indoor localization methods based on computer vision. Sensors **20**(9), 1–36 (2020)
7. Kim, Y.G., Park, T.H.: Localization of mobile robots from full detection of ceiling outlines. In: Proceedings of the IEEE International Conference on Information and Automation, Ningbo, China, pp. 1515–1520, August 2016
8. Lan, G., Wang, J., Chen, W.: An improved indoor localization system for mobile robots based on landmarks on the ceiling. In: IEEE International Conference on Robotics and Biomimetics (ROBIO), 3–7 December, Qingdao, China, pp. 1395–1399 (2016)

9. Vidal, J., Lin, C-Y.: Simple and robust localization system using ceiling landmarks and infrared light. In: IEEE International Conference on Control and Automation, Kathmandu, Nepal, pp. 583–587, 1–3 June (2016)
10. Wang, W., Luo, Z., Song, P., Sheng, S., Rao, Y., Soo, Y.G., Yeong, C.F., Duan, F.: A ceiling feature-based vision control system for a service robot. In: Proceedings of the 36th Chinese Control Conference, Dalian, China, pp. 6614–6619, 26–28 July 2017
11. Ribacki, A., Jorge, V.A.M., Mantelli, M., Maffei, R., Prestes, E.: Vision-based global localization using ceiling space density. In: IEEE International Conference on Robotics and Automation (ICRA), Brisbane, Australia, pp. 3502–3507, 21–25 May (2018)
12. Carreira, F., Calado, J., Cardeira, C., Oliveira, P.: Navigation system for mobile robots using PCA-based localization from ceiling depth images: experimental validation. In: 13th APCA International Conference on Automatic Control and Soft Computing, Ponta Delgada, Azores, Portugal, pp. 159–164, 4–6 June 2018
13. Carreira, F., Calado, J.M.F., Cardeira, C.B., Oliveira, P.J.C.R.: Enhanced PCA-based localization using depth maps with missing data. J. Intell. Robot. Syst. February 2015
14. Chatzimilioudis, G., Cuzzocrea, A., Gunopulos, D., Mamoulis, N.: A novel distributed framework for optimizing query routing trees in wireless sensor networks via optimal operator placement. J. Comput. Syst. Sci. **79**(3) (2013)
15. Cuzzocrea, A., Mansmann, S.: OLAP visualization: models, issues, and techniques, 2nd edn. In: Encyclopedia of Data Warehousing and Mining (2009)
16. Cuzzocrea, A., De Maio, C., Fenza, G., Loia, V., Parente, M.: OLAP analysis of multidimensional tweet streams for supporting advanced analytics. In: Proceedings of the 31st Annual ACM Symposium on Applied Computing (2016)
17. Lin, Q., Liu, X., Zhang, Z.: Mobile robot self-localizationusing visual odometry based on ceiling vision. In: Symposium Series on Computational Intelligence, Xiamen, China, pp. 1435–1439, 6–9 December 2019
18. Avgeris, M., Spatharakis, D., Athanasopoulos, N., Dechouniotis, D., Papavassiliou, S.: Single vision-based self-localization for autonomous robotic agents. In: 7th International Conference on Future Internet of Things and Cloud Workshops, Istanbul, Turkey, pp.123–129, 26–28 August 2019
19. Zhang, C., Zhang, X.: Visible light localization using conventional light fixtures and smartphones. IEEE Trans. Mob. Comput. **18**, 2968–2983 (2018)
20. Ogawa, Y., Lee, J-H., Mori, S., Takagi, A., Kasuga, C., Hashimoto, H.: The positioning system using the digital mark pattern-the method of measurement of a horizontal distance. In: IEEE SMC 1999 Conference Proceedings, Tokyo, Japan, pp. 731–741 (1999)
21. Wang, W., Luo, Z., Song , P., Sheng, S., Rao, Y., Soo, Y.G., Yeong, C.F., Duan, F.: A ceiling feature-based vision control system for a service robot. In: 136th Chinese Control Conference (CCC), Dalian, pp. 6614–6619 (2017)
22. Cuzzocrea, A., Mumolo, E., Pirro, D., Vercelli, G.: An efficient CUDA-based approximate two-dimensional dynamic programming algorithm for advanced computer vision applications. In: IEEE International Conference on Systems, Man, and Cybernetics, Budapest, Hungary, pp.4251–4258, 9–12 October 2016
23. He, M., Zhu, C., Huang, Q., Ren, B., Liu, J.: A review of monocular visual odometry. Vis. Comput. **36**(5), 1053–1065 (2020)
24. Uchida, S., Sakoe, H.: A monotonic and continuous two-dimensional warping based on dynamic programming. Fourteenth Int. Conf. Pattern Recogn. **16–20**(August), 521–524 (1998)

25. Levin, E., Pieraccini, R.: Dynamic planar warping for optical character recognition. In: 1992 IEEE International Conference on Acoustics, Speech, and Signal Processing, San Francisco, California, USA, pp. 149–152, 23–26 March 1992
26. Lei, H., Govindaraju, V.: Direct image matching by dynamic warping. In: IEEE Conference on Computer Vision and Pattern Recognition Workshops, Washington, DC, USA, 27 June-2 July (2004)
27. Vinotha, K.: Bellman equation in dynamic programming. Int. J. Comput. Algorithm, **05**(1), 35–37 (2016)
28. Becker, C., Joaquin, S., Tokusei, K., Latombe, J.-C.: Reliable navigation using landmarks. IEEE Conf. Robot. Autom. **2**, 401–406 (1995)
29. Keysers, D., Unger, W.: Elastic image matching is NP-complete. Pattern Recognit. Lett. **24**(1–3), 445–453 (2003)
30. Cannataro, M., Cuzzocrea, A., Pugliese, A.: XAHM: an adaptive hypermedia model based on XML. In: Proceedings of the 14th SEKE International Conference (2002)
31. Cuzzocrea, A.: Combining multidimensional user models and knowledge representation and management techniques for making web services knowledge-aware. Web Intell. Agent Syst. Int. J. **4**(3), 289–312 (2006)

Analysis of Visualized Bioindicators Related to Activities of Daily Living

Tomokazu Matsui[1(✉)], Kosei Onishi[1], Shinya Misaki[1], Hirohiko Suwa[2],
Manato Fujimoto[1], Teruhiro Mizumoto[3], Wataru Sasaki[4], Aki Kimura[4],
Kiyoyasu Maruyama[4], and Keiichi Yasumoto[1]

[1] Nara Institute of Science and Technology, Ikoma, Japan
matsui.tomokazu.mo4@is.naist.jp
[2] Nara Institute of Science and Technology/RIKEN, Ikoma, Japan
h-suwa@is.naist.jp
[3] Osaka University, Suita, Japan
mizumoto@ist.osaka-u.ac.jp
[4] Mitsubishi Electric Corporation, Tokyo, Japan
Kimura.Aki@cb.mitsubishielectric.co.jp

Abstract. Estimating the stress associated with activities of daily living in the home is important for understanding and improving residents' quality of life. However, conventional stress estimation methods are mostly used in specific situations, and stress has not been estimated for long-term activities of daily living. In this paper, we visualize changes in stress caused by activities such as household chores by measuring bioindicators associated with daily living activities in the home, and to obtain knowledge for establishing a method for estimating long-term daily quality of life. To collect lifestyle and stress data, two subjects completed a lifestyle and stress measurement questionnaire daily in a smart home for approximately two weeks. We visualized the obtained data and confirmed that the results suggest several relationships between daily living activities and bioindicators of stress level, for instance, cleaning the room and dishwashing resulted in more stress compared to smartphone use and sleep.

1 Introduction

One of the recent research challenges in the field of information science is to improve peoples' quality of life (QoL) using the Internet of Things (IoT) and information and communication technology (ICT). In particular, it is desirable to have people work on improving their own QoL in their home, which is a private space. Therefore, there has been a great deal of research on in-home living to improve residents' QoL [1–4].

A promising approach to improve QoL is to reduce stress in daily life. However, although it is possible to measure stress in daily life with simple stress checkers and questionnaires, it is difficult to identify the causes of stress. To reduce stress in daily life, we need to understand the stressors in our lives.

Performing household chores is an essential activity that residents must perform in their dwellings. Neglecting household chores has a negative impact on the resident's

© The Author(s), under exclusive license to Springer Nature Switzerland AG 2021
L. Barolli et al. (Eds.): AINA 2021, LNNS 225, pp. 731–744, 2021.
https://doi.org/10.1007/978-3-030-75100-5_62

hygiene and environment [5]. However, excessive household chores can cause a lot of stress to the occupants. If we record the stresses that residents feel during household chores, we can use this as a basis for providing an in-home service that reduces stress and improves QoL, with a focus on household chores, which takes up most of the time in daily life. For example, if someone has to get up early the next morning after work and go out, he or she could perform only light household chores to prepare for the next day's work while still being able to complete them. It also provides an opportunity to rethink the way we approach household chores. Specifically, to reduce the burden of household chores, residents can improve their household chore methods, introduce new appliances and household chore services, and review the division of household chores between family members and housemates. However, not enough research has been done to measure the stress (burden) of household chores.

In this paper, we propose a method for estimating QoL by measuring bioindicators of people performing chores in the home and using these indicators to estimate stress. We measured the life activity data of two subjects for two weeks in a smart home at the Nara Institute of Science and Technology, and evaluated their subjective impressions using a questionnaire. The results indicate that several relationships exist between daily activities and bioindicators, such as the tendency of increased stress while performing various household chores and the tendency of relaxation during rest and entertainment.

2   Related Work

Here, we describe existing studies related to QoL improvement and stress estimation. Quality of life is a measure of satisfaction with one's living conditions. The study of QoL [6] started originally as a concept to discuss a patient's QoL after medical treatment. Nowadays, it is used not only in the medical field, but also as a concept related to general QoL, such as work–life balance and happiness. Work–life balance describes the harmony (or disharmony) between work and daily life when not working, and the idea is to improve and enhance QoL by balancing work and home life and avoiding mental stress from work and depression from physical fatigue.

In particular, QoL that is directly related to a person's health is termed health-related QoL (HRQoL), and is evaluated across various domains such as physical state, psychological state, social interaction, economic and occupational state, and religious and spiritual state [7]. The World Health Organization (WHO) is trying to quantitatively evaluate HRQoL with their own metric, called the WHOQOL [8,9] and short form WHOQOL (SF) [10]. These indices are evaluated using paper questionnaires, but the WHOQOL-100 form [8] requires answers for 100 items in six domains, and the SF-36 form [10] requires answers for 36 items in eight domains. Because of the heavy burden of completing these long questionnaires, routine assessment of the QoL is a difficult task.

Amenomori et al. [11] proposed a method to continuously measure HRQoL with a smaller questionnaire by exploiting tracking and biometric data from smartphones and smartwatches. In addition, Asma et al. [12] used the Hamilton Depression Rating Scale to identify negative psychological states; Garcia et al. [3] proposed the Oldenburg Burnout Inventory to grasp working stress; and Natasha et al. [4] proposed a method to quantitatively measure and estimate the next day's health, stress, and happiness using a smart device. These studies used smart devices to improve QoL by measuring and estimating the user's state to provide feedback to the user. However, the causes of decreased QoL in daily life have yet to be determined.

One possible reason for decreased QoL is the stress of daily life. There are many ways to estimate the stress a person is feeling [13, 14]. Here, we describe a method for estimating stress from biometric information acquired by sensors.

Heart rate variability can be an indicator of various physiological functions, including autonomic functions [15, 16]. Heart rate variability is measured non-invasively using small sensors and thus can be used to estimate stress without significant inconvenience. However, without information related to the activities of daily life, identifying the activities that cause stress is impossible, making it difficult to make adjustments that could reduce stress levels.

In this study, we visualized changes in stress caused by household chores using bioindicators associated with activities of daily living in the home. Our purpose is to obtain knowledge for establishing a method for estimating QoL in long-term daily life. In the experiment, two participants completed an initial questionnaire and a pre-bedtime questionnaire during the experiment, and data collection was conducted in a smart home testbed of the Nara Institute of Science and Technology.

3   Proposed Method

To correlate life activities with stress, we measured stress-related bioindicators, collected subjective assessments, and tracked life activities. We describe how they were measured in the following sections.

3.1   Stress-Related Bioindicators

The heartbeat has periodic variation called heart rate variability. Analysis of this variability can be used to estimate the stress state [15, 16].

Various methods have been proposed for heart rate variability analysis. Frequency-domain analysis uses the low-frequency/high-frequency (LF/HF) ratio. LF is the power spectrum in the low-frequency range of 0.04 to 0.15 Hz and represents the activity of sympathetic and parasympathetic nerve activity associated with blood pressure variation. HF is the power spectrum in the high-frequency range of 0.15 to 0.4 Hz and represents the activity of the sympathetic and parasympathetic nerve activity associated with respiratory variation. Thus, the LF/HF ratio indicates the activity of the sympathetic nervous system. In a relaxed state, both LF and HF signals appear, but the HF component is relatively high and the LF/HF ratio is low. In a stressful state, the LF/HF ratio is higher because the HF component is less prominent. Therefore, if an individual has a high LF/HF ratio during a particular activity, it is presumed that he or she is feeling stress. Thus, heart rate variability may be an important indicator in stress estimation.

The heart rate interval between electrocardiogram R peaks, or the RR interval (RRI), and the standard deviation (SD) of the RRI are also important indices for estimating stress [17, 18]. Toyohuku et al. [19] proposed a simple method for estimating parasympathetic nerve activity using a Lorenz plot (also called a Poincaré plot) of the RRI. In this method, the n-th value of the RRI is plotted on a graph with the n-th value on the horizontal axis and the n + i-th value on the vertical axis. Specifically, all the plot points are projected onto the $y = x$ and $y = -x$ axes, the SD of the distance from the origin $(0, 0)$ of the $y = x$ axis is given as σ_x, the SD of the distance from the origin $(0, 0)$ of the $y = -x$ axis is given as σ_{-x}, and the ellipse output from the following equation is the area:

$$S = \pi \times \sigma_x \times \sigma_{-x}, \tag{1}$$

where the area of the output ellipse represents the magnitude of the RRI fluctuation. This method has been used to judge whether the parasympathetic nervous system has been activated [20–22]. In this study, the LF/HF ratio and RRI were measured, visualized, and analyzed for stress estimation.

This was accomplished with a chest-mounted WHS-3 heart rate sensor [23], which can measure the LF/HF ratio and transmit the data with Bluetooth communication. Then, the LF/HF ratio can be displayed and recorded on a tablet device.

3.2 Subjective Assessments Related to Stress

Questionnaires were administered before the experiment and before bedtime each day to collect the subjects' subjective evaluations. Subjects reported their preferences for performing household chores and their level of fatigue. Table 1 lists the questions about the preferences and strengths of each household chore performed before the experiment, where each item was rated on a 5-point scale. Table 2 shows the pre-bedtime questionnaire (7-point scale except for the first three questions) about the stress and fatigue of each household chore performed during the day.

The pre-bedtime questionnaire in Table 2 is based on the NASA Task Load Index (NASA-TLX) [24], which was developed to visualize the factors of workload, and designed as a questionnaire about all the chores performed throughout the day. There are six NASA-TLX rating scales: Mental Demand, Physical Demand, Temporal Demand,

Table 1. Content of pre-experiment questionnaire.

Item	Description
Washing preferences (up to 5)	Psychological factors such as likes and dislikes
Cleaning preferences (up to 5)	
Cooking preferences (up to 5)	
Dishwashing preferences (up to 5)	
Washing skill (up to 5)	A high degree of skill, such as being familiar
Cleaning skill (up to 5)	
Cooking skill (up to 5)	
Dishwashing skill (up to 5)	
Frequency of doing household chores (description)	How to do your daily chores, and how to order them?
Housework needs (up to 5)	

Table 2. Content of pre-bedtime questionnaire.

Item	Description
Current physical condition (up to 5)	Current status grasping
Current fatigue Level (up to 5)	
Satisfaction or sense of accomplishment in accomplishing household chores (up to 5)	
The most stressful chores	Psychological factors due to the stress of household tasks (one of "laundry, cleaning, cooking, and washing dishes")
Laundry stress (up to 7)	
Cleaning stress (up to 7)	
Cooking Stress (up to 7)	
Dishwashing stress (up to 7)	
Household chores that caused the most fatigue (physically)	Physical elements of household tasks due to fatigue (one of "laundry, cleaning, cooking, and dishwashing")
Laundry fatigue (up to 7)	
Cleaning fatigue (up to 7)	
Cooking fatigue (up to 7)	
Dishwashing fatigue (up to 7)	
"Mental Demand: How much mental and perceptual activity was needed? (up to 7)	Workload factor identification
Criterion: choose the one with a larger load if you feel that the task demands more or the content is more complex"	
"Physical Demand: How much physical activity did you need? (up to 7) Criteria: if you feel that the amount of work required by the task is too much, choose the one with a larger load"	
"Temporal Demand: How much time pressure did you feel for the progress of the task, the rate of progress, and the timing of the task occurring? (up to 7) Criterion: if you are pressed for time and in a hurry to do the work, choose the one with the greater load"	
"Performance: How well do you think you are doing in setting tasks in accomplishing your goals? (up to 7) Criterion: in this case, if I felt I could perform the household tasks well, I would choose satisfaction"	
"Effort: How hard did you work on the required task? (up to 7) Criterion: if you are forced to use a lot of ingenuity in your own way to do the work, choose the one with a larger load"	
"Frustration: How did you feel frustrated, stressed, satisfied, or relaxed while performing the task? (up to 7) Criterion: choose the one with a larger load if you were feeling frustrated when performing the task"	

Performance, Effort, and Frustration. Although NASA-TLX is mainly used to identify workload factors, it was used to measure the various demands of the overall housework activity in this study. The questionnaire was administered at the end of each day, and each item is rated on a 7-point scale.

3.3 Daily Activities

Participants recorded the start and end times of each daily activity using an activity recording application on a tablet device. Figure 1 shows the operation screen of the

Fig. 1. Activity annotation application.

Activity Recording Application for Everyday Life. Each activity has an icon that can be tapped to switch from one activity to another. The current activities are indicated by the icon turning blue; icons representing all other activities remain gray.

4   Experiment

This section describes the experiment we conducted to collect activity data and bioindicator data on daily life, and visualize the data to clarify the relationship between activity and bioindicators. This experiment was approved by the Ethics Review Board of the Nara Institute of Science and Technology (Approval Number 2019-I-9). The RRI and LF/HF ratio values in the data were preprocessed using an interquartile range to remove outliers. Note that the data on bathing activity were measured during the time just before bathing to the removal of the measurement device and from the time of wearing the device again after bathing to the operation of the application to select another activity.

4.1   Experimental Methods

Two participants (students) lived in the one-bedroom smart home testbed at the Nara Institute of Science and Technology during the data collection experiment, and data were collected for 13 days from December 27, 2019, to January 8, 2020. The experiment was conducted for the first 7 d for subject ID01 and for the later 6 days for subject ID02. Table 3 lists definitions of the target activities and definitions of the target activities' start and end. We asked the participants to fill out a questionnaire before the experiment and to complete a bedtime questionnaire every day during the experiment term. The participants were instructed to wear the WHS-3 at all times during the experiment, except when bathing. We also instructed them to record the start time and end time of each activity of daily living by using the application, which was developed for this experiment for mobile devices. These were the only restrictions on the participants.

Table 3. Target activities.

Activity	Definition of activity	Start	End
StayingAtHome	While at home	Going in through the front door	Stepping out of the doorway
PersonalHygiene	Hygiene activities such as hand washing and tooth brushing	Entering the bathroom	Leaving the bathroom
Cooking	Cooking ingredients in the kitchen	Entering the kitchen	Cooking finish
Eating	Eating a food	Start eating the food	Finish eating the food
WashingDishes	Washing dishes	Start washing the dishes	Dishwashing is over
UsingToilet	Going to the bathroom	Stepping out of the doorway	Going in through the front door
Bathing	Washing the body in the bathroom and soaking in the bathtub	Entering the bathroom	Getting out of the bathroom
CleaningRoom	Vacuuming the room	Turn on the vacuum cleaner	Turn off the vacuum cleaner
CleaningBathroom	Cleaning the bathroom	Entering the bathroom	Getting out of the bathroom
Laundry	Putting the laundry in the washing machine	Finish putting in the laundry, start folding the laundry	Finish folding the laundry
WorkingOnPC	Working with a PC	Open PC	Close the PC
WatchingTelevision	Watching TV	Turn on the TV	Turning off the TV
PlayingGames	Playing video games	Turn on the game console	Turning off the game console
ReadingBook	Reading books	Open the book	Close the book
UsingSmartphones	Using smartphones	Operating with a smart phone	Put your smartphone
Sleeping	Sleeping	Getting into bed	Get out of bed
Dressing	Preparing for an outing, such as makeup and changing clothes	Start preparing	Ready to go

4.2 Analysis Results

In this section, we first present the results of the analysis of the overall tendency by participant, followed by the results of analysis by activity and Lorenz plots.

4.2.1 Results of Overall Tendency Analysis

Tables 4 and 5 show results of the pre-bedtime questionnaire and the pre-experiment questionnaire, respectively. For the pre-bedtime questionnaire, Table 4 gives the average

Table 4. Results for pre-bedtime questionnaire.

Items	ID01	ID02
Current physical condition (up to 5)	3.71	2.83
Current fatigue level (up to 5)	3.14	3.00
Satisfaction or sense of accomplishment in accomplishing household chores. (up to 5)	3.00	3.17
Chores that were the most stressful	Cooking	Cooking
Laundry stress (up to 7)	2.14	3.50
Cleaning stress (up to 7)	1.71	2.83
Cooking stress (up to 7)	3.29	3.4
Dishwashing stress (up to 7)	2.57	2.67
Housework that was the most exhausting. (physically and mentally)	Cooking	Cooking
Laundry fatigue (up to 7)	2.00	3.20
Cleaning fatigue (up to 7)	2.00	3.20
Cooking fatigue (up to 7)	3.57	3.17
Dishwashing fatigue (up to 7)	2.86	2.33
Mental demand (up to 7)	4.14	4.67
Physical demand (up to 7)	3.57	2.17
Temporal demand (up to 7)	2.57	2.67
Performance (up to 7)	4.71	2.17
Effort (up to 7)	3.43	3.33
Frustration (up to 7)	3.14	3.17

responses to the questions, most of which were rated on a scale of 1 to 7, and the questions requiring a textual response (to select household chores) show the most frequent responses. The RRI histograms for each participant obtained throughout the experiment

Table 5. Results for pre-experiment questionnaires. (up to 5)

Items	ID01	ID02
Laundry preferences	5	3
Cleaning preferences	2	5
Cooking preferences	3	4
Dishwashing preferences	4	2
Laundry skill	5	2
Cleaning skill	2	4
Cooking skill	3	2
Dishwashing skill	4	4
Frequency of doing household chores	Often do housework after accumulating	Perform chores when needed
The need for household chores	5	4

are shown in Fig. 2, and the LF/HF ratio histograms for each participant are shown in Fig. 3.

Figure 2 shows that the shape and location of the frequency distribution of RRI differed between participants. The histogram of participant ID01 has a multimodal distribution, suggesting that the variation of RRI between activities is relatively large, and many RRI values are overall in a low range. In contrast, the histogram of ID02 is a normal distribution, and the RRI variation between activities is relatively small. From the results of the bedtime questionnaire, ID01 rated him/herself less stressed and fatigued from housework than did ID02. In contrast, ID01 is evaluated as being overloaded, as indicated by the NASA-TLX-referenced questions, such as various requirements and work generation, which may affect the RRI.

Figure 3 shows that the LF/HF ratio histograms do not show as clear a distributional difference as the RRI histograms. The ID01 distribution has more frequencies in the LF/HF ratio region than does the ID02 distribution. In other words, ID01 is more susceptible to stress due to daily activities than ID02. Regarding ID02, the frequency decrease is larger in the region around the LF/HF ratio of 3, compared with ID01. Thus, the figures indicate that activities can be divided into those with larger LF/HF and those with lower LF/HF.

4.2.2 Analysis Results by Activity

Table 6 shows the total duration of each activity for each participant, as well as the corresponding means and SDs of the RRI and LF/HF ratio for each activity. In this experiment, participants were encouraged to perform natural activities as usual, and therefore, the duration of each activity varied from subject to subject. Both participants tended to spend less time cleaning the room and dishwashing compared to sleeping and PC work. Also, a large difference appears between participants in the durations of activities such as cooking and TV watching. Thus, the participants engaging in stressful activities attained a higher index of stress during the experimental period, which may affect the shape of the distribution of bioindicator histograms. Some activities may be performed in parallel, such as watching TV during household chores. In this case, the collected biometric indices are used to calculate the activities' RRIs, LF/HF, and Lorenz plots as though they affect all activities taking place at that time.

The standard deviation of the RRI (hereafter, SDNN) is lower for both participants for all household activities (room cleaning, cooking, laundry, and dishwashing) compared to the other activities. The SDNN is known to be lower in a stressed state and higher in a relaxed state, indicating higher stress in the overall activity. The results of the pre-experiment questionnaire revealed that participants ID01 and ID02 do not prefer cleaning and dishwashing, respectively, as their household chores. Each participant's SDNN is relatively low for each activity, suggesting an association between household chore preferences and biometric indices.

RRI boxplots for each participant's activities are shown in Figs. 4a and 4b, and the corresponding LF/HF ratio boxplots are shown in Figs. 5a and 5b. From Figs. 4a and 4b, the data for household chores activities show relatively low RRI variability, indicating a stressed state. In contrast, data for sleep and smartphone use show relatively high RRI variability, indicating a relaxed state. Figures 5a and 5b show that watching television

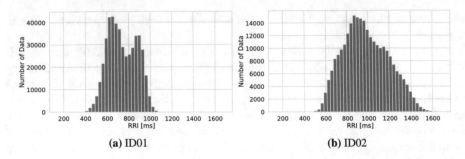

(a) ID01   **(b)** ID02

Fig. 2. RRI histograms.

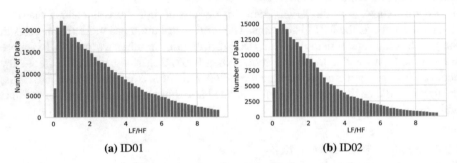

(a) ID01   **(b)** ID02

Fig. 3. LF/HF ratio histograms.

Table 6. Statistics for each activity.

Activity	ID01					ID02				
	Time [min]	RRI mean [ms]	RRI SD [ms]	LF/HF mean	LF/HF SD	Time [min]	RRI mean [ms]	RRI SD [ms]	LF/HF mean	LF/HF SD
Bathing	140.1	554.9	202.7	1.015	0.993	109.2	643.2	112.9	3.883	2.070
CleaningBathroom	15.6	460.2	52.1	1.758	1.337	0.0	0.0	0.0	0.0	0.0
CleaningRoom	25.4	455.5	29.3	1.237	0.959	29.6	724.6	72.6	3.835	1.773
Cooking	100.6	563.2	37.9	3.816	2.450	222.5	734.6	87.2	4.448	2.243
Dressing	15.0	578.0	137.3	2.492	1.714	26.6	658.7	72.3	4.053	2.161
Eating	163.0	555.2	54.2	4.175	2.291	162.1	748.8	118.2	4.431	2.201
Laundry	41.8	560.6	48.2	3.259	2.119	6.4	756.0	76.0	2.695	1.618
PersonalHygiene	23.1	614.8	84.0	5.086	2.059	22.5	766.4	87.4	3.284	1.650
ReadingBook	143.2	652.9	52.4	3.388	1.998	291.4	883.2	103.8	3.422	2.056
Sleeping	4180.6	846.8	95.6	2.284	2.112	3276.9	1106.6	199.4	1.764	1.710
StayingAtHome	8634.9	737.4	133.8	2.965	2.235	6024.6	979.8	210.4	2.458	2.013
UsingSmartphone	1016.1	745.5	100.3	4.092	2.268	199.4	936.7	162.6	3.120	2.078
UsingToilet	135.0	672.8	113.3	2.995	2.171	91.5	728.4	85.8	2.379	2.369
WashingDishes	43.3	586.1	52.9	3.832	2.664	40.2	654.3	67.2	4.591	1.970
WatchingTelevision	2073.5	617.2	70.2	3.976	2.211	177.9	868.0	122.7	2.337	1.647
Working on PC	3013.0	655.1	62.1	3.557	2.096	1226.2	919.0	109.6	2.699	1.827

and personal hygiene have different stress tendencies (high for ID01 and low for ID02) between participants, showing the effects of individual differences, the type of content watched on television, and different physical conditions of the participants.

According to the results of the pre-bedtime questionnaire, the most frequent stressful activity of each participant is cooking; however, according to each boxplot, the RRI tended to be relatively low and the LF/HF relatively high during this activity, reflecting subjective evaluations. Comparison with the results of the pre-experiment questionnaire shows that the LF/HF values of ID02's cooking and dishwashing activities are higher

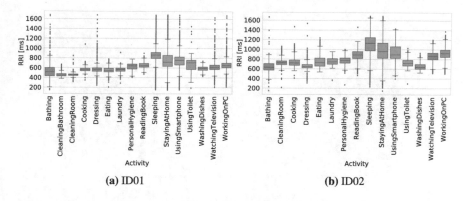

Fig. 4. RRI boxplots.

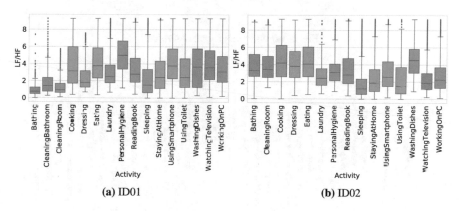

Fig. 5. LF/HF ratio boxplots.

when he or she had a high preference for the activity. In contrast, the self-reported preferences and strengths for activities may not always correspond to the LF/HF ratio, as seen in the washing and cleaning activities of ID01. This result is affected by the influence of the devices, each bioindicator, and other mental or physical factors not considered. In particular, bioindicators such as LF/HF ratio and RRI are affected by various mental and physical factors, such as psychological stress, individual differences, and chronic stress from the environment. Therefore, while the results obtained by this experiment can be regarded as a guide for estimating QoL for general living activities in the home, they are not absolute.

4.2.3 Analysis Results with Lorenz Plots

The Lorenz plot for each participant is shown in Figs. 6a and 6b. The Lorenz plot of ID02 showed that the data tended to be more scattered in the high RRI region and denser in the low RRI region. This indicates that fluctuations in the RRI are large, and the parasympathetic nervous system is activated during sleep and relaxation. In contrast, the Lorenz plot of ID01 shows that the data spread in the high RRI region is slightly

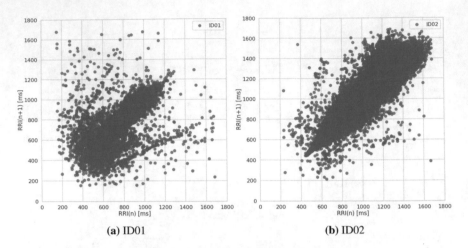

(a) ID01   **(b)** ID02

Fig. 6. Lorenz plots for all activities.

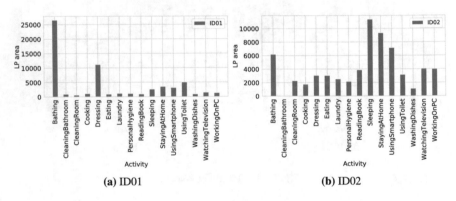

(a) ID01   **(b)** ID02

Fig. 7. Areas of lorenz plots.

wider than that in the low RRI region, and that the variability of data in the high RRI region is smaller. In other words, the fluctuation of the heart rate interval is small during sleep and relaxation, suggesting the possibility that the participants were not able to rest sufficiently. There is no significant difference in the shape of the Lorenz plots for each activity. Overall, the activities with a small area on the Lorenz plots show a wide distribution only in the $y = x$ direction with small differences in shape and clustering in the low RRI region. For activities with large areas, the distribution is wide not only in the $y = x$ direction but also in the $y = -x$ direction, and extends into the high RRI region.

Figures 7a and 7b show the area of the Lorenz plot for each participant's activity. The figure shows that, for both participants, sleeping, smartphone use, and watching TV have a large Lorenz plot area and activated the parasympathetic nervous system. The area is small for various household tasks, and the parasympathetic nervous system is inactive, indicating that they are under stress. The area of Lorenz plots corresponding to each participant's activity is also small, although participants ID01 and ID02 indicated cleaning and dishwashing as their least preferred activities in the pre-experiment questionnaire. These results reveal that household chore preferences and loads can be estimated from the area information of Lorenz plots calculated from the RRI. The Lorenz plot area of ID01's bathing activity is considered to be abnormally high. This is the data obtained within the range while sensor attachment and detachment happened.

5   Conclusion

In this paper, we visualized the subjective evaluation of stress and bioindicators for each activity for estimating the QoL variability in daily life activities. The results showed a tendency of increased stress during various household chores and a tendency of relaxation during rest and recreation, although the subjective evaluations did not necessarily coincide with the stress indicated by the bioindicators. A more generalizable evaluation requires a longer-term experiment with more participants. As future work, we will explore the possibility of providing services, such as optimal timing estimation for housekeeping and relaxation activity recommendations based on daily life activities, by accounting for the chronic stress to which each subject is subjected and personal characteristics, such as the ability to recover from stress. Generally, parasympathetic activity varies greatly among individuals, and real-time stress assessment by Lorenz plot area is difficult because of the need to normalize the area for each person. However, if we generate Lorenz plots in advance and calculate the area at rest or during each activity, real-time stress assessment for each individual is possible by calculating the difference between the usual area and the area at a specific time.

Acknowledgements. This study was supported in part by the Japan Society for the Promotion of Science, Grants-in-Aid for Scientific Research numbers JP20H04177 and JP19K11924.

References

1. Al-Surimi, K., Al-Harbi, I., El-Metwally, A., Badri, M.: Quality of life among home healthcare patients in Saudi Arabia: household-based survey. Health Qual. Life Outcomes **17**(1), 1–9 (2019)
2. Imanishi, M., Tomohisa, H., Higaki, K.: Impact of continuous in-home rehabilitation on quality of life and activities of daily living in elderly clients over 1 year. Geriatr. Gerontol. Int. **17**(11), 1866–1872 (2017)
3. Garcia-Ceja, E., Osmani, V., Mayora, O.: Automatic stress detection in working environments from smartphones' accelerometer data: a first step. IEEE J. Biomed. Health Inform. **20**(4), 1053–1060 (2015)
4. Jaques, N., Taylor, S., Nosakhare, E., Sano, A., Picard, R.: Multi-task learning for predicting health, stress, and happiness. In: NIPS Workshop on Machine Learning for Healthcare (2016)
5. Cerrato, J., Cifre, E.: Gender inequality in household chores and work-family conflict. Front. Psychol. **9**, 1330 (2018)
6. Group, W., et al.: The development of the world health organization quality of life assessment instrument (the whoqol). In: Quality of Life Assessment: International Perspectives, pp. 41–57. Springer, Heidelberg (1994)

7. McKenna, S.: Quality of life and pharmacoeconomics in clinical trials (1997)
8. Group, W.: The world health organization quality of life assessment (whoqol): development and general psychometric properties. Soc. Sci. Med. **46**(12), 1569–1585 (1998)
9. Group, W., et al.: Development of the world health organization whoqol-bref quality of life assessment. Psychol. Med. **28**(3), 551–558 (1998)
10. Ware Jr., J.E.: Sf-36 health survey update. Spine **25**(24), 3130–3139 (2000)
11. Amenomori, C., Mizumoto, T., Suwa, H., Arakawa, Y., Yasumoto, K.: A method for simplified hrqol measurement by smart devices. In: International Conference on Wireless Mobile Communication and Healthcare, pp. 91–98. Springer (2017)
12. Ghandeharioun, A., Fedor, S., Sangermano, L., Ionescu, D., Alpert, J., Dale, C., Sontag, D., Picard, R. : Objective assessment of depressive symptoms with machine learning and wearable sensors data. In: 2017 Seventh International Conference on Affective Computing and Intelligent Interaction (ACII) (2017)
13. Lovibond, P.F., Lovibond, S.H.: The structure of negative emotional states: comparison of the depression anxiety stress scales (DASS) with the beck depression and anxiety inventories. Behav. Res. Ther. **33**(3), 335–343 (1995)
14. Tennant, R., Hiller, L., Fishwick, R., Platt, S., Joseph, S., Weich, S., Parkinson, J., Secker, J., Stewart-Brown, S.: The Warwick-Edinburgh mental well-being scale (WEMWBS): development and UK validation. Health Qual. Outcomes **5**(1), 63 (2007)
15. Kim, H.G., Cheon, E.J., Bai, D.S., Lee, Y.H., Koo, B.H.: Stress and heart rate variability: a meta-analysis and review of the literature. Psychiatry Invest. **15**(3), 235 (2018)
16. Sztajzel, J., et al.: Heart rate variability: a noninvasive electrocardiographic method to measure the autonomic nervous system. Swiss Med. Wkly. **134**(35–36), 514–522 (2004)
17. Baek, H.J., Lee, H.B., Kim, J.S., Choi, J.M., Kim K.K., Park, K.S.: Nonintrusive biological signal monitoring in a car to evaluate a driver's stress and health state. Telemedicine e-Health, **15**(2), 182–189 (2009)
18. Yamamoto, K., Toyoda, K., Ohtsuki, T.: Spectrogram-based non-contact RRI estimation by accurate peak detection algorithm. IEEE Access **6**, 60369–60379 (2018)
19. Toyofuku, F., Yamaguchi, K., Hagiwara, H.: Simplified method for estimating parasympathetic nerves activity by Lorenz plot of ECG RR intervals. Jpn. J. Ergon. **42**(Supplement), 512–515 (2006)
20. Brennan, M., Palaniswami, M., Kamen, P.: Poincare plot interpretation using a physiological model of HRV based on a network of oscillators. Am. J. Physiol.-Heart Circulatory Physiol. **283**(5), H1873–H1886 (2002)
21. Kamen, P.W., Krum, H., Tonkin, A.M.: Poincare plot of heart rate variability allows quantitative display of parasympathetic nervous activity in humans. Clin. Sci. **91**(2), 201–208 (1996)
22. Mourot, L., Bouhaddi, M., Perrey, S., Rouillon, J.-D., Regnard, J.: Quantitative poincare plot analysis of heart rate variability: effect of endurance training. Euro. J. Appl. Physiol. **91**(1), 79–87 (2004)
23. Union Tool Co. Sensor products. https://www.uniontool.co.jp/en/product/sensor/. Accessed 31 Jan 2021
24. Hart, S.G.: Nasa task load index (TLX). volume 1.0; computerized version (1986)

Non-contact Person Identification by Piezoelectric-Based Gait Vibration Sensing

Keisuke Umakoshi[1], Tomokazu Matsui[1], Makoto Yoshida[1,2], Hyuckjin Choi[1],

Manato Fujimoto[1(✉)], Hirohiko Suwa[1,3], and Keiichi Yasumoto[1]

[1] Nara Institute of Science and Technology, Ikoma, Japan
{manato,h-suwa}@is.naist.jp
[2] Onkyo Corporation, Higashiosaka, Japan
makoto.yoshida@jp.onkyo.com
[3] RIKEN, Wako, Japan

Abstract. Floor vibrations caused by walking person (hereinafter, gait vibrations) have recently been explored as a way to determine their locations and identities, and the technologies for measuring such gaits may enable low-cost elderly monitoring services and crime prevention systems. In this paper, we report on the development of a system that can accurately capture both high- and low-level signal gait vibrations using a piezoelectric sensor, and then propose a new system that can identify a walking person based on a small number of footsteps. Our proposed system uses two key approaches to accurately obtain such gait vibrations. The first uses a combination of a source follower circuit in parallel and a piezoelectric sensor. The second involves widening the dynamic range by the use of a dual power supply drive. We then show how we can increase the accuracy of our system by combining multiple footsteps rather than using a single footstep, and thus achieve a more robust system regardless of the distance between the sensor and the target. In experiments comparing five different machine learning (ML) models conducted with six test participants to evaluate our system, person identification results obtained using only a single footstep showed accuracy levels up to 70.8% of average F-measure when using the Light Gradient Boosting Machine (LightGBM) classifier, while for other methods, the average F-measures were 63.1%, 75.9%, and 87.1% in cases of using the first, first and second, and from first to third footsteps from each back-and-forth walk test, respectively.

1 Introduction

Amidst the ongoing improvements to sensing technologies, various companies and research institutes have been working on estimating human activities and their contexts as a means to providing more sophisticated services. When attempting to grasp the human context, it is essential to answering a number of questions (Where is it happening? When did it happen? Who is involved? What are they doing and why?) to understand a person's activities in detail. Most existing studies related to person identification methods seek ways to determine such information, even though it is often difficult to identify. However, the research continues for effective methods because inexpensive

© The Author(s), under exclusive license to Springer Nature Switzerland AG 2021
L. Barolli et al. (Eds.): AINA 2021, LNNS 225, pp. 745–757, 2021.
https://doi.org/10.1007/978-3-030-75100-5_63

and accurate person identification procedures have the potential to make excellent contributions to elderly monitoring services and crime prevention systems.

Many previous methods have been developed to help identify people [1–8]. Of these, biometrics [1,2] are among the best-known technologies. However, face recognition [1] and fingerprint recognition [2] systems are often burdensome because they require the user to perform certain predetermined tasks, such as looking at the camera for authentication. Other methods involve extracting personal movements for identification with sensors such as cameras, Wi-Fi Channel State Information (CSI), and backscatter communication. However, it is difficult for camera-based methods [3] to distinguish identities if the subject is not clearly visible in the camera view field due to bright lighting, obstacles, etc. Also, Wi-Fi-based walking identification methods [4,9] require observations of a large number of footsteps, while backscatter communication-based methods [5] require very strong radio power, which limits its availability. Currently, the person identification technology using gait data (gait vibrations) [6–8] has been attracting attention as a way to resolve these problems. However, while the identification accuracy achieved with this technology is high, there are still unsolved problems. For example, the method is not always usable in narrow environments and it is often expensive to install.

In response to these issues, we herein propose a new system that efficiently detects low-level signal gait vibrations with a single piezoelectric sensor and can distinguish individuals with high accuracy levels using data from, e.g., three footsteps or fewer. The major contributions of this paper are summarized as follows.

First, we designed a circuit to detect both high- and low-level gait vibration signals by (1) widening the output signal level of a piezoelectric sensor using a field-effect transistor (FET), (2) designing a low-noise circuit using a circuit simulator, and (3) widening the dynamic range of the newly designed circuit by adopting a dual supply battery drive for voltage amplification. The resulting system can obtain accurate gait vibration data even when the gait vibration signal level is low. Second, we carried out evaluation experiments utilizing the proposed system with the designed low-noise circuit. Specifically, we evaluated the personal identification accuracy utilizing five different machine learning models (SVM, Random Forest, Xgboost, LightGBM, DNN). As a result, the average F-measure for the person identification using only a single footstep was up to 70.8% when using the LightGBM classifier. Finally, to identify individuals with just a few footsteps, we evaluated our system in three ways: using the first footstep, the first and second footsteps, and from the first to third footsteps, of each back-and-forth walking test, which will be described in detail later. The obtained results showed that the average F-measures when using the LightGBM classifier were 63.1%, 75.9%, and 87.1%, respectively.

2 Related Work

Recently, various studies utilizing gait vibrations generated in buildings have been conducted [10]. In this section, we describe the existing research related to our work.

2.1 Geophones/Seismic Sensor-Based System

Zhang et al. [11] proposed a method for estimating the number of people in a building. To accomplish this, they installed geophones at five locations under floors in a building and counted people if their gait vibrations exceeded a preset threshold. When evaluating the effectiveness of their proposed method, they compared the predicted number of people with the log data of the room access control system and obtained a correlation coefficient of 0.836. Separately, Mirshekari et al. [12] detected events such as door opening, closing, and walking persons by acquiring floor vibrations and using machine learning (ML) to analyze those events. That work showed that adapting the training model generated at a specific location to different locations improved the event detection rate by up to 10 times compared to other methods. However, these studies focused on estimating the number of people or detecting events without considering the issue of person identification.

Meanwhile, Pan et al. [6] proposed a person identification system called FootprintID that combines supervised learning and iterative transductive learning models in which they applied ML to gait data acquired in the case of an average walking speed. They then annotate that gait data with various walking speed data using iterative transductive learning to generate accurate training models. This method reduced the amount of labeled data and achieved an accuracy rate of up to 96% for 10 participants. However, this system did not use low-level signal gait vibrations that are sensitive to noise, requires seven footsteps of data to achieve sufficiently high accuracy, and could only identify individuals in wide open space. In other words, this system can only be used under specific conditions, such as within large rooms.

Clemente et al. [7] constructed a highly accurate person identification system by adding new features based on [13]. In their system, four seismic sensors are installed at the four corners of a 2.5 m × 3.0 m area and used to identify individuals using a simple majority voting method. In that system, F-measure of 63% was obtained for each individual sensor, while F-measure of 71.9% was obtained using a voting method based on the weighting of each sensor. However, this system is expensive to install and requires at least four sensors to identify an individual with one footstep.

2.2 Piezoelectric Sensor-Based System

To resolve the problems described in Sect. 2.1, we decided to base our system on piezoelectric sensors, which have been attracting attention due to their low cost and ease of installation in furniture and floors. Another attractive characteristic is that piezoelectric sensors are mostly insensitive to sounds from speakers and human voices, and are thus relatively resistant to environmental noises, which is a difference from a normal microphone. Hence, even with obstacles in the vicinity, vibrations transmitted to the floor or furniture are less affected by the noise than wireless communication-based systems such as Wi-Fi.

Kashimoto et al. [14] proposed a system that estimated a user's location by using various vibrations generated by the user's activity. However, this study focused solely on location estimation rather than individual identification. In addition, this system is based on a circuit with insufficient voltage amplification for obtaining fine-granularity floor vibration data.

Separately, Akiyama et al. [8] proposed a method of estimating a user's walking direction using two horizontally installed piezoelectric sensors. Their system generated differential data from the acquired gait vibrations and used linear discriminant analysis to estimate the user's walking direction. Our system uses parallel source-follower circuits. Source-follower circuit is a circuit that lowers the output impedance while keeping the voltage constant. This significantly increases the current amplification, and thus allows detailed walking vibrations to be obtained. From the acquired results, it was possible to classify the walking direction of a person with about 83% accuracy. However, this system only estimates the walking direction without identifying the individual. Additionally, despite the use of a single power supply consisting of two AA batteries, the dynamic range is relatively narrow and low-level signal gait vibrations cannot be correctly acquired due to noise. Furthermore, when the amplifier circuit gain is adjusted to obtain low-level signal gait vibrations, the original dynamic range is narrow and the high-level signal gait vibration becomes saturated (unusable) in the sensor vicinity.

2.3   Our Approach

To resolve the problems highlighted in Sects. 2.1 and 2.2, we constructed a new system that can efficiently detect low-level signal gait vibrations using a single piezoelectric sensor, and thus identify individuals with high levels of accuracy using as few footsteps as possible. In tests conducted to evaluate our system, we begin by assuming that a single user is walking within the detection range of our piezoelectric sensor.

3   System Requirements

To realize person identification using gait vibrations, we set the following four requirements for our proposed system:

Req. 1: The system should be able to accurately acquire low-level signal gait vibrations.
Req. 2: The system should be unaffected by internal and natural noise.
Req. 3: The system should be capable of detecting a wide range of gait vibrations.
Req. 4: The system should be able to identify the individual in the fewest footsteps possible.

Suppose the output impedance of the piezoelectric sensor is higher than the input impedance of the amplifier circuit. In that case, an output voltage drop occurs, which makes it difficult to detect vibrations generated at a distance from the sensor, which in turn means that it is necessary to devise a way to reduce the output impedance of the piezoelectric sensor. In the circuit shown in Fig. 1, the output impedance is reduced by introducing a source-follower circuit equipped with a field-effect transistor (FET), thus fulfilling Req. 1.

Since walking gait vibration characteristics are mainly expressed in terms of their low-frequency component, to accurately capture those characteristics to the greatest extent possible, it is vital to design a circuit that is not susceptible to low-frequency noise. Hence, to fulfill Req. 2, we used an LTspice[1] circuit simulator to produce a design that is unaffected by such noise.

For Req. 3, since it is important to widen the dynamic range to detect a wide range of gait vibrations, we obtained increased voltage by adopting a dual power supply drive. Here, the dynamic range refers to the ratio of the maximum and minimum signals. As a result, in addition to low-level signals, unsaturated high-level signal gait vibrations can be acquired without saturation, as well as low-level signal, thus fulfilling Req. 4.

4   Proposed System

In this section, we describe our proposed person identification system, which is based on a piezoelectric sensor, as well as an effective method for using this system.

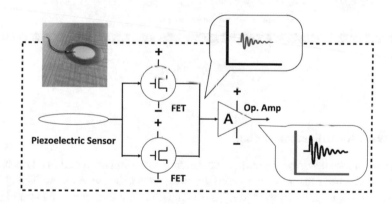

Fig. 1. Circuit diagram.

4.1   Circuit Structure

The circuit in Fig. 1 consists of a piezoelectric sensor, the FET, and an operational amplifier. The 7BB-41-2L0 piezoelectric sensor developed by Murata Manufacturing Co., Ltd. is used in our system. This circuit uses a FET source follower circuit to prevent the voltage drop caused by a case of the output impedance of the piezoelectric sensor from being higher than the input impedance of the amplifier circuit. As shown in Fig. 1, the current amplification ratio was increased by parallel source-follower circuit, which made it possible to acquire unsaturated low-level gait signal vibrations.

The source-follower circuit only amplifies the voltage current, which is essential to obtaining accurate low-level signal gait vibration data. To accomplish this, we amplify

---

[1] https://www.analog.com/en/design-center/design-tools-and-calculators/ltspice-simulator. html.

the signal by using a non-inverting amplifier circuit with an amplification gain of 10x. To minimize the noise effects, circuit elements were selected using the LTspice circuit simulator.

However, there is still a problem because high-level signal vibrations become saturated in the vicinity of the piezoelectric sensor, which means they cannot be acquired even if a gain adjustment is applied to acquire the low-level signal vibrations. In this research, we address this issue by supplying a dual power supply to both the source-follower and non-inverting amplifier circuits, thereby achieving a much wider dynamic range than that would be possible with conventional systems.

Finally, since humming noise is a serious problem in gait vibration feature extraction, we use AA batteries (1.2 V × 4 batteries) to remove that effect.

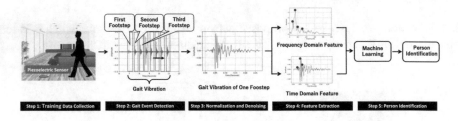

Fig. 2. System flow.

4.2   Person Identification Method

In this section, we describe the person identification method designed to take advantage of our piezoelectric sensor-based system. Figure 2 shows the system flow. The objective of our method is to use ML to identify individuals via the five steps of 1) training data collection, 2) gait event detection, 3) normalization and denoising, 4) feature extraction, and 5) person identification model construction. Each step is described below.

4.2.1   Training Data Collection
In Step 1, we collected gait vibrations to be used as training data. As shown in Fig. 2, we asked participants to walk normally along a section of flooring to obtain gait vibration data. Since private homes are the target environment of this study, participants were asked to walk without shoes but while wearing socks. Our system collected data using a data logger capable of sampling at 10 kHz.

4.2.2   Gait Event Detection
In Step 2, after collecting the training data, we detected single footstep gait events from the measured vibration waveforms. In this study, we use a change point detection algorithm that uses background noise variance as a threshold. To detect gait events, we obtained gait vibration variance values separated by two millisecond intervals and

shifted by one millisecond. The starting point of the gait event is when the variance value exceeds 10 times the threshold, and the end point of the gait event is when it falls below 1.5 times the threshold.

4.2.3 Normalization and Denoising

In Step 3, we normalized the gait events and removed the background noise. Since differences in signal levels due to the distance relationship between the piezoelectric sensor location and the gait vibration location could cause a decrease in identification accuracy, we resolved this problem by dividing the gait signal by the sum of the squares of that signal. This process normalizes the gait event and prevents the deterioration of identification accuracy. Additionally, since it is possible that gait vibration characteristics could be affected by building vibrations, we use the wavelet transform to remove such noise.

Table 1. Extracted features.

Domain	Feature names
Time domain	Event duration, Standard deviation, Entropy, First five peak values, Maximum peak value, Maximum peak location, Five values before the maximum peak, Five values after the maximum peak
Frequency domain	Spectra, Centroid, First five peak values, First five peak locations, Peak number

4.2.4 Feature Extraction

In Step 4, after normalizing the gait events and removing background noise, we extracted the features needed to identify individuals from their gait events. Table 1 shows the list of features. This system utilizes 13 features that have been shown to be highly significant for identifying individuals, as reported in [7].

4.2.5 Person Identification Model Construction

Finally, we used the features described in the previous section to identify individuals. To construct our person identification model, we used five ML models: support-vector machine (SVM), Random Forest, XGboost, Light Gradient Boosting Machine (Light-GBM), and deep neural network (DNN), which are considered to be representative classification models. In this study, gait data were collected under the condition that no vibrations other than the walking vibration occurred. For each training model, the hyper- parameters were adjusted using the grid search method.

5   Experiment

In this section, we describe evaluation experiments conducted using our proposed system. First, the objective and overview of the experiments are described, followed by the experimental conditions and evaluation methods.

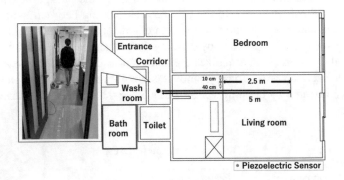

Fig. 3. Experimental environment.

5.1   Experimental Objective and Overview

To evaluate the effectiveness of our piezoelectric sensor-based person identification system, we conducted an evaluation experiment in a smart home environment of the Nara Institute of Science and Technology (NAIST), shown in Fig. 3. Initially, we asked our test participants to walk along a specific path and collected his/her gait vibrations by using our system. Figure 4 shows the measurement devices used in this experiment. As can be seen in the figure, one piezoelectric sensor is placed on the floor at the midpoint of the walking path 10 cm away from the wall and 40 cm from the walking path. A 215 g brass weight is placed on top of that piezoelectric sensor to make it easier to measure gait vibrations. The sampling rate is 10 kHz, and the data are logged on the laptop personal computer (PC) using an Analog Discovery 2 measurement device. Since this experiment tested a scenario in which a person enters and leaves in a room, we instructed the participants to enter the room, walk naturally on the path, turn around, and then exit at their own pace. All of the experiments in this study were conducted with the approval of the NAIST Ethics Review Committee (Approval Number: 2020-I-16).

5.2   Experimental Conditions

At the initial stage of our experiment, which was conducted with the cooperation of six participants (five males and one female). Each participant was asked to walk in their normal manner along the path shown in Fig. 3 while wearing socks. More specifically, each participant was instructed to walk forward 5 m from the starting point, turn around, and then return to the original position. In this experiment, we collected data of three

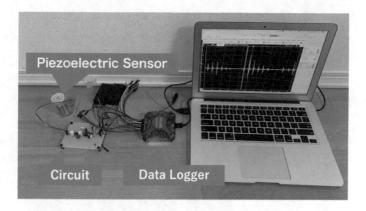

Fig. 4. Measurement devices.

sessions, each of which consisted of 10 consecutive round-trips, from each participant. To avoid collecting data consecutively from the same participant, we changed participants or inserted breaks between sessions.

5.3 Evaluation Method

Four evaluation methods were used in these experiments. First of all, we generated an identification model for each participant using just a single footstep. For each training data set, we constructed a person identification model using the five ML models described in Sect. 4.2.5, and then evaluated the resulting identification accuracy by leave-one-session-out cross-validation. The most accurate person identification model acquired was then used for the three other evaluation methods. More specifically, to determine whether individuals could be identified via a small number of footsteps, we evaluated the individual identification accuracy using the first, first and second, and first to third footsteps of each back-and-forth walking trip. In this experiment, precision, recall, and F-measure (indicated by F1) were used to evaluate the proposed system.

5.3.1 Person Identification Using a Single Footstep
For person identification using a single footstep, each ML model returns a confidence level that is the probability of belonging to each class. The class with the highest confidence value is considered the identification result. In our evaluation, one round-trip consists of approximately 18 footsteps and each participant performs 10 round-trip walks, which means that one session generates approximately 180 footsteps of data of a person.

5.3.2 Person Identification Using Multiple Footsteps
In cases where the identification result is decided by majority vote for each footstep, split votes may result and the person may not be correctly identified. For example, if the first footstep is "label A," the second footstep is "label B," and the third footstep is "label C," it would be impossible to determine the person by a simple majority vote.

Table 2. Person identification results using a single footstep.

	SVM			Random forest			XGBoost			LightGBM			DNN		
	Recall	Precision	F1	Recall	Precision	F1	Recall	Precision	F1	Recall	Precision	F1	Recall	Precision	F1
A	58.8%	54.3%	56.5%	62.8%	62.4%	62.4%	61.3%	60.5%	60.8%	63.4%	64.9%	64.1%	69.7%	57.2%	62.8%
B	72.8%	67.9%	70.3%	81.6%	60.4%	69.4%	79.9%	69.2%	74.0%	82.0%	69.5%	75.1%	76.3%	72.5%	74.3%
C	75.4%	66.8%	70.8%	65.5%	77.6%	70.3%	69.5%	76.6%	72.1%	72.0%	77.1%	73.8%	66.8%	76.4%	70.8%
D	54.8%	70.1%	61.4%	63.2%	63.8%	63.5%	67.3%	68.6%	67.8%	69.0%	71.5%	70.0%	62.9%	71.0%	66.7%
E	69.3%	72.2%	70.7%	60.2%	82.2%	69.5%	71.0%	78.0%	74.1%	74.1%	80.8%	77.1%	70.1%	76.5%	72.6%
F	59.1%	61.9%	60.4%	61.3%	58.5%	59.9%	63.9%	63.2%	63.5%	65.3%	64.5%	64.8%	67.5%	67.4%	67.4%
Ave.	65.0%	65.5%	65.0%	65.7%	67.4%	65.8%	68.8%	69.3%	68.7%	70.9%	71.3%	70.8%	68.9%	70.1%	67.4%
Acc.	64.8%			65.5%			68.4%			70.5%			68.8%		

Accordingly, this study uses the confidence value and sums up this value for each class. As a result, the class with the highest confidence value becomes the identification result for all of the footsteps. As mentioned earlier, each session generates approximately 180 footsteps of data per person. From these data, we constructed test data in three different ways. Since the first method is for an evaluation using only the first footstep, we extracted all the first footsteps from each back-and-forth walk of each participant. Consequently, this test data contain data of 20 first footsteps. Similarly, we construct evaluation test data using the first two footsteps, and then the first three footsteps. Hence, for the first two and first three footsteps, we collect data of 40 and 60 footsteps, respectively. During all experiments, if some gait events are omitted due to the low signal level, we consider the following measurement detected by the sensor as the next footstep's data.

6   Results

In this section, we describe the results of our experiments. First, we show the results of person identification using only a single footstep. Then, we show the results of the person identification from each back-and-forth walk in three ways: the first footstep, the first and second footsteps, and the first to third footsteps.

6.1   Person Identification Using a Single Footstep

Table 2 shows the results of person identification using a single footstep. These results show that the accuracy of LightGBM is the highest, with an average F-measure of 70.8%. Moreover, the other training models identified the participants with an average F-measure of 60% or better accuracy. Participant E, who shows the highest accuracy, has a mean F-measure of 77.1%, and Participant A, who has the lowest accuracy, has a mean F-measure of 64.1%, thereby indicating that our proposed system is robust and can identify individuals with relatively high accuracy. However, this result does not guarantee good identification accuracy in cases involving low-level signal gait vibrations, such as situations where gait vibrations are generated at an excessive distance from the sensor. Therefore, we conducted experiments using LightGBM, which showed the highest recognition accuracy, for three cases of identification data: the first footstep, the first and second footsteps, and the first to third footsteps of each back-and-forth walk.

Table 3. Results using multiple footsteps.

	Recall	Precision	F1	Accuracy
First footstep only	62.9%	69.9%	63.1%	63.0%
First and second footsteps	76.2%	78.5%	75.9%	76.1%
First to third footsteps	87.2%	88.1%	87.1%	87.3%

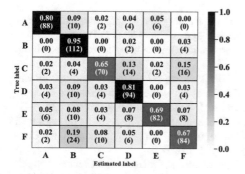

Fig. 5. Confusion matrix (using the first footstep only).

6.2 Person Identification Using Multiple Footsteps

Table 3 shows the person identification results for the first and multiple footsteps from each back-and-forth walk, while the confusion matrices for these results are shown in Figs. 5 to 7. The average F-measure for the first footstep detected by the system from the walking start was 63.1%, which indicates that our system is able to acquire gait vibrations generated at an expected distance from the sensor. The average F-measure using the first and second footsteps is 75.9%, thus achieving a 12.8% accuracy improvement compared to the case of using only the first footstep. The reason for this is that the confidence of each class is higher because of the observation of two footsteps of gait

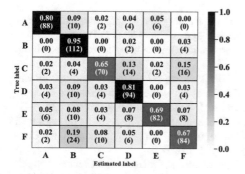

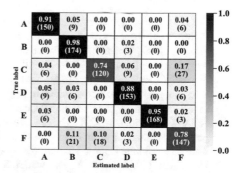

Fig. 6. Confusion matrix (using the first and second footsteps).

Fig. 7. Confusion matrix (using the first to third footsteps).

vibration. Furthermore, the mean F-measure of the person identification results when using the first to third footsteps of the walking test was 87.1%, which was 24% more accurate than using only the first footstep, and 11.2% more accurate than using the first and second footsteps. These results show that the proposed system can identify individuals with an average F-measure of more than 80% by observing gait vibrations from their first three footsteps.

7   Conclusion

In this paper, we proposed a new system that efficiently detects low-level signal gait vibrations with a single piezoelectric sensor and a ML-based method that can identify an individual with high accuracy from a small number of footsteps. To show the effectiveness of our proposed system, we conducted a person identification experiment in a smart home at the NAIST. As a result, we obtained two findings. First, the obtained accuracy using five different ML models showed that the average F-measure for person identification using only a single footstep was up to 70.8% when using LightGBM. Second, in order to investigate whether individuals could be identified with just a few footsteps, we evaluated the accuracy of individual identification in three ways of using the first footstep, the first and second footsteps, and the first to third step footsteps, from each back-and-forth walk. The mean F-measures of those cases were 63.1%, 75.9%, and 87.1%, respectively. These results showed that the proposed system achieves a recognition accuracy comparable to existing systems that can identify an individual with an average of 83% with five footsteps [13] and an average accuracy of 96% with seven footsteps [6]. In the future, we plan to improve the circuit to acquire gait vibrations at lower signal levels. In addition, we plan to select additional features to improve the system accuracy and to make it possible to more accurately identify individuals with fewer footsteps.

Acknowledgement. This study was supported in part by the Japan Society for the Promotion of Science, Grants-in-Aid for Scientific Research number JP20H04177.

References

1. Sun, Y., Chen, Y., Wang X., Xiaoou, T.: Deep learning face representation by joint identification-verification. Adv. Neural Inf. Process. Syst. 1988–1996 (2014)
2. Hrechak, A.K., McHugh, J.A.: Automated fingerprint recognition using structural matching. Pattern Recogn. **23**(8), 893–904 (1990)
3. Jiewen, Z., Ruize, H., Yiyang, G., Liang, W., Wei, F., Song, W.: Human identification and interaction detection in cross-view multi-person videos with wearable cameras. In: Proceedings of the 28th ACM International Conference on Multimedia, pp. 2608–2616 (2020)
4. Korany, B., Karanam, C.R., Cai, H., Mostofi, Y.: Xmodal-id: Using wifi for through-wall person identification from candidate video footage. In: The 25th Annual International Conference on Mobile Computing and Networking, pp. 1–15 (2019)
5. Higashino, T., Uchiyama, A., Saruwatari, S., Yamaguchi, H., Watanabe, T.: Context recognition of humans and objects by distributed zero-energy IoT devices. In: 2019 IEEE 39th International Conference on Distributed Computing Systems (ICDCS), pp. 1787–1796. IEEE (2019)

6. Pan, S., Yu, T., Mirshekari, M., Fagert, J., Bonde, A., Mengshoel, O.J., Noh, H.Y., Zhang, P.: Footprintid: indoor pedestrian identification through ambient structural vibration sensing. Proc. ACM Interact. Mob. Wear. Ubiquitous Technol. 1(3), 1–31 (2017)
7. Clemente, J., Song, W.Z., Valero, M., Li, F., Liy, X.: Indoor person identification and fall detection through non-intrusive floor seismic sensing. In: 2019 IEEE International Conference on Smart Computing (SMARTCOMP), pp. 417–424. IEEE (2019)
8. Akiyama, S., Yoshida, M., Moriyama, Y., Suwa, H., Yasumoto, K.: Estimation of walking direction with vibration sensor based on piezoelectric device. In: 2020 IEEE International Conference on Pervasive Computing and Communications Workshops (PerCom Workshops), pp. 1–6. IEEE (2020)
9. Zeng, Y., Pathak, P., Mohapatra, P.: Wiwho: wifi-based person identification in smart spaces. In: 2016 15th ACM/IEEE International Conference on Information Processing in Sensor Networks (IPSN), pp. 1–12 (2016)
10. Wan, C., Wang, L., Phoha, V.V.: A survey on gait recognition. ACM Comput. Surv. (CSUR) 51(5), 1–35 (2018)
11. Zhang, Y., Pan, S., Fagert, J., Mirshekari, M., Noh, H.Y., Zhang, P., Zhang, L.: Occupant activity level estimation using floor vibration. In: Proceedings of the 2018 ACM International Joint Conference and 2018 International Symposium on Pervasive and Ubiquitous Computing and Wearable Computers, pp. 1355–1363 (2018)
12. Mirshekari, M., Fagert, J., Bonde, A., Zhang, P., Noh, H.Y.: Human gait monitoring using footstep-induced floor vibrations across different structures. In: Proceedings of the 2018 ACM International Joint Conference and 2018 International Symposium on Pervasive and Ubiquitous Computing and Wearable Computers, pp. 1382–1391 (2018)
13. Pan, S., Wang, N., Qian, Y., Velibeyoglu, I., Noh, H.Y., Zhang, P.: Indoor person identification through footstep induced structural vibration. In: Proceedings of the 16th International Workshop on Mobile Computing Systems and Applications, pp. 81–86 (2015)
14. Kashimoto, Y., Fujimoto, M., Suwa, H., Arakawa, Y., Yasumoto, K.: Floor vibration type estimation with piezo sensor toward indoor positioning system. In: 2016 International Conference on Indoor Positioning and Indoor Navigation (IPIN), pp. 1–6. IEEE (2016)

Design of Healthcare Information Sharing Methods Using Range-Based Information Disclosure Incentives

Miku Honda[1]([✉]), Joji Toshima[2], Takuo Suganuma[3], and Akiko Takahashi[1,2]

[1] National Institute of Technology, Sendai College, 4-16-1 Ayashi-Chuo,
Aoba-ku, Sendai 989-3128, Japan
a1911530@sendai-nct.jp, akiko@sendai-nct.ac.jp
[2] Graduate School of Information Sciences, Tohoku University, 6-3-09 Aoba, Aramaki-aza
Aoba-ku, Sendai 980-8579, Japan
toshima@ci.cc.tohoku.ac.jp
[3] Cyberscience Center, Tohoku University, 6-3 Aoba, Aramaki-aza Aoba-ku,
Sendai 980-8578, Japan
suganuma@tohoku.ac.jp

Abstract. The means by which healthcare information is collected and utilized are progressing. Collecting and analyzing such information on a large scale can result in the extraction of useful information and further service development. However, the provision of healthcare information has privacy considerations because healthcare information contains personal information. Therefore, in this paper, we propose a healthcare information sharing method that promotes the sharing of useful healthcare information with ample privacy safeguards using auto-negotiation considerations and user/system intentions. In addition, we evaluate the effectiveness of the proposed method in simulation experiments.

1 Introduction

With the widespread use of wearable devices that continuously collect sensor data, the collection and utilization of user health data (e.g., sleep time and walking time) is increasing [1, 2]. However, although collecting and analyzing such information on a large scale may yield useful information, such information is generally managed and utilized by companies that provide devices. Here, the services the user can access with a device (information provider) while providing information are limited to those provided by the company. Thus, there are methods to appropriately collect information based on the consent of the information provider [3]. These methods are based on the notion that information owned by an individual belongs to that individual, and they should have the right to use it freely.

The goal is to widely use and distribute the information to business operators (information collectors), e.g., companies, and not limit the spread of information to companies that provide the devices. In such an information sharing context, public healthcare information provision imposes a burden on users due to the personal nature of healthcare information. As a result, users may be reluctant to provide healthcare information, which may limit the amount of data gathered for analysis. Thus, it is difficult to collect large volumes of high-quality healthcare information. Therefore, a new framework is needed to promote the use of healthcare information among a wide range of information collectors. The new framework will prioritize the collection of large amounts of high-quality healthcare information while considering privacy. To provide healthcare information on an information provider's own initiative, it is necessary to have a motivation to consider the risks incurred by the information provider and assume that the information provider may provide the information.

For this problem, we proposed a method to promote information provision by giving incentives while considering the protection of information. Using auto-negotiation methods, we proposed a way for determining the required level of incentives and information protection level. It is necessary to gain user buy-in while considering the sensitive nature of healthcare information, objective information provider/information collector evaluations, and a combination of the two.

We propose a healthcare information sharing method that promotes the provision of useful healthcare information with appropriate privacy safeguards without burdening information providers and collectors. In addition, we apply the proposed healthcare information sharing method to a healthcare information sharing system and evaluate its effectiveness through simulation experiments that assume different types of information providers.

2 Related Work and Proposed Method

2.1 Related Work

As the populations in Japan, Northern Europe, and the United States of America continue to rapidly age, the medical care and welfare burden continues to increase. As a result, it is becoming increasingly difficult to respond to conventional medical and insurance systems [4]. There are some methods to sharing information that promote adherence to a healthy diet and individualized physical activity [5]. Sharing healthcare information is effective relative to improving both health and lifestyle and sharing healthcare information is required to create personalized healthcare services [6].

To promote the provision of healthcare information, it is necessary to have a motivation to provide healthcare information. The conditions under which personal information may be provided are described as follows. The information provider receives economic benefits, and the provision of information improves the convenience of services. Incentives are an effective way to influence human behavior [7], and money is the most effective incentive, but it can be replaced with a voucher. It is effective to give a small incentive to promote changes in individual behavior; however, given privacy considerations, information providers are often reluctant to make their own personal information available.

Healthcare information contains personal information, and the handling risk of such information is high. Therefore, it is important to utilize such information with care.

However, the utilization of healthcare information has not progressed [8]. Regarding utilization of personal information on the Internet, e.g., browsing history and location information, various privacy infringement risk indicators and personal information protection methods have been proposed. However, it is possible that certain individual privacies will be abused as a result of access to such information because it is impossible to completely quantify privacy risk using the existing method [9]. Thus, it is important to develop a sense of satisfaction from the individual to promote information provision. One effective way to increase survey cooperation rates is through financial incentives [10, 11]; however, the extent to which incentive requirements vary is related to each individual's degree of privacy awareness due to considering the psychological burden of the information provider [12]. In addition, personal information may be abused, and disclosure of information that users wish to keep private is becoming problematic.

Therefore, it is becoming increasingly important to control the information disclosed by the information providers [13]. Such information providers must disclose personal information by adjusting the types of information provided based on data collectors' intentions. However, it is difficult for general users to determine appropriate personal information to provide.

2.2 Technical Problems and the Proposal

As discussed in Sect. 2.1, methods for sharing personal health information are available; however, providing healthcare information is a concern for users owing to privacy problems. To address this issue, some methods have previously been proposed to promote information provision through incentives. However, the amount of incentive required to promote healthcare information provision differs for each diffident individual. In addition, it is difficult for general users to determine appropriate healthcare information to provide. Thus, it is a burden for both the information provider and information collector to negotiate a benefit for both parties while considering such factors. There are technical problems involved in promoting healthcare information sharing through incentives, particularly when considering the information protection and usage intent.

(P1)   It is difficult to provide users sufficient incentives to motivate them to provide healthcare information.

Different people require different amounts of incentives to provide such sensitive information. In addition, it is difficult to determine the requisite level of incentive because user-provided healthcare information is used in different ways.

(P2)   It is difficult to reasonably protect sensitive user-provided information.

This difficulty exists because sharing sensitive information is mutually beneficial, but it is possible that personal information may be abused, which could lead to leakage of information the user wishes to keep private.

Therefore, it is necessary to determine the information to be disclosed based on the user's unique intentions. Therefore, this study proposes the following to solve the above technical issues.

(S1) Healthcare information should be shared using an auto-negotiation method between software agents.

We propose the promotion of useful healthcare information sharing via consensus building using auto-negotiation to consider the intentions of both information providers and information collectors. Given that agents reflect intentions between information providers and collectors, the level of incentives and information protection level can be determined by automated negotiation between agents. Thus, this method is expected to encourage users to provide confidential healthcare information.

3   Proposed Healthcare Information Sharing Method

3.1   Immediate Evaluation of Healthcare Information

	1	2	3	4	⋯	n
Steps	2300	2200	2100	1100	⋯	2000
Body weight	33	58	90	44	⋯	2
Exercise time	5.3	6.4	3.2	7.8	⋯	0
Heart rate	200	150	100	80	⋯	50
Water intake	1500	900	2000	600	⋯	400
Calorie intake	2000	3000	5000	600	⋯	800
Sleep time	9.5	8.2	7.7	6.0	⋯	4.5

Fig. 1. Example of healthcare information

The information-provider agent obtains healthcare information data that can be acquired on a daily basis from smart devices, including wearable devices, equipped with a wide variety of sensors. Healthcare information is sent to information-collector agents owned by people seeking data such as companies, developers, and organizations using network technology. The information-collector agent immediately evaluates the user-provided healthcare information according to several factors, e.g., quality and sensitivity. In this study, among smart devices, wearable devices are treated as testbeds. The evaluation point is data that can be obtained from a wearable device worn by an information provider. In this evaluation, only numerical data, e.g., blood pressure and sleep time, are considered because it is easy to uniquely determine the evaluation compared with verbal data. Healthcare information is evaluated quantitatively using the Mahalanobis distance measure [14]. The Mahalanobis distance is calculated using a group of high-quality healthcare information established by an information-collector agent, which relies on immediate evaluation of the healthcare data. We consider that the preference information of healthcare information differs depending on the given information collector. Therefore, good healthcare information should be defined individually by each information collector. Here, we categorize healthcare information into seven classes of numerical data (Fig. 1) that can be acquired from wearable devices.

Considering the above, the Mahalanobis distance used for immediate evaluation is defined as follows:

$$d = \sqrt{(\vec{x} - \vec{\mu}) \sum{}^{-1} (\vec{x} - \vec{\mu})} \tag{1}$$

where d denotes the Mahalanobis distance, x denotes the data provided by information provider, and μ denotes the mean vector calculated from the set of high-quality healthcare information established by the information-collector agent.

The closer the data are to a set of high-quality healthcare information, the smaller the Mahalanobis distance. Thus, it is possible to perform high-level evaluation of high-quality healthcare information. Here, the Mahalanobis distance is normalized to [0, 1] and used to immediately evaluate the healthcare information.

3.2 Utility Functions

Auto-negotiation is performed between the information-collector agent and information-provider agent. Here, the information-provider agent reflects the will of the information provider, and the information-collector agent reflects the will of the information collector. The level of incentives and amount of information disclosed is determined via auto-negotiation between these agents. This auto-negotiation is performed based on the "alternating offers" negotiation protocol [15], which is frequently used in bilateral negotiations. In the alternating offers negotiation protocol, each agent selects one of the three actions (Accept, Offer, or End Negotiation). Each agent will select an "Offer" until "Accept" or "End negotiation" is selected. This behavior alternates between the information-collector agent and information-provider agent. The negotiation algorithm is shown in Algorithm 1. When selecting an "offer, each agent offers a proposal in the range of [0, 1], with their utility function considered as a probability distribution.

(a) Accept.

    The agent accepts the bid presented by the other agent. The bid's incentive level and amount of information disclosed are determined as a result of negotiation, and the negotiation is completed.

```
Algorithm 1
Require:
  Maximum number of negotiations lim
  Information provider agent PA
  Information collector agent CA
Ensure:
  Agreement
t ← 0
while t<lim do
    offer ← Offer by PA
    if CA accepts offer then
      return offer
    end if
    offer ← Offer by CA
    if PA accepts offer then
      return offer
    end if
    t ← t+1
end while
return End Negotiation .
```

(b) Offer.
   The agent rejects the bid presented by the other agent. The agent presents a new
   bid to the other agent.

(c) End Negotiation.
   The agent rejects the bid presented by the other agent. The agent does not present
   a new bid and ends the negotiation.

Each agent determines its conduct according to each utility function. The utility
function quantifies the degree of satisfaction, i.e., the utility, a person feels as a result
of hypothetical profit or loss. Therefore, it is necessary to properly express how utilities
differ based on the intentions of the individuals who experience the profits and losses.
The utility value is defined in the range [0, 1]. Here, if the utility is 0, the person is
very dissatisfied with the profit or loss, and, if the utility is 1, the person is completely
satisfied with the profit or loss.

Considering the above, the utility function is defined as follows:

$$U = \sum_{i=1}^{2} u_i \times w_i \tag{2}$$

where U denotes the utility function, u_1 denotes the utility of information protection,
w_1 denotes the intentional weight of information protection, u_2 denotes the utility of the
incentive, and w_2 denotes the intentional weight of the incentive.

Here, intentional weights w_1 and w_2 are given in the range [0, 1]. When summed, w_1
and w_2 should equal 1. Parameters w_1 and w_2 express the difference between the user's
and system's intentions relative to incentives and information protection.

Utility u_i is expressed by a different equation between the system agents and user agents. Here, utility u_{u1}, u_{u2} is a value that expresses the degree of user satisfaction, and the value of utility u_{s1}, u_{s2} expresses the system's degree of satisfaction. u_{u1}, u_{u2}, and u_{s1}, u_{s2} are expressed as follows:

$$U_{ui} = (x - I_p + 1)^r (i = 1, 2), \tag{3}$$

$$U_{si} = I_e - (x(1 - I_e))^r (i = 1, 2), \tag{4}$$

where I_p denotes the parameter of the information provider's sensitive evaluation of healthcare information. I_p is defined in the parameter [0, 1]. I_e denotes the immediate evaluation of the healthcare information (Sect. 3.1). In addition, parameter r denotes the degree of risk preference.

The utility function is outlined in Fig. 2(a) and Fig. 2(b). Note that the information provider's utility function is a monotonically increasing function, and the information collector's utility function is a monotonically decreasing type function. Here, utility function U takes the form of a risk-averse function if the risk preference degree r becomes greater than 1 and becomes a risk-loving function if r is less than 1.

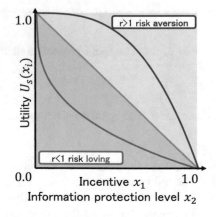 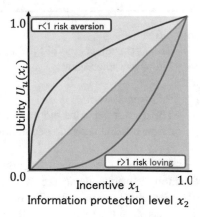

Fig. 2. (a) Utility function of information collector (b) Utility function of information provider

3.3   Design of Healthcare Information Sharing Model

This model comprises an information-provider side agent that operates on behalf of the information provider, an information-providing agent (PA), an information-collector agent that operates on behalf of the information collector, an information-collector agent (CA), and a healthcare information sharing system (HIS) that manages the collected healthcare information. The model is shown in Fig. 3. The processing flow of this model is summarized as follows. Note that each step outlines the equivalent process shown in Fig. 3.

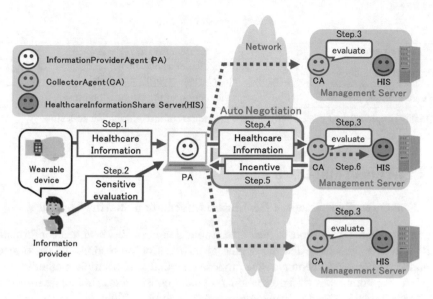

Fig. 3. Healthcare information-sharing model

Step 1. The PA obtains the information provider's healthcare information.
Step 2. The PA obtains the information provider's sensitive evaluation of the healthcare information
Step 3. The CA evaluates the healthcare information based on the quality and sensitivity of the information.
Step 4. The PA selects the CA with the highest evaluation of the value of the information and performs an automated negotiation. The agent determines the information to be disclosed and the level of incentive.
Step 5. The CA gives the PA the incentive established based on the valuation (Step 4).
Step 6. The CA sends the anonymized healthcare information to the HIS.

Table 1. Types of information providers

	Sensitive evaluation	*Immediate evaluation of healthcare information*
A	High [0.7–1.0]	Low [0.0–0.3]
B	High [0.7–1.0]	High [0.7–1.0]
C	Low [0.0–0.3]	High [0.7–1.0]
D	Low [0.0–0.3]	Low [0.0–0.3]

This method promotes information sharing by providing an incentive to the user according to the quality of the shared information in consideration of the extent to which the information is protected.

Table 2. Types of methods

Method name	Explanation of the method
Proposed method	Perform auto negotiations and calculate the amount of incentives and information protection level
Require simple method	Fully accept the request of the information providers
Quality simple method	Fully accept the request of the information collectors

4   Experiment and Discussion

4.1   Evaluation Experiment of Healthcare Information Sharing Method

To confirm whether the proposed method can collect high-quality healthcare information while reducing incentives and the degree of information protection, we performed a simulation experiment to compare the proposed method to two simple methods.

Here, the four types of information providers considered in this experiment are detailed in Table 1. These information providers assuming 500 people each, who have different qualities and sensitive evaluation of healthcare information. In this experiment, the utility function parameters (Sect. 3.2) were set to $w_1 = 0.5$, $w_2 = 0.5$, and $r = 2$. The behaviors of the information providers and information collectors followed the processing flow presented in Sect. 3.3.

Here, each information provider provided healthcare information every day for seven days. We assumed three information collectors. The information-collector agent generated and evaluated random numbers from the range of the immediate evaluation of healthcare information (shown in Table 1). The information-provider agent selected the information-collector agent who highly evaluated the healthcare information, and they automatically negotiated with the collector who highly evaluated the healthcare information. In the auto-negotiation, the upper limit of the negotiation was 30 times, and, if the upper limit was exceeded, the utility, amount of incentive obtained, and degree of information protection was 0. If the auto-negotiation succeeded, the information provider provided healthcare information using the utility calculated by the utility function as a probability. The experimental results were taken as the average results of auto-negotiation between an information provider's agent and the system's agent relative to the requisite amount of incentive, the information protection levels, the quality of the collected healthcare information, the number of collected healthcare information, and the utility. The amount of incentive, information protection levels, utility, and quality of the collected healthcare information were calculated as the average value of the experimental results for the three types of information providers. The proposed method yields excellent results for various types of information providers. Therefore, we confirm that the proposed method is excellent for any type of information provider. As a result, it is desirable to reduce incentives and the information protection level and increase the utility, the quality of the collected information, and the number of collected healthcare information. We confirmed the effectiveness of the proposed method by comparing the results of these experiments to two simple methods (Table 2). The experimental results are shown in Fig. 4. Regarding incentive, the proposed method scored −62% compared to the require simple method and +47% compared to the quality simple method. Here, regarding the degree of information protection, the same tendency for the amount of incentive was observed. For the information protection level, the proposed method scored −63% compared to the require simple method and +56% compared to the quality simple method. For utility, the proposed method scored −32% compared to the require simple method and +24% compared to the quality simple method. For the amount of healthcare information, the proposed method scored −15% compared to the require simple method and +19% compared to the quality simple method. In addition, for information quality, the proposed method scored +9% compared to the require simple method and +9% compared to the quality simple method.

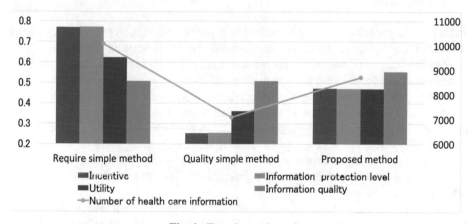

Fig. 4. Experimental results

4.2 Discussion

The experimental results demonstrate that the proposed method reduced the amount of incentive and information protection compared to the require simple method but increased these factors compared to the quality simple method. In the quality simple method, the amount of incentive and information protection level presented by the information collector is the experimental results. Information collectors seek to reduce incentives and the information protection level as much as possible; therefore, we considered that the incentive and information protection level was the lowest compared to any other method. We found that the amount of incentives and degree of information protection level were minimized. In addition, utility was the lowest among all methods, and the amount of healthcare information provided was also the smallest because the quality simple method only prioritizes the intent of the information collector. Thus, the quality simple method does not provide healthcare information as the utility felt by the information provider decreases. Since the require simple method accepts all information provider requests, the utility felt was high, and the probability of providing health care information increased. Therefore, the require simple method can collect the most health care information. However, the amount of incentive and information protection level was the highest among all methods, and the sustainability of services is considered in the follow. The proposed method collected higher quality health care information compared to the simple methods because healthcare information was collected by considering the intentions of both the information provider and information collector through auto-negotiations. Therefore, the information quality obtained by the proposed method was greater than that of the simple methods; thus, it is possible to collect a large amount of useful healthcare information using the proposed method.

Thus, we consider that the experimental results confirmed the effectiveness of the proposed healthcare information collection method.

5   Conclusion

In this paper, we have proposed a method to promote information provision by giving incentives while considering the protection of healthcare information. We proposed promoting the sharing of useful healthcare information by consensus building using the auto-negotiation considering the intentions of information provider/information collector. In the proposed method, agents reflect the intentions of information providers and information collectors, and the amount of incentive and items to provide information are determined automatically via auto-negotiation between the agents. In addition, to evaluate the effectiveness of the proposed healthcare information sharing method, we performed simulation experiments. The experimental results confirmed that the proposed method outperforms the simple method and that it enables dynamic evaluation of healthcare information.

In future, we plan to evaluate the effectiveness of the proposed method by considering various other conditions, e.g., the probability of a user providing healthcare information and continuation rate. We also plan to conduct an experiment in which healthcare information is acquired automatically from smart devices, e.g., wearable devices.

Acknowledgments. This work was supported by JSPS KAKENHI Grant Number 19K04895.

References

1. Itao, K., Komazawa, M., Katada, Y., Itao, K., Kobayashi, H., Luo, Z.W.: Age-related change of the activity of autonomic nervous system measured by wearable heart rate sensor for long period of time. In: Cipresso, P., Matic, A., Lopez, G. (eds.): Pervasive Computing Paradigms for Mental Health, Springer International Publishing, pp. 33–38 (2014)
2. Swan, M.: Sensor mania! the internet of things, wearable computing, objective metrics, and the quantified self 2.0. J. Sens. Actuator Netw. **1**(3) 217–253 (2012)
3. Mitchell, A.: Right Side Up: Building Brands in the Age of the Organized Consumer. Harper Collins Business, US (2001)
4. Dall, T.M., et al.: An aging population and growing disease burden will require alarge and specialized health care workforce by 2025. Health Aff. **32**, 2013–2020 (2013)
5. DeVahl, J., King, R., Williamson, J.W.: Academic incentives for students can increase participation in and effectiveness of a physical activity program. J. Am. Coll. Health, **53**(6), 295–298 (2005)

6. Pickard, K., Swan, M.: Big desire to share big health data: a shift in consumer attitudes toward personal health information (2014)
7. Farooqui, M.A., Tan, Y.T., Bilger, M., Finkelstein, E.A.: Effects of financial incentives on motivating physical activity among older adults: results from a discrete choice experiment. BMC Public Health **14**(1), 1–9 (2014)
8. Perera, G., Holbrook, A., Thabane, L., Foster, G., Willison, D.J.: Views on health information sharing and privacy from primary care practices using electronic medical records. Int. J. Med. Inform. **80**(2), 94–101 (2011)
9. Machanavajjhala, A., et al.: ℓ-diversity: privacy beyond k-anonymity. ACM Trans. Knowl. Discov. Data (TKDD) **1**(1) (2007)
10. Hann, I.H., Hui, K.L., Lee, T.S., Png, I.P.: The value of online information privacy: an empirical investigation. Ind. Organ. 0304001, University Library of Munich, Germany (2003)
11. Danezis, G., Lewis, S., Anderson, R.: How much is location privacy worth? In: Proceedings of WEIS (2005)
12. Sasank, R., Deborah, E., Mark, H., Mani, S.: Examining micro-payments for participatory sensing data collections. In Proceedings of the 12th ACM International Conference on Ubiquitous Computing (2010)
13. Westin, A.F.: Privacy and freedom. Wash. Lee Law Rev. **25**(1),166 (1968)
14. Qunigob, M.: An introduction to mahalanobis distance for MTS methods. Qual. Eng. **9**(1), 13–21 (2001)
15. Rubinstein, A.: Perfect equilibrium in a bargaining model. Econometrica, 97–109 (1982)

Author Index

Printed in the United States
by Baker & Taylor Publisher Services